Advanced Wireless Communications

Advanced Wireless Communications

4G Technologies

Savo Glisic
University of Oulu, Finland

John Wiley & Sons, Ltd

Other Wiley Editorial Offices

John Wiley & Sons Inc., 111 River Street, Hoboken, NJ 07030, USA

Jossey-Bass, 989 Market Street, San Francisco, CA 94103-1741, USA

Wiley-VCH Verlag GmbH, Boschstr. 12, D-69469 Weinheim, Germany

John Wiley & Sons Australia Ltd, 33 Park Road, Milton, Queensland 4064, Australia

John Wiley & Sons (Asia) Pte Ltd, 2 Clementi Loop #02-01, Jin Xing Distripark, Singapore 129809

John Wiley & Sons Canada Ltd, 22 Worcester Road, Etobicoke, Ontario, Canada M9W 1L1

Library of Congress Cataloging-in-Publication Data

British Library Cataloguing in Publication Data

A catalogue record for this book is available from the British Library

ISBN 0-470-86776-0

Typeset in 10/12pt Times by TechBooks, New Delhi, India
Printed and bound in Great Britain by Biddles Ltd, Kings Lynn, Norfolk
This book is printed on acid-free paper responsibly manufactured from sustainable forestry
in which at least two trees are planted for each one used for paper production.

To my family

Contents

Preface

At the present time, the wireless communications research community and industry are about to start discussions on standardization activities for the fourth generation (4G) of these systems. In the meantime, the research community has generated a number of promising solutions for significant improvements in system performance. Adaptive coding and modulation, iterative (turbo) decoding algorithms, space–time coding, multiple antennas and MIMO channels, multicarrier modulation, multiuser detection and ultra wideband radio are examples of enabling technologies for 4G. Significant new results have also been reported in the field of channel modeling and measurements for these applications, including new frequency bands. It is now up to the industry to implement these solutions and make appropriate selections from the variety of possible options. In order to really benefit from this opportunity, the industry will have to accept higher complexity and perfection in the implementation, otherwise the accumulated implementation losses may degrade significantly the system performance.

In the past, the term 'new generation' referred to mobile cellular communications. At this stage in the evolution of wireless communications, there is a tendency to agree that 4G will integrate mobile communications specified by International Mobile Telecommunications (IMT) standards and Wireless Local Area Networks (WLAN) or, in general, Broadband Radio Access Networks (BRAN). The expected data rates in the segment of mobile communications (cellular networks) are in the range of 50–100 Mbits/s and in WLANs up to 1 Gbit/s. The core network will be based on the Public Switched Telecommunications Network (PSTN), ATM (WATM) and Public Land Mobile Networks (PLMN) based on Internet Protocol (IP). Each of the segments of the system will be further enhanced in the future. Energy aware and, more generally, context aware, design in each layer of the network, including cross-layer optimization, will be an important segment of the overall system optimization philosophy. The inter-technology roaming of mobile terminals will be based on a reconfigurable software radio concept.

The focus of the book is on the system elements that provide adaptability and reconfigurability, and discussion of how much these features can improve the system performance. At the same time, the book provides a solid overview of the conventional technologies, although this is not the focus of our discussion. The measure of performance is the reconfiguration efficiency defined as a ratio between the reconfiguration (signal to noise ratio) gain and the relative increase in complexity.

The content of this book is based on the results of the research community in the field of wireless communications across the globe, and a number of authors have been explicitly cited in the text. This by no means represents an exhaustive list of those whose work is building up the enabling technology for 4G.

I would like to thank all those who have helped in the technical preparation of the manuscript.

S. Glisic
Naperville, Chicago, IL
December 2003

I

Adaptive Modulation and Coding

1

Fundamentals

1.1 4G AND THE BOOK LAYOUT

At the present time, the research community and industry in the field of telecommunications are considering possible choices for solutions in the fourth generation (4G) of wireless communications. This chapter will start with a generic 4G system concept that integrates available advanced technologies and will then focus on system adaptability and reconfigurability as a possible option to meet a variety of service requirements, available resources and channel conditions. The elements of such a concept can be found in [1–51]. The chapter will also try to offer a vision beyond the state of the art with the emphasis on how the advanced technologies can be used for an efficient 4G multiple access. Among a number of relevant issues, the focus will be on:

- adaptive and reconfigurable coding and modulation;

- adaptive and reconfigurable space–time coding;

- Advanced Time Division Multiple Access (ATDMA), Code Division Multiple Access (CDMA), Orthogonal Frequency Division Multiple Access (OFDMA), Multicarrier CDMA (MC CDMA) and Ultra Wide Band (UWB) radio;

- antenna array signal processing;

- adaptive and reconfigurable software radio, including discussion on the strategic difference between macro and micro reconfigurability;

- user location in 4G, including network synchronization;

- channel modeling;

- cross-layer optimization, including adaptive and power-efficient MAC layer design, adaptive and power-efficient routing on IP and TCP layer and the concept of a green wireless network.

Advanced Wireless Communications. S. Glisic
© 2004 John Wiley & Sons, Ltd. ISBN: 0-470-86776-0

As the complexity of the advanced algorithms and required processing power increases, there is more and more need for a systematic approach in the analysis of the implementation losses in the system, depending on the precision (price) of the technology used. The chapter will offer a possible approach based on system sensitivity function and show a few examples from practice.

Another important aspect of wireless system design is power consumption. This will also be covered in the chapter, including several layers in the network.

At this stage of the evolution of wireless communications, there is a tendency to agree that 4G will integrate mobile communications specified by International Mobile Telecommunications (IMT) standards and Wireless Local Area Networks (WLAN), or, in general, Broadband Radio Access Networks (BRAN). The core network will be based on the Public Switched Telecommunications Network (PSTN) and Public Land Mobile Networks (PLMN) based on Internet Protocol (IP) [13, 16, 19, 24, 32, 41, 51]. This concept is summarized in Figure 1.1. Each of the segments of the system will be further enhanced in the future. The inter-technology roaming of the mobile terminal willbe based on a reconfigurable radio concept presented in its generic form in Figure 1.2. Based on this concept,

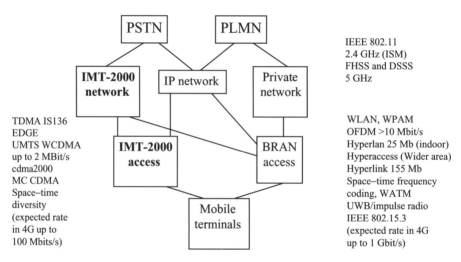

Figure 1.1 IMT-2000 and WLAN convergence.

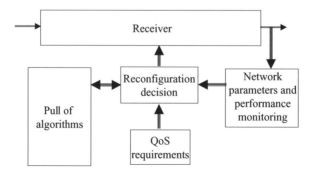

Figure 1.2 Reconfigurable radio concept inter-system roaming and QoS provisioning.

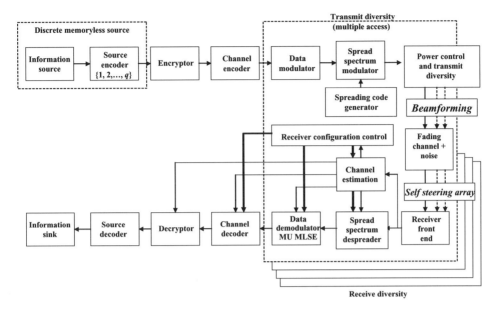

Figure 1.3 Generic block diagram of a digital communication system.

a generic block diagram of the 4G communication system is presented in Figure 1.3. For any specific configuration of the system, the blocks in the figure will have a specific form. The relation between the system presented in Figure 1.3 and the book layout is shown in Figure 1.4. The numbers in different blocks refer to the corresponding chapter in the book.

The remainder of this chapter considers the general structure of 4G signals, mainly Advanced Time Division Multiple Access (ATDMA), Code Division Multiple Access (CDMA), Orthogonal Frequency Division Multiplexing (OFDM), Multicarrier CDMA (MC CDMA) and Ultra Wide Band (UWB) signals. These signals will be elaborated upon later in the book.

Chapter 2 introduces adaptive coding. The book has no intention to cover all details of coding, but rather to focus on those components that enable code adaptability and re-configurability. Within this concept the chapter covers: adaptive and reconfigurable block and convolutional codes, punctured convolutional codes/code reconfigurability, maximum likelihood decoding/Viterbi algorithm, systematic recursive convolutional codes, concatenated codes with interleavers, the iterative (turbo) decoding algorithm and a discussion on adaptive coding practice and prospects.

Chapter 3 covers adaptive and reconfigurable modulation. This includes coded modulation, Trellis Coded Modulation (TCM) with examples of TCM schemes such as two-, four- and eight-state trellises, and QAM with three bits per symbol transmission. The chapter further discusses signal set partitioning, equivalent representation of TCM, TCM with multidimensional constellations, adaptive coded modulation for fading channels and adaptation to maintain fixed distance in the constellation.

Chapter 4 introduces space–time coding. It starts with a discussion on diversity gain, the encoding and transmission sequence, the combining scheme and ML decision rule for the two-branch transmit diversity scheme with one and M receivers. In the next step, it introduces a general discussion on space–time coding within a concept of space–time trellis modulation. The discussion is then extended to introduce space–time block codes

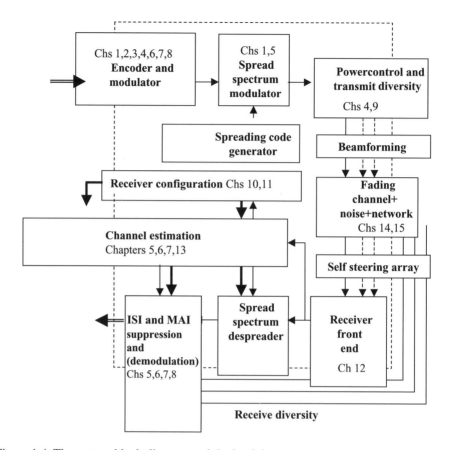

Figure 1.4 The system block diagram and the book layout.

from *orthogonal designs*, mainly linear processing orthogonal designs and generalized real orthogonal designs. The chapter also covers channel estimation imperfections. It continues with quasi-orthogonal space–time block codes, space–time convolutional codes and algebraic space–time codes. It also includes differential space–time modulation with a number of examples.

Layered space–time coding and *concatenated* space–time block coding are also discussed. Estimation of MIMO channels and space–time codes for frequency selective channels are discussed in detail. MIMO system optimization, including gain optimization by singular value decomposition (svd), are also discussed.

Chapter 5 covers Code Division Multiple Access – CDMA. It includes topics like pseudorandom sequences, multiuser CDMA receivers and signal subspace-based channel estimation.

Chapter 6 deals with Time Division Multiple Access, mainly equalization, detection for a statistically known, time-varying channel, adaptive MLSE equalization, adaptive joint channel identification and data demodulation, turbo equalization, Kalman filter based joint channel estimation and equalization using higher order signal statistics

Chapter 7 covers Orthogonal Frequency Division Multiplexing (OFDM) and MC CDMA. The following topics are discussed: timing and frequency offset in OFDM, fading channel estimation for OFDM systems, 64-DAPSK and 64-QAM modulated OFDM signals, space–time coding with OFDM signals, layered space–time coding for MIMO-OFDM, space–time coded TDMA/OFDM reconfiguration efficiency, multicarrier CDMA systems, multicarrier DS-CDMA broadcast systems, frame by frame adaptive rate coded multicarrier DS-CDMA systems, intermodulation interference suppression in multicarrier DS-CDMA systems, successive interference cancellation in multicarrier DS-CDMA systems, MMSE detection of multicarrier CDMA and multiuser receiver for space–time coded multicarrier CDMA systems.

Chapter 8 introduces *Ultra Wide Band Radio*. It covers topics like: UWB multiple access in Gaussian channels, the UWB channel, a UWB system with M-ary modulation, M-ary PPM UWB multiple access, coded UWB schemes, multiuser detection in UWB radio, UWB with space–time processing and beamforming for UWB radio.

Chapter 9 covers antenna array signal processing, with the focus on space–time receivers for CDMA communications, MUSIC and ESPRIT DOA estimation, joint array combining and MLSE receivers, joint combiner and channel response estimation and complexity reduction in wideband beamforming.

Chapter 10 discusses adaptive/reconfigurable software radio. The focus is on energy-efficient adaptive radio, frame length adaptation, energy-efficient adaptive error control, processing gain adaptation, trellis-based processing/adaptive maximum likelihood sequence equalization, a software radio architecture for linear multiuser detection and reconfigurable ASIC architecture.

Chapter 11 provides examples of software radio architectures. These examples include: a low power DSP core-based software radio architecture, software radio architecture with smart antennas, the software realization of a GSM base station, software realization of WCDMA (FDD) downlink in base station systems and designing a DS-CDMA indoor system over FPGA platforms.

Chapter 12 covers some selected issues for network overlay in 4G, mainly adaptive self-reconfigurable interference suppression schemes for CDMA wireless networks, such as multilevel ξ structures and multilevel G (global) structures.

Chapter 13 presents the problem of user location in 4G networks. It covers basic location technologies and then focuses on a few key topics such as frame detection and network synchronization.

Chapter 14 discusses channel modeling and measurements for 4G. It includes macrocellular environments (1.8 GHz), urban spatial radio channels in macro/microcell (2.154 GHz), MIMO channels in micro and picocell environments (1.71/2.05 GHz), outdoor mobile channel (5.3 GHz), microcell channel (8.45 GHz), wireless MIMO LAN environments (5.2 GHz), indoor WLAN channel (17 GHz), indoor WLAN channel (60 GHz) and a UWB channel model.

Chapter 15 includes discussion on adaptive 4G networks. It covers the adaptive MAC layer, minimum energy routing in peer-to-peer mobile wireless networks, least resistance routing in wireless networks and power optimal routing in wireless networks for guaranteed TCP layer QoS.

1.2 GENERAL STRUCTURE OF 4G SIGNALS

In this section we will summarize the signal formats used in the existing wireless systems and point out possible evolutionary routes towards the 4G system. The focus will be on ATDMA, WCDMA, OFDMA, MC CDMA and UWB signals.

Table 1.1 TDMA system parameters

	Europe (ETSI)	North America(TIA)	Japan (MPT)
Access method	TDMA	TDMA	TDMA
Carrier spacing	200 kHz	30 kHz	25 kHz
Users per carrier	8 (16)	3 (6)	3 (tbd)
Modulation	GMSK	$\pi/4$ DPSK	$\pi/4$ DPSK
Voice codec	RPE 13 kb/s	VSELP 8 kb/s	tbd
Voice frame	20 ms	20 ms	20 ms
Channel code	convolutional	convolutional	convolutional
Codec bit rate	22.8 kb/s	13 kb/s	11.2 kb/s
TDMA frame duration	4.6 ms	20 ms	20 ms
Interleaving	$\approx$40 ms	27 ms	27 ms
Associated control channel	extra slot	in slot	in slot
Handoff method	MAHO	MAHO	MAHO

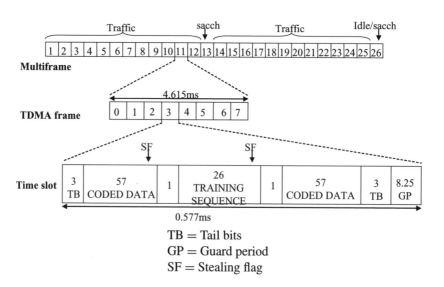

TB = Tail bits
GP = Guard period
SF = Stealing flag

Figure 1.5 Digital cellular TDMA systems: GSM slot and frame structure showing 130.25 b per time slot (0.577 ms), eight time slots/TDMA frame (full rate), and 13 TDMA frames/multiframe.

1.2.1 Advanced time division multiple access (ATDMA)

In a TDMA system each user is using a dedicated time slot within a TDMA frame, as shown in Figure 1.5 for GSM (Global System of Mobile Communications) and in Figure 1.6 for ADC (American Digital Cellular System). Additional data about the signal format and system capacity are given in Tables 1.1 and 1.2.

The evolution of the ADC system resulted in TIA (Telecommunications Industry Association) Universal Wireless Communications (UWC) standard 136. The basic system parameters are summarized in Table 1.3. The evolution of GSM resulted in a system known

Table 1.2   Approximate capacity in Erlang per km² assuming a cell radius of 1 km (site distance of 3 km) in all cases and three sectors per site. The Lee merit gain is the number of channels per site assuming an optimal reuse plan

	Analog FM	GSM pessimistic optimistic		ADC pessimistic optimistic		JDC pessimistic optimistic	
Bandwidth	25 MHz	25 MHz		25 MHz		25 MHz	
# of voice channels	833	1000		2500		3000	
Reuse plan	7	4	3	7	4	7	4
Channels per site	119	250	333	357	625	429	750
Erlang/km²	11.9	27.7	40.0	41.0	74.8	50.0	90.8
Capacity gain	1.0	2.3	3.4	3.5	6.3	4.2	7.6
(Lee merit gain)	(1.0)	(2.7)	(3.4)	(3.8)	(6.0)	(4.0)	(7.2)

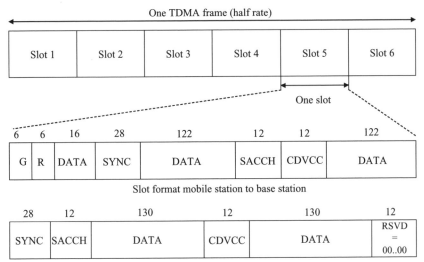

G = Guard time
R = Ramp up time
RSVD = Reserved bits

Figure 1.6 ADC slot and frame structure for down- and uplink with 324 bits per time slot (6.67 ms) and 3(6) time slots/TDMA frame for full rate (half rate).

as Enhanced Data rates for GSM Evolution (EDGE) with parameters also summarized in Table 1.3.

If TDMA is chosen for 4G the signal formats will be further enhanced by using multidimensional trellis (space–time–frequency) coding and advanced signal processing, as described in Chapters 2–4 and 6. This will also be combined with OFDM and MC CDMA signal formats, described in Chapter 7.

Table 1.3   Parameters of UWC-136 and EDGE signals

Key characteristic	TIA UWC-136	GSM radio interface (for reference only)
Multiple access	TDMA	TDMA
Bandwidth	30/200/1600 kHz	200 kHz
Bit rate	48.6 kbps	270.8 kbps
	72.9 kbps	for EDGE 812.5 kbps
	270.8 kbps	
	361.1 kbps	
	722.2 kbps	
	2.6 Mbps	
	5.2 Mbps	
Duplexing	FDD/TDD	FDD
Carrier spacing	30/200/1600 kHz	200 kHz
Inter BS timing	Asynchronous (sync. possible)	Asynchronous (sync. possible)
Inter-cell synchronization	not required	not required
Base station synchronization	not required	not required
Cell search scheme	L1 power based, L2 parameter based, L3 service/network/ operator based	L1 power based, L2 parameter based, L3 service/network/ operator based
Frame length	40/40/4.6/4.6 ms	4.6 ms
HO	HHO	HHO
Downlink data modulation	$\pi/4$ DPSK $\pi/4$ coherent QPSK 8PSK GMSK Q-O-QAM B-O-QAM	GMSK 8PSK
Downlink power control	Per slot and per carrier	Per slot
Downlink variable rate accommodation	slot aggregation	slot aggregation
Uplink data modulation	$\pi/4$ DPSK $\pi/4$ coherent QPSK 8PSK GMSK Q-O-QAM B-O-QAM	GMSK 8PSK
Uplink power control	Per slot and per carrier	BS directed MS power control
Uplink variable rate accommodation	slot aggregation	slot aggregation

Table 1.3   (*Cont.*).

Key characteristic	TIA UWC-136	GSM radio interface (for reference only)
Channel coding	Punctured convolutional code ($R = 1/2, 2/3, 3/4, 1/1$) Soft or hard decision coding	Convolutional coding. Rate dependent on service
Interleaving periods	0/20/40/140/240 ms	Dependent on service
Rate detection	Via L3 signaling	Via stealing flags
Other features	Space and frequency diversity; MRC / 'MRC-like' Support for hierarchical structures	MRC
Random access mechanism	Random access with shared control feedback (SCF), also reserved access	Random
Power control steps	4 dB	2 dB
Super frame length	720 ms/640 ms (hyperframe is 1280 ms)	720 ms
Slots/frame	6 per 30 kHz carrier 8 per 200 kHz carrier 16–64 per 1.6 MHz carrier	8
Focus of backward compatibility	AMPS/IS54/136/GSM	GSM

1.2.2   Code division multiple access (CDMA)

The CDMA technique is based on spreading the spectra of the relatively narrow information signal S_n by a code c, generated by a much higher clock (chip) rate. Different users are separated by using different uncorrelated codes. As an example, the narrowband signal in this case can be a PSK signal of the form:

$$S_n = b(t, T_m) \cos \omega t \tag{1.1}$$

where $1/T_m$ is the bit rate and $b = \pm 1$ is the information. The baseband equivalent of Equation (1.1) is:

$$S_n^b = b(t, T_m) \tag{1.1a}$$

A spreading operation, presented symbolically by operator $\varepsilon(\ )$, is obtained if we multiply the narrowband signal by a pseudonoise (PN) sequence (code) $c(t, T_c) = \pm 1$. The bits of the sequence are called chips and the chip rate $1/T_c \gg 1/T_m$. The wideband signal can be represented as:

$$S_w = \varepsilon(S_n) = c S_n = c(t, T_c) b(t, T_m) \cos \omega t \tag{1.2}$$

The baseband equivalent of Equation (1.2) is:

$$S_w^b = c(t, T_c) b(t, T_m) \tag{1.2a}$$

Despreading, represented by operator $D(\ )$, is performed if we use $\varepsilon(\ )$ once again and bandpass filtering, with the bandwidth proportional to $2/T_m$, represented by operator $\text{BPF}(\ )$, resulting in:

$$D(S_w) = \text{BPF}\left(\varepsilon(S_w)\right) = \text{BPF}\left(cc\,b\cos\omega t\right) = \text{BPF}\left(c2\,b\cos\omega t\right) = b\cos\omega t \quad (1.3)$$

The baseband equivalent of Equation (1.3) is:

$$D\left(S_w^b\right) = \text{LPF}\left(\varepsilon\left(S_w^b\right)\right) = \text{LPF}\left(c(t,T_c)c(t,T_c)b(t,T_m)\right) = \text{LPF}\left(b(t,T_m)\right) = b(t,T_m) \tag{1.3a}$$

where $\text{LPF}(\ )$ stands for low pass filtering. This approximates the operation of correlating the input signal with the locally generated replica of the code $\text{Cor}(c, S_w)$. Non-synchronized despreading would result in:

$$D_\tau(\ );\ \text{Cor}(c_\tau, S_w) = \text{BPF}\left(\varepsilon_\tau(S_w)\right) = \text{BPF}\left(c_\tau\,c\,b\cos\omega t\right) = \rho(\tau)b\cos\omega t \quad (1.4)$$

In Equation (1.4) BPF would average out the signal envelope $c_\tau c$, resulting in $E(c_\tau c) = \rho(\tau)$. The baseband equivalent of Equation (1.4) is:

$$D_\tau(\ );\ \text{Cor}\left(c_\tau, S_w^b\right) = \int_0^{T_m} c_\tau S_w^b\,dt = b(t,T_m)\int_0^{T_m} c_\tau c\,dt = d\rho(\tau) \tag{1.4a}$$

This operation would extract the useful signal b as long as $\tau \cong 0$, otherwise the signal would be suppressed because $\rho(\tau) \cong 0$ for $\tau \geq T_c$. Separation of multipath components in a RAKE receiver is based on this effect. In other words, if the received signal consists of two delayed replicas of the form:

$$r = S_w^b(t) + S_w^b(t - \tau)$$

the despreading process defined by Equation (1.4a) would result in:

$$D_\tau(\ );\text{Cor}(c,r) = \int_0^{T_m} cr\,dt = b(t,T_m)\int_0^{T_m} c(c+c_\tau)\,dt = b\rho(0) + b\rho(\tau)$$

Now, if $\rho(\tau) \cong 0$ for $\tau \geq T_c$ all multipath components reaching the receiver with a delay larger then the chip interval will be suppressed. If the signal transmitted by user y is despread in receiver x the result is:

$$D_{xy}(\ );\ \text{BPF}\left(\varepsilon_{xy}(S_w)\right) = \text{BPF}\left(c_x\,c_y\,b_y\cos\omega t\right) = \rho_{xy}(t)b_y\cos\omega t \quad (1.5)$$

So, in order to suppress the signals belonging to other users (Multiple Access Interference – MAI), the crosscorrelation functions should be low. In other words, if the received signal consists of the useful signal plus the interfering signal from the other user:

$$r = S_{wx}^b(t) + S_{wy}^b(t) = b_x c_x + b_y c_y \tag{1.6}$$

the despreading process at the receiver of user x would produce:

$$D_{xy}(\ );\ \text{Cor}(c_x, r) = \int_0^{T_m} c_x r\,dt$$

$$= b_x\int_0^{T_m} c_x c_x\,dt + b_y\int_0^{T_m} c_x c_y\,dt = b_x\rho_x(0) + b_y\rho_{xy}(0) \quad (1.7)$$

When the system is properly synchronized $\rho_x(0) \cong 1$, and if $\rho_{xy}(0) \cong 0$ the second component representing MAI will be suppressed. This simple principle is elaborated in the WCDMA standard, resulting in a collection of transport and control channels. The system is based on 3.84 Mchips rate and up to 2 Mbits/s data rate. In a special downlink high data rate shared channel, the data rate and signal format are adaptive. There shall be mandatory support for QPSK and 16-QAM and optional support for 64-QAM based on UE capability which will proportionally increase the data rate. For details see www.3gpp.com. CDMA is discussed in detail in Chapter 5.

1.2.3 Orthogonal frequency division multiplexing (OFDM)

In wireless communications, the channel imposes a limit on data rates in the system. One way to increase the overall data rate is to split the data stream into a number of parallel channels and use different subcarriers for each channel. The concept is presented in Figure 1.7 and represents the basic idea of an OFDM system. The overall signal can be represented as:

$$x(t) = \sum_{n=0}^{N-1} \left\{ D_n e^{j2\pi \frac{n}{N} f_s t} \right\}; \quad -\frac{k_1}{f_s} < t < \frac{N + k_2}{f_s} \tag{1.8}$$

In other words, complex data symbols $[D_0, D_1, \ldots, D_{N-1}]$ are mapped in OFDM symbols $[d_0, d_1, \ldots, d_{N-1}]$ such that:

$$d_k = \sum_{n=0}^{N-1} D_n e^{j2\pi \frac{kn}{N}} \tag{1.9}$$

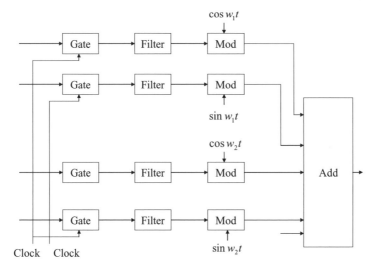

Figure 1.7 An early version of OFDM.

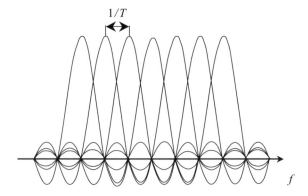

Figure 1.8 Spectrum overlap in OFDM.

The output of the FFT block at the receiver produces data per channel. This can be represented as:

$$\tilde{D}_m = \frac{1}{N} \sum_{k=0}^{N-1} r_k e^{-j2\pi m \frac{k}{2N}}$$

$$r_k = \sum_{n=0}^{N-1} H_n D_n e^{j2\pi \frac{n}{2N} k} + n(k) \qquad (1.10)$$

$$\tilde{D}_m = \begin{cases} H_n D_n + N(n), n = m \\ N(n), n \neq m \end{cases}$$

The signal spectra are given in Figure 1.8, while the two forms of system block diagram are given in Figures 1.9(a) and 1.9(b).

In order to eliminate residual inter-symbol interference, a guard interval after each symbol is used, as shown in Figure 1.10.

An example of an OFDM signal specified by the IEEE 802.11a standard is shown in Figure 1.11. The signal parameters are: 64 points FFT, 48 data subcarriers, 4 pilots, 12 virtual subcarriers, DC component 0, Guard interval 800 ns. A discussion of OFDM and an extensive number of references on the topic are included in Chapter 7.

1.2.4 Multicarrier CDMA (MC CDMA)

Good performance and the flexibility to accommodate multimedia traffic are incorporated in MC CDMA, which is obtained by combining the CDMA and OFDM signal formats.

Figure 1.12 shows the DS-CDMA transmitter of the jth user for the binary phase shift keying/coherent detection (CBPSK) scheme and the power spectrum of the transmitted signal, respectively, where $G_{DS} = T_m/T_c$ denotes the processing gain and $C^j(t) = [C_1^j \ C_2^j \cdots C_{G_{DS}}^j]$ the spreading code of the jth user.

Figure 1.13 shows the MC CDMA transmitter of the jth user for the CBPSK scheme and the power spectrum of the transmitted signal, respectively, where G_{MC} denotes the processing gain, N_C the number of subcarriers, and $C^j(t) = [C_1^j \ C_2^j \cdots C_{G_{MC}}^j]$ the spreading code of the jth user. The MC CDMA scheme is discussed assuming that the number of

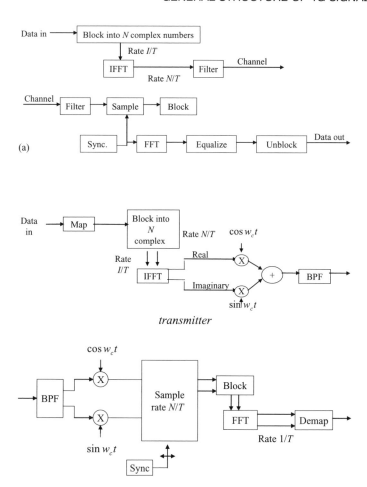

transmitter

receiver

Figure 1.9 (a) Basic OFDM system; (b) system with complex transmission.

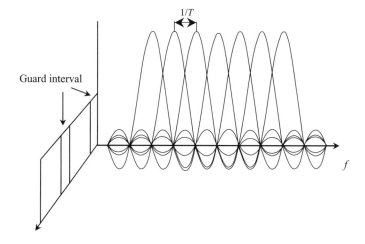

Figure 1.10 OFDM time and frequency span.

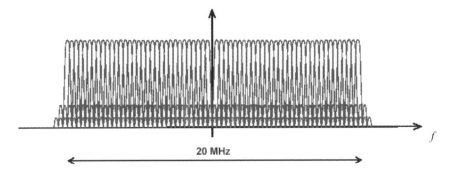

Figure 1.11 802.11a/HYPERLAN OFDM.

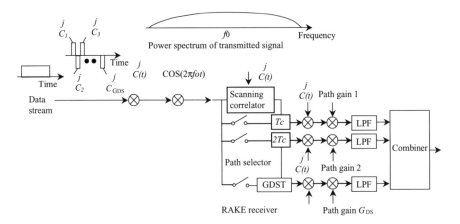

Figure 1.12 DS-CDMA scheme.

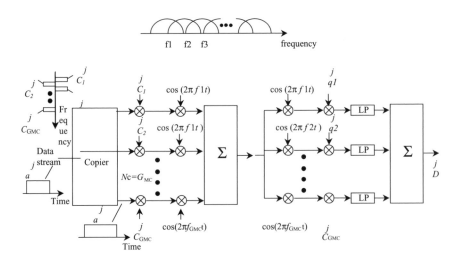

Figure 1.13 MC CDMA scheme.

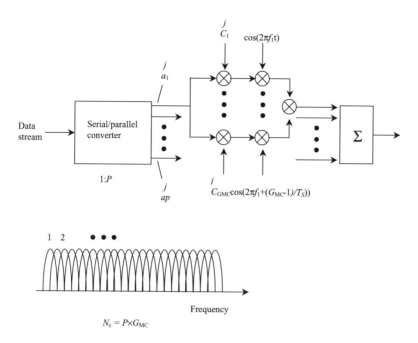

Figure 1.14 Modification of the MC CDMA scheme: spectrum of its transmitted signal.

subcarriers and the processing gain are all the same. However, we do not have to choose $N_C = G_{MC}$, and, actually, if the original symbol rate is high enough to become subject to frequency selective fading, the signal needs to be first S/P-converted before spreading over the frequency domain. This is because it is crucial for multicarrier transmission to have frequency non-selective fading over each subcarrier.

Figure 1.14 shows the modification to ensure frequency non-selective fading, where T_S denotes the original symbol duration, and the original data sequence is first converted into P parallel sequences, and then each sequence is mapped onto G_{MC} subcarriers ($N_C = P \times G_{MC}$). The multicarrier DS-CDMA transmitter spreads the S/P-converted data streams using a given spreading code in the time domain so that the resulting spectrum of each subcarrier can satisfy the orthogonality condition with the minimum frequency separation. This scheme is originally proposed for an uplink communication channel, because the introduction of OFDM signaling into the DS-CDMA scheme is effective for the establishment of a quasi-synchronous channel.

Figure 1.15 shows the multicarrier DS-CDMA transmitter of the jth user and the power spectrum of the transmitted signal, respectively, where G_{MD} denotes the processing gain, N_C the number of subcarriers, and $C^j(t) = [C_1^j \ C_2^j \cdots C_{G_{MD}}^j]$ the spreading code of the jth user.

The multitone MT CDMA transmitter spreads the S/P-converted data streams using a given spreading code in the time domain so that the spectrum of each subcarrier prior to the spreading operation can satisfy the orthogonality condition with the minimum frequency separation. Therefore, the resulting spectrum of each subcarrier no longer satisfies the orthogonality condition. The MT CDMA scheme uses longer spreading codes in proportion to the number of subcarriers, as compared with a normal (single carrier) DS-CDMA scheme, therefore, the system can accommodate more users than the DS-CDMA scheme.

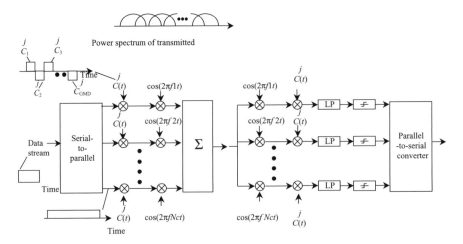

Figure 1.15 Multicarrier DS-CDMA scheme.

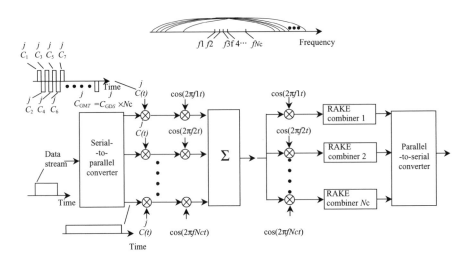

Figure 1.16 MT CDMA scheme.

Figure 1.16 shows the MT CDMA transmitter of the jth user for the CBPSK scheme and the power spectrum of the transmitted signal, respectively, where G_{MT} denotes the processing gain, N_C the number of subcarriers, and $C^j(t) = [C_1^j \ C_2^j \cdots C_{G_{MT}}^j]$ the spreading code of the jth user.

All of these schemes will be discussed in detail in Chapter 7.

1.2.5 Ultra wide band (UWB) signals

For the multipath resolution in indoor environments a chip interval of the order of a few nanoseconds is needed. This results in a spread spectrum signal with bandwidth in the order of a few GHz. Such a signal can also be used with no carrier, resulting in what is called impulse radio (IR) or Ultra Wide Band (UWB) radio. A typical form of the signal used

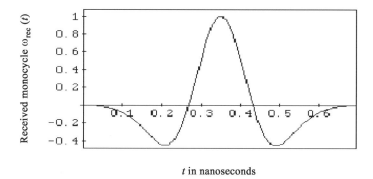

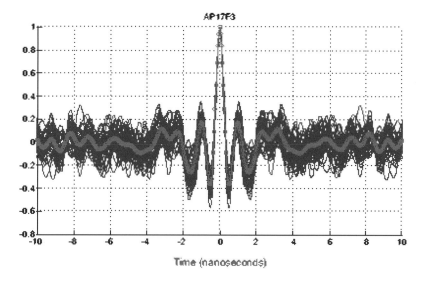

Figure 1.17 A typical ideal received monocycle $\omega_{\text{rec}}(t)$ at the output of the antenna subsystem as a function of time in *nanoseconds*.

Figure 1.18 A collection of received pulses in different locations [52].

in this case is shown in Figure 1.17. A collection of pulses received on different locations within the indoor environment is shown in Figure 1.18. The Ultra Wide Band radio will be discussed in detail in Chapter 8. In this section we will initially define only a possible signal format.

A typical time-hopping format used in this case can be represented as:

$$s_{\text{tr}}^{(k)}\left(t^{(k)}\right) = \sum_{j=-\infty}^{\infty} \omega_{\text{tr}}\left(t^{(k)} - jT_f - c_j^{(k)}T_c - \delta d_{[j/N_s]}^{(k)}\right) \tag{1.11}$$

where $t^{(k)}$ is the kth transmitter's clock time and T_f is the *pulse repetition time*. The transmitted pulse waveform ω_{tr} is referred to as a *monocycle*. To eliminate collisions due to multiple access, each user (indexed by k) is assigned a distinctive time shift pattern $\{c_j^{(k)}\}$

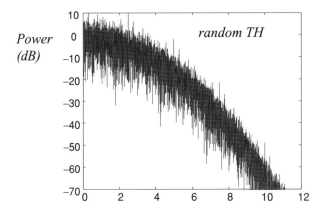

Figure 1.19 Spectrum of a TH signal.

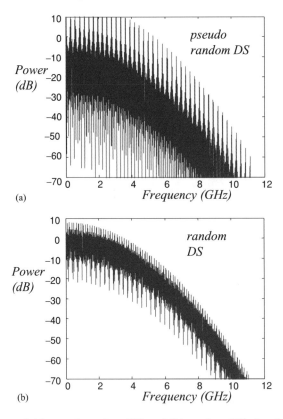

Figure 1.20 Spectra of (a) pseudorandom DS and (b) random DS signals.

called a *time-hopping sequence*. This provides an additional time shift of $c_j^{(k)}T_c$ seconds to the jth monocycle in the pulse train, where T_c is the duration of addressable time delay bins. For a fixed T_f, the *symbol rate*, R_s, determines the number, N_s, of monocycles that are modulated by a given binary symbol as $R_s = (1/N_sT_f)s^{-1}$. The modulation index δ is

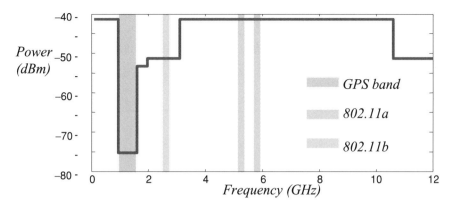

Figure 1.21 FCC frequency mask.

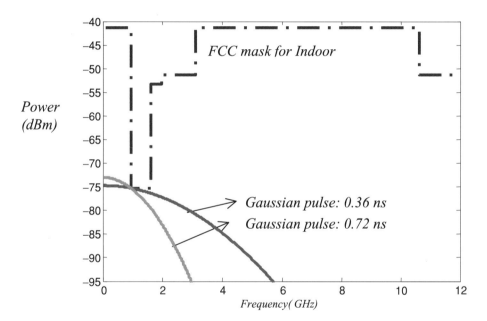

Figure 1.22 FCC mask and possible UWB signal spectra.

chosen to optimize performance. For performance prediction purposes, most of the time the data sequence $\{d_j^{(k)}\}_{j=-\infty}^{\infty}$ is modeled as a wide-sense stationary random process composed of equally likely symbols. For data, a pulse position data modulation is used.

When K users are active in the multiple access system, the composite received signal at the output of the receiver's antenna is modeled as:

$$r(t) = \sum_{k=1}^{K} A_k s_{\text{rec}}^{(k)}(t - \tau_k) + n(t) \tag{1.12}$$

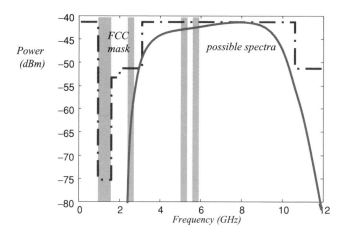

Figure 1.23 Single band UWB signal.

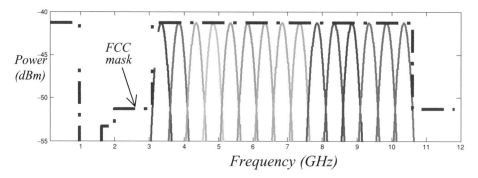

Figure 1.24 Multiband UWB signal.

The antenna/propagation system modifies the shape of the transmitted monocycle $\omega_{tr}(t)$ to $\omega_{rec}(t)$ on its output. An idealized received monocycle shape $\omega_{rec}(t)$ for a free space channel model with no fading is shown in Figure 1.17.

The optimum receiver for a single bit of a binary modulated impulse radio signal in additive white Gaussian noise (AWGN) is a correlation receiver:

$$\text{'decide } d_0^{(1)} = 0\text{' if}$$

$$\sum_{j=0}^{N_s-1} \overbrace{\int_{\tau_1+jT_f}^{\tau_1+(j+1)T_f} r(u,t)\upsilon\left(t - \tau_1 - jT_f - c_j^{(1)}T_c\right) \mathrm{d}t}^{\text{pulse correlator output } \triangleq \, \alpha_j(u)} > 0 \qquad (1.13)$$

$$\underbrace{\hphantom{\sum_{j=0}^{N_s-1} \int_{\tau_1+jT_f}^{\tau_1+(j+1)T_f} r(u,t)\upsilon\left(t - \tau_1 - jT_f - c_j^{(1)}T_c\right) \mathrm{d}t}}_{\text{test statistic } \triangleq \, \alpha(u)}$$

where $\upsilon(t) \triangleq \omega_{rec}(t) - \omega_{rec}(t - \delta)$.

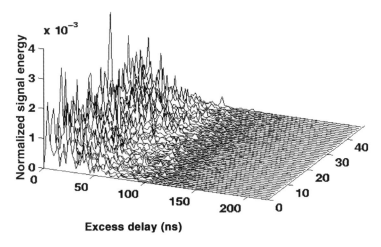

Figure 1.25 A collection of channel delay profiles [53].

The spectrum of a signal using TH is shown in Figure 1.19. If, instead of TH, a DS signal is used, the signal spectrum is shown in Figure 1.20(a) for a pseudorandom code and in Figure 1.20(b) for a random code. The FCC (Frequency Control Committee) mask for indoor communications is shown in Figure 1.21. Possible options for UWB signal spectra are given in Figures 1.22 and 1.23 for single band, and in Figure 1.24 for multiband signal formats. For a frequency selective fading channel, a collection of channel delay profiles is shown in Figure 1.25. For more details see http://www.uwb.org and http://www.uwbmultiband.org.

The *optimal detection* in a multiuser environment, with knowledge of all time-hopping sequences, leads to complex parallel receiver designs [2]. However, if the number of users is large and no such multiuser detector is feasible, then it is reasonable to approximate the combined effect of the other users' dehopped interfering signals as a Gaussian random process. All details regarding the system performance will be discussed in Chapter 8.

2

Adaptive Coding

Channel coding is a well established technical field that includes both strong theory and a variety of practical applications. Both theory and practice are well documented in open literature. In this chapter we provide a brief review of the basic principles and results in this field, with the emphasis on those parameters which are important for adaptability and reconfigurability of coding and decoding algorithms. This is an important characteristic for applications in wireless systems where a strong request for energy preservation suggests a system operation where quality of service (QoS) is met with minimum effort. In an environment with changing propagation conditions, this requires a possibility to adapt the complexity of the system. This is the focus of the presentation in this chapter and for the conventional details related to channel coding the reader is referred to the classical references in the field.

2.1 ADAPTIVE AND RECONFIGURABLE BLOCK CODING

The simplest way to improve the probability of correct detection of a bit is to repeat the transmission of the same bit (repetition code) and base the detection of the bit on so-called majority logic. As an example, if each bit is repeated three times, the decoder will base the decision on the observation of the three bits. The error will now occur if two or more bits are received incorrectly. This is a simple solution but rather inefficient from the point of view of bandwidth utilization. The next option is the family of codes based on the parity check principle. An oversimplified example is given in Figure 2.1. For every two input bits $\mathbf{u} = (u_1, u_2)$, a parity check bit, $x_3 = u_1 + u_2$, is created so that the transmitted bits are $\mathbf{x} = (x_1, x_2, x_3) = (u_1, u_2, x_3)$ as indicated in the figure. This simple example can be further expanded to include a number of parity check bits. For a number of input bits k, $n - k$ parity check bits are generated, resulting in a code word of length n. For this we use the notation (n, k) block codes. The art of block code construction consists of finding such parity check

---

Advanced Wireless Communications. S. Glisic
© 2004 John Wiley & Sons, Ltd. ISBN: 0-470-86776-0

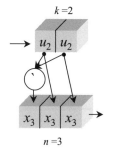

$$x_1 = u_1, x_2 = u_2, x_3 = u_1 + u_2$$

Data words	Code words
00	000
01	011
10	101
11	110

Figure 2.1 Encoder for the parity check code (3, 2).

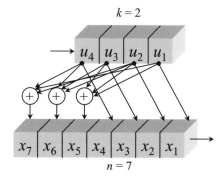

Figure 2.2 Encoder for the Hamming code (7, 4).

rules that would provide the best error correction capabilities with the minimum amount of redundant bits.

The ratio $R_c = k/n$ is called coding rate. An example of such a code is the Hamming code (7, 4) shown in Figure 2.2.

The Hamming code (7, 4) is defined by the relations:

$$
\begin{aligned}
x_i &= u_i, \; i = 1, 2, 3, 4 \\
x_5 &= u_1 + u_2 + u_3 \\
x_6 &= u_2 + u_3 + u_4 \\
x_7 &= u_1 + u_2 + u_4
\end{aligned}
\tag{2.1}
$$

The output *code words* $\mathbf{x}(x_1, x_2, x_3, x_4, x_5, x_6, x_7)$ for all possible *input words* $\mathbf{u}(u_1, u_2, u_3, u_4)$ are shown in Table 2.1.

One can see from Table 2.1 that out of 2^n possible code words, only 2^k are used in the encoder. A collection of these code words is called a code book. The decoder will decide in the favor of the code word from the code book that is the closest to the given received code word that may contain a certain number of errors. So, in order to minimize the bit error rate, the art of coding consists of choosing those code words for the code book that differ from each other as much as possible. This difference is quantified by Hamming distance d defined as the number of bit positions where the two words are different. A number of families of block codes are described in the literature. An example is cyclic codes, represented by Bose–Chaudhuri–

Table 2.1 Hamming code (7, 4)

Data words $\mathbf{u}(u_1, u_2, u_3, u_4)$	Code words $\mathbf{x}(x_1, x_2, x_3, x_4, x_5, x_6, x_7)$
0000	0000 000
0001	0001 011
0010	0010 110
0011	0011 101
0100	0100 111
0101	0101 100
0110	0110 001
0111	0111 010
1000	1000 101
1001	1001 110
1010	1010 011
1011	1011 000
1100	1100 010
1101	1101 001
1110	1110 100
1111	1111 111

Five code words

1	6	11	16	**21**	**26**	**31**	···	66	71
2	7	12	17	**22**	**27**	**32**	···	67	72
3	8	13	18	**23**	**28**	**33**	···	68	73
4	9	14	19	**24**	**29**	**34**	···	69	74
5	10	15	**20**	**25**	**30**	35	···	70	75

Information digits — Parity check digits

Figure 2.3 Interleaving.

Hocquengham (BCH) codes and its non-binary subclass known as Reed–Solomon (RS) codes. The RS codes operate with symbols created from m bits, which are elements in the extension Galois field GF(2^m) of GF(2). Now, an RS code is defined as a block of n m-ary symbols defined over GF(2^m), constructed from k input information symbols by adding an $n - k = 2t$ number of redundant symbols from the same extension field, giving an $n = k + 2t$ symbols code word. From now on we will use the notation for such codes: RS(n, k, t) over GF(2^m) or BCH(n, k, t) over GF(2). Such codes can correct t errors. For details of code construction and detection the reader is referred to the classical literature [54–71].

In the case of burst errors, bit interleaving, illustrated in Figure 2.3, is used. The figure represents a scheme for the interpretation of a (75,25) interleaved code derived from a (15, 5) BCH code. A burst of length $b = 15$ is spread into $t = 3$ error patterns in each of the five code words of the interleaved code.

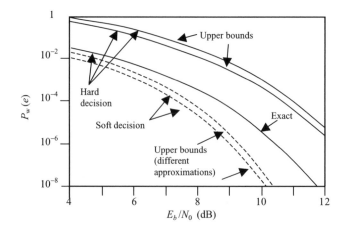

Figure 2.4 Word error probability for the (7, 4) Hamming code: hard decision and soft decision curves. Binary antipodal transmission.

In general, the decoding algorithms can be based on two different options. If a hard decision is performed for each bit separately and the detected word is compared with the possible candidates from the code book, the process is referred to as *hard decision* decoding. For such a decision the probability that a wrong code word is selected is given as [54–71]:

$$P_w(e) \leq (M - 1)\left[\sqrt{4p(1 - p)}\right]^{d_{\min}} \tag{2.2}$$

where $M = 2^k$ indicates the number of code words in the code book, assumed to be equally likely, and $d_{\min}$ is the minimum Hamming distance between the code words. Parameter p represents the bit error rate. The second option is to create a sum of analog signal samples at bit positions and to compare this sum (metrics) with the possible values created from the code book. This is referred to in the literature as *soft decision* decoding. The code word error rate (WER) in this case is given as [54–71]:

$$P_w(e) \leq \frac{(M - 1)}{2} \text{erfc}\left(\sqrt{\frac{d_{\min} R_c E_b}{N_0}}\right) \tag{2.3}$$

The word error rate for Hamming code (7,4) is given in Figure 2.4. One can see that soft decision decoding offers better performance.

If the word error occurs, then for a high signal to noise ratio, the decoder will choose the word with minimum distance from the correct one. It will make other choices with much lower probability. As a consequence, the decoded word will contain $2t + 1$ errors which can be anywhere in the n-bit word. So, the bit error probability can be approximated as:

$$P_b(e) \cong ((2t + 1)/n)P_w(e) \tag{2.4}$$

The bit error rate for some BCH codes is given in Figure 2.5. In general, from the figure, one can see that the longer the code the better the performance. One should be aware that this also means higher complexity. This represents the basis for code adaptability and reconfigurability. Figure 2.6 summarizes these results for a broader class of codes.

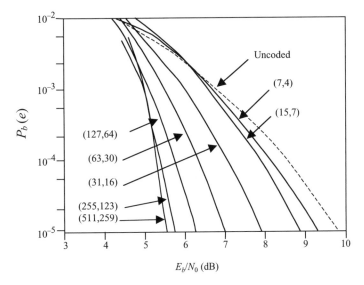

Figure 2.5 Bit error probability curves for some BCH codes, with rate R_c of about 0.5.

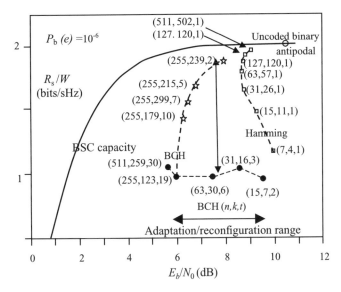

Figure 2.6 Performance of different BCH and Hamming codes. Each code is identified with three numbers: the block length n, the number of information bits k, and the number of corrected errors t.

For a Hamming code and available signal to noise ratio in the range $E_b/N_0 = 9$–10 dB, we can increase the data rate R_s/W approximately by a factor of two if we reconfigure the code from H(7, 4, 1) to H(511, 502, 1). A similar effect can be achieved by reconfiguration of the BCH codes in the range (255, 123, 19) to (255, 239, 2) if the available SNR changes

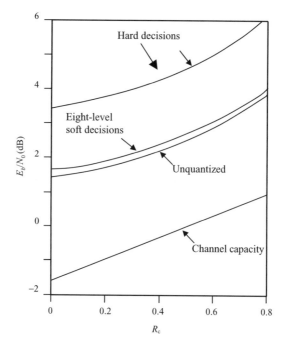

Figure 2.7 Cutoff rate-based bounds on the required signal to noise ratio as a function of the code rate R_c for hard and soft decisions and binary antipodal modulation over the AWGN channel.

in the range 6–8 dB. For a fixed data rate $R_s/W \cong 1$, we can reduce the required SNR for BER $= 10^{-6}$, from 9.5 to 5.5 dB (coding gain) by reconfiguring the BCH code from BCH(15, 7, 2) to BCH(255, 123, 19). The dependence of power consumption on the code reconfiguration range will be discussed in Chapter 10.

To relate these tradeoffs analytically we can use Relation (2.5), which represents a bound on word error probability as a function of coding rate, code length and required signal to noise ratio represented through bit error rate p.

$$P_w(e) \le 2^{-n(R_0 - R_c)}, \quad R_c \le R_0 \tag{2.5}$$

where, for the binary symmetric channel (BSC) with $p(0) = p(1) = 1/2$,

$$R_0 = 1 - \log_2 \left[1 + 2\sqrt{p(1-p)} \right] \tag{2.6}$$

These relations are illustrated in Figure 2.7.

Finally, adaptation/reconfiguration efficiency is defined as:

$$
\begin{aligned}
E_{ff} &= \frac{-\Delta \text{SNR}}{\Delta_r \text{complexity}} \\
&= \frac{\text{coding gain}}{\text{relative increase in complexity}} = \frac{10^{g_c(\text{dB})/10}}{D_r}
\end{aligned} \tag{2.7}
$$

This parameter will be used throughout the book to compare different schemes.

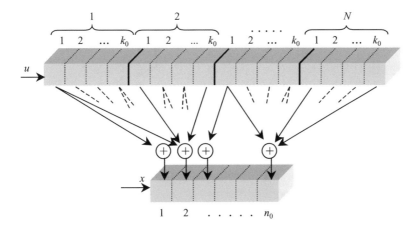

Figure 2.8 Convolutional encoder.

2.2 ADAPTIVE AND RECONFIGURABLE CONVOLUTIONAL CODES

In order to further increase the coding gain, a new class of codes, known as convolutional codes, is used [71–98]. A general block diagram of a convolutional encoder in serial form for an (n_0, k_0) code with constraint length N is given in Figure 2.8. N blocks, k_0 bits each, are used to produce a code word n_0 bits long. For every new input block a new code word is generated.

One can show that the encoder output can be represented as:

$$\mathbf{x} = \mathbf{u}\mathbf{G}_\infty \tag{2.8}$$

where

$$\mathbf{G}_\infty = \begin{bmatrix} G_1 & G_2 & \cdots & G_N & & & \\ & G_1 & G_2 & \cdots & G_N & & \\ & & G_1 & G_2 & \cdots & G_N & \\ & & & G_1 & G_2 & \cdots & G_N \\ & & & & \cdots & \cdots & \cdots & \cdots & \cdots \end{bmatrix} \tag{2.9}$$

Submatrices $\mathbf{G}_i$, containing k_0 rows and n_0 columns, define connectivity of the ith block of the input register with n_0 elements of the output register. In the notation, '1' means a connection and '0' no connection.

We will use the following notation: code rate $R_c = k_0/n_0$, memory $\nu = (N - 1)k_0$ and such a code will be denoted as an (n_0, k_0, N) convolutional code. An example of the encoder for $k_0 = 1$ is shown in Figure 2.9.

For the example in Fig. 2.9(a) we have:

$$\begin{aligned} \mathbf{G}_1 &= \begin{bmatrix} 1 & 1 & 1 \end{bmatrix} \\ \mathbf{G}_2 &= \begin{bmatrix} 0 & 1 & 1 \end{bmatrix} \\ \mathbf{G}_3 &= \begin{bmatrix} 0 & 0 & 1 \end{bmatrix} \end{aligned} \tag{2.10}$$

An equivalent representation is shown in Figure 2.9(b). For this representation we use

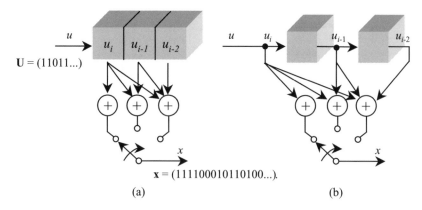

Figure 2.9 Two equivalent schemes for the convolutional encoder of the (3, 1, 3) code.

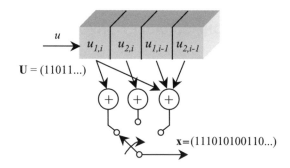

Figure 2.10 Convolutional encoder for the (3, 2, 2) code.

notation that might be more convenient:

$$\mathbf{G} = \begin{bmatrix} g_{1,1} & \cdots & g_{1,n_0} \\ & \vdots & \\ g_{k_0,1} & \cdots & g_{k_0,n_0} \end{bmatrix} \qquad (2.11)$$

with

$$\begin{aligned} g_{1,1} &= (100) \\ g_{1,2} &= (110) \text{ octal numbers}(110 \rightarrow 6) \\ g_{1,3} &= (111) \end{aligned} \qquad (2.12)$$

An example for $k_0 = 2$ is shown in Figure 2.10. In this case we have:

$$\mathbf{G}_1 = \begin{bmatrix} 1 & 0 & 1 \\ 0 & 1 & 0 \end{bmatrix}$$

$$\mathbf{G}_2 = \begin{bmatrix} 0 & 0 & 1 \\ 0 & 0 & 1 \end{bmatrix} \qquad (2.13)$$

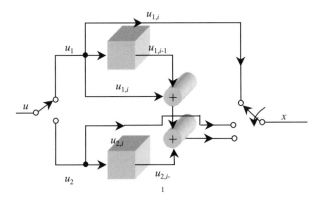

Figure 2.11 Parallel implementation of the same convolutional encoder shown in Figure 2.10.

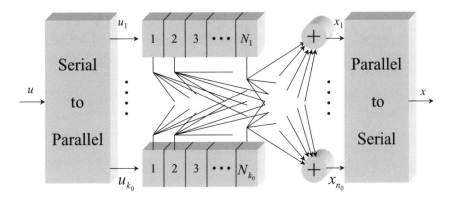

Figure 2.12 General block diagram of a convolutional encoder in parallel form for an (n_0, k_0, N) code.

One can show that the same encoder can be implemented in parallel form, as shown in Figure 2.11. The N general equivalent parallel presentation is shown in Figure 2.12, and two more examples are shown in Figures 2.13 and 2.14.

For the encoder from Figure 2.9 one can verify, and then generalize for any encoder with arbitrary n_0 and $k_0 = 1$, that, for each input digit u_i, n_0 digits will be generated in accordance with:

$$(x_{i1}, x_{i2}, x_{i3}, \ldots, x_{in_0}) = u_i \mathbf{G}_1 + u_{i-1} \mathbf{G}_2 + u_{i-3} \mathbf{G}_3 + \cdots + u_{i-N+1} \mathbf{G}_N$$

$$= \sum_{k=1}^{N} u_{i-N+1} \mathbf{G}_N \tag{2.14}$$

This relation has the form of convolution, hence the name *convolutional coding*.

For an efficient insight into the operation of the encoder, a state diagram, illustrated in Figure 2.15, is used.

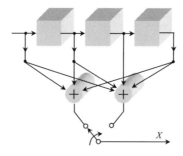

Figure 2.13 Encoder for the (2, 1, 4) convolutional code.

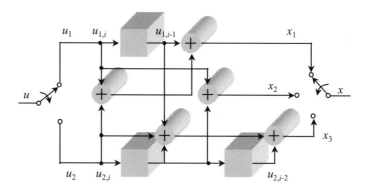

Figure 2.14 Encoder for the (3, 2, 2) convolutional code.

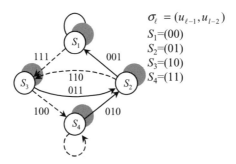

$$\sigma_\ell = (u_{l-1}, u_{l-2})$$
$$S_1 = (00)$$
$$S_2 = (01)$$
$$S_3 = (10)$$
$$S_4 = (11)$$

Figure 2.15 State diagram for the (3, 1, 3) convolutional code.

The (3, 1, 3) encoder, from Figure 2.9, has memory $v = N - 1 = 2$. We define the state σ_l of the encoder at discrete time l as the content of its memory at the same time, $\sigma_l \triangleq (u_{l-1}, u_{l-2})$. There are $N_\sigma = 2^v = 4$ possible states. That is, 00, 01, 10 and 11.

A solid arrow (edge) on Figure 2.15, represents a transition between two states forced by the input digit '0,' whereas a dashed arrow (edge) represents a transition forced by the input digit '1.' The label on each arrow represents the output digits corresponding to that

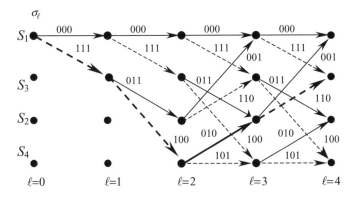

Figure 2.16 Trellis diagram for the (3, 1, 3) convolutional code. The boldface path corresponds to the input sequence 1101.

transition. As an example we have:

$$\begin{aligned}
\mathbf{u} &= (\quad\quad 1 \quad\quad 1 \quad\quad 0 \quad\quad 1 \quad\quad 1 \ldots) \\
path\ & S_1 \rightarrow S_3 \rightarrow S_4 \rightarrow S_2 \rightarrow S_3 \rightarrow S_4 \ldots \\
\mathbf{x} &= (\quad\quad 111 \quad 100 \quad 010 \quad 110 \quad 100 \ \ldots \ )
\end{aligned} \tag{2.15}$$

The concept of a state diagram can be applied to any (n_0, k_0, N) code with memory ν. The number of states is $N_\sigma = 2^\nu$. There are 2^{k_0} edges entering each state and 2^{k_0} edges leaving each state. The labels on each edge are sequences of length n_0. The equivalent representations using a trellis diagram or tree diagram are shown in Figure 2.16 and Figure 2.17 respectively.

An appropriate measure of the maximum likelihood decoder complexity for a convolutional code is the number of visited edges per decoded bit. Now, a rate k_0/n_0 code has 2^{k_0} edges leaving and entering each trellis state and a number of states $N_\sigma = 2^\nu$, where ν is the memory of the encoder. Thus, each trellis section, corresponding to k_0 input bits, has a total number of edges equal to $2^{k_0+\nu}$. As a consequence, an (n_0, k_0, N) code has a decoding complexity:

$$D = \frac{2^{k_0+\nu}}{k_0} \tag{2.16}$$

Reconfiguration from code 1 to code 2 can offer the coding gain g_{12}. This process will, in general, increase the complexity from D_1 to D_2 resulting in:

$$\begin{aligned}
D_r = D_2/D_1 &= \frac{k_{01}}{k_{02}} 2^{(k_{02}-k_{01})+(\nu_2-\nu_1)} \\
&= \frac{k_{01}}{k_{02}} 2^{\Delta k_0+\Delta \nu} = \frac{k_{01}}{k_{02}} 2^{\Delta(k_0+\nu)} = \frac{k_{01}}{k_{02}} 2^{\Delta(k_0 N)}
\end{aligned} \tag{2.17}$$

where Δ should be used as an operator. Equation (2.7) defining the reconfiguration efficiency now becomes:

$$E_{ff} = 10^{g_{12}/10} \frac{k_{02}}{k_{01}} 2^{-\Delta(k_0+\nu)} \tag{2.18}$$

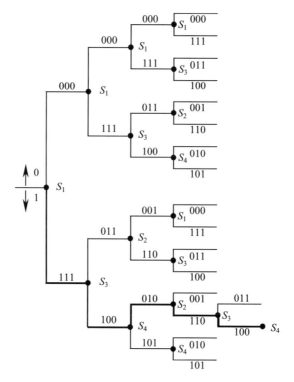

Figure 2.17 Tree diagram for the $(3, 1, 3)$ convolutional code. The boldface path corresponds to the input sequence 11011.

The maximum value of D_r will be referred to as the *reconfiguration range* and the maximum value of the gain g_{12} as the *adaptation range*.

2.2.1 Punctured convolutional codes /code reconfigurability

The increase of complexity inherent in passing from rate $1/n_0$ to rate k_0/n_0 codes can be mitigated using so-called *punctured convolutional codes*. A rate k_0/n_0 punctured convolutional code can be obtained by starting from a rate $1/n_0$ and deleting parity check symbols. An example is given in Figure 2.18.

Suppose now, that for every four parity check digits generated by the encoder, one (the last) is punctured, i.e. not transmitted. In this case, for every two input bits, three bits are generated by the encoder, thus producing a rate 2/3 code. The equivalent representation of the encoder is shown in Figure 2.19.

For more details, see [71, 99–101]

From the previous example, we can derive the conclusion that a rate k_0/n_0 convolutional code can be obtained considering k_0 trellis sections of a rate $1/2$ mother code. Measuring the decoding complexity as before for the punctured code we have:

$$D_{\text{punc}} = \frac{k_0 2^{\nu+1}}{k_0}$$

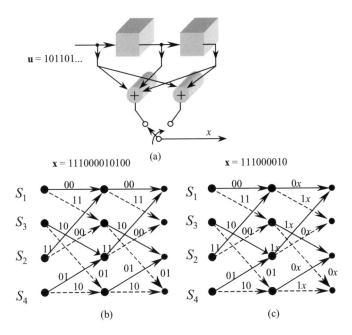

Figure 2.18 Encoder (a) and trellis (b) for a (2, 1, 3) convolutional code. The trellis (c) refers to the rate 2/3 punctured code.

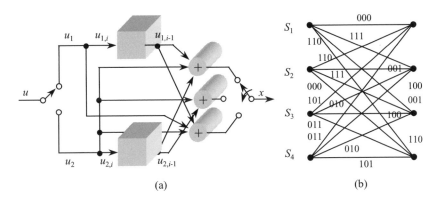

Figure 2.19 Encoder (a) and trellis (b) for the (3, 2, 3) convolutional code equivalent to the rate 2/3 punctured code described.

so that the ratio between the case of the unpunctured to the punctured solution yields:

$$\Delta D_r = \frac{D}{D_{\text{punc}}} = \frac{2^{k_0}}{2k_0}$$

which shows that, for $k_0 > 2$, there is an increasing complexity reduction. Unfortunately, in general, this would also reduce coding gain with respect to the mother code so that the relative increase in the reconfiguration efficiency can be represented as:

$$\Delta E_{ff} = \Delta D_r 10^{\Delta g_{12}/10} \tag{2.19}$$

where Δg_{12} is negative.

2.2.2 Maximum likelihood decoding/Viterbi algorithm

The maximum likelihood (ML) decoder will select the code word from the trellis, like the one in Figure 2.16, whose distance from the received sequence is minimal. Theoretically, the decoder must find the path through the trellis for which:

$$U(\sigma_{K-1}) \triangleq \max_r U^{(r)}(\sigma_{K-1}) \triangleq \max_r P(\mathbf{y} \mid \mathbf{x}^{(r)})$$

$$\equiv \max_r \left[\ln \prod_{l=0}^{K-1} P(\mathbf{y}_l \mid \mathbf{x}_l^{(r)}) \right] = \max_r \left[\sum_{l=0}^{K-1} \ln P(\mathbf{y}_l \mid \mathbf{x}_l^{(r)}) \right] \tag{2.20}$$

where $\mathbf{x}_l$ and $\mathbf{y}_l$ are n_0 binary transmitted and received digits respectively between discrete times l and $l+1$. The branch metric

$$V_l^{(r)}(\sigma_{l-1}, \sigma_l) \triangleq \ln P(\mathbf{y}_l \mid \mathbf{x}_l^{(r)}) \tag{2.21}$$

for antipodal modulation (with transmitted and received energy E) and assuming that the lth branch of the rth path has been transmitted, is:

$$y_{jl} = \sqrt{E}(2x_{jl}^{(r)} - 1) + v_j$$

So, we have:

$$P(\mathbf{y}_l \mid \mathbf{x}_l^{(r)}) = \prod_{j=1}^{n_0} P(y_{jl} \mid x_{jl}^{(r)})$$

$$= \prod_{j=1}^{n_0} \frac{1}{\sqrt{\pi N_0}} \exp\left\{ -\frac{\left[y_{jl} - \sqrt{E}(2x_{jl}^{(r)} - 1) \right]^2}{N_0} \right\} \tag{2.22}$$

$$V_l^{(r)}(\sigma_{l-1}, \sigma_l) = \sum_{j=1}^{n_0} y_{jl}(2x_{jl}^{(r)} - 1) \tag{2.23}$$

In order to reduce the number of trajectories for which the metric is calculated, the Viterbi algorithm accumulates the result only for those branches with the highest metric (survivors) that have the best chances to be selected at the end as the final choice [54–71]. A maximum *a posteriori* probability (MAP) detector is defined in Appendix 2.1.

In the remainder of this section we provide several examples of the performance results.

Figure 2.20 demonstrates that increasing the decoding delay (length of the observed trajectories) over 30 bit intervals cannot reduce BER significantly. In other words, for the given code, only trajectories up to 30 bits should be observed. From Figure 2.21 one can see that soft decision decoding provide a roughly 3 dB coding gain compared with hard decision decoding. Figures 2.22 and 2.23 present BER performance for coding rates 1/2 and 1/3 respectively. For these examples, $k_0 = 1$ and reconfiguration efficiency, defined by

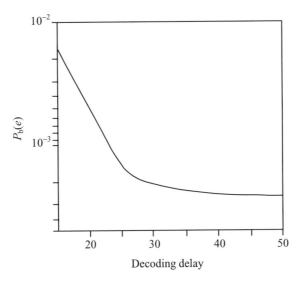

Figure 2.20 Simulated bit error probability versus the decoding delay for the decoding of the rate 1/2 (2, 1, 7) convolutional code. The signal to noise ratio E/N_0 is 3 dB.

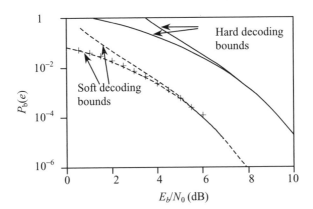

Figure 2.21 Performance bounds with hard and soft decision maximum likelihood decoding for the (3, 1, 3) convolutional code. The curve with '+' refers to simulation results obtained with the soft decision Viterbi algorithm.

Equation (2.18), becomes:

$$E_{ff} = 10^{g_{12}/10} 2^{-\Delta(1+(N-1))} = 10^{g_{12}/10} 2^{-\Delta N} \tag{2.24}$$

As an example, one can see from Figure 2.22 that for BER $= 10^{-6}$, the coding gain $g_{12} \cong 6.7 - 4 = 2.7$ dB can be achieved by reconfiguration of the code from $N = 3$ to $N = 9$. So the reconfiguration efficiency is $E_{ff}(R_c = 1/2) = 10^{0.27} 2^{-6}$. On the other hand, from Figure 2.23, one can see that approximately the same coding gain can be achieved by reconfiguration of the rate $R_c = 1/3$ code from $N = 3$ to $N = 8$, giving

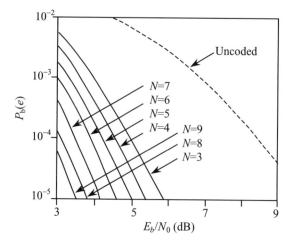

Figure 2.22 Upper bounds to the soft ML decoding bit error probability of different convolutional codes of rate $1/2$.

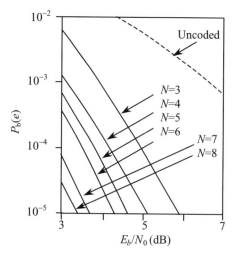

Figure 2.23 Upper bounds to the soft ML decoding bit error probability of different convolutional codes of rate $1/3$.

$E_{ff}(R_c = 1/3) = 10^{0.27}2^{-5}$. In other words, we find that the efficiency of the $R_c = 1/3$ code is higher by a factor of two.

Achievable coding gains for different codes are shown in Table 2.2. These results can be used for calculation of reconfiguration efficiency for different types of code.

2.2.3 Systematic recursive convolutional codes

Systematic convolutional codes generated by feed-forward encoders yield, in general, lower free distances than non-systematic codes. In this section, we will show how to derive

Table 2.2   Achievable coding gains with some convolutional codes of rate R_c and constraint length N at different values of the bit error probability. Eight-level quantization soft decision Viterbi decoding is used. The last line gives the asymptotic upper bound $10 \log_{10} d_f R_c$

E_b/N_0 for uncoded transmission (dB)	$P_b(e)$	$R_c = 1/3$ N		$R_c = 1/2$ N			$R_c = 2/3$ N		$R_c = 3/4$ N	
		7	8	5	6	7	6	8	6	9
6.8	10^{-3}	4.2	4.4	3.3	3.5	3.8	2.8	3.1	2.6	2.6
9.6	10^{-5}	5.7	5.9	4.3	4.6	5.1	4.2	4.6	3.6	4.2
11.3	10^{-7}	6.2	6.5	4.9	5.3	5.8	4.7	5.2	3.9	4.8
$\to \infty$	$\to 0$	7.0	7.3	5.4	6.0	7.0	5.2	6.7	4.8	5.7

a systematic encoder from every rate $1/n_0$ non-systematic encoder, which generates a systematic code with the same *weight enumerating function* as the non-systematic one which is relevant for the free distance [71–98]. The systematic codes are used for turbo code construction, to be discussed later.

Consider, for simplicity, a rate $1/2$ feed-forward encoder characterized by the two generators (in polynomial form) $g_{1,1}(Z)$ and $g_{1,2}(Z)$. Using the power series $u(Z)$ to denote the input sequence $\mathbf{u}$, and $x_1(Z)$ and $x_2(Z)$ to denote the two sequences $\mathbf{x}_1$ and $\mathbf{x}_2$ forming the code $\mathbf{x}$, we have the relationships:

$$x_1(Z) = u(Z)g_{1,1}(Z)$$
$$x_2(Z) = u(Z)g_{1,2}(Z)$$
(2.25)

To obtain a systematic code we need to have either $x_1(Z) = u(Z)$ or $x_2(Z) = u(Z)$. To obtain the first equality, let us divide both equations by $g_{1,1}(Z)$, so that:

$$\tilde{x}_1(Z) \triangleq \frac{x_1(Z)}{g_{1,1}(Z)} = u(Z)$$
$$\tilde{x}_2(Z) \triangleq \frac{x_2(Z)}{g_{1,1}(Z)} = \frac{u(Z)}{g_{1,1}(Z)}g_{1,2}(Z)$$
(2.26)

Defining now a new input sequence $\tilde{u}(Z)$ as:

$$\tilde{u}(Z) \triangleq \frac{u(Z)}{g_{1,1}(Z)}$$
(2.27)

the relations become:

$$\tilde{x}_1(Z) \triangleq \tilde{u}(Z)g_{1,1}(Z)$$
$$\tilde{x}_2(Z) \triangleq \tilde{u}(Z)g_{1,2}(Z)$$
(2.28)

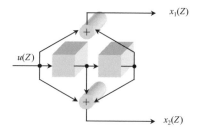

Figure 2.24 Rate 1/2, four-state feed-forward encoder generating the non-systematic code.

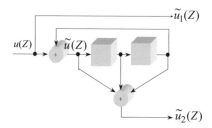

Figure 2.25 Rate 1/2, four-state recursive encoder generating the systematic code.

2.2.3.1 Example

Let us assume that the initial encoder is defined by the polynomials $g_1^A(Z) = 1 + Z^2$, $g_2^A(Z) = 1 + Z + Z^2$. Its equivalent recursive encoder is obtained as explained previously with generators $g_1^B(Z) = 1$, $g_2^B(Z) = (1 + Z + Z^2)/(1 + Z^2)$. These steps are illustrated in Figures. 2.24 and 2.25.

2.3 CONCATENATED CODES WITH INTERLEAVERS

Further improvements in performance can be achieved if two codes are combined (concatenated) to encode the same message, as illustrated in Figure 2.26 (parallel concatenation) and Figure 2.27 (serial concatenation).

For ML decoding we should consider an equivalent trellis characterized by $2^{v_1 + v_2}$ states. The complexity and reconfiguration efficiency are defined by the same equations as before, with $v \Rightarrow v_1 + v_2$. BER curves for the two types of concatenation are shown in Figures 2.28 and 2.29.

2.3.1 The iterative decoding algorithm

Due to a significant increase in the number of states in the equivalent trellis of the concatenated codes, the complexity of the ML decoder might become unacceptable. A suboptimal, iterative algorithm, known as a *turbo decoder*, provides a sufficiently good approximation of the ML decoder and offers performance approaching Shannon's limit (channel capacity). An heuristic explanation for the iterative decoder is based on the assumption that

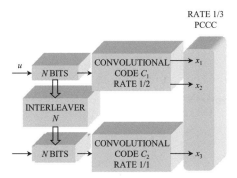

Figure 2.26 Block diagram of a rate $1/3$ parallel concatenated convolutional code (PCCC). Rate $1/1$ code could be obtained using the same rate $1/2$ systematic encoder that generates C_1 and transmitting only the parity check bit.

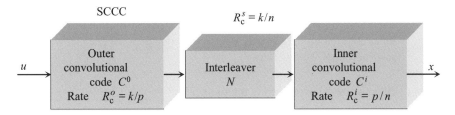

Figure 2.27 Block diagram of a rate k/n serially concatenated convolutional code (SCCC).

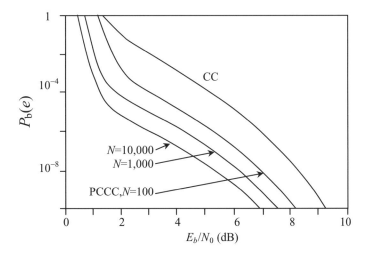

Figure 2.28 Average upper bounds to the bit error probability for a rate $1/3$ PCCC obtained by concatenating two rate $1/2$, four-state convolutional codes (CC), through a uniform interleaver of sizes $N = 100, 1000, 10000$.

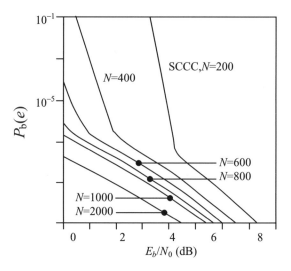

Figure 2.29 Average upper bound to the bit error probability for a rate 1/3 SCCC using as outer code a four-state, rate 1/2, non-recursive convolutional encoder, and as inner code a four-state, rate 2/3, recursive convolutional encoder with uniform interleavers of various sizes.

Equation (2.20) can now be expressed in the form [71]:

$$\hat{u}_k \triangleq \arg \max_i [\text{APP}(k, i)]$$

where the *a posteriori* probability APP (k, i) is defined as:

$$\text{APP}(k, i) \triangleq p(u_k = i, \; \mathbf{y}_1, \mathbf{y}_2)$$
$$= \sum_{\mathbf{u}:u_k=i} p(\mathbf{y}_1 \mid c_1(\mathbf{u})) p(\mathbf{y}_2 \mid c_2(\mathbf{u})) p_a(\mathbf{u}) \tag{2.29}$$

$$p(\mathbf{y}_1 \mid c_1(\mathbf{u})) = \prod_{j=1}^{N_1} p(y_{1j} \mid c_{1j}(\mathbf{u}))$$

$$p(\mathbf{y}_2 \mid c_2(\mathbf{u})) = \prod_{m=1}^{N_2} p(y_{2m} \mid c_{2m}(\mathbf{u}))$$

$$p_a(\mathbf{u}) = \prod_{l=1}^{K} p_a(u_l)$$

In other words, it is assumed that the interleaver will randomize the data stream so that the output of the two encoders can be considered independent. Based on this assumption we can also write:

$$p(\mathbf{y}_1 \mid c_1(\mathbf{u})) \cong \prod_l \tilde{P}_{1l}(u_l)$$
$$p(\mathbf{y}_2 \mid c_2(\mathbf{u})) \cong \prod_l \tilde{P}_{2l}(u_l) \tag{2.30}$$

and Equation (2.29) becomes:

$$
\text{APP}(k, i) = \left[\sum_{\mathbf{u}:u_k=i} p(\mathbf{y}_2 \Big| c_2(\mathbf{u})) \prod_{l \neq k} \tilde{P}_{1l}(u_l) p_a(u_l) \right] \tilde{P}_{1k}(i) p_a(i)
$$

$$
\text{APP}(k, i) = \left[\sum_{\mathbf{u}:u_k=i} p(\mathbf{y}_1 \Big| c_1(\mathbf{u})) \prod_{l \neq k} \tilde{P}_{2l}(u_l) p_a(u_l) \right] \tilde{P}_{2k}(i) p_a(i)
$$

(2.31)

The solution to Equation (2.31) can be represented as:

$$
\tilde{P}_{1k}(i) = \sum_{u:u_k=i} p(\mathbf{y}_1 \Big| c_1(\mathbf{u})) \prod_{l \neq k} \tilde{P}_{2l}(u_l) p_a(u_l)
$$

$$
\tilde{P}_{2k}(i) = \sum_{u:u_k=i} p(\mathbf{y}_2 \Big| c_2(\mathbf{u})) \prod_{l \neq k} \tilde{P}_{1l}(u_l) p_a(u_l)
$$

(2.32)

Finally, Equation (2.29) becomes:

$$
\text{APP}(k, i) = \tilde{P}_{1k}(i) \tilde{P}_{2k}(i) p_a(u_l)
$$

(2.33)

The maximization of Equation (2.33) can be performed in a number of iterations as:

$$
\tilde{P}_{2k}^{(0)} = 1, \quad k = 1, \dots, K
$$

$$\vdots$$

$$
\tilde{P}_{1k}^{(m)} = \sum_{\mathbf{u}:u_k=i} p(\mathbf{y}_1 \Big| c_1(\mathbf{u})) \prod_{l \neq k} \tilde{P}_{2l}^{(m-1)}(u_l) p_a(u_l), \quad k = 1, \dots, K
$$

(2.34)

$$
\tilde{P}_{2k}^{(0)} = \sum_{\mathbf{u}:u_k=i} p(\mathbf{y}_2 \Big| c_2(\mathbf{u})) \prod_{l \neq k} \tilde{P}_{1l}^{(m)}(u_l) p_a(u_l), \quad k = 1, \dots, K
$$

For binary signaling, $u_k = 0, 1$ and the signal probabilities can be replaced by using log likelihood ratios (LLR) defined as:

$$
\lambda_k(\text{APP}) \equiv \log \frac{\displaystyle\sum_{\mathbf{u}:u_k=0} p(\mathbf{y}_1 \mid c_1(\mathbf{u})) p(\mathbf{y}_2 \mid c_2(\mathbf{u})) p_a(\mathbf{u})}{\displaystyle\sum_{\mathbf{u}:u_k=1} p(\mathbf{y}_1 \mid c_1(\mathbf{u})) p(\mathbf{y}_2 \mid c_2(\mathbf{u})) p_a(\mathbf{u})}
$$

(2.35)

By introducing the following notation

$$
\lambda_k \equiv \log \frac{p(y_k \mid 0)}{p(y_k \mid 1)}
$$

$$
\lambda_{1j} \equiv \log \frac{p(y_{1j} \mid 0)}{p(y_{1j} \mid 1)} \qquad j = 1, \dots N_1
$$

$$
\lambda_{2m} \equiv \log \frac{p(y_{2m} \mid 0)}{p(y_{2m} \mid 1)} \qquad m = 1, \dots N_2
$$

$$
\lambda_a \equiv \log \frac{p_a(0)}{p_a(1)}
$$

(2.36)

$$\pi_{1l} \equiv \log \frac{\tilde{P}_{1l}(0)}{\tilde{P}_{1l}(1)}$$

$$\pi_{2l} \equiv \log \frac{\tilde{P}_{2l}(0)}{\tilde{P}_{2l}(1)}$$

and

$$\lambda_k(\text{APP}) \equiv \log \frac{\displaystyle\sum_{\mathbf{u}:u_k=0} \frac{p(\mathbf{y}_1 \mid c_1(\mathbf{u}))}{p(\mathbf{y}_1 \mid 0)} \frac{p(\mathbf{y}_2 \mid c_2(\mathbf{u}))}{p(\mathbf{y}_2 \mid 0)} \frac{p_a(\mathbf{u})}{p_a(\mathbf{0})}}{\displaystyle\sum_{\mathbf{u}:u_k=1} \frac{p(\mathbf{y}_1 \mid c_1(\mathbf{u}))}{p(\mathbf{y}_1 \mid 0)} \frac{p(\mathbf{y}_2 \mid c_2(\mathbf{u}))}{p(\mathbf{y}_2 \mid 0)} \frac{p_a(\mathbf{u})}{p_a(\mathbf{0})}} \qquad (2.37)$$

$$\frac{p(\mathbf{y}_1 \mid c_1(\mathbf{u}))}{p(\mathbf{y}_1 \mid 0)} = \prod_{j=1}^{N_1} \frac{p(y_{1j} \mid c_{1j}(\mathbf{u}))}{p(y_{1j} \mid 0)}$$

$$\frac{p(y_{1j} \mid c_{1j}(\mathbf{u}))}{p(y_{1j} \mid 0)} = \begin{cases} \exp(-\lambda_{1j}) & if\ c_{1j}(\mathbf{u}) = 1 \\ 0 & if\ c_{1j}(\mathbf{u}) = 0 \end{cases} = \exp\left[-\lambda_{1j})c_{1j}(\mathbf{u})\right]$$

$$\frac{p(y_{2m} \mid c_{2m}(\mathbf{u}))}{p(y_{2m} \mid 0)} = \exp\left[-\lambda_{2m})c_{2m}(\mathbf{u})\right]$$

$$\frac{p_a(u_l)}{p_a(0)} = \exp(-u_l\lambda_a)$$

Equation (2.35) becomes:

$$\lambda_k(\text{APP}) = \log\left\{ \sum_{u:k_k=0} \exp - \left[\sum_{j=1}^{N_1} c_{1j}(\mathbf{u})\lambda_{1j} + \sum_{m=1}^{N_2} c_{2m}(\mathbf{u})\lambda_{2m} + \sum_{l=1}^{K} u_l\lambda_a \right] \right\}$$

$$- \log\left\{ \sum_{u:k_k=1} \exp - \left[\sum_{j=1}^{N_1} c_{1j}(\mathbf{u})\lambda_{1j} + \sum_{m=1}^{N_2} c_{2m}(\mathbf{u})\lambda_{2m} + \sum_{l=1}^{K} u_l\lambda_a \right] \right\} \qquad (2.38)$$

and the iterative procedure (2.34) can be represented as:

$$\pi_{2k}^{(0)} = 0$$

$$\vdots$$

$$\pi_{1k}^{(m)} = \log\left\{ \sum_{u:k_k=0} \exp - \left[\sum_{j} c_{1j}(\mathbf{u})\lambda_{1j} + \sum_{l \neq k} u_l\left(\lambda_a + \pi_{2l}^{(m-1)}\right) \right] \right\}$$

$$- \log\left\{ \sum_{u:k_k=1} \exp - \left[\sum_{j} c_{1j}(\mathbf{u})\lambda_{1j} + \sum_{l \neq k} u_l\left(\lambda_a + \pi_{2l}^{(m-1)}\right) \right] \right\} \qquad (2.39)$$

$$\pi_{2k}^{(m)} = \log\left\{ \sum_{u:k_k=0} \exp - \left[\sum_{m} c_{2m}(\mathbf{u})\lambda_{2m} + \sum_{l \neq k} u_l\left(\lambda_a + \pi_{1l}^{(m)}\right) \right] \right\}$$

$$- \log\left\{ \sum_{u:k_k=1} \exp - \left[\sum_{m} c_{2m}(\mathbf{u})\lambda_{2m} + \sum_{l \neq k} u_l\left(\lambda_a + \pi_{1l}^{(m)}\right) \right] \right\}$$

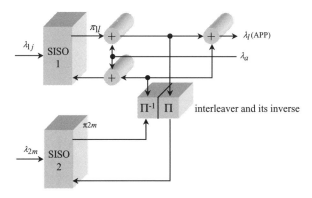

Figure 2.30 Block diagram of the iterative decoding scheme for binary convolutional codes.

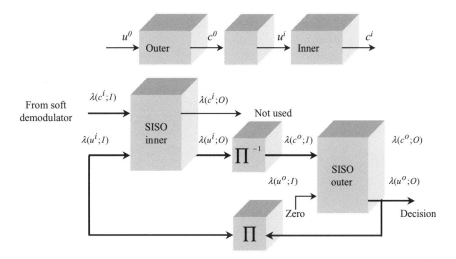

Figure 2.31 Block diagrams of the encoder and iterative decoder for serially concatenated convolutional codes.

The final value of λ_k (APP) is calculated as

$$\lambda_k(\text{APP}) = \pi_{1k} + \pi_{2k} + \lambda_a \qquad (2.40)$$

and the MAP decision is made according to the sign of λ_k. The details of the numerical evaluation of Equation (2.39) are available in standard literature [102–112], the most relevant being the original work by Berrou [113].

These iterations contain implicitly the trellis constraints imposed by the trellis structure. For these purposes forward and backward recursions are used [102–113].

In the block diagrams shown in Figures 2.30 and 2.31, these calculations are performed in soft input soft output (SISO) blocks [102–112]. Performance curves are given in Figures 2.32, 2.33 and 2.34. In order to avoid repetition, for details of iterative calculations, defined by Equation (2.39), the reader is referred to the classical literature [102–112]. Instead

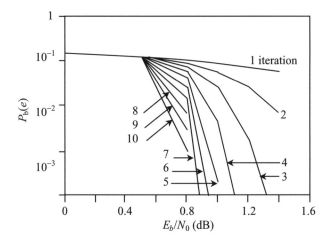

Figure 2.32 Bit error probability obtained by simulating the iterative decoding algorithm. Rate 1/2 PCCC based on 16-state rate 2/3 and 2/1 CCs, and interleaver with size $N = 8920$.

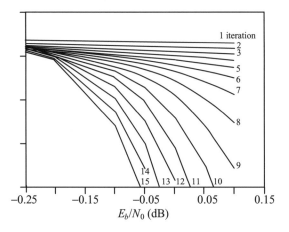

Figure 2.33 Simulated bit error probability performance of a rate 1/4 serially concatenated code obtained with two eight-state constituent codes and an interleaver yielding an input decoding delay equal to 16 384.

of going into these details we will get back, once again, to the issue of reconfiguration efficiency. As already mentioned, for two concatenated codes with $k_{01} = k_{02}$ and an ML decoder, Equation (2.18) becomes:

$$E_{ff} = 10^{g_{12}/10} 2^{-\Delta(k_0 v)}$$

So, for the two codes with $v_1 = v_2 = v$, the reconfiguration from a single convolutional

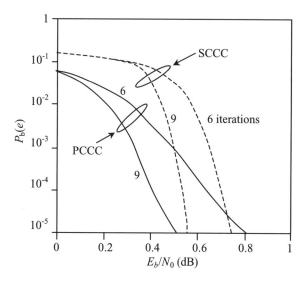

Figure 2.34 Comparison of rate 1/3 PCCC and SCCC. The PCCC is obtained concatenating two equal rate 1/2 four-state codes, whereas the SCCC concatenates two four-state rate 1/2 and rate 2/3 codes. The curves refer to six and nine iterations of the decoding algorithm and to an equal input decoding delay of 16 384.

code (CC) to parallel concatenated CC (PCCC) with $k_{01} = k_{02}$, gives:

$$E_{ff} = 10^{g_{12}/10} 2^{-v}$$

If a turbo decoder is used, the relative complexity changes and Equation (2.19) gives:

$$\Delta E_{ff} = 10^{-|g_{12}|/10} \frac{2^v}{2I}$$

where I is the number of iterations.

2.4 ADAPTIVE CODING, PRACTICE AND PROSPECTS

Efficient error control on time varying channels can be performed by implementing an adaptive control system where the optimum code is selected according to the actual channel conditions.

There is a number of burst error correcting codes that could be used in these adaptive schemes. Three major classes of burst error correcting code are binary fire block codes, binary Iwadare–Massey convolutional codes [114], and non-binary Reed–Solomon block codes. In practical communication systems these are decoded by hard decision decoding methods. Performance evaluation based on experimental data from satellite mobile communication channels [115] shows that the convolutional codes with the soft decision decoding Viterbi algorithm are superior to all of the above burst error correcting codes.

Superior error probability performance and availability of a wide range of code rates without changing the basic coded structure, motivate the use of punctured convolutional codes [80–83] with the soft decision Viterbi decoding algorithm in the proposed adaptive scheme. To obtain the full benefit of the Viterbi algorithm on bursty channels, ideal interleaving is assumed.

An adaptive coding scheme using incremental redundancy in a hybrid automatic repeat request (ARQ) error control system is reported by Wu *et al.* [116]. The channel model used is BSC with time variable bit error probability. The system state is chosen according to the channel bit error rate. The error correction is performed by shortened cyclic codes with variable degrees of shortening. When the channel bit error rate increases, the system generates additional parity bits for error correction.

An FEC adaptive scheme for matching code to the prevailing channel conditions was reported by Chase [117]. The method is based on convolutional codes with Viterbi decoding and consists of combining noisy packets to obtain a packet with a code rate low enough (less than $1/2$) to achieve the specified error rate. Other schemes that use a form of adaptive decoding are reported in [118–123]. Hybrid ARQ schemes based on convolutional codes with sequential decoding on a memoryless channel were reported by Drukarev and Costello [124,125] while a Type-II hybrid ARQ scheme formed by concatenation of convolutional codes with block codes was evaluated on a channel represented by two states [126].

In order to implement the adaptive coding scheme it is necessary again to use a return channel. The channel state estimator (CSE) determines the current channel state, based on counting the number of erroneous blocks. Once the channel state has been estimated, a decision is made by the *'reconfiguration block'* whether to change the code, and the corresponding messages are sent to the encoder and locally to the decoder.

In FEC schemes only error correction is performed, while in hybrid ARQ schemes retransmission of erroneous blocks is requested whenever the decoded data is labeled as unreliable.

The adaptive error protection is obtained by changing the code rates. For practical purposes it is desirable to modify the code rates without changing the basic structure of the encoder and decoder. Punctured convolutional codes are ideally suited for this application. They allow almost continuous change of the code rates, while decoding is done by the same decoder.

The encoded digits at the output of the encoder are periodically deleted according to the deleting map specified for each code. Changing the number of deleted digits varies the code rate. At the receiver end, the Viterbi decoder operates on the trellis of the parent code and uses the same deleting map as in the encoder in computing path metrics [100].

The Viterbi algorithm based on this metric is a maximum likelihood algorithm on channels with Gaussian noise, since on these channels the most probable errors occur between signals that are closest together in terms of squared Euclidean distance. However, this metric is not optimal for non-Gaussian channels. The Viterbi algorithm allows use of channel state information for fading channels [127].

However, a disadvantage of punctured convolutional codes compared to other convolutional codes with the same rate and memory order is that error paths are typically long. This requires quite long decision depths of the Viterbi decoder.

A scheme with ARQ rate compatible convolutional codes was reported by Hagenauer [128]. In this scheme, rate compatible codes are applied. The rate compatibility constraint increases the system throughput, since in transition from a higher to a lower rate code, only incremental redundancy digits are retransmitted. The error detection is performed by a cyclic redundancy check which introduces additional redundancy.

APPENDIX 2.1 MAXIMUM *A POSTERIORI* DETECTION

The BCJR algorithm

The maximum likelihood decoder defined by Equation (2.20) minimizes the probability that the whole detected sequence is in error. In an alternative concept presented in this appendix we will be interested in minimizing the symbol error probability. The starting point is the average symbol *a posteriori* probability [71, 57–70]:

$$\text{APP} = E\{p(x_k/\mathbf{y})\} \tag{A2.1}$$

that should be maximized. In other words, the detector will decide in favor of the symbol for which the probability of correct detection is maximized. For simplicity $\mathbf{y}$ will now be replaced by y. For binary transmission, this means finding out which one of the two probabilities $P(x_k = 0 \mid y)$ and $P(x_k = 1 \mid y)$ is larger. For this we have to compare:

$$\Lambda_k = \frac{P(x_k = 1 \mid y)}{P(x_k = 0 \mid y)}$$

with a unit threshold. The transmitted symbol x_k is associated with one or more branches of the trellis stage at time k, and each one of these branches can be characterized by the pair of states, say (δ_k, δ_{k+1}), that it joins. Thus, we can write:

$$\Lambda_k = \frac{\displaystyle\sum_{(\delta_k, \delta_{k+1}):x_k=1} P(y, \delta_k, \delta_{k+1})}{\displaystyle\sum_{(\delta_k, \delta_{k+1}):x_k=0} P(y, \delta_k, \delta_{k+1})}$$

where the two summations are over those pairs of states for which $x_k = 1$ and $x_k = 0$, respectively, and the conditional probabilities from the previous equation are replaced by joint probabilities after using the Bayesian rule and canceling out the pdf of y, common to numerator and denominator.

Next, we need to compute the pdf $P(y, \delta_k, \delta_{k+1})$. By defining y_k^-, as components of the received vector before time k, and y_k^+, as components of the received vector after time k, we can write:

$$y = (y_k^-, y_k, y_k^+)$$

which results in:

$$
\begin{aligned}
p(y, \delta_k, \delta_{k+1}) &= p(y_k^-, y_k, y_k^+, \delta_k, \delta_{k+1}) \\
&= p(y_k^-, y_k, \delta_k, \delta_{k+1}) p(y_k^+ \mid y_k^-, y_k, \delta_k, \delta_{k+1}) \\
&= p(y_k^-, \delta_k) p(y_k, \delta_{k+1} \mid y_k^-, \delta_k) p(y_k^+ \mid y_k^-, y_k, \delta_k, \delta_{k+1})
\end{aligned}
$$

Due to the dependences among observed variables and trellis states, reflected by the trellis structure or, equivalently, by the Markov chain property of the trellis states, y_k^+ depends on

δ_k, δ_{k+1}, y_k^-, and y_k only through δ_{k+1}, and, similarly, the pair y_k, δ_{k+1} depends on δ_k, y_k^-, only through δ_k. Thus, by defining the functions:

$$\alpha_k(\delta_k) \equiv p(y_k^-, \delta_k)$$

$$\beta_{k+1}(\delta_{k+1}) \equiv p(y_k^+ \mid \delta_{k+1})$$

$$\gamma_{k,k+1}(\delta_k, \delta_{k+1}) \equiv p(y_k, \delta_{k+1} \mid \delta_k) = p(y_k \mid \delta_k, \delta_{k+1}) p(\delta_{k+1} \mid \delta_k)$$

we may write

$$p(y, \delta_k, \delta_{k+1}) = \alpha_k(\delta_k) \gamma_{k,k+1}(\delta_k, \delta_{k+1}) \beta_{k+1}(\delta_{k+1})$$

So, the *a posteriori* probability ratio can be rewritten in the form:

$$\Lambda_k = \frac{\displaystyle\sum_{\delta_k,\delta_{k+1}:x_k=1} \alpha_k(\delta_k) \gamma_{k,k+1}(\delta_k, \delta_{k+1}) \beta_{k+1}(\delta_{k+1})}{\displaystyle\sum_{\delta_k,\delta_{k+1}:x_k=0} \alpha_k(\delta_k) \gamma_{k,k+1}(\delta_k, \delta_{k+1}) \beta_{k+1}(\delta_{k+1})} \tag{A2.2}$$

Finally, we now describe how the functions $\alpha_k(\delta_k)$ and $\beta_{k+1}(\delta_{k+1})$ can be evaluated recursively. We represent the forward recursion as:

$$\begin{aligned}
\alpha_{k+1}(\delta_{k+1}) &= p(y_{k+1}^-, \delta_{k+1}) \\
&= p(y_k^-, y_k, \delta_{k+1}) \\
&= \sum_{\delta_k} p(y_k^-, y_k, \delta_k, \delta_{k+1}) \\
&= \sum_{\delta_k} p(y_k^-, \delta_k) p(y_k, \delta_{k+1} \mid \delta_k) \\
&= \sum_{\delta_k} \alpha_k(\delta_k) \gamma_{k,k+1}(\delta_k, \delta_{k+1})
\end{aligned}$$

with the initial condition $\alpha_0(s_1) = 1$ (s_1 denotes the initial state of the trellis) and the backward recursion as:

$$\begin{aligned}
\beta_k(\delta_k) &= p(y_{k-1}^+ \mid \delta_k) \\
&= \sum_{\delta_{k+1}} p(y_k, y_k^+, \delta_{k+1} \mid \delta_k) \\
&= \sum_{\delta_{k+1}} p(y_k, \delta_{k+1} \mid \delta_k) p(y_k^+ \mid \delta_{k+1}) \\
&= \sum_{\delta_{k+1}} \gamma_{k,k+1}(\delta_k, \delta_{k+1}) \beta_{k+1}(\delta_{k+1})
\end{aligned}$$

with the final value $\beta_K(s_K) = 1$. The combination of the latter two recursions with Equation (A2.2) forms the BCJR algorithm, named after the authors who first derived it, Bahl, Cocke, Jelinek and Raviv [80]. Roughly speaking, we can state that the complexity of the BCJR algorithm is about three times that of the Viterbi algorithm.

3

Adaptive and Reconfigurable Modulation

3.1 CODED MODULATION

In general we can use an $M = 2^b$ point constellation to transmit b bits of information. An example for $b = 2$ is shown in Figure 3.1(a). For this example the output symbol rate is $R_s = R_b/2$. If we use coding, for example a rate $2/3$ convolutional encoder, and the same constellation as shown in Figure 3.1(b), the output symbol rate and the bandwidth required will be now higher, $R_s = (3/4)R_b$.

The third option is shown in Figure 3.1(c). Instead of 4PSK, 8PSK (8 points constellation) is used to transmit the encoded bits and the output symbol rate now remains the same. Because there are only $2^2 = 4$ possible code words and $2^3 = 8$ available constellation points, a proper choice of constellation points used in adjacent symbol intervals provides a way to encode the signal. This subset of signal trajectories, generated in K symbol intervals will again be referred to as a trellis in Euclidean space and the modulation is referred to as *Trellis Coded Modulation* (TCM) [129–166].

The above example illustrates a need to further elaborate the efficiency of reconfiguration in such a way as to explicitly incorporate constraints imposed by the limited available bandwidth. For these purposes, let us represent the coding gain g_{12} as the gain in energy per bit per noise density:

$$g_{12} = \Delta E/N_0 = \Delta(PT)/N_0$$

If there is no bandwidth limitation, the coding gain may be used in a number of ways.

1. Operate with reduced power and save the battery life.

2. Keep the same transmit power and data rate and increase the coverage of the network.

3. Increase bit rate (reduce the bit interval).

Advanced Wireless Communications. S. Glisic
© 2004 John Wiley & Sons, Ltd. ISBN: 0-470-86776-0

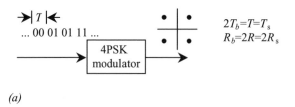

(a)

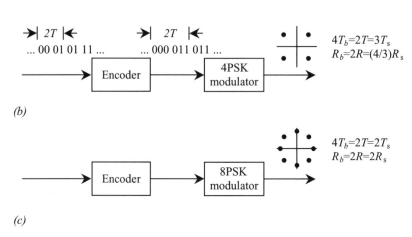

(b)

(c)

Figure 3.1 Three digital communication schemes transmitting two bits every T seconds. (a) Uncoded transmission with 4PSK; (b) 4PSK with a rate 2/3 encoder and bandwidth expansion; (c) 8PSK with a rate 2/3 encoder and no bandwidth expansion.

If the bandwidth is fixed and the coding gain is not available, we may have to reduce the data rate in order to maintain the required E_b/N_0 for a specified QoS. This suggests that the reconfiguration efficiency, defined by Equation (2.18), be further modified as follows:

$$E_{ff} = \frac{10^{g_{12}(\mathrm{dB})/10} b_r}{D_r} = \left(\frac{k_{02}}{k_{01}}\right)\frac{10^{g_{12}(\mathrm{dB})/10}}{D_r} \qquad (3.1)$$

where $b_r = k_{02}/k_{01}$ is the relative change in the number of bits per symbol for the same symbol period T. One should notice, that in Chapter 2, coding gain was defined by taking into account this effect through the coding rate R_c. In the next chapter, Equation (3.1) will be further modified to replace b_r by the relative change in the system capacity. This way we remove the portion of the gain due to bit rate reduction and take into account only those contributions due to increased efficiency of the decoding algorithm.

3.1.1 Euclidean distance

The demodulator will decide in favor of the trajectory in the trellis which is closest to the received signal (*minimum* distance from the received trajectory). This is characterized by the Euclidean distance, defined as:

$$\delta^2 = \sum_{i=0}^{K-1} |\mathbf{r}_i - \mathbf{x}_i|^2 \tag{3.2}$$

where $\mathbf{r}_i$ is the received signal and $\mathbf{x}_i$ the possible transmitted signal. In other words, the Euclidean distance is minimized by taking $\mathbf{x}_i = \hat{\mathbf{x}}_i, i = 0, \ldots, K - 1$, if the received sequence is closer to $\hat{\mathbf{x}}_i, \ldots, \hat{\mathbf{x}}_{K-1}$ than to any other allowable signal sequence.

By increasing the constellation size M' to $M > M'$, and selecting M'^K sequences as a subset of S^K, we can have sequences which are less tightly packed and hence increase the minimum distance among them. We obtain a minimum distance, δ_{free}, between any two sequences, which turns out to be greater than the minimum distance, δ_{min}, between signals in S'. Hence, use of maximum likelihood sequence detection will yield a 'distance gain' of a factor of $\delta_{\text{free}}^2/\delta_{\text{min}}^2$.

The free distance of a TCM scheme is the minimum Euclidean distance between two paths forming an error event.

3.1.2 Examples of TCM schemes

In order to analyze Equation (3.1) in more detail, let us assume transmission with two bits per symbol. For such transmission a 4PSK modulation ($M' = 4$) would be enough. We can expand the constellation to $M = 8$, as shown in Figure 3.2, and use the trellis with $S = 2$ states, as shown in Figure 3.3.

The asymptotic coding gain of a TCM scheme is defined as:

$$\gamma = \frac{\delta_{\text{free}}^2(M)/\varepsilon}{\delta_{\text{min}}^2(M')/\varepsilon'}$$

For PSK signals, $M' = 4$ and $\delta_{\text{min}}^2/\varepsilon' = 2$. For a TCM scheme based on the 8PSK constellation whose signals we label $\{0, 1, 2, \ldots, 7\}$, as shown in Figure 3.2, we have:

$$\varepsilon' = \frac{\delta'^2}{4\sin^2 \pi/8}$$

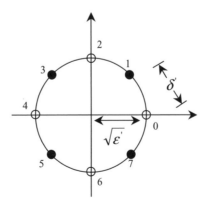

Figure 3.2 $M = 8$ point constellation.

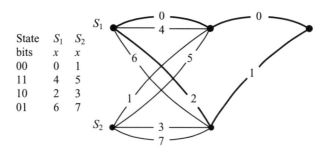

Figure 3.3 A TCM scheme based on a two-state trellis, $M' = 4$ and $M = 8$.

3.1.2.1 Two-state trellis

If the encoder is in state S_1, the subconstellation $\{0, 2, 4, 6\}$ is used. In state S_2, constellation $\{1, 3, 5, 7\}$ is used instead, as shown in Figure 3.3. The free distance of this TCM scheme is the smallest among the distances between signals associated with parallel transitions (error events of length 1) and the distances associated with a pair of paths in the trellis that originate from a common node and merge into a single node at a later time (error events of length greater than 1). The pair of paths yielding the free distance is shown in Figure 3.3. With $\delta(i, j)$ denoting the Euclidean distance between signals i and j, we have the following:

$$\frac{\delta_{free}^2}{\varepsilon} = \frac{1}{\varepsilon}[\delta^2(0, 2) + \delta^2(0, 1)] = 2 + 4\sin^2\frac{\pi}{8} = 2.586$$

The asymptotic coding gain over 4PSK is:

$$\gamma = \frac{2.586}{2} = 1.293 \Rightarrow 1.1 \text{ dB} \tag{3.3}$$

In this case, $\Delta k_0 = 0$ and $\Delta v = 1$ so that reconfiguration efficiency, defined by Equation (3.1), gives:

$$E_{ff} = \left(\frac{k_{02}}{k_{01}}\right)^2 \gamma 2^{-\Delta(k_0+v)} = 1.293/2 = 0.646 \tag{3.4}$$

For a given symbol error probability, the bit error probability (BER) will depend on the mapping of the source bits onto the signals in the modulator's constellation (see Figure 3.3). To minimize the BER, this mapping should be chosen in such a way that, whenever a symbol error occurs, the signal erroneously chosen by the demodulator differs from the transmitted one by the least number of bits. For high signal to noise ratios, most of the errors occur by mistaking a signal for one of its nearest neighbors. So, a reasonable choice is a mapping where neighboring signal points in the constellation correspond to binary sequences that differ in only one digit. This is called *Gray mapping*. In this case, the bit and symbol error probabilities are related as $P_s(e)/b = P_b(e)$. For the evaluation of the symbol error probability the reader is referred to the classical references [167–170].

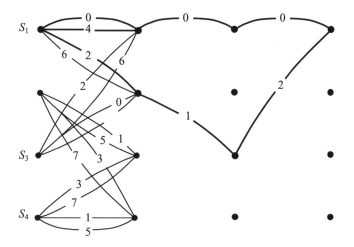

Figure 3.4 A TCM scheme based on a four-state trellis, $M' = 4$ and $M = 8$.

3.1.2.2 Four-state trellis

In this case the trellis is as given in Figure 3.4.

We associate the constellation $\{0, 2, 4, 6\}$ with states S_1 and S_3, and $\{1, 3, 5, 7\}$ with S_2 and S_4. In this case, the error event leading to δ_{free} has length 1 (a parallel transition).

$$\frac{\delta_{\text{free}}^2}{\varepsilon} = \delta^2(0, 4) = 4$$

$$\gamma = \frac{4}{2} = 2 \Rightarrow 3 \text{ dB} \tag{3.5}$$

In this case, $\Delta v = 2$ and Equation (3.4) gives $E_{ff} = 2/4 = 0.5$, which is lower than Equation (3.4). This means that effort invested is larger than the gain obtained.

3.1.2.3 Eight-state trellis

For the case of eight states, the trellis is shown in Figure 3.5. The four symbols associated with the branches emanating from each node are used as node labels. The first symbol in each node label is associated with the uppermost transition from the node, the second symbol with the transition immediately below it, etc.

The coding gain is calculated as:

$$\frac{\delta_{\text{free}}^2}{\varepsilon} = \frac{1}{\varepsilon}[\delta^2(0, 6) + \delta^2(0, 7) + \delta^2(0, 6)] = 2 + 4\sin^2\frac{\pi}{8} + 2 = 4.586$$

$$\gamma = \frac{4.586}{2} = 2.293 \Rightarrow 3.6 \text{ dB} \tag{3.6}$$

In this case, $\Delta v = 3$ and Equation (3.4) gives $E_{ff} = 2.293/8 = 0.286$, which is lower than in the previous case of the four-state trellis. This means again that effort invested is larger than the gain obtained.

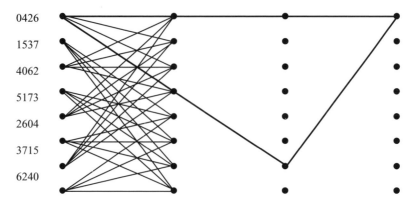

Figure 3.5 A TCM scheme based on an eight-state trellis, $M' = 4$ and $M = 8$.

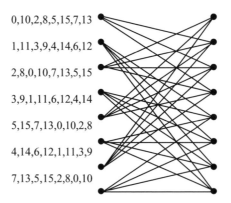

Figure 3.6 A TCM scheme based on an eight-state trellis, $M' = 8$ and $M = 16$.

3.1.2.4 QAM 3 bits per symbol

In this case, the trellis is as given in Figure 3.6 and the signal constellation as in Figure 3.7.

In this case, we have two subsets of points $\{0, 2, 5, 7, 8, 10, 13, 15\}$ and $\{1, 3, 4, 6, 9, 11, 12, 14\}$. For the basic 8QPSK constellation we have $\delta^2_{\min}/\varepsilon' = 0.8$, and coding gain can be represented as:

$$\frac{\delta^2_{\text{free}}}{\varepsilon} = \frac{1}{\varepsilon}[\delta^2(10, 13) + \delta^2(0, 1) + \delta^2(0, 5)] = \frac{1}{\varepsilon}[0.8\varepsilon + 0.4\varepsilon + 0.8\varepsilon] = 2 \tag{3.7}$$

$$\gamma = \frac{2}{0.8} = 2.5 \Rightarrow 3.98 \text{ dB}$$

In this case, $\Delta k_0 = 0$ and $\Delta v = 3$ so that reconfiguration efficiency, defined by Equation (3.1), gives:

$$E_{ff} = 2.5 \times 2^{-(0+3)} = 0.312 \tag{3.8}$$

which is higher than in the previous example. Additional results for free distances for a number of different schemes are given in Figure 3.8. These results can be used directly to evaluate the reconfiguration efficiency.

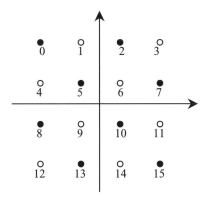

Figure 3.7 The 8 QAM constellation $\{0, 2, 5, 7, 8, 10, 13, 15\}$ and the 16 QAM constellation $\{0, 1, \ldots, 15\}$.

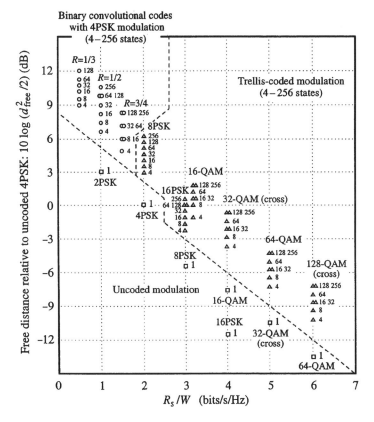

Figure 3.8 Free distance versus bandwidth efficiency of selected TCM schemes based on two-dimensional modulations. (Adapted from [130]) (PSK and QAM © 1987, IEEE).

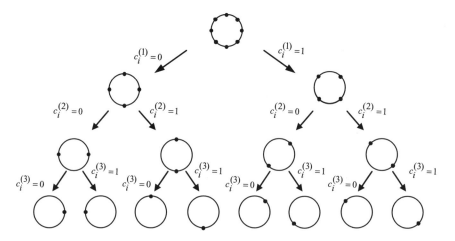

Figure 3.9 Set partition of an 8PSK constellation.

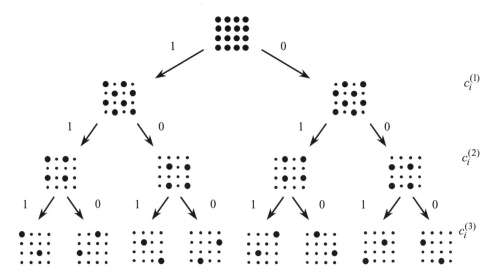

Figure 3.10 Set partition of a 16 QAM constellation.

3.1.3 Set partitioning

The M-ary constellation is successively partitioned into 2, 4, 8,..., subsets with size $M/2$, $M/4$, $M/8$,..., having progressively larger minimum Euclidean distances $\delta_{min}^{(1)}$, $\delta_{min}^{(2)}$, $\delta_{min}^{(3)}$,... as shown in Figure 3.9 and Figure 3.10.

Then, in accordance with Ungerboeck's rules, the following steps are taken:

1. Members of the same partition with the largest distance are assigned to parallel transitions.

2. Members of the next largest partition are assigned to 'adjacent' transitions, i.e. transitions stemming from, or merging into, the same node.

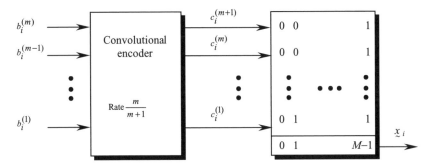

Figure 3.11 Representation of TCM.

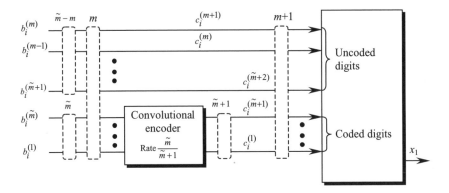

Figure 3.12 A TCM encoder where the bits that are left uncoded are shown explicitly.

3.1.4 Representation of TCM

A TCM encoder can be represented as a convolutional encoder encoding a block of m input bits $\mathbf{b}_i = (b_i^{(1)}, b_i^{(2)}, b_i^{(3)}, \ldots, b_i^{(m)})$ into a block of $m+1$ output bits $\mathbf{c}_i = (c_i^{(1)}, c_i^{(2)}, c_i^{(3)}, \ldots, c_i^{(m)})$, followed by a memoryless mapper into points of the constellation of size $M = 2^{m+1}$ (Figure 3.11). In the case when there are parallel transitions, not all bits are encoded, which is represented explicitly in Figure 3.12.

Uncoded digits cause parallel transitions; a branch in the trellis diagram of the code is now associated with $2^{m-\tilde{m}}$ signals. An example for $m = 2$ and $\tilde{m} = 1$ is shown in Figure 3.13. The trellis nodes are connected by parallel transitions associated with two signals each. The trellis has four states, as does the rate $1/2$ convolutional encoder, and its structure is determined by the latter.

3.1.5 TCM with multidimensional constellation

In general, we can use m channels for transmission and generate an m-dimensional trellis for the overall signal representation.

As an example of a TCM scheme based on multidimensional signals, consider the four-dimensional constellation obtained by pairing ($m = 2$) 4PSK signals. This is denoted as

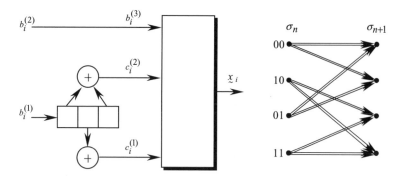

Figure 3.13 A TCM encoder with $m = 2$, $\tilde{m} = 1$ and the corresponding trellis.

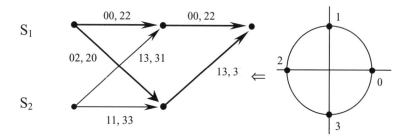

Figure 3.14 A two-state TCM scheme based on a 2×4PSK constellation. The error event providing the free Euclidean distance is also shown.

2×4PSK. With the signal labeling of Figure 3.14, the $4^2 = 16$ four-dimensional signals are:

$$\{00, 01, 02, 03, 10, 11, 12, 13, 20, 21, 22, 23, 30, 31, 32, 33\}$$

This constellation achieves the same minimum squared distance as two-dimensional 4PSK,

$$\delta^2_{\min} = \delta^2(00, 01) = \delta^2(0, 1) = 2$$

The following subconstellation has eight signals and a minimum squared distance four:

$$S = \{00, 02, 11, 13, 20, 22, 31, 33\}$$

With S partitioned into the four subsets:

$$\{00, 22\} \quad \{20, 02\} \quad \{13, 31\} \quad \{11, 33\}$$

the choice of a two-state trellis provides the TCM scheme shown in Figure 3.14. This has a squared free distance of 8.

If m channels are used independently, the overall data rate would be $k_{01} = mk_0$ and the complexity would be m times the complexity of the demodulation per trellis. So the normalized complexity per bit would be:

$$D_1 = \frac{m2^{k_0+v}}{mk_0} = \frac{2^{k_0+v}}{k_0}$$

For m dimensional trellises, only $k_{02} = k_0$ bits are transmitted:

$$D_2 = \frac{m2^{k_0+v}}{k_0} \quad \text{and} \quad D_r = \frac{D_2}{D_1} = m$$

Now we have:

$$E_{ff} = g_{12}\left(\frac{k_0}{mk_0}\right)\left(\frac{1}{m}\right) = \frac{g_{12}}{m^2} \tag{3.9}$$

For the previous example, $g_{12} = \gamma = 8/2 = 4$ and $m^2 = 4$ so that $E_{ff} = 1$.

For this reason, in the next chapter we will discuss multidimensional constellations obtained by using multiple antennas.

3.2 ADAPTIVE CODED MODULATION FOR FADING CHANNELS

In this section we describe the system which uses reconfiguration to improve performance in time varying fading channels [171–194]. Basically, for a better signal to noise ratio $\hat{\gamma}$, estimated at the receiver side, the higher constellation M is used at the receiver, as shown in Figure 3.15.

Let $\overline{\tau_j}$ be the average time that the adaptive modulation scheme continuously uses the constellation M_j. Since the constellation size is adapted to an estimate of the channel fade level (instantaneous signal to noise ratio), several symbol times may be required to obtain a good estimate. In addition, hardware and pulse shaping considerations generally dictate that the constellation size must remain constant over tens to hundreds of symbols. This results in the requirement that $\overline{\tau_j} \gg T \; \forall j$, where T is the symbol time.

Since each constellation M_j is associated with a range of fading values called the fading region R_j, $\overline{\tau_j}$ is the average time that the fading stays within the region R_j. The value of $\overline{\tau_j}$ is inversely proportional to the channel Doppler and also depends on the number and characteristics of the different fade regions. In Rayleigh fading with an average SNR of 20 dB and a channel Doppler of 100 Hz, $\overline{\tau_j}$ ranges from 0.7–3.9 ms, and thus, for a symbol rate of 100 ksymbols/s, the signal constellation remains constant over tens to hundreds of symbols. Similar results hold at other SNR values.

The flat fading assumption implies that the signal bandwidth B is much less than the channel coherence bandwidth $B_c = 1/T_M$, where T_M is the root mean square (rms) delay

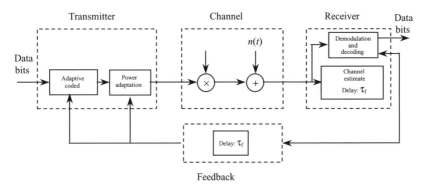

Figure 3.15 Block diagram of a system using adaptive modulation.

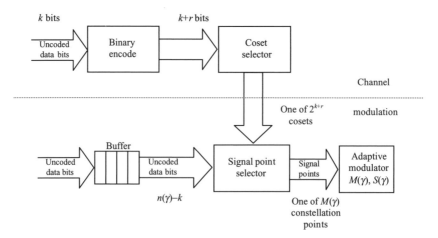

Figure 3.16 General structure for adaptive coded modulation.

spread of the channel. For Nyquist pulses $B = 1/T$, so flat fading occurs when $T \gg T_M$. Combining $T \gg T_M$ and $\overline{\tau}_j \gg T$ we get $\overline{\tau}_j \gg T \gg T_M$.

Wireless channels have rms delay spreads less than 30 µs in outdoor urban areas and less than around 1 µs in indoor environments. Taking the minimum $\overline{\tau}_j = 0.8$ ms, rates of the order of tens of ksymbols/s in outdoor channels and hundreds of ksymbols/s in indoor channels are practical for an adaptive scheme.

Modulation uses ideal Nyquist data pulses with a fixed symbol period $T = 1/B$. We also restrict $M(\gamma)$ to square M-QAM constellations of size $M_0 = 0$ and $M_j = 2^{2(j-1)}$, $j = 2, \ldots, J$. Thus, at each symbol time a constellation from the set $\{M_j : j = 0, 2, \ldots, J\}$ is used – the choice of constellation depends on the fade level γ over that symbol time. Choosing the M_0 constellation corresponds to no data transmission. Since the constellation set is finite, there will be a range of γ values over which a particular constellation M_j is used. Within that range the power must also be adapted to maintain the desired distance d_0 between the trajectories in the trellis. Thus, for each constellation M_j, the power adaptation $S_j(\gamma)$ associated with that constellation is a continuous function of γ. The basic premise for using adaptive modulation is to keep these distances constant by varying the size $M(\gamma)$, transmit power $S(\gamma)$, and/or symbol time $T(\gamma)$ of the transmitted signal constellation relative to γ, subject to an average transmit power constraint $\overline{S}$ on $S(\gamma)$. By maintaining $d_{\min}(t) = d_{\min}$ constant, the adaptive coded modulation exhibits the same coding gain as coded modulation designed for an AWGN channel with minimum code word distance $d_{\min}$. The detailed system block diagram is given in Figure 3.16.

The details about the coset codes used in the scheme can be found in [195, 196].

3.2.1 Maintaining a fixed distance

Define

$$M(\gamma) = \frac{\gamma}{\gamma_K^*} \tag{3.10}$$

where $\gamma_K^* \geq 0$ is a parameter which is optimized relative to the fade distribution to maximize spectral efficiency. For $\gamma < \gamma_K^* M_2$ the channel is not used. The constellation size M_j used for a given $\gamma \geq \gamma_K^* M_2$ is the largest M_j for which $M_j \leq M(\gamma)$. The range of γ values for which $M(\gamma) = M_j$ is thus $M_j \leq \gamma/\gamma_K^* M_{j+1}$, with $M_{J+1} \hat{=} \infty$. We call this range of fading values the fading region R_j associated with constellation M_j.

3.2.2 Information rate

For each γ, one redundant bit per symbol is used for the channel coding, so the number of information bits per symbol is $\log_2 M(\gamma) - 1$. Thus, the information rate for a single γ is $R_\gamma = [\log_2 M(\gamma) - 1]/T$ b/s, and the corresponding spectral efficiency is $R_\gamma/B = \log_2 M(\gamma) - 1$, since we use Nyquist pulses ($B = 1/T$).

Spectral efficiency is obtained by averaging the spectral efficiency for each γ weighted by its probability:

$$\frac{R}{B} = \sum_{j-2}^{J} (\log_2 M_j - 1) p(M_j \leq \gamma/\gamma_K < M_{j+1}) \tag{3.11}$$

where γ_K is picked to maximize Equation (3.11), subject to the average power constraint:

$$\sum_{j=2}^{J} \int_{\gamma_K M_j}^{\gamma_K M_{j+1}} S_j(\gamma) p(\gamma) \, d\gamma = \bar{S} \tag{3.12}$$

Similarly, the reconfiguration efficiency will now vary in time. Equation (3.1) can now be represented as:

$$E_{ff}(\gamma) = \left(\frac{k_{02}(\gamma)}{k_{01}} \right)^2 2^{-\Delta(k_0(\gamma) + v(\gamma))} g_{12}(\gamma) \tag{3.13}$$

and the average efficiency is obtained as:

$$E_{ff} = \int E_{ff}(\gamma) p(\gamma) \, d\gamma \tag{3.14}$$

Table 3.1 presents results of simulations for adaptive and non-adaptive systems with $M_j \in \{0, 4, 16, 64, 256\}$. One can see from the table that considerable SNR gains can be achieved with adaptive schemes.

Table 3.1 Comparison of adaptive and non-adaptive techniques

Spectral efficiency (bps/Hz)	BER	Trellis states	Average SNR (dB) Adaptive	Average SNR (dB) Non-adaptive
2	10^{-3}	4	10.5	18.5
		128	9.0	13.7
	10^{-6}	4	13	36.0
		128	11.3	21.0
3	10^{-3}	8	14.2	20.8
		128	13.1	16.8
	10^{-6}	8	16.4	36.5
		128	15.3	24.8

4

Space–Time Coding

4.1 DIVERSITY GAIN

In Chapters 2 and 3 the coding gain was used as a performance measure. Before we go into a detailed discussion on space–time coding, diversity gain will be defined and discussed. In Chapter 3 we briefly discussed the multidimensional trellis and pointed out the relatively high efficiency of such a concept. By using an additional dimension we provide a *diversity effect* which results in a considerable gain. These new dimensions may be additional frequency bands, different time slots or delayed replicas of the signal, or different antennas, resulting in frequency, time or space diversity respectively. In this section we elaborate the concept of diversity gain by using space diversity. In the subsequent sections of the chapter we will discuss space–time coding, where the concept of coding and diversity gain is combined into an integral performance measure.

A classical space diversity set-up with one transmitting and two receiving antennas is shown in Figure 4.1. The antenna diversity is realized in the receiver, hence the name *receiver diversity*. The following notation is used in the figure: the channel between the transmit antenna and the receiver antenna zero is denoted $\mathbf{h}_0$; that between the transmit antenna and the receiver antenna one is $\mathbf{h}_1$, where:

$$\mathbf{h}_0 = \alpha_0 e^{j\theta_0}$$
$$\mathbf{h}_1 = \alpha_1 e^{j\theta_1} \tag{4.1}$$

The resulting received baseband signals at antennas zero and one are:

$$\mathbf{r}_0 = \mathbf{h}_0 s_0 + \mathbf{n}_0$$
$$\mathbf{r}_1 = \mathbf{h}_1 s_0 + \mathbf{n}_1 \tag{4.2}$$

where $\mathbf{n}_0$ and $\mathbf{n}_1$ represent complex noise and interference.

In accordance with the discussion in Chapter 2, the ML decoder will choose signal $\mathbf{s}_i$ if and only if:

$$d^2(\mathbf{r}_0, \mathbf{h}_0\mathbf{s}_i) + d^2(\mathbf{r}_1, \mathbf{h}_1\mathbf{s}_i) \leq d^2(\mathbf{r}_0, \mathbf{h}_0\mathbf{s}_k) + d^2(\mathbf{r}_1, \mathbf{h}_1\mathbf{s}_k) \quad \forall i \neq k \tag{4.3}$$

Advanced Wireless Communications. S. Glisic
© 2004 John Wiley & Sons, Ltd. ISBN: 0-470-86776-0

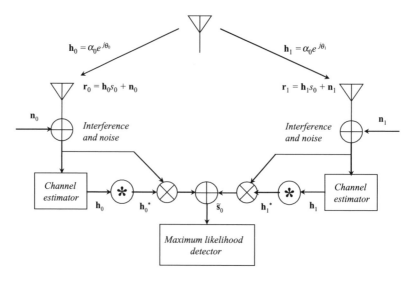

Figure 4.1 Two branch maximum ration receiver combiner (MRRC).

where $d^2(\mathbf{x}, \mathbf{y})$ is the squared Euclidian distance between $\mathbf{x}$ and $\mathbf{y}$:

$$d^2(\mathbf{x}, \mathbf{y}) = (\mathbf{x} - \mathbf{y})(\mathbf{x}^* - \mathbf{y}^*) \qquad (4.4)$$

A two branch *Maximum Ratio Receiver Combiner (MRRC)* would first create the signal:

$$\begin{aligned}\tilde{s}_0 &= \mathbf{h}_0^*\mathbf{r}_0 + \mathbf{h}_1^*\mathbf{r}_1 \\ &= \mathbf{h}_0^*(\mathbf{h}_0 s_0 + \mathbf{n}_0) + \mathbf{h}_1^*(\mathbf{h}_1 s_0 + \mathbf{n}_1) \\ &= (\alpha_0^2 + \alpha_1^2)s_0 + \mathbf{h}_0^*\mathbf{n}_0 + \mathbf{h}_1^*\mathbf{n}_1\end{aligned} \qquad (4.5)$$

with an equivalent distance defined as:

$$\mathbf{d}_i^2 = (\tilde{s}_0 - \beta s_i)(\tilde{s}_0 - \beta s_i)^*; \quad \beta = \alpha_0^2 + \alpha_1^2 \qquad (4.6)$$

and the ML detector would choose s_i if:

$$\left(\alpha_0^2 + \alpha_1^2\right)|s_i|^2 - \tilde{s}_0 s_i^* - \tilde{s}_0^* s_i \leq \left(\alpha_0^2 + \alpha_1^2\right)|s_k|^2 - \tilde{s}_0 s_k^* - \tilde{s}_0^* s_k \quad \forall i \neq k \qquad (4.7)$$

If Equation (4.6) is used in (4.7), the latter can also be represented in the following form. Choose s_i if:

$$\left(\alpha_0^2 + \alpha_1^2 - 1\right)|s_i|^2 + d^2(\tilde{s}_0, s_i) \leq \left(\alpha_0^2 + \alpha_1^2 - 1\right)|s_k|^2 + d^2(\tilde{s}_0, s_k) \quad \forall i \neq k \qquad (4.8)$$

For PSK signals (equal energy constellations):

$$|s_i|^2 = |s_k|^2 = E_s \quad \forall i, k \qquad (4.9)$$

where E_s is the energy of the signal. So, for PSK signals, the decision rule (4.8) may be simplified to:
Choose s_i if

$$d^2(\tilde{s}_0, s_i) \leq d^2(\tilde{s}_0, s_k) \quad \forall i \neq k \qquad (4.10)$$

Table 4.1 Space–time coding rules

	Antenna 0	Antenna 1
time t	s_0	s_1
time $t + T$	$-s_1^*$	s_0^*

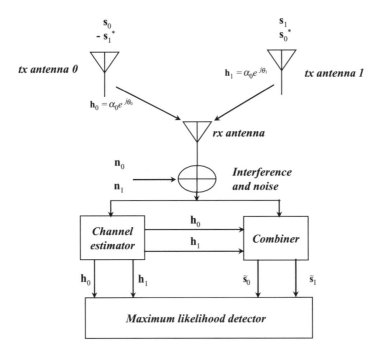

Figure 4.2 Two branch transmit diversity scheme with one receiver [197] © 1998, IEEE.

4.1.1 Two branch transmit diversity scheme with one receiver

Using two antennas in a mobile receiver might prove difficult in practice. For this reason, in this subsection we demonstrate how the same effect and diversity gain may be obtained by using two antennas at the transmitter and only one antenna at the receiver [197]. Implementing two antennas at the base station in mobile communication networks is a much simpler task. The system block diagram is shown in Figure 4.2.

4.1.1.1 The encoding and transmission sequence

The encoding is done in space and time (space–time (ST) coding) as defined in Table 4.1. The encoding may also be done in space and frequency. Instead of two adjacent symbol periods, two adjacent frequency subbands may be used (space–frequency coding).

4.1.1.2 The received signal

Assuming that the fading is constant across two consecutive symbols we have:

$$\mathbf{h}_0(t) = \mathbf{h}_0(t + T) = \mathbf{h}_0 = \alpha_0 e^{j\theta_0}$$
$$\mathbf{h}_1(t) = \mathbf{h}_1(t + T) = \mathbf{h}_1 = \alpha_1 e^{j\theta_1}$$

(4.11)

where T is the symbol interval. The received signals in two adjacent symbol intervals can be represented as:

$$\mathbf{r}_0 = \mathbf{r}(t) = \mathbf{h}_0 s_0 + \mathbf{h}_1 s_1 + \mathbf{n}_0$$
$$\mathbf{r}_1 = \mathbf{r}(t + T) = -\mathbf{h}_0 s_1^* + \mathbf{h}_1 s_0^* + \mathbf{n}_1$$

(4.12)

4.1.1.3 The combining scheme

The signals received in two adjacent symbol intervals are combined as follows:

$$\tilde{s}_0 = \mathbf{h}_0^* \mathbf{r}_0 + \mathbf{h}_1 \mathbf{r}_1^*$$
$$\tilde{s}_1 = \mathbf{h}_1^* \mathbf{r}_0 + \mathbf{h}_0 \mathbf{r}_1^*$$

(4.13)

$$\tilde{s}_0 = (\alpha_0^2 + \alpha_1^2) s_0 + \mathbf{h}_0^* \mathbf{n}_0 + \mathbf{h}_1 \mathbf{n}_1^*$$
$$\tilde{s}_1 = (\alpha_0^2 + \alpha_1^2) s_1 - \mathbf{h}_0 \mathbf{n}_1^* + \mathbf{h}_1^* \mathbf{n}_0$$

(4.14)

4.1.1.4 ML decision rule

For each of the signals s_0 and s_1 the decision rule is the same, Inequality (4.3) or (4.10). The resulting combined signals are equivalent to that obtained from two branch MRRC given by Equation (4.5). The only difference is phase rotations on the noise components which do not degrade the effective SNR. The resulting diversity order from the new branch transmit diversity scheme with one receiver is equal to that of two branch MRRC. This is known as Alamouti code [197].

4.1.2 Two transmitters and M receivers

There may be applications where a higher order of diversity is needed and multiple receive antennas are also feasible. In such cases, it is possible to provide a diversity order of $2M$, with two transmit and M receive antennas. Figure 4.3 represents an illustration of a special case of two transmit and two receive antennas. The generalization to N transmit and M receive antennas will be further elaborated in the subsequent sections. The notation is explained in Tables 4.2 and 4.3.

The space–time coding rule is given again by Table 4.1

Table 4.2   Channel notation

	rx antenna 0	*rx* antenna 1
tr antenna 0	$\mathbf{h}_0$	$\mathbf{h}_2$
tr antenna 1	$\mathbf{h}_1$	$\mathbf{h}_3$

Table 4.3   Signal notation

	rx antenna 0	*rx* antenna 1
time t	$\mathbf{r}_0$	$\mathbf{r}_2$
time $t + T$	$\mathbf{r}_1$	$\mathbf{r}_3$

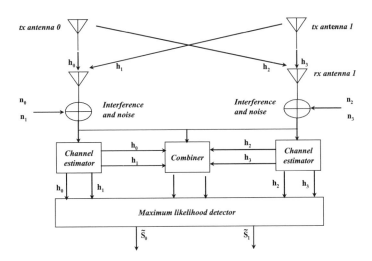

Figure 4.3 Two transmitters and two receivers [197] © 1998, IEEE.

4.1.2.1 *The received signal*

The received signals in two adjacent symbol intervals with notation given in Table 4.3 are:

$$\begin{aligned}
\mathbf{r}_0 &= \mathbf{h}_0 s_0 + \mathbf{h}_1 s_1 + \mathbf{n}_0 \\
\mathbf{r}_1 &= -\mathbf{h}_0 s_1^* + \mathbf{h}_1 s_0^* + \mathbf{n}_1 \\
\mathbf{r}_2 &= \mathbf{h}_2 s_0 + \mathbf{h}_3 s_1 + \mathbf{n}_2 \\
\mathbf{r}_3 &= -\mathbf{h}_2 s_1^* + \mathbf{h}_3 s_0^* + \mathbf{n}_3
\end{aligned} \tag{4.15}$$

where $\mathbf{n}_0$, $\mathbf{n}_1$, $\mathbf{n}_2$, and $\mathbf{n}_3$ are complex random variables representing receiver thermal noise and interference.

4.1.2.2 The combiner

The combiner is defined by the following rule:

$$\tilde{s}_0 = h_0^* r_0 + h_1 r_1^* + h_2^* r_2 + h_3 r_3^*$$
$$\tilde{s}_1 = h_1^* r_0 - h_0 r_1^* + h_3^* r_2 - h_2 r_3^* \tag{4.16}$$

$$\tilde{s}_0 = \left(\alpha_0^2 + \alpha_1^2 + \alpha_2^2 + \alpha_3^2\right)s_0 + h_0^* n_0 + h_1 n_1^* + h_2^* n_2 + h_3 n_3^*$$
$$\tilde{s}_1 = \left(\alpha_0^2 + \alpha_1^2 + \alpha_2^2 + \alpha_3^2\right)s_1 - h_0 n_1^* + h_1^* n_0 - h_2 n_3^* + h_3^* n_2 \tag{4.17}$$

So, the order of $2 \times 2 = 4$ diversity is achieved. For uncorrelated channel and noise, $\alpha_i^2 = \alpha^2$, and perfect channel estimation, the signal to noise ratio is given as:

$$(4\alpha^2)^2 / 4\alpha^2 N_0 = 4\alpha^2 / N_0 \tag{4.18}$$

4.1.2.3 ML decoder

The ML decoder will be operating as follows:
For s_0, choose s_i if:

$$\left(\alpha_0^2 + \alpha_1^2 + \alpha_2^2 + \alpha_3^2 - 1\right)|s_i|^2 + d^2(\tilde{s}_0, s_i) \leq \left(\alpha_0^2 + \alpha_1^2 + \alpha_2^2 + \alpha_3^2 - 1\right)|s_k|^2 + d^2(\tilde{s}_0, s_k) \tag{4.19}$$

For s_1, choose s_i if:

$$\left(\alpha_0^2 + \alpha_1^2 + \alpha_2^2 + \alpha_3^2 - 1\right)|s_i|^2 + d^2(\tilde{s}_1, s_i) \leq \left(\alpha_0^2 + \alpha_1^2 + \alpha_2^2 + \alpha_3^2 - 1\right)|s_k|^2 + d^2(\tilde{s}_1, s_k) \tag{4.20}$$

4.1.2.4 The BER performance

The BER results are shown in Figure 4.4. Significant diversity gain measured in SNR improvements is evident.

4.2 SPACE–TIME CODING

In the previous section we introduced the concept of space–time coding. We are now going to look more closely at the general model and bring in more detail in the analysis of such a system. In general, we will consider a system where data is encoded by a channel code and the encoded data is split into n streams that are simultaneously transmitted using n transmit antennas. The received signal at each receive antenna is a linear superposition of the n transmitted signals plus noise.

4.2.1 The system model

The system consists of n antennas in the base station and m antennas in the mobile. Data is encoded by the channel encoder, S/P converted, and divided into n streams of data. Each stream of data is used as the input to a pulse shaper. The output of each shaper is then

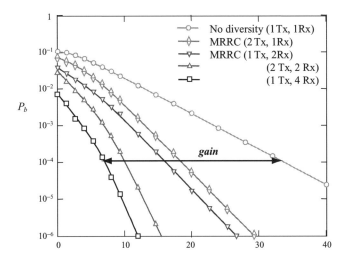

Figure 4.4 The BER performance comparison of coherent BPSK with MRRC and two branch transmit diversity in Rayleigh fading.

modulated. At each time slot with index l, the output of modulator i is a signal $\mathbf{c}_l^i$ that is transmitted using transmit antenna (Tx antenna) i for $1 < i < n$.

The n signals are transmitted simultaneously, each from a different transmit antenna, and all of these signals have the same transmission period T. The signal at each receive antenna is a noisy superposition of the n transmitted signals corrupted by Rayleigh or Rician fading.

Elements of the signal constellation are contracted by a factor of $\sqrt{E_s}$ chosen so that the average energy of the constellation is 1.

At the receiver, the demodulator computes a decision statistic based on the received signals arriving at each receive antenna $1 < j < m$. The signal $\mathbf{r}_l^j$ received by antenna j at discrete time l is given by:

$$\mathbf{r}_l^j = \sum_{i=1}^n \alpha_{i,j} \mathbf{c}_l^j \sqrt{E_s} + \eta_l^j \qquad (4.21)$$

where the noise η_l^j at discrete time l is modeled as independent samples of a zero mean complex Gaussian random variable with variance $N_0/2$ per dimension. The coefficient $\alpha_{i,j}$ is the path gain from transmit antenna i to receive antenna j. It is assumed that these path gains are constant during a frame of t symbol intervals and vary from one frame to another (quasistatic flat fading).

4.2.2 The case of independent fade coefficients

The ML receiver will decide erroneously in favor of a signal:

$$\mathbf{e} = e_1^1 e_1^2 \cdots e_1^n e_2^1 e_2^2 \cdots e_2^n \cdots e_l^1 e_l^2 \cdots e_l^n$$

assuming that

$$\mathbf{c} = c_1^1 c_1^2 \cdots c_1^n c_2^1 c_2^2 \cdots c_2^n \cdots c_l^1 c_l^2 \cdots c_l^n$$

was transmitted, with a probability that can be approximated by:

$$P(\mathbf{c} \to \mathbf{e}|\alpha_{i,j}, \ i = 1, \ 2, \ldots, \ n, \ j = 1, \ 2, \ldots, \ m) \leq \exp\left(-d^2(\mathbf{c}, \mathbf{e})E_s/4N_0\right) \quad (4.22)$$

where $N_0/2$ is the noise variance per dimension and

$$d^2(\mathbf{c}, \mathbf{e}) = \sum_{j=1}^{m} \sum_{l=1}^{t} \left| \sum_{i=1}^{n} \alpha_{i,j}\left(c_l^i - e_l^i\right) \right|^2 \quad (4.23)$$

is the distance between the two trajectories measured in a time frame of t symbol intervals. Setting $\Omega_j = (\alpha_{1,j}, \ldots, \alpha_{n,j})$, we rewrite Equation (4.23) as:

$$d^2(\mathbf{c}, \mathbf{e}) = \sum_{j=1}^{m} \sum_{l=1}^{t} \sum_{i'=1}^{n} \alpha_{i,j}\overline{\alpha_{i',j}} \sum_{t=1}^{l} \left(c_t^i - e_t^i\right)\overline{\left(c_t^{i'} - e_t^{i'}\right)}.$$

where $\bar{x}$ stands for the complex conjugate of x. After simple manipulations, we observe that:

$$d^2(\mathbf{c}, \mathbf{e}) = \sum_{j=1}^{m} \Omega_j \mathbf{A} \Omega_j^* \quad (4.24)$$

where $\mathbf{x}^*$ stands for the transpose conjugate, and elements of matrix $\mathbf{A}$ are defined as $A_{pq} = \mathbf{x}_p \cdot \mathbf{x}_q$ and $\mathbf{x}_p = (c_1^p - e_1^p, \ c_2^p - e_2^p, \ldots, \ c_l^p - e_l^p)$ for $1 \leq p, q \leq n$. Thus:

$$P(\mathbf{c} \to \mathbf{e}|\alpha_{i,j}, \ i = 1, \ 2, \ldots, \ n, \ j = 1, \ 2, \ldots, \ m) \leq \prod_{j=1}^{m} \exp\left(-\Omega_j \mathbf{A}(\mathbf{c}, \mathbf{e})\Omega_j^* E_s/4N_0\right) \quad (4.25)$$

where $A_{pq} = \sum_{t=1}^{l} (c_t^p - e_t^p)\overline{(c_t^q - e_t^q)}$. This can be also represented as:

$$\mathbf{A} = \mathbf{B}^* \mathbf{B}$$
$$\text{where} \quad \mathbf{B} = \{b_{it}\} = \{e_t^i - c_t^i\} \quad (4.26)$$

Matrix $\mathbf{B}$ is given in an explicit form later by Equation (4.48). From now on we will express $d^2(\mathbf{c}, \mathbf{e})$ in terms of the eigenvalues of the matrix $\mathbf{A}(\mathbf{c}, \mathbf{e})$ defined by $\mathbf{V}\mathbf{A}(\mathbf{c}, \mathbf{e})\mathbf{V}^* = \mathbf{D} = \text{diag}\{\lambda_i\}$. For details of eigenvalue decomposition, see Appendix 5.1.

If we define vector $(\beta_{1,j}, \ldots, \beta_{n,j}) = \Omega_j \mathbf{V}^*$, then we have:

$$\Omega_j \mathbf{A}(\mathbf{c}, \mathbf{e})\Omega_j^* = \sum_{i=1}^{n} \lambda_i |\beta_{i,j}|^2 \quad (4.27)$$

At this point recall that $\alpha_{i,j}$ are samples of a complex Gaussian random variable with mean $\bar{\alpha}_{ij}$. Let

$$\mathbf{K}^j = (\bar{\alpha}_{1j}, \bar{\alpha}_{2j}, \bar{\alpha}_{3j}, \cdots, \bar{\alpha}_{nj}) \quad (4.28)$$

Since $\mathbf{V}$ is unitary, $\{\mathbf{v}_1, \mathbf{v}_2, \ldots, \mathbf{v}_n\}$ is an orthonormal basis of C^n and $\beta_{i,j}$ are independent complex Gaussian random variables with variance 0.5 per dimension and mean $\mathbf{K}^j \cdot \mathbf{v}_i$. If $K_{i,j} = |\bar{\beta}_{i,j}|^2 = |\mathbf{K}^j \cdot v_i|^2$, then $|\beta_{i,j}|$ are independent Rician distributions with pdf:

$$p(|\beta_{i,j}|) = 2|\beta_{i,j}| \exp(-|\beta_{i,j}|^2 - K_{i,j})I_0(2|\beta_{i,j}|\sqrt{K_{i,j}}) \quad (4.29)$$

for $|\beta_{i,j}| \geq 0$, where $I_0(\cdot)$ is the zero order modified Bessel function of the first kind. To compute an upper bound on the average probability of error, we simply average:

$$\prod_{j=1}^{m} \exp\left(-(E_s/4N_0)\sum_{i=1}^{n}\lambda_i|\beta_{i,j}|^2\right) \tag{4.30}$$

with respect to independent Rician distributions of $|\beta_{i,j}|$ to arrive at:

$$P(\mathbf{c} \to \mathbf{e}) \leq \prod_{j=1}^{m}\left(\prod_{i=1}^{n}\frac{1}{1+\dfrac{E_s}{4N_0}\lambda_i}\exp\left(-\frac{K_{i,j}\dfrac{E_s}{4N_0}\lambda_i}{1+\dfrac{E_s}{4N_0}\lambda_i}\right)\right) \tag{4.31}$$

4.2.3 Rayleigh fading

In this case $\bar{\alpha} = 0$, giving $K_{i,j} = 0$ for all i and j. Thus Inequality (4.31) can be written as:

$$P(\mathbf{c} \to \mathbf{e}) \leq \left(\frac{1}{\displaystyle\prod_{i=1}^{n}(1+\lambda_i E_s/4N_0)}\right)^m \tag{4.32}$$

Let r denote the rank of matrix $\mathbf{A}$, then the kernel of $\mathbf{A}$ has dimension $n-r$ and exactly $n-r$ eigenvalues of $\mathbf{A}$ are zero. Say the non-zero eigenvalues of $\mathbf{A}$ are $\lambda_1, \lambda_2, \ldots, \lambda_r$, then it follows from Inequality (4.32) that:

$$P(\mathbf{c} \to \mathbf{e}) \leq \left(\prod_{i=1}^{r}\lambda_i\right)^{-m}(E_s/4N_0)^{-rm} \tag{4.33}$$

Thus, a diversity advantage of mr and a coding advantage of $(\lambda_1\lambda_2\cdots\lambda_r)^{1/r}$ is achieved. Recall that $\lambda_1\lambda_2\cdots\lambda_r$ is the absolute value of the sum of determinants of all the principal $r \times r$ cofactors of $\mathbf{A}$, Moreover, it is easy to see that the ranks of $\mathbf{A}(\mathbf{c}, \mathbf{e})$, and $\mathbf{B}(\mathbf{c}, \mathbf{e})$, defined as $\mathbf{A}(\mathbf{c}, \mathbf{e}) = \mathbf{B}(\mathbf{c}, \mathbf{e})*\mathbf{B}(\mathbf{c}, \mathbf{e})$, are equal.

4.2.4 Design criteria for Rayleigh space–time codes

The rank criterion. In order to achieve the maximum diversity mn, the matrix $\mathbf{B}(\mathbf{c}, \mathbf{e})$ has to be full rank for any code words $\mathbf{c}$ and $\mathbf{e}$. If $\mathbf{B}(\mathbf{c}, \mathbf{e})$ has minimum rank r over the set of two tuples of distinct code words, then a diversity of rm is achieved.

The determinant criterion. Suppose that a diversity benefit of rm is our target. The minimum of r roots of the sum of determinants of all $r \times r$ principal cofactors of $\mathbf{A}(\mathbf{c}, \mathbf{e}) = \mathbf{B}(\mathbf{c}, \mathbf{e})*\mathbf{B}(\mathbf{c}, \mathbf{e})$ taken over all pairs of distinct code words $\mathbf{e}$ and $\mathbf{c}$ corresponds to the coding advantage, where r is the rank of $\mathbf{A}(\mathbf{c}, \mathbf{e})$. Special attention in the design must be paid to this quantity for any code words $\mathbf{e}$ and $\mathbf{c}$. The design target is making this sum as large as possible. If a diversity of nm is the design target, then the minimum of the determinant of $\mathbf{A}(\mathbf{c}, \mathbf{e})$ taken over all pairs of distinct code words $\mathbf{e}$ and $\mathbf{c}$ must be maximized.

For large signal to noise ratios,

$$P(\mathbf{c} \rightarrow \mathbf{e}) \leq \left(\frac{E_s}{4N_0}\right)^{-rm} \left(\prod_{i=1}^{r} \lambda_i\right)^{-m} \left[\prod_{j=1}^{m} \prod_{i=1}^{r} \exp(-K_{i,j})\right] \tag{4.34}$$

Thus, a diversity of rm and a coding advantage of:

$$(\lambda_1 \lambda_2 \cdots \lambda_r)^{-1/r} \left[\prod_{j=1}^{m} \prod_{i=1}^{r} \exp(-K_{i,j})\right]^{1/rm} \tag{4.35}$$

is achievable. The derivation in this section is based on the original work presented in [198].

4.2.5 Code construction

In the presence of one receive antenna, little can be gained in terms of capacity increase by using more than four transmit antennas. Similarly, if there are two receive antennas, almost all the capacity increase can be obtained using $n = 6$ transmit antennas.

As has been indicated earlier, we can use multidimensional trellis codes for a wireless communication system that employs n transmit antennas and (optional) receive antenna diversity where the channel is a quasistatic flat fading channel. The encoding for these trellis codes is obvious, with the exception that *at the beginning and the end of each frame, the encoder is required to be in the zero state.* At each time t, depending on the state of the encoder and the input bits, a transition branch is chosen. If the label of this branch is $q_t^1 q_t^2 \cdots q_t^n$, then transmit antenna i is used to send constellation symbols q_t^i, $i = 1, 2, \cdots, n$ and all of these transmissions are simultaneous. Let us consider the 4PSK and 8PSK constellations as shown in Figure 4.5.

4.2.5.1 Examples

A number of results have been published on space–time code design [197–242]. We will present several illustrations.

In Figure 4.5 the first signal constellation is 4PSK, where the signal points are labeled by the elements of $\mathbb{Z}_4$, the ring of integers modulo 4. We consider the four-state trellis code shown in Figure 4.6 [198]. The edge label $\mathbf{x}_1 \mathbf{x}_2$ indicates that signal $\mathbf{x}_1$ is transmitted over the first antenna and that signal $\mathbf{x}_2$ is transmitted over the second antenna. This code has a very simple description in terms of a sequence (b_k, a_k) of binary inputs [198]. The output

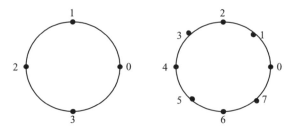

Figure 4.5 4PSK and 8PSK constellations.

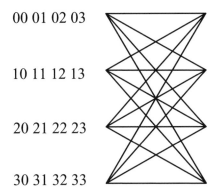

00 01 02 03

10 11 12 13

20 21 22 23

30 31 32 33

Figure 4.6 2-space–time code, 4PSK, four states, 2 b/s/Hz.

signal pair $\mathbf{x}_1^k \mathbf{x}_2^k$ at time k is given by:

$$\left(\mathbf{x}_1^k,\ \mathbf{x}_2^k\right) = b_{k-1}(2,0) + a_{k-1}(1,0) + b_k(0,2) + a_k(0,1) \tag{4.36}$$

where the addition takes place in $\mathbb{Z}_4$.

$$\left(\mathbf{x}_1^k,\ \mathbf{x}_2^k\right) = a_{k-2}(2,2) + b_{k-1}(2,0) + a_{k-1}(1,0) + b_k(0,2) + a_k(0,1)$$

$$\left(\mathbf{x}_1^k,\ \mathbf{x}_2^k\right) = b_{k-2}(0,2) + a_{k-2}(2,0) + b_{k-1}(2,0) + a_{k-1}(1,2) + b_k(0,2) + a_k(0,1)$$

Figure 4.7 represents 2-space–time codes for the 4PSK constellation and 8 and 16 state encoders for 2 b/s/Hz. The two output bits $(\mathbf{x}_1^k, \mathbf{x}_2^k)$ as a function of the input bits are also shown in the figure. Several additional examples are given in Figures 4.8–4.11 [198].

$$\begin{aligned}\left(\mathbf{x}_1^k,\ \mathbf{x}_2^k\right) = &\ a_{k-3}(2,2) + b_{k-2}(3,3) + a_{k-2}(2,0) \\ &+ b_{k-1}(2,2) + a_{k-1}(1,1) \\ &+ b_k(0,2) + a_k(0,1)\end{aligned} \tag{4.37}$$

Assuming that the input to the encoder at time k is the three input bits (d_k, b_k, a_k), the output of the encoder at time k is:

$$\begin{aligned}\left(\mathbf{x}_1^k,\ \mathbf{x}_2^k\right) = &\ d_{k-1}(4,0) + b_{k-1}(2,0) + a_{k-1}(5,0) \\ &+ d_k(0,4) + b_k(0,2) + a_k(0,1)\end{aligned} \tag{4.38}$$

where the computation is performed in $\mathbb{Z}_8$, the ring of integers modulo 8, and the elements of the 8PSK constellation have the labeling given in Figure 4.5.

If $\mathbf{r}_t^j$ is the received signal at receive antenna j at time t, the branch metric for a transition labeled $q_t^1 q_t^2 \cdots q_t^n$ is given by:

$$\sum_{j=1}^{m}\left| \mathbf{r}_t^j - \sum_{i=1}^{n} \alpha_{i,j} q_t^i \right|^2 \tag{4.39}$$

The Viterbi algorithm is then used to compute the path with the lowest accumulated metric. The frame error probability for four different examples of the coding is shown in Figures 4.12–4.14. The gain shown in these figures should be used in expressions for E_{ff} discussed

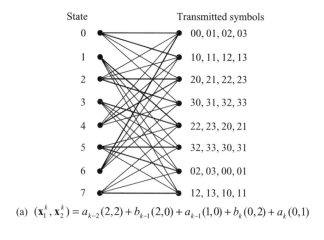

(a) $(\mathbf{x}_1^k, \mathbf{x}_2^k) = a_{k-2}(2,2) + b_{k-1}(2,0) + a_{k-1}(1,0) + b_k(0,2) + a_k(0,1)$

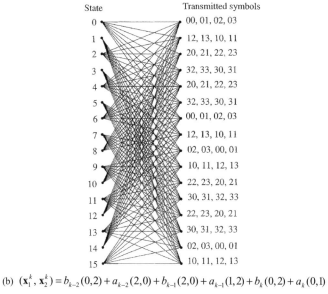

(b) $(\mathbf{x}_1^k, \mathbf{x}_2^k) = b_{k-2}(0,2) + a_{k-2}(2,0) + b_{k-1}(2,0) + a_{k-1}(1,2) + b_k(0,2) + a_k(0,1)$

Figure 4.7 2-space–time codes, 4PSK(a), 8 and (b)16 states, 2 b/s/Hz.

in Chapters 2 and 3 to evaluate the overall reconfiguration efficiency for different schemes. In general, the expression for efficiency should be further modified as follows.

4.2.6 Reconfiguration efficiency of space–time coding

Let *rm* be the diversity advantage of the system with *n* transmit and *m* receive antennas. For a block of *l* symbols, with constellation Q of 2^b elements, the equivalent rate of transmission *R* in a system with ST coding satisfies:

$$R \leq \frac{\log[A_{2^{bl}}(n, r)]}{l} \tag{4.40}$$

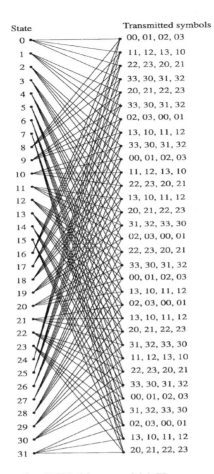

Figure 4.8 2-space–time code, 4PSK, 32 states, 2 b/s/Hz.

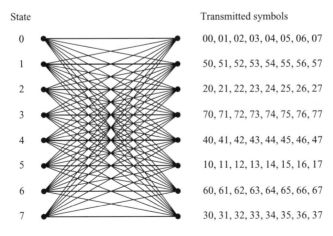

Figure 4.9 2-space–time code, 8PSK, eight states, 3 b/s/Hz.

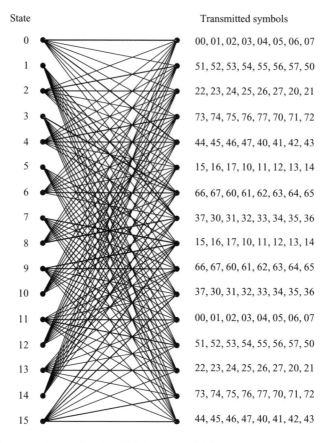

State   Transmitted symbols

0   00, 01, 02, 03, 04, 05, 06, 07

1   51, 52, 53, 54, 55, 56, 57, 50

2   22, 23, 24, 25, 26, 27, 20, 21

3   73, 74, 75, 76, 77, 70, 71, 72

4   44, 45, 46, 47, 40, 41, 42, 43

5   15, 16, 17, 10, 11, 12, 13, 14

6   66, 67, 60, 61, 62, 63, 64, 65

7   37, 30, 31, 32, 33, 34, 35, 36

8   15, 16, 17, 10, 11, 12, 13, 14

9   66, 67, 60, 61, 62, 63, 64, 65

10   37, 30, 31, 32, 33, 34, 35, 36

11   00, 01, 02, 03, 04, 05, 06, 07

12   51, 52, 53, 54, 55, 56, 57, 50

13   22, 23, 24, 25, 26, 27, 20, 21

14   73, 74, 75, 76, 77, 70, 71, 72

15   44, 45, 46, 47, 40, 41, 42, 43

Figure 4.10 2-space–time code, 8PSK, 16 states, 3 b/s/Hz.

in bits per second per Hertz, where $A_{2^{bl}}(n, r)$ is the maximum size of a code length n and minimum Hamming distance r defined over an alphabet of size 2^{bl} [198].

On the other hand, if b is the transmission rate of a multiple antenna system employed in conjunction with an r-space–time trellis code, the trellis complexity of the space–time code is at least $2^{b(r-1)}$ [198].

The reconfiguration gain is defined as the solution to:

$$P(\mathbf{c} \rightarrow \mathbf{e} \,|\, r, m, g(\text{ST})E_s/4T) = P(\mathbf{c} \rightarrow \mathbf{e} \,|\, r = 1, m = 1, E_s/4T) \qquad (4.41)$$

Using Inequality (4.33) for Rayleigh fading, Equation (4.41) gives:

$$\left(\prod_{i=1}^{r} \lambda_i \right)^{-m} (g(\text{ST})E_s/4N_0)^{-rm} = \lambda_1^{-1}(E_s/4N_0)$$

resulting in

$$g(\text{ST}) = \left[\lambda_1^{-1}(E_s/4N_0) \right]^{rm} \left\{ E_s/4N_0 \left(\prod_{i=1}^{r} \lambda_i \right)^{-1/r} \right\} \qquad (4.42)$$

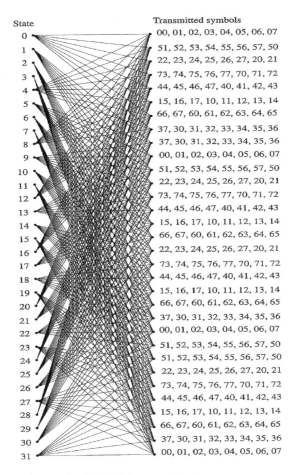

State	Transmitted symbols
0	00, 01, 02, 03, 04, 05, 06, 07
1	51, 52, 53, 54, 55, 56, 57, 50
2	22, 23, 24, 25, 26, 27, 20, 21
3	73, 74, 75, 76, 77, 70, 71, 72
4	44, 45, 46, 47, 40, 41, 42, 43
5	15, 16, 17, 10, 11, 12, 13, 14
6	66, 67, 60, 61, 62, 63, 64, 65
7	37, 30, 31, 32, 33, 34, 35, 36
8	37, 30, 31, 32, 33, 34, 35, 36
9	00, 01, 02, 03, 04, 05, 06, 07
10	51, 52, 53, 54, 55, 56, 57, 50
11	22, 23, 24, 25, 26, 27, 20, 21
12	73, 74, 75, 76, 77, 70, 71, 72
13	44, 45, 46, 47, 40, 41, 42, 43
14	15, 16, 17, 10, 11, 12, 13, 14
15	66, 67, 60, 61, 62, 63, 64, 65
16	22, 23, 24, 25, 26, 27, 20, 21
17	73, 74, 75, 76, 77, 70, 71, 72
18	44, 45, 46, 47, 40, 41, 42, 43
19	15, 16, 17, 10, 11, 12, 13, 14
20	66, 67, 60, 61, 62, 63, 64, 65
21	37, 30, 31, 32, 33, 34, 35, 36
22	00, 01, 02, 03, 04, 05, 06, 07
23	51, 52, 53, 54, 55, 56, 57, 50
24	51, 52, 53, 54, 55, 56, 57, 50
25	22, 23, 24, 25, 26, 27, 20, 21
26	73, 74, 75, 76, 77, 70, 71, 72
27	44, 45, 46, 47, 40, 41, 42, 43
28	15, 16, 17, 10, 11, 12, 13, 14
29	66, 67, 60, 61, 62, 63, 64, 65
30	37, 30, 31, 32, 33, 34, 35, 36
31	00, 01, 02, 03, 04, 05, 06, 07

Figure 4.11 2-space–time code, 8PSK, 32 states, 3 b/s/Hz.

The relative complexity over the signal constellation Q with 2^b elements is:

$$D_r(ST) > 2^{b(r-1)}/2^b = 2^{b(r-2)} \tag{4.43}$$

Now the reconfiguration efficiency defined by Equation (3.1) becomes:

$$E_{ff} = \left(\frac{R}{b}\right)\frac{g(ST)}{D_r} = [\lambda_1^{-1}(E_s/4N_0)]^{rm}\left\{E_s/4N_0\left(\prod_{i=1}^{r}\lambda_i\right)^{-1/r}\right\}\left(\frac{R}{b}\right)2^{-b(r-2)} \tag{4.44}$$

Now we continue with different examples of ST codes. Consider the 8PSK signal constellation, where the encoder maps a sequence of three bits $a_k b_k c_k$ at time k to ii with $i = 4a_k + 2b_k + c_k$. It is easy to show that the equivalent space–time code for this delay diversity code has the trellis representation given in Figure 4.15. The minimum determinant of this code is $(2 - \sqrt{2})^2$. As in the 4PSK case, one can improve the coding advantage of the above codes by constructing encoders with more states. An example is given in Figure 4.16 [198] with the constellation in Figure 4.17.

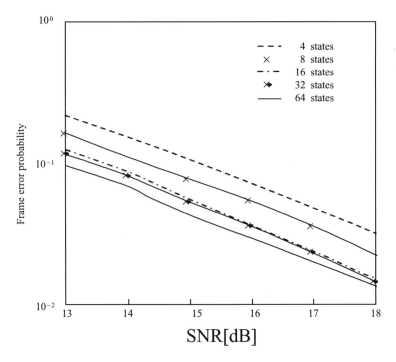

Figure 4.12 Codes for 4PSK with rate 2 b/s/Hz that achieve diversity 2 with one receive and two transmit antennas.

4.2.6.1 An r space–time trellis code for r > 2

As an example, a 4-space–time code for four transmit antennas is considered. The limit on the transmission rate is 2 b/s/Hz. Thus, the trellis complexity of the code is bounded below by 64. The input to the encoder is a block of length two of bits a_1, b_1, corresponding to an integer $i = 2a_1 + b_1 \in \mathbb{Z}_4$. The 64 states of the trellis correspond to the set of all three tuples (s_1, s_2, s_3) with $s_i \in \mathbb{Z}_4$ for $1 \leq i \leq 3$. At state (s_1, s_2, s_3) upon input data i, the encoder outputs (i, s_1, s_2, s_3) elements of the 4PSK constellation (see Figure 4.5) and moves to state (i, s_1, s_2). Given two code words **c** and **e**, the associated paths in the trellis diverge at time t_1 from a state and remerge in another state at a later time $t_2 \leq l$. It is easy to see that the t_1th, $(t_1 + l)$th, $(t_2 - 1)$th, and t_2th columns of the matrix $\mathbf{B}(\mathbf{c}, \mathbf{e})$ are independent. Thus, the above design gives a 4-space–time code.

4.2.7 Delay diversity

The encoder block diagram of a delay diversity transmitter is given in Figure 4.18, with

$$\begin{aligned} c_t^1 &= \tilde{c}_{t-1}^1 \\ c_t^2 &= \tilde{c}_t^2 \end{aligned}$$

(4.45)

where c_t^1 and c_t^2 are the symbols of the equivalent space–time code at time t and $c_t^1 c_t^2$ is the output of the encoder at time t.

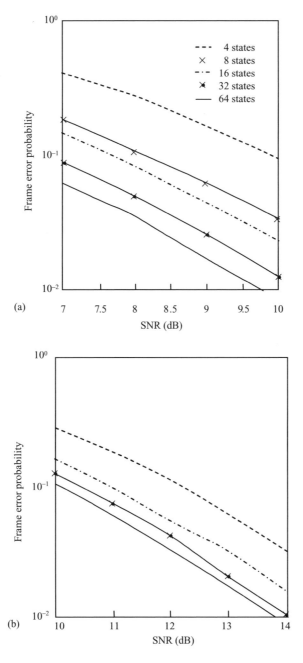

Figure 4.13 (a) Codes for 4PSK with rate 2 b/s/Hz that achieve diversity 4 with two receive and two transmit antennas; (b) codes for 8PSK with rate 3 b/s/Hz that achieve diversity 4 with two receive and two transmit antennas.

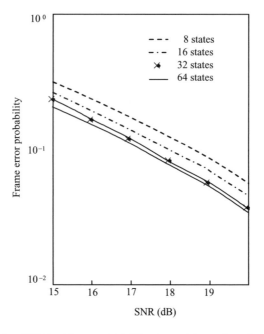

Figure 4.14 Codes for 8PSK with rate 3 b/s/Hz that achieve diversity 2 with one receive and two transmit antennas.

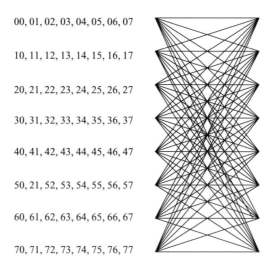

Figure 4.15 Space–time realization of a delay diversity 8PSK code constructed from a repetition code.

Next, we consider the block code [198]

$$\mathbf{C} = \{00, 15, 22, 37, 44, 51, 66, 73\}$$

of length two defined over the alphabet 8PSK instead of the repetition code. This block code is the best in the sense of product distance [206] among all the codes of cardinality eight

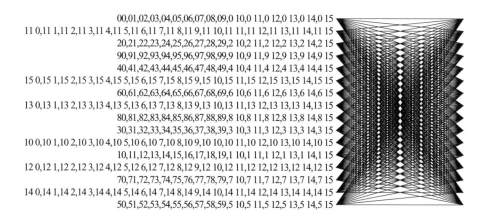

00,01,02,03,04,05,06,07,08,09,0 10,0 11,0 12,0 13,0 14,0 15
11 0,11 1,11 2,11 3,11 4,11 5,11 6,11 7,11 8,11 9,11 10,11 11,11 12,11 13,11 14,11 15
20,21,22,23,24,25,26,27,28,29,2 10,2 11,2 12,2 13,2 14,2 15
90,91,92,93,94,95,96,97,98,99,9 10,9 11,9 12,9 13,9 14,9 15
40,41,42,43,44,45,46,47,48,49,4 10,4 11,4 12,4 13,4 14,4 15
15 0,15 1,15 2,15 3,15 4,15 5,15 6,15 7,15 8,15 9,15 10,15 11,15 12,15 13,15 14,15 15
60,61,62,63,64,65,66,67,68,69,6 10,6 11,6 12,6 13,6 14,6 15
13 0,13 1,13 2,13 3,13 4,13 5,13 6,13 7,13 8,13 9,13 10,13 11,13 12,13 13,13 14,13 15
80,81,82,83,84,85,86,87,88,89,8 10,8 11,8 12,8 13,8 14,8 15
30,31,32,33,34,35,36,37,38,39,3 10,3 11,3 12,3 13,3 14,3 15
10 0,10 1,10 2,10 3,10 4,10 5,10 6,10 7,10 8,10 9,10 10,10 11,10 12,10 13,10 14,10 15
10,11,12,13,14,15,16,17,18,19,1 10,1 11,1 12,1 13,1 14,1 15
12 0,12 1,12 2,12 3,12 4,12 5,12 6,12 7,12 8,12 9,12 10,12 11,12 12,12 13,12 14,12 15
70,71,72,73,74,75,76,77,78,79,7 10,7 11,7 12,7 13,7 14,7 15
14 0,14 1,14 2,14 3,14 4,14 5,14 6,14 7,14 8,14 9,14 10,14 11,14 12,14 13,14 14,14 15
50,51,52,53,54,55,56,57,58,59,5 10,5 11,5 12,5 13,5 14,5 15

Figure 4.16 2-space–time 16 QAM code, 16 states, 4 b/s/Hz.

0	1	2	3
○	○	○	○
7	6	5	4
○	○	○	○
8	9	10	11
○	○	○	○
15	14	13	12
○	○	○	○

Figure 4.17 The 16 QAM constellation.

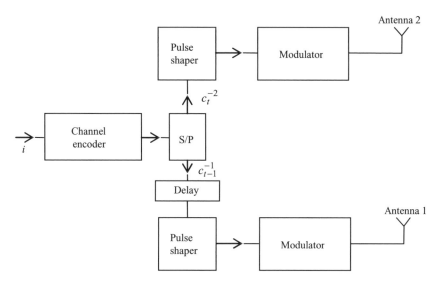

Figure 4.18 The block diagram of a delay diversity transmitter.

and of length two defined over the alphabet 8PSK. This means that the minimum of the product distance $|c_1 - e_1||c_2 - e_2|$ between pairs of distinct code words $\mathbf{c} = c_1 c_2 \in C$ and $\mathbf{e} = e_1 e_2 \in C$ is maximal among all such codes. The delay diversity code constructed from this block code is identical to the space–time code given by the trellis diagram of Figure 4.9. The minimum determinant of this delay diversity code is thus 2.

The 16-state code for the 16 QAM constellation given in Figure 4.16, is obtained from the block code

$$\{0\ 0, 1\ 11, 2\ 2, 3\ 9, 4\ 4, 5\ 15, 6\ 6, 7\ 13, 8\ 8, 9\ 3, 10\ 10, 11\ 1, 12\ 12, 13\ 7, 14\ 14, 15\ 5\}$$

using the same delay diversity construction. Again, this block code is optimal in the sense of product distance.

The delay diversity code construction can also be generalized to systems having more than two transmit antennas. For instance, the 4PSK 4-space–time code given before is a delay diversity code. The corresponding block code is the repetition code. By applying the delay diversity construction to the 4PSK block code

$$\{0\ 0\ 0\ 0, 1\ 2\ 3\ 1, 2\ 1\ 2\ 3, 3\ 3\ 1\ 2\}$$

one can obtain a more powerful 4PSK 4-space–time code having the same trellis complexity.

4.3 SPACE–TIME BLOCK CODES FROM ORTHOGONAL DESIGNS

ML decoding of a multidimensional trellis requires a large complexity. In this section we present a special class of space–time codes for which maximum likelihood decoding is achieved in a simple way through decoupling of the signals transmitted from different antennas rather than joint detection. This uses the orthogonal structure of the space–time block code and gives a maximum likelihood decoding algorithm which is based only on linear processing at the receiver. The presentation is based on [201, 235–242].

Space–time block codes are designed to achieve the maximum diversity order for a given number of transmit and receive antennas subject to the constraint of having a simple decoding algorithm. Unfortunately, space–time block codes constructed in this way only exist for few sporadic values of n.

4.3.1 The channel model and the diversity criterion

At time t the signal $\mathbf{r}_t^j$, received at antenna j, is given by Equation (4.21) which for $\sqrt{E_s} = 1$ becomes:

$$\mathbf{r}_t^j = \sum_{i=1}^{n} \alpha_{i,j} c_t^i + \eta_t^j \tag{4.46}$$

Assuming perfect channel state information is available, the receiver computes the decision metric for l symbols and Expression (4.39) gives:

$$\sum_{t=1}^{l} \sum_{j=1}^{m} \left| \mathbf{r}_t^j - \sum_{i=1}^{n} \alpha_{i,j} c_t^i \right|^2 \tag{4.47}$$

In order to achieve the maximum diversity mn, the matrix $\mathbf{A}(\mathbf{c}, \mathbf{e}) = \mathbf{B}(\mathbf{c}, \mathbf{e})^*\mathbf{B}(\mathbf{c}, \mathbf{e})$ with

$$
\mathbf{B}(\mathbf{c},\ \mathbf{e}) =
\begin{bmatrix}
e_1^1 - c_1^1 & e_2^1 - c_2^1 & \cdots & \cdots & e_l^1 - c_l^1 \\
e_1^2 - c_1^2 & e_2^2 - c_2^2 & \cdots & \cdots & e_l^2 - c_l^2 \\
e_1^3 - c_1^3 & e_2^3 - c_2^3 & \ddots & \vdots & e_l^3 - c_l^3 \\
\vdots & \vdots & \ddots & \ddots & \vdots \\
e_1^n - c_1^n & e_2^n - c_2^n & \cdots & \cdots & e_l^n - c_l^n
\end{bmatrix}
\tag{4.48}
$$

has to be full rank for any pair of distinct code words $\mathbf{c}$ and $\mathbf{e}$. If $\mathbf{B}(\mathbf{c}, \mathbf{e})$ has minimum rank r over the set of pairs of distinct code words, then a diversity of rm is achieved.

4.3.2 Real orthogonal designs

For $n = 2, 4$ or 8, mathematics literature offers orthogonal sets of signals defined as:

$$
\mathbf{O}_2(\mathbf{x}_1, \mathbf{x}_2) =
\begin{bmatrix}
x_1 & x_2 \\
-x_2 & x_1
\end{bmatrix}
\tag{4.49}
$$

The 4×4 design

$$
\mathbf{O}_4(\mathbf{x}_1, \mathbf{x}_2, \mathbf{x}_3, \mathbf{x}_4) =
\begin{bmatrix}
x_1 & x_2 & x_3 & x_4 \\
-x_2 & x_1 & -x_4 & x_3 \\
-x_3 & x_4 & x_1 & -x_2 \\
-x_4 & -x_3 & x_2 & x_1
\end{bmatrix}
\tag{4.50}
$$

and 8×8 design

$$
\mathbf{O}_8(\mathbf{x}_1, \mathbf{x}_2, \ldots \mathbf{x}_8) =
\begin{bmatrix}
x_1 & x_2 & x_3 & x_4 & x_5 & x_6 & x_7 & x_8 \\
-x_2 & x_1 & x_4 & -x_3 & x_6 & -x_5 & -x_8 & x_7 \\
-x_3 & -x_4 & x_1 & x_2 & x_7 & x_8 & -x_5 & -x_6 \\
-x_4 & x_3 & -x_2 & x_1 & x_8 & -x_7 & x_6 & -x_5 \\
-x_5 & -x_6 & -x_7 & -x_8 & x_1 & x_2 & x_3 & x_4 \\
-x_6 & x_5 & -x_8 & x_7 & -x_2 & x_1 & -x_4 & x_3 \\
-x_7 & x_8 & x_5 & -x_6 & -x_3 & x_4 & x_1 & -x_2 \\
-x_8 & -x_7 & x_6 & x_5 & -x_4 & -x_3 & x_2 & x_1
\end{bmatrix}
\tag{4.51}
$$

4.3.3 Space–time encoder

At time slot 1, nb bits arrive at the encoder and select constellation signals $s_1, \ldots, s_n$. Setting $x_i = s_i$ for $i = 1, 2, \ldots, n$, we arrive at a matrix $\mathbf{C} = \mathbf{O}(s_1, \ldots, s_n)$ with entries $\pm s_1, \pm s_2, \ldots, \pm s_n$. At each time slot $t = 1, 2, \ldots, n$, the entries $C_{ti}, i = 1, 2, \ldots, n$ are transmitted simultaneously from transmit antennas $1, 2, \ldots, n$. The rate of transmission is b bits/s/Hz.

4.3.4 The diversity order

The rank criterion requires that the matrix $\mathbf{O}(\tilde{s}_1, \ldots, \tilde{s}_n) - \mathbf{O}(s_1, \ldots, s_n)$ be non-singular for any two distinct code sequences $(\tilde{s}_1, \ldots, \tilde{s}_n) \neq (s_1, \ldots, s_n)$. Clearly, $\mathbf{O}(\tilde{s}_1 - s_1, \ldots, \tilde{s}_n -$

$s_n) = \mathbf{O}(\tilde{s}_1, \ldots, \tilde{s}_n) - \mathbf{O}(s_1, \ldots, s_n)$ where $\mathbf{O}(\tilde{s}_1 - s_1, \ldots, \tilde{s}_n - s_n)$ is the matrix constructed from $\mathbf{O}$ by replacing x_i with $\tilde{s}_i - s_i$ for all $i = 1, 2, \ldots, n$. The determinant of the orthogonal matrix $\mathbf{O}$ is easily seen to be:

$$\det(\mathbf{OO}^{\mathrm{T}})^{1/2} = \left[\sum_i x_i^2\right]^{n/2} \tag{4.52}$$

where $\mathbf{O}^{\mathrm{T}}$ is the transpose of $\mathbf{O}$. Hence:

$$\det[\mathbf{O}(\tilde{s}_1 - s_1, \ldots, \tilde{s}_n - s_n)] = \left[\sum_i |\tilde{s}_i - s_i|^2\right]^{n/2} \tag{4.53}$$

which is non-zero. It follows that $\mathbf{O}(\tilde{s}_1, \ldots, \tilde{s}_n) - \mathbf{O}(s_1, \ldots, s_n)$ is non-singular and the maximum diversity order nm is achieved.

4.3.5 The decoding algorithm

Rows of $\mathbf{O}$ are all permutations of the first row of $\mathbf{O}$ with possibly different signs. Let $e_1, \ldots, e_n$ denote the permutations corresponding to these rows and let $\delta_k(i)$ denote the sign of x_i in the kth row of $\mathbf{O}$. Then $e_k(p) = q$ means that x_p is up to a sign change the (k, q)th element of $\mathbf{O}$. Since the columns of $\mathbf{O}$ are pairwise-orthogonal, it turns out that minimizing the metric of Expression (4.47) amounts to minimizing:

$$\sum_{i=1}^n S_i \tag{4.54}$$

where

$$S_i = \left(\left|\left[\sum_{t=1}^n \sum_{j=1}^m r_t^j \alpha_{e_t(i),j}^* \delta_t(i)\right] - s_i\right|^2 + \left(-1 + \sum_{k,l} |\alpha_{k,l}|^2\right) |s_i|^2\right) \tag{4.55}$$

The value of S_i only depends on the code symbol s_i, the received symbols $\{r_t^j\}$, the path coefficients $\{\alpha_{i,j}\}$, and the structure of the orthogonal design $\mathbf{O}$. It follows that minimizing the sum given in Expression (4.54) amounts to minimizing Equation (4.55) for all $1 \le i \le n$. Thus, the maximum likelihood detection rule is to form the decision variables:

$$R_i = \sum_{t=1}^n \sum_{j=1}^m r_t^j \alpha_{e_t(i),j}^* \delta_t(i) \tag{4.56}$$

for all $i = 1, 2, \ldots, n$ and decide in favor of s_i among all the constellation symbols s if:

$$s_i = \arg \min_{s \in A} |R_i - s|^2 + \left(-1 + \sum_{k,l} |\alpha_{k,l}|^2\right) |s_i|^2 \tag{4.57}$$

This is a very simple decoding strategy that provides diversity.

4.3.6 Linear processing orthogonal designs

The above properties are preserved even if we allow linear processing at the transmitter. Therefore, we relax the definition of orthogonal designs to allow linear processing at the transmitter. Signals transmitted from different antennas will now be linear combinations of constellation symbols.

A linear processing orthogonal design in variables $x_1, x_2, \ldots, x_n$ is an $n \times n$ matrix $\mathbf{E}$ such that:

- the entries of $\mathbf{E}$ are real linear combinations of variables $x_1, x_2, \ldots, x_n$;
- $\mathbf{E}^{\mathrm{T}}\mathbf{E} = \mathbf{D}$, where $\mathbf{D}$ is a diagonal matrix with (i, i)th diagonal element of the form $(l_1^i x_1^2 + l_2^i x_2^2 + \cdots l_n^i x_n^2)$, with the coefficients $(l_1^i, l_2^i, \ldots, l_n^i)$ strictly positive numbers.

It is easy to show that transmission using a linear processing orthogonal design provides full diversity and a simplified decoding algorithm as above.

4.3.7 Generalized real orthogonal designs

Since the simple maximum likelihood decoding algorithm described above is achieved because of orthogonality of columns of the design matrix, we may generalize the definition of linear processing orthogonal designs. This creates new and simple transmission schemes for any number of transmit antennas.

A generalized orthogonal design $\mathbf{G}$ of size n is a $p \times n$ matrix with entries $0, \pm x_1, \pm x_2, \ldots, \pm x_k$, such that $\mathbf{G}^{\mathrm{T}}\mathbf{G} = \mathbf{D}$ where $\mathbf{D}$ is a diagonal matrix with diagonal $\mathbf{D}_{ii}, i = 1, 2, \ldots, n$ of the form $(l_1^i x_1^2 + l_2^i x_2^2 + \cdots l_n^i x_n^2)$ and coefficients $(l_1^i, l_2^i, \ldots, l_n^i)$ strictly positive numbers. The rate of $\mathbf{G}$ is $R = k/p$.

A full-rate generalized orthogonal design has entries of the form $\pm x_1, \pm x_2, \ldots, \pm x_p$.

The generalized orthogonal signal sets are:

$$
\mathbf{G}_3 = \begin{bmatrix} x_1 & x_2 & x_3 \\ -x_2 & x_1 & -x_4 \\ -x_3 & x_4 & x_1 \\ -x_4 & -x_3 & x_2 \end{bmatrix} \tag{4.58}
$$

$$
\mathbf{G}_5 = \begin{bmatrix} x_1 & x_2 & x_3 & x_4 & x_5 \\ -x_2 & x_1 & x_4 & -x_3 & x_6 \\ -x_3 & -x_4 & x_1 & x_2 & x_7 \\ -x_4 & x_3 & -x_2 & x_1 & x_8 \\ -x_5 & -x_6 & -x_7 & -x_8 & x_1 \\ -x_6 & x_5 & -x_8 & x_7 & -x_2 \\ -x_7 & x_8 & x_5 & -x_6 & -x_3 \\ -x_8 & -x_7 & x_6 & x_5 & -x_4 \end{bmatrix} \tag{4.59}
$$

$$\mathbf{G_6} = \begin{bmatrix} x_1 & x_2 & x_3 & x_4 & x_5 & x_6 \\ -x_2 & x_1 & x_4 & -x_3 & x_6 & -x_5 \\ -x_3 & -x_4 & x_1 & x_2 & x_7 & x_8 \\ -x_4 & x_3 & -x_2 & x_1 & x_8 & -x_7 \\ -x_5 & -x_6 & -x_7 & -x_8 & x_1 & x_2 \\ -x_6 & x_5 & -x_8 & x_7 & -x_2 & x_1 \\ -x_7 & x_8 & x_5 & -x_6 & -x_3 & x_4 \\ -x_8 & -x_7 & x_6 & x_5 & -x_4 & -x_3 \end{bmatrix} \tag{4.60}$$

$$\mathbf{G_7} = \begin{bmatrix} x_1 & x_2 & x_3 & x_4 & x_5 & x_6 & x_7 \\ -x_2 & x_1 & x_4 & -x_3 & x_6 & -x_5 & -x_8 \\ -x_3 & -x_4 & x_1 & x_2 & x_7 & x_8 & -x_5 \\ -x_4 & x_3 & -x_2 & x_1 & x_8 & -x_7 & x_6 \\ -x_5 & -x_6 & -x_7 & -x_8 & x_1 & x_2 & x_3 \\ -x_6 & x_5 & -x_8 & x_7 & -x_2 & x_1 & -x_4 \\ -x_7 & x_8 & x_5 & -x_6 & -x_3 & x_4 & x_1 \\ -x_8 & -x_7 & x_6 & x_5 & -x_4 & -x_3 & x_2 \end{bmatrix} \tag{4.61}$$

4.3.8 Encoding

At time slot 1, kb bits arrive at the encoder and select constellation symbols $s_1, s_2, \ldots, s_n$. The encoder populates the matrix by setting $x_i = s_i$ and at time $t = 1, 2, \ldots, p$ the signals $G_{t1}, \ldots, G_{tn}$ are transmitted simultaneously from antennas $1, 2, \ldots, n$. Thus, kb bits are sent during each p transmissions. It can be proved, as in Equation (4.53), that the diversity order is nm. It should be mentioned that the rate of a generalized orthogonal design is different from the throughput of the associated code. To motivate the definition of the rate, we note that the theory of space–time coding proves that for a diversity order of nm, it is possible to transmit b bits per time slot and this is the best possible. Therefore, the rate R of this coding scheme is defined to be kb/pb which is equal to k/p.

4.3.9 The Alamouti scheme

The space–time block code, already used in Section 4.1 was proposed by Alamouti [197]. The code uses the complex orthogonal signal set:

$$\begin{bmatrix} x_1 & x_2 \\ -x_2^* & x_1^* \end{bmatrix} \tag{4.62}$$

Suppose that there are 2^b signals in the constellation. At the first time slot, 2^b bits arrive at the encoder and select two complex symbols s_1 and s_2. These symbols are transmitted simultaneously from antennas one and two, respectively. At the second time slot, signals $-s_2^*$ and s_1^* are transmitted simultaneously from antennas one and two, respectively.

The ML detector will minimize the decision statistic:

$$\sum_{j=1}^{m} \left(\left| r_1^j - \alpha_{1,j} s_1 - \alpha_{2,j} s_2 \right|^2 + \left| r_2^j - \alpha_{1,j} s_1^* - \alpha_{2,j} s_2^* \right|^2 \right) \tag{4.63}$$

over all possible values of s_1 and s_2. The minimizing values are the receiver estimates of s_1

and s_2, respectively. This is equivalent to minimizing the decision statistics:

$$\left(\left|\left[\sum_{j=1}^{m} \left(r_1^j \alpha_{1,j}^* + \left(r_2^j\right)^* \alpha_{2,j}\right)\right] - s_1\right|^2 + \left(-1 + \sum_{j=1}^{m} \sum_{i=1}^{2} |\alpha_{i,j}|^2\right) |s_1|^2\right) \tag{4.64}$$

for detecting s_1 and the decision statistics:

$$\left(\left|\left[\sum_{j=1}^{m} \left(r_1^j \alpha_{2,j}^* + \left(r_2^j\right)^* \alpha_{1,j}\right)\right] - s_2\right|^2 + \left(-1 + \sum_{j=1}^{m} \sum_{i=1}^{2} |\alpha_{i,j}|^2\right) |s_2|^2\right) \tag{4.65}$$

for decoding s_2. The scheme provides full diversity of $2m$ using m receive antennas.

4.3.10 Complex orthogonal designs

Given a complex orthogonal signal set $\mathbf{O_c}$ of size n, we replace each complex variable $x_i = x_i^1 + x_i^2 \mathbf{i}$, $1 \le i \le n$ by the 2×2 real matrix:

$$\begin{bmatrix} x_i^1 & x_i^2 \\ -x_i^2 & x_i^1 \end{bmatrix} \tag{4.66}$$

In this way, x_i^* is represented by:

$$\begin{bmatrix} x_i^1 & -x_i^2 \\ x_i^2 & x_i^1 \end{bmatrix} \tag{4.67}$$

$\mathbf{i}x_i$ is represented by:

$$\begin{bmatrix} -x_i^2 & x_i^1 \\ -x_i^1 & -x_i^2 \end{bmatrix} \tag{4.68}$$

and so forth. It is easy to see that the $2n \times 2n$ matrix formed in this way is a real orthogonal design of size $2n$.

A complex orthogonal signal set of size n exists if and only if $n = 2$.

4.3.11 Generalized complex orthogonal designs

Let $\mathbf{G_c}$ be a $p \times n$ matrix whose entries are $0, \pm x_1, \pm x_1^*, \pm x_2, \pm x_2^*, \ldots, \pm x_k, \pm x_k^*$ or their product with $\mathbf{i}$. If $\mathbf{G_c^* G_c} = \mathbf{D_c}$ where $\mathbf{D_c}$ is a diagonal matrix with (i, i)th diagonal element of the form:

$$\left(l_1^i |x_1|^2 + l_2^i |x_2|^2 + \cdots l_k^i |x_k|^2\right)$$

and the coefficients $(l_1^i, l_2^i, \ldots, l_n^i)$ all strictly positive numbers, then $\mathbf{G_c}$ is referred to as a generalized orthogonal design of size n and rate $R = k/p$. For instance, rate $1/2$ codes for transmission using three and four transmit antennas are given by:

$$
G_c^3 =
\begin{bmatrix}
x_1 & x_2 & x_3 \\
-x_2 & x_1 & -x_4 \\
-x_3 & x_4 & x_1 \\
-x_4 & -x_3 & x_2 \\
x_1^* & x_2^* & x_3^* \\
-x_2^* & x_1^* & -x_4^* \\
-x_3^* & x_4^* & x_1^* \\
-x_4^* & -x_3^* & x_2^*
\end{bmatrix}
\tag{4.69}
$$

and

$$
G_c^4 =
\begin{bmatrix}
x_1 & x_2 & x_3 & x_4 \\
-x_2 & x_1 & -x_4 & x_3 \\
-x_3 & x_4 & x_1 & -x_2 \\
-x_4 & -x_3 & x_2 & x_1 \\
x_1^* & x_2^* & x_3^* & x_4^* \\
-x_2^* & x_1^* & -x_4^* & x_3^* \\
-x_3^* & x_4^* & x_1^* & -x_2^* \\
-x_4^* & -x_3^* & x_2^* & x_1^*
\end{bmatrix}
\tag{4.70}
$$

These transmission schemes and their analogs for higher n give full diversity but lose half of the theoretical bandwidth efficiency.

4.3.12 Special codes

It is natural to ask for higher rates than $1/2$ when designing generalized complex linear processing orthogonal designs for transmission with n multiple antennas. For $n = 2$, Alamouti's scheme gives a rate one design. For $n = 3$ and 4, rate $3/4$ generalized complex linear processing orthogonal designs are given by:

$$
H_3 =
\begin{bmatrix}
x_1 & x_2 & \frac{x_3}{\sqrt{2}} \\
-x_2^* & x_1^* & \frac{x_3}{\sqrt{2}} \\
\frac{x_3^*}{\sqrt{2}} & \frac{x_3^*}{\sqrt{2}} & \frac{(-x_1-x_1^*+x_2-x_2^*)}{2} \\
\frac{x_3^*}{\sqrt{2}} & -\frac{x_3^*}{\sqrt{2}} & \frac{(x_1-x_1^*+x_2+x_2^*)}{2}
\end{bmatrix}
$$

$$
H_4 =
\begin{bmatrix}
x_1 & x_2 & \frac{x_3}{\sqrt{2}} & \frac{x_3}{\sqrt{2}} \\
-x_2^* & x_1^* & \frac{x_3}{\sqrt{2}} & -\frac{x_3}{\sqrt{2}} \\
\frac{x_3^*}{\sqrt{2}} & \frac{x_3^*}{\sqrt{2}} & \frac{(-x_1-x_1^*+x_2-x_2^*)}{2} & \frac{(x_1-x_1^*-x_2-x_2^*)}{2} \\
\frac{x_3^*}{\sqrt{2}} & -\frac{x_3^*}{\sqrt{2}} & \frac{(x_1-x_1^*+x_2+x_2^*)}{2} & -\frac{(x_1+x_1^*+x_2-x_2^*)}{2}
\end{bmatrix}
\tag{4.71}
$$

4.3.13 Performance results

A collection of results is shown in Figures 4.19–4.26. The transmission using two transmit antennas employs the 8PSK constellation and the code $\mathbf{G}_2$. For three and four transmit antennas, the 16 QAM constellation and the codes $\mathbf{H}_3$ and $\mathbf{H}_4$, respectively, are used. Since $\mathbf{H}_3$ and $\mathbf{H}_4$ are rate 3/4 codes, the total transmission rate in each case is 3 bits/s/Hz. It is seen that at the bit error rate of 10^{-5} the rate 3/4 16 QAM code $\mathbf{H}_4$ gives about 7 dB gain over the use of an 8PSK $\mathbf{G}_2$ code.

Transmission using two transmit antennas employs the 4PSK constellation and the code $\mathbf{G}_2$. For three and four transmit antennas, the 16-QAM constellation and the codes $\mathbf{G}_3$ and $\mathbf{G}_4$, respectively, are used. Since $\mathbf{G}_3$ and $\mathbf{G}_4$ are rate 1/2 codes, the total transmission rate in each case is 2 bits/s/Hz. It is seen that at the bit error rate of 10^{-5} the rate 1/2 16-QAM code $\mathbf{G}_4$ gives about 5 dB gain over the use of a 4PSK $\mathbf{G}_2$ code.

The transmission using two transmit antennas employs the binary PSK (BPSK) constellation and the code $\mathbf{G}_2$. For three and four transmit antennas, the 4PSK constellation and the codes $\mathbf{G}_3$ and $\mathbf{G}_4$, respectively, are used. Since $\mathbf{G}_3$ and $\mathbf{G}_4$ are rate 1/2 codes, the total transmission rate in each case is 1 bit/s/Hz. It is seen that at the bit error rate of 10^{-5} the rate 1/2 4PSK code $\mathbf{G}_4$ gives about 7.5 dB gain over the use of a BPSK $\mathbf{G}_2$ code.

If number of receive antennas is increased, this gain reduces to 3.5 dB. The reason is that much of the diversity gain is already achieved using two transmit and two receive antennas.

4.4 CHANNEL ESTIMATION IMPERFECTIONS

So far we have been assuming that a perfect channel estimation is available for the operation of the ML decoder collecting statistics defined by Expression (4.39) or Expression (4.47).

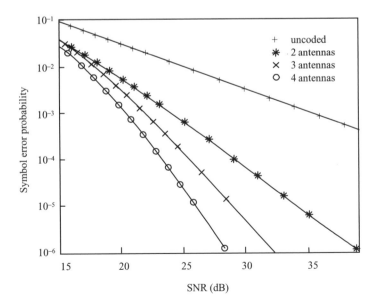

Figure 4.19 Symbol error probability versus SNR for space–time block codes at 3 bits/s/Hz; one receive antenna.

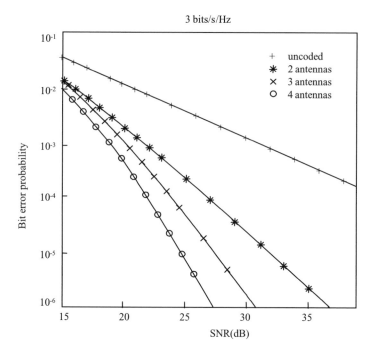

Figure 4.20 Bit error probability versus SNR for space–time block codes at 3 bits/s/Hz; one receive antenna.

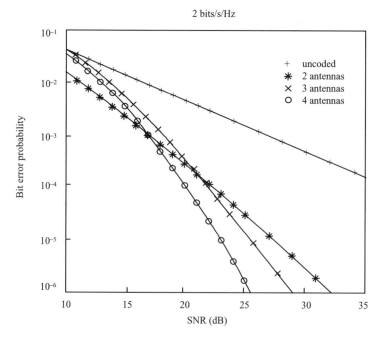

Figure 4.21 Bit error probability versus SNR for space–time block codes at 2 bits/s/Hz; one receive antenna.

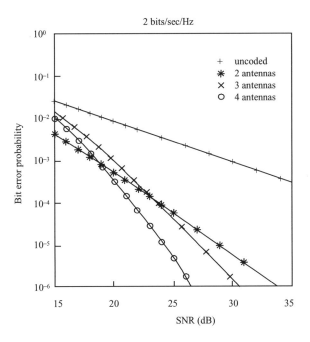

Figure 4.22 Symbol error probability versus SNR for space–time block codes at 2 bits/s/Hz; one receive antenna.

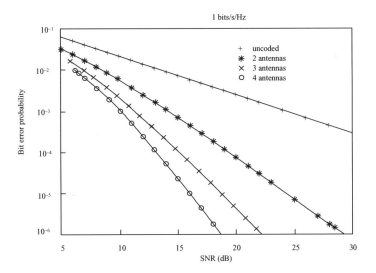

Figure 4.23 Bit error probability versus SNR for space–time block codes at 1 bit/s/Hz; one receive antenna.

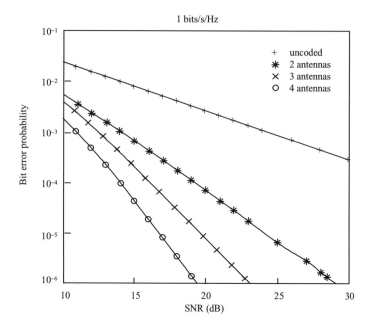

Figure 4.24 Symbol error probability versus SNR for space–time block codes at 1 bit/s/Hz; one receive antenna.

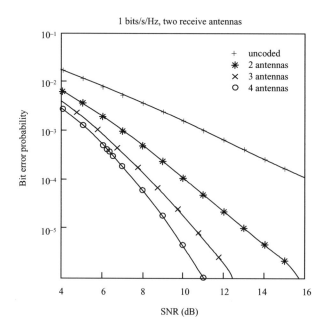

Figure 4.25 BER versus SNR for space–time block codes at. 1 bit/s/Hz; two receive antennas.

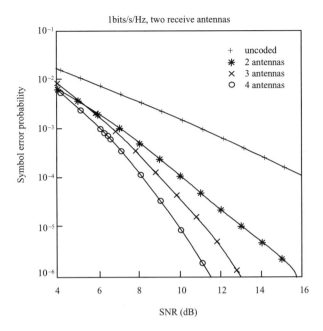

Figure 4.26 Symbol error probability versus SNR for space–time block codes at 1 bit/s/Hz; two receive antennas.

Let us look at this assumption more carefully. First of all, the errors in the channel estimation will depend on the estimator structure.

4.4.1 Channel estimator

At the beginning of each frame of symbols to be transmitted from transmit antenna i, a sequence $\mathbf{W_i}$ of length k pilot symbols:

$$\mathbf{W}_i = (W_{i,1}, W_{i,2}, \ldots, W_{i,k})$$

is appended. The sequences $\mathbf{W}_1, \mathbf{W}_2, \ldots, \mathbf{W}_n$ are designed to be orthogonal to each other:

$$\mathbf{W_p}\bar{\mathbf{W}}_\mathbf{q} = \sum_{j=1}^{k} W_{p,j} \bar{W}_{q,j} = 0$$

whenever $p \neq q$. In the previous expression $\bar{W}$ stands for the conjugate of W and $\bar{\mathbf{W}}$ is the conjugate transpose of $\mathbf{W}$. Let $\mathbf{r}^j = (r_1^j, r_2^j, \ldots, r_k^j)$ be the observed sequence of received signals at antenna j during the training period. Then:

$$r_t^j = \sum_{i=1}^{n} \alpha_{i,j} W_{i,t} + \eta_{t,j}, \quad 1 \leq j \leq m, \quad 1 \leq t \leq k$$

where the channel coefficients $\alpha_{i,j}$ are independent samples of a complex Gaussian random variable with mean zero and variance 0.5 per dimension, and $\eta_{t,j}$ are independent samples of a zero mean complex Gaussian random variable with variance $N_0/2$ per dimension. Let $\eta^j = (\eta_{1,j}, \eta_{2,j}, \ldots, \eta_{k,j})$. Our goal is to estimate $\alpha_{i,j}$, $i = 1, 2, \ldots, n$, $j = 1, 2, \ldots, m$ using the statistic $\mathbf{r}^j$.

The unbiased estimator $\beta_{i,j}$ having the least variance is given by the ratio of inner products $(\mathbf{r}^j \cdot \bar{\mathbf{W}}^i)/(\mathbf{W}^i \cdot \bar{\mathbf{W}}^i)$. Indeed, since $\mathbf{W}^p \bar{\mathbf{W}}^q = 0$ it is easy to see that:

$$\mathbf{r}^j \cdot \bar{\mathbf{W}}^i = \alpha_{i,j}(\mathbf{W}^i \cdot \bar{\mathbf{W}}^i) + \eta^j \cdot \bar{\mathbf{W}}^i$$

thus

$$\alpha_{i,j} = \frac{\mathbf{r}^j \cdot \bar{\mathbf{W}}^i}{\mathbf{W}^i \cdot \bar{\mathbf{W}}^i} - \frac{\eta^j \cdot \bar{\mathbf{W}}^i}{\mathbf{W}^i \cdot \bar{\mathbf{W}}^i}$$

In other words,

$$\beta_{i,j} = \alpha_{i,j} + \frac{\eta^j \cdot \bar{\mathbf{W}}^i}{\mathbf{W^i} \cdot \bar{\mathbf{W}}^i}$$

The random variable $\beta_{i,j}$ has zero mean. The variance of the estimation error is $N_0/2kE_s$ per dimension which is the minimum given by the Cramer–Rao bound. Simulation results for imperfect channel estimation (mismatch) with $n = 2$, $k = 8$, and the frame length 130 symbols are shown in Figure 4.27.

More details on system imperfections can be found in [203, 243–244].

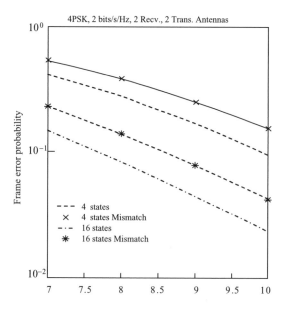

Figure 4.27 Performance of four and 16 state 4PSK codes in the presence of channel estimation error, 2 bits/s/Hz, two receive and two transmit antennas.

4.5 QUASI-ORTHOGONAL SPACE–TIME BLOCK CODES

It was shown in Section 4.3 that a complex orthogonal design that provides full diversity and full transmission rate for a space–time block code is not possible for more than two antennas. Previous attempts have been concentrated in generalizing orthogonal designs which provide space–time block codes with full diversity and a high transmission rate. In this section we discuss rate one codes which are quasi-orthogonal and provide partial diversity. The decoder for these codes works with pairs of transmitted symbols instead of single symbols.

An example of a full rate full diversity complex space–time block code is the Alamouti scheme already discussed in Section 4.3. In this section we will use the notation:

$$\mathcal{A}_{12} = \begin{bmatrix} x_1 & x_2 \\ -x_2^* & x_1^* \end{bmatrix} \tag{4.72}$$

Here we use the subscript 12 to represent the indeterminates x_1 and x_2 in the transmission matrix. Now, let us consider the following space–time block code with block length T symbol intervals where K bits are transmitted over N transmit antennas and received with M receive antennas, for $N = T = K = 4$:

$$\mathcal{A} = \begin{bmatrix} \mathcal{A}_{12} & \mathcal{A}_{34} \\ -\mathcal{A}_{34}^* & \mathcal{A}_{12}^* \end{bmatrix} = \begin{bmatrix} x_1 & x_2 & x_3 & x_4 \\ -x_2^* & x_1^* & -x_4^* & x_3^* \\ -x_3^* & -x_4^* & x_1^* & x_2^* \\ x_4 & -x_3 & -x_2 & x_1 \end{bmatrix} \tag{4.73}$$

It is easy to see that the minimum rank of matrix $\mathcal{A}(s_1 - \bar{s}_1, s_2 - \bar{s}_2, s_3 - \bar{s}_3, s_4 - \bar{s}_4)$, the matrix constructed from $\mathcal{A}$ by replacing x_i with $s_i - \bar{s}_i$, is 2. Therefore, a diversity of $2M$ is achieved while the rate of the code is one. Note that the maximum diversity of $4M$ for a rate one code is impossible in this case.

4.5.1 Decoding

Assuming perfect channel state information is available, the ML receiver computes the decision metric:

$$\sum_{m=1}^{M} \sum_{t=1}^{T} \left| r_{t,m} - \sum_{n=1}^{N} \alpha_{n,m} \mathcal{A}_{tn} \right|^2 \tag{4.74}$$

over all possible $x_k = s_k \in \mathcal{A}$ and decides in favor of the constellation symbols $s_1, \ldots, s_K$ that minimize this sum.

4.5.2 Decision metric

Now, if we define V_i, $i = 1, 2, 3, 4$, as the ith column of $\mathcal{A}$, it is easy to see that

$$\langle V_1, V_2 \rangle = \langle V_1, V_3 \rangle = \langle V_2, V_4 \rangle = \langle V_3, V_4 \rangle = 0 \tag{4.75}$$

where $\langle V_i, V_j \rangle = \sum_{l=1}^{4} (V_i)_l (V_j)_l^*$ is the inner product of vectors V_i and V_j. Therefore, the subspace created by V_1 and V_4 is orthogonal to the subspace created by V_2 and V_3.

Using this orthogonality, the maximum likelihood decision metric, Expression (4.74), can be calculated as the sum of two terms $f_{14}(x_1, x_4) + f_{23}(x_2, x_3)$, where f_{14} is independent of x_2 and x_3 and f_{23} is independent of x_1 and x_4. Thus, the minimization of Expression (4.74) is equivalent to minimizing these two terms independently. In other words, first the decoder finds the pair (s_1, s_4) that minimizes $f_{14}(x_1, x_4)$ among all possible (x_1, x_4) pairs. Then, or in parallel, the decoder selects the pair (s_2, s_3) which minimizes $f_{23}(x_2, x_3)$. This reduces the complexity of decoding without sacrificing the performance.

Simple manipulation of Expression (4.74) provides the following formulas for $f_{14}(.)$ and $f_{23}(.)$:

$$
\begin{aligned}
f_{14}(x_1, x_4) = \sum_{m=1}^{M} & \left(\left(\sum_{n=1}^{4} |\alpha_{n,m}|^2 \right) (|x_1|^2 + |x_4|^2) \right. \\
& + 2\mathrm{Re}\{(-\alpha_{1,m}r_{1,m}^* - \alpha_{2,m}^*r_{2,m} - \alpha_{3,m}^*r_{3,m} - \alpha_{4,m}r_{4,m}^*)x_1 \\
& + (-\alpha_{4,m}r_{1,m}^* + \alpha_{3,m}^*r_{2,m} + \alpha_{2,m}^*r_{3,m} - \alpha_{1,m}r_{4,m}^*)x_4 \\
& \left. + (\alpha_{1,m}\alpha_{4,m}^* - \alpha_{2,m}^*\alpha_{3,m} - \alpha_{2,m}\alpha_{3,m}^* + \alpha_{1,m}^*\alpha_{4,m})x_1x_4^*\} \right)
\end{aligned}
\tag{4.76}
$$

$$
\begin{aligned}
f_{23}(x_2, x_3) = \sum_{m=1}^{M} & \left(\left(\sum_{n=1}^{4} |\alpha_{n,m}|^2 \right) (|x_2|^2 + |x_3|^2) \right. \\
& + 2\mathrm{Re}\{(-\alpha_{2,m}r_{1,m}^* - \alpha_{1,m}^*r_{2,m} - \alpha_{4,m}^*r_{3,m} - \alpha_{3,m}r_{4,m}^*)x_2 \\
& + (-\alpha_{3,m}r_{1,m}^* + \alpha_{4,m}^*r_{2,m} + \alpha_{1,m}^*r_{3,m} - \alpha_{2,m}r_{4,m}^*)x_3 + (\alpha_{2,m}\alpha_{3,m}^* \\
& \left. - \alpha_{1,m}^*\alpha_{4,m} - \alpha_{1,m}\alpha_{4,m}^* + \alpha_{2,m}^*\alpha_{3,m})x_2x_3^*\} \right)
\end{aligned}
\tag{4.77}
$$

There are other structures which provide behaviors similar to those of Equation (4.73). A few examples are given below:

$$
\begin{bmatrix} A_{12} & A_{34} \\ -A_{34} & A_{12} \end{bmatrix} \quad \begin{bmatrix} A_{12} & A_{34} \\ A_{34} & -A_{12} \end{bmatrix} \quad \begin{bmatrix} A_{12} & A_{34} \\ A_{34}^* & -A_{12}^* \end{bmatrix}
\tag{4.78}
$$

A similar idea can be used to combine two rate 3/4 transmission matrices (4×4) to build a rate 3/4 transmission matrix (8×8) and so on. An example of an 8×8 matrix which provides a rate 3/4 code is given below:

$$
\begin{bmatrix}
x_1 & x_2 & x_3 & 0 & x_4 & x_5 & x_6 & 0 \\
-x_2^* & x_1^* & 0 & -x_3 & x_5^* & -x_4^* & 0 & x_6 \\
x_3^* & 0 & -x_1^* & -x_2 & -x_6^* & 0 & x_4^* & x_5 \\
0 & -x_3^* & x_2^* & -x_1 & 0 & x_6^* & -x_5^* & x_4 \\
-x_4 & -x_5 & -x_6 & 0 & x_1 & x_2 & x_3 & 0 \\
-x_5^* & x_4^* & 0 & x_6 & -x_2^* & x_1^* & 0 & x_3 \\
x_6^* & 0 & -x_4^* & x_5 & x_3^* & 0 & -x_1^* & x_2 \\
0 & x_6^* & -x_5^* & -x_4 & 0 & x_3^* & -x_2^* & -x_1
\end{bmatrix}
\tag{4.79}
$$

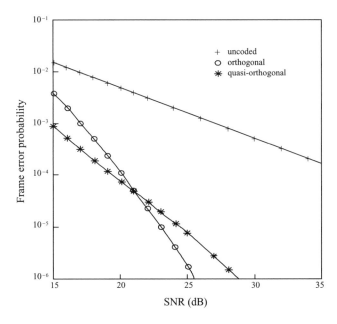

Figure 4.28 Bit error probability versus SNR for space–time block codes at 2 bits/s/Hz; 1 receive antenna.

Here, $n = 8$ antennas, $k = 6$ symbols, $p = 8$ transmissions. In this code, if we define $\mathcal{V}_i, i = 1, 2, \ldots, 8$, as the ith column, we have:

$$\langle \mathcal{V}_1, \mathcal{V}_i \rangle = 0, \ i \neq 5 \quad \langle \mathcal{V}_2, \mathcal{V}_i \rangle = 0, \ i \neq 6$$

$$\langle \mathcal{V}_3, \mathcal{V}_i \rangle = 0, \ i \neq 7 \quad \langle \mathcal{V}_4, \mathcal{V}_i \rangle = 0, \ i \neq 8$$

$$\langle \mathcal{V}_5, \mathcal{V}_i \rangle = 0, \ i \neq 1 \quad \langle \mathcal{V}_6, \mathcal{V}_i \rangle = 0, \ i \neq 2$$

$$\langle \mathcal{V}_7, \mathcal{V}_i \rangle = 0, \ i \neq 3 \quad \langle \mathcal{V}_8, \mathcal{V}_i \rangle = 0, \ i \neq 4 \tag{4.80}$$

The maximum likelihood decision metric, Expression (4.74), can be calculated as the sum of three terms $f_{14}(x_1, x_4) + f_{25}(x_2, x_5) + f_{36}(x_3, x_6)$ and similarly the decoding can be done using pairs of constellation symbols.

Figure 4.28 illustrates the performance of transmission of 2 bits/s/Hz using four transmit antennas and the rate one quasi-orthogonal code, the rate 1/2 full diversity orthogonal code and the uncoded 4PSK. The appropriate modulation schemes to provide the desired transmission rate for the space–time block codes, are 4PSK for the rate one code and 16 QAM for the rate 1/2 code.

Figure 4.29 presents the performance for four transmit antennas for space–time block codes at 3 bits/s/Hz. The rate one code and the uncoded system use 8PSK and the rate 3/4 code uses 16 QAM.

More details on the topic can be found in [238].

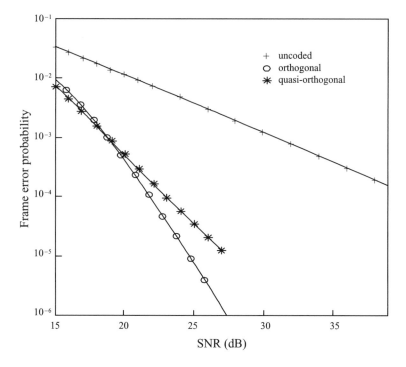

Figure 4.29 Bit error probability versus SNR for space–time block codes at 3 bits/s/Hz; 1 receive antenna.

4.6 SPACE–TIME CONVOLUTIONAL CODES

In Section 4.2 the coding gain was defined as:

$$\eta = \min_{c,e} \left(\prod_{i=1}^{r} \lambda_i \right)^{1/r}$$

over all code word pairs. Due to the similarity of Inequality (4.33) to an error bound for trellis coded modulation, rm is called the diversity gain (the slope of the pairwise error probability on a log–log plot) and η is called the coding gain (η^{-rm} is an offset on a log–log plot).

Consider only codes that achieve maximum diversity gain. Of these, we search for codes that give the largest possible coding gain.

Similarly to Equations (2.8)–(2.13) in Chapter 2, consider a set of convolutional codes whose output at time k is:

$$[x_1(k),\ x_2(k), \ldots, x_n(k)] = \mathbf{b}_r(kR)\mathbf{G}$$
$$= \mathbf{b}_r(kR)[G_1 G_2 \cdots G_n] \tag{4.81}$$

where $G_i, i = 1, \ldots, n$ is the ith column of the matrix $\mathbf{G}$ which is given by:

$$\mathbf{G} = \begin{bmatrix} g_{11} & g_{12} & \cdots & g_{1n} \\ g_{21} & g_{22} & \cdots & g_{2n} \\ g_{31} & g_{32} & \cdots & g_{3n} \\ g_{41} & g_{42} & \cdots & g_{4n} \\ \cdots & \cdots & \cdots & \cdots \\ g_{QR,1} & g_{QR,2} & \cdots & g_{QR,n} \end{bmatrix} \tag{4.82}$$

and $\mathbf{b}_r(kR) = [b_1, \ldots, b_{QR}]$ is a length QR binary row vector, $b_j \in \{0, 1\}$ for $i = 1, \ldots, n$ are taken to be in an alphabet of size s so that $g_{ij} \in \{0, 1, \ldots s - 1\}$. The arithmetic in Equation (4.81) is mod-s so $x_i(j) \in \{0, 1, \ldots s - 1\}$ also. Thus, the output code word is $(c_1(k), \ldots, c_n(k)) = (z[x_1(k))], \ldots, z[x_n(k))])$ where $z[x] = \exp(j(2\pi x/s + \phi))$ and $0 \leq \phi \leq 2\pi$ allows arbitrary rotation.

At each time slot, R bits are input into the convolutional encoder and the state is determined by $(Q - 1)R$ bits. For space–time convolutional coding using Equation (4.81), the outputs $x_1(k), \ldots, x_n(k)$ are each mapped into a constellation of size s and transmitted simultaneously from n antennas. The input to the encoder at time slot k is $b_1, \ldots, b_R$. The state is given by $b_{R+1}, \ldots, b_{QR}$. For the next time slot, new input bits are received and the old input bits are shifted to the right into the state which explains the subscript on $\mathbf{b}_r(kR)$.

The results of the search for the good codes are presented in Tables 4.4–4.9 with the following notation:

η_t (4.2)	– coding gain for the trellis space–time codes defined in Section 4.2.
η_c (4.6)	– coding gain for the convolutional space–time codes defined in Section 4.6.
$e_{p\,\min}$	– minimum effective product distance.
L	– total number of time slots considered in the bound calculation.
$\eta_{AP}(L)$	– considers all trellis paths describing error events of length L or longer.
$\eta_{CP}(L)$	– considers sets of code word pairs (continuous error paths) and searches for smallest upper bound.
$\tilde{Q}$	– the minimum length error event of $\tilde{Q}$ slots.

Performance results are shown in Figures 4.30–4.34.
More details on the topic can be found in [231, 245–251].

4.7 ALGEBRAIC SPACE–TIME CODES

Let C be a binary convolutional code of rate $1/L_t$ with the transfer function encoder $\mathbf{Y}(D) = X(D)\mathbf{G}(D)$, where $\mathbf{G}(D) = [G_1(D)G_2(D)\ldots G_{Lt}(D)]$. In the natural space–time formatting of C for BPSK transmission, the output sequence corresponding to $Y_i(D) = X(D)G_i(D)$ is assigned to the ith transmit antenna. The number of transmit antennas is L_t. The resulting space–time code C satisfies the binary rank criterion under relatively mild conditions on the connection polynomials $G_i(D)$.

Table 4.4   Optimum q-state 2 b/s/Hz 4PSK space–time codes [231] © 2002, IEEE

q	$\eta_t(4.2)$	$\eta_c(4.6)$	$\bar{\eta}_{AP}(\tilde{Q})$	$\bar{\eta}_{CP}(\tilde{Q})$	$\mathbf{G}^{T}$
4	2	$\sqrt{8}$	3.54	3.80	$\begin{bmatrix} 2 & 0 & 1 & 2 \\ 2 & 2 & 2 & 1 \end{bmatrix}$
8	$\sqrt{12}$	4	3.42	4.00	$\begin{bmatrix} 0 & 2 & 1 & 0 & 2 \\ 2 & 1 & 0 & 2 & 2 \end{bmatrix}$
16	$\sqrt{12}$	$\sqrt{32}$	5.76	6.24	$\begin{bmatrix} 0 & 2 & 1 & 1 & 2 & 0 \\ 2 & 2 & 1 & 2 & 0 & 2 \end{bmatrix}$
32	$\sqrt{12}$	6	5.63	6.33	$\begin{bmatrix} 2 & 0 & 1 & 2 & 1 & 2 & 2 \\ 2 & 2 & 0 & 1 & 2 & 0 & 2 \end{bmatrix}$

Table 4.5   Optimum q-state 2 b/s/Hz 4PSK space–time codes [231] © 2002, IEEE

q	η_t	η_c	$e_{p\,min}$	$\bar{\eta}_{AP}(L)$	$\bar{\eta}_{CP}(L)$	$\mathbf{G}^{T}$
4	2	$\sqrt{8}$	8	3.54	3.80	$\begin{bmatrix} 2 & 0 & 1 & 2 \\ 2 & 2 & 2 & 1 \end{bmatrix}$
8	$\sqrt{12}$	4	16	3.42	4.00	$\begin{bmatrix} 0 & 2 & 1 & 0 & 2 \\ 2 & 1 & 0 & 2 & 2 \end{bmatrix}$
16	$\sqrt{12}$	$\sqrt{32}$	32	5.76	6.24	$\begin{bmatrix} 0 & 2 & 1 & 1 & 2 & 0 \\ 2 & 2 & 1 & 2 & 0 & 2 \end{bmatrix}$
32	$\sqrt{12}$	6	36	5.63	6.33	$\begin{bmatrix} 2 & 0 & 1 & 2 & 1 & 2 & 2 \\ 2 & 2 & 0 & 1 & 2 & 0 & 2 \end{bmatrix}$

Table 4.6   Optimum q-state 1 b/s/Hz BSK space–time codes [231] © 2002, IEEE

q	η_c	$e_{p\,min}$	$\bar{\eta}_{AP}(L)$	$\bar{\eta}_{CP}(L)$	$\mathbf{G}^{T}$
2	4	4	4.00	4.00	$\begin{bmatrix} 0 & 1 \\ 1 & 1 \end{bmatrix}$
4	$\sqrt{48}$	12	7.33	6.93	$\begin{bmatrix} 0 & 1 & 1 \\ 1 & 0 & 1 \end{bmatrix}^{\dagger}$
4	$\sqrt{32}$	8	7.33	7.73	$\begin{bmatrix} 0 & 1 & 1 \\ 1 & 1 & 1 \end{bmatrix}$
8	$\sqrt{80}$	20	10.06	10.47	$\begin{bmatrix} 1 & 0 & 1 & 1 \\ 1 & 1 & 0 & 1 \end{bmatrix}$
16	$\sqrt{112}$	28	11.38	14.27	$\begin{bmatrix} 1 & 1 & 0 & 1 & 1 \\ 0 & 1 & 1 & 1 & 1 \end{bmatrix}$

† denotes that the code is catastrophic

Table 4.7 Optimum q-state 1 b/s/Hz BSK 3-space–time codes [231] © 2002, IEEE

q	η_c	$e_{p\,min}$	$\bar{\eta}_{AP}(L)$	$\bar{\eta}_{CP}(L)$	$\mathbf{G}^T$
4	4	$\dfrac{64}{27}$	5.08	5.08	$\begin{bmatrix} 0 & 1 & 1 \\ 1 & 0 & 1 \\ 1 & 1 & 1 \end{bmatrix}$
8	$256^{1/3}$	$\dfrac{256}{27}$	7.65	7.05	$\begin{bmatrix} 1 & 0 & 0 & 1 \\ 1 & 0 & 1 & 0 \\ 1 & 1 & 1 & 1 \end{bmatrix}^{\dagger}$
8	$192^{1/3}$	$\dfrac{64}{9}$	7.65	8.32	$\begin{bmatrix} 1 & 0 & 1 & 1 \\ 1 & 1 & 0 & 1 \\ 1 & 0 & 1 & 1 \end{bmatrix}$
16	8	$\dfrac{512}{27}$	10.00	10.18	$\begin{bmatrix} 1 & 0 & 0 & 1 & 1 \\ 1 & 1 & 0 & 1 & 0 \\ 1 & 1 & 1 & 0 & 1 \end{bmatrix}$

† denotes that the code is catastrophic

Table 4.8 Optimum q-state 1b/s/Hz BSK 4-space–time codes [231] © 2002, IEEE

q	η_c	$e_{p\,min}$	$\bar{\eta}_{AP}(L)$	$\bar{\eta}_{CP}(L)$	$\mathbf{G}^T$
8	4	1	5.97	5.97	$\begin{bmatrix} 0 & 1 & 0 & 1 \\ 0 & 1 & 1 & 1 \\ 1 & 0 & 1 & 0 \\ 1 & 1 & 1 & 0 \end{bmatrix}$
16	$1280^{1/4}$	5	8.11	8.37	$\begin{bmatrix} 1 & 0 & 0 & 0 & 1 \\ 1 & 0 & 1 & 1 & 1 \\ 1 & 1 & 0 & 1 & 1 \\ 1 & 1 & 1 & 1 & 0 \end{bmatrix}^{\dagger}$
16	$1024^{1/4}$	4	7.99	9.32	$\begin{bmatrix} 0 & 1 & 1 & 0 & 1 \\ 1 & 1 & 0 & 0 & 1 \\ 1 & 1 & 1 & 1 & 0 \\ 1 & 1 & 1 & 1 & 1 \end{bmatrix}$
32	$4352^{1/4}$	17	9.80	10.38	$\begin{bmatrix} 1 & 0 & 0 & 0 & 0 & 1 \\ 1 & 0 & 1 & 1 & 1 & 1 \\ 1 & 1 & 1 & 0 & 1 & 0 \\ 1 & 1 & 1 & 1 & 0 & 0 \end{bmatrix}$

† denotes that the code is catastrophic

Table 4.9   Optimum q-state 2 b/s/Hz 4 PSK space–time code using three transmit antennas [231] © 2002, IEEE

q	η_c	$e_{p\ \min}$	$\overline{\eta}_{AP}(L)$	$\overline{\eta}_{CP}(L)$	$\mathbf{G}^T$					
16	$32^{1/3}$	$\dfrac{256}{27}$	3.90	4.72	$\begin{bmatrix} 0 & 2 & 1 & 2 & 2 & 0 \\ 1 & 2 & 2 & 0 & 0 & 2 \\ 2 & 2 & 0 & 2 & 1 & 2 \end{bmatrix}$					

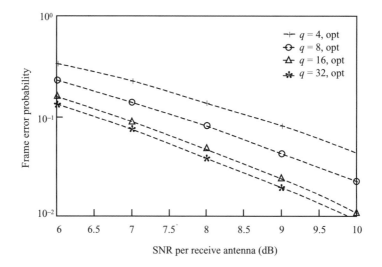

Figure 4.30   Performance comparison of some best 2 b/s/Hz, QPSK, q-state STCs with two transmit and two receive antennas (SNR per receive antenna $= nE_s/N_0$).

4.7.1 Full spatial diversity

Code C , associated with the $1/L_t$ convolutional code C, satisfies the binary rank criterion, and thus achieves full spatial diversity for BPSK transmission if and only if the transfer function matrix $\mathbf{G}(D)$ of C has full rank L_t as a matrix of coefficients over the binary field $\mathbb{F}$. This result follows directly from the stacking construction and can be easily generalized to recursive convolutional codes.

It is straightforward to see that the zeros symmetry codes satisfy the stacking construction conditions. However, the set of binary rate $1/L_t$ convolutional codes with the optimal free distance d_{free} offers a richer class of space–time codes as their associated natural space–time codes usually achieve full spatial diversity. Furthermore, these codes outperform the zeros symmetry codes uniformly. A collection of these codes is given in Table 4.10.

4.7.2 QPSK modulation

Experimentally, codes obtained by replacing each zero in the binary generator matrix by a two yield the best simulated frame error rate performance in most cases. These codes are

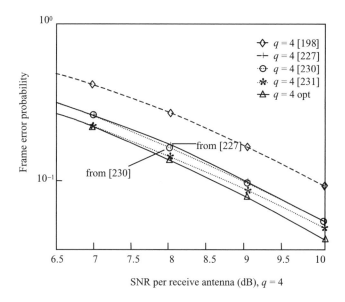

Figure 4.31 Performance comparisons of some 2 b/s/Hz, QPSK, four-state STCs with two transmit and two receive antennas.

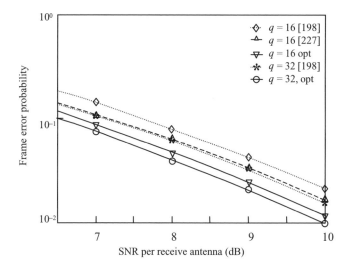

Figure 4.32 Performance comparisons of some 2 b/s/Hz, QPSK, 16- and 32-state STCs with two transmit and two receive antennas.

reported in Table 4.11 for different numbers of transmit antennas and constraint lengths v and are used in generating the simulation results in the next section.

A space–time code has zeros symmetry if every baseband code word difference $f(\mathbf{c}) - f(\mathbf{e})$ is upper and lower triangular and has appropriate non-zero entries to ensure full rank. The zeros symmetry property is sufficient for full rank but not necessary. Some simulation results for the system performance are shown in Figures 4.35–4.40.

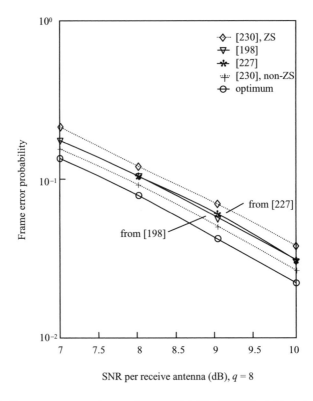

Figure 4.33 Performance comparisons of some 2 b/s/Hz, QPSK, eight-state STCs with two transmit and two receive antennas.

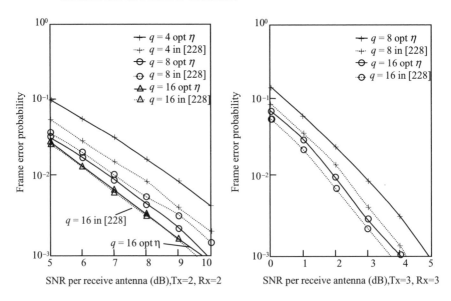

Figure 4.34 Performance of some 1 b/s/Hz, BPSK, q-state STCs with two transmit and two receive or three transmit and three receive antennas (performance of the best noncatastrophic codes similar).

Table 4.10 Natural full diversity space–time convolutional codes with optimal free distance for BPSK modulation [252] © 2002, IEEE

L_t	v	Connection polynomials
2	2	5, 7
	3	64, 74
	4	46, 72
	5	65, 57
	6	554, 744
3	3	54, 64, 74
	4	52, 66, 76
	5	47, 53, 75
	6	554, 624, 764
4	4	52, 56, 66, 76
	5	53, 67, 71, 75
5	5	75, 71, 73, 65, 57

Table 4.11 Linear $\mathbb{Z}_4$ space–time codes for QPSK modulation [252] © 2002, IEEE

L	v	Connection polynomials
2	1	$1 + 2D, 2 + D.$
	2	$1 + 2D + D^2, 1 + D + D^2.$
	3	$1 + D + 2D^2 + D^3, 1 + D + D^2 + D^3.$
	4	$1 + 2D + 2D^2 + D^3 + D^4, 1 + D + D^2 + 2D^3 + D^4.$
	5	$1 + D + 2D^2 + D^3 + 2D^4 + D^5, 1 + 2D + D^2 + D^3 + D^4 + D^5.$
3	2	$1 + 2D + 2D^2, 2 + D + 2D^2, 2 + 2D + D^2.$
	3	$1 + 2D + D^2 + D^3, 1 + D + 2D^2 + D^3, 1 + D + D^2 + D^3.$
	4	$1 + 2D + D^2 + 2D^3 + D^4, 1 + D + 2D^2 + D^3 + D^4, 1 + D + D^2 + D^3 + D^4.$
	5	$1 + 2D + 2D^2 + D^3 + D^4 + D^5, 1 + 2D + D^2 + 2D^3 ++ D^4 + D^5,$ $1 + D + D^2 + D^3 + 2D^4 + D^5.$
4	3	$1 + 2D + 2D^2 + 2D^3, 2 + D + 2D^2 + 2^3, 2 + 2D + D^2 + 2D^3, 2 + 2D + 2D^2 + D^3$
	4	$1 + 2D + D^2 + 2D^3 + D^4, 1 + 2D + D^2 + D^3 + D^4, 1 + D + 2D^2 + D^3 + D^4,$ $1 + D + D^2 + D^3 + D^4.$
	5	$1 + 2D + D^2 + 2D^3 + D^4 + D^5, 1 + D + 2D^2 + D^3 + D^4 + D^5,$ $1 + D + D^2 + 2D^3 + 2D^4 + D^5, 1 + D + D^2 + D^3 + 2D^4 + D^5.$
5	4	$1 + 2D + 2D^2 + 2D^3 + 2D^4, 2 + D + 2D^2 + 2D^3 + 2D^4, 2 + 2D + D^2 + 2D^3 + 2D^4,$ $2 + 2D + 2D^2 + D^3 + 2D^4, 2 + 2D + 2D^2 + 2D^3 + D^4,$
	5	$1 + D + D^2 + D^3 + 2D^4 + D^5, 1 + D + D^2 + D^3 + 2D^4 + D^5,$ $1 + D + D^2 + 2D^3 + D^4 + D^5, 1 + D + 2D^2 + D^3 + 2D^4 + D^5,$ $1 + 2D + D^2 + D^3 + D^4 + D^5.$

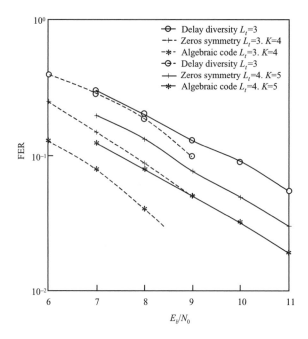

Figure 4.35 Performance of BPSK space–time codes with $L_r = 1$. The number of transmit antennas is represented by L_t, receive antennas by L_r, and constraint lengths by K. The number of bits per frame is 100.

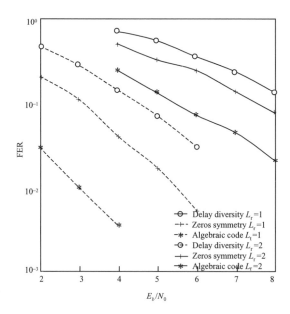

Figure 4.36 FER for BPSK space–time codes with $L_t = 5$ and $K = 6$.

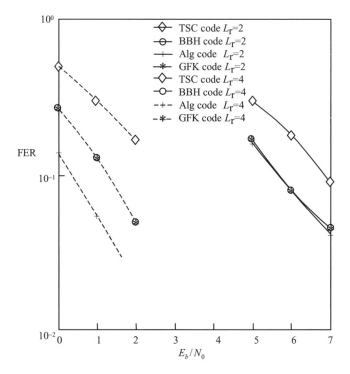

Figure 4.37 Performance of four-state QPSK space–time codes with $L_t = 2$. Four-state codes due to Tarokh, Seshadri, and Calderbank (TSC) [198], Baro, Bauch, and Hansmann (BBH) [227], Grimm, Fitz, and Krogmeier (GFK) [229], and the new linear $\mathbb{Z}_4$ code in Table 4.11 with two and four receive antennas.

4.8 DIFFERENTIAL SPACE–TIME MODULATION

When channel estimation is not available, a differential modulation and detection might be a solution. We start with a simple transmission scheme [261] for exploiting diversity given by two transmit antennas when neither the transmitter nor the receiver has access to channel state information. At the receiver, decoding is achieved with low decoding complexity. The transmission provides full spatial diversity and requires no channel state side information at the receiver. The scheme can be considered as the extension of conventional differential detection schemes to two transmit antennas.

In traditional differential phase shift keying (DPSK) the data is encoded in the difference of the phase of two consecutive symbols. For b bits per symbol, the symbol would usually have the form:

$$\Delta(l) = e^{i\theta(l)} = e^{2\pi i l/L}, \quad l = 0, 1, \ldots L - 1 \tag{4.83}$$

where $i = \sqrt{-1}$ and $L = 2^b$. If the data sequence is $z_1, z_2, z_3, \ldots$ the transmitted symbol sequences would be $s_0, s_1, s_2, \ldots$ where, at time t,

$$s_t = \Delta(z_t)s_{t-1} \tag{4.84}$$

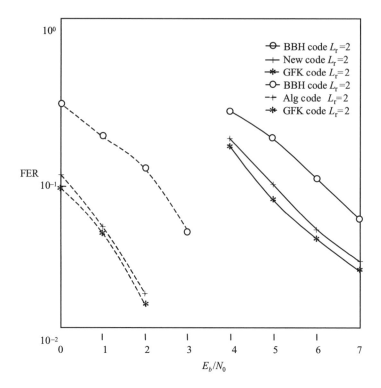

Figure 4.38 Performance of eight-state QPSK space–time codes with $L_t = 2$.

Bits are mapped into the symbols by using *Gray mapping*, e.g. for $b = 2$:

$$M(z) \Rightarrow \Delta(z) = \Delta(l)$$

$$\Delta(00) = \Delta(0) = e^{2\pi i \times 0/4} = 1$$

$$\Delta(01) = \Delta(1) = e^{2\pi i \times 1/4} = e^{i\pi/2}$$

$$\Delta(10) = \Delta(3) = e^{2\pi i \times 3/4} = e^{-i\pi/2} \qquad (4.85)$$

$$\Delta(11) = \Delta(2) = e^{2\pi i \times 2/4} = -1$$

This ensures that the most probable symbol errors cause the minimum number of bit errors.

Signal r_t from the received signal sequence $r_0, r_1, r_2, \ldots$ can be represented as:

$$r_t = h_t s_t + n_t \qquad (4.86)$$

where h_t is the channel gain. The receiver would extract the information by processing the received signal samples as follows:

$$\hat{\theta}_t = \arg r_t r_{t-1}^* \qquad (4.87)$$

which, for $h_t \cong h_{t-1}$, $|h_t| = 1$ and no noise, gives:

$$\hat{\theta}_t = \arg h_t s_t h_{t-1}^* s_{t-1}^* = \arg |h_t|^2 \Delta(z_t) |s_{t-1}|^2 = \arg \Delta(z_t) \qquad (4.88)$$

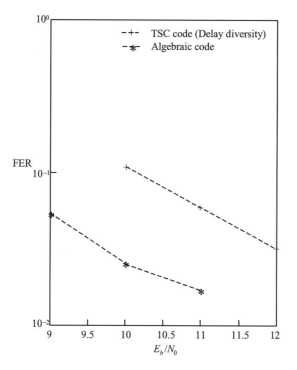

Figure 4.39 FER for QPSK space–time codes with $L_t = 4$ and $L_r = 1$. The new 64-state space–time code for four transmit antennas and the TSC 64-state code (i.e. delay diversity).

By using analogy with the previous discussion we will now extend this principle to two-dimensional constellations. Later in the chapter we will generalize these schemes to multi-dimensional constellations.

Let us restrict the constellation $\mathcal{A}$ to 2^b-PSK for some $b = 1, 2, 3, \ldots$, but in reality only BPSK, QPSK, and 8PSK are of interest. Thus,

$$\mathcal{A} = \left\{ \frac{e^{2\pi ki/2^b}}{\sqrt{2}} \middle| k = 0, 1, \ldots, 2^b - 1 \right\} \tag{4.89}$$

In order to implementl Equation (4.84) in a two-dimensional constellation, let us first consider the following relation between the vectors.

Given a pair of 2^b-PSK constellation symbols, x_1 and x_2, we first observe that the complex vectors $(x_1 x_2)$ and $(-x_2^* x_1^*)$ are orthogonal to each other and have unit lengths defined by $\sum_i x_i x_i^* = 1$. Any two-dimensional vector $\mathcal{X} = (x_3 x_4)$ can be uniquely represented in the orthonormal basis given by these vectors. In other words, there exists a unique complex vector $\mathbf{P}_{\mathcal{X}} = (\mathbf{A}_{\mathcal{X}} \mathbf{B}_{\mathcal{X}})$ such that $\mathbf{A}_{\mathcal{X}}$ and $\mathbf{B}_{\mathcal{X}}$ satisfy the vector equation:

$$(x_3 x_4) = \mathbf{A}_{\mathcal{X}}(x_1 x_2) + \mathbf{B}_{\mathcal{X}}(-x_2^* x_1^*) \tag{4.90}$$

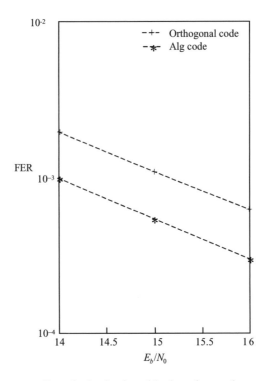

Figure 4.40 Performance of an algebraic short block and an orthogonal code with $L_t = 3$ and $L_r = 1$. The new algebraic code is used with QPSK modulation and the orthogonal code is used with 16 QAM. The increased size of the constellation in the case of the orthogonal code is necessary to support the same throughput [202]. More details on the topic can be found in [252–260].

or

$$x_3 = \mathbf{A}_\mathcal{X} x_1 - \mathbf{B}_\mathcal{X} x_2^*$$
$$x_4 = \mathbf{A}_\mathcal{X} x_2 + \mathbf{B}_\mathcal{X} x_1^*$$
$$x_3 x_1^* = \mathbf{A}_\mathcal{X} x_1 x_1^* - \mathbf{B}_\mathcal{X} x_2^* x_1^*$$
$$x_4 x_2^* = \mathbf{A}_\mathcal{X} x_2 x_2^* + \mathbf{B}_\mathcal{X} x_1^* x_2^*$$

giving $\mathbf{A}_\mathcal{X}$ and $\mathbf{B}_\mathcal{X}$ as:

$$\mathbf{A}_\mathcal{X} = x_3 x_1^* + x_4 x_2^* \tag{4.91}$$
$$\mathbf{B}_\mathcal{X} = -x_3 x_2 + x_4 x_1$$

These relations will be used later to implement Equation (4.84). For an equivalent implementation of Equation (4.85) we will need a few additional definitions.

We define the set $\mathcal{V}_\mathcal{X}$ to consist of all the vectors $\mathbf{P}_\mathcal{X}$, $\mathcal{X} \notin \mathcal{A} \times \mathcal{A}$. The set $\mathcal{V}_\mathcal{X}$ has the following properties:

- *Property A*: It has 2^{2b} elements corresponding to the pairs $(x_3 x_4)$ of constellation symbols.

- *Property B*: All elements of $\mathcal{V}_\mathcal{X}$ have unit length.

- *Property C*: For any two distinct elements $\mathcal{X} = (x_1 x_2)$ and $\mathcal{Y} = (y_1 y_2)$ of $\mathcal{A} \times \mathcal{A}$

$$||P_{\mathcal{X}} - P_{\mathcal{Y}}|| = ||(x_1 x_2) - (y_1 y_2)||.$$

- *Property D*: The minimum distance between any two distinct elements of $V_{\mathcal{X}}$ is equal to the minimum distance of the 2^b-PSK constellation $\mathcal{A}$.

The above properties hold because the mapping $\mathcal{X} \to P_{\mathcal{X}}$ is just a change of basis from the standard basis given by vectors $\{(1\ 0), (0\ 1)\}$ to the orthonormal basis given by $\{(x_1\ x_2), (-x_2^*\ x_1^*)\}$, which preserves the distances between the points of the two-dimensional complex space.

The first ingredient of construction is the choice of an arbitrary set V having Properties A and B. It is also handy if V has Properties C and D as well. As a natural choice for such a set V, we may fix an arbitrary pair $\mathcal{X} \in \mathcal{A} \times \mathcal{A}$ and let $V = V_{\mathcal{X}}$. Because the 2^b-PSK constellation $\mathcal{A}$ always contains the signal point $1/\sqrt{2}$, we choose to fix $\mathcal{X} = ((1/\sqrt{2})\ (1/\sqrt{2}))$.

We also need an arbitrary bijective mapping $\mathcal{M}$ of blocks of $2b$ bits onto V. Among all the possibilities for $\mathcal{M}$, we choose the mapping analogous to Equation (4.85). Given a block $\mathcal{B}$ of $2b$ bits, the first b bits are mapped into a constellation symbol a_3 and the second b bits are mapped into a constellation symbol a_4 using *Gray mapping*.

Let $a_1 = a_2 = 1/\sqrt{2}$, then $\mathcal{M}(\mathcal{B}) = (A(\mathcal{B})\ B(\mathcal{B}))$ is defined by Equation (4.91):

$$A(\mathcal{B}) = a_3 a_1^* + a_4 a_2^*$$
$$B(\mathcal{B}) = -a_3 a_2 + a_4 a_1 \tag{4.92}$$

Clearly, $\mathcal{M}$ maps any $2b$ bits onto V. Conversely, given $(A(\mathcal{B})\ B(\mathcal{B}))$, the pair $(a_3\ a_4)$ is recovered by Equation (4.90):

$$(a_3\ a_4) = A(\mathcal{B})(a_1\ a_2) + B(\mathcal{B})(-a_2^*\ a_1^*) \tag{4.93}$$

The block $\mathcal{B}$ is then constructed by inverse Gray mapping of a_3 and a_4.

4.8.1 The encoding algorithm

The transmitter begins the transmission by sending arbitrary symbols s_1 and s_2 from transmit antennas one and two respectively at time one, and symbols $-s_2^*$ and s_1^* at time two unknown to the receiver. These two transmissions do not convey any information. The transmitter then encodes the rest of the data in an inductive manner. Suppose that s_{2t-1} and s_{2t} are sent, respectively, from transmit antennas one and two at time $2t - 1$, and that $-s_{2t}^*$, s_{2t-1}^* are sent, respectively, from antennas one and two at time $2t$. At time $2t + 1$, a block of $2b$ bits $\mathcal{B}_{2t+1}$ arrives at the encoder. The transmitter uses the mapping $\mathcal{M}$ given by Equation (4.92) and computes $\mathcal{M}(\mathcal{B}_{2t+1}) = (A(\mathcal{B}_{2t+1})\ B(\mathcal{B}_{2t+1}))$. Then, in accordance with Equation (4.92) it computes:

$$(s_{2t+1}\ s_{2t+2}) = A(\mathcal{B}_{2t+1})(s_{2t-1}\ s_{2t}) + B(\mathcal{B}_{2t+1})(-s_{2t}^*\ s_{2t-2}^*) \tag{4.94}$$

The transmitter then sends s_{2t+1} and s_{2t+2}, respectively, from transmit antennas one and two at time $2t + 1$, and $-s_{2t+2}^*$, s_{2t+1}^* from antennas one and two at time $2t + 2$. This process is inductively repeated until the end of the frame (or end of the transmission). The block diagram of the encoder is given in Figure 4.41.

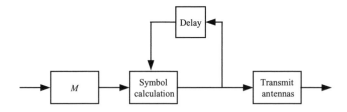

Figure 4.41 Transmitter block diagram.

4.8.1.1 *Example*

We assume that the constellation is BPSK ($b = 1$) consisting of the points $-1/\sqrt{2}$ and $1/\sqrt{2}$. Then the set $\mathcal{V} = \{(1\ 0), (0\ 1), (-1\ 0), (0\ -1)\}$. Recall that the Gray mapping maps a bit $i = 0, 1$ to $(-1)^i/\sqrt{2}$. We set $a_1 = a_2 = 1/\sqrt{2}$. Then the mapping $\mathcal{M}$ maps two bits onto $\mathcal{V}$ and is given by:

$$\mathcal{M}(00) = (1\ 0)$$
$$\mathcal{M}(10) = (0\ 1)$$
$$\mathcal{M}(01) = (0\ -1) \tag{4.95}$$
$$\mathcal{M}(11) = (-1\ 0)$$

Now suppose that at time $2t - 1$, $s_{2t-1} = 1/\sqrt{2}$ and $s_{2t} = -1/\sqrt{2}$ are sent, respectively, from antennas one and two, and at time $2t$, $-s_{2t}^* = 1/\sqrt{2}$ and $s_{2t-1}^* = 1/\sqrt{2}$ are sent. Suppose that the input to the encoder at time $2t + 1$ is the block of bits 10. Since $\mathcal{M}(10) = (0\ 1)$, we have $A(10) = 0$ and $B(10) = 1$. Then the values s_{2t+1} and s_{2t+2} corresponding to input bits 10 are computed as follows:

$$(s_{2t+1}\ s_{2t+2}) = 0 \cdot \left(\frac{1}{\sqrt{2}}\ \frac{-1}{\sqrt{2}}\right) + 1 \cdot \left(\frac{1}{\sqrt{2}}\ \frac{1}{\sqrt{2}}\right)$$
$$= \left(\frac{1}{\sqrt{2}}\ \frac{1}{\sqrt{2}}\right) \tag{4.96}$$

Thus, at time $2t + 1$, $s_{2t+1} = 1/\sqrt{2}$ and $s_{2t+2} = 1/\sqrt{2}$ are sent, respectively, from antennas one and two, and at time $2t + 2$, $-s_{2t+2}^* = -1/\sqrt{2}$ and $s_{2t+1}^* = 1/\sqrt{2}$ are sent, respectively, from antennas one and two. Transmitted symbols at time $2t + 1$ and $2t + 2$ correspond to the input bits 00, 10, 01 and 11 for this scenario. The results are summarized in Tables 4.12 and 4.13.

4.8.2 Differential decoding

For notational simplicity, we first present the results for one receive antenna. Write r_t for r_t^1, η_t for η_t^1 and α_1, α_2, respectively, for $\alpha_{1,1}$, $\alpha_{2,1}$, knowing that this can cause no confusion since there is only one receive antenna.

$r_{2t-1}, r_{2t}, r_{2t+1}$, and r_{2t+2} are received signal samples. Let

$$\Lambda(\alpha_1, \alpha_2) = \begin{bmatrix} \alpha_1 & \alpha_2^* \\ \alpha_2 & -\alpha_1^* \end{bmatrix} \tag{4.97}$$

Table 4.12   Transmitted symbols at time $2t + 1$ [261]
© 2002, IEEE

Input bits at time $2t + 1$	Antenna 1	Antenna 2
00	$\frac{1}{\sqrt{2}}$	$\frac{-1}{\sqrt{2}}$
10	$\frac{1}{\sqrt{2}}$	$\frac{1}{\sqrt{2}}$
01	$\frac{-1}{\sqrt{2}}$	$\frac{-1}{\sqrt{2}}$
11	$\frac{-1}{\sqrt{2}}$	$\frac{1}{\sqrt{2}}$

Table 4.13   Transmitted symbols at time $2t + 2$ [261]

Input bits at time $2t + 1$	Antenna 1	Antenna 2
00	$\frac{1}{\sqrt{2}}$	$\frac{1}{\sqrt{2}}$
10	$\frac{-1}{\sqrt{2}}$	$\frac{1}{\sqrt{2}}$
01	$\frac{1}{\sqrt{2}}$	$\frac{-1}{\sqrt{2}}$
11	$\frac{-1}{\sqrt{2}}$	$\frac{-1}{\sqrt{2}}$

and

$$N_{2t-1} = (\eta_{2t-1} \; \eta_{2t}^*) \tag{4.98}$$

Then the received signal can be represented as:

$$(r_{2t-1} \; r_{2t}^*) = (s_{2t-1} \; s_{2t})\mathbf{\Lambda}(\alpha_1, \alpha_2) + N_{2t-1} \tag{4.99}$$

and

$$(r_{2t+1} \; r_{2t+2}^*) = (s_{2t+1} \; s_{2t+2})\mathbf{\Lambda}(\alpha_1, \alpha_2) + N_{2t+1} \tag{4.100}$$

The receiver will create

$$
\begin{aligned}
(r_{2t+1} \; r_{2t+2}^*) \cdot (r_{2t-1} \; r_{2t}^*) = {} & (s_{2t+1} \; s_{2t+2})\mathbf{\Lambda}(\alpha_1, \alpha_2)\mathbf{\Lambda}^*(\alpha_1, \alpha_2)(s_{2t-1}^* \; s_{2t}^*) \\
& + (s_{2t+1} \; s_{2t+2})\mathbf{\Lambda}(\alpha_1, \alpha_2)N_{2t-1}^* \\
& + N_{2t+1}\mathbf{\Lambda}^*(\alpha_1, \alpha_2)(s_{2t-1} \; s_{2t})^* + N_{2t+1}N_{2t-1}^*
\end{aligned} \tag{4.101}
$$

which gives:

$$
\begin{aligned}
r_{2t+1}r_{2t-1}^* + r_{2t+2}^*r_{2t} = {} & (|\alpha_1|^2 + |\alpha_2|^2)(s_{2t+1}s_{2t-1}^* + s_{2t+1}s_{2t}^*) \\
& + (s_{2t+1} \; s_{2t+2})\mathbf{\Lambda}(\alpha_1, \alpha_2)N_{2t-1}^* \\
& + N_{2t+1}\mathbf{\Lambda}^*(\alpha_1, \alpha_2)(s_{2t-1} \; s_{2t})^* + N_{2t+1}N_{2t-1}^*
\end{aligned}
$$

For a more compact representation we use the following notation:

$$\mathcal{R}_1 = r_{2t+1}r_{2t-1}^* + r_{2t+2}^*r_{2t}$$

$$\mathcal{N}_1 = (s_{2t+1} \; s_{2t+2})\mathbf{\Lambda}(\alpha_1, \alpha_2)N_{2t-1}^* + N_{2t+1}\mathbf{\Lambda}^*(\alpha_1, \alpha_2)(s_{2t-1} \; s_{2t})^* + N_{2t+1}N_{2t-1}^*$$

which gives

$$\mathcal{R}_1 = (|\alpha_1|^2 + |\alpha_2|^2)A(\mathcal{B}_{2t-1}) + \mathcal{N}_1 \tag{4.102}$$

For the second vector term in the right side of Equation (4.94) we have:

$$(r_{2t} - r_{2t-1}^*) = (-s_{2t}^*\ s_{2t-1}^*)\mathbf{\Lambda}(\alpha_1, \alpha_2) + N_{2t}$$

where

$$N_{2t} = (\eta_{2t} - \eta_{2t-1}^*)$$

The receiver will now create

$$(r_{2t+1}\ r_{2t+2}^*) \cdot (r_{2t} - r_{2t-1}^*) = (s_{2t+1}\ s_{2t+2})\mathbf{\Lambda}(\alpha_1, \alpha_2)\mathbf{\Lambda}^*(\alpha_1, \alpha_2)(-s_{2t}\ s_{2t-1})$$
$$+ (s_{2t+1}\ s_{2t+2})\mathbf{\Lambda}(\alpha_1, \alpha_2)N_{2t}^*$$
$$+ N_{2t+1}\mathbf{\Lambda}^*(\alpha_1, \alpha_2)(-s_{2t}^*\ s_{2t-1}^*)^* + N_{2t+1}N_{2t}^*$$

giving:

$$r_{2t+1}r_{2t}^* - r_{2t+2}^*r_{2t-1} = (|\alpha_1|^2 + |\alpha_2|^2)(-s_{2t+1}s_{2t}^* + s_{2t+2}s_{2t-1}^*)$$
$$+ (s_{2t+1}\ s_{2t+2})\mathbf{\Lambda}(\alpha_1, \alpha_2)N_{2t}^*$$
$$+ N_{2t+1}\mathbf{\Lambda}^*(\alpha_1, \alpha_2)(-s_{2t}^*\ s_{2t-1}^*)^* + N_{2t+1}N_{2t}^*$$

With the notation

$$\mathcal{R}_2 = r_{2t+1}r_{2t}^* - r_{2t+2}^*r_{2t-1}$$

$$\mathcal{N}_2 = +(s_{2t+1}\ s_{2t+2})\mathbf{\Lambda}(\alpha_1, \alpha_2)N_{2t}^* + N_{2t+1}\mathbf{\Lambda}^*(\alpha_1, \alpha_2)(-s_{2t}^*\ s_{2t-1}^*)^* + N_{2t+1}N_{2t}^*$$

we have

$$\mathcal{R}_2 = (|\alpha_1|^2 + |\alpha_2|^2)B(\mathcal{B}_{2t-1}) + \mathcal{N}_2 \tag{4.103}$$

The final result can be represented as:

$$(\mathcal{R}_1\mathcal{R}_2) = (|\alpha_1|^2 + |\alpha_2|^2)(A(\mathcal{B}_{2t-1})\ B(\mathcal{B}_{2t-1})) + (\mathcal{N}_1\mathcal{N}_2) \tag{4.104}$$

Because the elements of $\mathcal{V}$ have equal length, to compute $(A(\mathcal{B}_{2t-1})B(\mathcal{B}_{2t-1}))$, the receiver now computes the closest vector of $\mathcal{V}$ to $(\mathcal{R}_1\mathcal{R}_2)$. Once this vector is computed, the inverse mapping of $\mathcal{M}$ is applied and the transmitted bits are recovered (see Table 4.12 and Table 4.13). The receiver block diagram and BER curves are given in Figures 4.42 and 4.43 respectively. As in the traditional, one dimensional DPSK system, there is loss in the performance (about 3 dB at BER $= 0.001$).

More results on the topic can be found in [261–272].

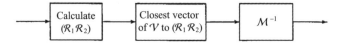

Figure 4.42 Receiver block diagram.

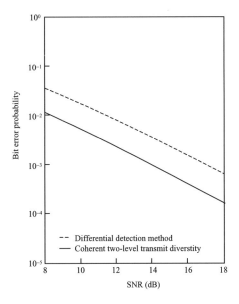

Figure 4.43 Performance of the differential detection and coherent detection: two-level transmit diversity scheme for BPSK constellation.

4.9 MULTIPLE TRANSMIT ANTENNA DIFFERENTIAL DETECTION FROM GENERALIZED ORTHOGONAL DESIGNS

In this section we explicitly construct multiple transmit antenna differential encoding/decoding schemes based on generalized orthogonal designs. These constructions generalize the two transmit antenna differential detection scheme discussed in the previous section. The presentation is based on [273, 274].

4.9.1 Differential encoding

We consider the specific example code, Γ_{84}, and only one receive antenna, $M = 1$. The generalization to other codes is straightforward. The code set is defined as:

$$
\Gamma_{84} = \begin{bmatrix}
x_1 & x_2 & x_3 & x_4 \\
-x_2 & x_1 & -x_4 & x_3 \\
-x_3 & x_4 & x_1 & -x_2 \\
-x_4 & -x_3 & x_2 & x_1 \\
x_1^* & x_2^* & x_3^* & x_4^* \\
-x_2^* & x_1^* & -x_4^* & x_3^* \\
-x_3^* & x_4^* & x_1^* & -x_2^* \\
-x_4^* & -x_3^* & x_2^* & x_1^*
\end{bmatrix}
\tag{4.105}
$$

4.9.2 Received signal

At each time, there is only one received signal, $r_{t,1}$, which can be noted by r_t and this is related to the constellation symbols s_1, s_2, s_3, s_4 by:

$$
\begin{aligned}
r_1 &= \alpha_1 s_1 + \alpha_2 s_2 + \alpha_3 s_3 + \alpha_4 s_4 + \eta_1 \\
r_2 &= -\alpha_1 s_2 + \alpha_2 s_1 - \alpha_3 s_4 + \alpha_4 s_3 + \eta_2 \\
r_3 &= -\alpha_1 s_3 + \alpha_2 s_4 + \alpha_3 s_1 - \alpha_4 s_2 + \eta_3 \\
r_4 &= -\alpha_1 s_4 - \alpha_2 s_3 + \alpha_3 s_2 + \alpha_4 s_1 + \eta_4 \\
r_1 &= \alpha_1 s_1 + \alpha_2 s_2 + \alpha_3 s_3 + \alpha_4 s_4 + \eta_1 \\
r_2 &= -\alpha_1 s_2 + \alpha_2 s_1 - \alpha_3 s_4 + \alpha_4 s_3 + \eta_2 \\
r_3 &= -\alpha_1 s_3 + \alpha_2 s_4 + \alpha_3 s_1 - \alpha_4 s_2 + \eta_3 \\
r_4 &= -\alpha_1 s_4 - \alpha_2 s_3 + \alpha_3 s_2 + \alpha_4 s_1 + \eta_4
\end{aligned}
\tag{4.106}
$$

By using Equation (4.165), one can show that the following relations are valid:

$$
(r_1, r_2, r_3, r_4, r_5^*, r_6^*, r_7^*, r_8^*) = (s_1, s_2, s_3, s_4)\Omega + (\eta_1, \eta_2, \eta_3, \eta_4, \eta_5^*, \eta_6^*, \eta_7^*, \eta_8^*) \tag{4.107}
$$

where

$$
\Omega = \begin{bmatrix}
\alpha_1 & \alpha_2 & \alpha_3 & \alpha_4 & \alpha_1^* & \alpha_2^* & \alpha_3^* & \alpha_4^* \\
\alpha_2 & -\alpha_1 & -\alpha_4 & \alpha_3 & \alpha_2^* & -\alpha_1^* & -\alpha_4^* & \alpha_3^* \\
\alpha_3 & \alpha_4 & -\alpha_1 & -\alpha_2 & \alpha_3^* & \alpha_4^* & -\alpha_1^* & -\alpha_2^* \\
\alpha_4 & -\alpha_3 & \alpha_2 & -\alpha_1 & \alpha_4^* & -\alpha_3^* & \alpha_2^* & -\alpha_1^*
\end{bmatrix}
\tag{4.108}
$$

If, instead of $\mathbf{S} = (s_1, s_2, s_3, s_4)$, other orthogonal vectors are used, we have:

$$
\begin{aligned}
(-r_2, r_1, r_4, -r_3, -r_6^*, r_5^*, r_8^*, -r_7^*) =&\ (s_2, -s_1, s_4, -s_3)\Omega \\
&+ (-\eta_2, \eta_1, \eta_4, -\eta_3, -\eta_6^*, \eta_5^*, \eta_8^*, -\eta_7^*) \\
(-r_3, -r_4, r_1, r_2, -r_7^*, -r_8^*, r_5^*, r_6^*) =&\ (s_3, -s_4, -s_1, s_2)\Omega \\
&+ (-\eta_3, -\eta_4, \eta_1, \eta_2, -\eta_7^*, -\eta_8^*, \eta_5^*, \eta_6^*) \\
(-r_4, r_3, -r_2, r_1, -r_8^*, r_7^*, -r_6^*, r_5^*) =&\ (s_4, s_3, -s_2, -s_1)\Omega \\
&+ (-\eta_4, \eta_3, -\eta_2, \eta_1, -\eta_8^*, \eta_7^*, -\eta_6^*, \eta_5^*)
\end{aligned}
\tag{4.109}
$$

4.9.3 Orthogonality

Note that, in general, for $\mathbf{S} = (s_1, s_2, s_3, s_4)^{\mathrm{T}}$, vectors

$$
\begin{aligned}
\mathbf{V}_1(\mathbf{S}) &= (s_1, s_2, s_3, s_4, s_1^*, s_2^*, s_3^*, s_4^*)^{\mathrm{T}} \\
\mathbf{V}_2(\mathbf{S}) &= (s_2, -s_1, s_4, -s_3, s_2^*, -s_1^*, s_4^*, -s_3^*)^{\mathrm{T}} \\
\mathbf{V}_3(\mathbf{S}) &= (s_3, -s_4, -s_1, s_2, s_3^*, -s_4^*, -s_1^*, s_2^*)^{\mathrm{T}} \\
\mathbf{V}_4(\mathbf{S}) &= (s_4, s_3, -s_2, -s_1, s_4^*, s_3^*, -s_2^*, -s_1^*)^{\mathrm{T}}
\end{aligned}
\tag{4.110}
$$

are orthogonal to each other. Therefore, for specific constellation symbols $\mathbf{S}$, vectors $\mathbf{V}_1(\mathbf{S}), \mathbf{V}_2(\mathbf{S}), \mathbf{V}_3(\mathbf{S}), \mathbf{V}_4(\mathbf{S})$ can create a basis for the four-dimensional subspace of any arbitrary four-dimensional constellation symbols and their conjugates in an eight-dimensional space. If the constellation symbols are real numbers, vectors

$$\mathbf{V}_1'(\mathbf{S}) = (s_1, s_2, s_3, s_4)^{\mathrm{T}}$$
$$\mathbf{V}_2'(\mathbf{S}) = (s_2, -s_1, s_4, -s_3)^{\mathrm{T}}$$
$$\mathbf{V}_3'(\mathbf{S}) = (s_3, -s_4, -s_1, s_2)^{\mathrm{T}} \quad (4.111)$$
$$\mathbf{V}_4'(\mathbf{S}) = (s_4, s_3, -s_2, -s_1)^{\mathrm{T}}$$

which only contain the first four elements of vectors $\mathbf{V}_1(\mathbf{S})$, $\mathbf{V}_2(\mathbf{S})$, $\mathbf{V}_3(\mathbf{S})$, $\mathbf{V}_4(\mathbf{S})$, create a basis for the space of any arbitrary four-dimensional real constellation symbols.

4.9.4 Encoding

Step 1. Let us assume that we use a signal constellation with 2^b elements. For each block of Kb bits, the encoding is done by first calculating the K-dimensional vector of symbols $\mathbf{S} = (s_1, s_2, \dots, s_K)^{\mathrm{T}}$.

Step 2. Indeterminates $x_1, x_2, \dots, x_K$ in Γ are replaced by $s_1, s_2, \dots, s_K$ to establish the matrix $\mathbf{X}$ which is used for transmission in a manner similar to a regular space–time block code. The main issue here is how to calculate $\mathbf{S} = (s_1, s_2, \dots, s_K)^{\mathrm{T}}$ such that non-coherent detection is possible.

Step 3. Let us assume that $\mathbf{S}_v$ is the vector which is used for the vth block of Kb bits. Also, $\mathbf{X}(\mathbf{S}_v)$ defines what to transmit from each antenna during the transmission of the vth block. In other words, $\mathbf{X}_n(\mathbf{S}_v), n = 1, 2, \dots, N$ is the nth column of $\mathbf{X}(\mathbf{S}_v)$ which contains T symbols which are transmitted from the nth antenna sequentially.

We fix a set V which consists of 2^{Kb} unit length vectors $\mathbf{P}_1, \mathbf{P}_2, \dots, \mathbf{P}_{2^{Kb}}$ where each vector $\mathbf{P}_l$ is a $K \times 1$ vector of real numbers $\mathbf{P}_l = (P_{l1}, P_{l2}, \dots, P_{lK})^{\mathrm{T}}$. We define an arbitrary one-to-one mapping β which maps Kb bits onto V. We start with an arbitrary vector $\mathbf{S}_1$. Then, let us assdume that $\mathbf{S}_v$ is used for the vth block. For the $(v + 1)$th block, we use the Kb input bits to pick the corresponding vector $\mathbf{P}_l$ in V using the one-to-one mapping β. Then, we calculate:

$$\mathbf{S}_{v+1} = \sum_{k=1}^{K} P_{lk}\mathbf{V}_k'(\mathbf{S}_v) \quad (4.112)$$

where $\mathbf{V}_k'(\mathbf{S}_v)$ is a K-dimensional vector which includes the first K elements of $\mathbf{V}_k(\mathbf{S}_v)$. We use $\mathbf{X}(\mathbf{S}_{v+1})$ for transmission at the following T time slots.

4.9.4.1 Example

We pick four BPSK symbols, i.e. symbol $s_k = 1$ if the corresponding bit is zero and $s_k = -1$ otherwise. Then, $\mathbf{P}_l$ is defined as the projection of the vector (s_1, s_2, s_3, s_4) onto $\mathbf{V}_1'(\mathbf{S})$, $\mathbf{V}_2'(\mathbf{S})$, $\mathbf{V}_3'(\mathbf{S})$, $\mathbf{V}_4'(\mathbf{S})$, where $\mathbf{S} = (1, 1, 1, 1)^{\mathrm{T}}$. In fact, the elements of $\mathbf{P}_l$ can be calculated using the following equations:

$$P_{l1} = (s_1 + s_2 + s_3 + s_4)/4$$
$$P_{l2} = (s_1 - s_2 + s_3 - s_4)/4$$
$$P_{l3} = (s_1 - s_2 - s_3 + s_4)/4 \quad (4.113)$$
$$P_{l4} = (s_1 + s_2 - s_3 - s_4)/4$$

This completes the one-to-one mapping β which is needed for encoding and decoding.

4.9.5 Differential decoding

Let us recall that the received signal r_t is related to the transmitted signals by:

$$r_t = \sum_{n=1}^{N} \alpha_n \mathbf{X}_{t,n} + \eta_t \tag{4.114}$$

Assuming $T = 2K$, which results in a rate $1/2$ code, and defining

$$\mathbf{R} = \left(r_1, r_2, \ldots, r_K, r_{K+1}^*, r_{K+2}^*, \ldots, r_{2K}^* \right),$$

we have

$$\mathbf{R} = \mathbf{S}^{\mathrm{T}} \mathbf{\Omega} \left(\alpha_1, \alpha_2, \ldots, \alpha_N \right) + \mathbf{N} \tag{4.115}$$

where $\mathbf{N} = (\eta_1, \eta_2, \ldots, \eta_K, \eta_{K+1}^*, \eta_{K+2}^*, \ldots, \eta_{2K}^*)$ and

$$\mathbf{\Omega}(\alpha_1, \alpha_2, \ldots, \alpha_N) = (\mathbf{\Lambda}(\alpha_1, \alpha_2, \ldots, \alpha_N)\mathbf{\Lambda}(\alpha_1^*, \alpha_2^*, \ldots, \alpha_N^*)) \tag{4.116}$$

where $\mathbf{\Lambda}(\alpha_1, \alpha_2, \ldots, \alpha_N)$ is the $K \times K$ matrix defined by

$$\mathbf{\Lambda}_{i,j} = \delta(i, j)\alpha_{\varepsilon(i,j)}, \quad i = 1, 2, \ldots, K, \quad j = 1, 2, \ldots, K \tag{4.117}$$

where $\varepsilon (i, j) = l \Leftrightarrow B_{j,l} = s_i$ or $B_{j,l} = -s_i$, and

$$\delta (i, j) = \begin{cases} 1, & \text{if } B_{j,l} = s_i \\ -1, & \text{if } B_{j,l} = -s_i \end{cases}$$

4.9.5.1 Example

For the rate $1/2$ code defined by $\mathbf{\Gamma}_{84}$ in Equation (4.105), $\mathbf{\Lambda}(\alpha_1, \alpha_2, \alpha_3, \alpha_4)$ is a 4×4 matrix defined by:

$$\mathbf{\Lambda} = \begin{bmatrix} \alpha_1 & \alpha_2 & \alpha_3 & \alpha_4 \\ \alpha_2 & -\alpha_1 & -\alpha_4 & \alpha_3 \\ \alpha_3 & \alpha_4 & -\alpha_1 & -\alpha_2 \\ \alpha_4 & -\alpha_3 & \alpha_2 & -\alpha_1 \end{bmatrix} \tag{4.118}$$

Differential decoding is enabled by the fact that $\mathbf{\Omega}\mathbf{\Omega}^* = 2 \sum_{n=1}^{N} |\alpha_n|^2 I_K$. To prove it, let us set $N = 0$ in Equations (4.114) and (4.115). Then, we have $\mathbf{R} = \mathbf{S}^{\mathrm{T}} \mathbf{\Omega}$ which results in:

$$\mathbf{R}\mathbf{R}^* = \mathbf{S}^{\mathrm{T}} \, \mathbf{\Omega}\mathbf{\Omega}^* \, \mathbf{S}^{*\mathrm{T}} \tag{4.119}$$

On the other hand

$$\mathbf{R}\mathbf{R}^* = \sum_{t=1}^{T} |r_t|^2 = \sum_{t=1}^{T} \left(\sum_{n=1}^{N} \alpha_n \mathbf{X}_{t,n} \right) \left(\sum_{n=1}^{N} \alpha_n \mathbf{X}_{t,n} \right)^* \tag{4.120}$$

We rewrite the above formulas using matrix $\mathbf{C}$ as follows:

$$
\begin{aligned}
\mathbf{RR}^* &= (\alpha_1, \alpha_2, \ldots, \alpha_N)\mathbf{X}^T\mathbf{X}^{*T}(\alpha_1, \alpha_2, \ldots, \alpha_N)^* \\
&= (\alpha_1, \alpha_2, \ldots, \alpha_N)(\mathbf{X}^*\mathbf{X})^T(\alpha_1, \alpha_2, \ldots, \alpha_N)^* \\
&= (\alpha_1, \alpha_2, \ldots, \alpha_N)2\sum_{k=1}^{K}|s_k|^2 I_K(\alpha_1, \alpha_2, \ldots, \alpha_N)^* \\
&= 2\sum_{k=1}^{K}|s_k|^2(\alpha_1, \alpha_2, \ldots, \alpha_N)(\alpha_1, \alpha_2, \ldots, \alpha_N)^*
\end{aligned}
$$

(4.121)

Therefore, we have:

$$
\mathbf{RR}^* = \mathbf{S}^T\mathbf{\Omega}\mathbf{\Omega}^*\mathbf{S}^{*T} = 2\sum_{k=1}^{K}|s_k|^2\sum_{n=1}^{N}|\alpha_n|^2 \tag{4.122}
$$

Since Equation (4.122) is true for any $\mathbf{S} = (s_1, s_2, \ldots, s_K)^T$, we should have

$$
\mathbf{\Omega}\mathbf{\Omega}^* = 2\sum_{n=1}^{N}|\alpha_n|^2 I_K
$$

4.9.6 Received signal

Let us recall that $\mathbf{S}_v$ and $\mathbf{S}_{v+1}$ are used for the vth and $(v+1)$th blocks of Kb bits, respectively. Using Γ_{84}, for each block of data we receive eight signals. To simplify the notation, we denote the received signals corresponding to the vth block by $r_1^v, r_2^v, \ldots, r_8^v$ and the received signals corresponding to the $(v+l)$th block by $r_1^{v+1}, r_2^{v+1}, \ldots, r_8^{v+1}$. Let us construct the following vectors (see Equation (4.109)).

$$
\begin{aligned}
\mathbf{R}_v^1 &= \left(r_1^v, r_2^v, r_3^v, r_4^v, r_5^{v*}, r_6^{v*}, r_7^{v*}, r_8^{v*}\right) \\
\mathbf{R}_v^2 &= \left(-r_2^v, r_1^v, r_4^v, -r_3^v, -r_6^{v*}, r_5^{v*}, r_8^{v*}, -r_7^{v*}\right) \\
\mathbf{R}_v^3 &= \left(-r_3^v, -r_4^v, r_1^v, r_2^v, -r_7^{v*}, -r_8^{v*}, r_5^{v*}, r_6^{v*}\right) \\
\mathbf{R}_v^4 &= \left(-r_4^v, r_3^v, -r_2^v, r_1^v, -r_8^{v*}, r_7^{v*}, -r_6^{v*}, r_5^{v*}\right) \\
\mathbf{R}_{v+1} &= \left(r_1^{v+1}, r_2^{v+1}, r_3^{v+1}, r_4^{v+1}, r_5^{(v+1)*}, r_6^{(v+1)*}, r_7^{(v+1)*}, r_8^{(v+1)*}\right)
\end{aligned}
$$

(4.123)

By using Equations (4.107) and (4.109), we have:

$$
\begin{aligned}
\mathbf{R}_{v+1}\mathbf{R}_v^{k*} &= \mathbf{S}_{v+1}^T\mathbf{\Omega}\mathbf{\Omega}^*\,\mathbf{V}_k'\,(\mathbf{S}_v)^{*T} + \mathbf{N}_k \\
&= 2\sum_{n=1}^{4}|\alpha_n|^2\,\mathbf{S}_{v+1}^T\mathbf{V}_k'\,(\mathbf{S}_v)^{*T} + \mathbf{N}_k \\
&= 2\sum_{n=1}^{4}|\alpha_n|^2\,\mathbf{P}_{lk} + \mathbf{N}_k
\end{aligned}
$$

(4.124)

Therefore, we can write:

$$
\begin{aligned}
\mathbf{R} &= \left(R_{v+1}R_v^{1*},\ R_{v+1}R_v^{2*},\ R_{v+1}R_v^{3*},\ R_{v+1}R_v^{4*}\right) \\
&= \left(2\sum_{n=1}^{4}|\alpha_n|^2\right)\mathbf{P}_l + \mathbf{N}'
\end{aligned}
$$

(4.125)

where $\mathbf{N}' = (\eta_1, \eta_2, \ldots, \eta_K)$.

4.9.7 Demodulation

Because the elements of $\mathbf{R}$ have equal lengths, to compute $\mathbf{P}_l$, the receiver can compute the closest vector of $\mathbf{V}$ to $\mathbf{R}$. Once this vector is computed, the inverse mapping of β is applied and the transmitted bits are recovered.

4.9.8 Multiple receive antennas

If there is more than one receive antenna, the same procedure can be used. In this case, first assuming that only the receive antenna m exists, we compute $\mathbf{P}^m$ using the same method for $\mathbf{P}$ given above. Then, after calculating M vectors $\mathbf{P}^m$, $m = 1, 2, \ldots, M$, the closest vector of $\mathbf{V}$ to $\sum_{m=1}^{M} \mathbf{P}^m$ is computed. The inverse mapping of β is applied to the closest vector to compute the transmitted bits. It is easy to show that $4M$-level diversity is achieved by using this method.

4.9.9 The number of transmit antennas lower than the number of symbols

So far, we have considered an example where $N = K$, i.e. the real matrix which creates Γ is a square matrix. When the number of transmit antennas is less than the number of symbols, $N < K$, the same approach works. Let us use the following example:

$$\Gamma_{83} = \begin{bmatrix} x_1 & x_2 & x_3 \\ -x_2 & x_1 & -x_4 \\ -x_3 & x_4 & x_1 \\ -x_4 & -x_3 & x_2 \\ x_1^* & x_2^* & x_3^* \\ -x_2^* & x_1^* & -x_4^* \\ -x_3^* & x_4^* & x_1^* \\ -x_4^* & -x_3^* & x_2^* \end{bmatrix} \tag{4.126}$$

When there is only one receive antenna, $M = 1$, the received signals are related to the constellation symbols s_1, s_2, s_3, s_4 by:

$$\begin{aligned} r_1 &= \alpha_1 s_1 + \alpha_2 s_2 + \alpha_3 s_3 + \eta_1 \\ r_2 &= -\alpha_1 s_2 + \alpha_2 s_1 - \alpha_3 s_4 + \eta_2 \\ r_3 &= -\alpha_1 s_3 + \alpha_2 s_4 + \alpha_3 s_1 + \eta_3 \\ r_4 &= -\alpha_1 s_4 - \alpha_2 s_3 + \alpha_3 s_2 + \eta_4 \\ r_5 &= \alpha_1 s_1^* + \alpha_2 s_2^* + \alpha_3 s_3^* + \eta_5 \\ r_6 &= -\alpha_1 s_2^* + \alpha_2 s_1^* - \alpha_3 s_4 + \eta_6 \\ r_7 &= -\alpha_1 s_3^* + \alpha_2 s_4 + \alpha_3 s_1^* + \eta_7 \\ r_8 &= -\alpha_1 s_4 - \alpha_2 s_3^* + \alpha_3 s_2^* + \eta_8. \end{aligned} \tag{4.127}$$

Equation (4.107) now becomes:

$$(r_1, r_2, r_3, r_4, r_5^*, r_6^*, r_7^*, r_8^*) = (s_1, s_2, s_3, s_4)\, \boldsymbol{\Omega} + (\eta_1, \eta_2, \eta_3, \eta_4, \eta_5^*, \eta_6^*, \eta_7^*, \eta_8^*) \tag{4.128}$$

where

$$
\Omega = \begin{bmatrix}
\alpha_1 & \alpha_2 & \alpha_3 & 0 & \alpha_1^* & \alpha_2^* & \alpha_3^* & 0 \\
\alpha_2 & -\alpha_1 & 0 & \alpha_3 & \alpha_2^* & -\alpha_1^* & 0 & \alpha_3^* \\
\alpha_3 & 0 & -\alpha_1 & -\alpha_2 & \alpha_3^* & 0 & -\alpha_1^* & -\alpha_2^* \\
0 & -\alpha_3 & \alpha_2 & -\alpha_1 & 0 & -\alpha_3^* & \alpha_2^* & -\alpha_1^*
\end{bmatrix}
\tag{4.129}
$$

The matrix Ω for Γ_{83} can be calculated from Ω for Γ_{84}, and Equation (4.108) by setting $\alpha_4 = 0$. Therefore, again for specific constellation symbols S, vectors $\mathbf{V}_1(\mathbf{S}), \mathbf{V}_2(\mathbf{S}), \mathbf{V}_3(\mathbf{S}), \mathbf{V}_4(\mathbf{S})$ can create a basis for the four-dimensional subspace of any arbitrary four-dimensional constellation symbols and their conjugates in an eight-dimensional space and the same encoding and decoding schemes are applicable. The only difference in the final result is that $\sum_{n=1}^{4} |\alpha_n|^2$ is replaced by $\sum_{n=1}^{3} |\alpha_n|^2$.

4.9.10 Final result

Instead of Equation (4.125) we now have:

$$
\begin{aligned}
\mathbf{R} &= \left(\mathbf{R}_{v+1}\mathbf{R}_v^{1*}, \mathbf{R}_{v+1}\mathbf{R}_v^{2*}, \mathbf{R}_{v+1}\mathbf{R}_v^{3*}, \mathbf{R}_{v+1}\mathbf{R}_v^{4*} \right) \\
&= \left(2\sum_{n=1}^{3} |\alpha_n|^2 \right) \mathbf{P}_l + \mathbf{N}'
\end{aligned}
\tag{4.130}
$$

4.9.11 Real constellation set

The above rate $1/2$ space–time block codes can be applied to any complex constellation set. If the constellation set is real, rate 1 space–time block codes are available and the same approach works. For example, in the case of $T = K = 4$, the following space–time block code exists for $N = 4$:

$$
\Gamma = \begin{bmatrix}
x_1 & x_2 & x_3 & x_4 \\
-x_2 & x_1 & -x_4 & x_3 \\
-x_3 & x_4 & x_1 & -x_2 \\
-x_4 & -x_3 & x_2 & x_1
\end{bmatrix}
\tag{4.131}
$$

It is easy to show that:

$$
(r_1, r_2, r_3, r_4, r_1^*, r_2^*, r_3^*, r_4^*) = (s_1, s_2, s_3, s_4)\Omega + (\eta_1, \eta_2, \eta_3, \eta_4, \eta_1^*, \eta_2^*, \eta_3^*, \eta_4^*)
\tag{4.132}
$$

where Ω is defined by Equation (4.108). Similar differential encoding and decoding are possible if we use the following vectors for $\mathbf{R}_v^i$, $i = 1, 2, 3, 4$, and $\mathbf{R}_{v+1}$:

$$
\begin{aligned}
\mathbf{R}_v^1 &= \left(r_1^v, r_2^v, r_3^v, r_4^v, r_1^{v*}, r_2^{v*}, r_3^{v*}, r_4^{v*} \right) \\
\mathbf{R}_v^2 &= \left(-r_2^v, r_1^v, r_4^v, -r_3^v, -r_2^{v*}, r_1^{v*}, r_4^{v*}, -r_3^{v*} \right) \\
\mathbf{R}_v^3 &= \left(-r_3^v, -r_4^v, r_1^v, r_2^v, -r_3^{v*}, -r_4^{v*}, r_1^{v*}, r_2^{v*} \right) \\
\mathbf{R}_v^4 &= \left(-r_4^v, r_3^v, -r_2^v, r_1^v, -r_4^{v*}, r_3^{v*}, -r_2^{v*}, r_1^{v*} \right) \\
\mathbf{R}_{v+1} &= \left(r_1^{v+1}, r_2^{v+1}, r_3^{v+1}, r_4^{v+1}, r_1^{(v+1)*}, r_2^{(v+1)*}, r_3^{(v+1)*}, r_4^{(v+1)*} \right)
\end{aligned}
\tag{4.133}
$$

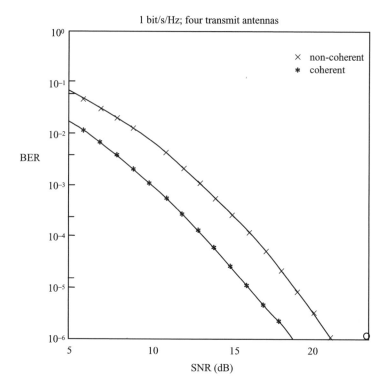

Figure 4.44 Performance results for coherent and non-coherent detection schemes; four transmit antennas, one receive antenna, BPSK.

This results in a full diversity, full rate scheme for differential detection. For example, as defined by Equation (4.133), performance results are shown in Figures 4.44 and 4.45 for four and three transmit antennas respectively.

More details on the topic can be found in [273–274].

4.10 LAYERED SPACE–TIME CODING

In this section we discuss a possibility of partitioning antennas at the transmitter into small groups, and using individual space–time codes, called component codes, to transmit information from each group of antennas. At the receiver, an individual space–time code is decoded by a linear processing technique that suppresses signals transmitted by other groups of antennas by treating them as interference. This receiver structure provides diversity and coding gain over uncoded systems. This combination of array processing at the receiver and coding techniques for multiple transmit antennas can provide reliable and very high data rate communication over narrowband wireless channels. A refinement of this basic structure gives rise to a multilayered space–time architecture that both generalizes and improves upon the layered space–time architecture proposed by Foschini, which is known in the literature as Bell Lab Layered space time (BLAST) coding [206, 273–284].

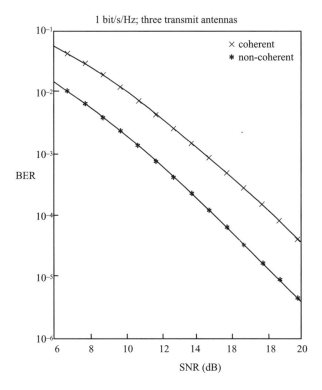

Figure 4.45 Performance results for coherent and non-coherent detection schemes; three transmit antennas, one receive antenna, BPSK.

For $1 \leq j \leq m$ the signal r_t^j received by antenna j at time t is given by:

$$r_t^j = \sum_{i=1}^{n} \alpha_{i,j} c_t^i + \eta_t^j \qquad (4.134)$$

where c_t^i is the encoded signal transmitted from transmit antenna i.

For any vector $\mathbf{x}$, let $\mathbf{x}^T$ denote the transpose of $\mathbf{x}$. We can now write Equation (4.134) in the vector form given by:

$$\mathbf{r}_t = \Omega \mathbf{c}_t + \eta_t \qquad (4.135)$$

where

$$
\begin{aligned}
\mathbf{c}_t &= \left(c_t^1, c_t^2, \ldots, c_t^n\right)^T \\
\mathbf{r}_t &= \left(r_t^1, r_t^2, \ldots, r_t^m\right)^T \\
\eta_t &= \left(\eta_t^1, \eta_t^2, \ldots, \eta_t^m\right)^T
\end{aligned}
\qquad (4.136)
$$

and

$$
\Omega = \begin{bmatrix}
\alpha_{1,1} & \alpha_{2,1} & \cdots & \cdots & \alpha_{n,1} \\
\alpha_{1,2} & \alpha_{2,2} & \cdots & \cdots & \alpha_{n,2} \\
\alpha_{1,3} & \alpha_{2,3} & \cdots & \cdots & \alpha_{n,3} \\
\vdots & \vdots & \ddots & \ddots & \vdots \\
\alpha_{1,m} & \alpha_{2,m} & \cdots & \cdots & \alpha_{n,m}
\end{bmatrix}
\tag{4.137}
$$

A *space–time product encoder* accepts a block of B input bits in each time slot t and these bits are divided into q strings of lengths $B_1, B_2, \ldots, B_q$ with $B_1 + B_2 + \cdots + B_q = B$. At the base station, n antennas are partitioned into q groups $G_1, G_2, \ldots, G_q$, respectively, comprising $n_1, n_2, \ldots, n_q$ antennas with $n_1 + n_2 + \ldots + n_q = n$. Each block $B_j, 1 \leq j \leq q$ is then encoded by a space–time encoder $\mathbf{X}_j$. The output of $\mathbf{X}_j$ goes through a serial to parallel converter and provides n_j sequences of constellation symbols for $1 \leq j \leq q$ which are simultaneously transmitted from the antennas of the group G_j. This gives a total of n sequences of constellation symbols that are transmitted simultaneously from antennas $1, 2, \ldots, n$.

A space–time product encoder can be considered as a set of q space–time encoders, called the *component codes*, operating in parallel on the same wireless communication channel, with each encoder using n_j transmit and m receive antennas for $1 \leq j \leq q$. It will be denoted an $\mathbf{X}_1 \times \mathbf{X}_2 \times \cdots \mathbf{X}_q$ encoder.

4.10.1 Receiver complexity

One approach to recovering the transmitted data at the receiver is to jointly decode the transmitted code word, but the difficulty here is decoding complexity. Indeed, if we require a diversity of rm, where $r \leq \min_j (n_j)$, then (see Section 4.2) the complexity of the trellis of $\mathbf{X}_j$ is at least $2^{B_j(r-1)}$ states and the complexity of the product code is at least $2^{B(r-1)}$ states. This means that if B is very large, the scheme may be too complex to implement.

4.10.2 Group interference suppression

The idea is to decode each code $\mathbf{X}_j$ separately while suppressing signals from other component codes. This approach has a much lower complexity but achieves a lower diversity order than the full diversity order nm, which is the product of the numbers of transmit and receive antennas.

4.10.3 Suppression method

Without any loss of generality, we take $j = 1$ and look to decode $\mathbf{X}_1$. There are $n - n_1$ interfering signals. We assume that there are $m \geq n - n_1 + 1$ receive antennas and that the receiver knows the matrix Ω (channel state information). The matrix

$$\Lambda(\mathbf{X}_1) = \begin{bmatrix} \alpha_{n_1+1,\,1} & \alpha_{n_1+2,\,1} & \cdots & \cdots & \alpha_{n,\,1} \\ \alpha_{n_1+1,\,2} & \alpha_{n_1+2,\,2} & \cdots & \cdots & \alpha_{n,\,2} \\ \alpha_{n_1+1,\,3} & \alpha_{n_1+2,\,3} & \cdots & \cdots & \alpha_{n,\,3} \\ \vdots & \vdots & \ddots & \ddots & \vdots \\ \alpha_{n_1+1,\,m} & \alpha_{n_1+2,\,m} & \cdots & \cdots & \alpha_{n,\,m} \end{bmatrix} \qquad (4.138)$$

has rank less than or equal to the number of its columns. Thus, rank $[\Lambda(\mathbf{X}_1)] \leq n - n_1$.

4.10.4 The null space

The *null space* N of this matrix is the set of all row vectors $\mathbf{x}$ such that $\mathbf{x}\Lambda(\mathbf{X}_1) = (0, 0, \ldots, 0)$. Furthermore,

$$\dim(N) + \text{rank}\,[\Lambda(\mathbf{X}_1)] = m$$

Since rank$[\Lambda(\mathbf{X}_1)] \leq n - n_1$, it follows that dim $(N) \geq m - n + n_1$. Hence we can compute a (not necessarily unique) *set of orthonormal vectors*:

$$\{\mathbf{v}_1, \mathbf{v}_2, \ldots, \mathbf{v}_{m-n+n_1}\}$$

in N. We let $\Theta(\mathbf{X}_1)$ denote the $(m - n + n_1) \times m$ matrix whose jth row is $\mathbf{v}_j$. Clearly, $\Theta(\mathbf{X}_1)\Theta(\mathbf{X}_1)^* = \mathbf{I}_{n_1-n+m}$, where $\Theta(\mathbf{X}_1)^*$ is the Hermitian of $\Theta(\mathbf{X}_1)$ and $\mathbf{I}_{n_1-n+m}$ is the $(m - n + n_1) \times (m - n + n_1)$ identity matrix. We multiply both sides of Equation (4.135) by $\Theta(\mathbf{X}_1)$ to arrive at:

$$\Theta(\mathbf{X}_1)\mathbf{r}_t = \Theta(\mathbf{X}_1)\Omega\mathbf{c}_t + \Theta(\mathbf{X}_1)\eta_t \qquad (4.139)$$

where

$$\Omega(\mathbf{X}_1) = \begin{bmatrix} \alpha_{1,1} & \alpha_{2,1} & \cdots & \cdots & \alpha_{n_1,1} \\ \alpha_{1,2} & \alpha_{2,2} & \cdots & \cdots & \alpha_{n_1,2} \\ \alpha_{1,3} & \alpha_{2,3} & \cdots & \cdots & \alpha_{n_1,3} \\ \vdots & \vdots & \ddots & \ddots & \vdots \\ \alpha_{1,m} & \alpha_{2,m} & \cdots & \cdots & \alpha_{n_1,m} \end{bmatrix} \qquad (4.140)$$

Since $\Theta(\mathbf{X}_1)\Lambda(\mathbf{X}_1) = \mathbf{0}$ is the all zero matrix, Equation (4.139) can be written as

$$\Theta(\mathbf{X}_1)\mathbf{r}_t = \Theta(\mathbf{X}_1)\Omega(\mathbf{X}_1)\mathbf{c}_t^1 + \Theta(\mathbf{X}_1)\eta_t \qquad (4.141)$$

where $\mathbf{c}_t^1 = (c_t^1, c_t^2, \ldots, c_t^{n_1})^{\mathrm{T}}$. Setting

$$\begin{aligned} \tilde{\mathbf{r}}_t &= \Theta(\mathbf{C}_1)\mathbf{r}_t \\ \tilde{\Omega} &= \Theta(\mathbf{C}_1)\Omega \\ \tilde{\eta}_t &= \Theta(\mathbf{C}_1)\eta_t \end{aligned} \qquad (4.142)$$

we arrive at the equation

$$\tilde{\mathbf{r}}_t = \tilde{\Omega}\mathbf{c}_t^1 + \tilde{\eta}_t \qquad (4.143)$$

This is an equation where all the signal streams out of antennas $n_1 + 1, n_1 + 2, \ldots, n$ are suppressed.

4.10.5 Receiver

Let us look at the structure of the decoder for the code $\mathbf{X}_1$, given that group interference suppression is performed to suppress all the signal streams out of antennas $n_1 + 1, n_1 + 2, \ldots, n$. To this end, suppose that $\mathbf{\Lambda}(\mathbf{X}_1)$ is given. The receiver computes a set of orthonormal vectors

$$\{\mathbf{v}_1, \mathbf{v}_2, \ldots, \mathbf{v}_{m-n+n_1}\}$$

and the matrix $\mathbf{\Theta}(\mathbf{X}_1)$ as described in the previous section. Let $\tilde{\Omega}_{ij}$ and $\tilde{\Omega}_{lk}$ denote the (i, j)th and (l, k)th elements of $\tilde{\mathbf{\Omega}}$. By definition $\tilde{\Omega}_{ij} = \mathbf{v}_i \boldsymbol{\omega}_j$ and $\tilde{\Omega}_{lk} = \mathbf{v}_l \boldsymbol{\omega}_k$ where $\boldsymbol{\omega}_j$ and $\boldsymbol{\omega}_k$ are, respectively, the jth and kth columns of $\mathbf{\Omega}(\mathbf{X}_1)$. The random variables $\tilde{\Omega}_{ij}$ and $\tilde{\Omega}_{lk}$ have zero means given $\mathbf{\Lambda}(\mathbf{X}_1)$. Moreover,

$$\mathrm{E}\left[\tilde{\Omega}_{ij}\tilde{\Omega}_{lk}^*\right] = \mathrm{E}\left[\mathbf{v}_i \boldsymbol{\omega}_j \boldsymbol{\omega}_k^* \mathbf{v}_l^*\right] = \mathbf{v}_i \, \mathrm{E}\left[\boldsymbol{\omega}_j \boldsymbol{\omega}_k^*\right] \mathbf{v}_l^* = \delta_{jk} \mathbf{v}_i \mathbf{v}_l^* = \delta_{jk} \delta_{il} \quad (4.144)$$

where δ is the Kronecker delta function given by $\delta_{rs} = 0$ if $r \neq s$ and $\delta_{rs} = 1$ if $r = s$. We conclude that the elements of the $(m - n + n_1) \times n_1$ matrix $\tilde{\mathbf{\Omega}}$ are independent complex Gaussian random variables of variance 0.5 per real dimension. Similarly, the components of the noise vector $\tilde{\boldsymbol{\eta}}_t$, $t = 1, 2, \ldots l$ are independent Gaussian random variables of variance $N_0/2$ per real dimension.

4.10.6 Decision metric

Assuming that all the code words of $\mathbf{X}_1$ are equiprobable, and given that group interference suppression is performed, the ML receiver for $\mathbf{X}_1$ decides in favor of the code word

$$c_1^1 c_1^2 \cdots c_1^{n_1} c_2^1 c_2^2 \cdots c_2^{n_1} \cdots c_l^1 c_l^2 \cdots c_l^{n_1}$$

if it minimizes the decision metric

$$\sum_{t=1}^{l} \left|\tilde{\mathbf{r}}_t - \tilde{\mathbf{\Omega}} \mathbf{c}_t^1\right|^2 \quad (4.145)$$

4.10.7 Multilayered space–time coded modulation

The previous discussion reduces code design for a multiple antenna communication system with n transmit and m receive antennas to that of designing codes for communication systems with n_i transmit and $n_i + m - n$, $i = 1, 2, \ldots, q$, receive antennas where $\sum_{i=1}^{q} n_i = n$ and $n_i \geq n - m + 1$. Using this insight, we may design a multilayered space–time coded modulation scheme. The idea behind such a system is multistage detection and cancellation.

4.10.8 Diversity gain

Suppose that $\mathbf{X}_1$ is decoded correctly using combined array processing and space–time coding. The space–time code $\mathbf{X}_1$ affords a diversity gain of $n_1 \times (n_1 + m - n)$. After decoding

$\mathbf{X}_1$ we may subtract the contribution of these code words to signals received at different antennas. This gives a communication system with $n - n_1$ transmit and m receive antennas.

In the next step the receiver uses combined array processing and space–time coding to decode $\mathbf{X}_2$. The space–time code $\mathbf{X}_2$ affords a diversity gain of $n_2 \times (n_2 + n_1 + m - n)$. Proceeding in this manner, we observe that by subtracting the contribution of previously decoded code streams $\mathbf{X}_j$, $j \leq k - 1$ to the received signals at different antennas, the space–time code $\mathbf{X}_k$ affords a diversity gain of $n_k \times (n_1 + \cdots + n_k + m - n)$.

We can choose space–time codes $\mathbf{X}_i$, $1 \leq i \leq q$ to provide these diversity gains and such that the sequence

$$n_1 (n_1 + m - n), n_2 (n_2 + n_1 + m - n), \ldots, n_k (n_1 + \cdots + n_k + m - n), \ldots, n_q m$$

is an increasing sequence. Assuming there was no decoding error in steps $1, 2, \ldots, k - 1$, then at decoding step k, the probability of error for the component code $\mathbf{X}_k$ is equal to the probability of error for $\mathbf{X}_k$ when employed in a communication system using n_k transmit and $(n_1 + \cdots + n_k + m - n)$ receive antennas.

4.10.9 Adaptive reconfigurable transmit power allocation

Since the diversity in each decoding stage k is more than that of the previous decoding stage $k - 1$, the transmit power out of each antenna at level k can be substantially less than that of the previous layer. Thus the transmitter should divide the available transmit power among different antennas in an unequal manner. Power allocation for this scenario is straightforward. In fact, powers at different levels could be allocated based on the diversity gains. In this way, the allocated powers may decrease geometrically in terms of the diversity gains. Other approaches are also possible.

4.10.9.1 Example 1

Transmitter. Here, four transmit and four receive antennas are used. The transmission rate is 4 bits/s/Hz. Let $\mathbf{X}$ denote the 32-state 4PSK (see Figure 4.46) space–time trellis code given in Figure 4.47. The product code $\mathbf{X}_1 \times \mathbf{X}_2$ where $\mathbf{X}_1 = \mathbf{X}_2 = \mathbf{X}$ will be used for transmission of 4 bits/s/Hz. At each time slot, upon the arrival of the four bits of the input data, the first two bits are used as the input to the encoder of $\mathbf{X}_1$ and the encoded symbols are transmitted by antennas one and two. The second two bits are used as the input to the encoder of $\mathbf{X}_2$ and the encoded signals are transmitted by antennas three and four. We assume that the average powers radiated from antennas one and two are equal but each is twice as much as the average power radiated from antennas three and four.

Receiver. Interference suppression is used to suppress $\mathbf{X}_2$ and decode $\mathbf{X}_1$. Upon decoding $\mathbf{X}_1$ the contributions of the code words transmitted from antennas one and two are subtracted from the received signals. Finally, $\mathbf{X}_2$ is decoded.

4.10.9.2 Simulation results

Figure 4.48 demonstrates the performance of this multilayered space–time coded architecture. Each frame consists of 130 transmissions from each transmit antenna. It is assumed that the channel matrix is perfectly known at the receiver. The horizontal axis shows the

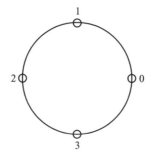

Figure 4.46 The 4PSK constellation.

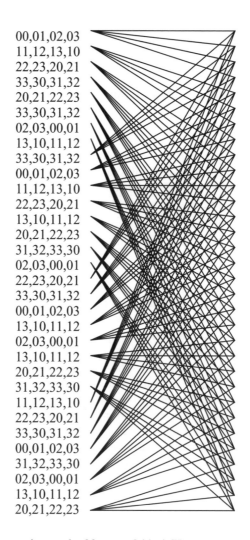

Figure 4.47 4PSK space–time code, 32 states 2 bits/s/Hz.

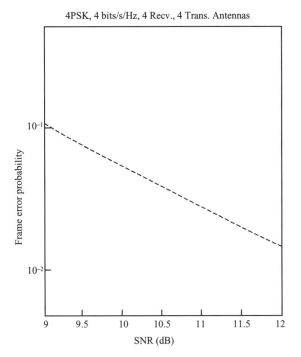

Figure 4.48 The performance of the scheme in Example 1.

receive signal to noise ratio per transmission time. Each transmission time corresponds to the transmission of four bits. Thus, the horizontal axis denotes the receive signal to noise ratio per four bits. For comparison, the graph of the outage capacity versus the signal to noise ratio for four transmit and four receive antennas is presented in Figure 4.49. The outage capacity is defined as the achievable capacity C_{out} for which the outage probability $P_{out} = P(C < C_{out}) < \varepsilon$. More details on MIMO channel capacity will be presented in Section 4.12. One can see that for a frame error probability of 10^{-1}, the system is about 6 dB away from the capacity.

4.10.9.3 Example 2

Here, eight transmit and eight receive antennas are used. The transmission rate is 8 bits/s/Hz. Let $\mathbf{X}$ denote the code given in Figure 4.47. We use the product code $\mathbf{X}_1 \times \mathbf{X}_2 \times \mathbf{X}_3 \times \mathbf{X}_4$ where $\mathbf{X}_1 = \mathbf{X}_2 = \mathbf{X}_3 = \mathbf{X}_4 = \mathbf{X}$ for transmission of 8 bits/s/Hz. At each time instance, upon the arrival of the eight bits of the input data, the first, second, third, and fourth blocks of length two of the input bits are respectively used as the input to encoders of $\mathbf{X}_1, \mathbf{X}_2, \mathbf{X}_3$ and $\mathbf{X}_4$. The output of encoders of $\mathbf{X}_i$, $1 \le i \le 4$ are, respectively, transmitted by antennas $2i - 1$ and $2i$. We assume that the average power radiated from antennas one and two is E_s, the average power radiated from antennas three and four is $E_s/2$, the average power radiated from antennas five and six is $E_s/4$, and the average power radiated from antennas seven and eight is $E_s/8$. Thus, the total signal to noise ratio at each receive antenna is $15\, E_s/4N_0$.

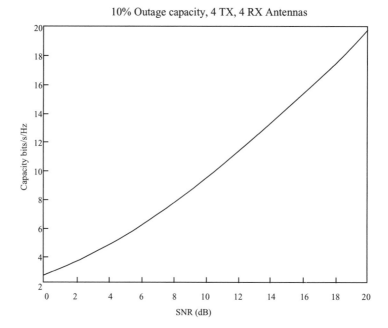

Figure 4.49 Outage capacity for four transmit and four receive antennas.

Decoder. Group interference suppression is used to decode $\mathbf{X}_1$. Upon decoding $\mathbf{X}_1$, the contributions of the code words transmitted from antennas one and two are subtracted from the received signals. Using this, $\mathbf{X}_2$ is decoded next and so forth. In Figure 4.50, we provide simulation results to demonstrate the performance of this multilayered space–time coded architecture. Each frame consists of 130 transmissions from each transmit antenna. It is assumed that the channel matrix is perfectly known at the receiver. The horizontal axis shows the receive signal to noise ratio per transmission time. Each transmission time corresponds to the transmission of eight bits. Thus, the horizontal axis denotes the receive signal to noise ratio per eight bits.

For comparison, we provide in Figure 4.51 the graph of the outage capacity versus the signal to noise ratio for eight transmit and eight receive antennas, as computed by Foschini and Gans [285]. We observe that for a frame error probability of 10^{-1}, we are about 9 dB away from the capacity.

More details on this topic can be found in [206, 273—284].

4.11 CONCATENATED SPACE–TIME BLOCK CODING

4.11.1 System model

The system model is given in Figure 4.52.

We begin by performance analysis and design criteria in quasi-static fading, where the fading coefficients are constant during a frame of length $2L$ and vary from one frame to another.

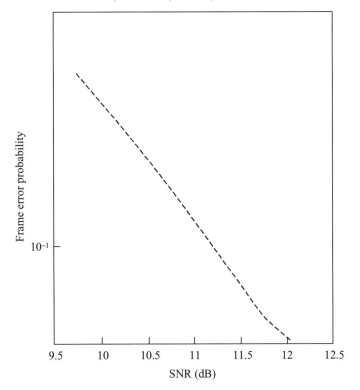

Figure 4.50 The performance of the scheme in Example 2.

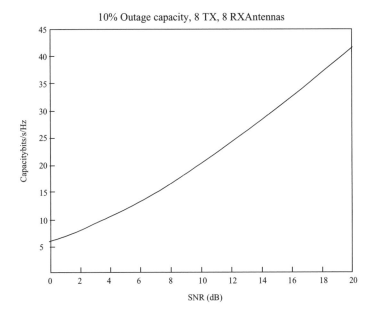

Figure 4.51 Outage capacity for eight transmit and eight receive antennas.

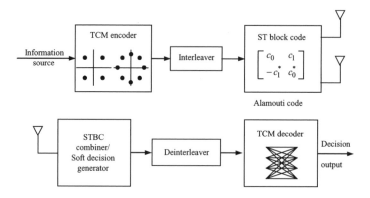

Figure 4.52 Block diagram of a space–time coded system concatenating STBC and TCM, $N = 2, M = 1$.

4.11.2 Product sum distance

Let η be the set of all i for which $c_i \neq e_i$ or $c_{i+1} \neq e_{i+1}$. Denote the number of elements in η by l_η. Then, at high signal to noise ratios (SNRs) [286],

$$P\left(\mathbf{C} \rightarrow \mathbf{E}\right) \leq \frac{1}{\left[\left(\dfrac{E_s}{4N_0}\right)^{l_\eta} d_P\left(l_\eta\right)\right]^2} \tag{4.146}$$

where

$$d_P(l_\eta) = \prod_{i \in \eta} \left\lfloor |c_i - e_i|^2 + |c_{i+1} - e_{i+1}|^2 \right\rfloor$$

is the product of Euclidean distances associated with two consecutive symbols along the error event path $(\mathbf{C} \rightarrow \mathbf{E})$. Parameter $d_P(l_\eta)$ is referred to as the *product–sum distance* over span 2. In addition, l_η is referred to as the *effective length* of this error event over span 2.

Let P_e denote the error event probability, then by using the union bound, an upper bound can be obtained as:

$$P_e \leq \sum_{L=1}^{\infty} \sum_{\mathbf{C}} \sum_{\mathbf{E} \neq \mathbf{C}} P\left(\mathbf{C}\right) P\left(\mathbf{C} \rightarrow \mathbf{E}\right) \tag{4.147}$$

where $P\left(\mathbf{C}\right)$ is the *a priori* probability of transmitting the symbol sequence $\mathbf{C}$ with length $2L$. By summing over all possible l_η and all possible $d_P(l_\eta)$, the error event probability can be further written as:

$$P_e \leq \sum_{l_\eta} \sum_{d_P(l_\eta)} \Xi(l_\eta, d_P(l_\eta)) \left[\left(\frac{E_s}{4N_0}\right)^{l_\eta} d_P(l_\eta)\right]^{-2} \tag{4.148}$$

where $\Xi(l_\eta, d_P(l_\eta))$ is the average number of error events having the span 2 effective length l_η and the product–sum distance $d_P(l_\eta)$.

4.11.3 Error rate bound

The smallest l_η and the smallest $d_P(l_\eta)$ dominate the error event probability at high SNRs. Denoting $R = \min(l_\eta)$ and $d_{\min}(R) = \min(d_P(R))$, then the error event probability is asymptotically approximated as:

$$P_e \cong \Xi(R, d_{\min}(R)) \frac{1}{\left(\dfrac{E_s}{4N_0}\right)^{2R} [d_{\min}(R)]^2} \qquad (4.149)$$

From Equation (4.149), we observe that the error event probability asymptotically varies with $2R$-power of SNR, so a diversity order of $2R$ is achieved. We further refer to R as the *built-in time diversity* or effective length of the concatenated space–time code.

The design criteria, in this case, involve the maximization of both the built-in-time diversity and the minimum product–sum distance of the trellis code at high SNRs for Rayleigh fading. This conclusion is, therefore, different from that of conventional TCM where the minimum product distance needs to be maximized. Thus, new optimal codes can be found, based upon these new criteria.

4.11.4 The case of low SNR

For low SNR we have [286]:

$$P(\mathbf{C} \to \mathbf{E}) \leq \left[1 + \frac{E_s}{4N_0} \sum_{i=0}^{2L-1} |c_i - e_i|^2 + o\left(\frac{E_s}{4N_0}\right) \right]^{-2} \qquad (4.150)$$

where $o(E_s/4N_0)$ denotes the summation of all the terms which include higher order quantities of $(E_s/4N_0)$. Equation (4.150) indicates that the squared Euclidean distance becomes the main factor. Thus, the dominant factor affecting the performance of trellis coded modulation for use with space–time block coding at low SNRs is the free Euclidean distance rather than the product–sum distance and built-in time diversity.

4.11.5 Code design

Here we explain the code design rules, by using the example with the four-state rate 2/3 8PSK trellis code for use with transmit diversity as it was presented in [286]. It is noted that the built-in time diversity (R) of a four-state code is equal to one and, therefore, optimized in this simplest case. In order to increase the product–sum distance of the code, parallel transitions in the trellis diagram are avoided. Thus, we can only consider the error events of actual length two to maximize the minimum product–sum distance of the underlying code. The signal transitions between states of consecutive stages can be represented by a 4×4 matrix $\mathbf{G}$, where the ijth element represents the signal transmitted from state i to state j between consecutive stages in the trellis diagram. Using set partitioning, the 8PSK signal set shown in Figure 4.53 is partitioned into two subsets $A_0 = (0, 2, 4, 6)$ and $A_1 = (1, 3, 5, 7)$. The design rules are given as follows.

Rule 1. Elements of each row of the matrix $\mathbf{G}$ are associated with signals from subsets A_0 or A_1. Specifically, signals with distance δ_1 and δ_3 are associated with branches diverging from one state to two adjacent states, with a state difference of two and one, respectively,

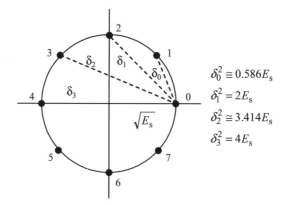

Figure 4.53 8 PSK signal set [286] © 2003, IEEE.

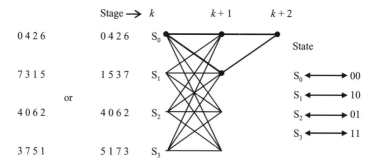

Figure 4.54 Trellis diagram of the four-state 2/3 8PSK trellis code [286] © 2003, IEEE

where the state difference is defined as the number of bits in which two states differ (see Figure 4.54).

Rule 2. The distance between branches remerging at one state from two adjacent states with a state difference of two is δ_2. The pair of signals remerging from two states with a state difference of one is associated with distance δ_0 or δ_3.

According to Equation (4.149), the minimum product–sum distance should be maximized. Rule 1 associates distance δ_1 to signals diverging from one state to two adjacent states with a state difference of two (two states with state difference of two are always adjacent) and guarantees that the distance between any two signals diverging from one state is at least δ_1. Thus, if we assign δ_2 to signals remerging at one state from two states with a state difference of two (Rule 2), the minimum product–sum distance will be:

$$d_{\min}(R) = \min\left(\delta_1^2 + \delta_2^2, \delta_0^2 + \delta_3^2\right) = \delta_0^2 + \delta_3^2 = 4.586 E_s$$

which is greater than that of the optimal single antenna four-state 2/3 8PSK code.

Using the above code design rules, the best four-state 2/3 8PSK trellis code for use with transmit diversity when perfect interleavers are assumed is shown in Figure 4.54. An equivalent code constructed using the design tools is also shown in the figure.

The eight-state trellis code can also be constructed. Due to the constraint of the trellis structure, the eight-state code becomes the Ungerboeck code [287]. Its minimum

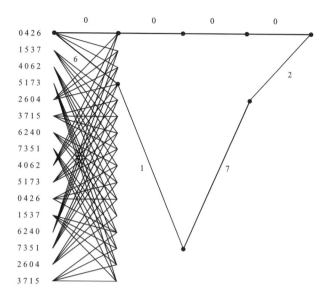

Figure 4.55 Trellis diagram of the 16-state 2/3 8PSK trellis code [286].

product–sum distance is $6E_S$. An eight-state code with a larger product–sum distance exists, but it is catastrophic. Obviously, the total diversity $(2R)$ of both the four and eight-state codes is equal to two, but it will increase to four when the number of states is increased to 16, since R is increased from one to two. After experimentation with various trellis structures and signal assignments, *the 16-state code* is also found based on the design criteria, and is shown in Figure 4.55. The minimum product–sum distance of this code is equal to:

$$d_{\min}(R) = \left(\delta_0^2 \delta_0^1\right)\left(\delta_0^2 \delta_0^1\right) = (2 + 0.586)\,E_s \times (2 + 0.586)\,E_s = 6.69E_s^2$$

The constructed 16-state code has an Ungerboeck representation, which means that it can be generated by a feedback-free convolutional encoder followed by the natural mapping. Performance results for concatenated code obtained using Rules 1 and 2 (R1 and 2 code) and traditional trellis codes are compared in Figures 4.56–4.63 with transmission matrix Γ of the STBC by Alamouti.

More details on the topic can be found in [286–294].

4.12 ESTIMATION OF MIMO CHANNEL

Channel estimation using training sequences is required for coherent detection in BLAST. In this section we present the maximum likelihood channel estimator and the optimal training sequences for block flat fading channels and analyze the estimation error. The optimal training length and training interval that maximize the throughput for a given target bit error rate are presented as functions of the Doppler frequency and the number of antennas.

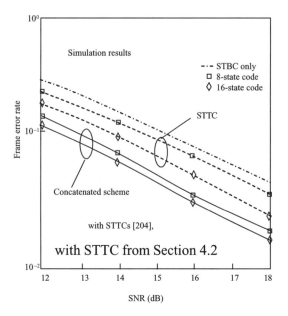

Figure 4.56 Quasi-static fading, $M = 1$.

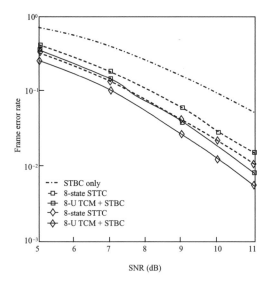

Figure 4.57 Quasi-static fading, $M = 2$. The dashed dotted line denotes the STBC only, and the solid lines denote the concatenated scheme.

4.12.1 System model

The system consists of M transmitting antennas and N receiving antennas. The vector of signals at the output of N receive antennas can be represented as:

$$\mathbf{y}_i = \sqrt{\frac{\rho}{M}} \mathbf{H}_i \mathbf{s}_i + \mathbf{w}_i \qquad (4.151)$$

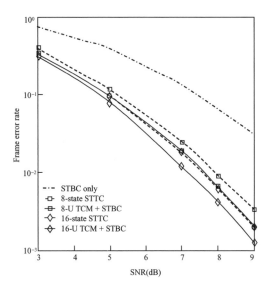

Figure 4.58 Quasi-static fading, $M = 3$. The dashed dotted line denotes the STBC only, and the solid lines denote the concatenated scheme.

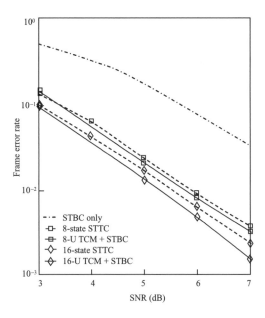

Figure 4.59 Quasi-static fading, $M = 4$. The dashed dotted line denotes the STBC only, and the solid lines denote the concatenated scheme.

where $\mathbf{H}_i$ is the $N \times M$ channel matrix, $\mathbf{s}_i$ is the $M \times 1$ transmitted signal vector and $\mathbf{w}_i$ is the $N \times 1$ vector of complex additive white Gaussian noise with zero mean and unit variance at time instant i. The average power of the components in $\mathbf{H}_i$ and $\mathbf{s}_i$ are normalized to unity, so the average signal to noise ratio (SNR) at each receiving antenna is ρ, independent of the number of transmitting antennas.

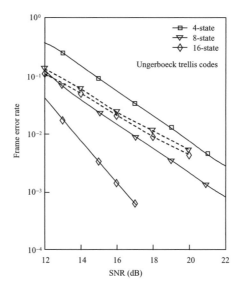

Figure 4.60 Perfect interleaving, $M = 1$. Performances without interleaving (dashed lines) are also shown for comparison.

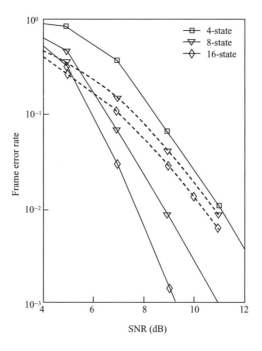

Figure 4.61 Perfect interleaving, $M = 2$. Performances without interleaving (dashed lines) are also shown for comparison.

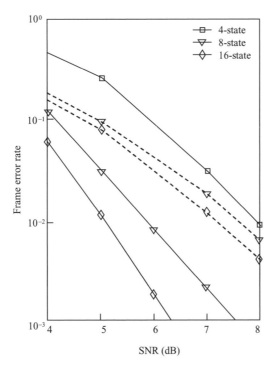

Figure 4.62 Perfect interleaving, $M = 3$. Performances without interleaving (dashed lines) are also shown for comparison.

An alternative presentation without normalization gives:

$$\mathbf{y} = \mathbf{Hs} + \mathbf{w} \tag{4151.a}$$

In this case, the signal power is constrained by $E(\mathbf{ss}^*) \leq P$ so that P/M is the maximum average power transmitted by each antenna, where ( )* stands for the Hermitian matrix/vector transpose conjugate. Starting from the definition of the mutual information exchanged in the MIMO channel, the capacity (maximum mutual information) $C(\mathbf{H})$ can be represented as [295, 296]

$$C(\mathbf{H}) = \log \det \left(\mathbf{I}_N + \frac{P}{M} \mathbf{HH}^* \right) \tag{4.151.b}$$

4.12.2 Training

During the training phase, training sequences of L_t symbols long are transmitted from all the transmitting antennas. An estimate of the channel, $\hat{\mathbf{H}}$, is obtained at the end. During the payload phase, data sequences of L_d symbols long are transmitted and jointly detected. We define L_t as the *training length* and $L = L_d + L_t$ as the *training interval*. The *duty cycle factor* $\eta = 1 - L_t/L$, is the fraction of time spent in data transmission.

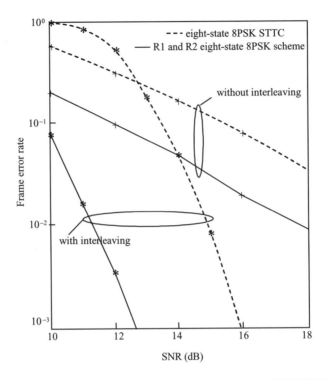

Figure 4.63 R1 and 2 eight-state 8PSK scheme versus eight-state 8PSK STTC at a spectral efficiency of 1.5 bit/s/Hz, $M = 1$.

4.12.3 Performance measure

Since the channel is continuously fading, the actual channel will deviate progressively from the channel estimate obtained at time $i = L_t$. The BER performance will be dominated by the worst channel estimation error. Therefore, we consider $\hat{\mathbf{H}} - \mathbf{H}_L$, as a measure of the channel estimation error, where $\mathbf{H}_L$ is the channel at the end of the training period.

4.12.4 Definitions

Define the difference between the channel at time i and at time L as:

$$\mathbf{\Delta H}_i = \mathbf{H}_i - \mathbf{H}_L \tag{4.152}$$

then, we can rewrite Equation (4.151) as follows:

$$\mathbf{y}_i = \sqrt{\frac{\rho}{M}}\mathbf{H}_L\mathbf{s}_i + \sqrt{\frac{\rho}{M}}\mathbf{\Delta H}_i\mathbf{s}_i + \mathbf{w}_i \tag{4.153}$$

Let $\mathbf{S}$ be the matrix of training symbols, $\mathbf{S} = \begin{bmatrix} \mathbf{s}_1 & \mathbf{s}_2 & \cdots & \mathbf{s}_{L_t} \end{bmatrix}$, where $\mathbf{s}_i$ for $1 \leq i \leq L_t$ is the $M \times 1$ training symbol vector at time i. Let the matrix of received signals be

$\mathbf{Y} = [\mathbf{y}_1\ \mathbf{y}_2 \cdots \mathbf{y}_{L_t}]$, and the matrix of noise be $\mathbf{W} = [\mathbf{w}_1\ \mathbf{w}_2 \cdots \mathbf{w}_{L_t}]$. Then:

$$\mathbf{Y} = \sqrt{\frac{\rho}{M}}\mathbf{H}_L\mathbf{S} + \mathbf{W} \times \mathbf{S}^* + \sqrt{\frac{\rho}{M}}[\boldsymbol{\Delta}\mathbf{H}_1\mathbf{s}_1\ \boldsymbol{\Delta}\mathbf{H}_2\mathbf{s}_2 \cdots \boldsymbol{\Delta}\mathbf{H}_{L_t}\mathbf{s}_{L_t}] \qquad (4.154)$$

4.12.5 Channel estimation error

For *block fading* channels where the channel realization remains constant within a block of certain length and then changes to an independent realization for the next block, the ML channel estimator is:

$$\hat{\mathbf{H}} = \sqrt{\frac{M}{\rho}}\mathbf{Y} \cdot \mathbf{S}^* \cdot (\mathbf{S}\mathbf{S}^*)^{-1} \qquad (4.155)$$

and the optimal training sequences which minimize the mean square estimation error are orthogonal across all transmitting antennas, i.e.

$$\mathbf{S}\mathbf{S}^* = L_t\mathbf{I}_M \qquad (4.156)$$

where $\mathbf{I}_M$ is the $M \times M$ identity matrix. A necessary condition for the matrix inversion $(\mathbf{S}\mathbf{S}^*)^{-1}$ to exist is $L_t \geq M$.

Equations (4.155) and (4.156) are suboptimal but practically appealing for continuous fading channels too. By applying these to Equation (4.154), we obtain:

$$\hat{\mathbf{H}} = \mathbf{H}_L + \boldsymbol{\Delta}\mathbf{H}_{\text{noise}} + \boldsymbol{\Delta}\mathbf{H}_{\text{Doppler}} \qquad (4.157)$$

where

$$\boldsymbol{\Delta}\mathbf{H}_{\text{noise}} = \frac{1}{L_t}\sqrt{\frac{M}{\rho}}\mathbf{W}\mathbf{S}^* \qquad (4.158)$$

is the estimation error due to noise and

$$\boldsymbol{\Delta}\mathbf{H}_{\text{Doppler}} = \frac{1}{L_t}\sum_{i=1}^{L_t}\boldsymbol{\Delta}\mathbf{H}_i \cdot (\mathbf{s}_i\mathbf{s}_i^*) \qquad (4.159)$$

is the estimation error due to the temporal variation of the channel.

4.12.6 Error statistic

It is easy to show that $\boldsymbol{\Delta}\mathbf{H}_{\text{noise}}$ has i.i.d. complex Gaussian entries of zero mean and variance of $M/(\rho L_t)$. We assume the components of $\mathbf{H}_i$ are uncorrelated with each other (rich scattering) and Rayleigh fading with respect to i. Let $\boldsymbol{\Delta}\mathbf{h}_n^T$ represent the nth row of $\boldsymbol{\Delta}\mathbf{H}_{\text{Doppler}}$.

$$E\left\{\left(\boldsymbol{\Delta}\mathbf{h}_{n_1}^T\right)^* \boldsymbol{\Delta}\mathbf{h}_{n_2}^T\right\} =$$

$$\delta_{n_1n_2} \cdot \frac{1}{L_t^2}\sum_{i_1=1}^{L_t}\sum_{i_2=1}^{L_t}\mathbf{s}_{i_1}\mathbf{s}_{i_1}^* \cdot [\xi(i_1 - i_2) - \xi(i_1 - L) - \xi(i_2 - L) + 1]\mathbf{s}_{i_2}\mathbf{s}_{i_2}^*$$

$$(4.160)$$

where δ_{jk} is the discrete Dirac delta function. $\xi(x) = J_0(2\pi f_{d\,max}T \cdot x)$, where $J_0(\cdot)$ is the zeroth order Bessel function of the first kind, $f_{d\,max}$ is the maximum Doppler frequency and T is the symbol period.

For channel estimation tracking, it is reasonable to assume that the phase change during one training period is small, i.e. $2\pi f_{d\,max}TL \ll 1$. Then Equation (4.160) can be simplified as:

$$E\{(\Delta \mathbf{h}_{n_1}^T)^* \Delta \mathbf{h}_{n_2}^T\} = \delta_{n_1 n_2} \cdot 2 \left(\frac{\pi f_{d\,max}T}{L_t}\right)^2 \cdot \left(\sum_{i=1}^{L_t}(L-i)\,\mathbf{s}_i \mathbf{s}_i^*\right)^2 \tag{4.161}$$

using $J_o(x) \approx 1 - x^2/4$ for small x.

4.12.7 Results

The result indicates that the estimation error due to temporal variation increases quadratically with the Doppler frequency. The error also depends on the training length L_t, the training interval L and the training sequences $\mathbf{s}_i$.

In the simple case where there is only one transmitting antenna and one receiving antenna, i.e. $M = N = 1$, the variance of the estimation error in Equation (4.161) can be computed directly:

$$\sigma_{Doppler}^2 = 2\left[\pi f_{d\,max}T\left(L - \frac{L_t+1}{2}\right)\right]^2 \tag{4.162}$$

We can see that if L is fixed and L_t increases, the error decreases. If L_t is fixed and L increases, the error increases. If both L_t and L increase at a fixed ratio L_t/L, the error increases. For $M, N > 1$, the expression for the mean square estimation error is generally very complicated and it depends on the exact training sequences. A possible choice of orthogonal training sequences is the FFT matrix, i.e.

$$S_{m,i} = e^{-j2\pi(m-1)(i-1)/L_t} \tag{4.163}$$

where $S_{m,i}$ is the (m, i)th component of the training matrix $\mathbf{S}$, $1 \le m \le M$, $1 \le i \le L_t$. It can be shown that with such training sequences, the leading component of the variance is the same as Equation (4.162). Therefore, the earlier observations also apply to multiple antenna systems.

4.12.7.1 Example M, N = 4

In this example, $M = 4$ transmitting antennas and $N = 4$ receiving antennas are used. The training sequences are the fast Fourier transform (FFT) sequences in Equation (4.163). The average receiving SNR is $\rho = 15$ dB. The carrier frequency is $f_c = 2$ GHz and the maximum Doppler frequency is $f_{d\,max} = 10$ Hz, which corresponds to a pedestrian speed. The symbol period is $T = 41\,\mu s$, corresponding to the IS-136 standard (see Chapter 1). The channel coefficients are generated using the Jakes model and continuously fading. Figure 4.64 shows the mean square error (MSE) of the channel estimation as a function of the training length L_t. Both L_t and the training interval L increase at a fixed ratio, $L_t/L = 20\,\%$.

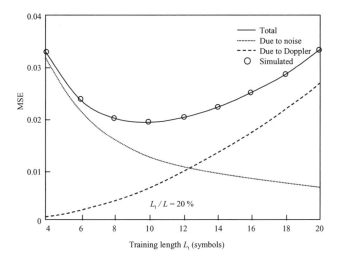

Figure 4.64 Channel estimation MSE versus the training length L_t. Four transmitting antennas and four receiving antennas. $L_t/L = 20\%$, ($\rho = 15$ dB, $f_{d\,max} = 10$ Hz).

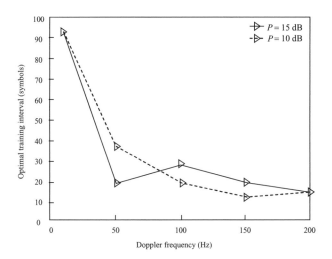

Figure 4.65 Optimal training interval versus Doppler frequency ($M, N = 4$, BER $= 3\%$).

The MSE due to noise decreases with L_t but the MSE due to temporal variation increases with L_t and L. As a result, the overall MSE first decreases and then increases.

Note here that as long as the flat fading model holds, the above results will apply to systems with different symbol period, T, if we scale the maximum Doppler frequency $f_{d\,max}$ appropriately. This is valid because the estimation error due to temporal variation depends only on the product $f_{d\,max}T$. Additional results are given in Figures 4.65 and 4.66.

More details on the topic can be found in [297–308].

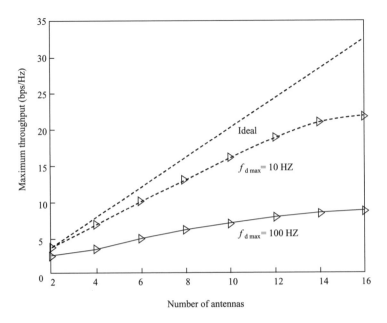

Figure 4.66 Maximum throughput versus the number of antennas resulting from the optimal training interval. 'Ideal' indicates the throughput with ideal channel estimation. ($\rho = 15$ dB, BER $= 3\%$).

4.13 SPACE–TIME CODES FOR FREQUENCY SELECTIVE CHANNELS

The presentation in this section is based on [309]. If a frequency selective channel is modeled as a symbol spaced, tap delay line of length L, the sampled version of one frame ($K + L - 1$ time slots) of the received signal, at antenna r, after matched filtering can be represented as

$$y_k^r = \sum_{l=0}^{L-1} \sum_{t=1}^{M_T} h_t^r(l) c_{k-l}^t + n_k^r \quad k = 1, \ldots, K + L - 1 \qquad (4.164)$$

In Equation (4.164) y_k^r is the received signal at antenna r and time slot k, n_k^r is a complex white Gaussian random noise sample at antenna r and time slot k with variance N_0 and $h_t^r(l)$ is a circularly symmetric complex Gaussian random variable with zero mean describing the lth tap gain. The variance of $h_t^r(l)$ is denoted as $\sigma^2(l)$ and $h_t^r(l)$ is normalized so that we have:

$$\sum_{l=0}^{L-1} \sigma^2(l) = 1 \qquad (4.165)$$

No channel knowledge at the transmitter and a coherent receiver with perfect channel state information are assumed. The vector $\mathbf{v} = (\sigma^2(0), \sigma^2(l), \ldots, \sigma^2(L-1))$ will be referred to as *the power delay profile vector* and is common for all subchannels. If the antennas are spaced sufficiently far apart, then $h_t^r(l)$ and $h_{t'}^{r'}(l')$ are independent if $t \neq t'$ or $r \neq r'$, which is referred to as *spatial independence*. The case when $h_t^r(l)$ and $h_{t'}^r(l')$ are independent for $l \neq l'$, is referred to as the *uncorrelated tap* case. Channels are said to have *uniform power delay profiles* if all components in the power delay profile vector are equal. Otherwise we have a *non-uniform power delay profile* [307–310]. Equation (4.164) can be represented in vector form as:

$$\mathbf{y}_r = \sum_{l=0}^{L-1} \mathbf{h}^r(l)\mathbf{C}(l) + \mathbf{n}_r \tag{4.166}$$

where

$$
\begin{aligned}
\mathbf{y}_r &= \left(y_1^r, \ldots, y_{K+L-1}^r\right) \\
\mathbf{n}_r &= \left(n_1^r, \ldots, n_{K+L-1}^r\right) \\
\mathbf{h}^r(l) &= \left(h_1^r(l), \ldots, y_{M_T}^r(l)\right) \\
\mathbf{C}(l) &= \left(\mathbf{0}_{M_T \times l} \quad \mathbf{C} \quad \mathbf{0}_{M_T \times (L-1-l)}\right)
\end{aligned}
\tag{4.167}
$$

with

$$
\mathbf{C} = \begin{bmatrix}
c_1^1 & c_2^1 & \cdots & c_K^1 \\
c_1^2 & c_2^2 & \cdots & c_K^2 \\
\vdots & \vdots & \vdots & \vdots \\
c_1^{M_T} & c_2^{M_T} & \cdots & c_K^{M_T}
\end{bmatrix}
\tag{4.168}
$$

which is the traditional code word matrix from the flat fading channel analysis. In fact, the $l = 0$ term in Equation (4.166) is exactly the flat fading signal term when M_T transmit antennas are employed. Due to the similarity of the other terms, the lth term in Equation (4.166) for $l > 0$ can be thought of as coming from an lth set of M_T virtual transmit antennas.

The tth row of the matrix $\mathbf{C}(l)$ represents the modulated output symbols transmitted from the tth transmit antenna over $K + L - 1$ time periods. Combining these rows together for $l = 0, \ldots, L - 1$ in the order of increasing l gives:

$$
\mathbf{C}_t = \begin{bmatrix}
c_1^t & c_2^t & \cdots & c_K^t & 0 & 0 & \cdots & 0 \\
0 & c_1^t & c_2^t & \cdots & c_K^t & 0 & \cdots & 0 \\
\vdots & \vdots & \vdots & \vdots & \vdots & \vdots & \vdots & \vdots \\
0 & 0 & \cdots & 0 & c_1^t & c_2^t & \cdots & c_K^t
\end{bmatrix}
\tag{4.169}
$$

Then, Equation (4.166) is also equal to:

$$\mathbf{y}_r = \sum_{l=1}^{M_T} \mathbf{h}_t^r \mathbf{C}_t + \mathbf{n}_r \tag{4.170}$$

where $\mathbf{h}_t^r = (h_t^r(0), h_t^r(1), \ldots, h_t^r(L-1))$ is a subchannel impulse response vector between transmit antenna t and receive antenna r.

We stack up all the signals received by the M_R receive antennas to get:

$$Y = HC_s + N \tag{4.171}$$

where

$$Y = \begin{bmatrix} y_1 \\ y_2 \\ \vdots \\ y_{M_R} \end{bmatrix} \quad N = \begin{bmatrix} n_1 \\ n_2 \\ \vdots \\ n_{M_R} \end{bmatrix} \quad C_s = \begin{bmatrix} C_1 \\ C_2 \\ \vdots \\ C_{M_R} \end{bmatrix}$$

$$H = \begin{bmatrix} h_1^1 & h_2^1 & \cdots & h_{M_T}^1 \\ h_1^2 & h_2^2 & \cdots & h_{M_T}^2 \\ \vdots & \vdots & \vdots & \vdots \\ h_1^{M_R} & h_2^{M_R} & \cdots & h_{M_T}^{M_R} \end{bmatrix}$$

Assume that the transmitted code word is C_s and the erroneously decoded code word is E_s. Define the codeword difference matrix as $B_s = C_s - E_s$. Define the non-negative definite Hermitian matrix $A_s = B_s B_s^H$, where H represents the conjugate transpose. Further, consider the $M_T L M_R \times M_T L M_R$ matrix $D_s = I_{M_R} \otimes A_s$, where I_{M_R}, is an $M_R \times M_R$ identity matrix and $\otimes$ is the Kronecker product. Now vectorize the channel matrix H^T, where T represents the transpose operation, to define the channel vector $h = \text{vec}(H^T)^T$. Let $K = E(h^H h)$ denote the correlation matrix of h. We only consider the case where K is full rank. Since K is a positive definite matrix, Cholesky factorization yields $K = F^H F$, where F is a lower triangular matrix. Using arguments from Section 4.2, the pairwise error probability is upper bounded by:

$$P(C_s \to E_s) \le \frac{1}{\det(I + \gamma FD_s F^H)} \tag{4.172}$$

where $\gamma = E_s/4N_0$, E_s is the average energy per symbol at each transmit antenna. At high SNR, Inequality (4.172) reduces to:

$$P(C_s \to E_s) \le \gamma^{-qM_R} \left(\prod_{i=1}^{qM_R} \lambda_i \right) \tag{4.173}$$

where q is the rank of the matrix A_s and λ_i, is the ith non-zero eigenvalue of $FD_s F^H$. An equation similar to Inequality (4.172) can be developed by using Equation (4.166) instead of Equation (4.170).

The maximum rank of matrix A_s is $M_T L$. Thus, the maximum diversity gain of an STC employed in an frequency selective channel is $M_T M_R L$, L times greater than that of the same STC for flat fading channels, which is only $M_T M_R$.

4.13.1 Diversity gain properties

Similarly to arguments used in Section 4.2, let C and E be two code word matrices from Equation (4.168). Let $C(l)$ and $E(l)$ denote the corresponding matrices from

Equation (4.167). Let $\mathbf{B}(l) = \mathbf{C}(l) - \mathbf{E}(l)$, then define $\mathbf{B}_{ds}$ as:

$$\mathbf{B}_{ds} = \begin{bmatrix} \mathbf{B}(0) \\ \mathbf{B}(1) \\ \vdots \\ \mathbf{B}(L-1) \end{bmatrix} \tag{4.174}$$

which can be easily derived from $\mathbf{B}_s$ and vice versa. Let $\mathbf{B} = \mathbf{C} - \mathbf{E}$ be the flat fading code word difference matrix and assume the rank of the matrix $\mathbf{B}$ is r_b. The matrix $\mathbf{B}$ is similar to an upper triangular matrix $\mathbf{T}$ whose first r_b diagonal elements are non-zero and other $M_T - r_b$ diagonal elements are zero. Likewise, the matrix $\mathbf{B}(l)$ is similar to $\mathbf{T}(l) = \mathbf{0}_{M_T \times l} \, \mathbf{T} \, \mathbf{0}_{M_T \times (L-1-l)}$. Thus, the difference matrix $\mathbf{B}_{ds}$ is similar to the following matrix:

$$\mathbf{T}_{ds} = \begin{bmatrix} \mathbf{T}(0) \\ \mathbf{T}(1) \\ \vdots \\ \mathbf{T}(L-1) \end{bmatrix} \tag{4.175}$$

It is apparent that the collection of vectors consisting of all r_b rows of the matrix $\mathbf{T}(0)$ and the r_bth row of each of the matrices $\mathbf{T}(l)$, $t = 1, \ldots, L - 1$ will be linearly independent, so $\mathbf{T}_{ds}$ has rank $r_b + L - 1$ or larger. Therefore, the minimum rank of difference matrix $\mathbf{B}_{ds}$ is $r_b + L - 1$.

4.13.2 Coding gain properties

Inequality (4.172) shows that the coding gain depends not only on the matrix $\mathbf{A}_s$, but on the channel correlation matrix $\mathbf{K}$ as well. With the assumption that the STC provides maximum diversity gain, $\mathbf{K}$ is full rank, and SNR is large, Inequality (4.172) is well approximated by:

$$P(\mathbf{C}_s \rightarrow \mathbf{E}_s) \leq \frac{\gamma^{-M_T M_R L}}{\det(\mathbf{K}) \det(\mathbf{D}_s)} \tag{4.176}$$

Denote $\mathbf{K}$ as a partitioned matrix with $E(\mathbf{h}_{t_1}^{r_1 H} \mathbf{h}_{t_2}^{r_2})$ as its $(M_T \times (r_1 - 1) + t_1, M_T \times (r_2 - 1) + t_2)$th block entry $(t_1, t_2 = 1, \ldots, M_T$ for each $r_1, r_2 = 1, \ldots, M_R)$, where $E(\mathbf{h}_{t_1}^{r_1 H} \mathbf{h}_{t_2}^{r_2})$ is an $L \times L$ correlation matrix.

Channel 1. In this case the taps are uncorrelated with spatial independence. Based on spatial independence, $E(\mathbf{h}_{t_1}^{r_1 H} \mathbf{h}_{t_2}^{r_2})$ reduces to the $\mathbf{0}$ matrix when $r_1 \neq r_2$ or $t_1 \neq t_2$. Furthermore, from the uncorrelated tap assumption, $E(\mathbf{h}_{t}^{r H} \mathbf{h}_{t}^{r})$ simplifies to a diagonal matrix with $\sigma^2(0), \ldots, \sigma^2(L-1)$ along the diagonal. Then, in this case, the correlation matrix $\mathbf{K}$ becomes a diagonal matrix, whose determinant is:

$$\det(\mathbf{K}) = \left(\prod_{l=0}^{L-1} \sigma^2(l) \right)^{M_T M_R} \tag{4.177}$$

For a specific code and *Channel 1* we want to know conditions for the best coding gain. Due to the arithmetic mean and geometric mean inequality and $\sum_{l=0}^{L-1} \sigma^2(l) = 1$, $\prod_{l=0}^{L-1} \sigma^2(l)$ will achieve the maximum L^{-L} if and only if each $\sigma^2(l)$ is equal to each other, which implies the channel has uniform power delay profile.

Channel 2. In this case the taps are correlated with spatial independence. Again $E(\mathbf{h}_{t_1}^{r_1\,\mathrm{H}}\mathbf{h}_{t_2}^{r_2})$ reduces to a $\mathbf{0}$ matrix when $r_1 \neq r_2$ or $t_1 \neq t_2$. However, $\mathbf{K}_t^r = E(\mathbf{h}_t^{r\,\mathrm{H}}\mathbf{h}_t^r)$ cannot be simplified to a diagonal matrix as in the case of *Channel 1*, but $\mathbf{K}_t^r$ still has exactly the same entries along the diagonal as those in *Channel 1*. More precisely, $\mathbf{K}$ can be described as the matrix formed by arranging a set of non-diagonal matrices along the partition diagonal of a larger partitioned matrix with the appropriate zero padding. According to the property of the determinant of a partitioned matrix, we have:

$$\det(\mathbf{K}) = \prod_{r=1}^{M_R}\prod_{t=1}^{M_T} \det\left(\mathbf{K}_t^r\right) \tag{4.178}$$

According to Hadamard's inequality that the determinant of a non-negative definite square matrix is not greater than the product of all its diagonal elements, we have:

$$\det\left(\mathbf{K}_t^r\right) \leq \prod_{l=0}^{L-1} \sigma^2(l) \tag{4.179}$$

Combining Equation (4.178) and Inequality (4.179), results in:

$$\det(\mathbf{K}) \leq \left(\prod_{l=0}^{L-1} \sigma^2(l)\right)^{M_T M_R} \tag{4.180}$$

Comparing Equation (4.177) and Inequality (4.180), we can see that with the spatial independence assumption, the coding gain of a specific STC for the uncorrelated tap assumption is always larger than or equal to that under a correlated tap assumption. In other words, the tap correlation will generally degrade the coding gain.

Channel 3. In this case the STC is applied to a spatially correlated frequency selective fading channel with correlated taps. Now, in general, none of the submatrices $E(\mathbf{h}_{t_1}^{r_1\,\mathrm{H}}\mathbf{h}_{t_2}^{r_2})$ can be reduced to a $\mathbf{0}$ matrix. In this case we will start from the following result in matrix calculus. Given three matrices $\mathbf{M}_1(m \times m)$, $\mathbf{M}_2(m \times n)$, $\mathbf{M}_3(n \times n)$, if

$$\mathbf{M} = \begin{bmatrix} \mathbf{M}_1 & \mathbf{M}_2 \\ \mathbf{M}_2^{\mathrm{H}} & \mathbf{M}_3 \end{bmatrix}$$

is positive definite, then $\det(\mathbf{M}) \leq \det(\mathbf{M}_1)\det(\mathbf{M}_3)$. This is known as Fischer's inequality. The correlation matrix $\mathbf{K}$ is a positive definite Hermitian matrix. So, by successively using Fischer's inequality, we find:

$$\det(\mathbf{K}) \leq \prod_{r=1}^{M_R}\prod_{t=1}^{M_T} \det\left(\mathbf{K}_t^r\right) \tag{4.181}$$

4.13.3   Space–time trellis code design

A systematic design procedure for space–time trellis codes (STTCs), presented in Section 4.2 for flat fading channels, will now be modified to handle frequency selective channels. The relationship between the symbols transmitted by real antennas and those transmitted by the virtual antennas is used. Define an R-bit binary input vector $\mathbf{a}_k = (a_{k,1}, a_{k,2}, \ldots, a_{k,R})$ and concatenate this with a $(Q-1)R$-bit current state vector to get $\bar{\mathbf{a}}_k = (\mathbf{a}_k, \mathbf{a}_{k-1}, \ldots, \mathbf{a}_{k-Q+1})$. From Section 4.2 an STTC can be represented by:

$$\bar{\mathbf{x}}_k = \left(x_k^1, x_k^2, \ldots, x_k^{M_T}\right) = \bar{\mathbf{a}}_k \mathbf{G} \tag{4.182}$$

where

$$\mathbf{G} = \begin{bmatrix} g_{11} & g_{12} & \cdots & g_{1,M_T} \\ g_{21} & g_{22} & \cdots & g_{2,M_T} \\ \cdots & \cdots & \cdots & \cdots \\ g_{QR,1} & g_{QR,2} & \cdots & g_{QR,M_T} \end{bmatrix}$$

At each time slot, each component of the length M_T output vector from Equation (4.182) is mapped into a constellation symbol and these symbols are transmitted simultaneously from M_T antennas. In this case the g_{ij}, $i = 1, \ldots, QR$, $j = 1, \ldots, M_T$ can be taken from an alphabet whose size is equal to the constellation size s. The trellis encoder starts and ends in state zero at the beginning and end of each frame.

Consider the case with $M_T = 2$, $R = 1$, $Q = 5$, $s = 2$ (BPSK) and $L = 3$. This results in a 16-state 1 b/s/Hz BPSK STTC for a frequency selective channel with three taps. Let $\mathcal{L}$ denote the shortest length error event, as in Section 4.2, for all such codes. First, all codes with maximum diversity gain and maximum coding gain of $\eta = 5.532$ are found [309]. There are many such codes. Among them, those with larger $\bar{\eta}_{CP}(\mathcal{L})$ are chosen [309]. In this case, there are only two codes yielding maximum $\bar{\eta}_{CP}(\mathcal{L}) = 9.357$. They are $\mathbf{G}_{11} = (11101, 11011)^T$ and $\mathbf{G}_{12} = (11011, 10111)^T$ respectively,[1] as listed in Table 4.14. Monte Carlo simulation is used to evaluate the code performance. Figure 4.67 shows the frame error rate (FER) versus signal to noise ratio (SNR $= M_T E_s/N_0$) per receive antenna. The frame size is 130 symbol intervals. A simulated channel with spatial independence and uncorrelated taps is assumed. Maximum likelihood decoding is performed at the receiver. An additional selection of good codes is listed in Table 4.15 [309].

The latest results in the field are presented in [311–352].

4.14 OPTIMIZATION OF A MIMO SYSTEM

4.14.1 The channel model

As before, the model consists of M transmit and N receive antenna elements. We assume that $M > N$. The antenna weights on the receive side are described as a column vector $\mathbf{U}$ with elements $U_1, U_2, \ldots$, and the vector is normalized so the norm is unity ($\mathbf{U}'\mathbf{U} = 1$). Similarly, $\mathbf{V}$ denotes the transmit weight vector. For notation, $*$ signifies complex conjugation, and $\prime$ transpose and conjugate. (M, N) refers to M antennas at the transmitter end and N antennas at the receiver end.

An example of a (3,2) system is shown in Figure 4.68. Similarly to Equation (4.151a) the transfer matrix from the transmit antennas to the receive antennas is described by

---

[1] $(11101, 11011)^T$ denotes $\begin{bmatrix} 1 & 1 & 1 & 0 & 1 \\ 1 & 1 & 0 & 1 & 1 \end{bmatrix}^T$ for simplicity.

Table 4.14 16-state BPSK STTCs with maximum η and different $\bar{\eta}_{CP}(\mathcal{L})$ [309].

No.	$\mathbf{G}^{\mathrm{T}}$	η	$\bar{\eta}_{CP}(\mathcal{L})$
$\mathbf{G}_{11}$	$\begin{bmatrix} 1 & 1 & 1 & 0 & 1 \\ 1 & 1 & 0 & 1 & 1 \end{bmatrix}$	5.532	9.357
$\mathbf{G}_{12}$	$\begin{bmatrix} 1 & 1 & 0 & 1 & 1 \\ 1 & 0 & 1 & 1 & 1 \end{bmatrix}$	5.532	9.357
$\mathbf{G}_{13}$	$\begin{bmatrix} 1 & 1 & 1 & 1 & 0 \\ 1 & 1 & 1 & 0 & 1 \end{bmatrix}$	5.532	8.512
$\mathbf{G}_{14}$	$\begin{bmatrix} 1 & 1 & 1 & 0 & 1 \\ 1 & 0 & 0 & 1 & 0 \end{bmatrix}$	5.532	7.008
$\mathbf{G}_{15}$	$\begin{bmatrix} 1 & 1 & 1 & 0 & 1 \\ 0 & 0 & 1 & 0 & 1 \end{bmatrix}$	5.532	6.465
$\mathbf{G}_{16}$	$\begin{bmatrix} 0 & 0 & 0 & 1 & 1 \\ 1 & 0 & 0 & 0 & 1 \end{bmatrix}$	5.532	5.532

Table 4.15 q-state BPSK STTCs for channels with two taps [309]

q	No.	$\mathbf{G}^{\mathrm{T}}$	η	$\bar{\eta}_{CP}(\mathcal{L})$
4	$\mathbf{G}_{21}$	$\begin{bmatrix} 1 & 1 & 1 \\ 1 & 0 & 1 \end{bmatrix}$	4.000	5.968
8	$\mathbf{G}_{22}$	$\begin{bmatrix} 1 & 1 & 1 & 1 \\ 1 & 0 & 0 & 1 \end{bmatrix}$	5.981	7.825
16	$\mathbf{G}_{23}$	$\begin{bmatrix} 1 & 1 & 0 & 1 & 1 \\ 1 & 0 & 1 & 0 & 1 \end{bmatrix}$	7.445	9.973
32	$\mathbf{G}_{24}$	$\begin{bmatrix} 1 & 1 & 1 & 0 & 0 & 1 \\ 0 & 0 & 1 & 1 & 0 & 1 \end{bmatrix}$	9.514	10.942

transmission matrix $\mathbf{H}$ with elements H_{ik}. They are random complex Gaussian quantities. A normalization:

$$E\langle |H_{ik}|^2 \rangle = 1 \tag{4.183}$$

is used. It is assumed that the angular spreads seen from both sides are so large that the antenna signals are spatially uncorrelated.

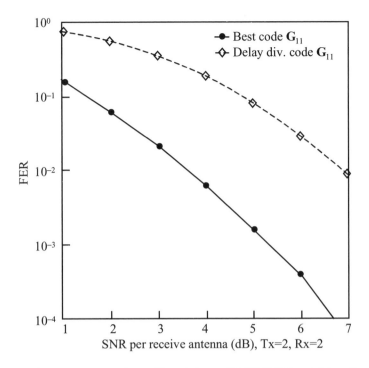

Figure 4.67 Performance comparison of best 16-state BPSK STTC and delay diversity code with two transmit and two receive antennas for a channel with uncorrelated power delay profile vector (1/3, 1/3, 1/3).

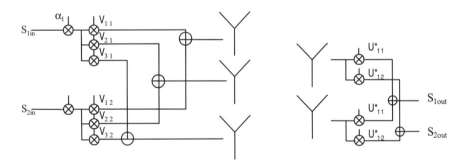

Figure 4.68 Transmission from three (M) transmit antennas to two (N) receive antennas [353].

4.14.2 Gain optimization by singular value decomposition (SVD)

The matrix $\mathbf{H}$ will, in general, be rectangular with N rows and M columns. An SVD expansion of $\mathbf{H}$ can be represented as (see Appendix 5.1)

$$\mathbf{H} = \mathbf{U}_\lambda \cdot \mathbf{D} \cdot \mathbf{V}'_\lambda \qquad (4.184)$$

where $\mathbf{D}$ is a diagonal matrix of real, non-negative singular values, the square roots of the eigenvalues of $\mathbf{G}$, where $\mathbf{G} = \mathbf{H}' \cdot \mathbf{H}$ is an $M \times M$ Hermitian matrix. The columns of the unitary matrices $\mathbf{U}_\lambda$ and $\mathbf{V}_\lambda$ are the corresponding singular vectors. Thus, Equation (4.184) is just a compact way of writing the set of independent channels [353]:

$$\mathbf{HV}_1 = \sqrt{\lambda_1}\mathbf{U}_1$$
$$\mathbf{HV}_2 = \sqrt{\lambda_2}\mathbf{U}_2$$
$$\vdots$$
$$\mathbf{HV}_N = \sqrt{\lambda_N}\mathbf{U}_N$$

$$(4.185)$$

The SVD is particularly useful for interpretation in the antenna context. For one particular eigenvalue, one can see that $\mathbf{V}_i$ is the transmit weight factor for excitation of the singular value $\sqrt{\lambda_i}$.

A receive weight factor of $\mathbf{U}'_i$, a conjugate match, gives the receive voltage S_r, and the square of that the received power, P_r:

$$S_r = \mathbf{U}'_i \mathbf{U}_i \sqrt{\lambda_i} = \sqrt{\lambda_i}$$
$$P_r = |S_r|^2 = \lambda_i$$

$$(4.186)$$

This clearly shows that the matrix $\mathbf{H}$ of transmission coefficients may be diagonalized, leading to a number of independent orthogonal modes of excitation, where the power gain of the ith mode or channel is λ_i. The weights applied to the arrays are given directly from the columns of the $\mathbf{U}_\lambda$ and $\mathbf{V}_\lambda$ matrices. Thus, the eigenvalues and their distributions are important properties of the arrays and the medium, and the maximum gain is of course given by the maximum eigenvalue. The number of non-zero eigenvalues may be shown to be the minimum value of M and N. The situation is illustrated in Figure 4.68, where the total power is distributed among the N parallel channels by weight factors α. These coefficients are discussed later in more detail. An important parameter is the trace of $\mathbf{G}$, i.e. the sum of the eigenvalues

$$\text{Trace} = \sum_i \lambda_i$$

$$(4.187)$$

which may be shown to have a mean value of MN. We illustrate the above relations by an example [353].

4.14.2.1 Example: (2, 2) system

For the $(M, N) = (2, 2)$ example, matrix $\mathbf{G}$ is given by:

$$\mathbf{G} = \begin{bmatrix} |H_{11}|^2 + |H_{12}|^2 & H_{11}H_{21}^* + H_{12}H_{22}^* \\ H_{11}^*H_{21} + H_{12}^*H_{22} & |H_{22}|^2 + |H_{21}|^2 \end{bmatrix}$$
$$= \begin{bmatrix} a & c \\ c^* & b \end{bmatrix}$$

$$(4.188)$$

and the two eigenvalues are (see Appendix 5.1)

$$\lambda_{\max} = \frac{1}{2}(a + b + \sqrt{(a-b)^2 + 4|c|^2})$$ (4.189)

and

$$\lambda_{\min} = \frac{1}{2}(a + b - \sqrt{(a-b)^2 + 4|c|^2})$$ (4.190)

Note that

$$\text{Trace} = \sum_i \lambda_i = a + b$$
$$= |H_{11}|^2 + |H_{12}|^2 + |H_{21}|^2 + |H_{22}|^2$$ (4.191)

so the sum of the eigenvalues displays the full fourth-order diversity.

The distribution of ordered eigenvalues may be found in [205], from which the distributions for $\lambda_{\min}$ and $\lambda_{\max}$ may be derived:

$$p(\lambda_{\min}) = 2e^{-2\lambda}$$ (4.192)
$$p(\lambda_{\max}) = e^{-\lambda}(\lambda^2 - 2\lambda + 2) - 2e^{-2\lambda}$$ (4.193)

In this particular case, it may be shown that the mean values are:

$$E\langle\lambda_{\max}\rangle = 3.5 \quad E\langle\lambda_{\min}\rangle = 0.5$$ (4.194)

The minimum eigenvalue is Rayleigh distributed with mean power 0.5.
The cumulative probability distribution for $\lambda_{\max}$ is:

$$\Pr(\lambda_{\max} < x) = 1 - e^{-x}(x^2 + 2) + e^{-2x}$$
$$\approx x^4/12 \quad x \ll 1$$ (4.195)

One can show that for the case of standard diversity $(M, N) = (1, 4)$:

$$\Pr(P < x) = 1 - e^{-x}(1 + x + x^2/2 + x^3/6)$$
$$\approx x^4/24 \quad x \ll 1$$ (4.196)

so the (2, 2) case displays full fourth-order diversity but with twice the cumulative probability for the same power level.

The cumulative probability distributions are shown in Figure 4.69, where the maximum eigenvalue (the array gain) follows the fourth-order maximum ratio diversity distribution quite closely.

One should be aware that in order to make full benefit of the maximum eigenvalue, the full knowledge of the channel at the transmitter is required, otherwise the eigenvectors cannot be found.

4.14.3 The general (M, N) case

For the (4, 4) case in Figure 4.70, the two arrays have 16 different uncorrelated transmission coefficients, so the diversity order is 16. In the asymptotic limit when N is large, it may be shown [354], [355] that the largest eigenvalue is bounded above by:

$$\lambda_{\max} < (\sqrt{c} + 1)^2 N; \quad c = M/N \geq 1$$ (4.197)

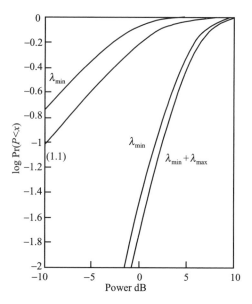

Figure 4.69 Cumulative probability distribution of eigenvalues for a $(T, R) = (2, 2)$ array with four uncorrelated paths. The maximum eigenvalue follows closely the fourth-order diversity with a shift of 0.75 dB.

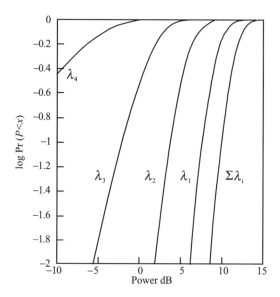

Figure 4.70 Cumulative probability distribution of eigenvalues (power) for two arrays of four elements each, including the sum of eigenvalues corresponding to a (1, 16) case.

whereas the smallest eigenvalue is bounded below by:

$$\lambda_{min} > (\sqrt{c} - 1)^2 N \quad c \geq 1 \tag{4.198}$$

In the previous examples, $c = 1$, and the upper asymptotic bound for this case is $4N$. These bounds should not be understood as absolute bounds, but rather as limits approached as N tends to infinity for a fixed c.

The mean array gains (mean of the maximum eigenvalues) are shown in Figure 4.71, together with the upper bound and the gain for the correlated, free space case, N^2. For $N = 10$, the true mean gain is just 1 dB below the upper bound. For a partly correlated case, we can expect the gain to lie between the $\rho = 0$ and the $\rho = 1$ cases, where ρ is the spatial correlation coefficient between the elements.

In some situations, it might be advantageous to have more antennas on one side than on the other, especially for asymmetric situations with heavy downloading of data from a base station. Again, the asymptotic upper bound for the largest eigenvalue is useful, Inequality (4.197). Introducing $M = cN$ directly we find:

$$G_{\text{upper bound}} = (\sqrt{M} + \sqrt{N})^2 \tag{4.199}$$

which, asymptotically, will approach M for large values of M and fixed N. This clearly illustrates that the composite gain of the link cannot be factored into one belonging to the transmitter and one belonging to the receiver.

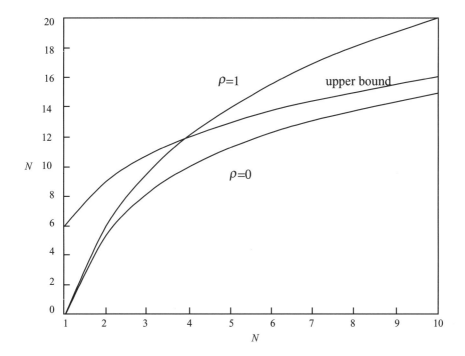

Figure 4.71 The gain relative to one element of (N, N) arrays in a correlated situation ($\rho = 1$), and in an uncorrelated case ($\rho = 0$). The upper bound equals $4N$, and is the asymptotic upper bound for the maximum eigenvalue for N tending to infinity.

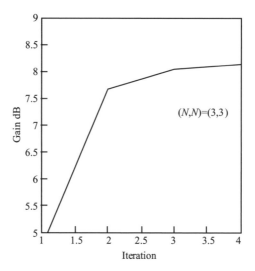

Figure 4.72 Convergence of gain by iterative transmissions between receiver and transmitter for a (3, 3) case. The gain values are mean values.

4.14.4 Gain optimization by iteration for a reciprocal channel

Since the channel is reciprocal, exactly the same weights may be used for transmission as for reception. Consider now the situation with transmission from M transmit antennas to N receive antennas (or vice versa). The iteration starts with an arbitrary $\mathbf{V}_1$, in the numerical calculations, chosen as a unit vector with equal elements. At the receive side, the weights are adjusted for maximum gain, and the same weights are then used for transmit since the channel is reciprocal. This may then be repeated a number of times. In principle, the process might converge to an eigenvalue different from the maximum one, but experience shows excellent performance [353].

An example of the convergence in the mean is shown for a (3, 3) case in Figure 4.72. After a few iterations, the gain has converged to the steady state. This might actually be a computationally efficient way of finding the maximum gain solution in practice without going through the trouble of finding the eigenvectors. For details regarding eigenvalue decomposition see Appendix 5.1.

4.14.5 Spectral efficiency of parallel channels

From Figure 4.70 with four independent channels one can see that there are other options for using the eigenvalues than using the largest for maximum gain. Another option is to keep them as parallel channels with independent information, as discussed in the previous sections of this chapter. The knowledge about the distribution of the eigenvalues and the upper and lower bounds may now be used for evaluating bounds on the theoretical capacity of the link. Shannon's capacity measure gives an upper bound on the realizable information rates through parallel channels, and how the power should be distributed over the channels to achieve maximum capacity through 'water filling' [356]. From Equation (4.151b), the basic expression for the spectral efficiency measured in bits/s/Hz for one Gaussian channel is given by:

$$C = \log_2(1 + P) \quad \text{bits / s / Hz} \tag{4.200}$$

where P is the signal to noise ratio, SNR, for one channel.

Assuming all noise powers to be the same, the 'water filling' concept is the solution to the maximum capacity, where each channel is filled up to a common level D:

$$\frac{1}{\lambda_1} + P_1 = \frac{1}{\lambda_2} + P_2 = \frac{1}{\lambda_3} + P_3 = \cdots = D \tag{4.201}$$

Thus, the channel with the highest gain receives the largest share of the power. The constraint on the powers is that:

$$\sum_i P_i = P \tag{4.202}$$

The weight factors α_i in Figure 4.68 equal P_i / P. In the case where level D drops below a certain $1/\lambda_i$, that power is set to zero. In the limit where the SNR is small ($P < 1/\lambda_2 - 1/\lambda_1$), only one eigenvalue, the largest, is left, and we are back to the maximum gain solution of the previous section. For the case of $(M, N) = (2, 2)$ (Figure 4.69) for $P < 2 - 2/7 = 2.34$ dB, only the largest eigenvalue is active, using the mean values from Equation (4.192). The capacity equals

$$C = \sum_{N'} \log_2(1 + \lambda_i P_i) = \sum_{N'} \log_2(\lambda_i D) \tag{4.203}$$

where the summation is over all channels with non-zero powers. The water filling is of course dependent on the knowledge of the channels on the transmit side. In the case where the channel is unknown at the transmitter, the only reasonable division of power is a uniform distribution over the antennas, i.e.

$$P_i = \frac{P}{M} \tag{4.204}$$

M being the number of transmit antennas [296]. It may also be argued that the transmit antenna 'sees' M eigenvalues, not taking into account that there are only N non-zero eigenvalues. Thus, power is lost by allocating power to the zero-valued eigenvalues.

It also follows that when $M = N$, the difference between the capacity for known and unknown channels is small for large P.

4.14.6 Capacity of the (*M, N*) array

It follows from Inequalities (4.197) and (4.198) that the instantaneous eigenvalues are limited by:

$$(\sqrt{M} - \sqrt{N})^2 < \lambda_i < (\sqrt{M} + \sqrt{N})^2 \tag{4.205}$$

So, for M much larger than N, all the eigenvalues tend to cluster around M. Furthermore, each of them will be non-fading due to the high MNth order diversity. Thus, the uncorrelated asymmetric channel with many antennas has a very large theoretical capacity of N equal, constant channels with high gains of M. The above illustrates in a mathematical sense the observation of Winters [357] that M should be of the order $2N$. In the limit of large M and

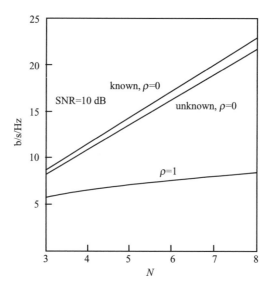

Figure 4.73 Mean capacity for two arrays of each N elements. The capacity grows linearly with the number of elements and is approximately the same for the known and the unknown channel. The total transmitted power is constant.

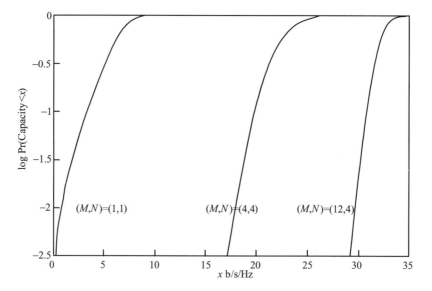

Figure 4.74 The cumulative probability distribution of capacity on a log scale for the (M, N) = (1, 1), (4, 4), and (12, 4) cases. The basic signal to noise ratio is 20 dB. The total radiated power is the same in all cases.

N, with M much larger than N, the capacity is easily found to be:

$$C = N \log_2 \left(1 + \frac{P}{N} M \right) \tag{4.206}$$

with the result that the theoretical capacity grows linearly with the number of elements N [356, 357] for M/N fixed. The result may conveniently be interpreted as N parallel channels, each with $1/N$ of the power and each having a gain of M. Note that this capacity is higher than the one used in [356, 357], where the power is divided between the M antennas instead of the N channels, given as:

$$C_{\text{unknown}} = N \log_2(1 + P) \tag{4.207}$$

The numerical results shown in Figure 4.73 support this approximate analysis for the mean values. It should be remembered that the potential gains are higher when a certain outage probability is studied due to the high order diversity effects. This is illustrated in Figure 4.74, which shows the cumulative probability distribution (on a log scale) of the capacity for the case of four receiving elements, and four and twelve transmitting elements. The signal to noise ratio is 20 dB for the (1,1) case, and it is worth emphasizing that the total power radiated remains constant. The improvement going from four to twelve transmitting antennas is mainly due to the improved gain of the smallest eigenvalues as indicated by Inequality (4.205). Using Equation (4.206) in the (12, 4) case gives 32.9 b/s/Hz.

II
Multiple Access

5

Code Division Multiple Access – CDMA

The basic principles of CDMA were discussed in Chapter 1. Here we focus on a few critical components of the system, mainly code generation, synchronization, power control and multiuser detection. More details can be found in the recent book on WCDMA [358].

5.1 PSEUDORANDOM SEQUENCES

5.1.1 Binary shift register sequences

Let us define a polynomial

$$h(x) = h_0 x^n + h_1 x^{n-1} + \cdots + h_{n-1} x + h_n \tag{5.1}$$

in the discrete field with two elements $h_i \in (0, 1)$ and $h_0 = h_n = 1$. An example polynomial could be $x^4 + x + 1$ or $x^5 + x^2 + 1$. The coefficients h_i of the polynomial can be represented by binary vectors 10011 and 100101, or in octal notation 23 and 45 (every group of three bits is represented by a number between 0 and 7). A binary sequence u is said to be a *sequence generated by* $h(x)$ if, for all integers j,

$$h_0 u_j \oplus h_1 u_{j-1} \oplus h_2 u_{j-2} \oplus \cdots \oplus h_n u_{j-n} = 0 \tag{5.2}$$

where $\oplus$ = addition modulo 2.

If we formally change the variables

$$j \to j + n, \quad \text{and} \quad h_0 = 1 \tag{5.3}$$

Advanced Wireless Communications. S. Glisic
© 2004 John Wiley & Sons, Ltd. ISBN: 0-470-86776-0

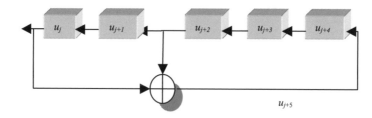

Figure 5.1 Sequence generator for the polynomial (45).

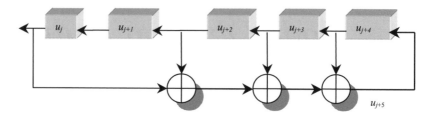

Figure 5.2 Sequence generator for polynomial (75).

then Equation (5.2) becomes:

$$u_{j+n} = h_n u_j \oplus h_{n-1} u_{j+1} \oplus \cdots h_1 u_{j+n-1} \tag{5.4}$$

In this notation, u_j is the jth bit (called chip) of the sequence u. Equation (5.4) suggests that the sequence u can be generated by an n-stage binary linear feedback shift register which has a feedback tap connected to the ith cell if $h_i = 1, 0 < i \leq n$. As an example, for $n = 5$, Equation (5.4) becomes:

$$u_{j+5} = h_5 u_j \oplus h_4 u_{j+1} \oplus h_3 u_{j+2} \oplus h_2 u_{j+3} \oplus h_1 u_{j+4} \tag{5.5}$$

For $x^5 + x^2 + 1$, octal representation (45), the coefficients h_i are:

h_0	h_1	h_2	h_3	h_4	h_5	
1	0	0	1	0	1	(octal representation 45)

and the block diagram of the circuit is shown in Figure 5.1.

Similarly, for the polynomial $x^5 + x^4 + x^3 + x^2 + 1$, the coefficients h_i are given as:

h_0	h_1	h_2	h_3	h_4	h_5	
1	1	1	1	0	1	(octal representation 75)

and by using Equation (5.4) one can get the generator shown in Figure 5.2.

Some of the properties of these sequences and definitions are listed below. Details can be found in standard literature listed at the end of the book, especially [359–379].

If u and v are generated by $h(x)$, then so is $u \oplus v$, where $u \oplus v$ denotes the sequence whose ith element is $u_i \oplus v_i$. The all zero state of the shift register is not allowed because, for this initial state, Equation (5.5) would continue to generate zero chips. For this reason, the period of u is at most $2^n - 1$, where n is the number of cells in the shift register, or equivalently, the degree of $h(x)$. If u denotes an arbitrary $\{0, 1\}$ valued sequence, then

$x(u)$ denotes the corresponding $\{+1, -1\}$ valued sequence, where the ith element of $x(u)$ is just $x(u_i)$:

$$x(u_i) = (-1)^{u_i} \qquad (5.6)$$

If T^i is a delay operator (delay for i chip periods) then we have:

$$T^i(x(u)) = x(T^i u) \quad \text{and} \quad \sum x(u) = x(u_0) + x(u_1) + \cdots + x(u_{N-1})$$
$$= N^+ - N^- = (N - N^-) - N^- \qquad (5.7)$$
$$= N - 2N^- = N - 2wt(u)$$

where $wt(u)$ denotes the Hamming weight of unipolar sequence u, that is, the number of ones in u, N is the sequence period and N^+ and N^- are the number of positive and negative chips in bipolar sequence $x(u)$. The crosscorrelation function between two bipolar sequences can be represented as:

$$\theta_{u,v}(l) \equiv \theta_{x(u),x(v)}(l) = \sum_{i=0}^{N-1} x(u_i)x(v_{i+l})$$
$$= \sum_{i=0}^{N-1} (-1)^{u_i}(-1)^{v_{i+l}} = \sum_{i=0}^{N-1} (-1)^{u_i \oplus v_{i+l}} = \sum_{i=0}^{N-1} x(u_i \oplus v_{i+l}) \qquad (5.8)$$

By using Equation (5.7) we have:

$$\theta_{u,v}(l) = N - 2wt(u \oplus T^l v) \qquad (5.9)$$

The periodic autocorrelation function $\theta_u(\cdot)$ is just $\theta_{u,u}(\cdot)$

$$\theta_u(l) = N - 2wt(u \oplus T^l u)$$
$$= N^+ - N^- = (N - N^-) - N^- = N - 2N^- \qquad (5.10)$$

5.1.2 Properties of binary maximal length sequences

As was mentioned earlier the all zero state of the shift register is not allowed because, based on Equation (5.4), the generator could not get out of this state. Bear in mind that the number of possible states of the shift register is 2^n. So, the period of a sequence u generated by the polynomial $h(x)$ cannot exceed $2^n - 1$, where n is the degree of $h(x)$. If u has this maximal period, $N = 2^n - 1$, it is called a maximal length sequence or m-sequence. To get such a sequence, $h(x)$ should be a primitive binary polynomial of degree n. There are exactly N non-zero sequences generated by $h(x)$, and they are just the N different phases of $u, Tu, T^2u, \ldots, T^{N-1}u$. Given distinct integers i and j, $0 \le i, j < N$, there is a unique integer, k, distinct from both i and j, such that $0 \le k < N$ and

$$T^i u \oplus T^j u = T^k u \qquad (5.11)$$

From the above discussion on the number of ones and zeros, $wt(u) = 2^{n-1} = 1/2(N + 1)$, so that from Equation (5.9) we have:

$$\theta_u(l) = \begin{cases} N, & \text{if} \quad l \equiv 0 \quad \mod N \\ -1, & \text{if} \quad l \neq 0 \quad \mod N \end{cases} \qquad (5.12)$$

From here on, $\tilde{u}$ will be called a characteristic m-sequence, or the characteristic phase of the m-sequence u if $\tilde{u}_i = \tilde{u}_{2i}$ for all $i \in \mathbb{Z}$.

Let q denote a positive integer, and consider the sequence v formed by taking every qth bit of u (i.e. $v_i = u_{qi}$ for all $i \in \mathbb{Z}$). The sequence v is said to be a *decimation* by q of u, and will be denoted by $u[q]$.

Assume that $u[q]$ is not identically zero. Then, $u[q]$ has period $N/\gcd(N, q)$, and is generated by the polynomial whose roots are the qth powers of the roots of $h(x)$, where $\gcd(N, q)$ is the greatest common divisor of the integers N and q. The tables of primitive polynomials are available in any book on coding theory. The reciprocal m-sequence v is generated by the reciprocal polynomial of $h(x)$, that is,

$$\hat{h}(x) = x^n h(x^{-1}) = h_n x^n + h_{n-1} x^{n-1} + \cdots + h_0. \tag{5.13}$$

5.1.3 Crosscorrelation spectra

Frequently, we do not need to know more than the set of crosscorrelation values together with the number of integers $l (0 \le l < N)$ for which $\theta_{u,v}(l) = c$ for each c in this set. Let u and v denote m-sequences of period $2^n - 1$. If $v = u[q]$, where either $q = 2^k + 1$ or $q = 2^{2k} - 2^k + 1$, and if $e = \gcd(n, k)$ is such that n/e is odd, then the spectrum of $\theta_{u,v}$ is three-valued [380–383] as:

$$\begin{array}{lll} -1 + 2^{(n+e)/2} & \text{occurs } 2^{n-e-1} + 2^{(n-e-2)/2} & \text{times} \\ -1 & \text{occurs } 2^n - 2^{n-e} - 1 & \text{times} \\ -1 - 2^{(n+e)/2} & \text{occurs } 2^{n-e-1} - 2^{(n-e-2)/2} & \text{times} \end{array} \tag{5.14}$$

The same spectrum is obtained if instead of $v = u[q]$, we let $u = v[q]$. Notice that if e is large, $\theta_{u,v}(l)$ takes on large values but only very few times, while if e is small, $\theta_{u,v}(l)$ takes on smaller values more frequently. In most instances, small values of e are desirable. If we wish to have $e = 1$, then clearly n must be odd in order that n/e be odd. When n is odd, we can take $k = 1$ or $k = 2$ (and possibly other values of k as well), and obtain $\theta(u, u[3])$, $\theta(u, u[5])$ and $\theta(u, u[13])$ all having the three-valued spectrum given by Expressions (5.14) (with $e = 1$). Suppose next that $n \equiv 2 \bmod 4$. Then, n/e is odd if e is even and a divisor of n. Letting $k = 2$, we obtain that $\theta(u, u[5])$ and $\theta(u, u[13])$ both have the three-valued spectrum given by Expressions (5.14) (with $e = 2$).

Let us define $t(n)$ as:

$$t(n) = 1 + 2[(n + 2)/2] \tag{5.15}$$

where $[\alpha]$ denotes the integer part of the real number α. Then if $n \ne 0 \bmod 4$, there exist pairs of m-sequences with three-valued crosscorrelation functions, where the three values are -1, $-t(n)$ and $t(n) - 2$. A crosscorrelation function taking on these values is called a *preferred three-valued crosscorrelation function* and the corresponding pair of m-sequences (polynomials) is called a *preferred pair of m-sequences* (polynomials).

Table 5.1 Set sizes and crosscorrelation bounds for the sets of
all m-sequences and for maximal connected sets [359]
© 1980, IEEE

n	$N = 2^n - 1$	Number of m-sequences	θ_c for set of all m-sequences	M_n	$t(n)$
3	7	2	5	2	5
4	15	2	9	0	9
5	31	6	11	3	9
6	63	6	23	2	17
7	127	18	41	6	17
8	255	16	95	0	33
9	511	48	113	2	33
10	1023	60	383	3	65
11	2047	176	287	4	65
12	4095	144	1407	0	129
13	8191	630	≥ 703	4	129
14	16 383	756	≥ 5631	3	257
15	32 767	1800	≥ 2047	2	257
16	65 535	2048	≥ 4095	0	513

Let u and v denote m-sequences of period $2^n - 1$, where n is a multiple of 4. If $v = u[-1 + 2^{(n+2)/2}] = u[t(n) - 2]$, then $\theta_{u,v}$ has a four-valued spectrum represented as:

$$
\begin{array}{lll}
-1 + 2^{(n+2)/2} & \text{occurs } (2^{n-1} - 2^{(n-2)/2})/3 & \text{times} \\
-1 + 2^{n/2} & \text{occurs } 2^{n/2} & \text{times} \\
-1 & \text{occurs } 2^{n-1} - 2^{(n-2)/2} - 1 & \text{times} \\
-1 - 2^{n/2} & \text{occurs } (2^n - 2^{n/2})/3 & \text{times}
\end{array}
\tag{5.16}
$$

5.1.4 Maximal connected sets of m-sequences

The preferred pair of m-sequences is a pair of m-sequences of period $N = 2^n - 1$, which has the preferred three-valued crosscorrelation function. The values taken on by the preferred three-valued crosscorrelation functions are -1, $-t(n)$ and $t(n) - 2$, where $t(n)$ is given by Equation (5.15). The pair of primitive polynomials that generate a preferred pair of m-sequences is called a preferred pair of polynomials. A connected set of m-sequences is a collection of m-sequences which has the property that each pair in the collection is a preferred pair. A largest possible connected set is called a *maximal connected set*, and the size of such a set is denoted by M_n. Some examples are given in Table 5.1.

5.1.5 Gold sequences

A set of Gold sequences of period $N = 2^n - 1$, consists of $N + 2$ sequences for which $\theta_c = \theta_a = t(n)$. A set of Gold sequences can be constructed from appropriately selected

m-sequences as described below. Suppose $f(x) = h(x)\hat{h}(x)$, where $h(x)$ and $\hat{h}(x)$ have no factors in common. The set of all sequences generated by $f(x)$ is of the form $a \oplus b$, where a is some sequence generated by $h(x)$, b is some sequence generated by $\hat{h}(x)$ and we do not make the usual restriction that a and b are non-zero sequences. We represent such a set by:

$$G(u, v) \triangleq \{u, v, u \oplus v, u \oplus Tv, u \oplus T^2v, \ldots, u \oplus T^{N-1}v\} \tag{5.17}$$

$G(u, v)$ contains $N + 2 = 2^n + 1$ sequences of period N. Let $\{u, v\}$ denote a preferred pair of m-sequences of period $N = 2^n - 1$ generated by the primitive binary polynomials $h(x)$ and $\hat{h}(x)$ respectively. Then set $G(u, v)$ is called a set of Gold sequences. For $y, z \in G(u, v)$, $\theta_{y,z}(l) \in \{-1, -t(n), t(n) - 2\}$ for all integers l, and $\theta_y(l) \in \{-1, -t(n), t(n) - 2\}$ for all $l \neq 0 \bmod N$. Every sequence in $G(u, v)$ can be generated by the polynomial $f(x) = h(x)\hat{h}(x)$. Note that the non-maximal length sequences belonging to $G(u, v)$ also can be generated by adding together (term by term, modulo 2) the outputs of the shift registers corresponding to $h(x)$ and $\hat{h}(x)$. The maximal length sequences belonging to $G(u, v)$ are, of course, the outputs of the individual shift registers. Let us compare the parameter $\theta_{\max} = \max\{\theta_a, \theta_c\}$ for a set of Gold sequences to a bound due to Sidelnikov, which states that for any set of N or more *binary* sequences of period N,

$$\theta_{\max} > (2N - 2)^{1/2} \tag{5.18}$$

For Gold sequences, they form an *optimal* set with respect to the bounds when n is odd. When n is even, Gold sequences are not optimal.

5.1.6 Gold-like and dual-BCH sequences

Let n be even and let q be an integer such that $\gcd(q, 2^n - 1) = 3$. Let u denote an m-sequence of period $N = 2^n - 1$ generated by $h(x)$, and let $v^{(k)}$, $k = 0, 1, 2$, denote the result of decimating $T^k u$ by q. The $v^{(k)}$ are sequences of period $N' = N/3$ which are generated by the polynomial $\hat{h}(x)$ whose roots are qth powers of the roots of $h(x)$. Gold-like sequences are defined as:

$$\begin{aligned} H_q(u) = \{&u, u \oplus v^{(0)}, u \oplus Tv^{(0)}, \ldots, u \oplus T^{N'-1}v^{(0)}, \\ &u \oplus v^{(1)}, u \oplus Tv^{(1)}, \ldots, u \oplus T^{N'-1}v^{(1)}, \\ &u \oplus v^{(2)}, u \oplus Tv^{(2)}, \ldots, u \oplus T^{N'-1}v^{(2)}\} \end{aligned} \tag{5.19}$$

Note that $H_q(u)$ contains $N + 1 = 2^n$ sequences of period N.

For $n \equiv 0 \bmod 4$, $\gcd(t(n), 2^n - 1) = 3$ vectors $v^{(k)}$ are taken to be of length N rather than $N/3$. Consequently, it can be shown that for the set $H_{t(n)}(u)$, $\theta_{\max} = t(n)$. We call $H_{t(n)}(u)$ a set of Gold-like sequences. The correlation functions for the sequences belonging to $H_{t(n)}(u)$ take on values in the set $\{-1, -t(n), t(n) - 2, -s(n), s(n) - 2\}$ where $s(n)$ is defined (for even n only) by:

$$s(n) = 1 + 2^{n/2} = \frac{1}{2}(t(n) + 1) \tag{5.20}$$

5.1.7 Kasami sequences

Let n be even and let u denote an m-sequence of period $N = 2^n - 1$ generated by $h(x)$. Consider the sequence $w = u[s(n)] = u[2^{n/2} + 1]$. w is a sequence of period $2^{n/2} + 1$ which is generated by the polynomial $h'(x)$, whose roots are the $s(n)$th powers of the roots of $h(x)$. Furthermore, since $h'(x)$ can be shown to be a polynomial of degree $n/2$, w is an m-sequence of period $2^{n/2} - 1$. Consider the sequences generated by $h(x)h'(x)$ of degree $3n/2$. Any such sequence must be of one of the forms $T^i u, T^j w, T^i u \oplus T^j w, 0 \le i < 2^n - 1, 0 \le j < 2^{n/2} - 1$. Thus, any sequence y of period $2^n - 1$ generated by $h(x)h'(x)$ is some phase of some sequence in the set $K_s(u)$ defined by:

$$K_s(u) \underset{=}{\Delta} \{u, u \oplus w, u \oplus Tw, \ldots, u \oplus T^{2^{n/2}-2}w\} \tag{5.21}$$

This set of sequences is called a small set of Kasami sequences with:

$$\theta = \{-1, -s(n), s(n) - 2\}$$
$$\theta_{max} = s(n) = 1 + 2^{n/2} \tag{5.22}$$

θ_{max} for the set $K_s(u)$ is approximately one half of the value of θ_{max} achieved by the sets of sequences discussed previously. $K_s(u)$ contains only $2^{n/2} = (N+1)1/2$ sequences, while the sets discussed previously contain $N+1$ or $N+2$ sequences.

Let n be even and let $h(x)$ denote a primitive binary polynomial of degree n that generates the m-sequence u. Let $w = u[s(n)]$ denote an m-sequence of period $2^{n/2} - 1$ generated by the primitive polynomial $h'(x)$ of degree $n/2$, and let $\hat{h}(x)$ denote the polynomial of degree n that generates $u[t(n)]$. Then, the set of sequences of period N generated by $h(x)\hat{h}(x)h'(x)$, called the *large set of Kasami sequences* and denoted by $K_L(u)$, is defined as follows:

1. If $n \equiv 2 \bmod 4$, then:

$$K_L(u) = G(u, v) \bigcup \left[\bigcup_{i=0}^{2^{n/2}-2} \{T^i w \oplus G(u, v)\} \right] \tag{5.23}$$

   where $v = u[t(n)]$, and $G(u, v)$ is defined in Equation (5.17).

2. If $n \equiv 0 \bmod 4$, then:

$$K_L(u) = H_{t(n)}(u) \bigcup \left[\bigcup_{i=0}^{2^{n/2}-2} \{T^i w \oplus H_{t(n)}(u)\} \right]$$
$$\times \bigcup \{v^{(j)} \oplus T^k w \colon 0 \le j \le 2, \ 0 \le k < (2^{n/2} - 1)/3\} \tag{5.24}$$

where $v^{(j)}$ is the result of decimating $T^j u$ by $t(n)$ and $H_{t(n)}(u)$ is defined earlier by Equation (5.19). In either case, the correlation functions for $K_L(u)$ take on values in the set $\{-1, -t(n), t(n) - 2, -s(n), s(n) - 2\}$ and $\theta_{max} = t(n)$. If $n \equiv 2 \bmod 4$, $K_L(u)$ contains $2^{n/2}(2^n + 1)$ sequences, while if $n \equiv 0 \bmod 4$, $K_L(u)$ contains $2^{n/2}(2^n + 1) - 1$ sequences. The large set of Kasami sequences contains both the small set of Kasami sequences and a set of Gold (or Gold-like) sequences as subsets. More interestingly, the correlation bound $\theta_{max} = t(n)$ is the *same* as that for the latter subsets. The previous discussion is summarized in Table 5.2 for some examples of codes.

Table 5.2   Polynomials generating various classes of sequences of periods 31, 63, 65, 127, and 255 [359] © 1980, IEEE

N	Polynomial	Construction	No.		Values taken on by the correlation functions										
31	3551	G	33				7		-1		-9					
	2373	G	33			11	7	3	-1	-5	-9					
63	14551	G	65		15				-1				-17			
	14343	G	65		15	11	7	3	-1	-5	-9	-13				
	12471	H_3	64		15		7		-1		-9		-17			
	1527	K_s	8				7		-1		-9					
	133605	K_L	520		15		7		-1		-9		-17			
65	10761		63		15	11	7	3	-1	-5	-9	-13				
127	41567	G	129		15				-1				-17			
255	231441	G	257	31	15				-1				-17			
	264455	G	257	31,...,	15	11	7	3	-1	-5	-9	-13	-17,...,	-29		
	326161	H_{33}	256	31	15				-1				-17		-33	
	267543	H_3	256	31	15				-1				-17		-33	
	11367	K_s	16		15				-1				-17			
	6031603	K_L	4111	31	15				-1				-17		-33	

5.1.8 JPL sequences

These sequences are constructed by combining sequence $S_1(t, T_c)$ of length L_1 and $S_2(t, T_c)$ of length L_2 with L_1, L_2 prime, as $S = S(t, T_c) = S_1(t, T_c) \oplus S_2(t, T_c)$ of length $L = L_1 \times L_2$. If the composite sequence is delayed for L_1 chips

$$S(t - L_1 T_c, T_c) = S_1(t - L_1 T_c, T_c) \oplus S_2(t - L_1 T_c, T_c)$$
$$= S_1(t, T_c) \oplus S_2(t - L_1 T_c, T_c) \tag{5.25}$$

and, summed up with its original version,

$$S(t, T_c) \oplus S(t - L_1 T_c, T_c) = S_1(t, T_c) \oplus S_2(t, T_c) \oplus S_1(t - L_1 T_c, T_c) \oplus S_2(t - L_1 T_c, T_c)$$
$$= S_1(t, T_c) \oplus S_1(t, T_c) \oplus S_2(t - L_1 T_c, T_c) \oplus S_2(t, T_c)$$
$$= S_2(t - L_3 T_c, T_c) \tag{5.26}$$

The result is only a component sequence S_2. In a similar way, by delaying the composite sequence for L_2 chips, a component sequence S_1 will be obtained. This can be used to synchronize sequence S of length $L_1 \times L_2$ by synchronizing separately component sequences S_1 and S_2 of length L_1 and L_2, which can be done much faster. The acquisition time is proportional to $T_{acq}(S) \sim \max[T_{acq}(S_1), T_{acq}(S_2)] \sim \max[L_1, L_2]$

5.1.9 Kronecker sequences

In this case the component sequences $S_1(t, T_{c1})$ of length L_1 and chip interval T_{c1}, and $S_2(t, T_{c2})$ with $L_2, T_{c2} = L_1 T_{c1}$ are combined as:

$$S(t, T_{c1}, T_{c2}) = S_1(t, T_{c1}) \oplus S_2(t, T_{c2}) \tag{5.27}$$

The composite sequence, S, synchronization is now performed in cascade, first S_1 with a much faster chip rate and then S_2. Correlation of S by S_1 gives:

$$F_2(S_1 \cdot S) = \rho_1 S_2 \tag{5.28}$$

and after that, this result is correlated with sequence S_2. The acquisition time is proportional to $T_{\text{acq}}(S) \sim T_{\text{acq}}(S_1) + T_{\text{acq}}(S_2) \sim L_1 + L_2$.

5.1.10 Walsh functions

A Walsh function of order n can be defined recursively as follows:

$$\mathbf{W}(n) = \begin{bmatrix} \mathbf{W}(n/2), & \mathbf{W}(n/2) \\ \mathbf{W}(n/2), & \mathbf{W}'(n/2) \end{bmatrix} \tag{5.29}$$

$\mathbf{W}'$ denotes the logical complement of $\mathbf{W}$, and $\mathbf{W}(1) = [0]$. Thus,

$$\mathbf{W}(2) = \begin{bmatrix} 0, & 0 \\ 0, & 1 \end{bmatrix} \quad \text{and} \quad \mathbf{W}(4) = \begin{bmatrix} 0, & 0, & 0, & 0 \\ 0, & 1, & 0, & 1 \\ 0, & 0, & 1, & 1 \\ 0, & 1, & 1, & 0 \end{bmatrix} \tag{5.30}$$

$\mathbf{W}(8)$ is as follows:

$$\mathbf{W}(8) = \begin{bmatrix} 0, & 0, & 0, & 0, & 0, & 0, & 0, & 0 \\ 0, & 1, & 0, & 1, & 0, & 1, & 0, & 1 \\ 0, & 0, & 1, & 1, & 0, & 0, & 1, & 1 \\ 0, & 1, & 1, & 0, & 0, & 1, & 1, & 0 \\ 0, & 0, & 0, & 0, & 1, & 1, & 1, & 1 \\ 0, & 1, & 0, & 1, & 1, & 0, & 1, & 0 \\ 0, & 0, & 1, & 1, & 1, & 1, & 0, & 0 \\ 0, & 1, & 1, & 0, & 1, & 0, & 0, & 1 \end{bmatrix} \tag{5.31}$$

One can see that any two rows from the matrix

$$w_k(n) = \{w_{k,j}(n)\}; j = 1, \ldots n$$
$$w_m(n) = \{w_{m,j}(n)\}$$

represent the sequences whose bipolar versions have crosscorrelation equal to zero (orthogonal codes). This is valid as long as the codes are aligned as in the matrix.

A modification of the previous construction rule is shown in Figure 5.3, producing orthogonal variable speeding factor (OVSF) sequences. At each node of the graph, a code $w_k(n/2)$ of length $n/2$ is producing two new codes of length n by a rule

$$w_k(n/2) \to w_{2k-1}(n) = \{w_k(n/2), w_k(n/2)\}$$
$$\to w_{2k}(n) = \{w_k(n/2), -w_k(n/2)\}$$

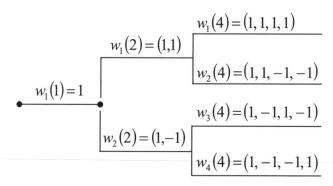

Figure 5.3 Flow graph generating OVSF codes of length 4.

5.1.11 Optimum PN sequences

If we represent the information bit stream as

$$\{b_n\} = \cdots, b_{-1}, b_0, b_1, b_2, \cdots; b_k = \pm 1 \tag{5.32}$$

and the sequence as a vector of chips

$$\mathbf{y} = (y_0, y_1, \ldots, y_{N-1}) \quad y_k = \pm 1 \tag{5.33}$$

then the product of these two streams would create

$$\hat{\mathbf{y}}_i = \cdots; b_{-1}\mathbf{y}; \quad b_0\mathbf{y}; \quad b_1\mathbf{y}; \cdots. \tag{5.34}$$

In other words, $\hat{\mathbf{y}}$ is the DSSS baseband signal which has as its ith element $\hat{y}_i = b_n y_k$ for all i such that $i = nN + k$ for k in the range $0 \leq k \leq N - 1$. A synchronous correlation receiver forms the inner product

$$\langle \hat{\mathbf{y}}_n, \mathbf{y} \rangle = b_n \langle \hat{\mathbf{y}}, \mathbf{y} \rangle = b_n \theta_y(0) \tag{5.35}$$

If the other signal is $\hat{\mathbf{x}}$ which is formed from the data sequence $\{b'_n\}$ and the signature sequence x (generated by a binary vector $\mathbf{x} = (x_0, x_1, \ldots, x_{N-1})$ in exactly the same manner as $\hat{\mathbf{y}}$ was formed from $\{b_n\}$ and $\mathbf{y}$, then we have for the overall received signal:

$$\hat{\mathbf{y}} + T^{-l}\hat{\mathbf{x}} \quad \text{where} \\ \hat{\mathbf{x}} = \cdots; \ b'_-\mathbf{x}; b'_0 \ \mathbf{x}; b'_1\mathbf{x}; \cdots \tag{5.36}$$

The output of a correlation receiver which is in synchronism with $\mathbf{y}$ is given by:

$$\mathbf{z}_n = \langle \hat{\mathbf{y}}_n, \mathbf{y} \rangle + \left[b'_{n-1} \sum_{i=0}^{l-1} x_{N-l+i} y_i + b'_n \sum_{i=l}^{N-1} x_{i-l} y_i \right] \tag{5.37}$$

Having in mind the following relations:

$$\sum_{i=0}^{l-1} x_{N-l+i} y_i = \sum_{i=0}^{N-1+m} x_{i-m} y_i$$

$$\sum_{i=l}^{N-1} x_{i-l} y_i = \sum_{j=0}^{N-1-l} x_j y_{j+l} \tag{5.38}$$

and the definition of aperiodic crosscorrelation function $C_{x,y}$

$$
C_{x,y}(l) = \begin{cases}
\sum_{j=0}^{N-1-l} x_j y^*_{j+l}, & 0 \le l \le N-1 \\[2ex]
\sum_{j=0}^{N-1+l} x_{j-l} y^*_j, & 1-N \le l < 0 \\[2ex]
0, & |l| \ge N
\end{cases}
\tag{5.39}
$$

Equation (5.37) becomes

$$
\mathbf{z}_n = b_n \theta_y(0) + [b'_{n-1} C_{x,y}(l-N) + b'_n C_{x,y}(l)]
\tag{5.40}
$$

The optimum sequences should minimize the interfering term for all values of l. Further details may be found in [371, 372, 384–397].

5.1.12 Golay code

As an appendix to this section, we will present Golay code [384] which is used in the WUMTS syncho channel due to its good aperiodic correlation properties. Additional information on the construction and implementation of these sequences can be found in [384–397].

In general for the two complementary sequences a and b of length N, with aperiodic autocorrelations $A(k)$ and $B(k)$ (delay k) we have:

$$
A(k) + B(k) = 0, k \ne 0 \quad \text{and} \quad A(0) + B(0) = 2N.
$$

The primary synchronization code (PSC), C_{psc} is constructed as a so-called generalized hierarchical Golay sequence. Define:

$$
a = \langle x_1, x_2, x_3, \dots, x_{16} \rangle = \langle 1, 1, 1, 1, 1, 1, -1, -1, 1, -1, 1, -1, 1, -1, -1, 1 \rangle
$$

The PSC is generated by repeating the sequence modulated by a Golay complementary sequence, and creating a complex-valued sequence with identical real and imaginary components. The PSC C_{psc} is defined as:

$$
C_{\text{psc}} = (1+j) \times \langle a, a, a, -a, -a, a, -a, -a, a, a, a, -a, a, -a, a, a \rangle;
$$

where the leftmost chip in the sequence corresponds to the chip transmitted first in time. The 16 secondary synchronization codes (SSCs), $\{C_{\text{ssc},1}, \dots, C_{\text{ssc},16}\}$, are complex-valued with identical real and imaginary components, and are constructed from position wise multiplication of a Hadamard sequence and a sequence z, defined as:

$$
z = \langle b, b, b, -b, b, b, -b, -b, b, -b, b, -b, -b, -b, -b, -b \rangle, \quad \text{where}
$$

$$
b = \langle x_1, x_2, x_3, x_4, x_5, x_6, x_7, x_8, -x_9, -x_{10}, -x_{11}, -x_{12}, -x_{13}, -x_{14}, -x_{15}, -x_{16} \rangle
$$

and $x_1, x_2, \dots, x_{15}, x_{16}$, are same as in the definition of the sequence a above.

The Hadamard sequences (see also Equations (5.29–5.31)) are obtained as the rows in a matrix $\mathbf{H}_8$ constructed recursively by:

$$
\mathbf{H}_0 = [1]
$$

$$
\mathbf{H}_k = \begin{bmatrix} \mathbf{H}_{k-1} & \mathbf{H}_{k-1} \\ \mathbf{H}_{k-1} & -\mathbf{H}_{k-1} \end{bmatrix}, \quad k \ge 1
$$

The rows are numbered from the top starting with row 0 (the all ones sequence). Denote the nth Hadamard sequence as a row of $\mathbf{H}_8$ numbered from the top, $n = 0, 1, 2, \ldots, 255$, subsequently. Furthermore, let $h_n(i)$ and $z(i)$ denote the ith symbol of the sequence h_n and z, respectively where $i = 0, 1, 2, \ldots, 255$ and $i = 0$ corresponds to the leftmost symbol. The kth SSC, $C_{\text{ssc},k}, k = 1, 2, 3, \ldots, 16$ is then defined as:

$$-C_{\text{ssc},k} = (1 + j) \times \langle h_m(0) \times z(0), h_m(1) \times z(1), h_m(2)$$
$$\times z(2), \ldots, h_m(255) \times z(255)\rangle;$$

where $m = 16 \times (k - 1)$ and the leftmost chip in the sequence corresponds to the chip transmitted first in time.

5.1.12.1 *Alternative generation*

The generalized hierarchical Golay sequences for the PSC described above may also be viewed as being generated (in real-valued representation) by the following methods:

Method 1: The sequence y is constructed from two constituent sequences x_1 and x_2 of length n_1 and n_2 respectively, using the following formula:

$$y(i) = x_2(i \bmod n_2) \, x_1(i \text{ div } n_2), i = 0 \cdots (n_1 n_2) - 1$$

The constituent sequences x_1 and x_2 are chosen to be the length 16 (i.e. $n_1 = n_2 = 16$) sequences:

- x_1 is defined to be the length $16 \, (N^{(1)} = 4)$ Golay complementary sequence [384] obtained by the delay matrix $\mathbf{D}^{(1)} = [8, 4, 1, 2]$ and weight matrix

$$\mathbf{W}^{(1)} = [1, \quad -1, \quad 1, \quad 1].$$

- x_2 is a generalized hierarchical sequence using the following formula, selecting $s = 2$ and using the two Golay complementary sequences x_3 and x_4 as constituent sequences. The lengths of the sequences x_3 and x_4 are called n_3 and n_4 respectively.

- $x_2(i) = x_4(i \bmod s + s(i \text{ div } sn_3)) \, x_3((i \text{ div } s) \bmod n_3), i = 0 \cdots (n_3 \, n_4) - 1.$

- x_3 and x_4 are defined to be identical and the length $4(N^{(3)} = N^{(4)} = 2)$ Golay complementary sequence obtained by the delay matrix $\mathbf{D}^{(3)} = \mathbf{D}^{(4)} = [1, 2]$ and weight matrix $\mathbf{W}^{(3)} = \mathbf{W}^{(4)} = [1, 1]$.

The Golay complementary sequences x_1, x_3 and x_4 are defined using the following recursive relation:

$$a_0(k) = \delta(k) \quad \text{and} \quad b_0(k) = \delta(k);$$
$$a_n(k) = a_{n-1}(k) + W_n^{(j)} \cdot b_{n-1}\big(k - D_n^{(j)}\big);$$
$$b_n(k) = a_{n-1}(k) - W_n^{(j)} \cdot b_{n-1}\big(k - D_n^{(j)}\big);$$
$$k = 0, 1, 2, \ldots, 2^{**}N^{(j)} - 1; n = 1, 2, \ldots, N^{(j)}.$$

The desired Golay complementary sequence x_j is defined by a_n assuming $n = N^{(j)}$. The Kronecker delta function is described by $\delta; k, j$ and n are integers.

Method 2: The sequence y can be viewed as a pruned Golay complementary sequence, generated using the following parameters which apply to the generator equations for a and b above:

- Let $j = 0$, $N^{(0)} = 8$.

- $[D_1^0, D_2^0, D_3^0, D_4^0, D_5^0, D_6^0, D_7^0, D_8^0] = [128, 64, 16, 32, 8, 1, 4, 2]$.

- $[W_1^0, W_2^0, W_3^0, W_4^0, W_5^0, W_6^0, W_7^0, W_8^0] = [1, -1, 1, 1, 1, 1, 1, 1]$.

- For $n = 4, 6$, set $b_4(k) = a_4(k)$, $b_6(k) = a_6(k)$.

5.2 MULTIUSER CDMA RECEIVERS

In this section we present a number of methods for CDMA multiple access interference cancellation. Multiple access interference is produced by the presence of the other users in the network which are located on the same bandwidth as our own signal. The common characteristic of all these schemes is some form of joint signal and parameter estimation for all signals present on the same bandwidth. It makes sense to implement this in a base station of a cellular system because all these signals are available there anyway. At the same time this concept will considerably increase the complexity of the receiver. Although very complex, these schemes are being standardized already because they offer significantly better performance.

If user k transmits bit stream b_k, with bit interval T, using spreading sequence s_k, then the low pass equivalent of the overall signal received in the base station can be represented as [398, 399]

$$r_t = S_t(\mathbf{b}) + \sigma n(t) \tag{5.41}$$

$$S_t(\mathbf{b}) = \sum_{i=-M}^{M} \sum_{k=1}^{K} b_k(i) s_k(t - iT - \tau_k) \tag{5.42}$$

where K is the number of users, $\mathbf{b} = (b_1, b_2, \cdots b_K)^T$ is the vector of bits of all users and the signal is observed in time interval $[-MT, MT]$. The noise component is represented by the second term of Equation (5.41) and τ_k is the delay of the signal from user k.

5.2.1 Synchronous CDMA channels

If the signals from different users are received synchronously, Equation (5.41) becomes:

$$r(t) = \sum_{k=1}^{K} b_k(j) s_k(t - jT) + \sigma n(t), \quad t \in [jT, jT + T] \tag{5.43}$$

If we use notation y_k for the output of the matched filter of user k then we have

$$y_k = \int_0^T r(t) s_k(t) \mathrm{d}t, \quad k = 1, \ldots, K \tag{5.44}$$

and we can write

$$y_1 = \sum_j b_k R_{1j} + n,$$

$$y_2 = \sum_j b_k R_{2j} + n_2$$

$$\vdots$$

$$y_k = \sum_j b_k R_{kj} + n_k$$

(5.45)

The vector form of these outputs can be presented as:

$$\mathbf{y} = \mathbf{Rb} + \mathbf{n}$$

(5.46)

where $\mathbf{R}$ is the non-negative definite matrix of crosscorrelations between the assigned waveforms:

$$R_{ij} = \int_0^T s_i(t)s_j(t)\,dt$$

(5.47)

Conventional single user detection can be represented as:

$$\hat{b}_k^c = \text{sgn}\ y_k$$

(5.48)

The optimum multiuser detector becomes:

$$\hat{b} \in \arg \min_{b\in\{-1,1\}^K} \int_0^T \left[r(t) - \sum_{k=1}^K b_k s_k(t) \right]^2 dt$$

$$= \arg \max_{b\in\{-1,1\}^K} 2\mathbf{y}^T\mathbf{b} - \mathbf{b}^T\mathbf{Rb}$$

(5.49)

5.2.2 The decorrelating detector

In the absence of noise, the matched filter output vector is $\mathbf{y} = \mathbf{Rb}$. This suggests that the detector should perform the following operation $\hat{b} = \text{sgn}\,\mathbf{R}^{-1}\mathbf{y}$. Note that the noise components in $\mathbf{R}^{-1}\mathbf{y}$ are correlated, and therefore $\text{sgn}\,\mathbf{R}^{-1}\mathbf{y}$ does not result in optimum decisions. It is interesting to point out that this detector does not require knowledge of the energies of any of the active users.

5.2.3 The optimum linear multiuser detector

The linear detector which minimizes the probability of bit error will be referred to as the optimum linear multiuser detector. Its operation can be represented as:

$$\hat{\mathbf{b}} = \text{sgn}\,(\mathbf{Ty}) = \text{sgn}\,(\mathbf{TRb} + \mathbf{Tn})$$

(5.50)

We will consider the set $I(\mathbf{R})$ of generalized inverses of the crosscorrelation matrix $\mathbf{R}$ and analyze the properties of the detector:

$$\hat{\mathbf{b}} = \text{sgn}\,\mathbf{R}^I\mathbf{y}$$

(5.51)

in Section 5.3. The special case $I(\mathbf{R}) = \mathbf{R}^{-1}$ is referred to as a decorrelating detector.

5.2.4   Multistage detection in asynchronous CDMA [400]

If the indexing of users is arranged in increasing order of their delays, then the output of the correlator of user k can be represented as:

$$z_k^{(i)}(0) = \int_{-\infty}^{\infty} r(t)s_k(t + iT - \tau_k)\,dt$$

$$= \eta_k^{(i)} + \sum_{l=k+1}^{K} R_{kl}(1)b_l^{(i-1)} + \sum_{l=1}^{K} R_{kl}(0)b_l^{(i)} + \sum_{l=1}^{k-1} R_{kl}(-1)b_l^{(i+1)} \quad (5.52)$$

where $\eta_k^{(i)}$ is the component of the statistic due to the additive channel noise. In vector notation, letting $\mathbf{z}^{(i)}(0) = [z_1^{(i)}(0), z_2^{(i)}(0), \ldots, z_K^{(k)}(0)]^{\mathrm{T}}$, we have:

$$\mathbf{z}^{(i)}(0) = \eta^{(i)} + \mathbf{R}(1)\mathbf{b}^{(i-1)} + \mathbf{R}(0)\mathbf{b}^{(i)} + \mathbf{R}(-1)\mathbf{b}^{(i+1)} \quad (5.53)$$

The multistage detector recreates the interfering term for each user based on bit estimations in the previous stage (iteration), subtracts the estimated MAI and then makes the new estimate of data which can be represented as:

$$\hat{b}_k^{(i)}(m+1) = \mathrm{sgn}\big[z_k^{(i)}(m)\big] \quad (5.54)$$

where

$$z_k^{(i)}(m) = z_k^{(i)}(0) - \sum_{l=k+1}^{K} h_{kl}(1)\hat{b}_l^{(i-1)}(m) - \sum_{l \neq k} h_{kl}(0)\hat{b}_l^{(i)}(m) - \sum_{l=1}^{k-1} h_{kl}(-1)\hat{b}_l^{(i+1)}(m)$$

$$(5.55)$$

Examples of probability of error curves are shown in Figure 5.4. All parameters are shown in the figure itself. One can see that even a two-stage detector may significantly improve the system performance. In order to further emphasize the role of MUD in the presence of the near–far effect, Figure 5.4 presents the BER for the case when the crosscorrelation is very high $r_{12} = 1/3$ (three chips long sequences). One can see that when the second user

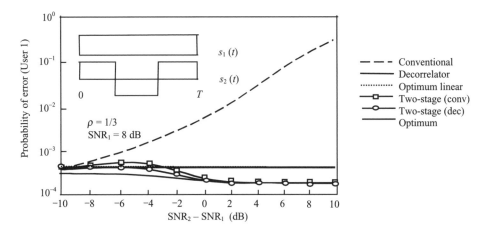

Figure 5.4   Error probability comparison of the linear, two-stage and optimum detectors for a two-user channel with $r_{12} = 1/3$ and signal to noise ratio of user 1 fixed at 8 dB [401] © 1991, IEEE.

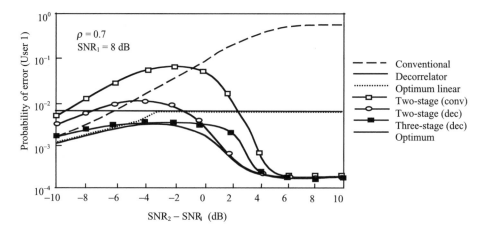

Figure 5.5 Error probability comparison for a two-user channel with $r_{12} = 0.7$ and signal to noise ratio of user 1 fixed at 8 dB [401] © 1991, IEEE.

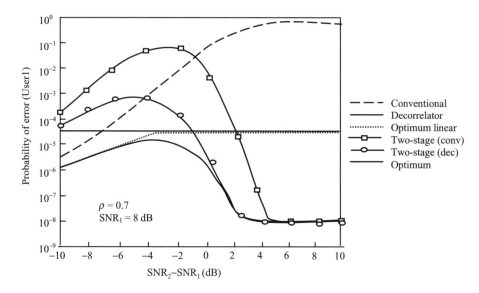

Figure 5.6 Error probability comparison of the linear, two-stage and optimum detectors for a two-user channel with $r_{12} = 0.7$ and signal to noise ratio of user 1 fixed at 12 dB.

becomes stronger and stronger, the improvement compared with a conventional detector is more significant.

This conclusion becomes more and more relevant if either r_{12} is increased, as in Figure 5.5, or signal to noise ratio is increased, as in Figure 5.6. Figure 5.7 demonstrates the same results for five users in the network.

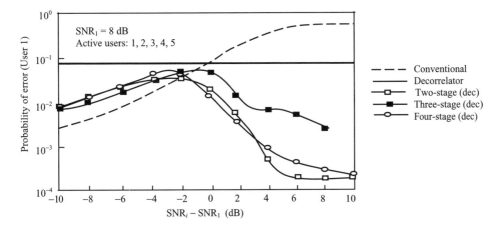

Figure 5.7 Probability of error, five users in the network [401] © 1991, IEEE.

5.2.5 Non-coherent detector

A conventional detector for differential phase keying signals is defined by the following equation:

$$\hat{b}_m = \text{sgn}\big[\text{Re}\{\overline{z_m(-1)}z_m(0)\}\big],$$
$$z_m(i) = \frac{1}{2}\int_{iT}^{(i+1)T} r(t)\overline{f_m(t-iT)}dt \tag{5.56}$$

where $f_m(t)$ is the signal matched filter function. In the trivial case it is the signal spreading code only.

In general, a non-coherent linear multiuser detector for the mth user, denoted by a non-zero transformation $\mathbf{h}^{(m)} \in C^K$, is defined by the decision:

$$\hat{b}_m = \text{sgn}\left[\overline{\text{Re}\left\{\sum_{k=1}^{K}\bar{h}_k^{(m)}z_k(-1)\sum_{l=1}^{K}\bar{h}_l^{(m)}z_l(0)\right\}}\right] \tag{5.57}$$

where K is the length of the code. A non-coherent decorrelating detector for user m is defined by the decision with the linear transformation $\mathbf{h} = \mathbf{d}$, where $\mathbf{d}$ denotes the complex conjugate of the mth column of a generalized inverse $\mathbf{R}^I$ of $\mathbf{R}$. If the mth user is linearly independent, it can be shown that $\mathbf{R}\bar{\mathbf{d}} = \mathbf{u}_m$, the mth unit vector. If all the signature signals are linearly independent, the $\mathbf{R}^{-1}$ exists and the decorrelating transformation $\mathbf{d}$ is uniquely characterized as the complex conjugate of the mth column of the inverse of $\mathbf{R}$. The receiver block diagram is shown in Figure 5.8.

5.2.6 Non-coherent detection in asynchronous multiuser channels [402]

The z-transform of Equation (5.53) gives:

$$\mathbf{Z}(z) = \mathbf{S}(z) \cdot \hat{\mathbf{D}}(z) + \mathbf{N}(z) \tag{5.58}$$

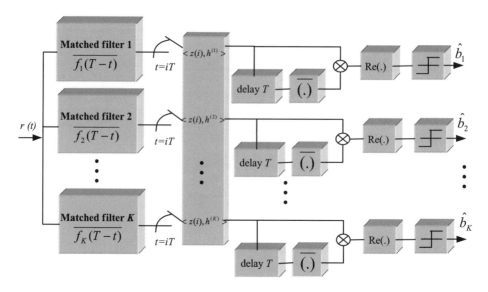

Figure 5.8 Linear multiuser DPSK detector.

where

$$\mathbf{S}(z) = \mathbf{R}(-1)z + \mathbf{R}(0) + \mathbf{R}(1)z^{-1}, \tag{5.59}$$

and $\mathbf{Z}(z)$, $\hat{\mathbf{D}}(z)$ and $\mathbf{N}(z)$ are the vector-valued z-transforms of the matched-filter output sequence, the sequence $\{\hat{d}(l) = A(l)d(l)\}$ and the noise sequence $\{n(l)\}$ at the output of the matched filters. If we define

$$\mathbf{G}(z) = [\mathbf{S}(z)]^{-1} = \frac{\text{adj } \mathbf{S}(z)}{\det \mathbf{S}(z)} \tag{5.60}$$

then we have

$$\hat{\mathbf{d}}(z) = \mathbf{G}(z)\mathbf{Z}(z) \tag{5.61}$$

and

$$\hat{\mathbf{b}}(i) = \text{sgn Re}[\tilde{\mathbf{d}}(i-1) \otimes \tilde{\mathbf{d}}^*(i)] \tag{5.62}$$

The system block diagram is shown in Figure 5.9.

5.2.7 Multiuser detection in frequency non-selective Rayleigh fading channels

Topics covered in the previous chapter are now repeated for the fading channel. Previously described algorithms are extended to the fading channel by using as much analogy as possible in the process of deriving the system transfer functions. In frequency selective channels, decorrelators are combined with the RAKE-type receiver in order to further improve the system performance. A number of simulation results are presented in order to illustrate the effectiveness of these schemes. The concept of this chapter is based on proper

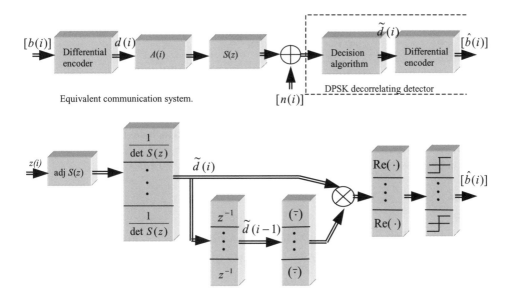

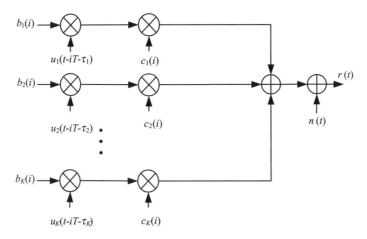

Figure 5.9 Non-coherent decorrelating detector.

Figure 5.10 Asynchronous CDMA flat Rayleigh fading channel model.

understanding of the channel model, covered in Chapter 14. The overall system model, including the channel model for frequency non-selective fading, is shown in Figure 5.10.

Parameters $c_k(i)$ are, for fixed i, independent, zero mean, complex-valued Gaussian random variables, with variances $\overline{|c_k|^2}$ with independent quadrature components. The time-varying nature of the channel is described via the spaced time correlation function of the kth channel $\Phi_k(\Delta t)$:

$$E\{c_k^*(i)c_k(j)\} = \Phi_k((j-i)T) \tag{5.63}$$

The received signal at the central receiver can be expressed as:

$$r(t) = S(t, \mathbf{b}) + n(t)$$

$$S(t, \mathbf{b}) = \sum_{i=-M}^{M} \sum_{k=1}^{K} b_k(i) c_k(i) u_k(t - iT - \tau_k) \tag{5.64}$$

$$u_k(t) = \sqrt{E_k} s_k(t) e^{j\phi_k}$$

where $u_k(t)$ is referred to as the user k signature sequence, and includes the signal amplitude (square root of signal energy), the code itself and the signal phase. By using proper notation, $r(t)$ can be represented as:

$$r(t) = \mathbf{b}^T \mathbf{C} \mathbf{u}_t + n(t) \tag{5.65}$$

where

$$\begin{aligned}
\mathbf{b}^T &[b_1(-M)b_2(-M) \cdots b_K(-M) \cdots b_1(M)b_2(M) \cdots b_K(M)] \\
\mathbf{u}_t &= [\mathbf{u}^T(t + MT) \cdots \mathbf{u}^T(t - MT)]^T \\
\mathbf{u}(t) &= [u_1(t - \tau_1) \cdots u_K(t - \tau_K)]^T \\
\mathbf{C} &= \mathrm{diag}(\mathbf{C}(-M) \cdots \mathbf{C}(M)) \\
\mathbf{C}(i) &= \mathrm{diag}(c_1(i) \cdots c_K(i))
\end{aligned} \tag{5.66}$$

5.2.7.1 Multiuser maximum likelihood sequence detection

By using analogy from the previous section, the likelihood function in this case can be represented as:

$$L(\mathbf{b}) = 2 \, \mathrm{Re}\{\mathbf{b}^H \mathbf{y}\} - \mathbf{b}^H \mathbf{C}^H \mathbf{R}_u \mathbf{C} \mathbf{b} \tag{5.67}$$

Upper index $()^H$ denotes the conjugate transpose and

$$\mathbf{y} = \int_{-\infty}^{+\infty} r(t) \mathbf{C}^H \mathbf{u}_t^* \, dt \tag{5.68}$$

represents the vector of matched filter outputs. The correlation matrix $\mathbf{R}_u$ can be represented as

$$\mathbf{R}_u = \int_{-\infty}^{+\infty} \mathbf{u}_t^* \mathbf{u}_t^T \, dt = \begin{bmatrix} \mathbf{R}_u(0) & \mathbf{R}_u(-1) & 0 & \cdots \\ \mathbf{R}_u(1) & \mathbf{R}_u(0) & \mathbf{R}_u(-1) & \cdots \\ & \ddots & \ddots & \\ \cdots & \mathbf{R}_u(1) & \mathbf{R}_u(0) & \mathbf{R}_u(-1) \\ \cdots & 0 & \mathbf{R}_u(1) & \mathbf{R}_u(0) \end{bmatrix} \tag{5.69}$$

with block elements of dimension $K \times K$

$$\mathbf{R}_u(i - j) = \int_{-\infty}^{+\infty} \mathbf{u}^*(t - iT) \mathbf{u}^T(t - jT) \, dt \tag{5.70}$$

and scalar elements

$$[\mathbf{R}_u(i - j)]_{mn} = \int_{-\infty}^{+\infty} u_m^*(t - iT - \tau_m) u_n(t - jT - \tau_n) \, dt \tag{5.71}$$

5.2.7.2 Decorrelating detector

If we slightly modify the vector notation, Equation (5.65) becomes:

$$r(t) = \sum_{i=-M}^{M} \mathbf{s}^{T}(t - iT)\mathbf{E}\boldsymbol{\Phi}\mathbf{C}(i)\mathbf{b}(i) + n(t) \tag{5.72}$$

with normalized signature waveform vector:

$$\mathbf{s}(t) = [s_1(t - \tau_1)s_2(t - \tau_2)\cdots s_K(t - \tau_K)]^{T} \tag{5.73}$$

$K \times K$ multichannel matrix

$$\mathbf{C}(i) = \text{diag}(c_1(i)c_2(i)\cdots c_K(i))$$
$$\mathbf{E} = \text{diag}\left(\sqrt{E_1}\sqrt{E_2}\cdots\sqrt{E_K}\right) \tag{5.74}$$

and matrix of carrier phases

$$\boldsymbol{\Phi} = \text{diag}\left(e^{j\phi_1}e^{j\phi_2}\cdots e^{j\phi_K}\right) \tag{5.75}$$

The $K \times K$ crosscorrelation matrix of normalized signature waveforms becomes

$$\mathbf{R}(\ell) = \int_{-\infty}^{+\infty} \mathbf{s}^*(t)\mathbf{s}^{T}(t + \ell T)\,dt, \tag{5.76}$$

The asynchronous nature of the channel is evident from the matrix elements

$$R_{mn}(\ell) = \int_{\ell T + \tau_m}^{(\ell+1)T + \tau_m} s_m^*(t - \tau_m)s_n(t + \ell T - \tau_n)\,dt \tag{5.77}$$

Since there is no inter-symbol interference, $\mathbf{R}(\ell) = 0$, $\forall |\ell| > 1$ and $\mathbf{R}(-1) = \mathbf{R}^{H}(1)$. Due to the ordering of the user $\mathbf{R}^{H}(1)$ is an upper triangular matrix with zero elements on the diagonal. The decorrelating detector front end consists of K filters matched to the normalized signature waveforms of the users. The output of this filter bank, sampled at the ℓ-*th* bit epoch is:

$$\mathbf{y}(\ell) = \int_{-\infty}^{+\infty} r(t)s(t - \ell T)\,dt \tag{5.78}$$

The vector of sufficient statistics can also be represented as:

$$\mathbf{y}(\ell) = \mathbf{R}(-1)\mathbf{E}\boldsymbol{\Phi} \quad \mathbf{C}(\ell+1)\mathbf{b}(\ell+1) + \mathbf{R}(0)\mathbf{E}\boldsymbol{\Phi} \quad \mathbf{C}(\ell)\mathbf{b}(\ell)$$
$$+ \mathbf{R}(1)\mathbf{E}\boldsymbol{\Phi}\mathbf{C}(\ell-1)\mathbf{b}(\ell-1) + \mathbf{n}_y(\ell) \tag{5.79}$$

The covariance matrix of the matched filter output noise vector sequence, $\{\mathbf{n}_y(\ell)\}$ is given by:

$$E\{\mathbf{n}_y^*(i)\mathbf{n}_y^{T}(j)\} = \sigma^2\mathbf{R}^*(i - j) \tag{5.80}$$

As in Equation (5.60) the decorrelator is a K-input K-output linear time-invariant filter with transfer function matrix:

$$\mathbf{G}(z) = [\mathbf{R}(-1)z + \mathbf{R}(0) + \mathbf{R}(1)z^{-1}]^{-1} \triangleq \mathbf{S}^{-1}(z) \tag{5.81}$$

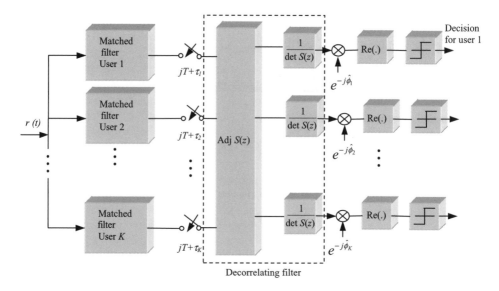

Figure 5.11 Coherent decorrelating multiuser detector.

The z-transform of the decorrelator output vector is:

$$\mathbf{P}(z) = \mathbf{E}\boldsymbol{\Phi}(\mathbf{Cb})(z) + N_p(z) \tag{5.82}$$

where $N_p(z)$ is the z-transform of the output noise vector sequence having power spectral density

$$\sigma^2 \mathbf{S}^{-1}(z) = \sigma^2 \sum_{m=-\infty}^{\infty} \mathbf{D}(m)z^{-m} \tag{5.83}$$

The receiver block diagram for coherent reception is shown in Figure 5.11. Performance results for the detector are shown in Figure 5.12. Significant improvement in the BER is evident.

5.2.8 Multiuser detection in frequency selective Rayleigh fading channels

By using analogy with Equation (5.64) the received signal in this case can be represented as:

$$r(t) = S(t, \mathbf{b}) + n(t)$$

$$S(t, \mathbf{b}) = \sum_{i=-M}^{M} \sum_{k=1}^{K} b_k(i)h_k(t - iT - \tau_k) \tag{5.84}$$

$$h_k(t) = c_k(t)^* u_k(t)$$

In Equation (5.84), $h_k(t)$ is the equivalent received symbol waveform of finite duration $[0, T_k]$ (convolution of equivalent low pass signature waveform $u_k(t)$ and the channel impulse response $c_k(t)$). We define the memory of this channel as v, the smallest integer such that $h_k(t) = 0$ for $t > (v + 1)T$, and all $k = 1 \cdots K$. The impulse response of the kth user

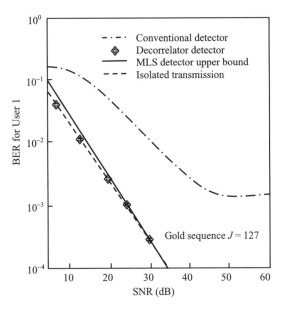

Figure 5.12 Bit error rate of user 1 for the two-user case with Rayleigh faded paths (same average path strength) and Gold sequences of period $J = 127$ [403].

channel is given by:

$$c_k(t) = \sum_{\ell=0}^{L-1} c_{k,l}(t)\delta(t - \tau_{k,\ell}) \tag{5.85}$$

When the signaling interval T is much smaller than the coherence time of the channel, the channel is characterized as slow fading, implying that the channel characteristics can be measured accurately. Since the channel is assumed to be Rayleigh fading, the coefficients $c_{k,\ell}(t)$ are modeled as independent zero mean complex-valued Gaussian random processes. We will use the following notation:

$$h_k(t) = \sum_{\ell=0}^{L-1} c_{k,l}(t)u_k(t - \tau_{k,l}) = \mathbf{c}_k^{\mathrm{T}}(t)\mathbf{u}_k(t) \tag{5.86}$$

For the single user vector of channel coefficients we use:

$$\mathbf{c}_k(t) = [c_{k,0}(t), c_{k,1}(t) \cdots c_{k,L-1}(t)]^{\mathrm{T}} \tag{5.87}$$

and for the signal vector of the delayed signature waveform:

$$\mathbf{u}_k(t) = [u_k(t - \tau_{k,0})u_k(t - \tau_{k,1}) \cdots u_k(t - \tau_{k,L-1})]^{\mathrm{T}} \tag{5.88}$$

The equivalent low pass signature waveform is represented as

$$u_k(t) = \sqrt{E_k}s_k(t)e^{j\phi_k} \tag{5.89}$$

where E_k is the energy, $s_k(t)$ is the real-valued, unit-energy signature waveform with period T and ϕ_k is the carrier phase. In this case the received signal given by Equation (5.84) becomes:

$$r(t) = S(t, \mathbf{b}) + n(t) = \mathbf{b}^{\mathrm{T}}\mathbf{h}_t + n(t) \tag{5.90}$$

The equivalent data sequence is as in Equation (5.66):

$$\mathbf{b} = [b_1(-M) \cdots b_K(-M) \cdots b_1(M) \cdots b_K(M)]^{\mathrm{T}} \tag{5.91}$$

The equivalent waveform vector of NK elements is:

$$\mathbf{h}_t = [\mathbf{h}^{\mathrm{T}}(t + MT) \cdots \mathbf{h}^{\mathrm{T}}(t - MT)]^{\mathrm{T}} \tag{5.92}$$

with

$$\mathbf{h}(t) = [h_1(t - \tau_1) \cdots h_K(t - \tau_K)]^{\mathrm{T}} = \mathbf{C}^{\mathrm{T}}(t)\mathbf{u}(t) \tag{5.93}$$

where

$$\mathbf{C}(t) = \begin{bmatrix} \mathbf{c}_1(t) & 0 & 0 & \cdots \\ & \mathbf{c}_2(t) & 0 & \cdots \\ & & \ddots & \\ \cdots & 0 & 0 & \mathbf{c}_K(t) \end{bmatrix} \tag{5.94}$$

is a $KL \times K$ multichannel matrix. KL is the total number of fading paths for all K users and:

$$\mathbf{u}(t) = [\mathbf{u}_1(t - \tau_1) \cdots \mathbf{u}_K(t - \tau_K)]^{\mathrm{T}} \tag{5.95}$$

is the equivalent signature vector of KL elements.

5.2.8.1 Multiuser maximum likelihood sequence detection

The log likelihood function in this case becomes:

$$L(\mathbf{b}) = 2\,\mathrm{Re}\{\mathbf{b}^{\mathrm{H}}\mathbf{y}\} - \mathbf{b}^{\mathrm{H}}\mathbf{H}\mathbf{b} \tag{5.96}$$

where superscript H denotes the conjugate transpose,

$$\mathbf{y} = \int_{-\infty}^{+\infty} r(t)\mathbf{h}_t^* \, \mathrm{d}t \tag{5.97}$$

is the output of the bank of matched filters sampled at the bit epoch of the users. Matrix $\mathbf{H}$ is an $N \times N$ block Toeplitz crosscorrelation waveform matrix with $K \times K$ block elements,

$$\mathbf{H}(i - j) = \int_{-\infty}^{+\infty} \mathbf{h}^*(t - iT)\mathbf{h}^{\mathrm{T}}(t - jT) \, \mathrm{d}t \tag{5.98}$$

5.2.8.2 Viterbi algorithm

Since every waveform $h_k(t)$ is time limited to $[0, T_k]$, $T_k < (v + 1)T$, it follows that $\mathbf{H}(l) = 0, \forall |l| > v + 1$ and $\mathbf{H}(j) = \mathbf{H}^{\mathrm{H}}(j)$ for $j = 1 \cdots v + 1$. Due to the ordering of the users $\mathbf{H}^{\mathrm{H}}(v + 1)$ is an upper triangular matrix with zero elements on the diagonal. Provided that knowledge of a channel is available, the MLS detector may be implemented as a dynamic programming algorithm of the Viterbi type. The vector Viterbi algorithm is the modification of the one introduced for M-input M-output linear channels where the dimensionality of the state space is $2^{(v+1)K}$. As in the case of the AWGN channel, a more efficient decomposition of the likelihood function results in an algorithm with a state space of dimension $2^{(v+1)K-1}$.

Frequency selective fading is described by the wide-sense stationary uncorrelated scattering model. The bandwidth of each signature waveform is much larger than the coherence bandwidth of the channel, $B_w \gg (\Delta f)_c$. The time varying frequency selective channel for each user can be represented as a tapped delay line with tap spacing $1/B_w$, so that Equation (5.86) becomes:

$$h_k(t) = \sum_{i=0}^{L-1} c_{k,i}(t) u_k \left(t - \frac{i}{B_w} \right)$$
$$= \mathbf{s}_k^T(t) \mathbf{E}_k \Phi_k \mathbf{c}_k(t) \tag{5.99}$$

The signature waveform vector may be described as:

$$\mathbf{s}_k(t) = \left[s_k(t), s_k \left(t - \frac{i}{B_w} \right) \cdots s_k \left(t - \frac{L-i}{B_w} \right) \right]^{\mathrm{T}} \tag{5.100}$$

and

$$\mathbf{E}_k = \sqrt{E_k} \mathbf{I}_L$$
$$\Phi_k = e^{j\phi_k} \mathbf{I}_L \tag{5.101}$$

For a data symbol duration much longer than the multipath delay spread, $T \gg T_m$, any inter-symbol interference due to channel dispersion can be neglected. Based on the above

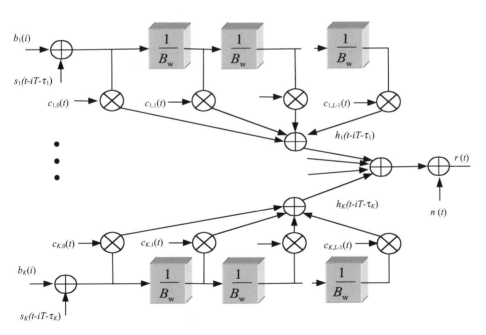

Figure 5.13 A synchronous CDMA frequency selective Rayleigh fading channel model.

discussion, the channel model is presented in Figure 5.13. If we use notation

$$
\begin{aligned}
\mathbf{b}(i) &= [b_1(i)b_2(i) \cdots b_K(i)]^{\mathrm{T}}, \quad i = -M \cdots M \\
\mathbf{s}(t) &= \left[\mathbf{s}_1^{\mathrm{T}}(t - \tau_1) \mathbf{s}_2^{\mathrm{T}}(t - \tau_2) \cdots \mathbf{s}_K^{\mathrm{T}}(t - \tau_K) \right]^{\mathrm{T}} \\
\mathbf{E} &= \operatorname{diag}(\mathbf{E}_1, \mathbf{E}_2, \cdots \mathbf{E}_K) \\
\mathbf{\Phi} &= \operatorname{diag}(\mathbf{\Phi}_1, \mathbf{\Phi}_2, \cdots \mathbf{\Phi}_K) \\
\mathbf{h}^{\mathrm{T}}(t) &= [h_1(t - \tau_1) \cdots h_K(t - \tau_K)] = \mathbf{s}^{\mathrm{T}}(t)\mathbf{E}\mathbf{\Phi}\mathbf{C}(t)
\end{aligned}
\tag{5.102}
$$

Equation (5.84) becomes:

$$
r(t) = \sum_{i=-M}^{M} \mathbf{h}^{\mathrm{T}}(t - iT)\mathbf{b}(i) + n(t) = \sum_{i=-M}^{M} \mathbf{s}^{\mathrm{T}}(t)\mathbf{E}\mathbf{\Phi}\mathbf{C}(t)\mathbf{b}(i) + n(t) \tag{5.103}
$$

We define a $KL \times KL$ crosscorrelation matrix of normalized signature waveforms,

$$
\mathbf{R}(l) = \int_{-\infty}^{+\infty} \mathbf{s}(t)\mathbf{s}^{\mathrm{T}}(t + lT) \, dt \tag{5.104}
$$

The asynchronous mode is evident from the structure of the $L \times L$ crosscorrelation matrix between the users m and n,

$$
\mathbf{R}_{mn}(l) = \int_{lT+\tau_m}^{(l+1)T+\tau_m} \mathbf{s}_m(t - \tau_m)\, \mathbf{s}_n^{\mathrm{T}}(t + lT - \tau_n) \, dt \tag{5.105}
$$

Since there is no inter-symbol interference, $\mathbf{R}(l) = 0, \forall |l| > 1$ and $\mathbf{R}(-1) = \mathbf{R}^{\mathrm{H}}(1)$. Due to the ordering of the users $\mathbf{R}^{\mathrm{H}}(1)$ is an upper triangular matrix with zero elements on the diagonal.

The front end of the multiuser detector consists of KL filters matched to the normalized properly delayed signature waveforms of the users, as shown in Figure 5.14. The output of this filter bank sampled at the bit epochs is given by the vector

$$
\mathbf{y}(l) = \int_{-\infty}^{+\infty} r(t)\mathbf{s}(t - lT) \, dt \tag{5.106}
$$

The vector of sufficient statistics can also be expressed as:

$$
\begin{aligned}
\mathbf{y}(l) &= \mathbf{R}(-1)\mathbf{E}\mathbf{\Phi}\mathbf{C}(l+1)\mathbf{b}(l+1) + \mathbf{R}(0)\mathbf{E}\mathbf{\Phi}\mathbf{C}(l)\mathbf{b}(l) \\
&\quad + \mathbf{R}(1)\mathbf{E}\mathbf{\Phi}\mathbf{C}(l-1)\mathbf{b}(l-1) + n(l)
\end{aligned}
\tag{5.107}
$$

The covariance matrix of the matched filter output noise vector is given by:

$$
E\{\mathbf{n}^*(i)\mathbf{n}^{\mathrm{T}}(j)\} = \sigma^2 \mathbf{R}(i - j)
$$

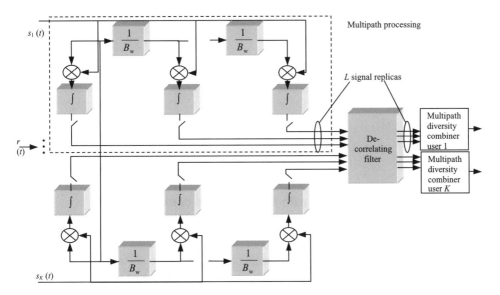

Figure 5.14 Multipath decorrelation.

Taking the z-transform gives

$$\mathbf{Y}(z) = \mathbf{S}(z)(\mathbf{E\Phi Cb})(z) + \mathbf{N}(z) \tag{5.108}$$

where $(\mathbf{E\Phi Cb})(z)$ is the transform of sequence

$$\{[\sqrt{E_1}e^{j\phi_1}c_{1,0}(i)b_1(i)\cdots\sqrt{E_1}e^{j\phi_1}c_{1,L-1}(i)b_1(i)\cdots\sqrt{E_K}e^{j\phi_K}c_{K,L-1}(i)b_K(i)]^{\mathrm{T}}\} \tag{5.109}$$

$\mathbf{S}(z)$ is the equivalent transfer function of the CDMA multipath channel which depends only on the signature waveforms of the users. The multipath decorrelating (MD) filter is a KL-input KL-output linear time-invariant filter with transfer function matrix:

$$\mathbf{G}(z)\underline{\underline{\Delta}}[\mathbf{S}(z)]^{-1} = \frac{\mathrm{adj}\mathbf{S}(z)}{\det\mathbf{S}(z)} = [\mathbf{R}(-1)z + \mathbf{R}(0) + \mathbf{R}(1)z^{-1}]^{-1} \tag{5.110}$$

The necessary and sufficient condition for the existence of a stable, but non-causal realization of the decorrelating filter is

$$\det[\mathbf{R}(-1)e^{jw} + \mathbf{R}(0) + \mathbf{R}(1)e^{-jw}]^{-1} \neq 0 \ \forall w \in [0, 2\pi] \tag{5.111}$$

The z-transform of the decorrelating detector outputs is

$$\mathbf{P}(z) = (\mathbf{E\Phi Cb})(z) + N_p(z) \tag{5.112}$$

$N_p(z)$ is the z-transform of a stationary, filtered Gaussian noise vector sequence. The z-transform of the noise covariance matrix sequence is equal to

$$\sigma^2[\mathbf{S}(z)]^{-1} = \sigma^2 \sum_{m=-\infty}^{\infty} \mathbf{D}(m)z^{-m} \tag{5.113}$$

The output of the decorrelating detector containing L signal replicas of user k may be expressed as:

$$\mathbf{p}_k(l) = \mathbf{c}_k \sqrt{E_k} e^{j\phi_k} b_k(l) + \mathbf{n}_k(l) \tag{5.114}$$

The noise covariance matrix is given by:

$$\sigma^2 [\mathbf{D}(0)]_{kk} = \sigma^2 \frac{1}{2\pi} \int\limits_{0}^{2\pi} [\mathbf{S}(e^{-jw})]_{kk}^{-1} dw \tag{5.115}$$

5.2.8.3 Coherent reception with maximal ratio combining

Since the front end of the coherent multiuser detector contains the decorrelating filter, the noise components in the L branches of the kth user are correlated. The usual approach prior to combining is to introduce the whitening operation, where whitening filter $(\mathbf{T}^H)^{-1}$ is obtained by Cholesky decomposition $[\mathbf{D}(0)]_{kk} = \mathbf{T}^T \mathbf{T}^*$. So, the output of the user of interest is given by

$$\begin{aligned} \mathbf{p}_{kw} &= \mathbf{f} \sqrt{E_k} e^{j\phi_k} b_k + \mathbf{n}_{kw} \\ &= \mathbf{p}'_{kw} b_k + \mathbf{n}_{kw} \end{aligned} \tag{5.116}$$

where

$$\mathbf{f} = (\mathbf{T}^H)^{-1} \mathbf{c}_k \tag{5.117}$$

and $\mathbf{n}_{kw}$ is a zero mean Gaussian white noise vector with covariance matrix $\sigma^2 \mathbf{I}_L$. The optimal combiner in this situation is the maximal ratio combiner (MRC). The receiver block diagram is shown in Figure 5.15. The output of the maximal ratio combiner can be represented as:

$$\hat{b}_k = \mathrm{sgn}\left(\mathbf{p}_{kw} \cdot \hat{\mathbf{p}}'_{kw}*\right) \tag{5.118}$$

For illustration purposes, a CDMA cellular mobile radio system with 1.25 MHz bandwidth and 9600 bps data rate is used. The multipath intensity profile of the mobile radio

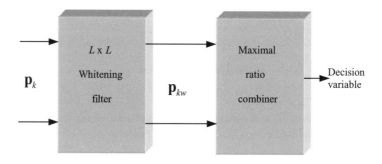

Figure 5.15 Maximal ratio combining after multipath decorrelation for coherent reception.

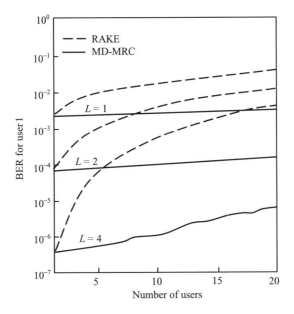

Figure 5.16 Error probability for a coherent RAKE and MD-MRC multiuser receiver for different multipath diversity order in a mobile radio channel using Gold signature sequences of length $J = 127$, average $i = 20$ dB [403].

channel is given by:

$$r(\tau) = \frac{P}{T_m} e^{-\frac{\tau}{T_m}} \tag{5.119}$$

where P is the total average received power and T_m is the multipath delay spread. Typical values of the multipath delay spread are $T_m = 0.5\,\mu$s for the suburban environment and $T_m = 3\,\mu$s for an urban environment. Therefore, we expect the multipath diversity reception with two branches in suburban areas and four to five branches in an urban setting. For the given parameters, inter-symbol interference is negligible, and the mobile radio channel can be described as a discrete multipath Rayleigh fading channel with mean square value of the path coefficients given by:

$$\overline{c_{k,l}^2} = \frac{1}{B_w} r\left(\frac{l}{B_w}\right) \quad l = 0 \cdots L - 1 \tag{5.120}$$

The BER versus the number of users is shown in Figures 5.16 and 5.17. One can see that for a large product KL, multiuser detector performance starts to degrade due to noise enhancement caused by matrix inversion.

More details on multiuser detection can be found in [404–414]. The latest results in this field including the systems using multiple antennas, can be found in [415–437], and [438–481].

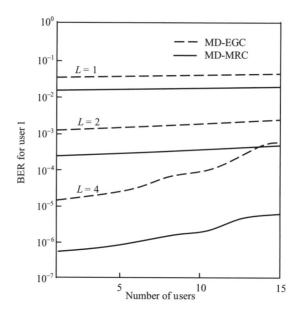

Figure 5.17 Multiuser receiver error probability for different multipath diversity orders in a mobile radio channel using Gold signature sequences of length $J = 127$, average SNR = 20 dB [403].

5.3 MINIMUM MEAN SQUARE ERROR (MMSE) LINEAR MULTIUSER DETECTION

If the amplitude of user k's signal in Equation (5.43) is A_k, then the vector of matched filter outputs $\mathbf{y}$ in Equation (5.46) can be represented as:

$$\mathbf{y} = \mathbf{RAb} + \mathbf{n} \tag{5.121}$$

where $\mathbf{A}$ is a diagonal matrix with elements A_k

$$\mathbf{A} = \text{diag}\|A_k\| \tag{5.122}$$

If the multiuser detector transfer function is denoted as M then the minimum mean square error (MMSE) detector is defined as

$$\min_{M \in R^{K \times K}} \mathbf{E}[\|\mathbf{b} - M\mathbf{y}\|^2] \tag{5.123}$$

One can show that the MMSE linear detector outputs the following decisions [441, 442, 445]:

$$\begin{aligned}
\hat{b}_k &= \text{sgn}\left(\frac{1}{A_k}([\mathbf{R} + \sigma^2 A^{-2}]^{-1}\mathbf{y})_k\right) \\
&= \text{sgn}(([\mathbf{R} + \sigma^2 A^{-2}]^{-1}\mathbf{y})_k)
\end{aligned} \tag{5.124}$$

Therefore, the MMSE linear detector replaces the transformation $\mathbf{R}^{-1}$ of the decorrelating detector by

$$[\mathbf{R} + \sigma^2 \mathbf{A}^{-2}]^{-1} \tag{5.125}$$

where

$$\sigma^2 \mathbf{A}^{-2} = \text{diag}\left\{\frac{\sigma^2}{A_1^2}, \cdots, \frac{\sigma^2}{A_K^2}\right\} \tag{5.126}$$

As an illustration, for the two users case we have:

$$[\mathbf{R} + \sigma^2 \mathbf{A}^{-2}]^{-1} = \left[\left(1 + \frac{\sigma^2}{A_1^2}\right)\left(1 + \frac{\sigma^2}{A_2^2}\right) - \rho^2\right]^{-1} \begin{bmatrix} 1 + \frac{\sigma^2}{A_2^2} & -\rho \\ -\rho & 1 + \frac{\sigma^2}{A_1^2} \end{bmatrix} \tag{5.127}$$

In the asynchronous case, similarly to the solution in Section 5.3 the MMSE linear detector is a K-input, K-output, linear, time-invariant filter with transfer function

$$[\mathbf{R}^{\mathrm{T}}[1]z + \mathbf{R}[0] + \sigma^2 \mathbf{A}^{-2} + \mathbf{R}[1]z^{-1}]^{-1} \tag{5.128}$$

In Figure 5.18, the BER is presented versus the near–far ratio for different detectors. One can see that MMSE shows better performance than the decorrelator. In the figure, the signal to noise ratio of the desired user is equal to 10 dB.

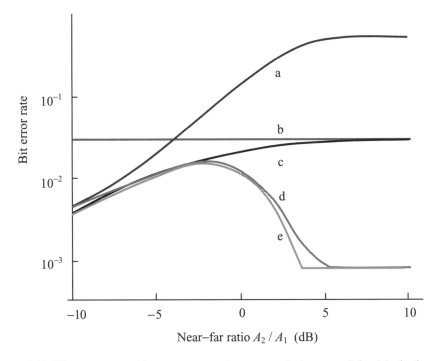

Figure 5.18 Bit error rate with two users and crosscorrelation $\rho = 0.8$: (a) single user matched filter; (b) decorrelator; (c) MMSE; (d) minimum (upper bound); (e) minimum (lower bound).

5.3.1   System model in multipath fading channels

In this section the channel impulse response and the received signal will be presented as

$$c_k(t) = \sum_{l=1}^{L_k} c_{k,l}^{(n)} \delta(t - \tau_{k,l}) \tag{5.129}$$

$$r(t) = \sum_{n=0}^{N_b-1} \sum_{k=1}^{K} \sum_{l=1}^{L} A_k b_k^{(n)} c_{k,l}^{(n)} s_k(t - nT - \tau_{k,l}) + n(t) \tag{5.130}$$

The received signal is time discretized, by antialias filtering and sampling $r(t)$ at the rate $1/Ts = S/T_c = SG/T$, where S is the number of samples per chip and $G = T/T_C$ is the processing gain. The received discrete time signal over a data block of N_b symbols is:

$$\mathbf{r} = \mathbf{SCAb} + \mathbf{n} \in C^{SGN_b} \tag{5.131}$$

where

$$\mathbf{r} = \left[\mathbf{r}^{\mathrm{T}(0)}, \dots, \mathbf{r}^{\mathrm{T}(N_b-1)} \right]^{\mathrm{T}} \in C^{SGN_b} \tag{5.132}$$

is the input sample vector with

$$\mathbf{r}^{\mathrm{T}(n)} = [r(T_s(nSG + 1)), \dots, r(T_s(n + 1)SG] \in C^{SG} \tag{5.133}$$

$$\mathbf{S} = [\mathbf{S}^{(0)}, \mathbf{S}^{(1)}, \dots, \mathbf{S}^{(N_b-1)}] \in R^{SGN_b \times KLN_b}$$

$$= \begin{bmatrix} \mathbf{S}^{(0)}(0) & \mathbf{0} & \cdots & \mathbf{0} \\ \vdots & \mathbf{S}^{(1)}(0) & \ddots & \vdots \\ \mathbf{S}^{(0)}(D) & \vdots & \ddots & \mathbf{0} \\ \mathbf{0} & \mathbf{S}^{(1)}(D) & \ddots & \mathbf{S}^{(N_b-1)}(0) \\ \vdots & \ddots & \ddots & \vdots \\ \mathbf{0} & \cdots & \mathbf{0} & \mathbf{S}^{(N_b-1)}(D) \end{bmatrix} \tag{5.134}$$

is the sampled spreading sequence matrix and $D = \lceil (T + T_m)/T \rceil$. In a single path channel, $D = 1$ due to the asynchronicity of users. In multipath channels, $D \geq 2$ due to the multipath spread. The code matrix is defined with several components $(S^{(n)}(0), \dots, S^{(n)}(D))$ for each symbol interval to simplify the presentation of the crosscorrelation matrix components. T_m is the maximum delay spread,

$$\mathbf{S}^{(n)} = \left[\mathbf{s}_{1,1}^{(n)}, \dots, \mathbf{s}_{1,L}^{(n)}, \dots, \mathbf{s}_{K,L}^{(n)} \right] \in R^{SGN_b \times KL} \tag{5.135}$$

where

$$
\mathbf{s}_{k,l}^{(n)} = \begin{cases}
\mathbf{0}_{SGN_b \times 1}^{T} & n = 0, \tau_{k,l} = 0 \\[2mm]
\left[[s_k(T_s(SG - \tau_{k,l} + 1)), \dots, s_k(T_s SG)]^{T}, \mathbf{0}_{(SGN_b - \tau_{k,l}) \times 1}^{T} \right]^{T} & n = 0, \tau_{k,l} > 0 \\[2mm]
\left[\mathbf{0}_{((n-1)SG + \tau_{k,l}) \times 1}^{T}, \mathbf{s}_k^{T}, \mathbf{0}_{(SG(N_b - n) - \tau_{k,l}) \times 1}^{T} \right]^{T} & 0 < n < N_b - 1 \\[2mm]
\left[\mathbf{0}_{(SG(N_b - 1) + \tau_{k,l}) \times 1}^{T}, [s_k(Ts), \dots, s_k(T_s(SG - \tau_{k,l}))] \right]^{T} & n = N_b - 1
\end{cases}
\tag{5.136}
$$

where $\tau_{k,l}$ is the time discretized delay in sample intervals and

$$
\mathbf{s}_k = \left[s_k(T_s), \dots, s_k(T_s SG) \right]^{T} \in R^{SG}
\tag{5.137}
$$

is the sampled signature sequence of the kth user. By analogy with Equation (5.94):

$$
\mathbf{C} = \mathrm{diag}\left[\mathbf{C}^{(0)}, \dots, \mathbf{C}^{(N_b - 1)} \right] \in C^{KLN_b \times KN_b}
\tag{5.138}
$$

is the channel coefficient matrix with

$$
\mathbf{C}^{(n)} = \mathrm{diag}\left[\mathbf{c}_1^{(n)}, \dots, \mathbf{c}_K^{(n)} \right] \in C^{KL \times K}
\tag{5.139}
$$

and

$$
\mathbf{c}_k^{(n)} = \left[c_{k,1}^{(n)}, \dots, c_{k,L}^{(n)} \right]^{T} \in C^{L}
\tag{5.140}
$$

Equation (5.122) now becomes:

$$
\mathbf{A} = \mathrm{diag}\left[\mathbf{A}^{(0)}, \dots, \mathbf{A}^{(N_b - 1)} \right] \in R^{KN_b \times KN_b}
\tag{5.141}
$$

the matrix of total received average amplitudes with

$$
\mathbf{A}^{(n)} = \mathrm{diag}[A_1, \dots, A_K] \in R^{K \times K}
\tag{5.142}
$$

The bit vector from Equation (5.91) becomes

$$
\mathbf{b} = \left[\mathbf{b}^{T(0)}, \dots, \mathbf{b}^{T(N_b - 1)} \right]^{T} \in \aleph^{KN_b}
\tag{5.143}
$$

with the modulation symbol alphabet $\aleph$ (with BPSK $\aleph = \{-1, 1\}$) and

$$
\mathbf{b}^{(n)} = \left[b_1^{(n)}, \dots, b_K^{(n)} \right] \in \aleph^{K}
\tag{5.144}
$$

and $\mathbf{n} \in C^{SGN_b}$ is the channel noise vector. It is assumed that the data bits are independent identically distributed random variables independent from the channel coefficients and the noise process.

The crosscorrelation matrix from Equation (5.104) for the spreading sequences can be formed as:

$$\mathbf{R} = \mathbf{S}^{\mathrm{T}}\mathbf{S} \in R^{KLN_b \times KLN_b} \tag{5.145}$$

$$= \begin{bmatrix} \mathbf{R}^{(0,0)} & \cdots & \mathbf{R}^{(0,D)} & \mathbf{0}_{KL} \cdots & & \mathbf{0}_{KL} \\ \vdots & \ddots & \ddots & \ddots & & \vdots \\ \mathbf{R}^{(D,0)} & \ddots & \ddots & \ddots & & \mathbf{0}_{KL} \\ \mathbf{0}_{KL} & \ddots & \ddots & \ddots & & \mathbf{R}^{(N_b-D,N_b-1)} \\ \vdots & \ddots & \ddots & \ddots & & \vdots \\ \mathbf{0}_{KL} & \cdots & \mathbf{0}_{KL} & \cdots & & \mathbf{R}^{(N_b-1,N_b-1)} \end{bmatrix}$$

where

$$\mathbf{R}^{(n,n-j)} = \sum_{i=0}^{D-j} \mathbf{S}^{\mathrm{T}(n)}(i)\mathbf{S}^{(n-j)}(i+j), \quad j \in \{0, \ldots, D\} \tag{5.146}$$

and $\mathbf{R}^{(n-j,n)} = \mathbf{R}^{\mathrm{T}(n,n-j)}$. The elements of the correlation matrix can be written as

$$\mathbf{R}^{(n,n')} = \begin{bmatrix} \mathbf{R}^{(n,n')}_{1,1} & \cdots & \mathbf{R}^{(n,n')}_{1,K} \\ \vdots & \ddots & \vdots \\ \mathbf{R}^{(n,n')}_{K,1} & & \mathbf{R}^{(n,n')}_{K,K} \end{bmatrix} \in R^{KL \times KL} \tag{5.147}$$

and

$$\mathbf{R}^{(n,n')}_{k,k'} = \begin{bmatrix} \mathbf{R}^{(n,n')}_{k1,k'1} & \cdots & R^{(n,n')}_{k1,k'L} \\ \vdots & \ddots & \vdots \\ \mathbf{R}^{(n,n')}_{kL,k'1} & \cdots & \mathbf{R}^{(n,n')}_{kL,k'L} \end{bmatrix} \in R^{L \times L} \tag{5.148}$$

with

$$\mathbf{R}^{(n,n')}_{kl,k'l'} = \sum_{j=\tau_{k,l}}^{SG-1+\tau_{k,l}} s_k(T_s(j - \tau_{k,l}))s_{k'}(T_s(j - \tau_{k'l'} + (n' - n)SG)) = \mathbf{s}^{\mathrm{T}(n)}_{k,l}\mathbf{s}^{(n')}_{k',l'} \tag{5.149}$$

which represents the correlation between users k and k', the lth and l'th paths, and between their nth and n'th symbol intervals.

5.3.2 MMSE detector structures

One of the conclusions in Section 5.2 was that noise enhancement in linear MUD causes system performance degradation for large products KL. In this section we consider a possibility for reducing the size of the matrix to be inverted by using multipath combining prior to MUD. The structure is called a postcombining detector and the basic block diagram of the receiver is shown in Figure 5.19.

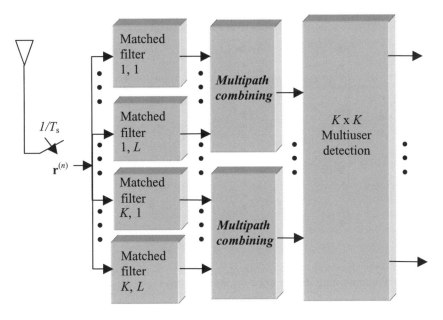

Figure 5.19 Postcombining interference suppression receiver.

The starting point in the derivation of the receiver structure is the cost function $E\{|\mathbf{b} - \hat{\mathbf{b}}|^2\}$ where

$$\hat{\mathbf{b}} = \mathbf{L}_{[post]}^{H}\mathbf{r} \tag{5.150}$$

The detector linear transform matrix is given as

$$\mathbf{L}_{[post]} = \mathbf{SCA}(\mathbf{AC}^{H}\mathbf{RCA} + \sigma^2\mathbf{I})^{-1} \in C^{SGN_b \times KN_b} \tag{5.151}$$

This result is obtained by minimizing the cost function and derivation details may be found in any standard textbook on signal processing. Here $\mathbf{R} = \mathbf{S}^T\mathbf{S}$ is the signature sequence crosscorrelation matrix defined by Equation (5.145). The output of the postcombining LMMSE receiver is

$$\mathbf{y}_{[post]} = (\mathbf{AC}^{H}\mathbf{RCA} + \sigma^2\mathbf{I})^{-1}(\mathbf{SCA})^{H}\mathbf{r} \in C^K \tag{5.152}$$

where $(\mathbf{SCA})^{H}\mathbf{r}$ is the multipath (MR) combined matched filter bank output. For non-fading AWGN:

$$\mathbf{L}_{[post]} = \mathbf{S}(\mathbf{R} + \sigma^2(\mathbf{A}^{H}\mathbf{A})^{-1})^{-1} \tag{5.153}$$

The postcombining LMMSE receiver in fading channels depends on the channel complex coefficients of all users and paths. If the channel is changing rapidly, the optimal LMMSE receiver changes continuously. The adaptive versions of LMMSE receivers have increasing convergence problems as the fading rate increases. The dependence on the fading channel state can be removed by applying a precombining interference suppression type of receiver. The receiver block diagram in this case is shown in Figure 5.20.

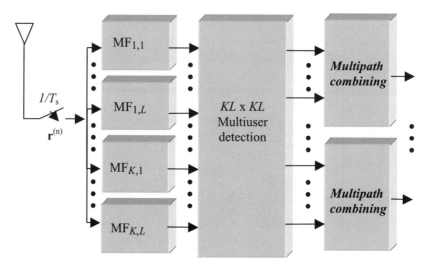

Figure 5.20 Precombining interference suppression receiver.

The transfer function of the detector is obtained by minimizing each element of the cost function

$$E\{|\mathbf{h} - \hat{\mathbf{h}}|^2\} \qquad (5.154)$$

where

$$\mathbf{h} = \mathbf{CAb} \qquad (5.155)$$

and

$$\hat{\mathbf{h}} = \mathbf{L}_{[\text{pre}]}^{\text{T}} \mathbf{r} \qquad (5.156)$$

is estimated.

The solution of this minimization is [482]:

$$\mathbf{L}_{[\text{pre}]} = \mathbf{S}\left(\mathbf{R} + \sigma^2 \mathbf{R}_h^{-1}\right)^{-1} \in \mathbf{R}^{SGN_b \times KLN_b} \qquad (5.157)$$

$$\mathbf{R}_h = \text{diag}\left[A_1^2 \mathbf{R}_{c_1}, \ldots, A_K^2 \mathbf{R}_{c_K}\right] \in \mathbf{R}^{KLN_b \times KLN_b} \qquad (5.158)$$

$$\mathbf{R}_{c_k} = \text{diag}[E[|c_{k,1}|^2], \ldots, E[|c_{k,L}|^2]] \in \mathbf{R}^{L \times L} \qquad (5.159)$$

$$\mathbf{y}_{[\text{pre}]} = \left(\mathbf{R} + \sigma^2 \mathbf{R}_h^{-1}\right)^{-1} \mathbf{S}^{\text{T}} \mathbf{r} \in C^{KL} \qquad (5.160)$$

The two detectors are compared in Figure 5.21. The postcombining scheme performs better. The illustration of LMMSE–RAKE receiver performance in the near–far environment is shown in Figure 5.22 [483]. Considerable improvement compared to conventional RAKE is evident.

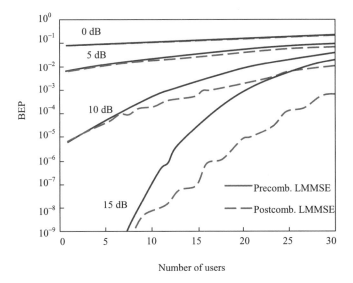

Figure 5.21 Bit error probabilities as a function of the number of users for the postcombining and precombining LMMSE detectors in an asynchronous two-path fixed channel with different SNRs and bit rate 16 kbit/s, Gold code of length 31, $td/T = 4.63 \times 10^{-3}$, maximum delay spread 10 chips [483].

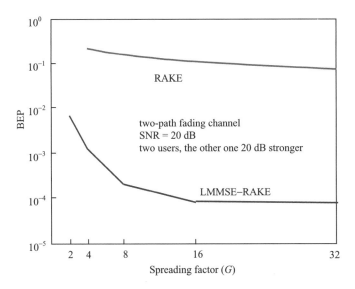

Figure 5.22 Bit error probabilities as a function of the near–far ratio for the conventional RAKE receiver and the precombining LMMSE (LMMSE–RAKE) receiver with different spreading factors (G) in a two-path Rayleigh fading channel with maximum delay spreads of 2 µs for $G = 4$, and 7 µs for other spreading factors. The average signal to noise ratio is 20 dB, the data modulation is BPSK, the number of users is 2, the other user has 20 dB higher power. Data rates vary from 128 kbit/s to 2.048 Mbit/s; no channel coding is assumed [483].

5.3.3   Spatial processing

When combined with multiple receive antennas the receiver structures may have one of the forms shown in Figure 5.23 [482–486].

The channel impulse response for the kth user's ith sensor can now be written as:

$$c_{k,i}(t) = \sum_{l=1}^{L_k} c_{k,l}^{(n)} e^{j2\pi \lambda^{-1} \langle e(\phi_{k,l}), \varepsilon_i \rangle} \delta(t - (\tau_{k,l,i})) \tag{5.161}$$

where L_k is the number of propagation paths (assumed to be the same for all users for simplicity; $L_k = L, \forall k$), $c_{k,l}^{(n)}$ is the complex attenuation factor of the kth user's lth path, $\tau_{k,l,i}$ is the propagation delay for the ith sensor, ε_i is the position vector of the ith sensor with respect to some arbitrarily chosen reference point, λ is the wavelength of the carrier, $e(\phi_{k,l})$ is a unit vector pointing to direction $\phi_{k,l}$ (direction of arrival), and $\langle ., . \rangle$ indicates the inner product.

Assuming that the number of propagation paths is the same for all users, the channel impulse response can be written as:

$$c_{k,i}(t) = \sum_{l=1}^{L} c_{k,l}^{(n)} e^{j2\pi \lambda^{-1} \langle e(\phi_{k,l}), \varepsilon_i \rangle} \delta(t - \tau_{k,l}) \tag{5.162}$$

The channel matrix for the ith sensor consists of two components

$$\mathbf{C}_i = \mathbf{C} \circ \mathbf{\Phi}_i \in C^{KLN_b \times KN_b} \tag{5.163}$$

where $\mathbf{C}$ is the channel matrix defined in Equation (5.139).$\circ$ is the Schur product defined as $\mathbf{Z} = \mathbf{X} \circ \mathbf{Y} \in C^{x \times y}$, i.e. all components of the matrix $\mathbf{X} \in C^{x \times y}$ are multiplied elementwise by the matrix $\mathbf{Y} \in C^{x \times y}$, $\mathbf{\Phi}_i = \mathrm{diag}(\tilde{\phi}_i) \otimes \mathbf{I}_{N_b}$ and $\tilde{\phi}_i = \mathrm{diag}(\varphi_1, \ldots, \varphi_K)$, $\varphi_k = [\phi_{k,1}, \ldots, \phi_{k,L}]^T$, is the matrix of the direction vectors

$$\varphi_i = \left[e^{j2\pi \lambda^{-1} \langle e(\phi_{1,1}), \varepsilon_i \rangle}, \ldots, e^{j2\pi \lambda^{-1} \langle e(\phi_{K,L}), \varepsilon_i \rangle} \right]^T \in C^{KL} \tag{5.164}$$

By using the previous notation one can show that the equivalent detector transform matrices are given as [482–485].

$$\mathbf{L}_{[STM]} = \sum_{i=1}^{I} \mathbf{S}(\mathbf{C} \circ \mathbf{\Phi}_i) \cdot \left(\sum_{i=1}^{I} \mathbf{A}^H (\mathbf{\Phi}_i^H \circ \mathbf{C}^H) \mathbf{R} (\mathbf{C} \circ \mathbf{\Phi}_i) \mathbf{A} + \sigma^2 \mathbf{I} \right)^{-1}$$

$$\mathbf{L}_{[SMT]} = \sum_{i=1}^{I} \mathbf{S} \mathbf{\Phi}_i \left(\sum_{i=1}^{I} \mathbf{\Phi}_i^H \mathbf{R} \mathbf{\Phi}_i + \sigma^2 \mathbf{R}_h^{-1} \right)^{-1} \tag{5.165}$$

$$\mathbf{L}_{[MST]i} = \mathbf{S} \left(\mathbf{R} + \sigma^2 \mathbf{R}_h^{-1} \right)^{-1}$$

$$\mathbf{L}_{[TMS]} = \mathbf{S} \mathbf{C} \mathbf{A} (\mathbf{A} \mathbf{C}^H \mathbf{R} \mathbf{C} \mathbf{A} + \sigma^2 \mathbf{I})^{-1}$$

5.4   SINGLE USER LMMSE RECEIVERS FOR FREQUENCY SELECTIVE FADING CHANNELS

5.4.1   Adaptive precombining LMMSE receivers

In this case the MSE criterion $E\{|\mathbf{h} - \hat{\mathbf{h}}|^2\}$ requires that the reference signal $\mathbf{h} = \mathbf{CAb}$ is available in adaptive implementations. For adaptive single user receivers, the optimization

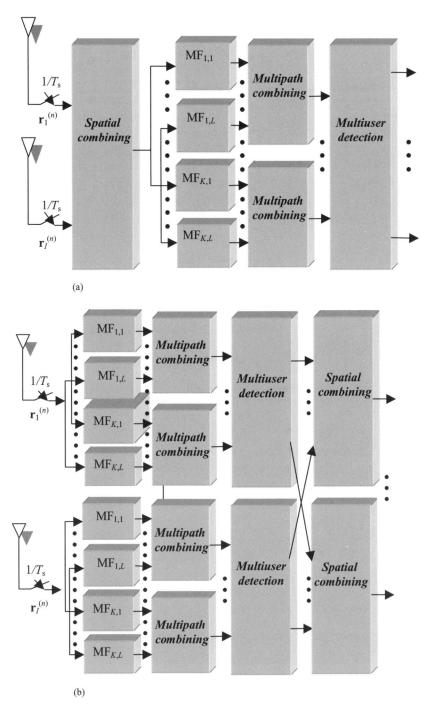

Figure 5.23 (a) The spatial–temporal multiuser (STM) receiver; (b) the TMS receive post combining interference suppression receiver with spatial signal processing (c) the SMT receiver; (d) the MST receiver. A precombining interference suppression receiver with spatial signal processing.

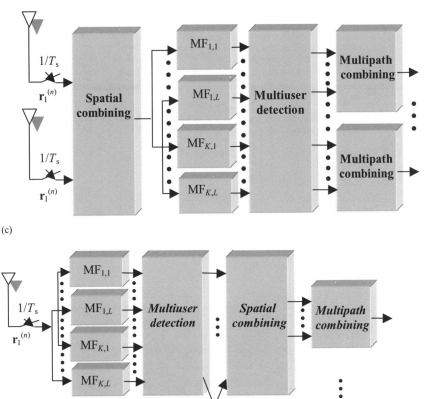

(c)

(d)

Figure 5.23 (*Cont.*).

criterion is presented for each path separately, i.e.

$$J_{k,l} = E\{|(\mathbf{h})_{k,l} - (\hat{\mathbf{h}})_{k,l}|^2\} \tag{5.166}$$

The receiver block diagram is given in Figure 5.24 [446–451].

By using notation

$$\bar{\mathbf{r}}^{(n)} = \left[\mathbf{r}^{\mathrm{T}(n-D)}, \ldots, \mathbf{r}^{\mathrm{T}(n)}, \ldots, \mathbf{r}^{\mathrm{T}(n+D)}\right]^{\mathrm{T}} \in \mathbf{C}^{MSG}$$

$$\mathbf{w}_{k,l}^{(n)} = \left[w_{k,l}^{(n)}(0), \ldots, w_{k,l}^{(n)}(MSG - 1)\right]^{\mathrm{T}} \in \mathbf{C}^{MSG} \tag{5.167}$$

$$y_{k,l}^{(n)} = \mathbf{w}_{k,l}^{\mathrm{H}(n)} \bar{\mathbf{r}}^{(n)}$$

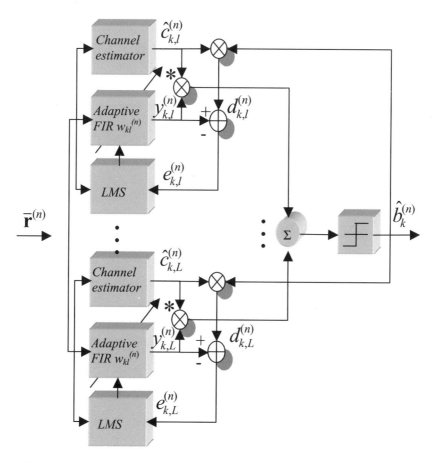

Figure 5.24 Block diagram of the adaptive LMMSE–RAKE receiver.

the bit estimation is defined as

$$\hat{b}_k^{(n)} = \mathrm{sgn}\left(\sum_{l=1}^{L} \hat{c}_{k,l}^{(n)} y_{k,l}^{(n)}\right) \tag{5.168}$$

The filter coefficients $\mathbf{w}$ are derived using the MSE criterion ($E[|e_{k,l}^{(n)}|^2]$). This leads to the optimal filter coefficients $\mathbf{w}_{[\mathrm{MSE}]k,l} = \mathbf{R}_{\bar{r}}^{-1}\mathbf{R}_{\bar{r}d_{k,l}}$ where $\mathbf{R}_{\bar{r}d_{k,l}}$ is the crosscorrelation vector between the input vector $\bar{\mathbf{r}}$ and the desired response $d_{k,l}$, and $\mathbf{R}_{\bar{r}}$ is the input signal crosscorrelation matrix. Adaptive filtering can be implemented by using a number of algorithms.

5.4.1.1 The steepest descent algorithm

In this case we have

$$\mathbf{w}_{k,l}^{(n+1)} = \mathbf{w}_{k,l}^{(n)} - \mu \nabla_{k,l} \tag{5.169}$$

where ∇ is the gradient of

$$\left(J_{k,l} = E\{\left|c_{k,l}A_kb_k - \mathbf{w}_{k,l}^{\mathrm{H}}\mathbf{r}\right|^2\}\right) \tag{5.170}$$

This can be represented as

$$\nabla_{k,l} = \frac{\partial J_{k,l}}{\partial \operatorname{Re}\{\mathbf{w}_{k,l}\}} + \mathrm{j}\frac{\partial J_{k,l}}{\partial \operatorname{Im}\{\mathbf{w}_{k,l}\}} = 2\frac{\partial J_{k,l}}{\partial \mathbf{w}_{k,l}^*} \tag{5.171}$$

If the processing window $M = 1$ we have $\bar{\mathbf{r}}^{(n)} = \mathbf{r}^{(n)} \triangleq \mathbf{r}$ and

$$\nabla_{k,l} = -2E[\mathbf{r}(c_{k,l}A_kb_k)^*] + 2E[\mathbf{r}\mathbf{r}^{\mathrm{H}}]\mathbf{w}_{k,l}$$
$$= -2\mathbf{R}_{rd_{k,l}} + 2\mathbf{R}_r\mathbf{w}_{k,l} \tag{5.172}$$

where $d_{k,l} = c_{k,l}A_kb_k$. If we assume that $A_k = 1, \forall k$

$$\mathbf{w}_{k,l}^{(n+1)} = \mathbf{w}_{k,l}^{(n)} - 2\mu\left(\mathbf{R}_{rd_{k,l}} - \mathbf{R}_r\mathbf{w}_{k,l}^{(n)}\right) \tag{5.173}$$

As a stochastic approximation, Equation (5.172) can be represented as

$$\nabla_{k,l} \approx -2\mathbf{r}(c_{k,l}b_k)^* + 2\mathbf{r}\mathbf{r}^{\mathrm{H}}\mathbf{w}_{k,l}^{(n)} = -2\mathbf{r}(c_{k,l}b_k)^* + 2\mathbf{r}y_{k,l}^*$$

From this equation, and assuming that $M > 1$, the LMS algorithm for updating the filter coefficients results in

$$\mathbf{w}_{k,l}^{(n+1)} = \mathbf{w}_{k,l}^{(n)} + 2\mu\bar{\mathbf{r}}^{(n)}\left(c_{k,l}^{(n)}b_k^{(n)} - y_{k,l}^{(n)}\right)^* \in \mathrm{C}^{MSG} \tag{5.174}$$

We decompose Equation (5.174) into adaptive and fixed components as:

$$\mathbf{w}_{k,l}^{(n)} = \bar{\mathbf{s}}_{k,l} + \mathbf{x}_{k,l}^{(n)} \in \mathrm{C}^{MSG} \tag{5.175}$$

where $\mathbf{x}_{k,l}^{(n)}$ is the adaptive filter component and

$$\bar{\mathbf{s}}_{k,l} = \left[\mathbf{0}_{(DSG+\tau_{k,l})\times 1}^{\mathrm{T}}, \mathbf{s}_k^{\mathrm{T}}, \mathbf{0}_{(DSG-\tau_{k,l})\times 1}^{\mathrm{T}}\right]^{\mathrm{T}} \tag{5.176}$$

is the fixed spreading sequence of the kth user with the delay $\tau_{k,l}$. In this case every branch from Figure 5.24 can be represented as shown in Figure 5.25.

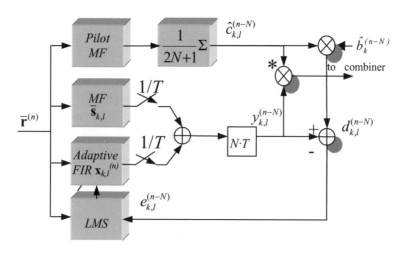

Figure 5.25 Block diagram of one receiver branch in the adaptive LMMSE–RAKE receiver.

In this case, Equation (5.47) gives:

$$\mathbf{x}_{k,l}^{(n+1)} = \mathbf{x}_{k,l}^{(n)} - 2\mu_{k,l}^{(n)}\left(c_{k,l}^{(n)}b_k^{(n)} - y_{k,l}^{(n)}\right)^*\bar{\mathbf{r}}^{(n)} = \mathbf{x}_{k,l}^{(n)} - 2\mu_{k,l}^{(n)}e_{k,l}^{*(n)}\bar{\mathbf{r}}^{(n)}$$

$$\mu_{k,l}^{(n)} = \mu/(\bar{\mathbf{r}}^{H(n)}\bar{\mathbf{r}}^{(n)}); \quad 0 < \mu < 1$$

$$e_{k,l}^{(n)} = d_{k,l}^{(n)} - y_{k,l}^{(n)} \tag{5.177}$$

The reference signal is

$$d_{k,l}^{(n)} = \hat{c}_{k,l}^{(n)}b_k^{(n)} \quad \text{or} \quad d_{k,l}^{(n)} = \hat{c}_{k,l}^{(n)}\hat{b}_k^{(n)} \tag{5.178}$$

and the channel estimator is using a pilot channel

$$\hat{c}_{k,l}^{(n)} = \frac{1}{2N+1}\sum_{i=-N}^{N}\bar{\mathbf{s}}_{p,l}^{T}\bar{\mathbf{r}}^{(n-i)} \tag{5.179}$$

To illustrate the system operation the following example is used [483]: carrier frequency 2.0 GHz, symbol rate 16 kbit/s, 31 chip Gold code and rectangular chip waveform. A synchronous downlink with equal energy two-path ($L = 2$) Rayleigh fading channel with vehicle speeds of 40 km/h (which results in the maximum normalized Doppler shift of 4.36×10^{-3}) and maximum delay spread of ten chip intervals. The number of users examined was 1–30 including the unmodulated pilot channel. The average energy was the same for the pilot channel and user data channels. A simple moving average smoother of length eleven symbols was used in a conventional channel estimator. Perfect channel estimation and ideal truncated precombining LMMSE receivers were used in the analysis to obtain the bit error probability lower bounds. The receiver processing window is three symbols ($M = 3$) unless otherwise stated. The adaptive algorithm used in the simulations was normalized LMS with

$$\mu_{k,l}^{(n)} = \frac{1}{100 \cdot (2D+1)SG}\left(\bar{\mathbf{r}}_{k,l}^{H(n)}\bar{\mathbf{r}}_{k,l}^{(n)}\right)^{-1}. \tag{5.180}$$

The simulation results were produced by averaging over the BERs of randomly selected users with different delay spreads.

The simulation results are shown in Figure 5.26. In general, one can notice that the improvement gains are lower than in the case of multiuser detectors. In the presence of a strong near–far effect, the improvements should be more evident.

5.4.1.2 Blind adaptive LMMSE–RAKE

In this case, in Equation (5.170) we use estimates of bits $\hat{b}_{k,l}$ instead of $b_{k,l}$ [452–455]:

$$\mathbf{x}_{k,l}^{(n+1)} = \mathbf{x}_{k,l}^{(n)} + 2\mu_{k,l}^{(n)}\left(c_{k,l}^{(n)}\hat{b}_{k,l}^{(n)} - y_{k,l}^{(n)}\right)^*\bar{\mathbf{r}}^{(n)} \tag{5.181}$$

The MSE criterion now gives

$$\mathbf{w}_{[\text{MSE}]k,l} = \mathbf{R}_{\bar{r}}^{-1}\mathbf{R}_{\bar{r}d_{k,l}} = \mathbf{R}_{\bar{r}}^{-1}\bar{s}_{k,l}E\lfloor|c_{k,l}|^2\rfloor \tag{5.182}$$

Similarly, the minimum output energy criteria defined as

$$\text{MOE}(E\lfloor|y_{k,l}|^2\rfloor) \tag{5.183}$$

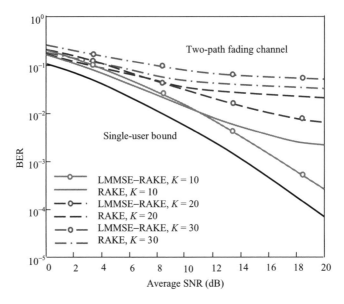

Figure 5.26 Simulated bit error rates as a function of the average SNR for the conventional RAKE and the adaptive LMMSE–RAKE in a two-path fading channel for vehicle speeds of 40 km/h with different numbers of users [483].

gives

$$\mathbf{w}_{[\text{MOE}]k,l} = \mathbf{R}_{\bar{r}}^{-1} \bar{s}_{k,l} \Big/ \left(\bar{\mathbf{s}}_{k,l}^{\mathrm{T}} \mathbf{R}_{\bar{r}}^{-1} \bar{s}_{k,l} \right) \tag{5.184}$$

An implementation example can be seen in [487]. The stochastic approximation of the gradient of Equation (5.172) for the MOE criterion gives

$$\nabla_{k,l} = \bar{\mathbf{r}}^{(n)} \bar{\mathbf{r}}^{\mathrm{H}(n)} \mathbf{w}_{k,l} \tag{5.185}$$

If we want to keep the useful signal autocorrelation unchanged, Equation (5.184) should be constrained to satisfy $\bar{\mathbf{s}}_{k,l}^{\mathrm{T}} \mathbf{x}_{k,l}^{(n)} = 0$. The orthogonality condition is maintained at each step of the algorithm by projecting the gradient onto the linear subspace orthogonal to $\bar{\mathbf{s}}_{k,l}^{\mathrm{T}}$. In practice, this is accomplished by subtracting an estimate of the desired signal component from the received signal vector. An implementation can be seen in [488]. So, we have:

$$\mathbf{x}_{k,l}^{(n+1)} = \mathbf{x}_{k,l}^{(n)} - 2\mu_{k,l}^{(n)} \bar{\mathbf{r}}^{\mathrm{H}(n)} \left(\bar{\mathbf{s}}_{k,l} + \mathbf{x}_{k,l}^{(n)} \right) \left(\bar{\mathbf{r}}^{(n)} - \mathbf{F}_{k,l} \left(\mathbf{F}_{k,l}^{\mathrm{T}} \bar{\mathbf{r}}^{(n)} \right) \right) \tag{5.186}$$

where

$$\mathbf{F}_{k,l} = \begin{bmatrix} \mathbf{0}_{\tau_{k,l} \times 1}^{\mathrm{T}}, \ \mathbf{s}_k^{\mathrm{T}}, & \mathbf{0}_{(2DSG-\tau_{k,l}) \times 1}^{\mathrm{T}} \\ \mathbf{0}_{(SG-\tau_{k,l}) \times 1}^{\mathrm{T}}, & \mathbf{s}_k^{\mathrm{T}}, \ \mathbf{0}_{((2D-1)SG-\tau_{k,l}) \times 1}^{\mathrm{T}} \\ \mathbf{0}_{(2DSG+\tau_{k,l}) \times 1}^{\mathrm{T}}, & [s_k(T_{\mathrm{s}}), \cdots, s_k(T_{\mathrm{s}}(SG - \tau_{k,l}))] \end{bmatrix}^{\mathrm{T}} \in \mathrm{R}^{MSG \times M} \tag{5.187}$$

is a block diagonal matrix of sampled spreading sequence vectors. Effectively, M separate filters are adapted.

5.4.1.3 Griffiths's algorithm

In this case instead of assuming that vector $\mathbf{R}_{\bar{r}d_{k,l}}$ is known, the instantaneous estimate for the covariance is used, i.e.

$$\mathbf{R}_{\bar{r}} \approx \bar{\mathbf{r}}^{(n)}\bar{\mathbf{r}}^{H(n)} \tag{5.188}$$

In this case the crosscorrelation is $\mathbf{R}_{\bar{r}d_{k,l}} = E[|c_{k,l}|^2]\bar{\mathbf{s}}_{k,l}$, and Griffiths's algorithm results in:

$$\mathbf{x}_{k,l}^{(n+1)} = \mathbf{x}_{k,l}^{(n)} + 2\mu_{k,l}^{(n)}\left(E[|c_{k,l}|^2]\mathbf{F}_{k,l}\mathbf{1}_M - \bar{\mathbf{r}}_{k,l}^{*(n)}\left(\bar{\mathbf{s}}_{k,l} + \mathbf{x}_{k,l}^{(n)}\right)^H\bar{\mathbf{r}}^{(n)}\right) \tag{5.189}$$

In practice, the energy of multipath components $(E[|c_{k,l}|^2])$ is not known and must be estimated.

5.4.1.4 Constant modulus algorithm

In this case the optimization criterion is $E[(|y_{k,l}|^2 - \omega)^2]$ where ω is the so-called constant modulus (CM), set according to the received signal power, i.e $\omega = E[|c_{k,l}|^2]$ or $\omega^{(n)} = |c_{k,l}^{(n)}|^2$. By using the CM algorithm, it is possible to avoid the use of the data decisions in the reference signal in the adaptive LMMSE–RAKE receiver by taking the absolute value of the estimated channel coefficients ($|\hat{c}_{k,l}^{(n)}|$) in adapting the receiver. In the precombining LMMSE receiver framework, the cost function for the BPSK data modulation is

$$E[|\|\hat{\mathbf{h}}|^2 - |\mathbf{h}|^2|^2] \tag{5.190}$$

The stochastic approximation of the gradient for the CM criterion is

$$\nabla_{k,l}^{(n+1)} = \left(|y_{k,l}^{(n)}|^2 - |\hat{c}_{k,l}^{(n)}|^2\right)\bar{\mathbf{r}}^{(n)}\bar{\mathbf{r}}^{H(n)}\mathbf{w}_{k,l} \tag{5.191}$$

Hence, the constant modulus algorithm can be expressed as:

$$\mathbf{x}_{k,l}^{(n+1)} = \mathbf{x}_{k,l}^{(n)} - 2\mu_{k,l}^{(n)}y_{k,l}^{*(n)}\left(|y_{k,l}^{(n)}|^2 - |\hat{c}_{k,l}^{(n)}|^2\right)\bar{\mathbf{r}}^{(n)} \tag{5.192}$$

5.4.1.5 Constrained LMMSE–RAKE, Griffiths's and constant modulus algorithms

The adaptive LMMSE–RAKE, the Griffiths's, and the constant modulus algorithm contain no constraints. By applying the orthogonality constraint $\bar{\mathbf{s}}_{k,l}^T\mathbf{x}_{k,l}^{(n)} = 0$ to each of these algorithms, an additional term $\bar{\mathbf{s}}_{k,l}^T\mathbf{x}_{k,l}^{(n)}\bar{\mathbf{s}}_{k,l}$ is subtracted from the new $\mathbf{x}_{k,l}^{(n+1)}$ update at every iteration. The constrained LMMSE–RAKE receiver becomes [385, 391]:

$$\mathbf{x}_{k,l}^{(n+1)} = \mathbf{x}_{k,l}^{(n)} + 2\mu_{k,l}^{(n)}\left(\hat{c}_{k,l}^{(n)}\hat{b}_k^{(n)} - y_{k,l}^{(n)}\right)^*\bar{\mathbf{r}}^{(n)} - \bar{\mathbf{s}}_{k,l}^T\mathbf{x}_{k,l}^{(n)}\bar{\mathbf{s}}_{k,l} \tag{5.193}$$

The Griffiths's and the constant modulus algorithms can also be defined in a similar way.

5.4.2 Blind least squares receivers

All blind adaptive algorithms described in the previous section are based on the gradient of the cost function. In practical adaptive algorithms, the gradient is estimated, i.e. the expectation in the optimization criterion is not taken but is replaced in most cases by some stochastic approximation. In fact, the stochastic approximation used in LMS algorithms is accurate only for small step sizes μ. This results in rather slow convergence, which may be intolerable in practical applications.

Another drawback with the blind adaptive receivers presented above is the delay estimation. Those receiver structures as such support only conventional delay estimation based on matched filtering (MF). The MF-based delay estimation is sufficient for the downlink receivers in systems with an unmodulated pilot channel, since the zero mean MAI can be averaged out if the rate of fading is low enough. If CDMA systems do not have the pilot channel, it would be beneficial to use some near–far resistant delay estimators.

5.4.3 Least squares (LS) receiver

One possible solution to both the convergence and the synchronization problems is based on blind linear least squares (LS) receivers. The cost function in this case is:

$$J_{[LS]k,l} = \sum_{j=n-N+1}^{n} \left(c_{k,l}^{(j)} b_k^{(j)} - \mathbf{w}_{k,l}^{H(n)} \bar{\mathbf{r}}^{(j)} \right)^2 \tag{5.194}$$

where N is the observation window in symbol intervals. Filter weights are given as

$$\mathbf{w}_{k,l}^{(n)} = \hat{\mathbf{R}}_{\bar{\mathbf{r}}}^{-1(n)} \bar{\mathbf{s}}_{k,l} \tag{5.195}$$

$\hat{\mathbf{R}}_{\bar{\mathbf{r}}}^{-1(n)}$ denotes the estimated covariance matrix over a finite data block called the sample covariance matrix. This matrix can be expressed as

$$\hat{\mathbf{R}}_{\bar{\mathbf{r}}}^{(n)} = \sum_{j=n-N+1}^{n} \bar{\mathbf{r}}^{(j)} \bar{\mathbf{r}}^{H(j)} \tag{5.196}$$

Analogous to the MOE criterion, the LS criterion can be modified as:

$$J_{[LS']k,l} = \sum_{j=n-N+1}^{n} \left(\mathbf{w}_{k,l}^{H(n)} \bar{\mathbf{r}}^{(j)} \right)^2, \quad \text{subject to } \mathbf{w}_{k,l}^{T} \bar{\mathbf{s}}_{k,l} = 1 \tag{5.197}$$

which results in

$$\mathbf{w}_{k,l}^{(n)} = \frac{\hat{\mathbf{R}}_{\bar{\mathbf{r}}}^{-1(n)} \bar{\mathbf{s}}_{k,l}}{\bar{\mathbf{s}}_{k,l}^{T} \hat{\mathbf{R}}_{\bar{\mathbf{r}}}^{-1(n)} \bar{\mathbf{s}}_{k,l}} \tag{5.198}$$

The adaptation of the blind LS receiver means updating the inverse of the sample covariance. The blind adaptive LS receiver is significantly more complex than the stochastic gradient based blind adaptive receivers. Recursive methods, such as the recursive least squares (RLS) algorithm, for updating the inverse and iteratively finding the filter weights are known. Also, the methods based on eigendecomposition of the covariance matrix have been proposed to avoid explicit matrix inversion.

5.4.4 Method based on the matrix inversion lemma

The general relation

$$(\mathbf{A} + \mathbf{BCD})^{-1} = \mathbf{A}^{-1} - \mathbf{A}^{-1}\mathbf{B}(\mathbf{DA}^{-1}\mathbf{B} + \mathbf{C}^{-1})^{-1}\mathbf{DA}^{-1} \qquad (5.199)$$

becomes

$$\hat{\mathbf{R}}_{\tilde{\mathbf{r}}}^{-1(n)} = \left(\hat{\mathbf{R}}_{\tilde{\mathbf{r}}}^{-1(n-1)} + \tilde{\mathbf{r}}^{(n)}\tilde{\mathbf{r}}^{H(n)}\right)^{-1}$$

$$= R_{\tilde{\mathbf{r}}}^{-1(n-1)} - \frac{\mathbf{R}_{\tilde{\mathbf{r}}}^{-1(n-1)}\tilde{\mathbf{r}}^{(n)}\tilde{\mathbf{r}}^{H(n)}\mathbf{R}_{\tilde{\mathbf{r}}}^{-1(n-1)}}{1 + \tilde{\mathbf{r}}^{H(n)}\mathbf{R}_{\tilde{\mathbf{r}}}^{-1(n-1)}\tilde{\mathbf{r}}^{(n)}} \qquad (5.200)$$

In time-variant channels, the old values of the inverses must be weighted by the so-called forgetting factor $(0 < \gamma < 1)$, which results in:

$$\hat{\mathbf{R}}_{\tilde{\mathbf{r}}}^{-1(n)} = \frac{1}{\gamma}\left(\hat{\mathbf{R}}_{\tilde{\mathbf{r}}}^{-1(n-1)} - \frac{\hat{\mathbf{R}}_{\tilde{\mathbf{r}}}^{-1(n-1)}\tilde{\mathbf{r}}^{(n)}\tilde{\mathbf{r}}^{H(n)}\hat{\mathbf{R}}_{\tilde{\mathbf{r}}}^{-1(n-1)}}{\gamma + \tilde{\mathbf{r}}^{H(n)}\hat{\mathbf{R}}_{\tilde{\mathbf{r}}}^{-1(n-1)}\tilde{\mathbf{r}}^{(n)}}\right) \qquad (5.201)$$

It is sufficient to initialize the algorithm as $\hat{\mathbf{R}}_{\tilde{\mathbf{r}}}^{-1(0)} = \mathbf{I}$. For illustration purposes a numerical example is shown in Figure 5.27 [483] and Table 5.3. System parameters are shown in the figure. In general, one can see that the blind algorithms are inferior when compared with LMMSE–RAKE using pilot symbols.

More information on the topic can be found in [456–463] and especially in the late publications that include MIMO channels [464–470].

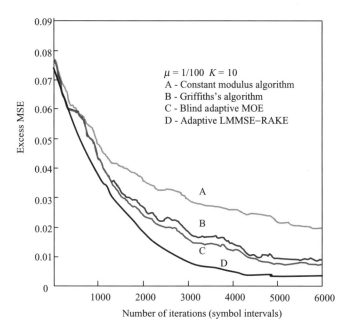

Figure 5.27 Excess mean squared error as a function of the number of iterations for different blind adaptive receivers in a two-path fading channel with vehicle speeds of 40 km/h, the number of active users $K = 10$, SNR $= 20$ dB, $\mu = 100^{-1}$ [483].

Table 5.3   The BERs of different blind adaptive receivers at an SNR of 20 dB in a two-path
Rayleigh fading channel at vehicle speeds of 40 km/h. The acronyms used are:
adaptive LMMSE–RAKE (LR), adaptive MOE (MOE), Griffiths's algorithm
(GRA), constant modulus algorithm with average channel tap powers (CMA2),
constrained adaptive LMMSE–RAKE (C-LR), constrained constant modulus
algorithm (C-CMA), constrained Griffiths's algorithm (C-GRA), constrained
constant modulus algorithm with average channel tap powers (C-CMA2) and
conventional RAKE (RAKE) [483]

Adaptive receiver	$K = 30$		$K = 15$		
	$\mu = 100^{-1}$	$\mu = 10^{-1}$	$\mu = 100^{-1}$	$\mu = 10^{-1}$	$\mu = 2^{-1}$
LR	4.5×10^{-2}	3.9×10^{-1}	6.3×10^{-4}	7.2×10^{-4}	3.0×10^{-2}
MOE	2.8×10^{-2}	4.2×10^{-2}	6.0×10^{-4}	2.1×10^{-3}	9.1×10^{-2}
GRA	2.8×10^{-2}	4.7×10^{-2}	6.4×10^{-4}	3.3×10^{-3}	1.2×10^{-1}
CMA	3.9×10^{-2}	4.0×10^{-1}	1.2×10^{-3}	2.1×10^{-2}	5.0×10^{-1}
CMA2	3.3×10^{-2}	4.0×10^{-1}	1.8×10^{-3}	2.1×10^{-2}	5.0×10^{-1}
C-LR	3.2×10^{-2}	4.2×10^{-2}	6.3×10^{-4}	6.4×10^{-4}	1.9×10^{-3}
C-CMA	3.3×10^{-2}	5.0×10^{-1}	6.1×10^{-4}	3.8×10^{-1}	5.0×10^{-1}
C-GRA	2.8×10^{-2}	4.2×10^{-2}	6.1×10^{-4}	2.3×10^{-3}	9.7×10^{-2}
C-CMA2	2.9×10^{-2}	5.0×10^{-1}	7.7×10^{-4}	2.7×10^{-1}	5.0×10^{-1}
RAKE	3.1×10^{-2}	3.1×10^{-2}	7.1×10^{-3}	7.1×10^{-3}	7.1×10^{-3}

5.5 SIGNAL SUBSPACE-BASED CHANNEL ESTIMATION FOR CDMA SYSTEMS

The practical CDMA systems use the pilot signals for channel estimation enabling even the
use of a smoother for these purposes. While the solution is simple, it is very inefficient in
the presence of Doppler or the near–far effect. The advanced channel estimation algorithms
in CDMA systems are based either on Kalman-type [471, 477–479] estimators for channels
with high dynamics, or a signal subspace-based approach in the presence of high levels
of multiple access interference [473, 489–498]. The best performance is obtained if a
joint detection of data and channel is used [474, 476]. Additional results including blind
estimation are given in [480, 481, 499]. In this section we present a multiuser channel
estimation problem through a signal subspace-based approach [473]. For these purposes
the received signal for K users will be presented as:

$$r(t) = \sum_{k=1}^{K} r_k(t) + \eta_t \quad -\infty < t < \infty \tag{5.202}$$

If the channel impulse response for user k is $h_k(t, \tau)$ we have

$$r_k(t) = h_k(t, \tau)^* s_k(t)$$
$$= \int_{-\infty}^{\infty} h_k(t, \alpha) s_k(\alpha) \, d\alpha \tag{5.203}$$

If phase shift keying (PSK) is used to modulate the data, then the baseband complex envelope representation of the kth user's transmitted signal is given by

$$s_k(t) = \sqrt{2P_k}e^{j\phi_k}\sum_i e^{j(2\pi M)m_k^{(i)}}a_k(t - iT) \tag{5.204}$$

where P_k is the transmitted power, ϕ_k is the carrier phase relative to the local oscillator at the receiver, M is the size of the symbol alphabet, $m_k^{(i)} \in \{0, 1, \ldots, M - 1\}$ is the transmitted symbol, $a_k(t)$ is the spreading waveform, and T is the symbol duration. The spreading waveform is given by

$$a_k(t) = \sum_{n=0}^{N-1} \Pi_{T_c}(t - nT_c)a_k^{(n)} \tag{5.205}$$

where $\Pi_{T_c}(t)$ is a rectangular pulse, T_c is the chip duration ($T_c = T/N$), and $\{a_k^{(n)}\}$ for $n = 0, 1, \ldots, N - 1$ is a signature sequence (possibly complex-valued since the signature alphabet need not be binary). The chip matched filter can be implemented as an integrate-and-dump circuit, and the discrete time signal is given by

$$r[n] = \frac{1}{T_c}\int_{nT_c}^{(n+1)T_c} r(t)\,dt \tag{5.206}$$

Thus, the received signal can be converted into a sequence of WSS random vectors by buffering $r[n]$ into blocks of length N

$$\mathbf{y}_i = [r[iN]\,r[1 + iN]\cdots r[N - 1 + iN]]^\mathrm{T} \in \mathbb{C}^N \tag{5.207}$$

where the nth element of the ith observation vector is given by $y_{i,n} = r[n + iN]$. Although each observation vector corresponds to one symbol interval, this buffering was done without regard to the actual symbol intervals of the users. Since the system is asynchronous, each observation vector will contain at least the end of the previous symbol (left) and the beginning of the current symbol (right) for each user. The factors due to the power, phase and transmitted symbols of the kth user may be collected into a single complex constant $c_k^{(i)}$, e.g. some constant times $\sqrt{2P_k}e^{j[\phi_k+(2\pi/M)m_k^{(i)}]}$, and Equation (5.207) becomes:

$$\mathbf{y}_i = \sum_{k=1}^{K}\left[c_k^{(i-1)}\mathbf{u}_k^\mathrm{r} + c_k^{(i)}\mathbf{u}_k^l\right] + \boldsymbol{\eta}_i = \mathbf{A}\mathbf{c}_i + \boldsymbol{\eta}_i \tag{5.208}$$

where $\boldsymbol{\eta}_i = [\eta_{i,0}, \ldots, \eta_{i,N-1}]^\mathrm{T} \in \mathbb{C}^N$ is a Gaussian random vector. Its elements are zero mean with variance $\sigma^2 = N_0/2T_c$ and are mutually independent. Vectors $\mathbf{u}_k^\mathrm{r}$ and $\mathbf{u}_k^l$ are the right side of the kth user's code vector followed by zeros, and zeros followed by the left side of the kth user's code vector, respectively. In addition, we have defined $\mathbf{c}_i = [c_1^{(i-1)}c_1^{(i)}\cdots c_K^{(i-1)}c_K^{(i)}]^\mathrm{T} \in \mathbb{C}^{2K}$ and the signal matrix. $\mathbf{A} = \lfloor\mathbf{u}_1^\mathrm{r}\mathbf{u}_1^l\cdots\mathbf{u}_K^\mathrm{r}\mathbf{u}_K^l\rfloor \in \mathbb{C}^{N\times 2K}$. We will start with the assumption that each user's signal goes through a single propagation path with an associated attenuation factor and propagation delay. We assume that these parameters vary slowly with time, so that for sufficiently short intervals the channel is approximately a linear time-invariant (LTI) system. The baseband channel impulse response can then be represented by a Dirac delta function as $h_k(t, \tau) = h_k(t) = \alpha_k\delta(t - \tau_k), \forall\tau$, where α_k is a complex-valued attenuation weight and τ_k is the propagation delay. Since there is just a single path, we assume that α_k is incorporated into $c_k^{(i)}$ and concentrate solely on the delay.

Let us define $v \in \{0, \ldots, N-1\}$ and $\gamma \in [0, 1)$ such that $(\tau_k/T_c) \bmod N = v + \gamma$. If $\gamma = 0$, i.e. the received signal is precisely aligned with the chip matched filter and only one chip will contribute to each sample, the signal vectors become:

$$
\begin{aligned}
\mathbf{u}_k^r &= \mathbf{a}_k^r(v) \equiv \left[a_k^{(N-v)} \cdots a_k^{(N-1)} 0 \cdots 0 \right]^T \\
\mathbf{u}_k^l &= \mathbf{a}_k^l(v) \equiv \left[0 \cdots 0 \quad a_k^{(0)} \cdots a_k^{(N-v-1)} \right]^T
\end{aligned}
\tag{5.209}
$$

Since the chip matched filter is just an integrator, the samples for a non-zero γ can be represented as:

$$
\begin{aligned}
\mathbf{u}_k^r &= (1-\gamma)\mathbf{a}_k^r(v) + \gamma \mathbf{a}_k^r(v+1) \\
\mathbf{u}_k^l &= (1-\gamma)\mathbf{a}_k^l(v) + \gamma \mathbf{a}_k^l(v+1)
\end{aligned}
\tag{5.210}
$$

For the more general case of a multipath transmission channel with L distinct propagation paths, the impulse response becomes a series of delta functions

$$
h_k(t, \tau) = h_k(t) = \sum_{p=1}^{L} \alpha_{k,p} \delta(t - \tau_{k,p})
\tag{5.211}
$$

The signal vectors can be represented as

$$
\mathbf{u}_k^r = \sum_{p=1}^{L} \alpha_{k,p} \left[(1 - \gamma_{k,p})\mathbf{a}_k^r(v_{k,p}) + \gamma_{k,p} \mathbf{a}_k^r(v_{k,p} + 1) \right]
$$

$$
\mathbf{u}_k^l = \sum_{p=1}^{L} \alpha_{k,p} \left[(1 - \gamma_{k,p})\mathbf{a}_k^l(v_{k,p}) + \gamma_{k,p} \mathbf{a}_k^l(v_{k,p} + 1) \right]
\tag{5.212}
$$

If we introduce the following notation

$$
\begin{aligned}
\mathbf{U}_k^r &= \left\lfloor \mathbf{a}_k^r(0) \cdots \mathbf{a}_k^r(N-1) \right\rfloor \in \mathbb{C}^{N \times N} \\
\mathbf{U}_k^l &= \left[\mathbf{a}_k^l(0) \cdots \mathbf{a}_k^l(N-1) \right] \in \mathbb{C}^{N \times N}
\end{aligned}
\tag{5.213}
$$

where the $\mathbf{a}_k^s$ are as defined in Equation (5.209), then the signal vectors may be expressed as a linear combination of the columns of these matrices

$$
\begin{aligned}
\mathbf{u}_k^r &= \mathbf{U}_k^r \mathbf{h}_k \\
\mathbf{u}_k^l &= \mathbf{U}_k^l \mathbf{h}_k
\end{aligned}
\tag{5.214}
$$

where $\mathbf{h}_k$ is the composite impulse response of the channel and the receiver front end, evaluated modulo the symbol period. Thus, the nth element of the impulse response is given by

$$
h_{k,n} = \sum_{j=0}^{\infty} \frac{1}{T_c} \int_{jT+nT_c}^{jT+(n+1)T_c} h_k(t)^* \prod_{T_c}(t) \, dt
\tag{5.215}
$$

For delay spread $T_m < T/2$, at most two terms in the summation will be non-zero.

5.5.1 Estimating the signal subspace

The correlation matrix of the observation vectors is given by:

$$\mathbf{R} = E\lfloor \mathbf{y}_i \mathbf{y}_i^\dagger \rfloor = \mathbf{ACA}^\dagger + \sigma^2 \mathbf{I} \tag{5.216}$$

where $\mathbf{C} = E\lfloor \mathbf{c}_i \mathbf{c}_i^\dagger \rfloor \in C^{2K \times 2K}$ is diagonal. The correlation matrix can also be expressed in terms of its eigenvector decomposition:

$$\mathbf{R} = \mathbf{VDV}^\dagger \tag{5.217}$$

where the columns of $\mathbf{V} \in C^{N \times N}$ are the eigenvectors of $\mathbf{R}$, and $\mathbf{D}$ is a diagonal matrix of the corresponding eigenvalues (λ_n). Details of eigenvector decomposition are given in Appendix 5.1. Furthermore,

$$\lambda_n = \begin{cases} d_n + \sigma^2, & \text{if } n \leq 2K \\ \sigma^2, & \text{otherwise} \end{cases} \tag{5.218}$$

where d_n is the variance of the signal vectors along the nth eigenvector and we assume that $2K < N$. Since the $2K$ largest eigenvalues of $\mathbf{R}$ correspond to the signal subspace, $\mathbf{V}$ can be partitioned as $\mathbf{V} = [\mathbf{V}_S \mathbf{V}_N]$, where the columns of $\mathbf{V}_S = [\mathbf{v}_{S,1} \cdots \mathbf{v}_{S,2K}] \in C^{N \times 2K}$ form a basis for the signal subspace S_Y, and $\mathbf{V}_N = [\mathbf{v}_{N,1} \cdots \mathbf{v}_{N,N-2K}] \in C^{N \times N-2K}$ spans the noise subspace N_Y. Readers less familiar with eigenvalue decomposition are referred to Appendix 5.1. Since we would like to track slowly varying parameters, we form a moving average or a Bartlett estimate of the correlation matrix based on the J most recent observations:

$$\hat{\mathbf{R}}_i = \frac{1}{J} \sum_{j=i-J+1}^{i} \mathbf{y}_j \mathbf{y}_j^\dagger \tag{5.219}$$

It is well known [500] that the maximum likelihood estimate of the eigenvalues and associated eigenvectors of $\mathbf{R}$ is just the eigenvector decomposition of $\hat{\mathbf{R}}_i$. Thus, we perform an eigenvalue decomposition of $\hat{\mathbf{R}}_i$ and select the eigenvectors corresponding to the $2K$ largest eigenvalues as a basis for $\hat{S}_Y$.

5.5.2 Channel estimation

Consider the projection of a given user's signal vectors into the estimated noise subspace

$$\left. \begin{array}{l} \mathbf{e}_k^r = \left(\mathbf{u}_k^{r\dagger} \hat{\mathbf{V}}_N\right)^{\mathrm{T}} \\ \mathbf{e}_k^l = \left(\mathbf{u}_k^{l\dagger} \hat{\mathbf{V}}_N\right)^{\mathrm{T}} \end{array} \right\} \in C^{N-2K} \tag{5.220}$$

If $\mathbf{u}_k^r$ and $\mathbf{u}_k^l$ both lie in the signal subspace, then their sum $\mathbf{u}_k = \mathbf{u}_k^r + \mathbf{u}_k^l$ must also be contained in $\mathbf{V}_S$. The projection of $\mathbf{u}_k$ into the estimated noise subspace

$$\tilde{\mathbf{e}}_k = (\mathbf{u}_k^\dagger \hat{\mathbf{V}}_N)^{\mathrm{T}} \tag{5.221}$$

is a Gaussian random vector [472] and thus has probability density function

$$p_{\tilde{\mathbf{e}}}(\tilde{\mathbf{e}}_k) = \frac{1}{\det[\pi \mathbf{K}]} \exp\{-\tilde{\mathbf{e}}_k^\dagger \mathbf{K}^{-1} \tilde{\mathbf{e}}_k\} \tag{5.222}$$

The covariance matrix $\mathbf{K}$ is a scalar multiple of the identity given by

$$\mathbf{K} = \frac{1}{J}\mathbf{u}_k^\dagger \mathbf{Q}\mathbf{u}_k \mathbf{I} \tag{5.223}$$

and

$$\mathbf{Q} = \sigma^2 \left[\sum_{k=1}^{2K} \frac{\lambda_k}{(\sigma^2 - \lambda_k)^2} \mathbf{v}_{S,k} \mathbf{v}_{S,k}^\dagger \right] \tag{5.224}$$

Therefore, within an additive constant, the log likelihood function of $\tilde{\mathbf{e}}_k$ is:

$$\Lambda(\tilde{\mathbf{e}}_k) = -(N - 2K)\ln(\mathbf{u}_k^\dagger \mathbf{Q}\mathbf{u}_k) - J\frac{\tilde{\mathbf{e}}_k^\dagger \tilde{\mathbf{e}}_k}{\mathbf{u}_k^\dagger \mathbf{Q}\mathbf{u}_k}$$

$$= -(N - 2K)\ln(\mathbf{u}_k^\dagger \mathbf{Q}\mathbf{u}_k) - J\frac{\mathbf{u}_k^\dagger \mathbf{V}_N \mathbf{V}_N^\dagger \mathbf{u}_k}{\mathbf{u}_k^\dagger \mathbf{Q}\mathbf{u}_k} \tag{5.225}$$

The exact $\mathbf{V}_N$ and $\mathbf{Q}$ are unknown, but we may replace them with their estimates. The best estimates will minimize $\tilde{\mathbf{e}}_k$, which will result in the maximum of the likelihood function.

Unfortunately, maximizing this likelihood function is prohibitively complex for a general multipath channel, so we will consider only a single propagation path. In this case, the vector $\mathbf{u}_k$ is a function of only one unknown parameter: the delay τ_k. To form the timing estimate, we must solve

$$\hat{\tau}_k = \arg \max_{\tau_k \in [0,T)} \Lambda(\mathbf{u}_k) \tag{5.226}$$

Ideally, we would like to differentiate the log likelihood function with respect to τ. However, the desired user's delay lies within an uncertainty region, $\tau_k \in [0, T]$, and $\mathbf{u}_k(\tau)$ is only piecewise continuous on this interval. To deal with these problems, we divide the uncertainty region into N cells of width T_c and consider a single cell, $c_v \equiv [vT_c, (v + 1)T_c)$. We again define $v \in \{0, \ldots, N - 1\}$ and $\gamma \in [0, 1)$ such that $(\tau/T_c) \bmod N = v + \gamma$, and for $\tau \in c_v$ the desired user's signal vector becomes

$$\mathbf{u}_k\{\tau\} = (1 - \gamma)\mathbf{u}_k(v) + \gamma\mathbf{u}_k(v + 1) \tag{5.227}$$

and

$$\frac{\mathrm{d}}{\mathrm{d}\tau}\mathbf{u}_k(\tau) = \mathbf{u}_k(v + 1) - \mathbf{u}_k(v)$$
$$= \text{a constant} \tag{5.228}$$

Thus, within a given cell, we can differentiate the log likelihood function and solve for the maximum in closed form. We then choose whichever of the N solutions yields the largest value for Equation (5.226). Details can be found in [473].

Under certain conditions, it may be possible to simplify this algorithm. Note that maximizing the log likelihood function (5.226) is equivalent to maximizing

$$\Lambda(\tilde{\mathbf{e}}_k) = -\frac{N - 2K}{J}\ln(\mathbf{u}_k^\dagger \mathbf{Q}\mathbf{u}_k) - \frac{\mathbf{u}_k^\dagger \mathbf{V}_N \mathbf{V}_N^\dagger \mathbf{u}_k}{\mathbf{u}_k^\dagger \mathbf{Q}\mathbf{u}_k} \tag{5.229}$$

As $J \to \infty$, the leading term goes to zero; thus, for large observation windows, we can use the following approximation:

$$\Lambda(\tilde{\mathbf{e}}_k) \approx -\frac{\mathbf{u}_k^\dagger \mathbf{V}_N \mathbf{V}_N^\dagger \mathbf{u}_k}{\mathbf{u}_k^\dagger \mathbf{Q} \mathbf{u}_k} \tag{5.230}$$

This yields a much simpler expression for the stationary points [473]. The MUSIC (Multiple Signal Classification) algorithm is equivalent to Equation (5.29) when one only maximizes the numerator and ignores the denominator, i.e. one assumes $\mathbf{u}_k^\dagger \mathbf{Q} \mathbf{u}_k$ is equal to one in Equation (5.28) or Equation (5.29). This yields an even simpler approximation for the log likelihood function

$$\Lambda(\tilde{\mathbf{e}}_k) \approx -\mathbf{u}_k^\dagger \mathbf{V}_N \mathbf{V}_N^\dagger \mathbf{u}_k \tag{5.231}$$

which further simplifies the solution for the stationary points [473].

For illustration purposes, the simulation results for five users with length 31 Gold codes are presented in Figures 5.28–5.30.

A single desired user was acquired and tracked in the presence of strong multiple access interference (MAI). The power ratio between each of the four interfering users and the desired user is designated the MAI level.

We first compare the true log likelihood estimate, Equation (5.225), with the large observation window approximation, Equation (5.29), and the MUSIC algorithm, Equation (5.30). This is done for a window size of 200 symbols and with a varying SNR. Figure 5.28(a) shows the probability of acquisition for each method, where acquisition is defined as $|\tau_k - \hat{\tau}_k| < 1/2T_c$. Using the approximate log likelihood function results in almost no drop in performance. Furthermore, when the SNR is poor, both probabilistic approaches considerably outperform the MUSIC algorithm. In Figure 5.28(b), we compare the RMSE of the delay estimate once acquisition has occurred, i.e. after processing enough symbols to reach within half of one chip. The approximate log-likelihood function experiences a slight increase in error at low SNR, but again both probabilistic methods do better than MUSIC.

The same parameters as a function of the window size are shown in Figure 5.29. One can say that for $J > 100$ the performance curve settles down to steady state values. The RMSE versus MAI and SNR are shown in Figure 5.30. One can see that for an extremely wide ranging near–far effect the performance is good.

APPENDIX 5.1 LINEAR AND MATRIX ALGEBRA

Definitions

Consider an $m \times n$ matrix $\mathbf{R}$ with elements $r_{ij}, i = 1, 2, \ldots, m; j = 1, 2, \ldots, n$. A shorthand notation for describing $\mathbf{R}$ is

$$[\mathbf{R}]_{ij} = r_{ij}$$

The *transpose* of $\mathbf{R}$, which is denoted by $\mathbf{R}^\mathrm{T}$, is defined as the $n \times m$ matrix with elements $\mathbf{r}_{ji}$ or

$$[\mathbf{R}^\mathrm{T}]_{ij} = r_{ji}$$

A square matrix is one for which $m = n$. A square matrix is symmetric if $\mathbf{R}^\mathrm{T} = \mathbf{R}$. The *rank* of a matrix is the number of linearly independent rows or columns, whichever is less.

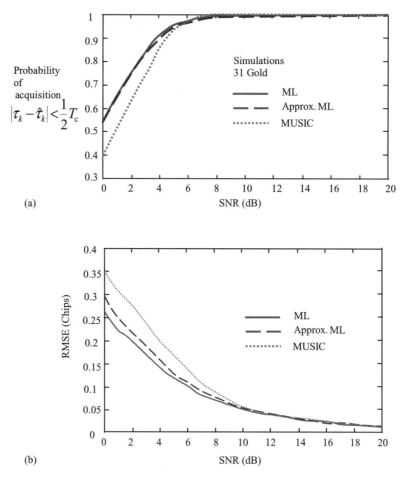

Figure 5.28 (a) Probability of acquisition for the maximum likelihood estimator, the approximate ML and the MUSIC algorithm [$K = 5$, $N = 31$, $J = 200$, MAI = 20 dB]; (b) Root mean squared error (RMSE) of the delay estimate in chips for the maximum likelihood (ML) estimator, the approximate ML and the MUSIC algorithm [$K = 5$, $N = 31$, $J = 200$, MAI = 20 dB].

The *inverse* of a square $n \times n$ matrix is the square $n \times n$ matrix $\mathbf{R}^{-1}$ for which

$$\mathbf{R}^{-1}\mathbf{R} = \mathbf{R}\mathbf{R}^{-1} = \mathbf{I}$$

where $\mathbf{I}$ is the $n \times n$ identity matrix. The inverse will exist if and only if the rank of $\mathbf{R}$ is n. If the inverse does not exist, then $\mathbf{R}$ is singular. The *determinant* of a square $n \times n$ matrix is denoted by det($\mathbf{R}$). It is computed as

$$\det(\mathbf{R}) = \sum_{j=1}^{n} r_{ij}C_{ij}$$

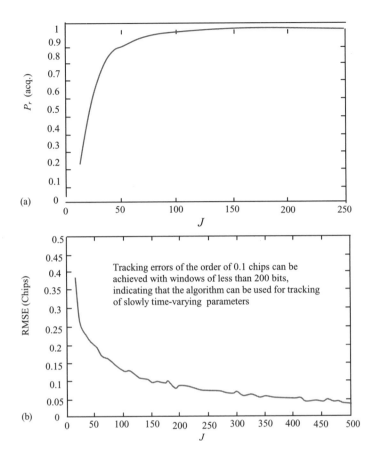

Figure 5.29 (a) Probability of acquisition and (b) root mean squared error (RMSE) of timing estimate in chips of the subspace-based maximum likelihood estimator for varying window size [$N = 31$, SNR $= 8$ dB, $K = 5$, MAI $= 20$ dB] [473] © 1996, IEEE.

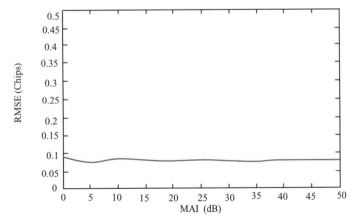

Figure 5.30 RMSE of the subspace-based maximum likelihood estimator for varying MAI level [$K = 5$, $N = 31$, $J = 200$, SNR $= 8$ dB].

where

$$C_{ij} = (-1)^{i+j} M_{ij}$$

M_{ij} is the determinant of the submatrix of **R** obtained by deleting the ith row and jth colunm and is termed the *minor* of r_{ij}. C_{ij} is the cofactor of r_{ij}. Note that any choice of i for $i = 1, 2, \ldots, n$ will yield the same value for det(**R**).

A *quadratic form Q* is defined as

$$Q = \sum_{i=1}^{n} \sum_{j=1}^{n} r_{ij} x_i x_j$$

In defining the quadratic form it is assumed that $r_{ji} = r_{ij}$. This entails no loss in generality since any quadratic function may be expressed in this manner. Q may also be expressed as

$$Q = \mathbf{x}^T \mathbf{R} \mathbf{x}$$

where $\mathbf{x} = [x_1 x_2 \cdots x_n]^T$ and **R** is a square $n \times n$ matrix with $r_{ji} = r_{ij}$ or **R** is a symmetric matrix.

A square $n \times n$ matrix **R** is *positive semidefinite* if **R** is symmetric and

$$\mathbf{x}^T \mathbf{R} \mathbf{x} \geq 0$$

for all $\mathbf{x} \neq \mathbf{0}$. If the quadratic form is strictly positive, then **R** is *positive definite*. When referring to a matrix as *positive definite* or positive semidefinite, it is always assumed that the matrix is symmetric. The *trace* of a square $n \times n$ matrix is the sum of its diagonal elements or

$$\text{tr}(\mathbf{R}) = \sum_{i=1}^{n} r_{ii}$$

A *partitioned m × n* matrix **R** is one that is expressed in terms of its submatrices. An example is the 2×2 partitioning

$$\mathbf{R} = \begin{bmatrix} \mathbf{R}_{11} & \mathbf{R}_{12} \\ \mathbf{R}_{21} & \mathbf{R}_{22} \end{bmatrix}$$

Each 'element' $\mathbf{R}_{ij}$ is a submatrix of **R**. The dimensions of the partitions are given as

$$\begin{bmatrix} k \times l & k \times (n - l) \\ (m - k) \times l & (m - k) \times (n - l) \end{bmatrix}$$

Special Matrices

A *diagonal* matrix is a square $n \times n$ matrix with $r_{ij} = 0$ for $i \neq j$, in other words, all elements off the principal diagonal are zero. A diagonal matrix appears as:

$$\mathbf{R} = \begin{bmatrix} r_{11} & 0 & \cdots & 0 \\ 0 & r_{22} & \cdots & 0 \\ \vdots & \vdots & \ddots & \vdots \\ 0 & 0 & \cdots & r_{nn} \end{bmatrix}$$

A diagonal matrix will sometimes be denoted by $\text{diag}(r_{11}, r_{22}, \ldots, r_{nn})$. Theinverse of a diagonal matrix is found by simply inverting each element on the principal diagonal. A generalization of the diagonal matrix is the square $n \times n$ block diagonal matrix

$$
\mathbf{R} = \begin{bmatrix} \mathbf{R}_{11} & \mathbf{0} & \cdots\cdots\cdots\cdots & \mathbf{0} \\ \mathbf{0} & \mathbf{R}_{22} & \cdots\cdots\cdots & \mathbf{0} \\ \vdots & & & \vdots \\ \mathbf{0} & \mathbf{0} & \cdots\cdots\cdots & \mathbf{R}_{kk} \end{bmatrix}
$$

in which all submatrices $\mathbf{R}_{ii}$ are square and the other submatrices are identically zero. The dimensions of the submatrices need not be identical. For instance, if $k = 2$, $\mathbf{R}_{11}$ might have dimension 2×2 while $\mathbf{R}_{22}$ might be a scalar. If all $\mathbf{R}_{ii}$ are nonsingular, then the inverse is easily found as:

$$
\mathbf{R}^{-1} = \begin{bmatrix} \mathbf{R}_{11}^{-1} & \mathbf{0} & \cdots\cdots\cdots & \mathbf{0} \\ \mathbf{0} & \mathbf{R}_{22}^{-1} & \cdots\cdots\cdots & \mathbf{0} \\ \vdots & & & \vdots \\ \mathbf{0} & \mathbf{0} & \cdots\cdots\cdots & \mathbf{R}_{kk}^{-1} \end{bmatrix}
$$

Also, the determinant is:

$$
\det(\mathbf{R}) = \prod_{i=1}^{n} \det(\mathbf{R}_{ii})
$$

A square $n \times n$ matrix is *orthogonal* if

$$
\mathbf{R}^{-1} = \mathbf{R}^{\mathrm{T}}
$$

For a matrix to be orthogonal the columns (and rows) must be orthonormal or, if

$$
\mathbf{R} = [\mathbf{r}_1 \mathbf{r}_2 \ldots \mathbf{r}_n]
$$

where $\mathbf{r}_i$ denotes the ith column, the conditions

$$
\mathbf{r}_i^{\mathrm{T}} \mathbf{r}_j = \begin{cases} 0 & \text{for } i \neq j \\ 1 & \text{for } i = j \end{cases}
$$

must be satisfied.

An *idempotent* matrix is a square $n \times n$ matrix which satisfies

$$
\mathbf{R}^2 = \mathbf{R}
$$

This condition implies that $\mathbf{R}^l = \mathbf{R}$ for $l \geq 1$. An example is the projection matrix

$$
\mathbf{R} = \mathbf{H}(\mathbf{H}^{\mathrm{T}}\mathbf{H})^{-1}\mathbf{H}^{\mathrm{T}}
$$

where $\mathbf{H}$ is an $m \times n$ full rank matrix with $m > n$.

A square $n \times n$ *Toeplitz* matrix is defined as

$$
[\mathbf{R}]_{ij} = r_{i-j}
$$

or

$$\mathbf{R} = \begin{bmatrix} r_0 & r_{-1} & r_{-2} & \cdots & r_{-(n-1)} \\ r_1 & r_0 & r_{-1} & \cdots & r_{-(n-2)} \\ \vdots & \vdots & \vdots & \vdots & \vdots \\ r_{n-1} & r_{n-2} & r_{n-3} & \cdots & r_0 \end{bmatrix}$$

Each element along a northwest–southeast diagonal is the same. If, in addition, $r_{-k} = r_k$, then $\mathbf{R}$ is symmetric Toeplitz.

Matrix manipulation and formulas

Some useful formulas for the algebraic manipulation of matrices are summarized in this section. For $n \times n$ matrices $\mathbf{R}$ and $\mathbf{P}$ the following relationships are useful.

$$(\mathbf{RP})^{\mathrm{T}} = \mathbf{P}^{\mathrm{T}}\mathbf{R}^{\mathrm{T}}$$
$$(\mathbf{R}^{\mathrm{T}})^{-1} = (\mathbf{R}^{-1})^{\mathrm{T}}$$
$$(\mathbf{RP})^{-1} = \mathbf{P}^{-1}\mathbf{R}^{-1}$$
$$\det(\mathbf{R}^{\mathrm{T}}) = \det(\mathbf{R})$$
$$\det(c\mathbf{R}) = c^n \det(\mathbf{R}) \quad (c \text{ a scalar})$$
$$\det(\mathbf{RP}) = \det(\mathbf{R})\det(\mathbf{P})$$
$$\det(\mathbf{R}^{-1}) = \frac{1}{\det(\mathbf{R})}$$
$$\mathrm{tr}(\mathbf{RP}) = \mathrm{tr}(\mathbf{PR})$$
$$\mathrm{tr}(\mathbf{R}^{\mathrm{T}}\mathbf{P}) = \sum_{i=1}^{n}\sum_{j=1}^{n}[\mathbf{R}]_{ij}[\mathbf{P}]_{ij}$$

For vectors $\mathbf{x}$ and $\mathbf{y}$ we have

$$\mathbf{y}^{\mathrm{T}}\mathbf{x} = \mathrm{tr}(\mathbf{xy}^{\mathrm{T}})$$

It is frequently necessary to determine the inverse of a matrix analytically. To do so one can make use of the following formula. The inverse of a square $n \times n$ matrix is

$$\mathbf{R}^{-1} = \frac{\mathbf{C}^{\mathrm{T}}}{\det(\mathbf{R})}$$

where $\mathbf{C}$ is the square $n \times n$ matrix of cofactors $\mathbf{R}$. The cofactor matrix is defined by

$$[\mathbf{C}]_{ij} = (-1)^{i+j} M_{ij}$$

where M_{ij} is the minor of r_{ij} obtained by deleting the ith row and jth column of $\mathbf{R}$. Another formula which is quite useful is the matrix inversion lemma

$$(\mathbf{R} + \mathbf{PCD})^{-1} = \mathbf{R}^{-1} - \mathbf{R}^{-1}\mathbf{P}(\mathbf{DR}^{-1}\mathbf{P} + \mathbf{C}^{-1})^{-1}\mathbf{DR}^{-1}$$

where it is assumed that $\mathbf{R}$ is $n \times n$, $\mathbf{P}$ is $n \times m$, $\mathbf{C}$ is $m \times m$, and $\mathbf{D}$ is $m \times n$ and that the indicated inverses exist. A special case known as *Woodbury's identity* results when $\mathbf{P}$ is an

$n \times 1$ column vector $\mathbf{u}$, $\mathbf{C}$ a scalar of unity, and $\mathbf{D}$ a $1 \times n$ row vector $\mathbf{u}^T$. Then

$$(\mathbf{R} + \mathbf{u}\mathbf{u}^T)^{-1} = \mathbf{R}^{-1} - \frac{\mathbf{R}^{-1}\mathbf{u}\mathbf{u}^T\mathbf{R}^{-1}}{1 + \mathbf{u}^T\mathbf{R}^{-1}\mathbf{u}}$$

Partitioned matrices may be manipulated according to the usual rules of matrix algebra by considering each submatrix as an element. For multiplication of partitioned matrices, the submatrices which are multiplied together must be conformable. As an illustration, for 2×2 partitioned matrices

$$\mathbf{RP} = \begin{bmatrix} \mathbf{R}_{11} & \mathbf{R}_{12} \\ \mathbf{R}_{21} & \mathbf{R}_{22} \end{bmatrix} \begin{bmatrix} \mathbf{P}_{11} & \mathbf{P}_{12} \\ \mathbf{P}_{21} & \mathbf{P}_{22} \end{bmatrix} = \begin{bmatrix} \mathbf{R}_{11}\mathbf{P}_{11} + \mathbf{R}_{12}\mathbf{P}_{21} & \mathbf{R}_{11}\mathbf{P}_{12} + \mathbf{R}_{12}\mathbf{P}_{22} \\ \mathbf{R}_{21}\mathbf{P}_{11} + \mathbf{R}_{22}\mathbf{P}_{21} & \mathbf{R}_{21}\mathbf{P}_{12} + \mathbf{R}_{22}\mathbf{P}_{22} \end{bmatrix}$$

The transposition of a partitioned matrix is formed by transposing the submatrices of the matrix and applying T to each submatrix. For a 2×2 partitioned matrix

$$\begin{bmatrix} \mathbf{R}_{11} & \mathbf{R}_{12} \\ \mathbf{R}_{21} & \mathbf{R}_{22} \end{bmatrix}^T = \begin{bmatrix} \mathbf{P}_{11}^T & \mathbf{P}_{21}^T \\ \mathbf{P}_{12}^T & \mathbf{P}_{22}^T \end{bmatrix}$$

The extension of these properties to arbitrary partitioning is straightforward. Determination of the inverses and determinants of partitioned matrices is facilitated by employing the following formulas. Let $\mathbf{R}$ be a square $n \times n$ matrix partitioned as

$$\mathbf{R} = \begin{bmatrix} \mathbf{R}_{11} & \mathbf{R}_{12} \\ \mathbf{R}_{21} & \mathbf{R}_{22} \end{bmatrix} = \begin{bmatrix} k \times k & k \times (n-k) \\ (n-k) \times k & (n-k) \times (n-k) \end{bmatrix}$$

Then,

$$\mathbf{R}^{-1} = \begin{bmatrix} \left(\mathbf{R}_{11} - \mathbf{R}_{12}\mathbf{R}_{22}^{-1}\mathbf{R}_{21}\right)^{-1} & -\left(\mathbf{R}_{11} - \mathbf{R}_{12}\mathbf{R}_{22}^{-1}\mathbf{R}_{21}\right)^{-1}\mathbf{R}_{12}\mathbf{R}_{22}^{-1} \\ -\left(\mathbf{R}_{22} - \mathbf{R}_{21}\mathbf{R}_{11}^{-1}\mathbf{R}_{12}\right)^{-1}\mathbf{R}_{21}\mathbf{R}_{11}^{-1} & \left(\mathbf{R}_{22} - \mathbf{R}_{21}\mathbf{R}_{11}^{-1}\mathbf{R}_{12}\right)^{-1} \end{bmatrix}$$

and

$$\begin{aligned} \det(\mathbf{R}) &= \det\left(\mathbf{R}_{22}\right)\det(\mathbf{R}_{11} - \mathbf{R}_{12}\mathbf{R}_{22}^{-1}\mathbf{R}_{21}) \\ &= \det(\mathbf{R}_{11})\det\left(\mathbf{R}_{22} - \mathbf{R}_{21}\mathbf{R}_{11}^{-1}\mathbf{R}_{12}\right) \end{aligned}$$

where the inverses of $\mathbf{R}_{11}$ and $\mathbf{R}_{22}$ are assumed to exist.

Theorems

Some important theorems are summarized in this section.

1. A square $n \times n$ matrix $\mathbf{R}$ is invertible (non-singular) if and only if its columns (or rows) are linearly independent or, equivalently, if its determinant is non-zero. In such a case, $\mathbf{R}$ is full rank. Otherwise, it is singular.

2. A square $n \times n$ matrix $\mathbf{R}$ is positive definite if and only if
   – it can be written as

$$\mathbf{R} = \mathbf{C}\mathbf{C}^T$$

   where $\mathbf{C}$ is also $n \times n$ and is full rank and hence invertible, or

– the principal minors are all positive. (The ith principal minor is the determinant of the submatrix formed by deleting all rows and columns with an index greater than i). If $\mathbf{R}$ can be written as in the previous equation, but $\mathbf{C}$ is not full rank or the principal minors are only non-negative, then $\mathbf{R}$ is positive semidefinite.

3. If $\mathbf{R}$ is positive definite, then the inverse exists and may be found from the previous equation as

$$\mathbf{R}^{-1} = (\mathbf{C}^{-1})^{T} (\mathbf{C}^{-1})$$

4. Let $\mathbf{R}$ be positive definite. If $\mathbf{P}$ is an $m \times n$ matrix of full rank with $m \leq n$, then $\mathbf{PRP}^{T}$ is also positive definite.

5. If $\mathbf{R}$ is positive definite (positive semidefinite), then
   – the diagonal elements are positive (non-negative)
   – the determinant of $\mathbf{R}$, which is a principal minor, is positive (non-negative).

Eigendecompostion of matrices

An *eigenvector* of a square $n \times n$ matrix $\mathbf{R}$ is an $n \times 1$ vector $\mathbf{v}$ satisfying

$$\mathbf{Rv} - \lambda\mathbf{v} \tag{A5.1}$$

for some scalar λ, which may be complex. λ is the eigenvalue of $\mathbf{R}$ corresponding to the eigenvector $\mathbf{v}$. It is assumed that the eigenvector is normalized to have unit length or $\mathbf{v}^T\mathbf{v} = 1$. If $\mathbf{R}$ is symmetric, then one can always find n linearly independent eigenvectors, although they will not in general be unique. An example is the identity matrix for which any vector is an eigenvector with eigenvalue 1. If $\mathbf{R}$ is symmetric, then the eigenvectors corresponding to distinct eigenvalues are orthonormal or $\mathbf{v}_i^T\mathbf{v}_j = \delta_{ij}$ and the eigenvalues are real. If, furthermore, the matrix is positive definite (positive semidefinite), then the eigenvalues are positive (non-negative). For a positive semidefinite matrix the rank is equal to the number of non-zero eigenvalues.

The defining previous relation can also be written as

$$\mathbf{R}[\mathbf{v}_1 \quad \mathbf{v}_2 \cdots \mathbf{v}_n] = [\lambda_1\mathbf{v}_1 \quad \lambda_2\mathbf{v}_2 \quad \cdots \quad \lambda_n\mathbf{v}_n]$$

or

$$\mathbf{RV} = \mathbf{V\Lambda}$$

where

$$\mathbf{V} = [\mathbf{v}_1 \quad \mathbf{v}_2 \quad \cdots \quad \mathbf{v}_n]$$
$$\mathbf{\Lambda} = \text{diag}(\lambda_1, \lambda_2, \cdots, \lambda_n)$$

If $\mathbf{R}$ is symmetric so that the eigenvectors corresponding to distinct eigenvalues are orthonormal and the remaining eigenvectors are chosen to yield an orthonormal eigenvector set, $\mathbf{V}$ is an orthonormal matrix. As such, its inverse is $\mathbf{V}^T$, so that the previous equation becomes

$$\mathbf{R} = \mathbf{V}\Lambda\mathbf{V}^{\mathrm{T}}$$
$$= \sum_{i=1}^{n} \lambda_i \mathbf{v}_i \mathbf{v}_i^{\mathrm{T}}$$

Also, the inverse is easily determined as

$$\mathbf{R}^{-1} = \mathbf{V}^{\mathrm{T}-1}\Lambda^{-1}\mathbf{V}^{-1} = \mathbf{V}\Lambda^{-1}\mathbf{V}^{\mathrm{T}} = \sum_{i=1}^{n} \frac{1}{\lambda_i} \mathbf{v}_i \mathbf{v}_i^{\mathrm{T}}$$

A final useful relationship follows as

$$\det(\mathbf{R}) = \det(\mathbf{V})\det(\Lambda)\det(\mathbf{V}^{-1}) = \det(\Lambda) = \prod_{i=1}^{n} \lambda_i$$

Calculation of eigenvalues and eigenvectors

We can write Equation (A5.1) as

$$\mathbf{R}\mathbf{v} - \lambda\mathbf{v} = 0 \Rightarrow (\mathbf{R} - \lambda\mathbf{I})\mathbf{v} = 0$$

These are n linear algebraic equations in the n unknowns v_1, v_2, v_3, ..v_n (the components of $\mathbf{v}$). For these equations to have a solution $\mathbf{v} \neq \mathbf{0}$, the determinant of the coefficient matrix $\mathbf{R} - \lambda\mathbf{I}$ must be zero. As an example, if $n = 2$, we have

$$\begin{bmatrix} r_{11} - \lambda & r_{12} \\ r_{21} & r_{22} - \lambda \end{bmatrix}\begin{bmatrix} v_1 \\ v_2 \end{bmatrix} = \begin{bmatrix} 0 \\ 0 \end{bmatrix}$$

which can be written as

$$(r_{11} - \lambda)v_1 + r_{12}v_2 = 0$$
$$r_{21}v_1 + (r_{22} - \lambda)v_2 = 0$$

(A5.2)

Now, $\mathbf{R} - \lambda\mathbf{I}$ is singular if and only if its determinant $\det(\mathbf{R} - \lambda\mathbf{I})$ is zero. This can be written as

$$\det(\mathbf{R} - \lambda\mathbf{I}) = \begin{vmatrix} r_{11} - \lambda & r_{12} \\ r_{21} & r_{22} - \lambda \end{vmatrix} = (r_{11} - \lambda)(r_{22} - \lambda) - r_{12}r_{21}$$
$$= \lambda^2 - (r_{11} + r_{22})\lambda + r_{11}r_{22} - r_{12}r_{21} = 0$$

This quadratic equation in λ is called the *characteristic equation* of $\mathbf{R}$. Its solutions are the eigenvalues λ_1 and λ_2 of $\mathbf{R}$. First determine these. Then use Equation (A5.2) with $\lambda = \lambda_1$ to determine the eigenvector $\mathbf{v}_1$ of $\mathbf{R}$ corresponding to λ_1. Finally, use Equation (A5.2) with $\lambda = \lambda_2$ to find an eigenvector $\mathbf{v}_2$ of $\mathbf{R}$ corresponding to λ_2. Note that if $\mathbf{v}$ is an eigenvector of $\mathbf{R}$ so is $k\mathbf{v}$ for any $k \neq 0$. As an example, suppose that

$$\mathbf{R} = \begin{bmatrix} -4.0 & 4.0 \\ -1.6 & 1.2 \end{bmatrix}$$

then we have

$$\det(\mathbf{R} - \lambda \mathbf{I}) = \begin{vmatrix} -4 - \lambda & 4 \\ -1.6 & 1.2 - \lambda \end{vmatrix} = \lambda^2 + 2.8\,\lambda + 1.6 = 0$$

It has the solutions $\lambda_1 = -2$ and $\lambda_2 = -0.8$. These are the eigenvalues of $\mathbf{R}$. Eigenvectors are obtained from Equation (A5.2). For $\lambda = \lambda_1 = -2$ we have, from Equation (A5.2):

$$(-4.0 + 2.0)v_1 + 4.0v_2 = 0$$
$$-1.6v_1 + (1.2 + 2.0)v_2 = 0$$

A solution of the equation is $v_1 = 2$, $v_2 = 1$. Hence, an eigenvector of $\mathbf{R}$ corresponding to $\lambda = \lambda_1 = -2$ is

$$\mathbf{v}_1 = \begin{bmatrix} 2 \\ 1 \end{bmatrix}$$

Similarly, for $\lambda = \lambda_2 = -0.8$

$$\mathbf{v}_2 = \begin{bmatrix} 1 \\ 0.8 \end{bmatrix}$$

So, for this example:

$$\mathbf{V} = [\mathbf{v}_1 \quad \mathbf{v}_2] = \begin{bmatrix} 2 & 1 \\ 1 & 0.8 \end{bmatrix}$$

6

Time Division Multiple Access – TDMA

The basic concept of time division multiple access (TDMA) has been discussed in Chapter 1. Within this chapter we cover the basic enabling technologies for TDMA. Coding and modulation are covered in Chapters 2, 3 and 4 so that in this chapter we focus on the remaining topics, mainly TDMA-specific channel estimation and equalization.

6.1 EQUALIZATION IN THE DIGITAL DATA TRANSMISSION SYSTEM

6.1.1 Zero-forcing equalizers

The basic problem of channel equalization is illustrated in Figure 6.1. Figure 6.1(a) presents the transmitted pulse $p(t)$, 6.1(b) shows the pulse after propagation through the channel $p_c(t)$, while Figure 6.1(c) shows the difference between the received $p_c(t)$ and the equalized pulse $p_{eq}(t) \cong p(t)$.

Figure 6.2 illustrates the mutual impact (inter-symbol interference) of the non-equalized and equalized signals. Finally, Figure 6.3 illustrates a general structure and function of the equalizer. To summarize, the transmitted signal for full response signaling is created in such a way that the pulse goes through zero at the time instances $\pm kT$, $k \neq 0$ from the pulse maximum (Nyquist signaling) $p(\pm kT) = 0$, $k \neq 0$. That way the adjacent symbols sampled at those instances will not be affected. The degradation caused by the channel will result in the pulse $p_c(\pm kT) \neq 0$, $k \neq 0$ which produces inter-symbol interference.

Advanced Wireless Communications. S. Glisic
© 2004 John Wiley & Sons, Ltd. ISBN: 0-470-86776-0

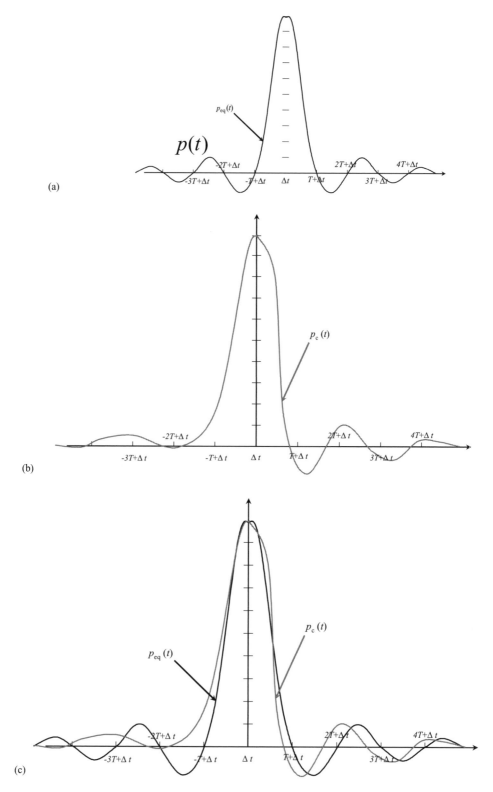

Figure 6.1 (a) Transmitted; (b) received; and (c) equalized pulses.

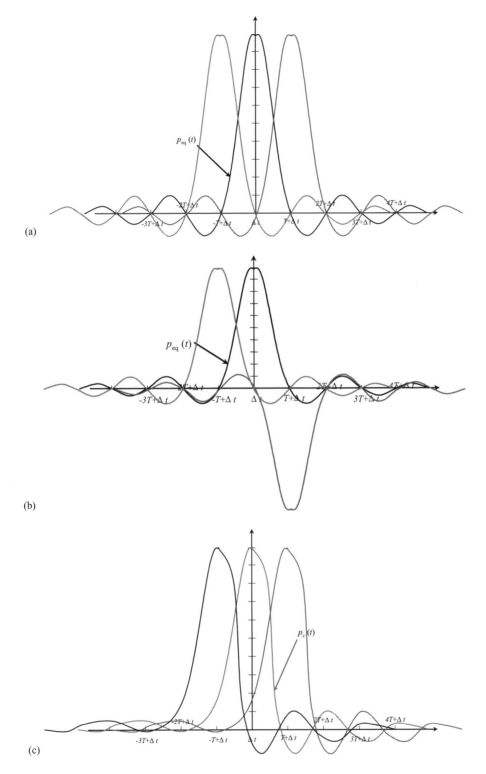

(a)

(b)

(c)

Figure 6.2 (a), (b) Transmitted signals and (c), (d) received signals.

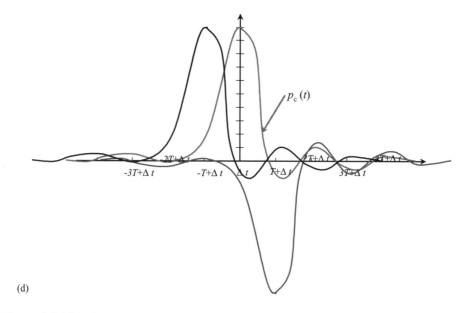

(d)

Figure 6.2 (*Cont.*)

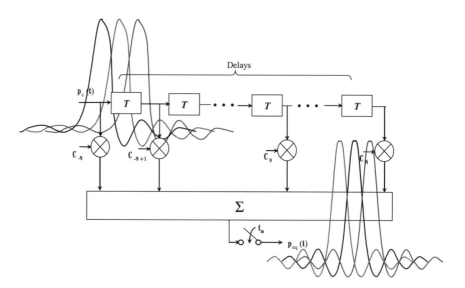

Figure 6.3 Transversal filter equalizer.

The equalizer is supposed to compensate for this degradation by regenerating a pulse $p_{eq}(\pm kT) \cong p(\pm kT) = 0,\ k \neq 0$.

This is represented by

$$p_{eq}(t) = \sum_{n=-N}^{N} C_n p_c(t - nT)$$

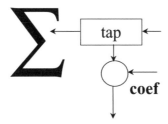

Figure 6.4 Transversal filter equalizer basic building block.

for $t = mT + \Delta t$

$$p_{eq}(mT + \Delta t) = \sum_{n=-N}^{N} C_n p_c((m-n)T + \Delta t)$$

$$= \begin{cases} 1 & m = 0 \\ 0 & m \neq 0 \end{cases} \quad m = 0, \pm 1, \pm 2, \ldots, \pm N \tag{6.1}$$

where $t = \Delta t$ is the sampling time for which $p_{eq}(t)$ is maximal. Equation (6.1) is implemented by the structure shown in Figure 6.3 with the basic building block shown in Figure 6.4. From time to time, throughout the rest of the book, we will explicitly emphasize the importance of this basic building block in order to be able to motivate some basic approaches in building up a common reconfigurable platform for different technologies.

Equation (6.1) for $m = 0, \pm 1, \pm 2, \ldots, \pm N$ can be written in matrix form as $\mathbf{P}_{eq} = \mathbf{P}_c \mathbf{C}$, where $\mathbf{P}_{eq}$ and $\mathbf{C}$ are vectors or column matrices given by

$$\mathbf{P}_{eq} = \begin{bmatrix} 0 \\ 0 \\ \cdot \\ \cdot \\ 0 \\ 1 \\ 0 \\ 0 \\ \cdot \\ \cdot \\ \cdot \\ 0 \end{bmatrix} \begin{array}{l} \left.\rule{0pt}{30pt}\right\} N \text{ zeros} \\ \\ \left.\rule{0pt}{30pt}\right\} N \text{ zeros} \end{array} \qquad \mathbf{C} = \begin{bmatrix} C_{-N} \\ C_{-N+1} \\ \cdot \\ \cdot \\ \cdot \\ C_0 \\ C_1 \\ \cdot \\ \cdot \\ \cdot \\ C_N \end{bmatrix} \tag{6.2}$$

and $\mathbf{P}_c$ is the $(2N+1) \times (2N+1)$ matrix of channel responses of the form

$$\mathbf{P}_c = \begin{bmatrix} p_c(0) & p_c(-1) & \cdots & p_c(-2N) \\ p_c(1) & p_c(0) & \cdots & p_c(-2N+1) \\ p_c(2) & p_c(1) & \cdots & p_c(-2N+2) \\ \cdot & \cdot & \cdot & \cdot \\ \cdot & \cdot & \cdot & \cdot \\ \cdot & \cdot & \cdot & \cdot \\ p_c(2N) & p_c(2N-1) & \cdots & p_c(0) \end{bmatrix} \tag{6.3}$$

The solution for the equalizer coefficients is

$$\mathbf{C} = \mathbf{P}_c^{-1}\mathbf{P}_{eq} \tag{6.4}$$

given $\mathbf{P}_c^{-1} \neq \mathbf{0}$. This way we force the $\mathbf{P}_{eq}(\pm kT) = \mathbf{0}$, hence the name zero-forcing equalizer. One should be aware that in a channel with noise $\mathbf{P}_c \Rightarrow \mathbf{P}_c + \mathbf{P}_n$, where $\mathbf{P}_n$ is the noise matrix given by (6.3) where channel samples are replaced with corresponding noise samples $p_c( ) \Rightarrow p_n( )$. Equation (6.4) now results in

$$\mathbf{P}_{eq} = \mathbf{P}_c\mathbf{C} + \mathbf{P}_n\mathbf{C} = \mathbf{P}_c\mathbf{C} + \mathbf{N}_{eq} \tag{6.5}$$

$$\mathbf{C} = \mathbf{P}_c^{-1}\mathbf{P}_{eq} - \mathbf{P}_c^{-1}\mathbf{N}_{eq} \tag{6.6}$$

The second term represents the noise enhancement factor and is a limiting factor in achieving good performance.

6.1.1.1 Example

Suppose that the sample values for the channel response are

$$
\begin{array}{lll}
p_c(-5) = 0.01 & p_c(-4) = -0.02 & p_c(-3) = 0.05 \\
p_c(-2) = -0.1 & p_c(-1) = 0.2 & p_c(0) = 1 \\
p_c(1) = -0.1 & p_c(2) = 0.1 & p_c(3) = -0.05 \\
p_c(4) = 0.02 & p_c(5) = 0.005 &
\end{array}
$$

The channel response matrix is

$$
\mathbf{P}_c = \begin{bmatrix}
1.0 & 0.2 & -0.1 & 0.05 & -0.02 \\
-0.1 & 1.0 & 0.2 & -0.1 & 0.05 \\
0.1 & -0.1 & 1.0 & 0.2 & -0.1 \\
-0.05 & 0.1 & -0.1 & 1.0 & 0.2 \\
0.02 & -0.05 & 0.1 & -0.1 & 1.0
\end{bmatrix}
$$

and its inverse

$$
\mathbf{P}_c^{-1} = \begin{bmatrix}
0.996 & -0.170 & 0.117 & -0.083 & 0.056 \\
0.118 & 0.945 & -0.158 & 0.112 & -0.083 \\
-0.091 & 0.133 & 0.937 & -0.158 & 0.117 \\
0.028 & -0.095 & 0.133 & 0.945 & -0.170 \\
-0.002 & 0.028 & -0.091 & 0.118 & 0.966
\end{bmatrix}
$$

The coefficient vector is the center column of $\mathbf{P}_c^{-1}$. Therefore,

$$C_{-2} = 0.117 \quad C_{-1} = -0.158 \quad C_0 = 0.937$$
$$C_1 = 0.133 \quad C_2 = -0.091$$

The sample values of the equalized pulse response are given as

$$p_{eq}(m) = \sum_{n=-2}^{2} C_n p_c(m - n)$$

where $\Delta t = 0$. For this example we have,

$$
\begin{aligned}
p_{eq}(0) &= (0.117)(0.1) + (-0.158)(-0.1) + (0.937)(1) \\
&\quad + (0.133)(0.2) + (-0.091)(-0.1) \\
&= 1.0
\end{aligned}
$$

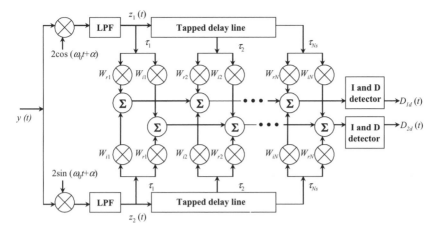

Figure 6.5 Transversal filter equalizer for QPSK.

which checks with the desired value of unity. Similarly, it can be verified that $p_{eq}(-2) = p_{eq}(-1) = p_{eq}(1) = p_{eq}(2) = 0$.

Values of $p_{eq}(n)$ for $n < -2$ or $n < 2$ are not zero. For example,

$$p_{eq}(3) = (0.117)(0.005) + (-0.158)(0.02) + (0.937)(-0.05)$$
$$+ (0.133)(0.1) + (-0.091)(-0.1)$$
$$= -0.027$$

$$p_{eq}(-3) = (0.117)(0.2) + (-0.158)(-0.1) + (0.937)(0.05)$$
$$+ (0.133)(-0.02) + (-0.091)(0.01)$$
$$= 0.082$$

6.2 LMS EQUALIZER

Let us now, in the next iteration, revisit the same problem of the channel equalization by introducing more details. We will be dealing with a QPSK signal and equalizer, as shown in Figure 6.5.

6.2.1 Signal model

The transmitted QPSK signal has the form

$$s_{tr}(t) = d_1(t) \cos \omega_0 t - d_2(t) \sin \omega_0 t \tag{6.7}$$

The received signal can be represented as

$$y(t) = s_{rec}(t) + n(t)$$
$$= s_{tr}(t) + \beta s_{tr}(t - \tau_m) + n_c(t) \cos(\omega_0 t + \alpha) - n_s(t) \sin(\omega_0 t + \alpha) \tag{6.7a}$$

We define the desired non-distorted complex signal at the receiver as

$$D(t) = (1 + \beta) [d_1(t) + j d_2(t)] \tag{6.8}$$

The cost function for the MMSE equalizer is defined as

$$C = E\lfloor |d_e(t) - D(t)|^2 \rfloor \tag{6.9}$$

If we define the complex vectors of equalizer coefficients **W** and LPF output signal samples **Z** (see Figure 6.5)

$$\mathbf{W} = \begin{bmatrix} W_1 \\ W_2 \\ \vdots \\ W_N \end{bmatrix} \qquad \mathbf{Z} = \begin{bmatrix} z(t) \\ z(t - \Delta) \\ \vdots \\ z(t - (N-1)\Delta) \end{bmatrix} \tag{6.10}$$

then the cost function becomes

$$C = E\lfloor |\mathbf{W}'\mathbf{Z} - D(t)|^2 \rfloor \tag{6.11}$$

By setting the gradient to zero we get

$$\mathbf{W}_{\text{opt}} = [E(\mathbf{Z}\mathbf{Z}')]^{-1} E(\mathbf{Z}D^*(t))$$

A sample of performance results is shown in Figure 6.6.

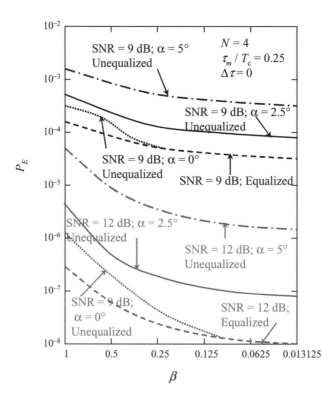

Figure 6.6 Error probability versus amplitude of a specular component for detection of equalized QPSK.

6.2.2 Adaptive weight adjustment

Setting the tap coefficients of the zero-forcing and LMS equalizers involves the solutions of a set of simultaneous equations. In the case of the zero-forcing equalizer, adjustment of the tap coefficients involves measuring the channel filter output at T second spaced sampling times in response to a test pulse, and solving for the tap gains. In the case of the LMS equalizer, these equations involve data, noise and multipath dependent parameters, which may be difficult to determine or may not be known at all. Known data assumes a kind of preamble (midamble) embedded into the data.

6.2.3 Automatic systems

So-called *preset* algorithms, use a training sequence. In *adaptive* algorithms, adjustment of the coefficients is performed continuously during data transmission.

A zero-forcing equalizer can be solved iteratively for the coefficient vector $\mathbf{C}$. Let $\mathbf{C}$ at the kth iteration be $\mathbf{C}^{(k)}$, then the error in the solution is

$$\mathbf{E}^{(k)} = \mathbf{P}_c\mathbf{C}^{(k)} - \mathbf{P}_{eq} \tag{6.12}$$

where $\mathbf{E}^{(k)}$ is a vector with $2N + 1$ components, each of which represent the error in a component of $\mathbf{C}^{(k)}$. Each component of $\mathbf{C}^{(k)}$ can be adjusted in accordance with the error in it.

6.2.4 Iterative algorithm

If A is a small positive constant, the adjustment algorithm for the jth component of $\mathbf{C}^{(k)}$ is

$$C_j^{(k+1)} = C_j^{(k)} - A\mathrm{sgn}\left(E_j^{(k)}\right) \tag{6.13}$$

The iteration process is continued until $C_j^{(k+1)}$ and $C_j^{(k)}$ differ by some suitable small increment. The algorithm converges under fairly broad restrictions.

6.2.5 The LMS algorithm

The gradient of the mean square error with respect to the jth tap gain is twice the *correlation* between the equalizer output and the error between actual and desired outputs. Two problems arise in applying it to adaptive weight adjustment. First, it requires the expectation or average to be taken. Since this is not available, the unbiased but noisy estimate $z^*(t - j\Delta) e(t)$ can be used. Secondly, the undistorted output $D(t)$ is not available unless a known data sequence is transmitted.

An alternative to sending a known data sequence is to assume that the detected data is correct (which is true even in a fairly bad channel giving an error probability of only 10^{-2}) and using the detected data to reconstruct an estimate of the undistorted output $D(t)$. Equalizers using this method of data estimation are called *decision-directed*.

A suitable decision-directed algorithm for weight adjustment of the LMS equalizer is

$$\mathbf{W}(n+1) = \mathbf{W}(n) - A\mathbf{Z}_T^* \left[d_e(t_n - \Delta) - D_d(t_n)\right] \tag{6.14}$$

More details on LMS-based equalizers can be found in [501–507].

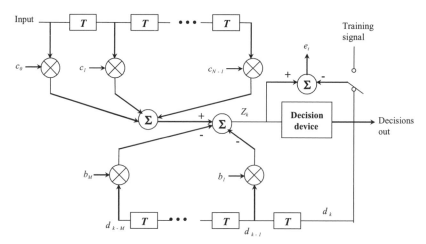

Figure 6.7 Decision feedback equalizer.

6.2.6 Decision feedback equalizer (DFE)

Non-linear equalizer structures may provide better performance under many circumstances. A simple non-linear equalizer is the decision feedback equalizer (DFE) which uses feedback of decisions on symbols already received to cancel the interference from symbols which have already been detected.

The basic idea is that, assuming past decisions are correct, the ISI contributed by these symbols can be canceled exactly by subtracting appropriately weighted past symbol values from the equalizer output. This is the purpose of the delay line with weights $b_1, b_2, \ldots, b_M$ shown in Figure 6.7. The forward transversal filter, with weights $c_0, c_1, \ldots, c_{N-1}$, then compensates for ISI over a smaller portion of the ISI-contaminated received signal. Both the feedback and feedforward coefficients can be adjusted simultaneously to minimize the mean square error.

Decision feedback equalizers are discussed in [508–541].

6.2.7 Blind equalizers

What we need for designing a blind equalizer is to *recover the transmitted message* (a_t) *from the received one* (x_t) *only, without any preamble for identification of the unknown channel.*

The equivalent system models are given in Figure 6.8 and Figure 6.9. The cost function defined by Equation (6.11) is now modified in such a way that it does not require knowledge of data. Options are presented by Equations (6.15) and (6.16). The remaining details are the same as in the previous discussion.

$$J(W) = E\left(\frac{1}{2}c_t^2(W) - \alpha\,|c_t(W)|\right)$$

$$\alpha = \frac{Ea_t^2}{E|a_t|}$$

(6.15)

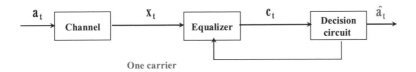

One carrier

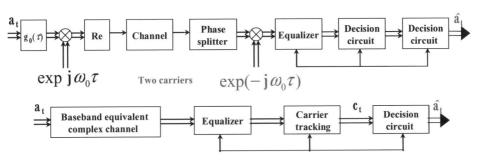

Two carriers

Baseband equivalent

Figure 6.8 System models.

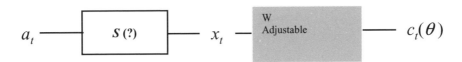

Figure 6.9 Equivalent model of the system from Figure 6.8.

For complex signals, parameters a_t, x_t, c_t, S and W in Figure 6.9 are complex. Re a_t and Im a_t are independent and the cost function is defined as

$$J(W) = E(\psi(\text{Re } c_t(W)) + \psi(\text{Im } c_t(W)))$$

$$\psi(x) = \frac{1}{2}x^2 - \alpha|x| \quad \text{(Sato function)} \tag{6.16}$$

$$\alpha = \left(\int x^2 v(\mathrm{d}x)\right)\left(\int |x| v(\mathrm{d}x)\right)^{-1}$$

Blind equalizers are discussed in [542–564].

6.3 DETECTION FOR A STATISTICALLY KNOWN, TIME VARYING CHANNEL

The system model is given in Figure 6.10. Parameters $h_T(\tau)$, $c(t, \tau)$ and $h_R(\tau)$ represent transmitter filter, channel and receive filter impulse responses respectively. The overall system pulse response is a convolution $f(\tau; t)$ of the three and is given by Equation (6.17).

$$f(\tau; t) = h_T(\tau) * c(\tau; t) * h_R(\tau)$$

$$r(kT_s) = r_k = \sum_n a_n f(kT_s - nT; kT_s) + w_k \tag{6.17}$$

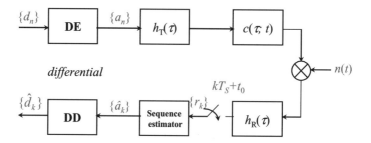

Figure 6.10 Block diagram of the communication system under consideration, in complex envelope form.

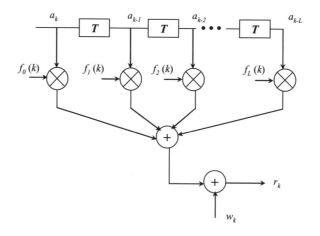

Figure 6.11 Tapped delay line model of the equivalent discrete time channel for the TS case. The blocks containing 'T' denote delays T_s.

Equation (6.17) also represents the samples of the overall received signal at sampling points kT_s. The sampling interval is T_s and the symbol interval is T.

6.3.1 Signal model

The convolution can also be presented as:

$$r_k = \sum_{n=0}^{L} a_{l-n} f_i(k) + w_k \tag{6.18}$$

For the T-spaced sampling (TS) case, we have $l = k$, and $i = n$; for the fractional spaced sampling (FS) case, $l = [k/2]$, and $i = 2n + m$, with $m = (k + 1)$ mod 2, i.e. $m = 1$ for k even, and $m = 0$ for k odd. Tapped delay line system models for TS and FS are given in Figures 6.11 and 6.12.

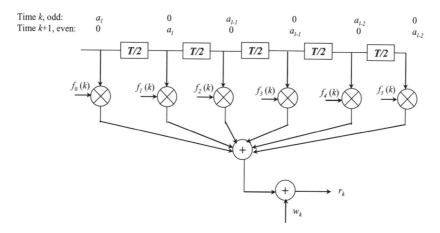

Figure 6.12 Tapped delay line model of the equivalent discrete time channel for the FS case and a channel impulse response length of $3T$. The blocks containing '$T/2$' denote delays of $T/2$ s, $l = [k/2]$, and $[x]$ denotes the smallest integer $\geq x$.

6.3.2 Channel model

For the channel model we assume:

- The worst case channel in terms of T_M and f_D.

- For the worst case value of T_M we use 20 μs.

- With the IS-54 symbol duration of $T = 1/24\,000 \approx 41.7$ μs, this corresponds to roughly $T_M = T/2$, i.e. the echo delay is half a symbol duration.

- For the impulse response shape, we use a 'double-spike' spaced by T_M : $c(\tau;t) = c_0(t)\delta(\tau) + c_1(t)\delta(\tau - T_M)$ where $E[|c_0(t)|^2] = E[|c_1(t)|^2] = 0.5$ so that the average energy of the channel is normalized to one.

- The response $f(\tau;t)$ is then

$$f(\tau;t) = c_0(t)h(\tau) + c_1(t)h(\tau - T_M)$$

- $h(\tau)$ is the full raised cosine response, equal to the convolution of $h_T(\tau)$ and $h_R(\tau)$.

- We set t equal to nT_s and t equal to kT_s (assuming the sampling phase $t_0 = 0$).

- For the TS case, we have, with $T_s = T$,

$$f_0^{TS}(k) = c_0(k)h(0) + c_1(k)h(T/2)$$
$$f_n^{TS}(k) = c_1(k)h((2n-1)T/2), \quad n = 1, 2, \dots$$

- The symmetry of $h(\tau)$ was used.

- For the TS case, we truncate the number of taps to $L + 1 = 3$. The normalized average tap energies are then $E[|f_0|^2] = 0.7717$, $E[|f_1|^2] = 0.2136$, and $E[|f_2|^2] = 0.0147$.

In the FS case, with $t = nT/2$ and $k = kT/2$

$$f_0^{\text{FS}}(k) = c_0(k)h(0) + c_1(k)h(T/2)$$
$$f_1^{\text{FS}}(k) = c_1(k)h(T/2) + c_1(k)h(0)$$
$$f_{2n}^{\text{FS}}(k) = c_1(k)h((2n-1)T/2) \quad n = \pm1, \pm2, \ldots$$
$$f_{2n+1}^{\text{FS}}(k) = c_0(k)h((2n+1)T/2) \quad n = \pm1, \pm2, \ldots$$

- We retain the most significant taps, yielding $2(L+1) = 6$.

- The average (normalized) tap energies here are $E[|f_{-2}|^2] = E[|f_3|^2] = 0.00735$, $E[|f_{-1}|^2] = E[|f_2|^2] = 0.1068$, and $E[|f_0|^2] = E[|f_1|^2] = 0.38585$.

- In both the TS and FS cases, we choose the minimum number of taps such that the truncated response contained at least 98 % of the energy of the untruncated response.

- The autocorrelation of the continuous time processes $c_0(t)$ and $c_1(t)$ is $r_c(\tau) = J_0(2\pi\tau f_D)$, where $J_0(x)$ is the zeroth order Bessel function of the first kind.

- For this model, for a vehicle speed of 30 m/s (67.1 mph) and a carrier frequency of 900 MHz, the maximum Doppler shift $f_D \approx 90$ Hz.

- For simulation purposes, the autocorrelation is approximated by something more easily synthesized, namely, the inverse Fourier transform of a Chebyshev Type I magnitude squared frequency response.

- For time separations $\tau \leq 50T$, the Chebyshev filter yields a very good approximation to the desired autocorrelation.

6.3.3 Statistical description of the received sequence

The following assumptions are made

- the received sequence $\mathbf{r}_N = (r_1, r_2, \ldots, r_N)^{\text{T}}$

- the transmitted sequence $\mathbf{a}_N = (a_1, a_2, \ldots, a_N)^{\text{T}}$

- the pdf is complex Gaussian

$$p(\mathbf{r}_N|\mathbf{a}_N) = \frac{1}{\pi \det|\mathbf{C}(\mathbf{a}_N)|} \cdot \exp\left(-\mathbf{r}_N^{\text{H}} \mathbf{C}^{-1}(\mathbf{a}_N) \mathbf{r}_N\right) \tag{6.19}$$

- $\mathbf{C}(\mathbf{a}_N)$ denotes the covariance matrix of $\mathbf{r}_N$ given $\mathbf{a}_N$.

- The superscript H denotes Hermitian (conjugate transpose).

- The elements of $\mathbf{C}(\mathbf{a}_N)$, abbreviated C_a, are obtained by taking the expectation $E[\mathbf{r}_N \cdot \mathbf{r}_N^{\text{H}}|\mathbf{a}_N]$:

$$c_{ij} = \sum_{n=0}^{L} \sum_{m=0}^{L} a_{i-n} a_{j-m}^* E[f_n(i) f_m^*(j)] + \sigma_w^2 \delta_{ij} \tag{6.20}$$

The assumption that the channel is statistically known means that $E\lfloor f_n(i) f_m^*(j)\rfloor$ are known.

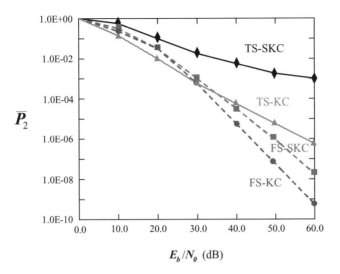

Figure 6.13 Plots of union upper bounds on the average sequence error probability versus E_b/N_0 for the TS and FS SKC detectors, and for the TS and FS known channel detector, for block length $N = 4$, using binary PSK modulation. Channel parameters are $f_D = 90$ Hz, $T_M = T/2$, and $t_0 = 0$.

6.3.4 The ML sequence (block) estimator for a statistically known channel

The maximum likelihood sequence estimator (MLSE) $\hat{\mathbf{a}}$ is defined as

$$\hat{\mathbf{a}} = \arg\min_{a} \Lambda(\mathbf{a}, \mathbf{r}) \tag{6.21}$$

where $\Lambda(\ )$ is the logarithm of Equation (6.19). The sequence metrics are

$$\Lambda(\mathbf{a}, \mathbf{r}) = \ln|C_{\mathbf{a}}| + \mathbf{r}^H C_{\mathbf{a}}^{-1} \mathbf{r} \tag{6.22}$$

Matrix inversion $\mathbf{C}^{-1}$ can be calculated by Cholesky decomposition (factorization). For a given vector of samples $\mathbf{r}$ and the known channel, the algorithm will search for $\mathbf{a}$ that minimizes Equation (6.21). Some results are given in Figures 6.13–6.17.

6.4 LMS-ADAPTIVE MLSE EQUALIZATION ON MULTIPATH FADING CHANNELS

Within this section we are going to drop the assumption that the channel is statistically known and deal with the problem where both data and channel have to be estimated simultaneously.

6.4.1 System and channel models

The system model is given in Figure 6.18.

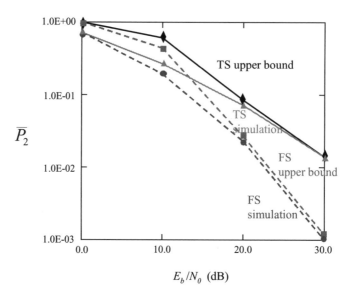

Figure 6.14 Plots of the union upper bounds and simulated average sequence error probability versus E_b/N_0 for the TS and FS SKC detectors, for block length $N = 4$, and binary PSK modulation. The channel parameters are $f_D = 90$ Hz, $T_M = T/2$, and $t_0 = 0$.

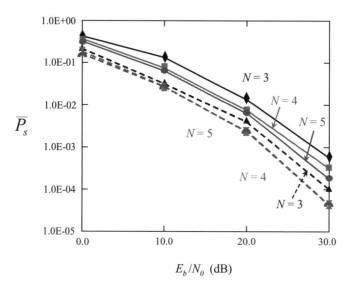

Figure 6.15 Plots of simulated average symbol error probability $\bar{P}_s$ versus E_b/N_0 for the FS SKC (solid lines) and FS KC (dashed lines) detectors, for block length $N = 3, 4, 5$ and binary PSK modulation. The channel parameters are $f_D = 90$ Hz, $T_M = T/2$, and $t_0 = 0$.

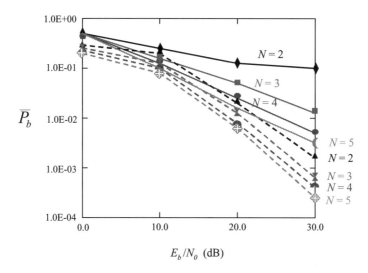

Figure 6.16 Plots of simulated average BEP $\bar{P}_b$ (after differential decoding) versus E_b/N_0 for the TS (solid lines) and FS (dashed lines) SKC detectors, for block length $N = 2$–5, and binary PSK modulation. The channel parameters are $f_D = 90$ Hz, $T_M = T/2$, and $t_0 = 0$.

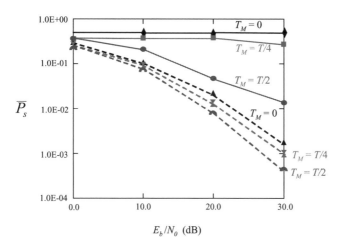

Figure 6.17 Plots of simulated average symbol error probability $\bar{P}_s$ versus E_b/N_0 for the TS and FS SKC detectors, for block length $N = 4$ and binary PSK modulation, showing the effect of mismatch between the estimated delay spread, $\hat{T}_M$, and the actual delay spread T_M. For all cases, $\hat{T}_M = T/2$, but the actual T_M is zero, $T/4$ and $T/2$. The other channel parameters are $f_D = 90$ Hz and $t_0 = 0$. Solid curves are TS results, dashed curves are FS results.

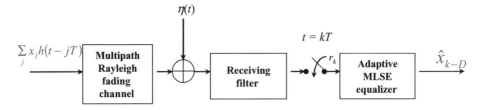

Figure 6.18 System model.

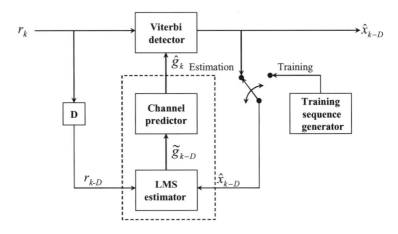

Figure 6.19 Adaptive MLSE equalizer.

The received signal can be represented as

$$r_k = \sum_{i=0}^{L} x_{k-i} g_{k,i}^* + \eta_k$$

$$r_k = \mathbf{g}_k^H \mathbf{x}_k + \eta_k \qquad (6.23)$$

where

$$\mathbf{g}_k = [g_{k,0}, \ g_{k,1}, \dots, g_{k,L}]^T, \quad \mathbf{x}_k = [x_k, \ x_{k-1}, \dots, x_{k-L}]^T$$

$$\frac{1}{2} E\{g_{k,i} \, g_{k-l,j}^*\} = R_g(l, i, j)$$

Parameter $g_{k,i}$ represents the samples of the equivalent channel impulse response. The operation of the system is presented in Figure 6.19. For data estimation by the Viterbi algorithm, the system uses the channel estimates. The Viterbi algorithm delay is DT. This class of algorithms is known as *delayed decision-directed equalization.*

The metric used in the Viterbi detector is $|r_k - \sum_{i=0}^{L} x_{k-i} \hat{g}_{k,i}^*|^2$

6.4.2 Adaptive channel estimator and LMS estimator model

Assuming that the feedback decisions are correct, i.e. $\hat{x}_{k-D} = x_{k-D}$ at the input of the LMS estimator, and the order of the LMS estimator is $L + 1$ the following step is implemented:

the LMS estimator updates $\tilde{g}_{k-D}$ each time by:

$$\tilde{g}_{k-D} = \tilde{g}_{k-D-1} + \mu \mathbf{x}_{k-D}\alpha^{*}_{k-D}$$

where

$$\alpha_{k-D} = r_{k-D} - \tilde{g}^{H}_{k-D-1}\mathbf{x}_{k-D} \tag{6.24}$$

6.4.3 The channel prediction algorithm

The linear prediction generates

$$\hat{\mathbf{g}}_k = \sum_{j=0}^{q} \alpha_j \tilde{g}_{k-D-j} \tag{6.25}$$

where $\alpha_0, \dots, \alpha_q$ are constants and q is a positive integer. Due to delay DT of the Viterbi algorithm, the straight line extrapolator is used

$$\hat{\mathbf{g}}_k = \tilde{g}_{k-D} + \frac{p}{q}(\tilde{g}_{k-D} - \tilde{g}_{k-D-q}) \tag{6.26}$$

where p is the prediction step. Although there exist many other channel predictions from simulation results and comparisons, the straight line extrapolator gives fair performance and very simple implementation. If the autocorrelation function $R_g(l, i, j)$ is known by the receiver, the prediction coefficients α_j can be obtained from the Wiener solution. Some performance results are given in Figures 6.20–6.24.

MLSE equalizers are discussed in [565–575].

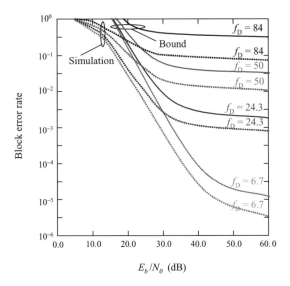

Figure 6.20 Analytical and simulation results for adaptive MLSE employing straight line extrapolation prediction on a two-tap Rayleigh fading channel with $f_D = 6.7$, 24.3, 50 and 84 Hz.

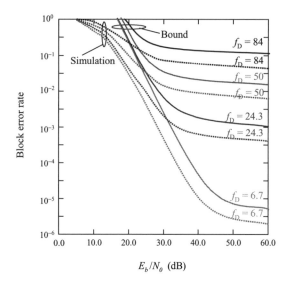

Figure 6.21 Analytical and simulation results for adaptive MLSE employing linear channel prediction on a two-tap Rayleigh fading channel with $f_D = 6.7, 24.3, 50$ and 84 Hz.

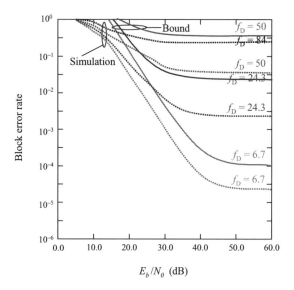

Figure 6.22 Analytical and simulation results for adaptive MLSE without channel prediction on a two-tap Rayleigh fading channel with $f_D = 6.7, 24.3, 50$ and 84 Hz. (Note that the analytical bound of the case of $f_D = 84$ Hz is greater than 1 and hence not shown.)

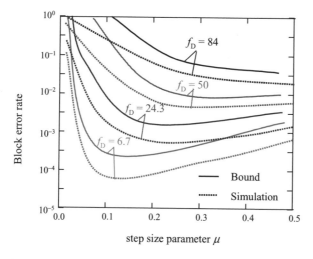

Figure 6.23 Analytical and simulation results obtained by varying the step size parameter μ at $E_b/N_0 = 35$ dB.

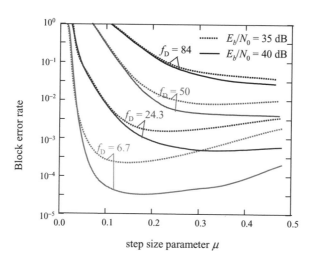

Figure 6.24 Analytical results obtained by varying the step size parameter μ at $E_b/N_0 = 35$ dB and 40 dB.

6.5 ADAPTIVE CHANNEL IDENTIFICATION AND DATA DEMODULATION

The estimation technique presented in Section 6.4 is rather inefficient when detection delay D or Doppler f_D increase. In such examples, a joint estimation of both data and channel would give better results. Unfortunately, pure joint estimation would be too complex. In this section we present an algorithm where the joint ML function is maximized by alternating the maximization process with respect to data (given the channel) and channel (given data from the previous iteration).

6.5.1 System model

The received signal is given as

$$r_k = \sum_{l=0}^{L} h_l(k)a_{k-l} + n_k, \quad k = 1, 2, \ldots, N$$

$$\mathbf{h}(k) = [h_0(k), \ h_1(k), \ldots, h_L(k)]^{\mathrm{T}}$$

(6.27)

The memory L of the channel is determined by the time spread $T_{\max}$ of the actual channel and the symbol period T_s:

$$L = \left[\frac{T_{\max}}{T_s}\right] + 1$$

6.5.2 Joint channel and data estimation

The approach used in this section is based on [576]. The process is described by the following steps.

6.5.2.1 *Block sequence estimation (BSE)*

- The received sequence is fed into the metric computation unit in blocks of N_b symbols at a time.

- The data demodulation is also performed in blocks, rather than symbol by symbol as in the conventional VA.

- At time k the receiver processes a block of N_b data symbols $(r_k, \ldots, r_{k+Nb-1})$.

- After computing the path metrics at time $k + N_b - 1$, the survivor paths of each state are traced backwards in the trellis in order to detect a merge.

- If a merge occurs within the block, at the time $k + N_b - 1 - \delta$ ($\delta < N_b$), decisions are made on the data sequence $\{a_k, \ldots, a_{k+Nb-1-\delta}\}$.

- The initial point of the next block is set at time $k + N_b - 1 - \delta$ and the next N_b data symbols are processed.

- This means that the portion of the received symbols $(r_{k+Nb-\delta}, \ldots, r_{k+Nb-1})$ is processed twice.

- The state metrics at time $k + N_b - 1 - \delta$ are reset to a large number except for the metric of the state at which the merge occurred. This metric is set to zero. This way the chances of errors associated with illegal state transitions at the beginning of the new block are reduced. It is equivalent to restarting the VA from a known initial state.

- If a merge is not detected within the block ($\delta < N_b$), the state with the minimum metric at the end of the block (at the time $k + N_b - 1$) is chosen and its survivor path is traced backwards in the trellis.

- A merge is assumed at time $k + N_b - \delta$ ($\delta = \lceil N_b/2 \rceil$) on the best survivor path.

- The state metrics at time $k + N_b - 1 - \delta$ are reset to a large number, except for the metric of the best survivor. This metric is set to zero.

6.5.2.2 Block adaptive channel estimation

In the derivation of the adaptive schemes for the CIR estimation we will temporarily assume that the channel is unknown but static. The discrete time index k will be dropped from Equation (6.27) to simplify the notation. We will revisit the time varying CIR scenario in subsequent sections. In vector notation, Equation (6.27) can be written as:

$$\mathbf{r} = \mathbf{A}\mathbf{h} + \mathbf{n} \tag{6.28}$$

where

$$\mathbf{r} = [r_1, \ldots, r_N]^\mathrm{T}, \mathbf{h} = [h_0, \ldots, h_L]^\mathrm{T}, \mathbf{n} = [n_1, \ldots, n_N]^\mathrm{T}, \quad \text{and}$$

$$\mathbf{A} = \begin{bmatrix} a_1 & a_0 & \cdots & a_{1-L} \\ a_2 & a_1 & \cdots & a_{2-L} \\ \vdots & \vdots & \vdots & \vdots \\ a_N & a_{N-1} & \cdots & a_{N-L} \end{bmatrix} \tag{6.29}$$

Let us use $\mathbf{a} = [a_{1-L}, \ldots, a_N]^\mathrm{T}$ to denote the transmitted data vector.

6.5.2.3 Maximum likelihood (ML)

ML estimates of the channel $\mathbf{h}_{\mathrm{ML}}$ and data $\mathbf{a}_{\mathrm{ML}}$ are those that maximize the conditional probability density function (pdf) $p(\mathbf{r}|\mathbf{a}, \mathbf{h})$ or, equivalently, minimize $|\mathbf{r} - \mathbf{A}\mathbf{h}|^2$. With respect to $\mathbf{A}$ and $\mathbf{h}$

$$(\mathbf{a}_{\mathrm{ML}}, \mathbf{h}_{\mathrm{ML}}) = \underset{a \in A^{N+L-1}, \mathbf{h} \in R^L}{\arg\min} \; |\mathbf{r} - \mathbf{A}\mathbf{h}|^2 \tag{6.30}$$

The joint minimization over $\mathbf{a}$ and $\mathbf{h}$ cannot be computed in closed form. When either $\mathbf{a}$ or $\mathbf{h}$ is fixed, this minimization is a well known problem. When the channel $\mathbf{h}$ is fixed, the ML estimate of $\mathbf{a}$ is computed via the VA. When the data $\mathbf{a}$ is given, the minimization of $C(\mathbf{A}, \mathbf{h}) = |\mathbf{r} - \mathbf{A}\mathbf{h}|^2$ over $\mathbf{h}$ is a standard least squares problem which results in

$$\mathbf{h}_{\mathrm{ML}} = (\mathbf{A}^H \mathbf{A})^{-1} \mathbf{A}^H \mathbf{r} \tag{6.31}$$

Thus, the joint minimization of $C(\mathbf{A}, \mathbf{h})$ over $\mathbf{A}$ and $\mathbf{h}$ can be viewed as an alternative minimization type of problem.

6.5.2.4 Iterative procedure

The above algorithms can be summarized as:

1. Start with an initial estimate $\hat{\mathbf{h}}(0)$ of the channel.

2. Minimize $C(\hat{\mathbf{A}}, \hat{\mathbf{h}})$ with respect to $\hat{\mathbf{A}}$ via the VA to obtain $\hat{\mathbf{A}}(l)$.

3. Minimize $C(\hat{\mathbf{A}}, \hat{\mathbf{h}})$ w.r.t. $\hat{\mathbf{h}}(l)$ to obtain

$$\hat{\mathbf{h}}(l+1) = [\hat{\mathbf{A}}^H(l)\hat{\mathbf{A}}(l)]^{-1}\hat{\mathbf{A}}^H(l)\mathbf{r} \tag{6.32}$$

4. Repeat Steps 2 and 3 until the algorithm converges.

Simulation studies have shown that this scheme can converge very fast to a global minimum (in approximately 5–6 iterations).

The above algorithm pays a heavy price as a tradeoff for its optimum performance. The receiver has to process a long received sequence (in the order of a few thousands of symbols) several times before it can provide the user with any reliable decisions about the transmitted data. This implies a significant decision delay. The algorithm is impractical for real-time implementation, especially when the channel is time varying. This is a strong motivation for constructing algorithms to perform the joint channel and data estimation recursively in time, which provides the user with reliable data estimates quickly enough to be able to adopt to channel variations too.

6.5.2.5 *Iterative CIR estimator*

In this case we have

$$\hat{\mathbf{h}}(l+1) = \hat{\mathbf{h}}(l) + \mu \left. \frac{\partial[\ln p(\mathbf{r}|\hat{\mathbf{A}}, \mathbf{h})]}{\partial \mathbf{h}} \right|_{\mathbf{h}=\hat{\mathbf{h}}(l)}$$
$$= \hat{\mathbf{h}}(l) + \mu\hat{\mathbf{A}}^H[\mathbf{r} - \hat{\mathbf{A}}\hat{\mathbf{h}}(l)] \tag{6.33}$$

If we can achieve perfect convergence, then

$$\hat{\mathbf{h}}(l+1) = \hat{\mathbf{h}}(l) \Leftrightarrow \hat{\mathbf{A}}^H[\mathbf{r} - \hat{\mathbf{A}}\hat{\mathbf{h}}(l)] = 0 \Leftrightarrow \mathbf{h}(l)$$
$$= (\hat{\mathbf{A}}^H\hat{\mathbf{A}})^{-1}\hat{\mathbf{A}}^H\mathbf{r} \tag{6.34}$$

which is exactly the ML estimate of the channel. In Equation (6.32), $\hat{\mathbf{A}}$ and $\mathbf{r}$ extend over the whole data record. If, instead, we consider only a portion of data, i.e. a block of N_b symbols, then each time we update the CIR estimate we can perform the joint ML estimation of data and channel recursively in time. Notice that Equation (6.33) has the form of a block least mean square (BLMS) equation. This is a consequence of the quadratic exponent of the Gaussian probability distribution of the noise vector. Given the CIR estimate $\hat{\mathbf{h}}(l)$ at the lth recursion, the corresponding block of data estimates can be provided by the BSE, as was developed earlier. Starting from an initial guess, $\hat{\mathbf{h}}(0)$, for the CIR we construct a joint data and channel estimation scheme which operates on successive blocks of data in a time recursive manner.

6.5.2.6 *BSE/BLMS algorithm*

In this case:

1. Start with an initial estimate $\hat{\mathbf{h}}(0)$ of the CIR. Initialize the block counter ($l \rightarrow 1$) and the symbol counter ($k \rightarrow 1$).

2. Receive the lth block of data $\mathbf{r}(l) = [r_k, \ldots, r_{k+N_b-1}]^T$, where N_b is the block length.

3. Apply the BSE on $\mathbf{r}(l)$, using $\hat{\mathbf{h}}(l-1)$ for metric computations, to obtain the lth block of data estimates:

$$\hat{\mathbf{a}}(l) = [\hat{a}_k, \ldots, \hat{a}_{k+N_b-1-\delta(l)}]^\mathsf{T}$$

where $\delta(l)$ is the merging delay of the BSE for the lth block.

4. Using the data estimates $\hat{\mathbf{a}}(l)$ and the first $N_b - \delta(l)$ entries of the lth block of received symbols $\mathbf{r}_\delta(l) = [r_k, \ldots, r_{k+N_b-1-\delta(l)}]^\mathsf{T}$, update the CIR estimate as follows:

$$\hat{\mathbf{h}}(l) = \hat{\mathbf{h}}(l-1) + \mu \hat{\mathbf{A}}^\mathsf{H}(l)[\mathbf{r}_\delta(l) - \hat{\mathbf{A}}(l)\hat{\mathbf{h}}(l-1)] \qquad (6.35)$$

The matrix $\hat{\mathbf{A}}(l)$ is given by:

$$\hat{\mathbf{A}}(l) = \begin{bmatrix} \hat{a}_k & \hat{a}_{k-1} & \cdots & \hat{a}_{k-L} \\ \hat{a}_{k+1} & \hat{a}_k & \cdots & \hat{a}_{k+1-L} \\ \vdots & \vdots & \vdots & \vdots \\ \hat{a}_{k+N_b-1-\delta(l)} & \hat{a}_{k+N_b-2-\delta(l)} & \cdots & \hat{a}_{k+N_b-1-\delta(l)-L} \end{bmatrix} \qquad (6.36)$$

and μ is a step size parameter.

5. Reset the block and symbol counters to $l \to l+1$ and $k \to k + N_b - \delta(l)$ and go to Step 2 until all data have been processed. An issue is the choice of the step size parameter μ. A large value of μ would provide faster adaptation but could lead to divergence of the channel estimate. On the other hand, if μ is too small, we will lose a lot of data while trying to acquire the channel.

6.5.2.7 *Recursive least squares (RLS) channel estimation*

In this case the CIR estimation is based on the minimization of the weighted sum of the squared errors

$$J(k) = \sum_{l=0}^{k} \lambda^{k-l} e^2(l) = \sum_{l=0}^{k} \lambda^{k-l} |r_l - \hat{\mathbf{h}}^\mathsf{T}(k) \cdot \hat{\mathbf{a}}(l)|^2 \qquad (6.37)$$

where $\hat{\mathbf{h}}(k) = \lfloor \hat{h}_0(k), \ldots, \hat{h}_L(k) \rfloor^\mathsf{T}$ is the CIR estimate at time k, $\hat{\mathbf{a}}(l) = [\hat{a}_l, \ldots, \hat{a}_{l-L}]^\mathsf{T}$ is the vector of the estimated data at times $l, \ldots, l-L$ and r_l is the received signal at time l. The parameter λ, $(0 < \lambda \le 1)$ gives more weight to recent errors. Thus, we allow for time varying channels. The RLS algorithm for the CIR estimation is described as [577–579]:

$$g(k) = \frac{\mathbf{P}(k)\hat{\mathbf{a}}^*(k)}{\lambda + \hat{\mathbf{a}}^\mathsf{T}(k)\mathbf{P}(k)\hat{\mathbf{a}}^*(k)}$$

$$\mathbf{P}(k+1) = \frac{1}{\lambda}[\mathbf{P}(k) - g(k)\hat{\mathbf{a}}^\mathsf{T}(k)\mathbf{P}(k)]$$

$$e(k) = r_k - \hat{\mathbf{h}}^\mathsf{T}(k)\hat{\mathbf{a}}(k) \qquad (6.38)$$

$$\hat{\mathbf{h}}(k+1) = \hat{\mathbf{h}}(k) + e(k)g(k)$$

$$k = 1, 2, \ldots$$

To initialize the algorithm we set $\mathbf{P}(0) = \varepsilon^{-1}\mathbf{I}$, where ε is a small positive constant.

6.5.2.8 BSE/RLS algorithm

Incorporating the CIR update into the recursive joint channel and data estimation algorithm gives:

1. Start with an initial estimate $\hat{\mathbf{h}}(0)$ of the CIR. Initialize the block counter ($l \to 1$) and the symbol counter ($k \to 1$).

2. Receive the lth block of data $\mathbf{r}(l) = [r_k, \ldots, r_{k+N_b-1}]^T$, where N_b is the block length.

3. Apply the BSE on $\mathbf{r}(l)$, using $\hat{\mathbf{h}}(l-1)$ for metric computations, to obtain the lth block of data estimates:

$$\hat{\mathbf{a}}(l) = [\hat{a}_k, \ldots, \hat{a}_{k+N_b-1-\delta(l)}]^T,$$

where $\delta(l)$ is the merging delay of the BSE for the lth block.

4. Using the data estimates $\hat{\mathbf{a}}(l)$ and the first $N_b - \delta(l)$ entries of the lth block of received symbols $\mathbf{r}_\delta(l) = [r_k, \ldots, r_{k+N_b-1-\delta(l)}]^T$, obtain the new CIR estimate $\hat{\mathbf{h}}(l)$ by executing the RLS recursion $N_b - \delta(l)$ times.

5. Reset the block and symbol counters to $l \to l+1$ and $k \to k + N_b - \delta(l)$ and go to Step 2 until all data have been processed.

A schematic diagram of adaptive BSE (BSE/BLMS or BSE/RLS) is shown in Figure 6.25.

The adaptive BSE derives its strength from the fact that, due to the variable decision delay, the data estimates used for CIR tracking are more likely to be taken from the ML path. The block processing of the data contributes to the averaging of the effects of symbol errors. Thus, the CIR estimator is fed with better data estimates, which improves its tracking capability compared to the conventional adaptive MLSE.

The possible data errors towards the end of each block, caused by computing the metrics for the whole block using the same CIR estimate, are alleviated by the fact that the last symbols of each block are processed again with the updated CIR estimate. If the fading rate gets higher, the block length N_b must be chosen carefully so that the CIR estimate is not outdated. A choice of $N_b \approx 5L$ is a good rule of thumb.

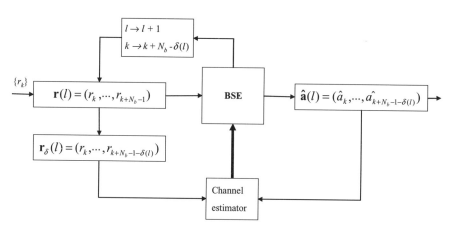

Figure 6.25 Schematic diagram for adaptive BSE [576] © 1998, IEEE.

6.5.3 Data estimation and tracking for a fading channel

In a static channel environment, the BSE/BLMS and BSE/RLS as described so far, operate starting from an initial guess $\hat{\mathbf{h}}(0)$ of the CIR. For a fading channel, the task of CIR acquisition is much heavier, especially when the fading rate becomes high. In such environments, the CIR acquisition is accomplished via a training sequence, which is periodically sent to the receiver as a portion of a fixed size data packet. This format is used in mobile communications. An example of such a data structure is the time division multiple access (TDMA) slot of the IS-54 North American Digital Cellular standard described briefly in Chapter 1.

6.5.3.1 Training

The BSE/BLMS, BSE/RLS can be applied in this signaling format with an appropriate adjustment in the CIR acquisition step. This is accomplished by feeding the above described adaptive algorithms (BLMS or LMS, RLS) with the training symbols at the header of the TDMA slot, producing the CIR estimate $\hat{\mathbf{h}}(0)$. As soon as the acquisition stage is over, the BSE/BLMS or BSE/RLS is activated for joint data estimation and channel tracking, as described in the previous subsections, for the duration of the entire TDMA slot. The performance of this scheme, obtained by simulation, is now presented.

6.5.3.2 Performance and computational complexity

In the evaluation of the system performance the same assumptions are used as in [576]:

- Static and fading channel environments.

- To cover both the static and time varying CIR, the SNR is defined as

$$\text{SNR} = 10 \log \left[\frac{\sigma_a^2}{\sigma^2} \sum_{i=0}^{L} E(|h_i|^2) \right]$$

 where σ_a^2 and σ^2 are the variances of the input data and additive noise, respectively.

- The above definition implies independence of the CIR variations from the transmitted data sequence, which is quite a reasonable assumption.

6.5.4 The static channel environment

- CIR is assumed constant for the duration of the entire received data record.

- In the simulation there is no training sequence, i.e. the BSE operates in a blind fashion.

- The modulation format is binary phase shift keying (BPSK).

- No coding is assumed.

- The input alphabet is $\{-1, 1\}$, i.e. the transmitted data sequence $\{a_k\}$ consists of real numbers.

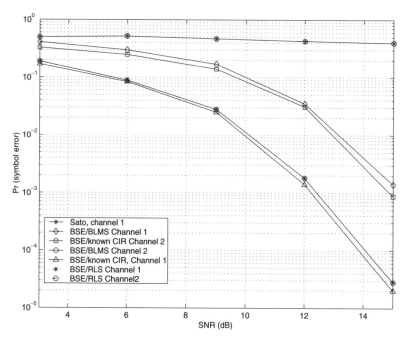

Figure 6.26 Probability of symbol error versus SNR using the BPSK waveform for the BSE/BLMS and the BSE/RLS algorithms over Channels 1 and 2.

- Two particular channel examples are used whose impulse responses are given by the vectors $\mathbf{h}_1 = [0.407, 0.815, 0.407]$ and $\mathbf{h}_2 = [0.1897, 0.5097, 0.6847, 0.46, 0.1545]$, respectively.

- Both channels exhibit deep nulls in their magnitude response within the frequency band of interest.

- Channel 2 has non-linear phase characteristics.

- As a result, linear equalization methods exhibit very poor performance on these channels.

- The MLSE for the corresponding *known* channel is used in Figure 6.26 for comparison

6.5.4.1 Simulations/BSE/BLMS

- The error rates are steady state after the channel acquisition has been completed.

- The acquisition (convergence) time is described independently.

- In both cases the block length is chosen to be $N_b = 8L$ and the step size parameter $\mu = 0.01$.

- The initial guesses for Channels 1 and 2 are $\hat{\mathbf{h}}(0) = [0, 1, 0]$ and $\hat{\mathbf{h}}(0) = [0, 0, 1, 0, 0]$, respectively.

- The probability of error is estimated over a data record of 100 000 symbols.

- The performance is very close to that of the known CIR environment.

- For comparison purposes, a linear blind equalizer (Sato's algorithm) described by Equation (6.16) is used. As expected, it fails to converge at all for this type of channel.

6.5.4.2 The BSE/RLS

- Is simulated for Channels 1 and 2, starting with the same initial guesses for the channels as in BSE/BMLS.

- For the initialization of the BSE/RLS, the parameters $\lambda = 1$ and $\varepsilon = 0.001$ are used in $P(0)$.

- The performance of the BSE/RLS in Figure 6.26 is also very close to that of the known channel.

- An improvement is observed for Channel 2 compared to the BSE/BLMS, especially at high SNR. This is no surprise, since acquisition capability (convergence speed) is better than the LMS at high SNR.

- The tradeoff for the improved performance of the BSE/RLS, however, is its increased computational complexity.

6.5.4.3 The effect of the block length N_b

BSE/BLMS is simulated on Channel 1 for $N_b = 8L, 5L, 4L$ and $3L$ symbols and the results are shown in Figure 6.27.

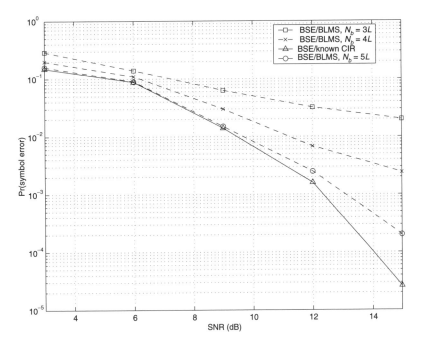

Figure 6.27 Probability of symbol error versus SNR using the BPSK waveform for the BSE/BLMS algorithm with various values of N_b over Channel 1.

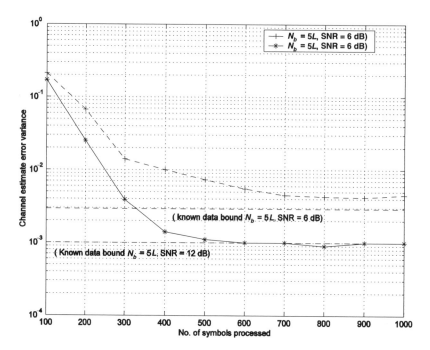

Figure 6.28   Channel estimate error variance for Channel 1 versus number of symbols processed using the BSE/BLMS algorithm at SNR equal to 6 and 12 dB.

- The performance degradation for smaller N_b must be attributed to the BSE algorithm.

- As the block length becomes smaller, the number of blocks in which no merge occurs increases.

- Bad data decisions are forced, reflecting upon the channel estimate too.

6.5.4.4 The convergence properties

- 300 independent Monte Carlo runs of BSE/BLMS are performed and the average squared error $|\hat{\mathbf{h}} - \mathbf{h}|^2$ as the number of processed data varies between 100 and 1000 symbols is computed.

- Simulations are performed for block sizes $N_b = 5L$, $10L$ and $15L$ points, and for SNR = 6 and 12 dB.

- The results are shown in Figure 6.28.

- For Channel 1, independent of the block size, we approach the steady state mean squared error within the first few hundred symbol periods.

- For SNR = 6 dB the steady state is reached after processing about 700 symbols, while for SNR = 12 dB we need approximately 400 symbols. This allows us to specify the *acquisition time* for the BSE/BLMS at about 400 symbol periods.

6.5.5 The time varying channel environment

In this segment a TDMA with the following parameters is simulated.

- The carrier frequency is assumed to be 900 MHz and the bit rate 48.6 kb/s.

- For QPSK modulation, that translates to a baud rate of 24.3 ksymbols/s, or a symbol period $T_s = 41$ μs.

- The mobile radio channel is assumed to be wideband with $L + 1$ taps.

- Each element of the CIR $\{h_i(k)\}_{i=0}^{L}$ is modeled as an independent low pass zero mean complex Gaussian random process with Rayleigh distributed amplitude and uniformly distributed phase in the interval $[-\pi, \pi]$.

- Each of the h_is is generated by passing a zero mean white complex Gaussian noise sequence through a digital second order Butterworth filter whose cutoff frequency f_d is determined by

$$f_d = T_s \left(f_c \frac{v}{v_c} \right)$$

where f_d is the normalized Doppler shift, f_c is the carrier frequency, and v and v_c are the vehicle speed and speed of light, respectively. This simulation procedure is widely used in the literature.

- For these system specifications the performance of all algorithms is investigated for vehicle speeds ranging from $v = 60$ mph (100 km/h) up to 150 mph (248 km/h).

- The corresponding normalized Doppler shift ranges from $f_d = 0.0034$ to 0.0085.

- A three-tap CIR ($L = 2$) is used. For the specifications of the IS-54 standard ($T_s = 41$ μs), this channel model covers a maximum time spread of $T_{max} = 123$ μs.

- The CIR elements $h_0(k), h_1(k), h_2(k)$ are constructed so as to have equal power $\sigma_h^2 = 1/3$ and to be independent of each other.

- This corresponds to a fading scenario where no *line of sight* is present (Rayleigh fading).

Figure 6.29 compares the performance of all algorithms tested: the BSE; the conventional adaptive MLSE; a variable D (delay in the CIR estimation loop) version of the conventional adaptive MLSE; and the Per Survivor Processing (PSP) known CIR environment. PSP estimates the channel only for the surviving trajectories in VA, and for those trajectories data is known. This corresponds to an ideal receiver capable of estimating the CIR perfectly, with zero delay, which is unrealistic when the channel is varying rapidly.

The adaptive D version of the conventional MLSE is derived by detecting a merge within N_b epochs from the current symbol and setting D equal to the merging delay. If no merge is detected, D is set to $N_b - 1$. The symbol error rate is computed as the fraction of data symbols that are in error over a time interval of 5000 consecutive TDMA slots, each packet of which has the structure shown in Chapter 1. For convenience, only the initial training symbols and the information symbols inside the TDMA slot were considered.

For LMS channel acquisition and tracking, the optimum value of the step size parameter was found to be $\mu = 0.06$ for $f_d = 0.0034$ and $\mu = 0.12$ for $f_d = 0.0085$, at SNR $= 20$ dB.

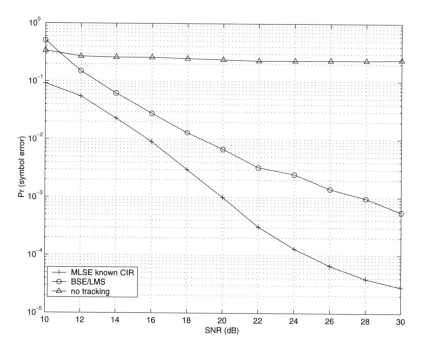

Figure 6.29 Symbol error rate performance of the various adaptive MLSE algorithms at a vehicle speed of 60 mph.

The optimum value of the weight parameter was found to be $\lambda = 0.73$ for $f_d = 0.0085$ at SNR = 20 dB. For all of the simulated algorithms presented, the initial CIR guess at the beginning of the first TDMA slot was set at $\hat{\mathbf{h}}(0) = [0, \ldots, 0]$. The last CIR estimate computed at the end of each subsequent slot is utilized to initialize the LMS or RLS at the beginning of the next slot.

6.5.5.1 Performance

In Figure 6.29, the performance of all the adaptive MLSE algorithms is presented for $f_d = 0.0034$ (82.6 Hz), which corresponds to a vehicle speed of 60 mph. The following parameters were used.

- An LMS update was used for the CIR acquisition and tracking. The step size parameter was set at $\mu = 0.06$ for all algorithms.

- For the conventional adaptive MLSE, the optimum performance was achieved for a delay in the CIR estimation loop equal to $D = 4$.

- The decision delay was set at $\delta = 5L$ for both the PSP, the conventional MLSE and the adaptive D MLSE.

- For the BSE/LMS a block length $N_b = 11$ was found to yield the best performance.

- Both the PSP and the BSE/LMS exhibit a clear advantage over the conventional adaptive MLSE at SNR above 20 dB.

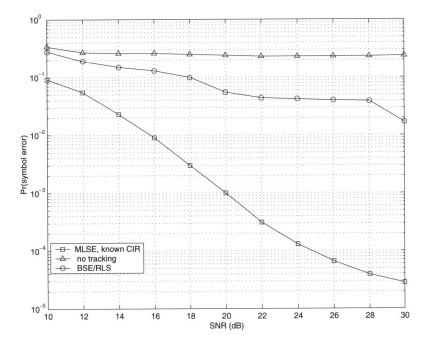

Figure 6.30 Symbol error rate performance of the various adaptive MLSE algorithms at a vehicle speed of 150 mph. The results from the use of either the LMS or the RLS channel estimates are shown.

- The adaptive D MLSE also exhibits a performance advantage over the conventional MLSE and gets closer to the BSE at high SNR.

- For comparison purposes a receiver which performs no tracking of the CIR after the initial acquisition mode was also simulated. It reaches a floor of about 2×10^{-1} at SNR = 20 dB. This clearly enforces the necessity of the use of CIR tracking algorithms even at this fading rate.

Figure 6.30 presents the performance of all algorithms for $f_d = 0.0085$ (206.5 Hz), and a speed of 150 mph. Both the LMS and RLS were used for CIR acquisition and tracking. For the LMS, the step size parameter was set at $\mu = 0.12$ and for the RLS, the weight parameter was set at $\lambda = 0.73$, for all algorithms. The optimum parameters were again found to be $D = 4$ and $N_b = 11$. The PSP and the BSE exhibit a clear advantage over the conventional and adaptive D MLSE at high SNR.

A much poorer performance was exhibited by all algorithms compared to the smaller vehicle speed in Figure 6.29. The performance degradation must be attributed to the higher fading rate making it very difficult to track the channel variations. The RLS exhibited slightly better performance at high SNR, which was anticipated. Absence of CIR tracking yields a floor of 0.5 probability of symbol error at SNR = 10 dB.

Figure 6.31 presents the error rates for the information symbols inside the TDMA slot for $f_d = 0.0085$ at SNR = 20 dB with LMS update. In the figure, only every other information symbol is plotted. The error rates tend to increase toward the end of the slot, even with CIR

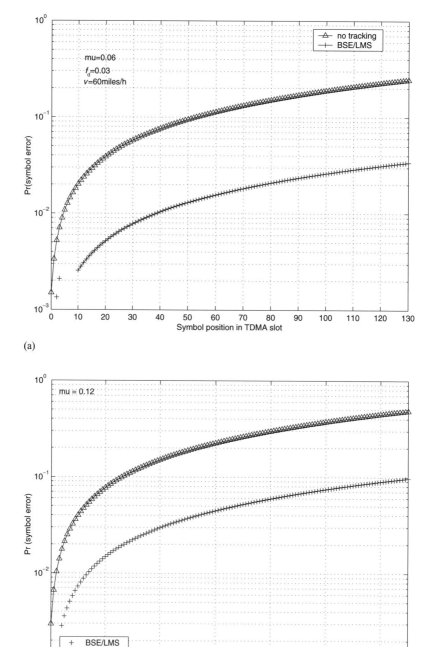

Figure 6.31 Data symbol error rate versus symbol position in the TDMA slot at SNR =
20 dB using the LMS channel acquisition and tracking algorithms (a) $v =$
60 mph; (b) $v = 150$ mph.

tracking algorithms. If no CIR tracking is employed, the error rates become unacceptable after the first 10–20 symbols within the TDMA slot. For BSE/LMS at 60 mph and error rate 10^{-3} there is a 10 dB worse performance from the lower bound of a known channel. PSP has manifested a similar performance. At 150 mph the error rate of all algorithms levels off at residual error rates larger than 10^{-2}, even at 30 dB. All algorithms fail at a velocity of 150 mph.

6.5.5.2 *Computational complexity and reconfiguration efficiency*

Computational complexity is measured by the required number of multiplications, N_{mul}, and additions, N_{add}, per data sample. The complexity is computed for a CIR of length $L + 1$ and an alphabet size M. The exact number of required operations per data symbol for the BSE cannot be expressed in closed form because of the variable merging delay inside each block. If $\bar{\delta}$ denotes the average merging delay over all blocks processed, then the required operations for metric computations per data symbol must be multiplied by a factor

$$f_{ex} = \frac{1}{[1 - \bar{\delta}/N_b]}$$

to account for processing the last part of each block twice. Parameter $\bar{\delta}$ is obtained from simulations.

The three-tap fading channel ($L = 2$) of the previous section is used and 5000 frames of PSK data ($M = 4$) are processed, amounting to 82 953 processed blocks, each of length $N_b = 11$ symbols. The fading rate is set at $f_d = 0.0034$ and the SNR is 30 dB. $\bar{\delta} = 2.9453$ symbols is found, which gives $f_{ex} = 1.365662$.

Table 6.1 presents the number of complex operations required per data symbol for metric calculations in the adaptive MLSE algorithm. Table 6.2 presents the required number of complex operations per symbol for this specific example, with both LMS and RLS updates for all channel estimators. For the Sato algorithm, discussed in Section 6.2, a tapped delay line equalizer with $L_{sa} = 16$ taps is assumed. From Tables 6.1 and 6.2, the conventional adaptive MLSE requires the smallest number of operations of all MLSE-based techniques, but significantly more operations than linear equalizers such as Sato's algorithm. The BSE requires an increased number of operations for the metric update compared to

Table 6.1   Complex operations per data symbol for adaptive MLSE algorithms (metric calculations)

		Conventional adaptive MLSE	BSE	PSP	Sato
Metric	N_{mul}	$M^{L+1}(L+1)$	$M^{L+1}(L+1)f_{ex}$	$M^{L+1}(L+1)$	0
Update	N_{add}	$M^{L+1}(L+1)$	$M^{L+1}(L+1)f_{ex}$	$M^{L+1}(L+1)$	0
CIR update	N_{mul}	$2(L+1)$	$2(L+1)$	$M^{L+1}(L+1)$	$2(L_{sa}+1)$
(LMS)	N_{add}	$2(L+1)$	$2(L+1)$	$M^{L+1}(L+1)$	$2(L_{sa}+1)$
CIR update	N_{mul}	$4(L^2+L)$	$4(L^2+L)$	$M^L 4(L^2+L)$	
(RLS)	N_{add}	$3(L^2+L)$	$3(L^2+L)$	$M^L 3(L^2+L)$	

Table 6.2   Complex operations per data symbol for $M = 4$ and $L = 2$ (channel estimations)

		Conventional adaptive MLSE	BSE	PSP	Sato
Metric	N_{mul}	192	262	192	0
Update	N_{add}	192	262	192	0
CIR update	N_{mul}	6	6	96	34
(LMS)	N_{add}	6	6	96	34
CIR update	N_{mul}	24	24	384	
(RLS)	N_{add}	24	24	384	

the conventional adaptive MLSE due to reprocessing the last part of each data block. A hardware implementation of the BSE would be more difficult due to the necessary checks for the detection of a merge inside each block. For the CIR estimation part, the PSP requires considerably more operations than all other techniques due to the multiple CIR estimates that it preserves (one for each state). This overhead becomes even larger if the RLS estimation algorithm is used for CIR tracking.

For the calculation of reconfiguration efficiency, relations defined in Chapters 2 and 3 are still valid. For practical applications the relative complexity D_r can be calculated by using data directly from Tables 6.1 and 6.2.

More details on the topic can be found in [580–593].

6.6   TURBO EQUALIZATION

In this section we discuss how the problem from Section 6.5 can be solved by using the turbo principle discussed in Chapter 2.

6.6.1   Signal format

The transmitted signal $s(t)$ is provided by the output of a filter whose impulse response is $h_e(t)$. The signal emitted can be expressed in the form:

$$s(t) = A \sum_k c_k h_e(t - kT) \exp j(2\pi f_0 t + \varphi_0) \tag{6.39}$$

In a multipath channel, the received signal $y(t)$, can be written as follows:

$$y(t) = \sum_{m=0}^{M-1} A_m(t) \sum_k c_k h_e(t - \tau_m - kT) \exp j(2\pi f_0 t + \varphi_0) + w(t) \tag{6.40}$$

where $A_m(t)$ are complex-valued independent multiplicative noise processes. The receiver matched filter output

$$R_n \triangleq R(nT) = \sum_{m=0}^{M-1} A_m(n) \sum_k c_{n-k} h_s(kT - \tau_m) + w_n \tag{6.41}$$

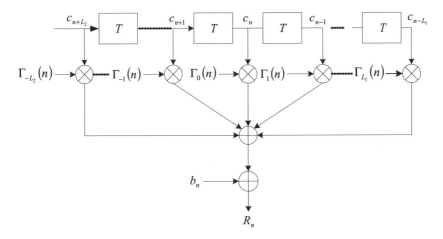

Figure 6.32 Equivalent discrete time model of a channel with inter-symbol interference.

where $A_m(n)$ equals $A_m(nT)$ by definition, w_n denotes the response of the receiving matched filter to the noise $w(t)$, sampled at time nT. $h_s(t)$ is defined by $h_s(t) = h_e(t) \times h_e^*(-t)$ and satisfies the Nyquist criterion.

$$\Gamma_k(n) = \sum_{m=0}^{M-1} A_m(n)h_s(kT - \tau_m) \qquad (6.42)$$

Let us suppose that the ISI is limited to $(L_1 + L_2)$ symbols. Equation (6.41) may be written in the form:

$$R_n = \sum_{k=-L_2}^{L_1} \Gamma_k(n)c_{n-k} + w_n \qquad (6.43)$$

6.6.2 Equivalent discrete time channel model

Quantities $\Gamma_k(n)$ are expressed as a linear combination of the multiplicative noises $A_m(n)$. Therefore, they are Gaussian in the case of a Rayleigh-type channel and constant in the case of a Gauss-type channel. Consequently, the set of modules made up of the modulator, the transmission channel and the demodulator can be represented by an equivalent discrete time channel as shown in Figure 6.32.

6.6.3 Equivalent system state representations

By new indexing in Equation (6.43) we have:

$$R_n = \sum_{k=0}^{L_1+L_2} \Gamma_{k-L_2}(n)c_{n+L_2-k} + b_n \qquad (6.44)$$

If we denote $S_n = (c_{n+L_2}, \ldots, c_{n-L_1+1})$ the state of the equivalent discrete time channel at time nT sample R_n depends on the channel state S_{n-1} and on the symbol c_{n+L_2}. Therefore,

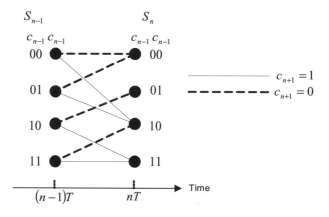

Figure 6.33   Trellis diagram for $L_1 = L_2 = 1$.

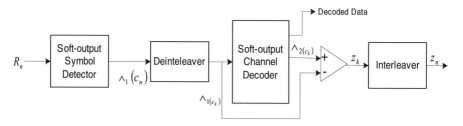

Figure 6.34   Extrinsic information deriving scheme.

the equivalent discrete time channel can be modeled as a Markov chain and its behavior can be represented by the trellis diagram shown in Figure 6.33.

6.6.4   Turbo equalization

In order to use a soft-input channel decoder, the symbol detector has to provide information about the reliability of the symbols estimated. This information may be obtained by using a soft-output Viterbi algorithm (SOVA) (the name often used for the algorithm described in Appendix 2.1), that associates an estimation of the logarithm of its likelihood ratio (LLR), $\Lambda_1(c_n)$, to each symbol c_n detected:

$$\Lambda_1(c_n) = \log \frac{\Pr\{c_n = +1|\mathbf{R}\}}{\Pr\{c_n = -1|\mathbf{R}\}} \tag{6.45}$$

where $\mathbf{R}$ denotes the vector of samples that constitutes the observation. After deinterleaving, the SOVA decoder provides a new LLR value of c_k, $\Lambda_2(c_k)$, that may be derived by analogy with the calculations used in Chapter 2 and expressed in the form:

$$\Lambda_2(c_k) = \Lambda_1(c_k) + z_k \tag{6.46}$$

where z_k is the extrinsic information associated with symbol c_k and provided by the channel decoder (Figure 6.34).

The extrinsic information z_k is another estimation of the LLR of symbol c_k conditioned on the decoding step:

$$z_k = \log \frac{\Pr\{c_k = +1 \mid \text{decoding}\}}{\Pr\{c_k = -1 \mid \text{decoding}\}} \tag{6.47}$$

Hence, z_k may be used through a feedback loop by the symbol detector after interleaving. This is the basis of the turbo equalization principle.

6.6.5 Viterbi algorithm

To evaluate the LLR of symbol c_{n-L_1}, the Viterbi algorithm used in the detector has to calculate a metric

$$\lambda_n^i = \left| R_n - r_n^i \right|^2 - 2\sigma_w^2 \log \Pr\{c_{n-L_1} = i\} \quad i = \pm 1 \tag{6.48}$$

where:

$$r_n^i = \sum_{k=0}^{L_1+L_2-1} \hat{\Gamma}_{k-L_2}(n) \cdot c_{n+L_2-k} + \hat{\Gamma}_{L_1}(n) \cdot i \quad i = \pm 1 \tag{6.49}$$

$\hat{\Gamma}_{k-L_2}(n)$, $0 \le k \le L_1 + L_2$ represents an estimation of quantity $\Gamma_{k-L_2}(n)$ and σ_w^2 denotes the variance of noise w_n, that is $\sigma_w^2 = E[|w_n|^2]$.

The *a priori* probabilities $\Pr\{c_{n-L_1} = i\}$ used in Equation (6.48) may be estimated from the extrinsic information z_{n-L_1}, if we assume that

$$z_{n-L_1} = \log \frac{\Pr\{c_{n-L_1} = +1\}}{\Pr\{c_{n-L_1} = -1\}} \tag{6.50}$$

From Equation (6.50)

$$\Pr\{c_{n-L_1} = +1\} \approx \frac{\exp z_{n-L_1}}{1 + \exp z_{n-L_1}}$$

$$\Pr\{c_{n-L_1} = -1\} \approx \frac{1}{1 + \exp z_{n-L_1}} \tag{6.51}$$

Using Equations (6.51) and (6.48), metrics λ_n^i are equal to:

$$\lambda_n^{+1} = \left| R_n - r_n^{+1} \right|^2 - \gamma z_{n-L_1}$$

$$\lambda_n^{-1} = \left| R_n - r_n^{-1} \right|^2 \tag{6.52}$$

Note that the common term $\log(1 + \exp z_{n-L_1})$ has been suppressed in Equation (6.52). Coefficient γ is a weight introduced to take into account variance σ_w^2 and the fact that the extrinsic information is only an *estimation* of the *a priori* probability. Its value depends on the signal to noise ratio, that is to say the reliability of the extrinsic information.

6.6.6 Iterative implementation of turbo equalization

The different processing stages in the turbo equalizer present a non-zero internal delay, so turbo equalizing can only be implemented in an iterative way. At each iteration q, a new value of extrinsic information is calculated and used by the symbol detector at the

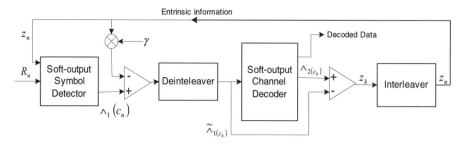

Figure 6.35 Principle of turbo equalization (under the zero internal delay assumption).

next iteration. Therefore, the turbo equalizer can be implemented in a modular pipelined structure, where each module is associated with one iteration. Then, performance in bit error rate (BER) terms is a function of the number of chained modules.

When extrinsic information is used by the symbol detector, it can be proved that, at iteration q, the LLR of symbol c_n, $\Lambda_1^q(c_n)$, may be expressed as:

$$\Lambda_1^q(c_n) = \hat{\Lambda}_1^q(c_n) + \gamma^q z_n^{q-1} \tag{6.53}$$

where $\hat{\Lambda}_1^q$ is a term depending on the samples of observation $\mathbf{R}$ and on z_n^{q-1}, $k \neq n$, and z_n^{q-1} denotes the extrinsic information of symbol c_n determined at iteration $q - 1$. If we apply the same approach as in turbo decoding described in Chapter 2, the quantity $\gamma^q z_n^{q-1}$ provided by the channel decoder at the previous iteration has to be subtracted from $\Lambda_1^q(c_n)$ (see Equation (6.52)), as illustrated in Figure 6.35. Hence, after deinterleaving, the channel decoder input is in fact equal to:

$$\tilde{\Lambda}_1^q(c_n) = \Lambda_1^q(c_n)\big|_{z_k^{q-1}=0} \tag{6.54}$$

At the channel decoder output, the extrinsic information z_k^q may also be written as follows, using Equation (6.46):

$$z_k^q = \Lambda_2^q(c_n)\big|_{\tilde{\Lambda}_1^q(c_n)=0} \tag{6.55}$$

6.6.7 Performance

Performance of this device has been evaluated for a rate $R = 1/2$ recursive systematic encoder with constraint length $K = 5$ and generators $G_l = 23, G_2 = 35$. Bits were interleaved in a non-uniform matrix whose dimensions are 64 by 64. The modulation used was a BPSK modulation, with a Nyquist filter whose transfer function $H_s(f)$ was a raised cosine with a rolloff $\alpha = 1$, on both Gaussian and Rayleigh channels.

For both channels, $M = 5$ independent paths were considered, each with a mean power $P_m = E[|A_m(n)|^2]$, so that the total mean power was normalized: $\sum_{m=0}^{M-1} P_m = 1$. The delays τ_m were chosen as multiples of T ($\tau_m = mT$, $\Gamma_k(n) = A_k(n)$ since $h_s[(k-m)T] = \delta_{k-m,0}$). The coefficients for the Gaussian channel were chosen equal to:

$$\Gamma_0(n) = \sqrt{0.45}, \quad \Gamma_1(n) = \sqrt{0.25}, \quad \Gamma_2(n) = \sqrt{0.15},$$
$$\Gamma_3(n) = \sqrt{0.10}, \quad \Gamma_4(n) = \sqrt{0.05}$$

For the Rayleigh channel, the five paths had equal mean power ($P_i = 1/M$, $\forall i \in$

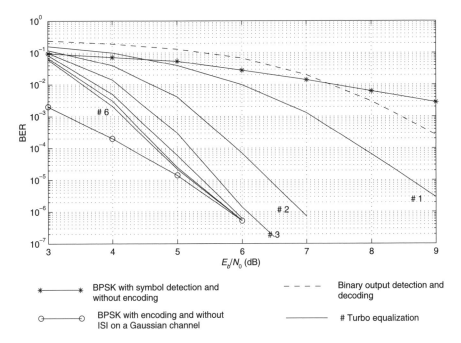

Figure 6.36 Performance of turbo equalization over a Gaussian channel (convolutional encoding with $K = 5$).

[1, M]}. A parameter BT, which is the product of the Doppler bandwidth and the symbol duration, fixes the variation velocity of the channel: the smaller BT is, the more slowly the channel parameters vary during a time interval symbol.

The discrete time equivalent channel was modeled by a 16-state trellis, and the symbol detector was working on the SOVA algorithm. The channel coefficients $\Gamma_k(n)$ were supposed perfectly known. After deinterleaving, the soft estimations provided by the SOVA detector were used by the decoder, which also worked on a 16-state trellis and the SOVA algorithm. The extrinsic information extracted from the decoder was used by the symbol detector according to the principle depicted in Figure 6.35.

The BER was computed as a function of signal to noise ratio E_b/N_0, where E_b is the mean energy received per information bit d_k and N_0 is the noise power bilateral spectral density. The signal to noise ratio E_b/N_0 may be expressed as:

$$\frac{E_b}{N_0} = \frac{\sum_{m=0}^{M-1} P_m}{\sigma_b^2} \tag{6.56}$$

The results are shown in Figures 6.36–6.38 for different numbers of iterations n. One can see that even for three iterations the performance curves approach the BER curve with no ISI.

More details on turbo equalization can be found in [594–603].

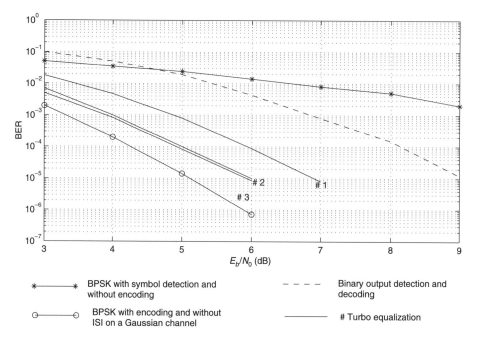

Figure 6.37 Performance of turbo equalization over a Rayleigh channel with $BT = 0.1$ (convolutional encoding with $K = 5$).

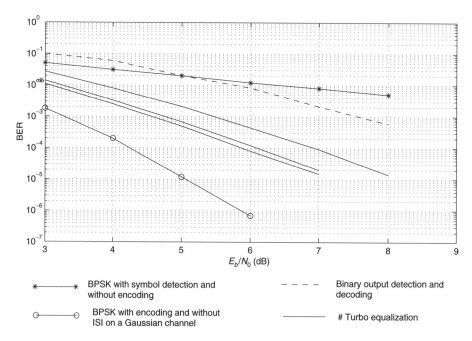

Figure 6.38 Performance of turbo equalization over a Rayleigh channel with $BT = 0.001$ (convolutional encoding with $K = 5$).

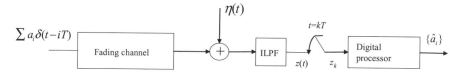

Figure 6.39 The signal model for the baseband communication system.

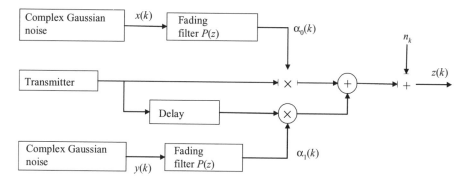

Figure 6.40 The fading channel model.

6.7 KALMAN FILTER BASED JOINT CHANNEL ESTIMATION AND DATA DETECTION OVER FADING CHANNELS

In this section we consider the problem of joint channel and data estimation in a fading channel based on using a Kalman-type estimator. The general system block diagram is shown in Figure 6.39. More details can be found in [604–607]. The presentation in this section is based on [604]. The channel model generator is shown in Figure 6.40.

6.7.1 Channel model

If the filter $P(z)$ in Figure 6.40 is modeled as

$$P(z) = \frac{D}{1 - Az^{-1} - Bz^{-2} - Cz^{-3}} \tag{6.57}$$

then at sampling time k, the CIR, $\mathbf{h}_k$, a complex Gaussian vector, is

$$\mathbf{h}_k = (h_{k,0}, h_{k,1}, \ldots, h_{k,\beta})^{\mathrm{T}} \tag{6.58}$$

truncated to a finite length of $(\beta + 1)$. By considering the third-order approximation of Equation (6.57), an autoregressive (AR) representation for the CIR can be introduced as

$$\mathbf{h}_k = A\mathbf{I}\mathbf{h}_{k-1} + B\mathbf{I}\mathbf{h}_{k-2} + C\mathbf{I}\mathbf{h}_{k-3} + D\mathbf{I}\mathbf{w}_k \tag{6.59}$$

where $\mathbf{I}$ is the identity matrix, $\mathbf{w}_k$ is a zero mean white complex circularly symmetric Gaussian process with the covariance matrix defined as $E(\mathbf{w}_k\mathbf{w}_l^{\mathrm{T}}) = \mathbf{Q}\delta_{kl}$ and $\mathbf{w}_l^{\mathrm{T}}$ is the conjugate transpose of $\mathbf{w}_l$. CIR at time k depends on its three consecutive previous values.

The state of such a system is a vector composed of three consecutive channel impulse responses as

$$\mathbf{x}_k = \left(\mathbf{h}_k^T, \mathbf{h}_{k-1}^T, \mathbf{h}_{k-2}^T\right)^T \tag{6.60}$$

Using Equations (6.60) and (6.59) gives

$$\mathbf{x}_{k+1} = \begin{bmatrix} A\mathbf{I} & B\mathbf{I} & C\mathbf{I} \\ \mathbf{I} & 0 & 0 \\ 0 & \mathbf{I} & 0 \end{bmatrix} \mathbf{x}_k + \begin{bmatrix} D\mathbf{I} \\ 0 \\ 0 \end{bmatrix} \mathbf{w}_k \tag{6.61}$$

$$\mathbf{x}_{k+1} = \mathbf{F}\mathbf{x}_k + \mathbf{G}\mathbf{w}_k \tag{6.62}$$

where $\mathbf{F}$ and $\mathbf{G}$ are $3(\beta + 1) \times 3(\beta + 1)$ and $3(\beta + 1) \times (\beta + 1)$ matrices given in Equation (6.61). $\mathbf{F}$ is called the state transition matrix and $\mathbf{G}$ is the process noise coupling matrix.

6.7.2 The received signal

If we introduce a $1 \times (\beta + 1)$ vector $\mathbf{H}_k$

$$\mathbf{H}_k = (a_k, a_{k-1}, a_{k-2}, \ldots, a_{k-\beta}, 0, \ldots, 0) \tag{6.63}$$

where a_k is the transmitted data sequence, then the received signal z_k becomes

$$z_k = \mathbf{H}_k \mathbf{x}_k + n_k \tag{6.64}$$

The Kalman filter is optimum for minimizing the mean square estimation error [608], in the above linear time varying system. The algorithm is complex and, in practice, suboptimal methods are more advantageous due to their implementation simplicity.

6.7.3 Channel estimation alternatives

In this segment the same assumptions are used as in [604]:

- To avoid the decision delay, the PSP method [609] is used. In this method, there is a channel estimate for every possible sequence.

- The estimated channel impulse response will be used to compute the branch metrics in the trellis diagram of the Viterbi algorithm.

- The number of required estimators is limited to the number of survivor branches (or the number of states) in the Viterbi algorithm trellis diagram.

- In PSP each surviving path keeps and updates its own channel estimate. This method eliminates the problem of decision delay, and in order to employ the best available information for data detection the data sequence of the shortest path is used for channel estimation along the same path.

- The data communication system is based on the IS-136 standard presented in Chapter 1.

- The modulation is QPSK with four possible symbols $(\pm 1 \pm j)$ and a symbol rate of 25 ksymbols/s.

- The differentially encoded data sequence is arranged into 162 symbol frames.

- The first 14 symbols of each frame make up a training preamble sequence to help the adaptation of the channel estimator.

- For the shaping filter at the transmitter, a finite impulse response (FIR) filter approximates a raised cosine frequency response with an excess bandwidth of 25 % (slightly different from the 35 % selected in IS-136).

- A two-ray fading channel model as described in Figure 6.40, where one ray has a fixed delay equal to one symbol period.

- The multiplicative coefficients of α_0 and α_1 are produced at the output of two fading filters, where the inputs are two independent zero mean complex Gaussian processes with equal variances.

- The length of the discrete impulse response of the shaping filter is set equal to the symbol interval so that the ISI at the receiver is only due to the multipath nature of the channel.

- The total length of the CIR is two symbol intervals, i.e. $\beta + 1 = 2$ if there is one sample per symbol interval.

- Therefore, there is ISI between two neighboring symbols and there are four possible states in the trellis diagram.

- The LMS algorithm, RLS algorithm or the Kalman filter can be used to estimate the channel impulse response.

6.7.4 Implementing the estimator

6.7.4.1 The Kalman filter algorithm [608]

The measurement update equations are:

$$
\begin{aligned}
\hat{\mathbf{x}}_{k|k} &= \hat{\mathbf{x}}_k + \mathbf{K}_k(z_k - \mathbf{H}_k\hat{\mathbf{x}}_k) \\
\mathbf{K}_k &= \mathbf{P}_k\mathbf{H}_k^{\mathrm{T}}\mathbf{R}_k^{-1} \\
R_k &= \mathbf{H}_k\mathbf{P}_k\mathbf{H}_k^{\mathrm{T}} + \mathrm{N}_0 \\
\mathbf{P}_{k|k} &= \mathbf{P}_k - \mathbf{K}_k\mathbf{H}_k\mathbf{P}_k
\end{aligned}
\tag{6.65}
$$

The time update equations are:

$$
\begin{aligned}
\hat{\mathbf{x}}_{k+1} &= \mathbf{F}\hat{\mathbf{x}}_{k|k} \\
\mathbf{P}_{k+1} &= \mathbf{F}\mathbf{P}_{k|k}\mathbf{F}^T + \mathbf{GQG}^{\mathrm{T}}
\end{aligned}
\tag{6.66}
$$

6.7.4.2 The RLS algorithm [110]

$$
\begin{aligned}
\hat{\mathbf{x}}_{k+1} &= \hat{\mathbf{x}}_k + \mathbf{K}_k(z_k - \mathbf{H}_k\hat{\mathbf{x}}_k) \\
\mathbf{K}_k &= \mathbf{P}_k\mathbf{H}_k^{\mathrm{T}}\mathbf{R}_k^{-1} \\
R_k &= \mathbf{H}_k\mathbf{P}_k\mathbf{H}_k^{\mathrm{T}} + \lambda \\
\mathbf{P}_{k+1} &= \lambda^{-1}(\mathbf{P}_k - \mathbf{K}_k\mathbf{H}_k\mathbf{P}_k)
\end{aligned}
\tag{6.67}
$$

In the above equations the measurement updated estimate $\hat{\mathbf{x}}_{k|k}$ is the linear least squares estimate of x_k given observations $\{z_0, z_1, \ldots, z_k\}$. $\hat{\mathbf{x}}_{k+1}$ is the time updated estimate of $\mathbf{x}_k$ given observations $\{z_0, z_1, \ldots, z_k\}$. The corresponding error covariance matrices of these estimations are

$$\mathbf{P}_{k|k} = E\lfloor(\mathbf{x}_k - \hat{\mathbf{x}}_{k|k})(\mathbf{x}_k - \hat{\mathbf{x}}_{k|k})^{\mathrm{T}}\rfloor$$

$$\mathbf{P}_k = E\lfloor(\mathbf{x}_k - \hat{\mathbf{x}}_k)(\mathbf{x}_k - \hat{\mathbf{x}}_k)^{\mathrm{T}}\rfloor$$

(6.68)

In the RLS algorithm, λ is a parameter called the forgetting factor.

6.7.5 The Kalman filter

The estimator is based on two parts: *measurement update equations* and *time update equations*. The RLS algorithm is essentially identical to the measurement update equations of the Kalman filter. The Kalman filter can be used for channel estimation when some *a priori* information about the channel is available at the receiver (i.e. the **F** and **G** matrices). The RLS algorithm, which is a suboptimal method, does not require this *a priori* information and its computational complexity is less than the Kalman filter.

6.7.6 Implementation issues

At the same precision, mathematically equivalent implementations can have different numerical stabilities, and some methods of implementation are more robust against roundoff errors.

In the Kalman filter and the RLS algorithm the estimation depends on the correct computation of the error covariance matrix. In an ill-conditioned problem, the solution will not be equal to the covariance matrix of the actual estimation uncertainty. Factors contributing to this problem are: large matrix dimensions, a growing number of arithmetic operations and poor machine precision. Solutions to these problems are factorization methods and square root filtering [610, 611].

6.7.6.1 Square root filtering

Some implementations are more robust against roundoff errors and ill-conditioned problems.The so-called square root filter implementations have generally better error propagation bounds than the conventional Kalman filter equations [612]. In the square root forms of the Kalman filter, matrices are factorized and *triangular square roots* are propagated in the recursive algorithm to preserve the symmetry of the covariance (information) matrices in the presence of roundoff errors.

Different techniques are used for changing the dependent variable of the recursive estimation algorithm to factors of the covariance matrix. A Cholesky factor of a symmetric non-negative definite matrix **M** is a matrix **C** such that $\mathbf{CC}^{\mathrm{T}} = \mathbf{M}$. Cholesky decomposition algorithms solve for a diagonal factor and either a lower triangular factor **L** or an upper triangular factor **U** such that $\mathbf{M} = \mathbf{UD}_{\mathrm{U}}\mathbf{U}^{\mathrm{T}} = \mathbf{LD}_{\mathrm{L}}\mathbf{L}^{\mathrm{T}}$, where $\mathbf{D}_{\mathrm{L}}$ and $\mathbf{D}_{\mathrm{U}}$ are diagonal factors with non-negative diagonal elements.

The square root methods propagate the L-D or U-D factors of the covariance matrix rather than the covariance matrix. The propagation of square root matrices implicitly preserves the Hermitian symmetry and non-negative definiteness of the computed covariance matrix.

The condition number $\kappa(\mathbf{P}) = [\text{eigenvalue}_{\text{max}}(\mathbf{P})/\text{eigenvalue}_{\text{min}}(\mathbf{P})]$ of the covariance matrix $\mathbf{P}$ can be written as

$$\kappa(\mathbf{P}) = \kappa(\mathbf{LDL}^{\text{T}}) = \kappa(\mathbf{BB}^{\text{T}}) = [\kappa(\mathbf{B})]^2$$

where $\mathbf{B} = \mathbf{LD}^{1/2}$. The condition number of $\mathbf{B}$ used in the square root method is much smaller than the condition number of $\mathbf{P}$ and this leads to improved numerical robustness of the algorithm.

In the square root method, the dynamic range of the numbers entering into computations will be reduced. Loosely speaking, we can say that the computations which involve numbers ranging between 2^{-N} to 2^{+N} will be reduced to ranges between $2^{-N/2}$ to $2^{+N/2}$, which would halve the length of required mantissa used in signal processing. All of these will directly affect the accuracy of computer computations.

Performance results are shown in Figures 6.41 and 6.42. Different Kalman algorithms will demonstrate superior performance only if the precision in the calculation of Equation (6.65) is high enough. From Figures 6.41 and 6.42, this means if the quantization is precise enough, requiring a mantissa length of 20 bits or higher.

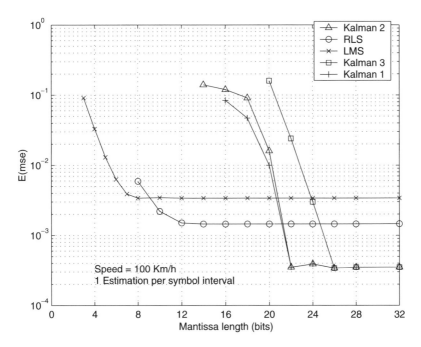

Figure 6.41 The effects of changing the word length on estimation error E (mse) ($E_b/N_0 = 15$ dB). Kalman 1 represents the WGS method, Kalman 2 is for the correlation method and Kalman 3 is the direct method, the PSP method is employed for detection.

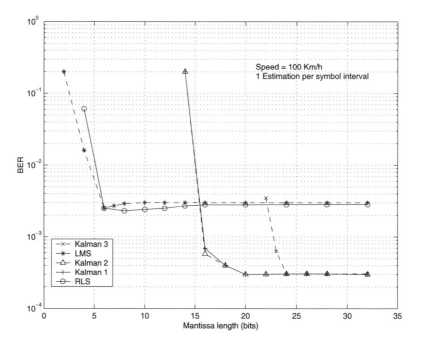

Figure 6.42 The effects of changing the word length on BER ($E_b/N_0 = 15$ dB). Kalman 1 represents the WGS method, Kalman 2 is for the correlation method and Kalman 3 is the direct method, the PSP method is employed for detection.

6.8 EQUALIZATION USING HIGHER ORDER SIGNAL STATISTICS

In order to speed up the algorithms, equalization based on higher order signal statistics may be used [613–619]. In this section we discuss a group of such algorithms derived from the cost function defined as average entropy of the set of the signal samples [613]. Maximization of the average entropy is obtained by minimizing the mutual information of the set of signal samples which result in minimum ISI. The block diagram of the equalizer based on joint entropy minimization (JEM) principles is shown in Figure 6.43.

6.8.1 Problem statement

For input data symbols $s(n)$, discrete time channel output $x(n)$ and the channel response $c(n)$, the received baseband symbol can be represented as

$$x(n) = \sum_{k=0}^{N} c(k)s(n-k) \tag{6.69}$$

We will assume that input symbols are independently identically distributed (i.i.d.) and zero mean, which is true in a digital communication system, and that $c(0) = 1$.

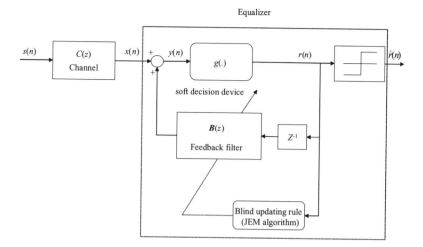

Figure 6.43 Block diagram of channel and JEM-DFE structure.

Signal $y(n)$ at the output of the equalizer (see Figure 6.43) can be presented as

$$y(n) = x(n) + \sum_{k=1}^{N} b(k)r(n-k)$$

$$= s(n) + \sum_{k=1}^{N} c(k)s(n-k) + \sum_{k=1}^{N} b(k)r(n-k) \tag{6.70}$$

$$= s(n) + \sum_{k=1}^{N} c(k)s(n-k) + \sum_{k=1}^{N} b(k)g(y(n-k))$$

Non-linearity $g(\cdot)$ is a strictly monotone (increasing or decreasing) differentiable function. A more general DFE structure usually consists of a feedforward finite impulse response (FIR) filter followed by the feedback FIR filter given in Figure 6.43. As discussed earlier in the chapter, the main purpose of the feedforward filter is to eliminate the precursor ISI and the feedback filter cancels the postcursor ISI. For simplicity, only the feedback part is considered. This is not a hurdle since it is possible to separate the adaptation of the feedback part from the feedforward part.

6.8.2 Signal model

By introducing auxiliary inputs and outputs we have

$$\mathbf{y}(n) = \mathbf{B} \cdot \mathbf{x}(n) \tag{6.71}$$

where

$$\mathbf{x}(n) = [x(n) \quad r(n-1) \quad \ldots \quad r(n-N)]^{\mathrm{T}}$$
$$\mathbf{y}(n) = [y(n) \quad r(n-1) \quad \ldots \quad r(n-N)]^{\mathrm{T}} \tag{6.72}$$

and

$$
\mathbf{B} =
\begin{bmatrix}
1 & b(1) & \cdots & b(N) \\
0 & 1 & \ddots & 0 \\
\vdots & \ddots & \ddots & \ddots \\
0 & \ddots & \cdots & 1
\end{bmatrix}
\tag{6.73}
$$

In the presence of ISI (i.e. dependence), the entropy of the received symbols is smaller than when they are independent. Thus, maximizing the joint entropy of the equalizer outputs will result in removing the ISI.

6.8.3 Derivation of algorithms for DFE

So, the algorithm is based on maximizing the joint entropy of the equalizer output $\mathbf{r}(n)$. The joint entropy of $\mathbf{r}(n) = g(\mathbf{y}(n))$ denoted by $\mathbf{H}[r_1(n), \ldots, r_{N+1}(n)]$ is defined as [578, 620] (Chapter 6)]:

$$
\begin{aligned}
\mathbf{H}[r_1(n), \ldots, r_{N+1}(n)] &\triangleq - E\{\ln f_\mathbf{r}(\mathbf{r}(n))\} \\
&= E\{\ln |J|\} - E\{\ln f_\mathbf{y}(\mathbf{y}(n))\} \\
&= E\{\ln |J|\} - E\{\ln\{|B|^{-1} f_\mathbf{x}(\mathbf{x}(n))\}\} \\
&= E\{\ln |J|\} - E\{\ln f_\mathbf{x}(\mathbf{x}(n))\}
\end{aligned}
\tag{6.74}
$$

where $|J|$ is the absolute value of the Jacobian of the transformation

$$
J = \det
\begin{bmatrix}
\dfrac{\partial r_1(n)}{\partial y_1(n)} & \cdots & \dfrac{\partial r_1(n)}{\partial y_{N+1}(n)} \\
\ddots & \cdots & \ddots \\
\dfrac{\partial r_{N+1}(n)}{\partial y_1(n)} & \cdots & \dfrac{\partial r_{N+1}(n)}{\partial y_{N+1}(n)}
\end{bmatrix}
$$

$$
J = \det
\begin{bmatrix}
\dfrac{\partial r(n)}{\partial y(n)} & \cdots & \dfrac{\partial r(n)}{\partial r(n - N)} \\
\ddots & \cdots & \ddots \\
\dfrac{\partial r(n - N)}{\partial y(n)} & \cdots & \dfrac{\partial r(n - N)}{\partial r(n - N)}
\end{bmatrix}
\tag{6.75}
$$

$$
=
\begin{bmatrix}
\dfrac{\partial r(n)}{\partial y(n)} & \cdots & 0 \\
\ddots & \cdots & \ddots \\
\dfrac{\partial r(n - N)}{\partial y(n)} & \cdots & 1
\end{bmatrix}
= \dfrac{\partial r(n)}{\partial y(n)}
\tag{6.76}
$$

The quantity $r_i(n)[x_i(n)]$ is the ith component of the vector $\mathbf{r}(n)[\mathbf{x}(n)]$ and $f_r(\mathbf{r}(n))[f_x(\mathbf{x}(n))]$ is the joint density function of the input vector $\mathbf{r}(n)$ [$\mathbf{x}(n)$].

Assume that the previous decisions are correct, i.e.

$$\text{Assumption: } r(n - k) = s(n - k), \; k = 1, \ldots, N \tag{6.77}$$

Under Assumption (6.77), $r(n - i)$ are independent of $r(n - j)$, $i > j$ and thus Equation (6.75) follows from Equation (6.75). The joint entropy can also be expressed as [620, Chapter 15]:

$$\mathbf{H}[r_1(n), \ldots, r_{N+1}(n)] = \sum_{i=1}^{N+1} \mathbf{H}[r_i(n)] - \mathbf{I}[r_1(n), \ldots, r_{N+1}(n)] \tag{6.78}$$

Maximizing the joint entropy of $\mathbf{r}(n)$ is the same as maximizing the first term in Equation (6.78) while minimizing the mutual information $\mathbf{I}[r_1(n), \ldots, r_{N+1}(n)]$. Since the previous decisions that the outputs of the non-linear function $g(\cdot)$ are the same as the transmitted symbols $s(n - k)$, $k = 1, \ldots, N$, $E\{\ln f_x(\mathbf{x}(n))\}$ in Equation (6.74) can be considered to be a constant with respect to the feedback filter coefficients $b(k)$, $k = 1, \ldots, N$. Then, maximizing $E\{\ln |J|\}$ is equivalent to maximizing $\mathbf{H}[r_1(n), \ldots, r_{N+1}(n)]$. In doing so, the statistical dependence between the current output of the non-linear function $r(n)$ and the previous outputs $(r(n - k), k = 1, \ldots, N)$ can be reduced, which leads to ISI suppression.

6.8.4 The equalizer coefficients

The joint entropy $\mathbf{H}[r_1(n), \ldots, r_{N+1}(n)]$ is a non-linear function of the unknown equalizer coefficients and does not lead easily to a closed form solution. A gradient descent algorithm is used for maximizing Equation (6.74). The equalizer coefficients are then updated iteratively

$$b^{n+1}(k) = b^n(k) + \mu E \left\{ \frac{\partial \ln |J|}{\partial b^n(k)} \right\} \tag{6.79}$$

where μ is the positive step size. This equation can be further simplified based on the choice of non-linearity used in the decision device.

6.8.5 Stochastic gradient DFE adaptive algorithms

Despite its superiority over the linear equalizer and its popularity, a major drawback of the DFE is that it suffers from error propagation. Any previous decision errors at the output of the decision device will produce the symbol estimate $y(n)$, whose ISI is not completely eliminated by the feedback filter. This error in turn will affect future decisions. An intuitive choice is one where we replace the hard limiter with a softer function. It is hoped that the errors in $r(n)$ can be reduced by using soft decisions.

The update equation given in (6.79) depends on the mapping used in the soft decision device. Various non-linear functions may be used for the soft decision device (function $g(\cdot)$ shown in Figure 6.43). We use two different functions for $g(\cdot)$ to derive new blind algorithms for DFE.

6.8.5.1 Equivalence of JEM and ISIC

First choice for the non-linear function is

$$g(n) = \alpha \cdot \tanh[\beta \cdot n] \tag{6.80}$$

resulting in the following

$$\frac{\partial \ln |J|}{\partial b(k)} = \frac{\partial \ln \left| \dfrac{\partial r(n)}{\partial y(n)} \right|}{\partial b(k)}$$

$$= -2\beta \tanh[\beta \cdot y(n)]r(n-k) \tag{6.81}$$

$$= -\frac{2\beta}{\alpha} r(n)r(n-k)$$

The gradient is a function of the non-linearity used in the decision device that provides a tool via which the algorithm characteristics can be varied. A new adaptive blind algorithm for DFE based on JEM can be obtained as follows:

$$\text{JEM-1:} \qquad b^{n+1}(k) = b^n(k) - \mu r(n)r(n-k) \tag{6.82}$$

Taylor series (TS) expansion of $r(n)$ leads to

$$r(n) = \alpha \cdot \tanh[\beta \cdot y(n)]$$

$$= \alpha\beta y(n) - \frac{\alpha\beta^3}{3}y^3(n) + \frac{2\alpha\beta^5}{15}y^5(n) + O(y^6(n)) \tag{6.83}$$

For simplicity only the first few terms of the expansion are considered. Further variations of JEM-1 can be obtained by substituting Equation (6.83) into Equation (6.82). By using only the first term in Equation (6.83) with $\alpha = 3$ and $\beta = 1/3$, a simpler update equation for the feedback filter coefficients is

$$\text{JEM-2:} \qquad b^{n+1}(k) = b^n(k) - \mu y(n)r(n-k) \tag{6.84}$$

JEM-2 is exactly the same as the ISIC algorithm [621] which uses soft decision feedback. Here, the same update equation is arrived at via the motivation to maximize the entropy of the observations at the output of a soft decision device.

6.8.5.2 Equivalence of JEM and decorrelation criterion

Another possible variation of Equation (6.82) is to use the TS approximation for both $r(n)$ and $r(n-k)$.

$$r(n)r(n-k) = \alpha^2\beta^2 y(n)y(n-k) - \frac{\alpha^2\beta^4}{3}\{y^3(n)y(n-k)$$

$$+ y^3(n-k)y(n)\} + \cdots \tag{6.85}$$

By using only the first term in Equation (6.85) with $\alpha = 3$ and $\beta = 1/3$, we have

$$\text{JEM-3:} \qquad b^{n+1}(k) = b^n(k) - \mu y(n)y(n-k) \tag{6.86}$$

JEM-3 is exactly the same as that for the adaptive blind algorithm based on the *decorrelation criterion* (DECA) [622].

6.8.5.3 *JEM and CMA-DFE*

Another function that has transfer characteristics similar to the hyperbolic tangent is the function with a cubic non-linearity

$$g(y(n)) = \alpha \cdot y(n) + \beta \cdot y^3(n) \tag{6.87}$$

With the above mapping we have

$$\frac{\partial \ln |J|}{\partial b(k)} = \frac{6\beta y(n)}{\alpha + 3\beta y^2(n)} r(n-k) \tag{6.88}$$

$$\frac{\partial \ln |J|}{\partial b(k)} = \left[\frac{6\beta}{\alpha} y(n) \left\{ 1 - \frac{3\beta}{\alpha} y^2(n) + \frac{9\beta^2}{\alpha} y^4 \right\} + O(y^7(n)) \right] r(n-k) \tag{6.89}$$

Dropping terms of order greater than three gives:

$$\text{JEM-4:} \qquad b^{n+1}(k) = b^n(k) + \mu y(n) \left\{ 1 - \frac{3\beta}{\alpha} y^2(n) \right\} \cdot r(n-k) \tag{6.90}$$

With $\alpha = 3$ and $\beta = 1$, Equation (6.90) coincides with the CMA-DFE algorithm, provided the soft decisions given by Equation (6.87) are used in the CM algorithm for a DFE. The original CMA-DFE uses hard decisions [623, 624].

6.8.6 Convergence analysis

6.8.6.1 *Alternative JEM-DFE scheme and Bussgang-type algorithm*

Since soft decisions can possibly smooth the error surface and allow the algorithm to escape from local minima, the authors in [508] suggest the use of soft decisions at the start of equalization. These can then be replaced with a hard decision device after a few iterations. It is difficult to decide when one can switch from soft to hard decisions. A simple technique based on a modified JEM-DFE scheme is suggested to overcome this difficulty (Figure 6.44).

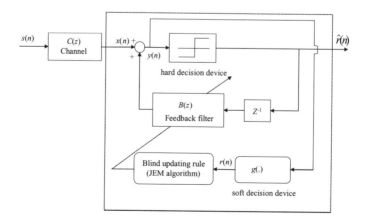

Figure 6.44 Block diagram of channel and alternative JEM-DFE structure.

For the update equations (for example, (6.82) and (6.84)), the soft decision device outputs, $r(n)$ and $r(n - k)$, are used. On the other hand, the hard decisions $\hat{r}(n - k)$ in Figure 6.43 are used as inputs to the feedback filter instead of $r(n - k)$; see Figure 6.43. In this way, soft and hard decision modes can be combined. A structure similar to JEM-DFE (see Figure 6.44) was used in [624] and will be used in what follows.

6.8.6.2 JEM-DFE structure in Figure 6.44 and classical Bussgang-type DFE

DFE can be considered as a linear equalizer [624]. The usual DFE structure consists of the feedforward filter followed by the feedback filter. The symbol estimate $y(n)$ can be expressed as:

$$y(n) = \sum_{k=-N_f}^{0} f(k)x(n - k) + \sum_{k=1}^{N_b} b(k)\hat{r}(n - k)$$

$$= \sum_{k=-N_f}^{N_b} w(k)\hat{x}(n - k) \tag{6.91}$$

The input to the equalizer is

$$\tilde{x}(n - k) = \begin{cases} x(n - k), & \text{when } -N_f \le k \le 0 \\ \hat{r}(n - k), & \text{when } 1 \le k \le N_b \end{cases} \tag{6.92}$$

The coefficients of the equalizer are

$$w(n - k) = \begin{cases} f(n - k), & \text{when } -N_f \le k \le 0 \\ b(n - k), & \text{when } 1 \le k \le N_b \end{cases} \tag{6.93}$$

and N_f and N_b are the feedforward and feedback filter lengths, respectively.

6.8.6.3 Bussgang-type algorithm [625 (Chapter 2)]

The algorithm is defined by

$$w^{n+1}(k) = w^n(k) + \mu\{\hat{g}(y(n)) - y(n)\}\tilde{x}(n - k) \tag{6.94}$$

where $\hat{g}(y(n))$ is some non-linear function of $y(n)$. Provided that N_f and N_b in Equation (6.94) are large enough, the symbol estimate $\{y(n)\}$ is *Bussgang* (see [626, Chapter 18], [627]). A stochastic process $\{y(n)\}$ is said to be a *Bussgang process* if it satisfies the condition $E\{y(n)y(n + k)\} = E\{y(n)g(y(n + k))\}$ where the function $g(\cdot)$ is a zero memory non-linearity.

Since the feedforward and feedback parts can be updated separately, consider only the feedback part of Equation (6.94). Then, the Bussgang-type DFE algorithm for the feedback part is

$$b^{n+1}(k) = b^n(k) + \mu\{\tilde{g}(y(n)) - y(n)\}r(n - k) \tag{6.95}$$

Comparing Equation (6.95) to Equations (6.82) and (6.90), $-r(n)$ in Equation (6.82), $-y(n)$ in Equation (6.84) and $y(n)\{1 - (3\beta/\alpha)y^2(n)\}$ in Equation (6.90) may be viewed to be equivalent to the error term $\{\hat{g}(y(n)) - y(n)\}$ in Equation (6.95). Thus, JEM-1, JEM-2 and JEM-4 can be considered to be special cases of the Bussgang-type DFE. By comparing Equation (6.95) to Equation (6.86), however, we note that JEM-3 cannot be categorized as a Bussgang-type algorithm.

6.8.6.4 Convergence of JEM algorithms

JEM-1 and JEM-2 are guaranteed to converge for super-Gaussian inputs but not for sub-Gaussian. The input $\{s(n)\}$ is sub-Gaussian (super-Gaussian) if the kurtosis $\gamma_s = E\{s(n)^4\} - 3[E\{s(n)^2\}]^2$ is less than zero (greater than zero). For example, for a BPSK signal, $\gamma_s = -2$ and hence it is sub-Gaussian.

Extensive simulations show, however, that these algorithms converge to the right solution in almost all the cases. Convergence of JEM-3 is not considered here since it cannot be categorized as a Bussgang-type algorithm. JEM-4 is guaranteed to converge. When $\alpha = \infty$ and $\beta = 1/\alpha = 0$, JEM-2 coincides with JEM-3.

6.8.7 Kurtosis-based algorithm

The JEM approach cannot guarantee that the maximization of the joint entropy leads to minimization of the mutual information. To compute the mutual information, the probability density function of $r(n)$, the output of the decision device, is required. The approximation of the mutual information (or contrast function) was derived by Comon (see [628]). Maximization of the contrast function, or minimization of the mutual information of the equalizer outputs, is equivalent to minimization of the sum of the autocumulants squared. This result was obtained by using the Edgeworth expansion to separate the constant mixtures of the independent input signals [628].

6.8.7.1 Modified signal model

Under Assumption (6.77), Equation (6.72) can be rewritten as:

$$\mathbf{x}(n) = [x(n) \quad s(n-1) \quad \cdots \quad s(n-N)]^T$$
$$\mathbf{y}(n) = [y(n) \quad s(n-1) \quad \cdots \quad s(n-N)]^T$$

(6.96)

The new equivalent channel model is

$$\mathbf{x}(n) = \mathbf{C} \cdot \mathbf{s}(n)$$

(6.97)

where

$$\mathbf{s}(n) = [s(n) \quad s(n-1) \quad \cdots \quad s(n-N)]^T$$

(6.98)

and

$$
\mathbf{C} = \begin{bmatrix} 1 & c(1) & \cdots & c(N) \\ 0 & 1 & \ddots & 0 \\ \vdots & \ddots & \ddots & \ddots \\ 0 & \ddots & \cdots & 1 \end{bmatrix}
\tag{6.99}
$$

The matrix $\mathbf{C}$ is full rank so there exists a unique solution to Equation (6.97). Here, the mixture matrix $\mathbf{C}$ and the source vector $\mathbf{s}(n)$ are partially unknown. Equalization is attained by estimating the matrix $\mathbf{B}$ in Equation (6.71). It can be easily shown that $\mathbf{B}$ is the inverse matrix of $\mathbf{C}$ when $c(n - k) = -b(n - k)$, $k = 1, \ldots, N$. To find the solution $\mathbf{B}$, several contrast functions can be used (see [628] and [629]).

6.8.7.2 Contrast function proposed in [629]

In this case we have

$$
J = \sum_{k=1}^{N+1} C_{4y_k}(0, 0, 0), = C_{4y}(0, 0, 0) + \text{Constant}
\tag{6.100}
$$

where $C_{4y_k}(0, 0, 0)$ is the kurtosis (all zero lag of the fourth-order cumulant) of $y(n)$. A criterion similar to one defined in Equation (6.100) for a single output FIR channel was originally proposed for multiinput multioutput (MIMO) channel equalization with MIMO-DFEs in [630]. The cost function for the present simplified case with only a single channel is

$$
J_{\text{AC}} = C_{4y}(0, 0, 0)
\tag{6.101}
$$

The update equation for the feedback filter coefficients based on the criterion in Equation (6.101) is:

$$
\begin{aligned}
b^{n+1}(k) = b^n(k) &+ \mu \text{sgn}\{\gamma_s\} \\
&\cdot [\, \hat{E}\{y(n)^3 r(n - k)\} - 3\hat{E}\{y(n)^2\} \\
&\cdot \hat{E}\{y(n) r(n - k)\}]
\end{aligned}
\tag{6.102}
$$

where $\hat{E}\{\cdot\}$ is the estimate of $E\{\cdot\}$. Note that the contrast-based cost function of Equation (6.100) is similar to the autocumulant (AC) criterion of Equation (6.101). Since the preprocessing stage (prewhitening of the observations) is missing in the AC approach, however, the proposed criterion can only be thought of as approximately minimizing the mutual information. Simulation examples indicate that the equalizer based on the AC criterion converges much faster than most existing algorithms.

6.8.7.3 Performance examples [613]

Example 1
An FIR non-minimum phase channel with impulse response given by

$$
c(n) = \delta[n] + 2\delta[n - 1]
$$

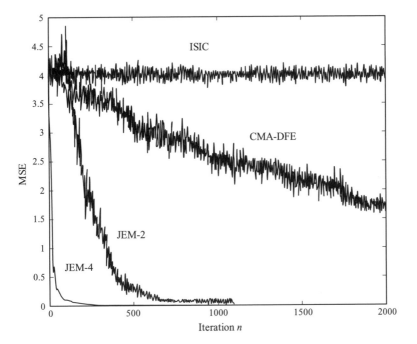

Figure 6.45 MSE comparison of ISIC, JEM-2, CMA-DFE and JEM-4: $(g(\cdot) = 3 \cdot \tanh[1/3y(n)])$ for JEM-2, $g(y(n)) = 3 \cdot y(n) + y^3(n)$ for JEM-4, $\mu_{ISIC} = \mu_{JEM-2} = \mu_{CMA-DFE} = 0.01, \mu_{JEM-4} = 0.001$ averaged over 100 Monte Carlo runs. SNR $= 20$ dB [613] © 1998, IEEE.

is used. The channel input $s(n)$ is a sequence of 2000 independent BPSK symbols, uniformly distributed over $\{1, -1\}$. The received symbols $x(n)$ consist of the distorted symbols $s(n)$ and additive noise at 20 dB. This data is used to update the one-tap feedback filter $(b(1))$, which is initialized with a zero. The results are shown in Figures 6.45 and 6.46.

Example 2
The performance of the adaptive blind algorithms was investigated with a multipath fading channel with impulse response

$$c(t) = \sum_{i=1}^{L_d} \varepsilon_i \, p(t - \tau_i)$$

where $p(t)$ is the pulse shape (raised cosine pulse). The delay $\{\tau_i\}$ is statistically independent and uniformly distributed in $[0, 3T]$ where T is the symbol interval. The attenuations $\{\varepsilon_i\}$ are independent zero mean Gaussian variables and the number of paths is six.

The equivalent discrete time baseband channel has a length of four symbol intervals $(c(k), \ k = 0, \ldots, 3)$. The channel coefficients $c(k)$ were normalized by $c(0)$. An array of three hundred sets of channels drawn from channel parameter distributions was generated. 10 000 independent uniformly distributed input symbol sequences $\{s(n)\}$ over $\{1, -1\}$ were used for each channel. For the sake of fair comparison, the step size was chosen to be the same for all the algorithms.

Table 6.3   SIR Comparison of (blind) DFE algorithms: $g(\cdot) = 3\tanh$
[$1/3y(n)$] for JEM-1 and JEM-2, $g(\cdot) = 3y(n) + y^3(n)$ for
JEM-4, and $\mu = 1e^{-4}$ for all algorithms, averaged over 185
simulated FIR channels: noise-free case [613] © 1998, IEEE

	SIR (dB)	Number of successes among a set of 300 channels
Unequalized	0.80	300
ISIC	4.87	300
JEM-2	5.70	300
CMA-DFE	10.88	295
JEM-4	58.92	185
DECA	6.16	300
AC	22.14	292
JEM-1	5.07	300

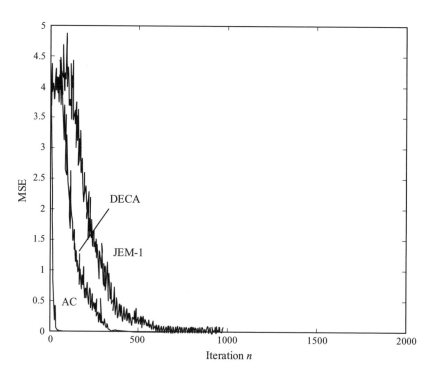

Figure 6.46 MSE comparison of AC, decorrelation algorithm (DECA) and JEM-1: $g(\cdot) =$
3 · tanh[$1/3y(n)$], $\mu_{AC} = \mu_{DECA} = \mu_{JEM-1} = 0.01$, averaged over 100 Monte
Carlo runs. SNR = 20 dB [613] © 1998, IEEE.

6.8.8 Performance results

Table 6.3 shows the results. Non-JEM-type algorithms yield about 5 dB–22 dB improvement in SIR. JEM-type algorithms exhibit an 11 dB–59 dB improvement. The JEM-4 algorithm is sensitive to step size. The third column of Table 6.3 shows the number of successful convergences. The JEM-4 algorithm became unstable with 115 simulated channels among 300. When the robustness and performance are considered simultaneously, JEM-1, JEM-2 or DECA are winners among the algorithms.

7

Orthogonal Frequency Division Multiplexing – OFDM and Multicarrier CDMA

The basic principles of the orthogonal frequency division multiplexing (OFDM) concept are presented in Chapter 1. Within this chapter we introduce further details with the main focus on synchronization, channel estimation, space–time frequency coding and some comparisons of efficiency in dealing with multipath propagation impacts by using equalization and TDMA or OFDM signal formats.

7.1 TIMING AND FREQUENCY OFFSET IN OFDM

As already indicated in Chapter 1, the transmitted OFDM signal can be represented as

$$x(t) = \sum_{k=-k_1}^{N+k_2+1} \sum_{n=0}^{N-1} D_k e^{j2\pi \frac{nk}{N}} w\left(t - \frac{k}{f_s}\right) \tag{7.1}$$

for $-k_1/f_s < t < (N + k_2)/f$, where k_1 and k_2 are pre- and postfix lengths and $w(t)$ is the time domain window function. The received signal $r(t)$ is filtered and sampled at the rate, or multiples of, $1/T$.

Advanced Wireless Communications. S. Glisic
© 2004 John Wiley & Sons, Ltd. ISBN: 0-470-86776-0

The sampled signal at the output of the receiver FFT with ideal channel can be represented as a convolution:

$$
y_n = \left[\sum_{k=-\infty}^{\infty} X_c \left(f + \frac{Nk}{T} \right) \right] \otimes T W(fT) \big|_{f=\frac{n}{T}}
\tag{7.2}
$$

where $X_c(f)$ is the Fourier transform of the periodically repeated analog equivalent of the signal generated by the transmitter's IFFT.

$X_c(f)$ is a line spectrum at k/T and $W(f)$ is the Fourier transform of window function $w(t)$.

Assuming that the sampling time has a relative phase offset of τ and that the offset does not change during one OFDM symbol, the sampled received signal for a non-dispersive channel can be simplified to:

$$
y_k = \sum_{n=0}^{N-1} D_n e^{j\varphi} e^{j2\pi \frac{n}{N} f_c t} \bigg|_{t=\dfrac{k+\tau}{f_s}}
\tag{7.3}
$$

where φ represents envelope delay distortion. After the Fourier transform at the receiver we have

$$
\begin{aligned}
\tilde{D}_m &= \frac{1}{N} \sum_{k=0}^{N-1} \sum_{n=0}^{N-1} D_n e^{j2\pi \frac{n}{N} k} e^{j\left(\varphi + 2\pi \frac{n}{N}\tau\right)} e^{-j2\pi \frac{m}{N} k} \\
&= \sum_{n=0}^{N-1} D_n e^{j\left(\varphi + 2\pi \frac{n}{N}\tau\right)} \sum_{k=0}^{N-1} e^{-j2\pi \frac{k}{N}(n-m)} = \begin{cases} 0, n \neq m \\ D_m e^{j(\varphi + 2\pi n\tau/N)}, n = m \end{cases}
\end{aligned}
\tag{7.4}
$$

The timing phase (error), or envelope delay distortion, does not violate the orthogonality of the subcarriers and the effect of the timing offset is a phase rotation which linearly changes with subcarriers' orders. On the other hand, envelope delay results in the same amount of rotation for all subcarriers.

It is straightforward to show that in a more general case, with pulse shaping filter with rolloff factor α and dispersive channel with impulse response $h(t)$, the detected data is attenuated and phase rotated such that $\tilde{D}_m = \gamma_m(\tau) D_m$, where:

$$
\gamma_m(\tau) = \begin{cases} H\left(\dfrac{m}{NT}\right) e^{j2\pi \frac{m}{NT}\tau}, 0 \le \dfrac{m}{N} \le \dfrac{1-\alpha}{2} \\[2ex] H\left(\dfrac{m}{NT}\right) e^{j2\pi \frac{m}{NT}\tau} + H\left(\dfrac{m-N}{NT}\right) e^{j2\pi \frac{m-N}{NT}\tau}, \dfrac{1-\alpha}{2} \le \dfrac{m}{N} \le \dfrac{1+\alpha}{2} \\[2ex] H\left(\dfrac{m-N}{NT}\right) e^{j2\pi \frac{m-N}{NT}\tau}, \dfrac{1+\alpha}{2} \le \dfrac{m}{N} \le 1 \end{cases}
\tag{7.5}
$$

where $H(m/NT)$ is the Fourier transform of $h(t)$ at frequency m/NT. The estimation of τ and φ will be discussed later in this section.

We will now see that frequency offset amounts to inter-channel interference which is similar to inter-symbol interference of a single carrier signal due to a timing jitter. In a non-dispersive channel with rectangular pulse shaping, the interference caused by frequency offset could be too constraining. The sampled signal is:

$$
y_k = \sum_{n=0}^{N-1} D_k e^{j2\pi t\left(\frac{n}{N} f_s + \delta f\right)} \bigg|_{t=\frac{k+\tau}{f_s}} = \sum_{n=0}^{N-1} D_k e^{j2\pi\left(\frac{n}{N} + \frac{\delta f}{f_s}\right)k}
\tag{7.6}
$$

After DFT we have

$$\tilde{D} = D_m \left(\frac{e^{j2\pi\Delta f} - 1}{e^{j2\pi\frac{\Delta f}{N}} - 1} \right) + \sum_{n=0}^{N-1} D_n \sum_{k=0}^{N-1} e^{j2\pi\frac{k}{N}(n-m+\Delta f)} + N_m \tag{7.7}$$

where $\Delta f = n\delta/f_s$. So, due to the frequency offset we have attenuation of the desired signal (the first term of Equation (7.7)), and an interference between different symbols of several subcarriers (the second term of Equation (7.7)).

In order to avoid frequency offset, the window function

$$w_n = w(t)|_{t=nT} \tag{7.8}$$

must be such that zero crossings of its Fourier transform are at multiples of symbol frequency

$$W_m = W(\omega)|_{\omega=2\pi m f_s} = \delta_m \tag{7.9}$$

A generalized *sinc* function of the form

$$\frac{\sin \omega n}{\omega n} \times g(n) \tag{7.10}$$

for any differentiable function, $g(t)$ satisfies the condition.

Another desired property of a sinc function is its low rate of change in the vicinity of in-frequency sampling points. A typical example is the raised cosine function in time:

$$w_{rc} = \begin{cases} T, & 0 \le |t| \le \frac{1-\beta}{2T} \\ \frac{\sin \pi \frac{t}{T}}{\pi \frac{t}{T}} \times \frac{\cos \beta\pi \frac{t}{T}}{1 - 4\beta^2 \frac{t^2}{T^2}}, & \frac{1-\beta}{2T} \le |t| \le \frac{1+\beta}{2T} \\ 0, & \text{elsewhere} \end{cases} \tag{7.11}$$

where β is the rolloff factor for time domain pulse shaping. A higher rolloff factor requires a longer cyclic extension and a guard interval, which consumes higher bandwidth. The results for adjacent channel interference (ACI) caused by frequency offset in systems with different windowing functions are shown in Figure 7.1.

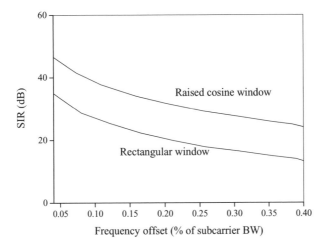

Figure 7.1 ACI caused by frequency offset.

7.1.1 Robust frequency and timing synchronization for OFDM

7.1.1.1 Symbol timing estimation algorithm

The symbol timing recovery relies on searching for a training symbol with two identical halves in the time. Consider the first training symbol where the first half is identical to the second half (in time order), except for a phase shift caused by the carrier frequency offset. If the conjugate of a sample from the first half is multiplied by the corresponding sample from the second half ($T/2$ seconds later), the effect of the channel should cancel, and the result will have a phase of approximately $\phi = \pi T \Delta f$. At the start of the frame, the products of each of these pairs of samples will have approximately the same phase, so the magnitude of the sum will be a large value. Let us use L complex samples in one half of the first training symbol (excluding the cyclic prefix), and let the sum of the pairs of products be

$$P(d) = \sum_{m=0}^{L-1} (r^*_{d+m} r_{d+m+L}) \tag{7.12}$$

This can be implemented with the iterative formula

$$P(d+1) = P(d) + (r^*_{d+L} r_{d+2L}) - (r^*_d r_{d+L}) \tag{7.13}$$

where d is a time index corresponding to the first sample in a window of $2L$ samples. This window slides along in time as the receiver searches for the first training symbol. The received energy for the second half symbol is defined by

$$R(d) = \sum_{m=0}^{L-1} |r_{d+m+L}|^2 \tag{7.14}$$

This can also be calculated iteratively. $R(d)$ may be used as part of an automatic gain control (AGC) loop. A timing metric can be defined as

$$M(d) = \frac{|P(d)|^2}{(R(d))^2}. \tag{7.15}$$

Equation (7.15) is shown in Figure 7.2 and Figure 7.3.

For the results in these figures, OFDM symbols are generated with 1000 frequencies, -500 to 499, and slightly oversampled at a rate of 1024 samples for the useful part of each symbol. In an actual hardware implementation, the ratio of the sampling rate to the number of frequencies would be higher to ease filtering requirements. The guard interval is set to about 10 % of the useful part, which is 102 samples.

7.1.1.2 Carrier frequency offset estimation algorithm

The main difference between the two halves of the first training symbol will be a phase difference of

$$\phi = \pi T \Delta f \tag{7.16}$$

which can be estimated by

$$\hat{\phi} = \text{angle}(P(d)) \tag{7.17}$$

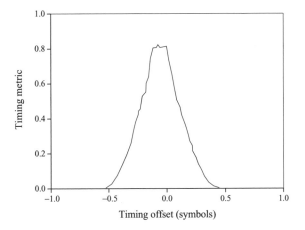

Figure 7.2 The timing metric for the AWGN channel (SNR = 10).

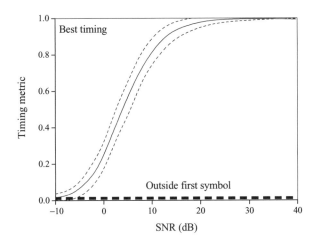

Figure 7.3 Expected value of timing metric with $L = 512$. Dashed lines indicate three standard deviations.

The second training symbol contains a PN sequence on the odd frequencies to measure these subchannels, and another PN sequence on the even frequencies to help determine frequency offset. If $|\hat{\phi}|$ can be guaranteed to be less than π, then the frequency offset estimate is

$$\Delta \hat{f} = \hat{\phi}/(\pi T) \tag{7.18}$$

and the even PN frequencies on the second training symbol would not be needed. Otherwise, the actual frequency offset would be

$$\frac{\phi}{\pi T} + \frac{2z}{T} \tag{7.19}$$

where z is an integer. By partially correcting the frequency offset, adjacent carrier interference (ACI) can be avoided, and then the remaining offset of $2z/T$ can be found. After the two training symbols are frequency corrected by $\hat{\phi}/(\pi T)$ (by multiplying the samples by $\exp(-j2t\hat{\phi}/T)$, let their FFTs be $x_{1,k}$ and $x_{2,k}$, and let the differentially-modulated PN sequence on the even frequencies of the second training symbol be v_k. The PN sequence v_k will appear at the output except it will be shifted by $2z$ positions because of the uncompensated frequency shift of $2z/T$. Note that because there is a guard interval and there is still a frequency offset, even if there were no differential modulation between training symbols 1 and 2 ($v_k = 1$) there would still be a phase shift between $x_{1,k}$ and $x_{2,k}$ of $2\pi(T + T_g)2z/T$. Since at this point the integer z is unknown, this additional phase shift is unknown. Since the phase shift is the same for each pair of frequencies, a metric similar to the previous one can be used.

Let X be the set of indices for the even frequency components,
$X = \{-W, -W + 2, \ldots, -4, -2, 2, 4, \ldots, W - 2, W\}$. The number of even positions shifted can be calculated by finding $\hat{g}$ to maximize

$$B(g) = \frac{\left|\sum_{k\in X} x^*_{1,k+2g} v^*_k x_{2,k+2g}\right|^2}{2\left(\sum_{k\in X} |x_{2,k}|^2\right)^2} \tag{7.20}$$

with integer g spanning the range of possible frequency offsets and W being the number of even frequencies with the PN sequence.

Then the frequency offset estimate would be:

$$\Delta\hat{f} = [\hat{\phi}/(\pi T)] + (2\hat{g}/T) \tag{7.21}$$

7.1.1.3 *Variance of carrier frequency offset estimator*

From Equations (7.18) and (7.20) we have:

$$\text{var}[\hat{\phi}/\pi] = \frac{1}{\pi^2 \cdot L \cdot \text{SNR}} \tag{7.22}$$

One should keep in mind the Cramer–Rao bound is

$$\text{var}[\hat{\phi}/\pi] \geq \frac{1}{\pi^2 \cdot L \cdot \text{SNR}} \tag{7.23}$$

At the correct frequency offset, all the signal products

$$s^*_{1,h+g_{\text{correct}}} v^*_h s_{2,h+g_{\text{correct}}}$$

have the same phase and

$$\mu_B = E[B(g_{\text{correct}})] = \frac{\sigma_s^4}{\left(\sigma_s^2 + \sigma_n^2\right)^2}$$

$$\text{var}[B(g_{\text{correct}})] = \frac{\sigma_s^4\left[(2 + 2\mu_B)\sigma_s^2\sigma_n^2 + (1 + 4\mu_B)\sigma_n^4\right]}{W\left(\sigma_s^2 + \sigma_n^2\right)^4} \tag{7.24}$$

At an incorrect frequency offset the signal products no longer add phase, and $B(g_{\text{incorrect}})$ has a chi-squared distribution with two degrees of freedom with

$$
\begin{aligned}
E[B(g_{\text{correct}})] &= \frac{1}{2W}\left(1 + \frac{\sigma_s^2}{\sigma_s^2 + \sigma_n^2}\right) < \frac{1}{W} \\
\text{var}[B(g_{\text{correct}})] &= \frac{1}{4W^2}\left(1 + \frac{3\sigma_s^4 + 2\sigma_s^2\sigma_n^2}{\left(\sigma_s^2 + \sigma_n^2\right)^2}\right) < \frac{1}{W^2}
\end{aligned}
\tag{7.25}
$$

7.2 FADING CHANNEL ESTIMATION FOR OFDM SYSTEMS

7.2.1 Statistics of mobile radio channels

The channel impulse response will be represented as

$$
h(t, \tau) = \sum_k \gamma_k(t)\delta(\tau - \tau_k)
\tag{7.26}
$$

Assume that $\gamma_k(t)$ has the same normalized time correlation function $r_t(\Delta t)$ for all k, and power spectrum $p_t(\Omega)$.

$$
r_{\gamma k}(\Delta t) \overset{\Delta}{=} E\{\gamma_k(t + \Delta t)\gamma_k^*(t)\} = \sigma_k^2 r_t(\Delta t)
\tag{7.27}
$$

where σ_k^2 is the average power of the kth path.

The frequency response of the time-varying radio channel at time t is

$$
H(t, f) \overset{\Delta}{=} \int_{-\infty}^{\infty} h(t, \tau)e^{-j2\pi f\tau}\,d\tau = \sum_k \gamma_k(t)e^{-j2\pi f\tau_k}
\tag{7.28}
$$

The time–frequency correlation is defined as

$$
\begin{aligned}
r_H(\Delta t, \Delta f) &\overset{\Delta}{=} E\{H(t + \Delta t, f + \Delta f)H^*(t, f)\} \\
&= \sum_k r_{\gamma k}(\Delta t)e^{-j2\pi\Delta f\tau_k} \\
&= r_t(\Delta t)\left(\sum_k \sigma_k^2 e^{-j2\pi\Delta f\tau_k}\right) \\
&= \sigma_H^2 r_t(\Delta t)r_f(\Delta f)
\end{aligned}
\tag{7.29}
$$

where $\sigma_H^2 \overset{\Delta}{=} \sum_k \sigma_k^2$ is the total average power of the channel impulse response. The frequency correlation is defined as

$$
r_f(\Delta f) = \sum_k \frac{\sigma_k^2}{\sigma_H^2}e^{-j2\pi\Delta f\tau_k}
\tag{7.30}
$$

where $r_t(0) = r_f(0) = 1$. Without loss of generality, we also assume that $\sigma_H^2 = 1$, so that it can be omitted. For block length T_f and tone spacing (subchannel spacing) Δf, the correlation function for different blocks and tones is

$$r_H[n, k] = r_t[n] r_f[k] \tag{7.31}$$

where

$$r_t[n] \triangleq r_t(nT_f) \quad r_f[k] \triangleq r_f(k\Delta f)$$

From Jakes's model of the channel we have:

$$r_t[n] = J_0(n\omega_d) \triangleq r_J[n]$$

$$p_J(\omega) = \begin{cases} \dfrac{2}{\omega_d} \dfrac{1}{\sqrt{1 - (\omega/\omega_d)^2}}, & if\,|\omega| < \omega_d \\ 0, & \text{otherwise} \end{cases} \tag{7.32}$$

As an example, for carrier frequency $f_c = 2\,\text{GHz}$, $f_d = 184\,\text{Hz}$ when the user is moving at 60 mph.

7.2.2 Diversity receiver

In the case of space (antenna) diversity, the signal from the mth antenna at the kth tone and the nth block can be represented as

$$x_m[n, k] = H_m[n, k]a[n, k] + \omega_m[n, k] \tag{7.33}$$

where $\omega_m[n, k]$ is additive Gaussian noise from the mth antenna at the kth tone and the nth block, with zero mean and variance ρ. Let us assume that $\omega_m[n, k]$ is independent for different ns, ks, or ms. $H_m[n, k]$, the frequency response at the kth tone and the nth block corresponding to the mth antenna, is assumed independent for different ms, but with the same statistics. $a[n, k]$ is the signal modulating the kth tone during the nth block and is assumed to have unit variance and be independent for different ks and ns.

With knowledge of the channel parameters, $a[n, k]$ can be estimated as $y[n, k]$ by an MMSE combiner

$$y[n, k] = \frac{\displaystyle\sum_{m=1}^{p} H_m^*[n, k]x_m[n, k]}{\displaystyle\sum_{m=1}^{p} |H_m[n, k]|^2} \tag{7.34}$$

The transceiver (transmitter/receiver) block diagram is given in Figure 7.4.

7.2.3 MMSE channel estimation

If the reference $a[n, k]$ is ideal (pilot symbols), a temporal estimation of $H[n, k]$ is

$$\tilde{H}[n, k] = x[n, k]a^*[n, k] \cong H[n, k] + \omega[n, k]a^*[n, k] \tag{7.35}$$

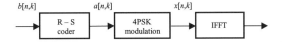

(a)

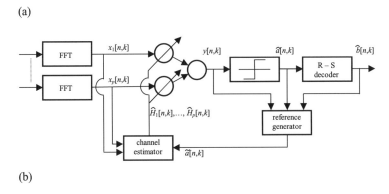

(b)

Figure 7.4 OFDM system with channel estimator. (a) Transmitter; (b) diversity receiver.

where * denotes the complex conjugate. $\tilde{H}[n, k]$s for different ns and ks are correlated; therefore, an MMSE channel estimator can be constructed as follows:

$$\hat{H}[n, k] = \sum_{l} \sum_{m=-\infty}^{0} c[m, l, k]\tilde{H}[n - m, k - l] \qquad (7.36)$$

where $c[m, l, k]$s are selected to minimize

$$\text{MSE}(\{c[m, l, k]\}) = E|\tilde{H}[n, k,] - H[n, k]|^2 \qquad (7.37)$$

If K is the number of tones in each OFDM block, then we will be using the following notation

$$c[m, k] \stackrel{\Delta}{=} \begin{pmatrix} c[m, k - 1, k] \\ \vdots \\ c[m, 0, k] \\ \vdots \\ c[m, -K + k, k] \end{pmatrix} \qquad \begin{aligned} c(\omega; k) &\stackrel{\Delta}{=} \sum_{n=-\infty}^{0} c[n, k]e^{-jn\omega} \\ \mathbf{C}(\omega) &\stackrel{\Delta}{=} (c(\omega; 1), c(\omega; 2), \cdots, c(\omega; K)) \end{aligned} \qquad (7.38)$$

Starting from the fact that the projection of the estimation error on $H[n, k]$ is zero, $E\{(\hat{H}[n, k] - H[n, k])\tilde{H}^*[n - m, k - l]\} = 0$ (for orthogonality principles see Chapter 5), the estimation coefficients are given by [631]:

$$\mathbf{C}(\omega) = \mathbf{U}^H \mathbf{\Phi}(\omega)\mathbf{U} \qquad (7.39)$$

where $\Phi(\omega)$ is a diagonal matrix with the lth element

$$\Phi_l(\omega) = 1 - \frac{1}{M_l(-\omega)\gamma_l[0]} \tag{7.40}$$

and $M_l(\omega)$ is a stable one-sided FT

$$M_l(\omega) = \sum_{n=0}^{\infty} \gamma_l[n]e^{-jn\omega} \tag{7.41}$$

$$M_l(\omega)M_l(-\omega) = \frac{d_l}{\rho} p_t(\omega) + 1 \tag{7.42}$$

The DC component $\gamma_l[0]$ in $M_l(\omega)$ can be found by

$$\gamma_l^2[0] = \exp\left\{\frac{1}{2\pi}\int_{-\pi}^{\pi}\ln\left[\frac{d_l}{\rho} p_t(\omega) + 1\right]d\omega\right\} \tag{7.43}$$

The d_ls and u_ls are the corresponding eigenvalues and eigenvectors of the frequency domain correlation matrix $\mathbf{R}_f$,

$$\mathbf{R}_f \overset{\Delta}{=} \begin{bmatrix} r_f[0] & r_f[1] & \cdots & r_f[K-1] \\ r_f[-1] & r_f[0] & \cdots & r_f[K-2] \\ \vdots & \vdots & \ddots & \vdots \\ r_f[-K+1] & r_f[-K+2] & \cdots & r_f[0] \end{bmatrix} \tag{7.44}$$

$\mathbf{U} = (u_1, \ldots, u_k)$ is a unitary matrix.

$$\mathbf{R}_f = \bar{\mathbf{U}}^{\mathrm{H}}\mathbf{D}\bar{\mathbf{U}} \quad \text{or} \quad \bar{\mathbf{U}}\mathbf{R}_f\bar{\mathbf{U}}^{\mathrm{H}} = \mathbf{D} \tag{7.45}$$

and $\mathbf{D} = \mathrm{diag}\{d_1, \ldots, d_K\}$ and $\sum_k d_k = K$. The processing is illustrated in Figure 7.5.

The unitary linear inverse transform $\mathbf{U}^{\mathrm{H}}$ and transform $\mathbf{U}$ in the figure perform the eigendecomposition of the frequency domain correlation. The estimator turns off the zero or small d_ls to reduce the estimation noise. For those large d_ls, linear filters are used to take advantage of the time domain correlation.

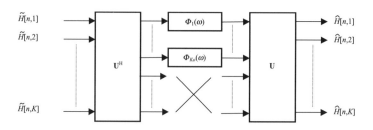

Figure 7.5 Channel estimator for OFDM systems.

One can show [631] that for *Jakes's model*:

$$\overline{\text{MMSE}}_J(\omega_d) = \frac{\rho}{K} \sum_{l-1}^{K} \left(1 - \left(\frac{\alpha_l}{2} \right)^{-(\omega_d)\pi} \exp\left\{ -\frac{\omega_d b(\alpha_l)}{\pi} \right\} \right) \tag{7.46}$$

$$\alpha_l \triangleq \frac{2d_l}{\omega_d \rho}$$

$$b(\alpha_l) \triangleq \begin{cases} \dfrac{\pi}{2}\alpha_l + \sqrt{1 - \alpha_l^2} \ln \dfrac{1 + \sqrt{1 - \alpha_l^2}}{\alpha_l}, & \text{if } \alpha_l < 1 \\[2ex] \dfrac{\pi}{2}\alpha_l - \sqrt{\alpha_l^2 - 1}\left(\dfrac{\pi}{2} - \arcsin \dfrac{1}{\alpha_l} \right), & \text{if } \alpha_l \geq 1 \end{cases}$$

7.2.4 FIR channel estimator

For a reader less familiar with eigenvalue decomposition, discussed in Chapter 5, we provide a simplified interpretation of the processing by using an approximation. Instead of proper eigenvectors we use DFT matrix $\mathbf{W}$ defined as

$$\mathbf{W} \triangleq \frac{1}{\sqrt{K}} \begin{pmatrix} 1 & 1 & \cdots & 1 \\ 1 & e^{j(2\pi/K)} & \cdots & e^{j(2\pi(K-1)/K)} \\ \vdots & \vdots & \ddots & \vdots \\ 1 & e^{j(2\pi(K-1)/K)} & \cdots & e^{j(2\pi(K-1)(K-1)/K)} \end{pmatrix} \tag{7.47}$$

and with notation

$$\bar{\mathbf{R}}_t = \begin{pmatrix} \bar{r}_t[0] & \bar{r}_t[1] & \cdots & \bar{r}_t[L-1] \\ \bar{r}_t[-1] & \bar{r}_t[0] & \cdots & \bar{r}_t[L-2] \\ \vdots & \vdots & \ddots & \vdots \\ \bar{r}_t[-L+1] & \bar{r}_t[-L+2] & \cdots & \bar{r}_t[0] \end{pmatrix} \tag{7.48}$$

$$\bar{\mathbf{r}}_t = (\bar{r}_t[0], \ \bar{r}_t[1], \cdots, \bar{r}_t[L-1])^{\text{T}}$$

$$\bar{r}_t[n] = \frac{\sin(n\omega_d)}{n\omega_d}$$

we have, for the coefficient matrix of the designed FIR channel estimator,

$$\bar{\mathbf{C}}(\omega) = \mathbf{W}^{\text{H}} \bar{\mathbf{\Phi}}(\omega) \mathbf{W} \tag{7.49}$$

with

$$\bar{\mathbf{\Phi}}(\omega) = \text{diag}\{\underbrace{c(\omega), \ldots, c(\omega)}_{K_0 \text{ elements}}, 0, \ldots 0\}$$

If the maximum delay spread is t_d, then for all $l \leq K_0$ ($K_0 = \lceil K t_d / T_s \rceil$), $d_l \approx 0$, where T_s is the symbol interval. In Equation (7.49), $c(\omega)$ is the FT of c_n given by

$$(c_0, c_1, c_2, \ldots, c_{L-1})^{\mathrm{T}} = \left(\bar{\mathbf{R}}_t + \frac{K_o \rho}{K} \mathbf{I} \right)^{-1} \bar{r}_t$$

and L is the length of the FIR estimator. The estimation error

$$\overline{\mathrm{MSE}} = \frac{K_o c_0 \rho}{K} \tag{7.50}$$

For the robust FIR channel estimator, the $\mathbf{U}$ in Figure 7.5 is the DFT matrix $\mathbf{W}$ and the $\Phi_k(\omega)$s for $k = 1, \ldots, K$ are $c(\omega)$.

7.2.5 System performance

The set of assumptions used to generate the performance curves is the same as in [631]: a two-way Rayleigh fading with delay from 0 to 40 µs and Doppler frequency from 10 to 200 Hz; the channels corresponding to different receivers have the same statistics; two antennas are used for receiver diversity. For the OFDM signal, it is assumed that the entire channel bandwidth, 800 kHz, is divided into 128 subchannels. The four subchannels on each end are used as guard tones and the rest (120 tones) are used to transmit data. To make the tones orthogonal to each other, the symbol duration is 160 µs. An additional 40µs guard interval is used to provide protection from ISI, the length $T_f = 200$ ms and a subchannel symbol rate $r_b = 5$ kBd.

To compare the performance of the OFDM system with and without the channel estimation, PSK modulation with coherent demodulation and differential PSK (DPSK) modulation with differential demodulation are used, respectively. The (40, 20) RS code, with each code symbol consisting of three quadrature PSK/differential quadrature PSK (QPSK/DQPSK) symbols grouped in frequency, is used in the system. Each OFDM block forms an RS code word. The RS decoder erases ten symbols, based on signal strength, and corrects five additional random errors. Hence, the simulated system can transmit data at 1.2 Mb/s before decoding or 600 kb/s after decoding, over an 800 kHz channel.

7.2.6 Reference generation

Four different ways to generate the reference signal are used:

1. *Undecoded/decoded dual mode reference*: If the RS decoder can successfully correct all errors in an OFDM block, the reference for the block can be generated by the decoded data; hence $\tilde{a}[n, k] = a[n, k]$. Otherwise, $\tilde{a}[n, k] = \hat{a}[n, k]$.

2. *Undecoded reference*: $\tilde{a}[n, k] = \hat{a}[n, k]$, no matter whether the RS decoder can successfully correct all errors in a block or not.

3. *Decoded/CMA dual mode reference*: The constant modulus algorithm (CMA) is used to generate a reference for the OFDM channel estimator. If the RS decoder can successfully correct all errors in a block, the reference for the block can be generated from the decoded data; hence $\tilde{a}[n, k] = a[n, k]$. Otherwise, the reference can use the projection of $y[n, k]$ on the unit circle, i.e. $\tilde{a}[n, k] = y[n, k]/\mathrm{mod}(y[n, k])$.

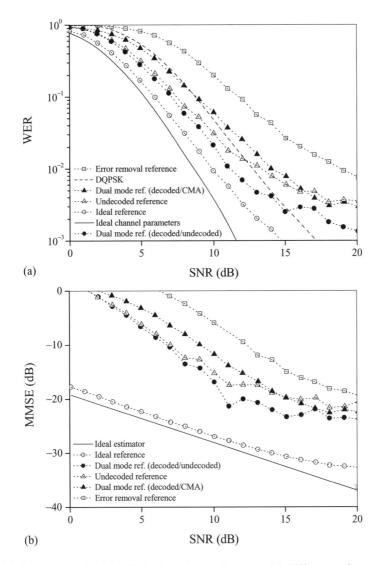

(a)

(b)

Figure 7.6 (a) WER and (b) MMSE of a robust estimator with different references versus
SNR for the system with 1 % training blocks when the 50-tap channel estimator
matches the channel with $f_d = 200$ Hz and $t_d = 5\mu$s.

4. *Error removal reference*: If the RS decoder can successfully correct all errors in a
   block, the reference for the block can be generated by the decoded data. Otherwise,
   the $\tilde{H}[n-1, k]$s are used instead of the $\tilde{H}[n, k]$s for $k = 1, \ldots, K$ respectively.

The results are shown in Figure 7.6.
In general a careful study of the results based on the previous discussion suggests the
following conclusions:

- If a channel estimator is designed to match the channel with 40 Hz maximum Doppler frequency and 20μs maximum delay spread, then for all channels with $f_d \leq 40$ Hz and $t_d \leq 20$ μs, the system performance is not worse than the channel with $f_d = 40$ Hz and $t_d = 20$ μs. For channels with $f_d > 40$ Hz or $t_d > 20$ μs, such as $f_d = 80$ Hz and $t_d = 20$ μs or $f_d = 40$ Hz and $t_d = 40$ μs, the system performance degrades dramatically.

- If the estimator is designed to match a Doppler frequency or delay spread larger than the actual ones, the system performance degrades only slightly compared with estimation that exactly matches the channel Doppler frequency and delay spread.

More details on channel estimation can be found in [632–641].

7.3   64 DAPSK AND 64 QAM MODULATED OFDM SIGNALS

In this section we discuss in more detail specific, high constellation, modulation schemes for OFDM signals. The transmission system and two options for the signal constellation are shown in Figures 7.7. and 7.8.

In this section we consider an OFDM system with $N = 1024$ subcarriers and parameters the same as in [642]. Each OFDM symbol transfers 6144 bits in total. The convolutional code has the memory length $m = 6$. The bit interleaver is a block interleaver with 83 rows and 74 columns, which means that only two bits of an OFDM symbol are not involved in the interleaving process. The performance of convolutional codes with code rates $1/2, 2/3$ and $3/4$ has been analyzed.

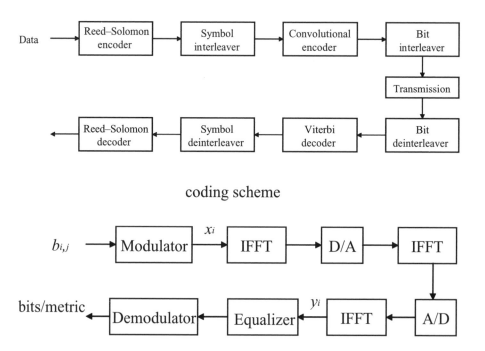

Figure 7.7 Transmission system.

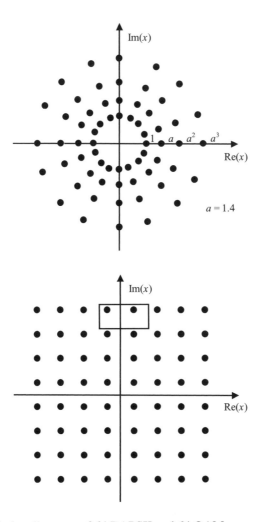

Figure 7.8 Constellation diagrams of 64 DAPSK and 64 QAM.

A (204, 188) RS code with $p = 8$ b/symbol has been chosen for the DTVB application. The objective of the concatenated code is to fulfil the requirement of a residual bit error rate (BER) of $10E^{-11}$ at the output of the RS decoder. For this, the BER at the output of the Viterbi decoder is required to be lower than $2 \times 10E^{-4}$ [643]. The above parameters are used in the system for digital terrestrial video broadcasting [644–647]. Such a system can transmit 34 Mbits/s over an 8 MHz radio channel.

The phase modulation is independent of the amplitude and identical to the well known 16 DPSK.

The input bits $b_{i,0}$, $b_{i,1}$, $b_{i,2}$, $b_{i,3}$ are used for this differential phase modulation in the 64 DAPSK scheme. The amplitude states $|x_i|$ are chosen from the constellation diagram depending on the previous amplitude state $|\tilde{x}_i|$ and the two information bits $b_{i,5}$ and $b_{i,4}$ according to Table 7.1. This means, e.g. for a subcarrier i if the amplitude in the previous OFDM symbol was $|\tilde{x}_i|$ and the input information bits $b_{i,5}$ and $b_{i,4}$ were both zero, then

Table 7.1 Differential amplitude modulation for 64 DAPSK. Choice of the current amplitude state $|x_i|$ depending on the previous state $|\tilde{x}_i|$ and the amplitude bits

	Amplitude bits $b_{i,4}b_{i,5}$			
$|\tilde{x}_i|$	00	01	11	10
1	1	a	a^2	a^3
a	a	a^2	a^3	1
a^2	a^2	a^3	1	a
a^3	a^3	1	a	a^2

the amplitude in the current OFDM symbol would be $|x_i| = |\tilde{x}_i|$ again. Table 7.1 shows the amplitude state $|x_i|$ for the subcarrier in the current OFDM symbol.

In each OFDM receiver after block synchronization, analog-to-digital (A/D) conversion, and removing of the guard interval, a fast Fourier transform (FFT) will produce the complex output states y_i for each subcarrier i. If the coherent 64 QAM is used, after the FFT a channel equalization must be performed.

This means that the channel transfer factor α_i is assumed to be known exactly for each subcarrier i. With this information, the transmitted state x_i of each subcarrier in the 64 QAM constellation diagram is evaluated by simple quotient

$$x_i \approx y_i/\alpha_i$$

in order to get the coded bit sequence for hard decision decoding or the metric increments for soft decision decoding.

For the non-coherent 64 DAPSK demodulation, first the quotient

$$r_i = y_i/\tilde{y}_i \approx x_i/\tilde{x}_i$$

of the currently received state y_i and the preceding state $\tilde{y}_i$ of the same subcarrier i in the receiver (FFT output) are calculated. The resulting complex quotient r_i is evaluated in order to get the phase and amplitude bits $b_{i,0}, b_{i,1}, \ldots, b_{i,5}$. This quotient r_i is nearly independent of channel transfer factor α_i if the radio channel does not change the transmission behavior too quickly. Therefore, pilot symbols, channel estimation and equalization are not needed in the 64 DAPSK receiver, which reduces the computation complexity.

The four bits $b_{i,0}, b_{i,1}, b_{i,2}, b_{i,3}$ are determined depending on the phase difference between y_i and $\tilde{y}_i$ only. This phase demodulation process is the same as the demodulation of 16 DPSK.

For the amplitude demodulation, Table 7.2 shows how the amplitude bits $b_{i,4}$ and $b_{i,5}$ are obtained. For the evaluation of $|r_i|$, simple amplitude thresholds are used.

Performance results are given in Figure 7.9 and Tables 7.3–7.5. On average, 4 dB SNR degradation must be accepted in order to have much simpler-to-implement DAPSK. More details on coherent APSK and DAPSK can be found in [642–652].

Table 7.2   Evaluation of the amplitude information bits $b_{i,5}$ and $b_{i,4}$ in the 64 DAPSK demodulation

				$|r_i| = |y_i/\tilde{y}_i$				
	a^{-3}	a^{-2}	a^{-1}	1	a^1	a^2	a^3
$b_{i,4}b_{i,5}$	01	11	10	00	01	11	10
thresholds		$a^{-2.5}$	$a^{-1.5}$	$a^{-0.5}$	$a^{-0.5}$	$a^{1.5}$	$a^{-2.5}$
	$a^{-2.5}$	$a^{-1.5}$	$a^{-0.5}$	$a^{0.5}$	$a^{1.5}$	$a^{2.5}$	$a^{3.5}$

Table 7.3   SNR [dB] required for BER $= 2 \times 10^{-4}$

		Ideal 64 QAM			64 DAPSK		
Code rate R	Metric	AWGN channel	Rice channel	Rayleigh channel	AWGN channel	Rice channel	Rayleigh channel
1/2	hard decision	17.3	19.6	24.1	21.4	23.7	28.6
2/3	hard decision	19.5	22.4	28.6	24.0	26.5	33.2
3/4	hard decision	20.9	23.9	32.8	25.7	28.2	37.4
1/2	soft decision	13.6	15.6	18.6	17.7	19.5	22.3
2/3	soft decision	16.8	19.1	22.4	20.6	22.8	26.5
3/4	soft decision	18.5	20.8	24.9	22.5	24.8	29.3

Table 7.4   Difference of required SNR [dB] for 64 QAM (ideal channel estimation) and 64 DAPSK modulation at BER $= 2 \times 10^{-4}$, fixed code rates

	Hard decision			Soft decision		
Code rate R	AWGN channel	Rice channel	Rayleigh channel	AWGN channel	Hire channel	Rayleigh channel
1/2	4.1	4.1	4.5	4.1	3.9	3.7
2/3	4.5	4.1	4.6	3.8	3.7	4.1
3/4	4.8	4.3	4.6	4.0	4.0	4.4

Table 7.5   Difference of required SNR [dB] for 64 QAM (ideal channel estimation) with $R = 3/4$ and 64 DAPSK modulation with $R = 2/3$ at BER $= 2 \times 10^{-4}$, fixed user data rate

	Hard decision			Soft decision		
Code rate R	AWGN channel	Rice channel	Rayleigh channel	AWGN channel	Rice channel	Rayleigh channel
QAM: 3/4 DAPSK: 2/3	3.1	2.6	0.4	2.1	2.0	1.6

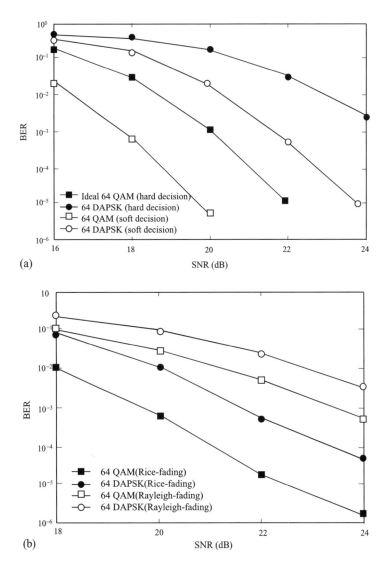

Figure 7.9 (a) Performance of ideal 64 QAM and 64 DAPSK (ring ratio $a = 1.4$) with a convolutional code $m = 6$, $R = 3/4$ for hard and soft decision decoding in an AWGN channel; (b) performance of ideal 64 QAM (convolutional code $R = 3/4$) and 64 DAPSK ($R = 2/3$) over the Rayleigh and Rice-fading channel (fixed user data rate).

7.4 SPACE–TIME CODING WITH OFDM SIGNALS

In this section we discuss space–time coding with an OFDM signal. The system block diagram is shown in Figure 7.10.

Table 7.6   Modulation parameters

Modulation scheme	Bits per symbol BPS	Decoding algorithm	No. of states	No. of transmitters	No. of termination symbols
4PSK	2	VA	4	2	1
			8	2	2
			16	2	2
			32	2	3
8PSK	3	VA	8	2	1
			16	2	2
			32	2	2

Table 7.7   The parameters associated with the turbo convolutional (n, k, K) TC(2, 1, 3) code

Code	Code rate R	Modulation mode	BPS	Random turbo interleaver depth	Random separation interleaver depth
TC(2,1, 3)	0.50			128 carriers	
		16 QAM	2	256	512
		64 QAM	3	384	768
				512 carriers	
		QPSK	1	512	1024
		16 QAM	2	1024	2048

In the system, Alamouti's $\mathbf{G}_2$ space–time block code is used

$$\mathbf{G}_2 = \begin{pmatrix} x_1 & x_2 \\ -\bar{x}_2 & \bar{x}_1 \end{pmatrix}$$

For comparison the system parameters used in this section are the same as in [653] and are specified in Tables 7.6–7.8.

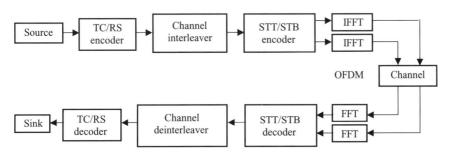

Figure 7.10 Turbo convolutional (TC) or Reed–Solomon (RS) code used with space–time trellis (STT) or space–time block (STB) encoder.

Table 7.8 The coding parameters of the Reed–Solomon codes

Code	Galois field	Rate	Correctable symbol errors
RS(105, 51)	2^{10}	0.49	27
RS(153, 102)	2^{10}	0.67	25

7.4.1 Signal and channel parameters

A two-ray channel impulse response having equal amplitudes and differential delay of 5 μs is used. The average signal power received from each transmitter antenna is the same. All multipath components undergo independent Rayleigh fading; Jakes's model. The receiver has a perfect knowledge of the CIR. A 128 subcarrier (160 μs) OFDM signal with a cyclic extension of 32 samples (40 μs) is used. The results are presented in Figure 7.11.

For increased delay spread and Doppler, the variation of the frequency domain fading envelope will eventually destroy the orthogonality of the space–time block code $\mathbf{G}_2$ (see Figure 7.13). For this reason the two transmission instants of the space–time block code $\mathbf{G}_2$ will have to be allocated to the same OFDM symbol. In the previous example they were allocated to the adjacent subcarriers. The transmission system for the time-varying channel is now modeled in Figure 7.12. The received signals are given by:

$$
\begin{aligned}
\tilde{x}_1 &= \bar{h}_{1,1}y_1 + h_{2,2}\bar{y}_2 \\
&= \bar{h}_{1,1}h_{1,1}x_1 + \bar{h}_{1,1}h_{2,1}x_2 + \bar{h}_{1,1}n_1 - h_{2,2}\bar{h}_{1,2}x_2 + h_{2,2}\bar{h}_{2,2}x_1 + h_{2,2}\bar{n}_2 \\
&= (|h_{1,1}|^2 + \{h_{2,2}|^2)x_1 + (\bar{h}_{1,1}h_{2,1} - h_{2,2}\bar{h}_{1,2})\,x_2 + \bar{h}_{1,1}n_1 + h_{2,2}\bar{n}_2 \\
\tilde{x}_2 &= \bar{h}_{2,1}y_1 + h_{1,2}\bar{y}_2 \\
&= \bar{h}_{2,1}h_{1,1}x_1 + \bar{h}_{2,1}h_{2,1}x_2 + \bar{h}_{2,1}n_1 - h_{1,2}\bar{h}_{1,2}1x_2 - h_{1,2}\bar{h}_{2,2}x_1 - h_{1,2}\bar{n}_2 \\
&= (|h_{2,1}|^2 + \{h_{1,2}|^2)x_2 + (\bar{h}_{2,1}h_{1,1} - h_{1,2}\bar{h}_{2,2})\,x_1 + \bar{h}_{2,1}n_1 - h_{1,2}\bar{n}_2 \quad (7.51)
\end{aligned}
$$

The signal to interference ratio (SIR) for signal x_1 is:

$$
\text{SIR} = \frac{|h_{1,1}|^2 + |h_{2,2}|^2}{\bar{h}_{1,1}h_{2,1} - h_{2,2}\bar{h}_{1,2}} \quad (7.52)
$$

and for signal x_2 is:

$$
\text{SIR} = \frac{|h_{2,1}|^2 + |h_{1,2}|^2}{\bar{h}_{2,1}h_{1,1} - h_{1,2}\bar{h}_{2,2}} \quad (7.53)
$$

The two transmission instants are no longer assumed to be associated with the same complex transfer function values. Performance curves are given in Figures 7.13 and 7.14.

The fading amplitude variation versus time is slower than that versus the subcarrier index within the OFDM symbols. This implies that the SIR attained would be higher, if we were to allocate the two transmission instants of the space–time block code $\mathbf{G}_2$ to the same subcarrier of consecutive OFDM symbols. This increase in SIR is achieved by doubling the delay of the system, since in this scenario two consecutive OFDM symbols have to be decoded.

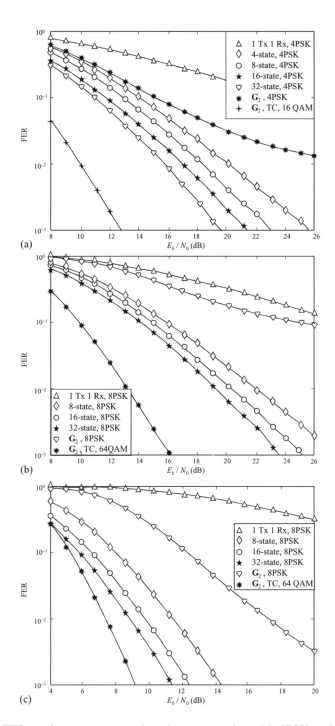

Figure 7.11 FER performance comparison between various (a) 4PSK and (b, c) 8PSK space–time trellis codes and the space–time block code $\mathbf{G}_2$ concatenated with the TC(2, 1, 3) code using (a, b) one or (c) two receivers and the 128 subcarrier OFDM modem over a channel having a CIR characterized by two equal-power rays separated by a delay spread of 5μs. The maximum Doppler frequency was 200 Hz. The effective throughput was (a) 2 BPS or (b, c) 3 BPS and the coding parameters are shown in Tables 7.6–7.7.

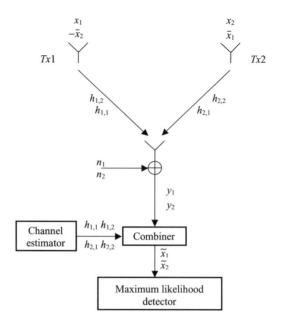

Figure 7.12 Baseband representation of the simple twin-transmitter space–time block code G_2 using one receiver over varying fading conditions.

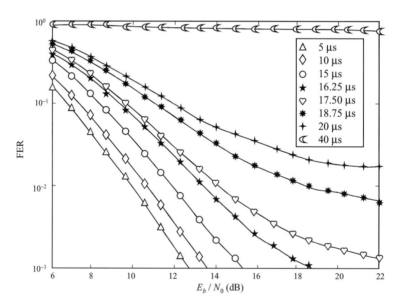

Figure 7.13 FER performance of the space–time block code G_2 concatenated with the TC(2, 1, 3) code using one receiver, the 128 subcarrier OFDM modem and 16 QAM. The CIR exhibits two equal-power rays separated by various delay spreads and a maximum Doppler frequency of 200 Hz. The coding parameters are shown in Tables 7.6–7.9.

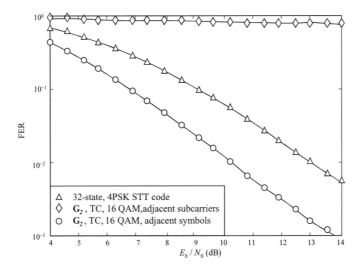

Figure 7.14 FER performance comparison between adjacent subcarriers and adjacent OFDM symbols allocation for the space–time block code G_2 concatenated with the TC(2, 1, 3) code using one receiver, the 128 subcarrier OFDM modem and 16 QAM over a channel having a CIR characterized by two equal-power rays separated by a delay spread of 40 μs. The maximum Doppler frequency was 100 Hz. The coding parameters are shown in Tables 7.6–7.9.

7.4.2 The wireless asynchronous transfer mode system

The CIR used in this case has three paths and is referred to as the shortened WATM CIR, as shown in Figure 7.15. A 512 subcarrier (2.2756 μs) OFDM time domain signal with a cyclic extension of 64 samples (0.2844 μs) is used. The sampling rate is 225 Msamples/s and the carrier frequency is 60 GHz. The results are shown in Figure 7.16.

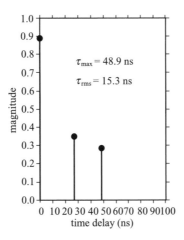

Figure 7.15 Short WATM channel impulse response.

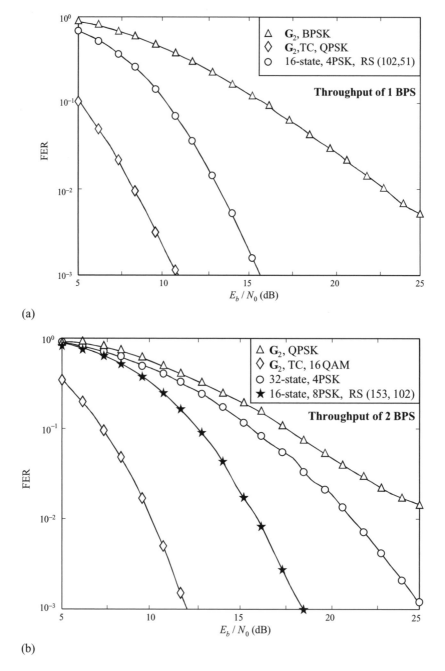

Figure 7.16 FER performance comparison between the TC(2, 1, 3) coded space–time block code **G**$_2$ and the RS(102, 51) GF(2^{10}) coded 16-state 4PSK space–time trellis code using one 512 subcarrier OFDM receiver over the shortened WATM channel at an effective throughput of (a) 1 BPS and (b) 2BPS. The coding parameters are shown in Tables 7.6–7.9.

Table 7.9   Switching SNR

System	NoTx	BPSK	QPSK	16 QAM	64 QAM
Speech	$-\infty$	3.31	6.48	11.61	17.64
Data	$-\infty$	7.98	10.42	16.76	26.33

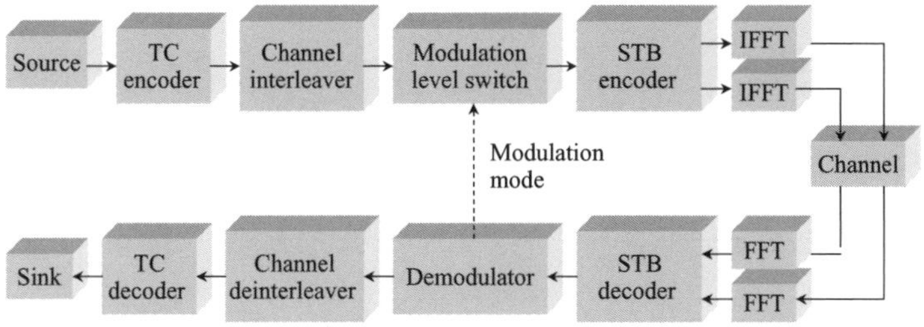

Figure 7.17 System overview of the turbo coded and space–time coded adaptive OFDM.

7.4.3   Space–time coded adaptive modulation for OFDM

All subcarriers in an adaptive OFDM (AOFDM) symbol are split into blocks of adjacent subcarriers, referred to as subbands. The same modulation scheme is employed for all subcarriers of the same subband. This substantially simplifies the task of signaling the modulation modes, since there are typically four modes and, for example, 32 subbands, requiring a total of 64 AOFDM mode signaling bits. The system is referred to as subband adaptive OFDM.

The system is presented in Figure 7.17. The choice of the modulation scheme to be used by the transmitter for its next OFDM symbol is determined by the channel quality estimate of the receiver, based on the current OFDM symbol. The following assumptions are used: the average signal power received from each transmitter antenna is the same; all multipath components undergo independent Rayleigh fading; the receiver has a perfect knowledge of the CIR and perfect signaling of the AOFDM modulation modes is available.

Optimized switching levels for adaptive modulation over a Rayleigh fading channel, shown in the instantaneous channel SNR (dB) are given in Table 7.9. The performance results are given in Figure 7.18.

7.4.4   Turbo and space–time coded adaptive OFDM

In this case the system parameters are as defined in Table 7.10. Performance results are shown in Figure 7.19.

For details on space–time coding with OFDMA signals and same set of parameters see [653–661].

7.5   LAYERED SPACE–TIME CODING FOR MIMO OFDM

In this section we extend the discussion on space–time coded OFDM to the case with $n_t = n_r = 4$ assuming the Jakes fading model and a layered architecture. The channel

Table 7.10   Coding rates and switching levels (dB) for TC(2, 1, 3) and space–time coded adaptive OFDM over the shortened WATM channel of Figure 7.15 for a target BER of 10^{-4}

	NoTx	BPSK	QPSK	16 QAM	64 QAM
		Half rate TC(2, 1, 3)			
Rate	–	0.50	0.50	0.50	0.50
Thresholds (dB)	$-\infty$	-4.0	-1.3	5.4	9.8
		Variable rate TC(2, 1, 3)			
Rate	–	0.50	0.67	0.75	0.90
Thresholds (dB)	$-\infty$	-4.0	2.0	9.70	21.50

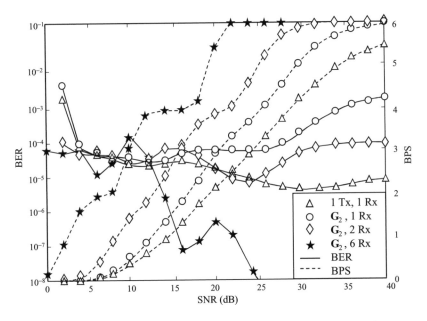

Figure 7.18 BER and BPS performance of 16 subband AOFDM employing the space–time block code $\mathbf{G}_2$ using multiple receivers for a target BER of 10^{-4} over the shortened WATM channel shown in Figure 7.15.

modeling is discussed in Chapter 14, but in this section we assume the channel estimation procedures as in [659] and [662] and the TU channel model considered in [662]. The OFDM signals assume a channel bandwidth of 1.25 MHz, which is divided into 256 subchannels. Two subchannels at each end of the band are used as guard tones, with the other 252 tones used to transmit data. The symbol duration is taken to be 204.8 μs so that the tones are orthogonal. A 20.2 μs guard interval is used to provide protection from inter-symbol interference, making the block duration $T_f = 225$ μs. The subchannel symbol rate is $r_b = 4.44$ kbaud. Systems with the same signal parameters are discussed, for example, in [662, 663].

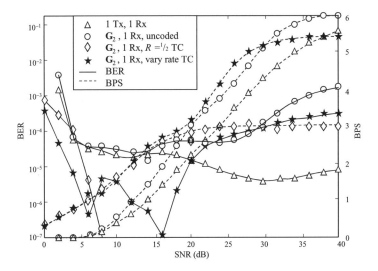

Figure 7.19 BER and BPS performance of 16 subband AOFDM employing the space–time block code $\mathbf{G}_2$ concatenated with both half rate and variable rate TC(2, 1, 3) at a target BER of 10^{-4} over the shortened WATM channel.

7.5.1 System model (two times two transmit antennas)

First we consider the $n_g = 2$ (groups of antennas) MIMO-OFDM implementation illustrated in Figure 7.20. In this case, two antenna space–time codes are employed that use 16 states and QPSK modulation. Data is grouped into blocks of 500 information bits, called words. Each word is coded into 252 symbols to form an OFDM block. Since this system uses $n_g = 2$, it can transmit two of these data blocks (1000 bits total) in parallel. Each time slot

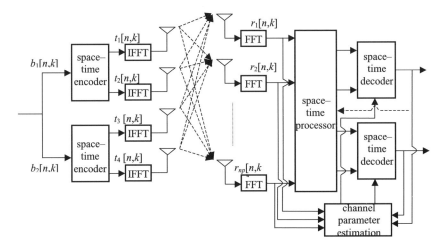

Figure 7.20 MIMO–OFDM using $n_g = 2$ individual space–time encoders, each using $n_t / n_g - 2$ transmit antennas.

consists of ten OFDM blocks with the first block used for training and the following nine blocks used for data transmission. This leads to a system capable of transmitting 4 Mbit/s using 1.25 MHz of bandwidth, so the transmission efficiency is 3.2 bit/s/Hz.

7.5.2 Interference cancellation

In general, all the interference cancellation schemes discussed in Chapter 5, modified for OFDM signals, can be used. One option has already been discussed in Chapter 4. An initial study of the system outlined in this section was provided in [664]. In [664] several interference cancellation approaches were described and performance was evaluated. Here, we focus on the successive interference cancellation approach based on signal quality (see Chapter 4). Basically, the strongest signal is detected first, subtracted from the input signal and then the procedure is repeated. We also assume that interleaving is employed. For the two-antenna, 16-state code given in Chapter 4 [3 (Figure 5)], the word error rate (WER) is

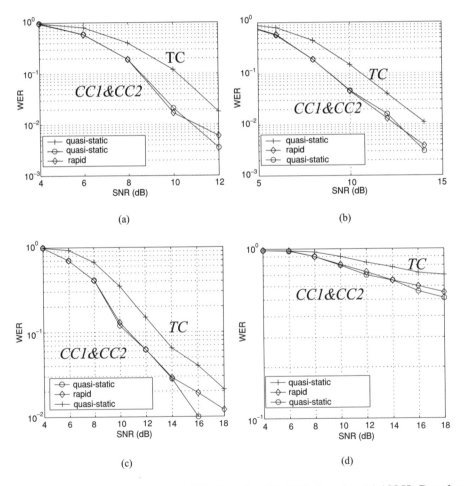

Figure 7.21 Performance curves. (a) 5 Hz Doppler; (b) 40 Hz Doppler; (c) 100 Hz Doppler; (d) 200 Hz Doppler.

Table 7.11 Generator matrices for the four transmit antenna codes used in figure 7.22. These codes are in GF(4) with elements denoted by $\{0, 1, a, 1+a\}$

$$\begin{bmatrix} (1+a)+D & a+(1+a)D & a+D & 1+(1+a)D \\ a+(1+a)D & a+D & 1+(1+a)D & 1+(1+a)D \\ a+D & 1+(1+a)D & 1+(1+a)D & (1+a)+aD \\ 1+(1+a)D & 1+(1+a)D & (1+a)+aD & 1+aD \end{bmatrix}$$

for the 16-state code

$$\begin{bmatrix} (1+a)+(1+a)D+aD^2 & (1+a)D+aD^2 & 1+D^2 & 1+D^2 \\ (1+a)D+aD^2 & 1+D^2 & 1+(1+a)D+(1+a)D^2 & (1+a)+aD+(1+a)D^2 \\ 1+D^2 & 1+(1+a)D+(1+a)D^2 & (1+a)+aD+(1+a)D^2 & (1+a)+aD \\ 1+(1+a)D+(1+a)D^2 & (1+a)+aD+(1+a)D^2 & (1+a)+aD & D+D^2 \end{bmatrix}$$

for the 256-state code

given in Figure 7.21 for the case where the channel has a TU delay profile and for Doppler frequencies of 5, 40, 100, and 200 Hz. The other two curves in Figure 7.21 illustrate the performance improvement that can be obtained using the improved convolutional space–time codes given in Section 4.6. One of these codes was designed to be optimal for the quasi-static fading model from Section 4.2 (CC1). The other code was designed to be optimal for the rapid fading model in Section 4.2 (CC2). The improved codes from Section 4.6 are optimal codes based on the criterion given in Section 4.2. All these details are available in Chapter 4. In summary:

$$\begin{bmatrix} D+2D^2 & 1+2D^2 \\ 2D & 2 \end{bmatrix} \text{ Figure 4.7 (TC)}$$

$$\begin{bmatrix} D+D^2 & 2+D \\ 2+D & 2+2D+2D^2 \end{bmatrix} \text{ quasi-static code Section 4.6 (CC1)}$$

$$\begin{bmatrix} 2D^2 & 2+D+2D^2 \\ 2+D & 2D+2D^2 \end{bmatrix} \text{ rapid fading code Section 4.6 (CC2)}$$

7.5.3 Four transmit antennas

Next we investigate the approach that uses four-antenna space–time codes. We consider 16-state and 256-state codes, designed using an *ad hoc* approach. The codes are presented in Table 7.11 and the performance results are given in Figure 7.22.

7.6 SPACE–TIME CODED TDMA/OFDM RECONFIGURATION EFFICIENCY

In this section we compare two options for combating the fading in digital wireless communications, mainly the TDMA-based concept with equalization versus the OFDM concept. The structure of the code and the equivalent model of the encoder are shown in Figures 7.23 and 7.24 respectively. This structure is considered for the EDGE system. The presentation in this section is based mainly on [665–667].

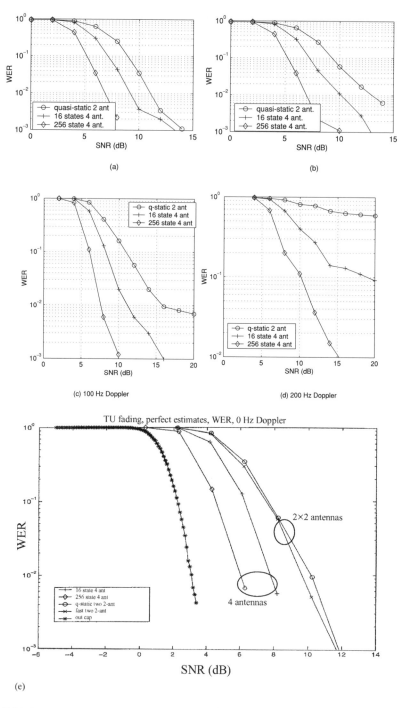

Figure 7.22 WER versus SNR of MIMO–OFDM systems with $n_t = n_r = 4$, TU channel with different Doppler frequencies. Here we compare the best code from the last figure with codes designed for four transmit antenna cases. See Table 7.11 and Figure 7.21 for details on the codes. (a) 5 Hz Doppler; (b) 40 Hz Doppler; (c) 100 Hz Doppler; (d) 200 Hz Doppler; (e) comparisons of WER for best MIMO–OFDM systems with perfect estimates and no Doppler.

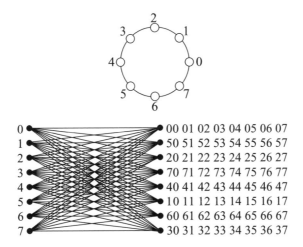

Figure 7.23 Eight-state 8PSK STTC with two transmit antennas and a spectral efficiency of 3 bits/s/Hz.

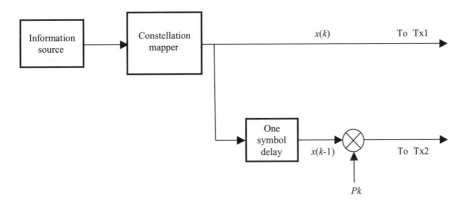

Figure 7.24 Equivalent encoder model for an eight-state 8PSK STTC with two transmit antennas.

7.6.1 Frequency selective channel model

The structure of this space–time trellis code (STTC) can be exploited to reduce the complexity of joint equalization and decoding in a frequency selective channel. This is achieved by embedding the space–time encoder in Figure 7.24 in the two channels $\mathbf{h}_1$ and $\mathbf{h}_2$, resulting in an equivalent single-input single-output (SISO) data-dependent channel impulse response (CIR) with memory $(v - 1)$, whose delay D-transform is given by

$$
\begin{aligned}
h_{\text{eqv}}^{\text{STTC}}(k, D) &= \mathbf{h}_1(0) + \sum_{m=1}^{v} (\mathbf{h}_1(m) + p_k \mathbf{h}_2(m - 1))D^m + p_k \mathbf{h}_2(v))D^{v+1} \\
&= h_1(D) + p_k D h_2(D)
\end{aligned}
\tag{7.54}
$$

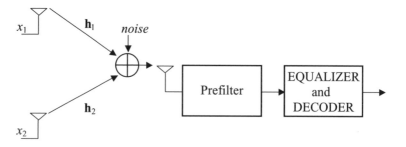

Figure 7.25 Receiver structure for STTC joint equalization/decoding with two transmit and one receive antennas.

where D is the delay operator and $p_k = \pm 1$ is data dependent. Therefore, trellis-based joint space–time equalization and decoding with 8^{v+1} states can be performed on this equivalent channel. The traditional trellis equalization would require 8^{2v} states, and STTC decoding requires eight states. The receiver block diagram is shown in Figure 7.25.

7.6.2 Front end prefilter

The objective of the prefilter is to shorten and shape the effective CIR seen by the equalizer to reduce its complexity (since the number of equalizer trellis states is exponential in the CIR memory).

7.6.3 Time-invariant channel

First, we describe the prefilter design problem for a time-invariant channel with memory v and then extend it to the data-dependent time-varying channel case. Assume that the FIR prefilter has N_f taps and denote its impulse response by the vector $\mathbf{w}$. Then, the impulse response of the effective channel at the output of the prefilter is given by $\mathbf{h}_{\text{eff}} = \mathbf{H}\mathbf{w}$, where $\mathbf{H}$ is the $(N_f + v) \times N_f$ Toeplitz convolution matrix. Let the vector $\mathbf{h}_{\text{win}}$ contain the $(N_b + 1)$ taps (where $N_b < v$) of $\mathbf{h}_{\text{eff}}$ to retain after shortening (whose energy is to be maximized), and let $\mathbf{h}_{\text{wall}}$ contain the remaining taps (whose energy is to be minimized). Then

$$\mathbf{h}_{\text{win}} = \mathbf{J}_{\text{win}}\mathbf{h}_{\text{eff}} = \underbrace{\mathbf{J}_{\text{win}}\mathbf{H}}_{\mathbf{H}_{\text{win}}}\mathbf{w} = \mathbf{H}_{\text{win}}\mathbf{w} \tag{7.55}$$

where the $(N_b + 1) \times (N_f + v)$-dimensional matrix $\mathbf{J}_{\text{win}}$ is constructed using columns of the identity matrix corresponding to tap positions of $\mathbf{h}_{\text{win}}$ within $\mathbf{h}_{\text{eff}}$. And

$$\mathbf{h}_{\text{wall}} = \mathbf{J}_{\text{wall}}\mathbf{h}_{\text{eff}} = \underbrace{\mathbf{J}_{\text{wall}}\mathbf{H}}_{\mathbf{H}_{\text{wall}}}\mathbf{w} = \mathbf{H}_{\text{wall}}\mathbf{w} \tag{7.56}$$

where the $(N_f + v - N_b - 1) \times (N_f + v)$-dimensional matrix $\mathbf{J}_{\text{wall}}$ is constructed from the columns of the identity matrix corresponding to tap positions of $\mathbf{h}_{\text{wall}}$ within $\mathbf{h}_{\text{eff}}$.

The prefilter design criterion maximizes the shortening signal to noise ratio (SSNR), the desired signal energy of the shortened channel contained in $\mathbf{h}_{\text{win}}$ divided by the residual ISI energy in $\mathbf{h}_{\text{wall}}$ plus the noise energy at the prefilter output.

7.6.4 Optimization problem

The problem reduces to the generalized eigenvector problem (for specific details see [666] and the references therein):

$$\max_{\mathbf{w}} \mathbf{w}^* \mathbf{B} \mathbf{w} \quad \text{subject to } \mathbf{w}^* \mathbf{A} \mathbf{w} = 1 \tag{7.57}$$

where $(\cdot)^*$ denotes the complex conjugate transpose operation, $\mathbf{B} = \mathbf{H}_{\text{win}}^* \mathbf{H}_{\text{win}}$, $\mathbf{A} = \mathbf{H}_{\text{wall}}^* \mathbf{H}_{\text{wall}} + \mathbf{R}_{zz}$, and $\mathbf{H}_{zz}$ is the noise autocorrelation matrix at the prefilter input. The solution has the form

$$\mathbf{w}_{\text{opt}} = (\mathbf{L}_{\mathbf{A}}^*)^{-1} \mathbf{u}_{\text{max}} \tag{7.58}$$

here, $\mathbf{A} = \mathbf{L}_{\mathbf{A}} \mathbf{L}_{\mathbf{A}}^*$ is the Cholesky factorization of the matrix $\mathbf{A}$, and $\mathbf{u}_{\text{max}}$ is the unitnorm eigenvector of matrix $(\mathbf{L}_{\mathbf{A}})^{-1} \mathbf{B} (\mathbf{L}_{\mathbf{A}})^{-1}$ that corresponds to its largest eigenvalue λ_{max}. The resulting optimal SSNR is given by

$$\text{SSNR}_{\text{opt}} = \frac{\mathbf{w}_{\text{opt}}^* \mathbf{B} \mathbf{w}_{\text{opt}}}{\mathbf{w}_{\text{opt}}^* \mathbf{A} \mathbf{w}_{\text{opt}}} = \lambda_{\text{max}} \tag{7.59}$$

Equation (7.59) provides the optimal prefilter for a time-invariant channel.

7.6.5 Average channel

For the eight-state 8PSK STTC with two transmit antennas, we can design the prefilter for the *average* of the equivalent channel given in Equation (7.54). It can be shown [666] that the matrices $\mathbf{A}$ and $\mathbf{B}$ in this case are modified to

$$\mathbf{B} = \left(\mathbf{H}_{\text{win}}^1\right)^* \mathbf{H}_{\text{win}}^1 + \left(\mathbf{H}_{\text{win}}^2\right)^* \mathbf{H}_{\text{win}}^2$$
$$\mathbf{A} = \left(\mathbf{H}_{\text{wall}}^1\right)^* \mathbf{H}_{\text{wall}}^1 + \left(\mathbf{H}_{\text{wall}}^2\right)^* \mathbf{H}_{\text{wall}}^2 + \mathbf{R}_{zz} \tag{7.60}$$

where $\mathbf{H}_{\text{win}}^i$ and $\mathbf{H}_{\text{wall}}^i$ $(i = 1, 2)$ are matrices corresponding to the constant channels $\mathbf{h}_1$ and $\mathbf{h}_2$ between the two transmit and the single receive antennas. The main attractive feature of the prefilter is that it is a single time-invariant (over a transmission block) FIR filter that shortens both channels $\mathbf{h}_1$ and $\mathbf{h}_2$ simultaneously without excessive noise enhancement.

7.6.6 Prefiltered *M*-BCJR equalizer

The algorithm as proposed in [668], is a reduced complexity version of the Bahl–Cocke–Jelinek–Raviv (BCJR) forward–backward algorithm presented in Appendix 2.1, also elaborated in [669], where at each trellis step, only the M active states associated with the highest metrics are retained. An improved version of the M-BCJR algorithm was proposed in [670] based on a log domain implementation of the BCJR algorithm and operates as follows [667].

The forward and backward recursions independently select trees of active nodes without restricting one to be a subtree of the other. To form soft decisions at any time instant, we use all edges with at least one active node.

Let L be the number of trellis steps; $Y_1 Y_2 \ldots Y_L$ the received outputs; s_t the trellis state at time t; S the number of trellis states and u_t the input at time t. The quantity calculated by the algorithm is not $Pr(u_t = u | Y_1 \ldots Y_L)$ as in BCJR, but an approximation, as detailed in what follows.

Using the channel observations and the channel description, calculate for each trellis step t the quantities

$$\gamma_t(i, j) = Pr(s_t = j; Y_t \mid s_{t-1} = i) \tag{7.61}$$

The forward recursion (see Appendix 2.1)

1. For $t = 0$ (initialization) $\alpha_0(0) = 1$, $\alpha_0(i) = 0$ for $i = 1 \ldots S$.
2. For $t = 1, \ldots, L - 1$
   - $\alpha_t(i) = \sum_j \alpha_{t-1}(j)\gamma_t(i, j)$
   - Let A_t denote the M largest αs at time t. Any sorting algorithm can be used to construct A_t. Set $\alpha_t(i) = 0$ if $\alpha_t(i) \notin A_t$.

The backward recursion

1. For $t = L$ (initialization) $\beta_L(0) = 1$, $\beta_L(i) = 0$ for $i = 1 \ldots S$.
2. For $t = L - 1, \ldots, 1$
   - $\beta_t(i) = \sum_j \beta_{t+1}(j)\gamma_{t+1}(i, j)$
   - Let B_t denote the M largest βs at time t. Any sorting algorithm can be used to construct B_t. Set $\beta_t(i) = 0$ if $\beta_t(i) \notin B_t$.

The probabilities of error E_α and E_β (in the sense that the correct state is not included in the M selected states) can be calculated as follows:

$$E_\alpha = Q\left(\sqrt{\frac{|\mathbf{h}(0)|^2 d_{\min}^2}{2\sigma_z^2}}\right) ; \quad E_\beta = Q\left(\sqrt{\frac{|\mathbf{h}(v)|^2 d_{\min}^2}{2\sigma_z^2}}\right) \tag{7.62}$$

where $Q(\cdot)$ is the standard Q function, $d_{\min}$ the minimum Euclidean distance between any two constellation points and σ_z^2 the noise variance.

7.6.7 Decision

To make a decision at step $0 < t < L$ on the input $u_t = u$, do the following:

1. Set $P(u_t = u) = 0$.
2. For all edges (i, j) that have input u
   - if $\beta_t(j) \neq 0$ and $\alpha_{t-1}(i) \neq 0$, then $P(u_t = u)+ = \alpha_{t-1}(i)\gamma(i, j)\beta_t(j)$;
   - if $\beta_t(j) \neq 0$ and $\alpha_{t-1}(i) = 0$, then $P(u_t = u)+ = E_\alpha\gamma(i, j)\beta_t(j)$;
   - if $\beta_t(j) = 0$ and $\alpha_{t-1}(i) \neq 0$, then $P(u_t = u)+ = \alpha_{t-1}(i)\gamma(i, j)E_\beta$.

The performance of the M-BCJR equalizer/decoder is further improved, especially for small M, by using the prefilter of the previous subsection to concentrate the channel energy in a smaller number of taps. In fact, two different prefilters should be used for the forward and backward recursions since the forward recursion favors a close to minimum phase channel, whereas the backward recursion favors a close to maximum phase channel [670]. The value of M and the number of prefilter taps can be jointly optimized to achieve the best performance complexity tradeoffs.

7.6.8 Prefiltered MLSE/DDFSE equalizer complexity

To evaluate reconfiguration efficiency we need to estimate the complexity of the algorithm. For a size 2^b signal constellation, n_i transmit antennas and MIMO channel memory of v, the MIMO MLSE equalizer has $2^{bn_i v}$ states in general. The number of equalizer states can be reduced by using the STTC trellis structure as shown in [671] or by a MIMO channel-shortening prefilter [672]. However, this complexity is still too high for large signal constellations, even for two transmit antennas and short-to-moderate MIMO channel memory. For example, for an 8PSK constellation and the EDGE TU channel (where $V = 3$), the number of full MLSE equalizer states is equal to $8^6 = 262\,144$ states.

7.6.9 Delayed decision feedback sequence estimation (DDFSE)

In order to reduce complexity, DDFSE, as discussed in Chapter 6, was introduced in [673]. This is a hybrid scheme between MLSE and decision feedback equalization (DFE) for channels with long memory. Basically, the CIR is divided into a leading part and a tail. Then, an MLSE equalizer is constructed based on the leading part, and the interfering effect of the CIR tail is canceled by feedback using previous (hard) decisions (assumed correct).

At time k, the branch metric $\xi(k)$ of the DDFSE equalizer/decoder is given by

$$\xi(k) = \left| y(k) - \sum_{i=0}^{n} \mathbf{h}_{\text{eqv}}^{\text{STTC}}(i)x(k-i) - \sum_{i=n+1}^{v+1} \mathbf{h}_{\text{eqv}}^{\text{STTC}}(i)\hat{x}(k-i) \right|^2 \qquad (7.63)$$

where

$y(k)$	is the kth received symbol;
n	is the design parameter ($0 \le n \le v$) that determines the number of DDFSE trellis states;
$\mathbf{h}_{\text{eqv}}^{\text{STTC}}$	is the impulse response vector of the equivalent channel;
$x(k)$	are all possible input symbols according to different transitions along the path history;
$\hat{x}(k)$	are the previous hard symbol decisions along the path history.

7.6.10 Equalization schemes for STBC

The focus is on the case of two transmit antennas where full-rate Alamouti-type space–time block codes can be constructed for any signal constellation.

7.6.10.1 Time-reversal space–time block coding (TR-STBC)

TR-STBC was introduced in [674] as an extension of the Alamouti STBC scheme to frequency selective channels by imposing the Alamouti orthogonal structure at a *block, not symbol*, level, as in the flat fading channel case. More specifically, the transmitted blocks from antennas one and two at time $(k + 1)$ (which were denoted by $\mathbf{x}_1^{(k+1)}$ and $\mathbf{x}_2^{(k+1)}$, respectively) are generated by the encoding rule (for $k = 0, 2, 4, \ldots$)

$$\mathbf{x}_1^{(k+1)} = -\mathbf{J}\bar{\mathbf{x}}_2^{(k)}; \quad \mathbf{x}_2^{(k+1)} = \mathbf{J}\bar{\mathbf{x}}_1^{(k)} \qquad (7.64)$$

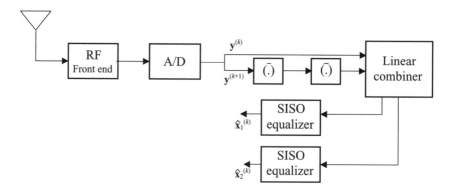

Figure 7.26 TR-STBC receiver block diagram. The operations $(\bar{\cdot})$ and $(\tilde{\cdot})$ denote complex conjugation and time reversal, respectively [667].

where $\mathbf{J}$ is the time reversal matrix that consists of ones on the main antidiagonal and zeros everywhere else. To eliminate inter-block interference (IBI) between adjacent blocks due to channel memory, length-v all-zero guard sequences are inserted between information blocks.

The TR-STBC receiver in Figure 7.26 employs linear combining techniques (a spatio-temporal matched filter) to eliminate the mutual interference effects between the two transmit antennas *while still achieving the maximum diversity gain* of $\|\mathbf{h}_1\|^2 + \|\mathbf{h}_2\|^2$ (where $\|\cdot\|$ denotes the norm of a vector).

TR-STBC uses a combination of complex conjugation, time reversal, and matched filtering operations, as described in detail in [659], to convert the two-input single-output channel to two single-input single-output (SISO) channels, each with an equivalent impulse response given by

$$h_{\text{eqv}}^{\text{TR-STBC}}(D) = h_1(D)\bar{h}_1(\bar{D}^{-1}) + h_2(D)\bar{h}_2(\bar{D}^{-1}) \tag{7.65}$$

to which standard SISO equalization schemes can be applied. In Equation (7.65), $h_i(D)$ is the D-transform of $\mathbf{h}_i(k)$, and $\bar{h}_i(\bar{D}^{-1})$ is the D-transform of $\mathbf{h}_i(-k)$ for $i = 1, 2$. A whitened matched filter (WMF) front end can be used to convert $h_{\text{eqv}}^{\text{TR-STBC}}(D)$ to its minimum phase equivalent followed by trellis or feedback equalization, as will be further discussed later in the section . TR-STBC assumes that the two channels $h_1(D)$ and $h_2(D)$ are fixed over two consecutive transmission blocks and perfectly known at the receiver. In the next two subsections, we describe two alternative STBC joint equalization/decoding schemes that use frequency domain processing.

7.6.10.2   OFDM-STBC

In this case at the receiver end in Figure 7.27, received blocks are processed in pairs where their FFTs are computed and linearly combined. Finally, gain and phase adjustment is performed using minimum mean square error frequency domain equalization (MMSE-FDE) with a single complex tap for each subchannel, followed by a decision device. While the use of the Alamouti STBC modifies the channel frequency gain at subchannel i from

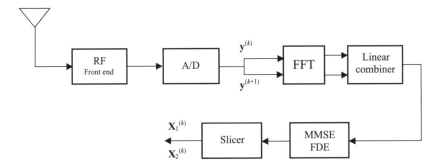

Figure 7.27 OFDM-STBC receiver block diagram [667].

$|H(i)|^2$ to $|H_1(i)|^2 + |H_2(i)|^2$, which provides increased immunity against fading, decision errors occurring at any subchannel result in an irreducible error floor.

OFDM has two main drawbacks, namely, a high peak to average ratio (PAR), which results in larger backoff with non-linear amplifiers, and high sensitivity to frequency errors and phase noise [675]. An alternative equalization scheme that overcomes these two drawbacks of OFDM while retaining its reduced implementation complexity advantage (due to the use of FFT) is the single-carrier (SC) FDE [676], which has been extended to receive diversity systems in [677] and to Alamouti-type STBC transmit diversity systems in [678], and is described next.

These schemes can be extended to more than two transmit antennas using the theory of *orthogonal designs* presented in Chapter 4.

7.6.11 Single-carrier frequency domain equalized space–time block coding SC FDE STBC

The SC FDE, shown in Figure 7.28, is distinct from OFDM in that the IFFT block is moved to the receiver end and placed before the decision device. As noted in [676], this causes the effects of deep nulls in the channel frequency response to be spread out, by the IFFT operation, over all symbols, thus reducing their effect and improving performance.

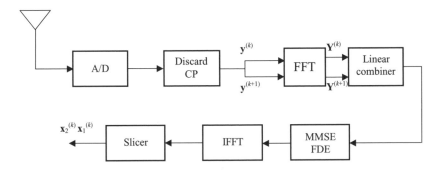

Figure 7.28 FDE-STBC receiver block diagram.

7.6.11.1 Encoder

Denote the nth symbol of the kth transmitted block (of length N) from antenna i by $\mathbf{x}_i^{(k)}(n)$. Then, the FDE-STBC encoding rule is given by [678]:

$$\mathbf{x}_1^{(k+1)}(n) = -\bar{\mathbf{x}}_2^{(k)}((-n)_N)$$
$$\mathbf{x}_2^{(k+1)}(n) = -\bar{\mathbf{x}}_1^{(k)}((-n)_N) \tag{7.66}$$
$$\text{for } n = 0, 1, \ldots, N - 1 \text{ and } k = 0, 2, 4, \ldots$$

where $(\cdot)_N$ denotes the modulo-N operation that distinguishes this encoding scheme from TR-STBC, Equation (7.64). In addition, a cyclic prefix (CP) is added to each transmitted block to eliminate IBI and make the two channel matrices circulant. Taking the discrete Fourier 66 transform (DFT) of Equation (7.66), we see

$$\mathbf{X}_1^{(k+1)}(m) = -\bar{\mathbf{X}}_2^{(k)}; \quad \mathbf{X}_2^{(k+1)}(m) = -\bar{\mathbf{X}}_1^{(k)}$$
$$\text{for } m = 0, 1, \ldots, N - 1 \text{ and } k = 0, 2, 4, \ldots \tag{7.67}$$

which reveals that this is also a *block-level* implementation of the symbol-level Alamouti encoding rule.

7.6.11.2 Receiver

After analog-to-digital (A/D) conversion, the CP part of each received block is discarded. Mathematically, we can express the input–output relationship over the *j*th received block as

$$\mathbf{y}^{(j)} = \mathbf{H}_1^{(j)}\mathbf{x}_1^{(j)} + \mathbf{H}_2^{(j)}\mathbf{x}_2^{(j)} + \mathbf{z}^{(j)} \tag{7.68}$$

where $\mathbf{H}_1^{(j)}$ and $\mathbf{H}_2^{(j)}$ are $N \times N$ circulant matrices whose first columns are equal to $\mathbf{h}_1^{(j)}$ and $\mathbf{h}_2^{(j)}$, respectively, appended by $(N - v - 1)$ zeros and $\mathbf{z}^{(j)}$ is the noise vector. Since $\mathbf{H}_1^{(j)}$ and $\mathbf{H}_2^{(j)}$ are circulant matrices, they admit the eigendecompositions

$$\mathbf{H}_1^{(j)} = \mathbf{Q}^*\mathbf{\Lambda}_1^{(j)}\mathbf{Q}; \quad \mathbf{H}_2^{(j)} = \mathbf{Q}^*\mathbf{\Lambda}_2^{(j)}\mathbf{Q} \tag{7.69}$$

where $\mathbf{Q}$ is the orthonormal FFT matrix and $\mathbf{\Lambda}_1^{(j)}$ (respectively $\mathbf{\Lambda}_2^{(j)}$) is a diagonal matrix whose (n, n) entry is equal to the nth FFT coefficient of $\mathbf{h}_1^{(j)}$ (resp. $\mathbf{h}_2^{(j)}$). Therefore, applying the FFT to $\mathbf{y}^{(j)}$, we get (for $j = k, k + 1$)

$$\mathbf{Y}^{(j)} = \mathbf{Q}\mathbf{y}^{(j)} = \mathbf{\Lambda}_1^{(j)}\mathbf{X}_1^{(j)} + \mathbf{\Lambda}_2^{(j)}\mathbf{X}_2^{(j)} + \mathbf{Z}^{(j)} \tag{7.70}$$

7.6.11.3 Processing

The length-N blocks at the FFT output are then processed in pairs, resulting in the two blocks (we drop the time index from the channel matrices since they are assumed fixed over the two blocks under consideration):

$$\underbrace{\begin{bmatrix} \mathbf{Y}^{(k)} \\ \bar{\mathbf{Y}}^{(k+1)} \end{bmatrix}}_{\mathbf{Y}} = \underbrace{\begin{bmatrix} \mathbf{\Lambda}_1 & \mathbf{\Lambda}_2 \\ \bar{\mathbf{\Lambda}}_2 & -\bar{\mathbf{\Lambda}}_1 \end{bmatrix}}_{\mathbf{\Lambda}} \underbrace{\begin{bmatrix} \mathbf{X}_1^{(k)} \\ \mathbf{X}_2^{(k)} \end{bmatrix}}_{\mathbf{X}} + \underbrace{\begin{bmatrix} \mathbf{Z}^{(k)} \\ \bar{\mathbf{Z}}^{(k+1)} \end{bmatrix}}_{\mathbf{Z}} \tag{7.71}$$

where $\mathbf{X}_1^{(k)}$ and $\mathbf{X}_2^{(k)}$ are the FFTs of the information blocks $\mathbf{x}_1^{(k)}$ and $\mathbf{x}_2^{(k)}$, respectively, and $\mathbf{Z}$ is the noise vector. To eliminate *inter-antenna interference*, the linear combiner $\mathbf{\Lambda}*$ is applied to $\mathbf{Y}$. Due to the orthogonal Alamouti-like structure of $\mathbf{\Lambda}$, a second-order diversity gain is achieved. By *Alamouti-like* we mean any 2×2 complex orthogonal matrix of the form

$$
\begin{bmatrix}
c_1 & c_2 \\
\pm \bar{c}_2 & \mp \bar{c}_1
\end{bmatrix}
$$

Then, the two decoupled blocks at the output of the linear combiner are equalized separately, using the MMSE-FDE [676], which consists of N complex taps per block that mitigate *inter-symbol interference*. Finally, the MMSE-FDE output is transformed back to the time domain using the inverse FFT where decisions are made. Note that the SC MMSE-FDE is equivalent to block MMSE linear equalization [679]; hence, its performance can be improved at the expense of increased complexity by adding a feedback section as discussed in [680].

7.6.11.4 *Channel estimator*

Formulate the channel estimation problem for the two-transmit one-receive scenario as in Chapter 4. The analysis can be easily generalized to multiple transmit/receive antennas. Transmit two training sequences $\mathbf{s}_1$ and $\mathbf{s}_2$ from the first and second antennas simultaneously in synchronized data blocks, where each block consists of N information symbols and N_t training symbols. For two transmit antennas, the receiver uses the $2N_t$ known training symbols to estimate the $2(v + 1)$ unknown channel coefficients. The observed training sequence output, which does not have interference from information or preamble symbols, can be expressed as

$$
\mathbf{y} = \begin{bmatrix} \mathbf{S}_1 & \mathbf{S}_2 \end{bmatrix} \begin{bmatrix} \mathbf{h}_1 \\ \mathbf{h}_2 \end{bmatrix} + \mathbf{z} = \mathbf{Sh} + \mathbf{z} \tag{7.72}
$$

where the column vectors $\mathbf{y}$ and $\mathbf{z}$ are of size $(N_t - v)$, $\mathbf{S}_1$, and $\mathbf{S}_2$, are Toeplitz matrices of size $(N_t - v) \times (v + 1)$ that contain training symbols. The MMSE channel estimate assuming that $\mathbf{S}$ has full column rank, is given by [681]:

$$
\hat{\mathbf{h}} = \begin{bmatrix} \hat{\mathbf{h}}_1 \\ \hat{\mathbf{h}}_2 \end{bmatrix} = (\mathbf{S}^*\mathbf{S})^{-1}\mathbf{S}^*\mathbf{y} \tag{7.73}
$$

where $(\cdot)^{-1}$ denotes the inverse. The estimation error (mean square error) is given as

$$
\text{MSE} = E\lfloor (\mathbf{h} - \hat{\mathbf{h}})^*(\mathbf{h} - \hat{\mathbf{h}}) \rfloor = \sigma_z^2 \text{tr}((\mathbf{S}^*\mathbf{S})^{-1}) \tag{7.74}
$$

where we assume that white noise with variance σ_z^2 and tr $(\cdot)$ denotes the trace of a matrix. The channel estimation MMSE is equal to

$$
\text{MMSE} = \frac{\sigma_z^2(v + 1)}{(N_t - v)} \tag{7.75}
$$

which is achieved if and only if

$$
\mathbf{S}^*\mathbf{S} = \begin{bmatrix} \mathbf{S}_1^*\mathbf{S}_1 & \mathbf{S}_2^*\mathbf{S}_1 \\ \mathbf{S}_1^*\mathbf{S}_2 & \mathbf{S}_2^*\mathbf{S}_2 \end{bmatrix} = (N_t - v)\mathbf{I}_{v+1} \tag{7.76}
$$

where $\mathbf{I}_{v+1}$ is the identity matrix of size $v + 1$. Two optimal training sequences that satisfy Equation (7.76) have an impulse-like autocorrelation sequence and zero cross-correlation. In this case, computing the channel estimates using Equation (7.73) reduces to a simple crosscorrelation (matrix–vector product).

7.6.11.5   Training sequences

For implementation purposes (to avoid non-linear amplifier distortion), it is desirable to use training sequences with constant amplitude. Optimal constant amplitude training sequences can be constructed from a Pth root-of-unity alphabet $A_p = \{e^{i2\pi k/P} | k = 0, 1, \ldots, P - 1\}$ (where $i = \sqrt{-1}$) without constraining the alphabet size P. Such sequences are the perfect roots-of-unity sequences (PRUS), which are also known as polyphase sequences. Chu [682] showed that for any training sequence length N_t, there exists a PRUS with alphabet size $P = 2N_t$. In [683] and [684] the interested reader can find details on how to design PSK-type training sequences for dual-antenna transmissions with negligible performance loss from PRUS.

7.6.11.6   Performance results

For the performance results generation signal and channel parameters as in [667] are used. An 8PSK modulation with two transmit and one receive antennas on the TU EDGE channel is used. The overall CIR length is effectively four symbol periods, i.e. $v = 3$.

   In EDGE, fading can be safely assumed to be quasi-static, i.e. the CIR can be assumed constant for the duration of a transmission block. This is due to the fact that the coherence time of the channel at around 1 GHz carrier frequency is much larger than the block duration of 577 μs, even for highway speeds. This eliminates the need for channel tracking at the receiver. In addition, assuming ideal frequency hopping, the fading process is independent from block to block. The noise samples are generated as independent samples of a zero mean complex Gaussian random variable with a variance of $1/$SNR per complex dimension. The reason for doubling the noise variance (compared with the single transmit antenna case) is that with two-antenna transmissions, we assume that the total transmitted power is the same as in the single antenna case and is divided equally between the two antennas. The average energy of the symbols transmitted from each antenna is normalized to one so that the signal to noise ratio is SNR. The results are shown in Figure 7.29.

   For reconfiguration efficiency, relations from Chapter 4 are applicable where values for D and g_{12} can be derived from results presented in Tables 7.12 and 7.13. For two configurations, g_{12} is obtained as a difference of the corresponding entries in column 2 and D_r as a ratio of the corresponding complexity numbers from column 3.

   In summary, for STTCs, the prefiltered M-BCJR equalizer/decoder outperforms the prefiltered DDFSE equalizer/decoder and has lower implementation complexity. For space–time block codes, TR-STBC achieves the best performance among the three investigated schemes. OFDM-STBC has the highest PAR and is the most sensitive to frequency errors but is also the most flexible among the three schemes in its support of multirate and multiQOS requirements. The three STBC schemes suffer the same amount of overhead (in the form of an all zero guard sequence for TR-STBC and a cyclic prefix guard sequence for FDE-STBC and OFDM-STBC).

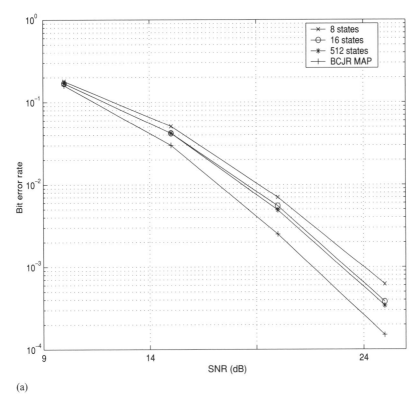

(a)

Figure 7.29 (a) BER performance of two-transmit one-receive eight-state 8PSK STTC with prefiltered M-BCJR equalizer as a function of M (the number of active states). BER of a 4096-state full BCJR-MAP equalizer is shown as a benchmark; (b) BER performance of two-transmit one-receive eight-state 8PSK STTC with prefiltered 64-state DDFSE, prefiltered 16-state M-BCJR, and full BCJR-MAP equalizers; (c) BER performance of two-transmit one-receive eight-state 8PSK STTC with prefiltered 16-state M-BCJR with perfect and estimated CSI. Full BCJR-MAP equalizer performance shown as BER lower bound; (d) BER performance of two-transmit one-receive OFDM-STBC, FDE-STBC, and TR-STBC. For OFDM and FDE-STBC, a size 64 FFT is assumed. For TR-STBC, an ideal whitened matched filter front end and a three-tap feedback filter are assumed; (e) effect of channel estimation on performance of SC FDE-STBC; (f) BER performance of two-transmit one-receive TR-STBC with 512-state full BCJR-MAP, eight-state M-BCJR, and SISO MMSE-DFE with $N_b = 3$ feedback taps. An ideal whitened matched filter front end is assumed.

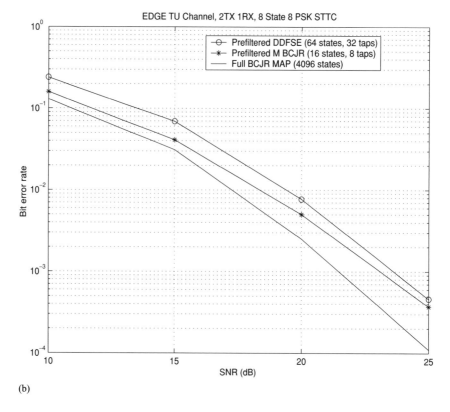

(b)

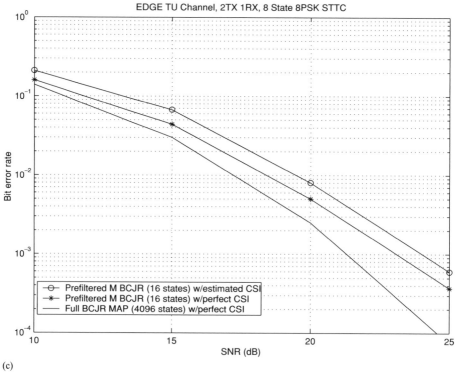

(c)

Figure 7.29 (*Cont.*).

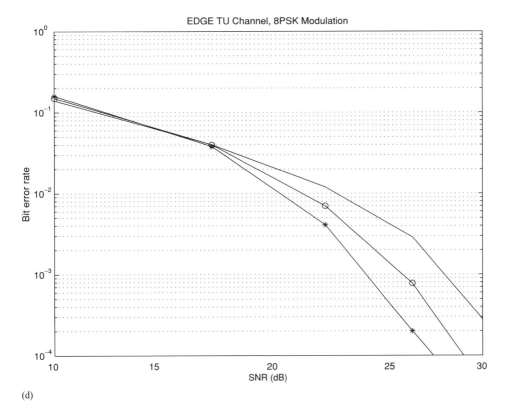

(d)

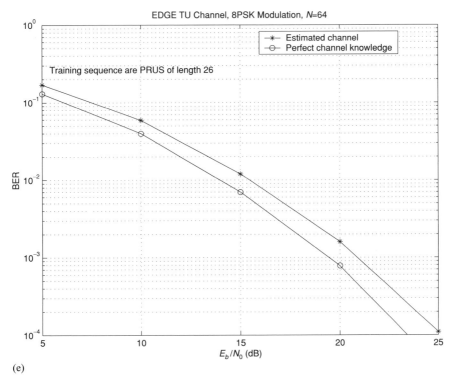

(e)

Figure 7.29 (*Cont.*).

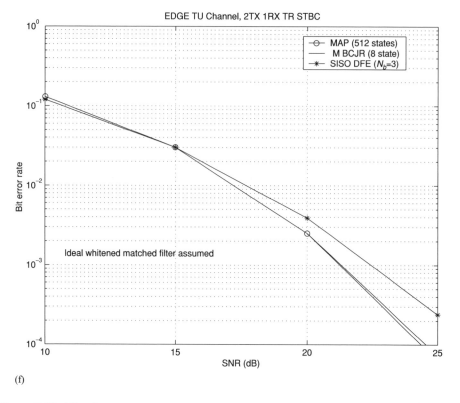

(f)

Figure 7.29 (*Cont.*).

Table 7.12   Performance and complexity comparison summary between the equalization schemes for the eight-state 8PSK STTC over the TU EDGE channel

Equalization scheme	SNR (dB) at BER $= 10^{-3}$	Receiver complexity (per block)
Full BCJR-MAP	21.3	4096 states (each direction)
Prefiltered M-BCJR	23.1	16 states (each direction) 8-tap prefilter (each direction)
Prefiltered DDFSE	23.6	64 states and 32-tap prefilter

Table 7.13   Performance and complexity comparison summary between the Alamouti-type STBC equalization schemes over the TU EDGE channel, assuming 8PSK modulation and block size of 64

Equalization scheme	SNR (dB) at BER $= 10^{-3}$	Receiver complexity (per block)
Full BCJR-MAP	21.3	512 states (each direction) and 20-tap WMF
TR-STBC	22.2	20-tap WMF and 3 feedback taps
FDE-STBC	24.2	Size 64 FFT/IFFT and 64-tap FDE
OFDM-STBC	26.5	Size 64 FFT and 64-tap FDE

7.7 MULTICARRIER CDMA SYSTEM

In Chapter 1 we discussed a variety of different structures for MC CDMA at the introductory level. In a number of the following sections we will provide more in-depth discussion on the performance of these systems. In the system shown in Figure 7.30 [685, 686] the MC-CDMA BPSK signal transmitted by the kth user is:

$$s_k(t) = \text{Re}\left\{ \sum_{n=-\infty}^{+\infty} u_k[n]h(t - nT_c - \tau_k) \sum_{l=0}^{L-1} e^{j(\omega_l t + \psi_{k,l})} \right\} \tag{7.77}$$

where

$$u_k[n] = AC_{pk}[n] + BC_{dk}[n]d_k[n] \tag{7.78}$$

A and B are the signal amplitudes of the pilot and the data channel respectively, $d_k[n]$ is binary data, $h(i)$ is the impulse response of the chip wave-shaping filter, ω_l and $\psi_{k,l}$ are the carrier frequency and carrier phase of the lth subcarrier, respectively, and L is the number of subcarriers.

The signal is transmitted through a fading channel. The bandwidth of the subcarriers in this section are selected such that each subcarrier experiences independent, slowly varying, flat Rayleigh fading. Assuming perfect average power control, the received signal is given by

$$r(t) = \text{Re}\left\{ \sum_{k=0}^{K-1} \sum_{n=-\infty}^{+\infty} u_k[n]h(t - nT_c - \tau_k) \sum_{l=0}^{L-1} \alpha_{k,l} e^{j(\omega_l t + \theta_{k,l})} \right\} + n_w(t) \tag{7.79}$$

where K is the total number of users, τ_k is the relative time delay of user k, $\alpha_{k,1}$ and $\theta_{k,1}$ are the fading amplitude and phase, respectively, of the lth path for the kth user, and $n_w(t)$ is zero mean white Gaussian noise with two-sided spectral density $\eta_0/2$.

The channel estimator based on a pilot signal is shown in Figure 7.31. Assuming

$$X(f) \equiv |H(f)|^2 = \begin{cases} \dfrac{1}{W}, & -\dfrac{W}{2} < f < \dfrac{W}{2} \\ 0, & \text{otherwise} \end{cases} \tag{7.80}$$

$$W = \frac{1}{T_c}$$

$$\mathcal{F}^{-1}|H(f)|^2 \equiv x(t)$$

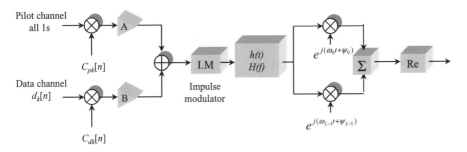

Figure 7.30 The complex transmitter block diagram [685].

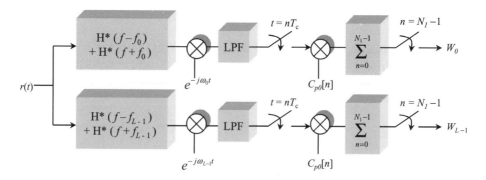

Figure 7.31 The complex channel estimator block diagram [685].

and

$$\int_{-\infty}^{+\infty} |H(f)|^2 \mathrm{d}f \equiv 1$$

the *l*th complex channel estimate is given by

$$W_l = \alpha_{0,l} e^{j\theta_{0,l}} A N_I + I_{p,l} + N_{p,l} \tag{7.81}$$

We have used the fact that the spreading sequences $C_{pk}[n]$ and $C_{dk}[n]$ are orthogonal in an estimation interval.

7.7.1 Data demodulation

Each subcarrier of the received signal is chip-matched filtered, demodulated, despread by the corresponding data spreading sequence, and then integrated over the bit interval of N chips (see Figure 7.32) . The *l*th demodulator output before combining is

$$Y_l = \alpha_{0,l} e^{j\theta_{0,l}} B d[0] N + I_{p,l} + N_{d,l} \tag{7.82}$$

where the interference term is

$$I_{d,l} = \sum_{n=0}^{N-1} C_{d0}[n] R_n \tag{7.83}$$

and the noise term is

$$N_{d,l} = \sum_{n=0}^{N-1} [n_w(t) * h^*(t)]_{t=nT_c} \cdot C_{d0}[n] \tag{7.84}$$

The combined signal is given by

$$Y = \sum_{l=0}^{L-1} W_l^* Y_l \tag{7.85}$$

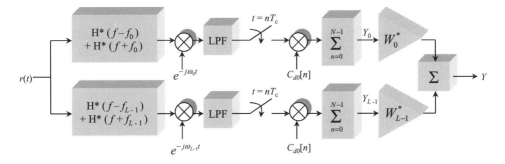

Figure 7.32 The complex data demodulation block diagram.

and the final decision statistic is

$$Z = \text{Re}\{Y\} = \sum_{l=0}^{L-1} \left[\tfrac{1}{2} W_l^* Y_l + \tfrac{1}{2} W_l Y_l^* \right] \qquad (7.86)$$

Bit error rate analysis for such a system can be found in [685].

7.7.2 Performance examples

In this example the signal parameters are the same as in [685]. The bandwidth of each subcarrier is fixed to be the coherence bandwidth of the channel. The total bandwidth is proportional to the number of subcarriers. To make the comparison between different bandwidths, the total transmit power is kept constant, that is, decreasing the transmit power per subcarrier as the number of subcarriers increases. Traditionally, assuming perfect channel estimation, the probability of error improves monotonically with the number of subcarriers [686]. However, when there is estimation error, the situation is different.

The probability of error is plotted against the number of subcarriers L in Figure 7.33(a) with ten total users, the estimation interval N_i equaling 64 chips, and E_b/η_0 of 4, 7, 10 and 13 dB. A processing gain of 64 is used, and there is equal power in the pilot and the data channel. As the number of subcarriers increases, the bit error rate first improves and then degrades. The increasing L helps performance by introducing diversity gain. At the same time, as L goes up, the transmit energy-per-band goes down; this causes more estimation error, and in turn, results in performance degradation. Thus, an optimal value of L exists. When we increase the E_b/η_0, the optimal L becomes larger, because the higher signal to noise ratio reduces the degradation due to the estimation error. Some additional results are shown in Figures 7.33 (b) and (c).

7.8 MULTICARRIER DS-CDMA BROADCAST SYSTEMS

As pointed out in Chapter 1, multicarrier direct sequence code division multiple access (DS-CDMA) systems can be classified into two categories: those with overlapping bandwidths [687, 688] and those with disjoint bandwidths [686]. In this section, the non-overlapping bandwidth system is considered, employed in the forward link (base-to-mobile link) of a

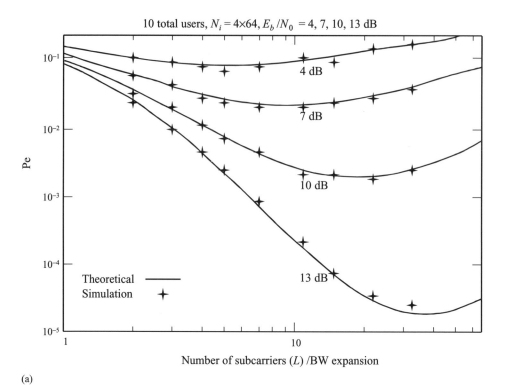

(a)

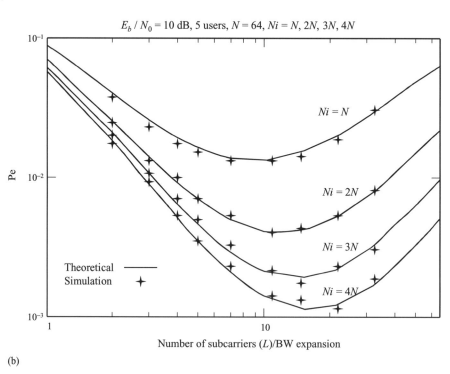

(b)

Figure 7.33 (a) Probability of error versus number of subcarriers for varying E_b/η_0; (b) probability of error versus number of subcarriers for different estimation intervals; (c) probability of error versus number of subcarriers for different numbers of users.

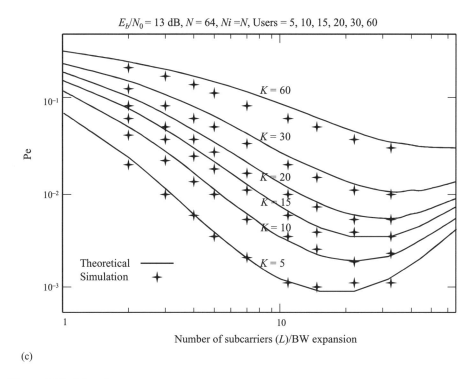

(c)

Figure 7.33 (*Cont.*).

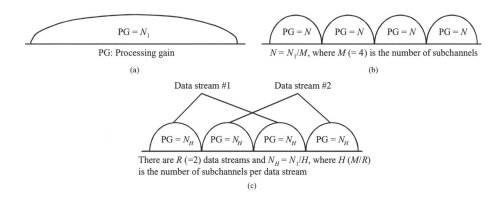

Figure 7.34 Spectra of (a) single-carrier CDMA; (b) multicarrier CDMA; and (c) hybrid multicarrier CDMA/FDM.

cellular system, wherein all user signals are synchronous. Using this type of multicarrier DS-CDMA to generate a wideband CDMA waveform, in particular by choosing the bandwidth of a subcarrier equal to that of a narrowband CDMA waveform, we can achieve some degree of compatibility between wideband (e.g. UMTS/WCDMA) and narrowband (e.g. UMTS/cdma2000) CDMA systems. The spectra of single-carrier and multicarrier CDMA are shown in Figures 7.34(a) and (b), respectively.

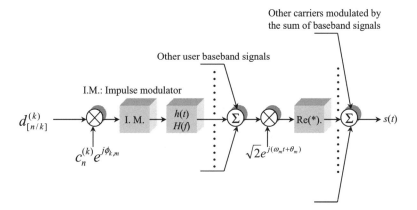

Figure 7.35 Block diagram of a multicarrier CDMA base station transmitter.

A base station transmitter using a multicarrier DS-CDMA system is shown in Figure 7.35. The transmitted signal is given by

$$s(t) = Re \left[\sum_{m=1}^{M} S_m(t) \sqrt{2} e^{j(\omega_m t + \theta_m)} \right] \qquad (7.87)$$

$S_m(t)$ is given by

$$S_m(t) = \sum_{k=1}^{K} \sqrt{E_c} \sum_{n=-\infty}^{+\infty} d_v^{(k)} c_n^{(k)} e^{j\phi_{k,m}} h(t - nT_c) \qquad (7.88)$$

where $d_v^{(k)}$ and $c_n^{(k)}$ are the data and the spreading sequences of the kth user, respectively, $v = \lfloor n/N \rfloor$ where $\lfloor x \rfloor$ is the largest integer less than or equal to x and N is the number of chips of the spreading sequence per data bit), $h(t)$ is the impulse response of the chip wave-shaping filter, $\phi_{k,m}$ is a carrier phase randomly chosen by a base station for the mth subchannel of the kth user, and $1/T_c$ is the chip rate. Both θ_m and $\phi_{k,m}$ are uniformly distributed over $[0, 2\pi)$. The energy per data bit is defined as $E_b \equiv E_c NM$. Note that the need for $\phi_{k,m}$ to be random for different users and different subchannels is to obtain a proper processing gain in the multicarrier system. The reason for this is as follows. A multicarrier system is known to provide an effective processing gain of MN [689] for an asynchronous communication link, where the carrier phase difference between different users' signals can be assumed to be random, and the phases of any given user are uncorrelated in different subchannels. Therefore, coherent combining of the despread signals from M correlators increases the processing gain by M times that obtained by despreading the signals in each subchannel. However, for the forward link, without the $\phi_{k,m}$, the interference components of the M correlator outputs are also combined coherently, and consequently, there is no increased processing gain achieved by the coherent combining of the M correlator outputs, only a diversity gain. With the random $\phi_{k,m}$, it is possible for the receivers to demodulate the data coherently by using a common pilot signal from a base station if each receiver is provided the values of the $\phi_{k,m}$ by the base station when it establishes a communication link.

It is assumed that $c_n^{(k)} = a_n b_n^{(k)}$, where a_n is a random sequence commonly used by all users and $b_n^{(k)}$ is a member of either an orthogonal or a quasi-orthogonal code set assigned to the kth user. We also have

$$E\{c_n^{(k)}c_{n+i}^{(k')}\} = 0 \quad \text{for } i \neq 0 \tag{7.89}$$

for all k and k'. It is also assumed that the $b_n^{(k)}$ satisfy the following relationship for all k:

$$\frac{1}{K-1} \sum_{\substack{k'=1 \\ k' \neq k}}^{k} \left(\sum_{n=0}^{N-1} b_n^{(k)} b_n^{(k')} \right)^2 = (1-q)N \tag{7.90}$$

where q is a measure of the orthogonality of the set of quasi-orthogonal codes. The quasi-orthogonal codes with these characteristics are standardized [690].

Further, we assume that the channel is a slowly varying frequency selective Rayleigh fading channel and is modeled as a finite length tapped delay line. The complex low-pass equivalent response for the mth subchannel is given by

$$c_m(t) = \sum_{i=0}^{L-1} \alpha_{m,i} e^{j\beta_{m,i}} \delta(t - iT_c) \tag{7.91}$$

where L is the number of resolvable paths, $\alpha_{m,i}$ are independent but not necessarily identically distributed Rayleigh random variables, and the $\beta_{m,i}$ are independently, identically distributed (i.i.d.), uniform random variables over $[0, 2\pi)$. A unit energy constraint is assumed, i.e. $\sum_{i=0}^{L-1} E\{\alpha_{m,i}^2\} = 1$. Then, for a constant multipath intensity profile (MIP), the second moment of each path of a subchannel is given by $E\{\alpha_{m,i}^2\} = 1/L$. For an exponential MIP, the second moments are assumed to be related to the second moment of the initial path strength by $E\{\alpha_{m,i}^2\} = E\{\alpha_{m,0}^2\} \exp(-ri)$, where r is the MIP decay factor. For comparison of the performance of single-carrier and multicarrier systems, we use the facts that $r = Mr_1$ and $L = L_1/M$, where r_1 and L_1 are the MIP decay factor and the number of resolvable paths, respectively, for the single-carrier system [691].

Since all signals that arrive from the base station at a given mobile unit propagate over the same path, they all fade in unison. The received signal at the desired mobile unit is then given by

$$r(t) = \text{Re}[R_c(t)] + n(t) \tag{7.92}$$

where $R_c(t)$ is the complex representation of the received signal given by:

$$R_c(t) = \sum_{m=1}^{M} \sum_{i=0}^{L-1} \alpha_{m,i} S_m(t - iT_c) \sqrt{2} e^{j(\omega_m t + \theta'_{m,i})} \tag{7.93}$$

$n(t)$ is additive white Gaussian noise (AWGN) with a two-sided power spectral density of $\eta_0/2$, and $\theta'_{m,i} = \theta_m + \beta_{m,i}$. The corresponding expression for the single-carrier system is obtained by setting $M = 1$ and L equal to L_1.

The receiver of the desired user ($k = 1$) is shown in Figure 7.36, where a RAKE receiver in each subchannel, and perfect phase recovery of each carrier from the pilot signal detector is assumed [689]. The chip wave-shaping filter given in [689] is assumed, where $X(f) \equiv |H(f)|^2$ is a raised cosine filter. The DS waveforms do not overlap and therefore, adjacent channel interference may be ignored.

Each data stream modulates $H(= M/R)$ disjoint carriers, and the number of chips of the spreading sequence per data bit is N_1/H. To make a fair comparison of the performance,

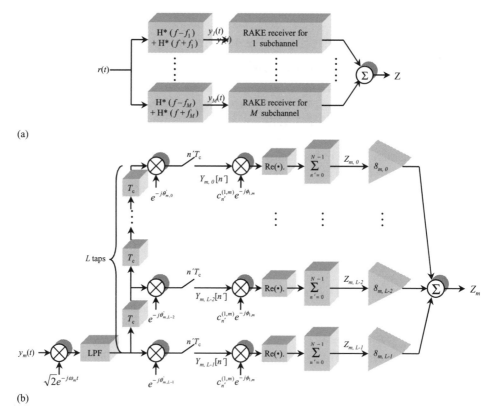

(a)

(b)

Figure 7.36 Block diagrams of (a) a multicarrier CDMA receiver and (b) a RAKE receiver of the mth subchannel for the first user.

given a fixed information rate and total bandwidth, the relationship $HRL = ML = L_1$ must be satisfied. Bit error rate analysis for such a system can be found in [692] and some results are shown in Figure 7.37.

7.9 FRAME BY FRAME ADAPTIVE RATE CODED MULTICARRIER DS-CDMA SYSTEM

In this section we discuss an adaptive rate convolutionally coded multicarrier direct sequence code division multiple access (DS-CDMA) system. In order to accommodate a number of coding rates easily and make the encoder and decoder structure simple, the rate compatible punctured convolutional (RCPC) code discussed in Chapter 2 is used. We choose the coding rate that has the highest data throughput in the signal to interference and noise ratio (SINR) sense. To achieve maximum data throughput, a rate adaptive system is used, based on the channel state information (the signal to interference to noise ratio, SINR, estimate). The SINR estimate is obtained by the soft decision Viterbi decoding metric. It will be demonstrated that the rate adaptive convolutionally coded multicarrier DS-CDMA system

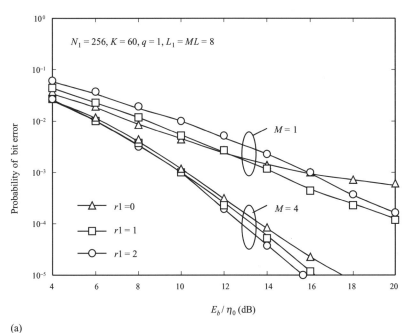

(a)

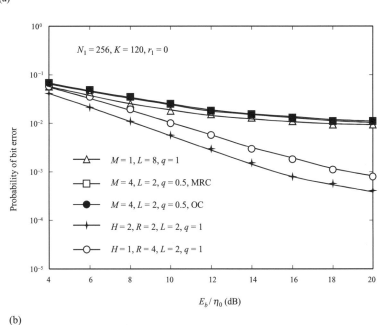

(b)

Figure 7.37 (a) Performance of a single-carrier and a multicarrier CDMA system in multipath fading channel when both systems employ an orthogonal code set; (b) performance comparison of a single-carrier CDMA, a multicarrier CDMA and a hybrid multicarrier CDMA/FDM system for $K = 120$ in a multipath fading channel with a constant MIP; (c) probability of bit error versus K for $E_b/\eta_0 = 15$ (decibels).

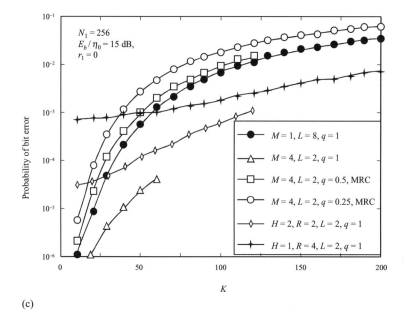

(c)

Figure 7.37 (*Cont.*).

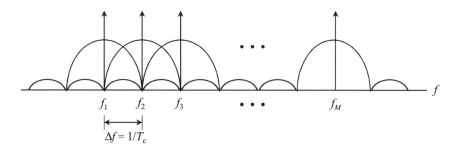

Figure 7.38 A typical CC-OM power spectral density.

can enhance spectral efficiency and provide frequency diversity.

7.9.1 Transmitter

The power spectral density of a convolutionally coded orthogonal multicarrier (CC-OM) signal and the transmitter for the rate adaptive CC-OM DS-CDMA system considered in this section are shown in Figure 7.38 and Figure 7.39 respectively. For user k, the (information) bits $\{b_k^i\}$, each with duration T_b, are encoded by the RCPC encoder of rate r. The relationship between T_b and the duration T_s of a coded binary symbol can be written as

$$T_s = r M T_b \tag{7.94}$$

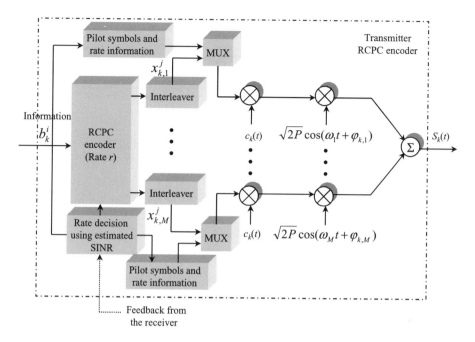

Figure 7.39 The transmitter model for user k in the adaptive rate CC-OM DS-CDMA system.

where M is the number of subchannels. The M coded binary symbols are allocated to M subchannels to get frequency diversity. They are interleaved to get time diversity as well as frequency diversity, and are spread by each user's pseudonoise (PN) signature waveform $c_k(t)$ with chip duration $T_c = T_s/N$, where N is the processing gain of the DS narrowband waveforms modulated by subcarriers. For CC-OM systems, we have $M = (2B_T/B_S) - 1$, where B_T and B_S are the total and subchannel bandwidths, respectively. Since we fix the subchannel bandwidth B_S (or equivalently, the symbol duration T_s) in this section, T_b varies according to Equation (7.94) when the code rate r changes.

The transmitted signal $s_k(t)$ of user k can be written as

$$s_k(t) = \sqrt{2P} \sum_{j=-\infty}^{\infty} \sum_{m=1}^{M} x_{k,m}^j c_k(t - jT_s) \cos(\omega_m t + \varphi_{k,m}) \tag{7.95}$$

The channel is assumed to be frequency selective Rayleigh fading and not to vary during one symbol duration. However, the subchannels are assumed to be non-selective by choosing the number of subcarriers appropriately as [686]:

$$MT_c \geq T \tag{7.96}$$

where T is the maximum delay spread of the channel. Then the complex low-pass impulse response of the subchannels of user k can be modeled as

$$h_{k,m}(t) = \alpha_{k,m} e^{j\beta_{k,m}} \delta(t) \tag{7.97}$$

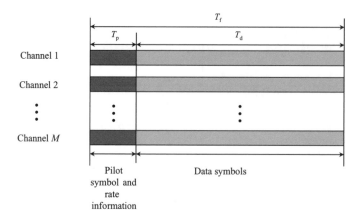

Figure 7.40 The frame structure.

where $\alpha_{k,m}$ is the fading amplitude and $\beta_{k,m}$ is the random phase of the mth subchannel, $m = 1, 2, \ldots, M$. The phases $\{\beta_{k,m}\}$ are i.i.d. uniform random variables on $[0, 2\pi)$. In general, the fading amplitudes $\{\alpha_{k,m}\}$ are correlated, but we can assume that they are i.i.d. Rayleigh random variables once the coded symbols are properly interleaved in the time domain.

Frame by frame transmission is assumed. Such a frame by frame transmission is typical of many cellular systems. The frame discussed in this section is shown in Figure 7.40. Each frame of duration T_f consists of a header of duration T_p and data symbols of duration T_d. We have $T_d = N_s T_s$ and $T_p = N_p T_s$, where N_s is the number of data symbols and N_p is normally 6–10. The header contains pilot symbols and information on the rate and channel state. The function of the MUX in Figure 7.39 is to combine the header and data symbols to make frames as shown in Figure 7.40.

7.9.2 Receiver

The receiver for the adaptive rate CC-OM DS-CDMA system in this section is shown in Figure 7.41. Let us assume that there are K users *in a cell* and power control is employed. Then the received signal *at the base station* can be written as

$$r(t) = \sqrt{2P} \sum_{j=-\infty}^{\infty} \sum_{k=1}^{K} \sum_{m=1}^{M} \alpha_{k,m} x_{k,m}^j c_k(t - \tau_k - jT_s) \cos(\omega_m t + \phi_{k,m}) + n(t) \quad (7.98)$$

7.9.3 Rate-compatible punctured convolutional (RCPC) codes

RCPC codes are discussed in Chapter 2. For this section, some additional details are specified. Let the code rate and constraint length of the parent code be $R = 1/n$ and L_c, respectively. The parent code is completely specified by the n generator polynomials $G^j(D) = g_0^j + g_1^j D + \cdots + g_{L_c-1}^j D^{L_c-1}$, $j = 1, 2, \ldots, n$, where $g_i^j \in \{0, 1\}$. The puncturing is done according to the rate compatibility criterion, which requires that lower rate codes use the same coded bits as the higher rate codes plus one or more additional bit(s).

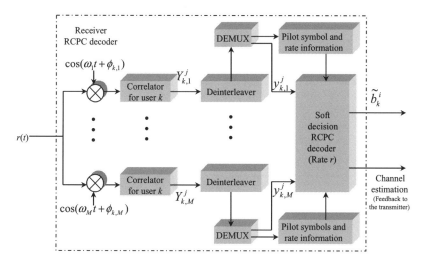

Figure 7.41 The receiver model for user k in the adaptive rate CC-OM DS-CDMA system.

The bits to be punctured are described by an $n \times p$ puncturing matrix **P** consisting of zeros and ones, where p is called the puncturing period. At time instant t, the output from each generator $G^j(D)$ is transmitted if $P(j, t \bmod p) = 1$ and punctured otherwise. Here, $\mathbf{P}(a, b)$ denotes the element on row a and column b in the matrix **P**. The number p of columns determines the number of code rates and the rate resolution that can be obtained. Generally, from a parent code of rate $1/n$, we can obtain a family of $(n - l)p$ different codes with rates

$$r = \frac{p}{np}, \frac{p}{np - 1}, \dots, \frac{p}{p + 1} \tag{7.99}$$

The code rate of RCPC codes can be changed during even one information bit transmission and, thus, unequal error protection can be obtained [638]. In this section, however, the code rate of RCPC codes is changed frame by frame, not bit by bit, because we assumed frame by frame transmission. An example of the RCPC encoder is shown in Figure 7.42.

7.9.4 Rate adaptation

A threshold-based adaptation scheme is used which adaptively changes the coding rate depending upon the SINR estimated. Let $\theta_0 = -\infty$, $\theta_1, \theta_2, \dots, \theta_Q = \infty$ be the SINR threshold values, which are chosen such that between θ_{j-1} and θ_j the channel coding rate r_j has the highest throughput. Here, Q is the number of possible code rates. Then, as discussed in Chapter 3, the transmitter mode (rate) adaptation scheme can be defined as follows:

$$\text{Choose } r_j \text{ if } \theta_{j-1} \le \text{SINR} < \theta_j, \quad j = 1, \dots, Q \tag{7.100}$$

In this method, bit by bit adaptation is not assumed due to the feedback delay. Instead, we choose the adaptation interval T_a in such a way that T_a is long enough to allow the transmission of at least one frame and short enough to react quickly to the possible change of the SINR. The transmitter can then adapt its data rate every $[T_a/T_f]T_f$. This allows efficient error recovery through ARQ mechanisms even with dynamic rate adaptation. The rate at

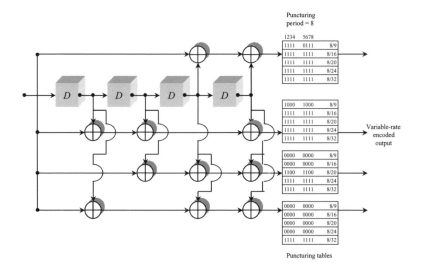

Figure 7.42 A five-rate RCPC encoder from a rate $1/4$ parent code [693] (D: delay of duration T_b).

which the transmitter reacts to the changes in the SINR depends on the SINR estimate and feedback delay in the system [694].

7.9.4.1 Example

RCPC codes with rate $1/4$ convolutional codes of constraint length $L_c = 5$ and 9 are used as the parent codes [694, 695], as the error control capability varies when the constraint length changes, we use two values of L_c. The number, M, of subcarriers is four and nine. The processing gain N is 192 and 96, for $M = 4$ and $M = 9$, respectively, when the total bandwidth

$$B_T = N(M + 1)/T_s \tag{7.101}$$

is fixed. We assume each frame contains 144 symbols with $T_f = 10$ ms and $T_p = 0$ (that is, we assume perfect feedback to simplify the simulations).

The adaptive rate CC-OM DS-CDMA system was implemented based on the above discussion in [696]. When $L_c = 5$, fixing the coding rate to $1/2$ allows us to get the highest throughput and the result is Figure 7.43(a). When $L_c = 9$, using the thresholds $\theta_1 = 2.5$ dB and $\theta_2 = 5.5$ dB, we get Figure 7.43(b). In these figures, the throughput of the conventional system (fixed rate with $1/M$) is also shown. It is clear that we can get much higher throughput with the adaptive system.

7.10 INTERMODULATION INTERFERENCE SUPPRESSION IN MULTICARRIER CDMA SYSTEMS

In this section, a coded multicarrier direct sequence code division multiple access (DS-CDMA) system is presented that, by the use of a minimum mean squared error receiver, achieves frequency diversity (instead of path diversity as in a conventional single carrier

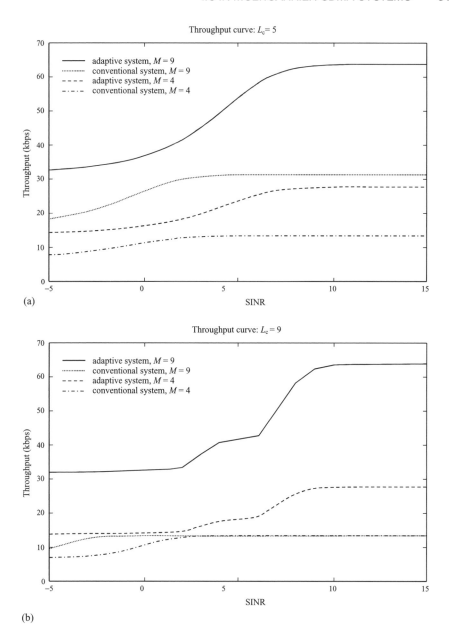

Figure 7.43 (a) The adaptive throughput curves and the conventional throughput curves when $L_c = 5$; (b) the adaptive throughput curves and the conventional throughput curves when $L_c = 9$.

(SC) RAKE DS-CDMA). It also has the ability to suppress the intermodulation distortion and partially compensate for the signal distortion introduced by a non-linear amplifier at the transmitter. A frequency selective Rayleigh fading channel is decomposed into M frequency non-selective channels, based on the channel coherence bandwidth. A rate $1/M$ convolutional code, after being interleaved, is used to modulate M different DS-CDMA waveforms. The system is shown to effectively combat intermodulation distortion in the presence of multiple access interference.

7.10.1 Transmitter

In this section, the transmitter shown in Figure 7.44 is considered. The input signal to the power amplifier for the kth user $s_k(t)$ is given by

$$s_k(t) = A_k \sum_{i=-\infty}^{\infty} \sum_{m=1}^{M} d_{k,m}^{(i)} p_{k,m}(t - iY - \tau_k) \cos(\omega_m t + \theta_{k,m}) \tag{7.102}$$

and $p_k(t)$ is a spreading (or signature) waveform given by

$$p_{k,m}(t) = \sum_{n=0}^{N-1} c_{k,m}^{(n)} h(t - nMT_c) \tag{7.103}$$

In Equation (7.103), $c_{k,m}^{(n)} \in \{\pm 1\}$ is the nth chip of the spreading sequence, N is the processing gain, which is taken to be equal to the period of the spreading sequence, $h(t)$ is the impulse response of the chip wave-shaping filter, $1/MT_c$ is the chip rate of the band-limited MC DS-CDMA system, and $1/T_c$ is the chip rate of a band-limited single carrier (SC) DS-CDMA system that occupies the same spread bandwidth as does the MC system. That is, $T = NMT_c$.

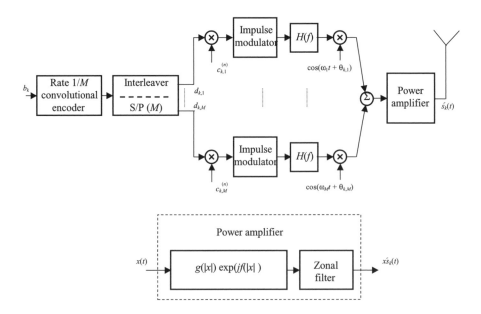

Figure 7.44 Transmitter block diagram for user k.

7.10.2 Non-linear power amplifier model

For an input signal formed by the sum of M subcarrier signals, i.e.

$$x(t) = Re \left\{ \sum_{m=1}^{M} a_m(t) \exp[j\omega_0 t + j\psi_m(t)] \right\}$$
$$= Re \left\{ A(t) \exp[j\omega_0 t + j\Psi(t)] \right\}$$

(7.104)

the output signal of a non-linear power amplifier can be represented by

$$\hat{x}(t) = Re \left\{ g[A(t)] \exp[j\omega_0 t + j\Psi(t) + jf[A(t)]] \right\}$$ (7.105)

where $g(A)$ and $f(A)$ represent the AM/AM (see Figure 7.45) and AM/PM conversion characteristics of the non-linear power amplifier.

7.10.3 MMSE receiver

As shown in Figure 7.46, M MMSE filters are combined with a soft-decision Viterbi decoder so that the soft outputs from the M MMSE filters are parallel-to-serial converted, deinterleaved, and decoded. The received DS-CDMA signal after the LPF for each subcarrier is despread (either partially or fully) over consecutive F chips, which is characterized by a parameter $N_t = \lceil N/F \rceil$. The decision symbols needed by the M MMSE filters can be obtained via an interleaver with a serial-to-parallel converter and an encoder which is identical to that used in the transmitter. The tap weight vector of the mth MMSE filter ω_m is chosen so as to minimize the conditional mean square error, conditioned on all

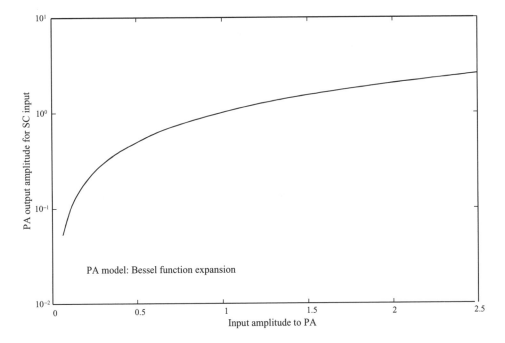

Figure 7.45 Power amplifier transfer function for single carrier.

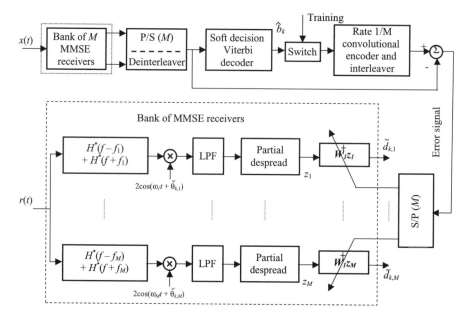

Figure 7.46 Receiver block diagram for user k.

parameters of the desired user (user 1) and certain parameters of the MAI and IM, i.e. with $q = \{\alpha_1, \check{\theta}_1, \rho^{\pm}_{k,I(m,h)}\}$, we have

$$MSE = E\{(\omega'_m \mathbf{z}_m - d_{1,m})^2 \mid \mathbf{q}\} \tag{7.106}$$

For the notation see Figure 7.46. Note that we omit the superscript i, which denotes the estimated bit, for notational simplicity. From Chapter 5, the optimum tap weight vector for Equation (7.106) is given by:

$$(\omega_m)_{\text{opt}} = \mathbf{R}_m^{-1} \mathbf{a}_m \tag{7.107}$$

where

$$\mathbf{R}_m = E\{z_m z'_m \mid \mathbf{q}\} \tag{7.108}$$

and

$$\mathbf{a}_m = E\{d_{1,m} z_m \mid \mathbf{q}\} \tag{7.109}$$

7.10.3.1 Example

For the numerical example, we use the same parameters as in [697], the PA model given in the previous section and $M = 4$, $N = 32$, $R = 1/4$ and constraint length $K = 3$ coded MC-CDMA system and an SC coded RAKE system with the same coding scheme and the same bandwidth as the coded MC-CDMA system. Note that there are not any IM terms in the SC system, but the non-linear distortion introduced by the PA is taken into account. The same PA output power is assumed for both the SC input signal and the MC input signal. The system performance is obtained by simulation. In Figures 4.47–48, the maximal number of resolvable paths for the SC is denoted L_p, and L_t is the actual number of RAKE taps.

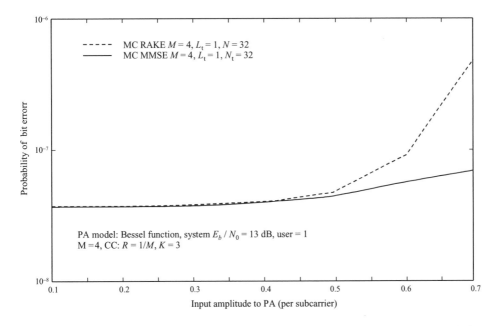

Figure 7.47 Probability of bit error in the presence of IMD (different PN (DPN) code for each subcarrier).

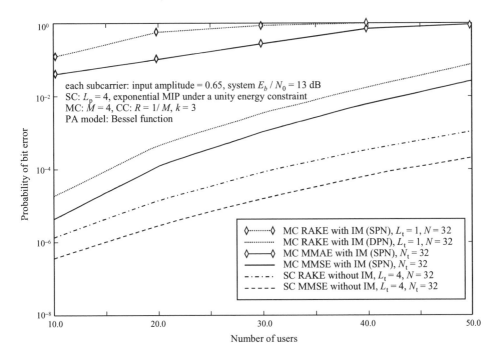

Figure 7.48 Comparison between the systems with the same (SPN) and the different (DPN) PN codes for each subcarrier.

7.11   SUCCESSIVE INTERFERENCE CANCELLATION IN MULTICARRIER DS-CDMA SYSTEMS

This section presents a successive interference cancellation (SIC) scheme for a multicarrier (MC) asynchronous DS-CDMA system, wherein the output of a convolutional encoder modulates band-limited spreading waveforms at different subcarrier frequencies. In every subband, the SIC receiver successively detects the interferers signals and subtracts them from that of the user of interest. The SIC receiver employs maximal ratio combining (SIC-MRC) for detection of the desired user, and feeds a soft decision Viterbi decoder.

7.11.1   System and channel model

A K-user asynchronous communication system is presented. We assume knowledge of the time delays and the spreading sequences of all the users, but no knowledge of the channel gains of the interferers. The user data symbols are input to a rate $1/M$ convolutional encoder. The output code symbols are interleaved and serial-to-parallel (S/P) converted such that M parallel code symbols may be transmitted simultaneously. Then, each of the M code symbols is replicated by a rate $1/R$ repetition code, and transmitted over MR subcarriers.

For example, if $M = 4$ and $R = 2$, then the $M = 4$ code symbols are mapped to a total of eight subcarriers in the following way: the first code symbol is transmitted on the first and the fifth subcarriers, the second code symbol is transmitted on the second and the sixth subcarriers, the third code symbol is transmitted on the third and the seventh subcarriers, and the fourth code symbol is transmitted on the fourth and the eighth subcarriers. Therefore, the minimum subcarrier distance for the same code symbol is maximized. The mapped code symbols are then multiplied by the spreading sequence assigned to the given user. The transmitted signal for the kth user is:

$$S_k(t) = \sqrt{2E_{ck}} \left\{ \sum_{j=-\infty}^{+\infty} a_k(t - jT - \tau_k) \times \sum_{m=1}^{MR} b_{k,[m]_M}^j \cos(\omega_m t + \theta_{k,m}) \right\} \qquad (7.110)$$

where

$$a_k(t) = \sum_{n=0}^{N-1} c_k^{(n)} h(t - nT_c)$$

We assume that the channel in each subband is a slow-varying frequency non-selective Rayleigh channel with transfer function

$$\zeta_{k,m} = \alpha_{k,m} \exp(j\beta_{k,m})$$

The received signal is given by

$$r(t) = \sum_{k=1}^{K} \sqrt{2E_{ck}} \sum_{j=-\infty}^{+\infty} a_k(t - jT - \tau_k) \times \sum_{m=1}^{MR} b_{k,[m]_M}^j \alpha_{k,m} \cos(\omega_m t + \theta_{k,m}') + n_w(t)$$

$$(7.111)$$

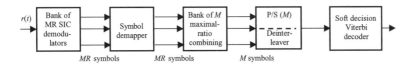

Figure 7.49 Receiver block diagram for the desired user 1.

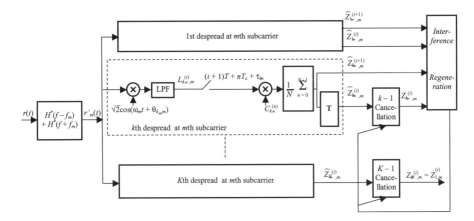

Figure 7.50 SIC demodulator at mth subcarrier for the desired user 1.

The receiver structure for the desired user, user 1, is shown in Figures 7.49–50. Interferers $2, 3, \ldots, K$ are renumbered as $1_m, 2_m, \ldots, (K-1)_m$, which defines the cancellation order such that in the m-th subband, interferer 1_m is the strongest, 2_m is the second strongest, and so on. The cancellation order in each subband is different, i.e. for every subband there is a distinct successive interference cancellation order.

7.11.1.1 *Example*

In the following numerical example the signal parameters are the same as in [698]. Gold sequences are used with $N = 31$ when convolutional coding is employed, and the constraint length of the convolutional code is three. Ideal power control is assumed, i.e. $E_{ck} \triangleq E_c$ for every user k. Also, $X(f)$ is a raised cosine function with a rolloff factor $\alpha = 0.5$, and all of the comparisons are based on the same transmitted data rate. The analytical results presented in [698] are averaged over 1000 realizations, and the cancellation order in the simulation of SIC on the mth subband is based upon the decreasing order of $|Z_{k_m,m}(i)|$, i.e. $|\hat{Z}_{1_m,m}(i)| \geq \cdots \geq |\hat{Z}_{(K-1)_m,m}(i)|$. To simulate the time-correlated Rayleigh fading channel, the Jakes model (see Chapter 14) is used with a bit rate of 20 000 bits/s and a maximum Doppler frequency of 100 Hz, while a block interleaver/deinterleaver is employed to separate adjacent data bits into a block size of 30×30, which results in a delay of 45 ms. Some results are given in Figure 7.51.

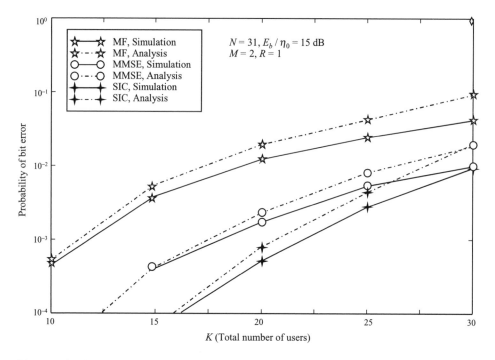

Figure 7.51 Comparisons of the analytical bounds with simulation results for the MF and SIC receivers in convolutionally coded MC CDMA, where perfect CSI is assumed.

7.12 MMSE DETECTION OF MULTICARRIER CDMA

The results from the previous section have already demonstrated that an MMSE detector performs better than a receiver with SIC. For this reason, in this section the minimum mean squared error (MMSE) detection of multicarrier code division multiple access (CDMA) signals is presented in more detail. The performance of two different design strategies for MMSE detection is compared. In one case, the MMSE filters are designed separately for each carrier, while in the other case the optimization of the filters is done jointly. Naturally, the joint optimization produces a better receiver, but the difference in performance is shown to be substantial. The multicarrier CDMA performance is then compared to that of a single-carrier CDMA system on a frequency selective fading channel. A mechanism to track the channel fading parameters for all the users' signals is presented which enables joint optimization of the receiver filters in a time-varying channel. Simulation results show that the performance of this receiver is close to the ideal theoretical results for moderate vehicle speeds. Performance begins to degrade when the normalized Doppler rate is higher than about 1 %.

The received signal on the system's reverse link on the mth carrier is given by

$$r_m(t) = \text{Re}\left\{ \sum_{i=-\infty}^{+\infty} \sum_{k=1}^{K} \sqrt{\frac{2P_k}{M}} \gamma_{k,m}(i) d_k(i) \times c_{k,m}(t - iT_b - \tau_k) \exp(j\omega_m t) \right\} + n_m(t)$$

$$(7.112)$$

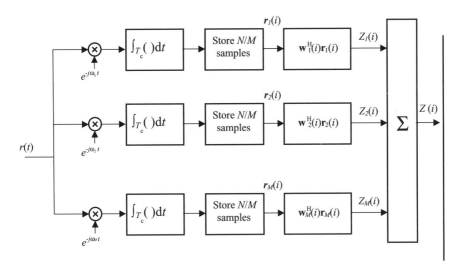

Figure 7.52 General receiver for multicarrier CDMA.

The received signal is processed with a chip-matched filter, which consists of an integrator with duration MT_c. The samples are stored for one bit interval, giving a column vector of length N/M:

$$\mathbf{r}_m(i) = \sum_{k=1}^{K} \sqrt{\frac{P_k}{P_1}} \gamma_{k,m}(i)[d_k(i)\mathbf{f}_{k,m} + d_k(i-1)\mathbf{g}_{k,m}] + \mathbf{n}_m(i) \qquad (7.113)$$

where $\mathbf{f}_{k,m}$ and $\mathbf{g}_{k,m}$ depend on the left- and right-cyclic shifts of $\mathbf{c}_{k,m}$, the spreading code of the kth user on the mth carrier.

A block diagram of a general linear receiver is shown in Figure 7.52. Each of the M received vectors is processed with a receiver filter $\mathbf{w}_m(i)$ to form a statistic $Z_m(i) = \mathbf{w}_m^{\mathrm{H}}(i)\mathbf{r}_m(i)$, for $m = 1, 2, \ldots, M$. Note the time dependence of the filters in the time-varying fading channel. The individual statistics are summed to form an overall decision statistic $Z(i) = \sum_{m=1}^{M} Z_m(i)$. Equivalently, we can define an overall receiver filter as

$$\mathbf{w}(i) = \left[\mathbf{w}_1^{\mathrm{T}}(i), \mathbf{w}_2^{\mathrm{T}}(i), \cdots, \mathbf{w}_M^{\mathrm{T}}(i)\right]^{\mathrm{T}} \qquad (7.114)$$

and an overall received vector as

$$\mathbf{r}(i) = \left[\mathbf{r}_1^{\mathrm{T}}(i), \mathbf{r}_2^{\mathrm{T}}(i), \ldots, \mathbf{r}_M^{\mathrm{T}}(i)\right]^{\mathrm{T}} \qquad (7.115)$$

which gives $Z(i) = \mathbf{w}^{\mathrm{H}}(i)\mathbf{r}(i)$.

We next consider two different design strategies for performing MMSE detection. The best performance is obtained when the filters $\mathbf{w}_1(i), \mathbf{w}_2(i), \cdots, \mathbf{w}_M(i)$ are designed jointly so as to minimize the composite mean squared error

$$J = E[|d_1(i) - \mathbf{w}^{\mathrm{H}}(i)\,\mathbf{r}\,(i)|^2] \qquad (7.116)$$

This gives the well-known Wiener solution $\mathbf{w}(i) = \mathbf{R}^{-1}(i)\,\mathbf{p}\,(i)$, with

$$\mathbf{R}(i) = E\lfloor\mathbf{r}(i)\mathbf{r}^{\mathrm{H}}(i)\rfloor \quad \text{and} \quad \mathbf{p}(i) = E\lfloor d_1^*(i)\,\mathbf{r}\,(i)\rfloor \qquad (7.117)$$

representing the correlation matrix and steering vector, respectively. These can be further decomposed as

$$\mathbf{R}(i) = \begin{bmatrix} \mathbf{R}_{1,1}(i) & \mathbf{R}_{1,2}(i) & \cdots & \mathbf{R}_{1,M}(i) \\ \mathbf{R}_{2,1}(i) & \mathbf{R}_{2,2}(i) & \cdots & \mathbf{R}_{2,M}(i) \\ \vdots & \vdots & \ddots & \vdots \\ \mathbf{R}_{M,1}(i) & \mathbf{R}_{M,2}(i) & \cdots & \mathbf{R}_{M,M}(i) \end{bmatrix} \tag{7.118}$$

where the individual submatrices are defined as

$$\mathbf{R}_{m,n}(i) = E\left\lfloor \mathbf{r}_m(i)\mathbf{r}_n^{\mathrm{H}}(i) \right\rfloor \tag{7.119}$$

and

$$\mathbf{p}(i) = \left[\mathbf{p}_1^{\mathrm{T}}(i), \mathbf{p}_2^{\mathrm{T}}(i), \ldots, \mathbf{p}_M^{\mathrm{T}}(i) \right]^{\mathrm{T}} \tag{7.120}$$

with

$$\mathbf{p}_m(i) = E\lfloor d_1^*(i)\mathbf{r}_m(i) \rfloor \tag{7.121}$$

An alternative suboptimal approach is to design the M filters separately by choosing each of the filters $\mathbf{w}_1(i), \mathbf{w}_2(i), \ldots, \mathbf{w}_M(i)$ to minimize the individual mean squared error quantities $J = E[|d_1(i) - \mathbf{w}_m^{\mathrm{H}}(i)\mathbf{r}_m(i)|^2]$, which leads to $\mathbf{w}_m(i) = \mathbf{R}_{m,m}^{-1}(i)\mathbf{p}_m(i)$. Together with the previous notation, an overall filter using this design strategy can then be written as $\mathbf{w}(i) = \mathbf{R}_B^{-1}(i)\mathbf{p}(i)$ where

$$\mathbf{R}_B(i) = \begin{bmatrix} \mathbf{R}_{1,1}(i) & 0 & \cdots & 0 \\ 0 & \mathbf{R}_{2,2}(i) & \cdots & 0 \\ \vdots & \vdots & \ddots & \vdots \\ 0 & 0 & \cdots & \mathbf{R}_{M,M}(i) \end{bmatrix} \tag{7.122}$$

An example of performance results is given in Figure 7.53. One can see that the joint detector demonstrates better performance.

Before proceeding to present a tracking algorithm, it is worthwhile to compare the performance of a multicarrier CDMA system to that of a single-carrier system which realizes diversity inherently by operating on a frequency selective fading channel. In order to get a fair comparison to the multicarrier case, the system will be assumed to employ spreading waveforms of length N chips/bit, where N in the multicarrier case is the composite processing gain. The received signal will consist of M resolvable components, delayed with respect to one another by a sufficient number of chips. Results for this case are shown for comparison to the multicarrier case in Figure 7.54.

7.12.1 Tracking the fading processes

The presentation in this section is based on [699–701]. The received vector on the mth carrier, from Equation (7.113), may be written as

$$\mathbf{r}_m(i) = (\mathbf{F}_m\mathbf{D}(i) + \mathbf{G}_m\mathbf{D}(i-1))\mathbf{P}\boldsymbol{\Gamma}_m(i) + \mathbf{n}_m(i) \tag{7.123}$$

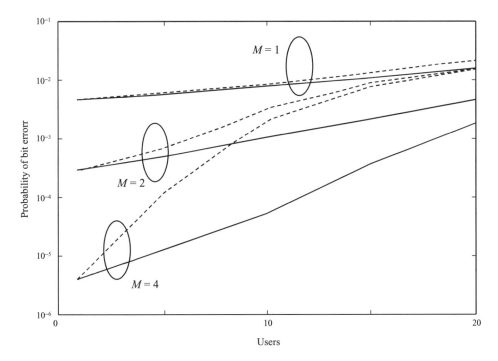

Figure 7.53 Probability of error versus number of users for the multicarrier CDMA system, with $E_b/N_0 = 17\,\mathrm{dB}$, composite processing gain of 32 chips/bit and M carriers. The solid lines are the Wiener solutions when all users' fading processes are tracked, the dashed lines are the Wiener solutions when only the desired user's fading processes are tracked.

where the matrices in this expression are defined as

$$\mathbf{F}_m = \left\lfloor \mathbf{f}_{1,m}, \mathbf{f}_{2,m}, \ldots, \mathbf{f}_{K,m} \right\rfloor$$

$$\mathbf{G}_m = [\mathbf{g}_{1,m}, \mathbf{g}_{2,m}, \ldots, \mathbf{g}_{K,m}]$$

$$\mathbf{D}(i) = \begin{bmatrix} d_1(i) & & & \\ & d_2(i) & & \\ & & \ddots & \\ & & & d_K(i) \end{bmatrix}$$

$$\mathbf{P}(i) = \begin{bmatrix} 1 & & & \\ & \sqrt{P_2/P_1} & & \\ & & \ddots & \\ & & & \sqrt{P_K/P_1} \end{bmatrix} \tag{7.124}$$

and the column vector of fading coefficients on the mth carrier, which is to be tracked, is

$$\boldsymbol{\Gamma}_m(i) = [\gamma_{1,m}(i), \gamma_{2,m}(i), \ldots, \gamma_{K,m}(i)]^{\mathrm{T}} \tag{7.125}$$

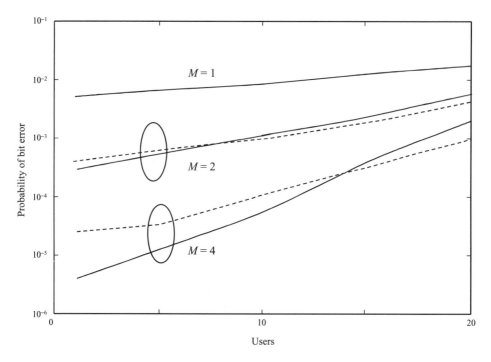

Figure 7.54 Probability of error versus number of users for the multicarrier CDMA system (solid curves) and the single-carrier system on a frequency selective
fading channel (dashed curves). For each $E_b/N_0 = 17$ dB, the composite
processing gain is 32 chips/bit, and M is the number of carriers used (for
multicarrier) or the number of resolvable paths present (for frequency selective). The Wiener solution is formed, with all users' fading processes tracked
perfectly.

Also, in Equation (7.123), recall that $\mathbf{n}_m(i)$ is a column vector of independent, complex
Gaussian noise samples, with the real and imaginary parts independent from each other and
each with variance $\sigma^2 = N/(2E_b/N_0)$. Then, Equation (7.123) can be rewritten as

$$\mathbf{r}_m(i) = \mathbf{\Lambda}_m(i)\mathbf{\Gamma}_m(i) + \mathbf{n}_m(i) \tag{7.126}$$

with

$$\mathbf{\Lambda}_m(i) = (\mathbf{F}_m\mathbf{D}(i) + \mathbf{G}_m\mathbf{D}(i-1))\mathbf{P} \tag{7.127}$$

For estimation purposes, we next assume that the fading processes are essentially constant
over a window of L bit intervals, which gives

$$\mathbf{r}_m^{(L)}(i) = \mathbf{\Lambda}_m^{(L)}(i)\mathbf{\Gamma}_m(i) + \mathbf{n}_m^L(i) \tag{7.128}$$

where matrices from L bit intervals have been concatenated to form

$$\mathbf{r}_m^{(L)}(i) = \left[\mathbf{r}_m^{\mathrm{T}}(i - (L - 1)), \ldots, \mathbf{r}_m^{\mathrm{T}}(i)\right]$$

$$\mathbf{\Lambda}_m^{(L)}(i) = \left[\mathbf{\Lambda}_m^{\mathrm{T}}(i - (L - 1)), \ldots, \mathbf{\Lambda}_m^{\mathrm{T}}(i)\right] \tag{7.129}$$

$$\mathbf{n}_m^{(L)}(i) = \left[\mathbf{n}_m^{\mathrm{T}}(i - (L - 1)), \ldots, \mathbf{n}_m^{\mathrm{T}}(i)\right]$$

We now assume that the receiver has knowledge of the data bits of all of the users. This would be reasonable either when the receiver is in training mode, or when decision feedback is used. In this case, the maximum likelihood estimate of the vector $\mathbf{\Gamma}_m(i)$ minimizes the quadratic cost function

$$C(\mathbf{\Gamma}_m(i)) = \left\|\mathbf{r}_m^{(L)}(i) - \mathbf{\Lambda}_m^{(L)}\mathbf{\Gamma}_m(i)\right\|^2$$

giving the least squares solution

$$\begin{aligned}\mathbf{\Gamma}_{m,ML}(i) &= \left(\left[\mathbf{\Lambda}_m^{(L)}(i)\right]^{\mathrm{T}}\left[\mathbf{\Lambda}_m^{(L)}(i)\right]\right)^{-1} \times \left[\mathbf{\Lambda}_m^{(L)}(i)\right]^{\mathrm{T}}\left[\mathbf{r}_m^{(L)}(i)\right] \\ &= \mathbf{Q}_m^{-1}(i)\mathbf{S}_m(i)\end{aligned} \tag{7.130}$$

where

$$\begin{aligned}\mathbf{Q}_m(i) &= \left[\mathbf{\Lambda}_m^{(L)}(i)\right]^{\mathrm{T}}\left[\mathbf{\Lambda}_m^{(L)}(i)\right] \\ &= \sum_{j=0}^{L-1}\mathbf{\Lambda}_m^{\mathrm{T}}(i - j)\mathbf{\Lambda}_m(i - j)\end{aligned}$$

$$\begin{aligned}\mathbf{S}_m(i) &= \left[\mathbf{\Lambda}_m^{(L)}(i)\right]^{\mathrm{T}}\left[\mathbf{r}_m^{(L)}(i)\right] \\ &= \sum_{j=0}^{L-1}\mathbf{\Lambda}_m^{\mathrm{T}}(i - j)\mathbf{r}_m(i - j)\end{aligned} \tag{7.131}$$

This form of the matrices suggests recursive estimates using exponentially weighted windows

$$\begin{aligned}\mathbf{Q}_m(i) &= \lambda\mathbf{Q}_m(i - 1) + \mathbf{\Lambda}_m^{\mathrm{T}}(i)\mathbf{\Lambda}_m(i) \\ \mathbf{S}_m(i) &= \lambda\mathbf{S}_m(i - 1) + \mathbf{\Lambda}_m^{\mathrm{T}}(i)\mathbf{r}_m(i)\end{aligned} \tag{7.132}$$

where $0 < \lambda < 1$ is the forgetting factor. Once the fading has been estimated according to this procedure, an estimate of the true Wiener solution of the tap weights may be formed.

To obtain some insight into the operation of this channel estimator, consider a single bit estimator, that is, let $\lambda = 0$. We have

$$\begin{aligned}\mathbf{\Gamma}_{m,ML}(i) &= \left[\mathbf{\Lambda}_m^{(L)}(i)\mathbf{\Lambda}_m^{(L)}(i)\right]^{-1}\left[\mathbf{\Lambda}_m^{(L)}(i)\mathbf{r}_m^{(L)}(i)\right] \\ &= \mathbf{\Gamma}_m(i) + \left[\mathbf{\Lambda}_m^{(L)}(i)\mathbf{\Lambda}_m^{(L)}(i)\right]^{-1}\left[\mathbf{\Lambda}_m^{(L)}(i)\mathbf{n}_m^{(L)}(i)\right]\end{aligned} \tag{7.133}$$

Thus, the estimate of $\mathbf{\Gamma}_m(i)$ is equal to the true value plus a term due only to the thermal noise, and independent of the multiaccess interference.

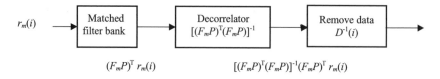

Figure 7.55 Block diagram of the estimator for fading processes using a single bit observation window.

Furthermore, if the system were synchronous, then the matrix $\mathbf{G}_m$ would be a zero matrix, and the estimate of $\mathbf{\Gamma}_m(i)$ could be written as

$$\mathbf{\Gamma}_{m,ML}(i) = \mathbf{D}^{-1}(i)[(\mathbf{F}_m\mathbf{P})^\mathrm{T}(\mathbf{F}_m\mathbf{P})]^{-1}[(\mathbf{F}_m\mathbf{P})^\mathrm{T}\mathbf{r}_m^{(L)}(i)] \tag{7.134}$$

The estimator could then be visualized as in Figure 7.55. The received signal is processed first with a matched filter bank. The MAI is then removed with a decorrelator. Finally, the data are removed, leaving an unbiased estimate of $\mathbf{\Gamma}_m(i)$.

One should keep in mind that the matrix $(\mathbf{F}_m\mathbf{P})^\mathrm{T}(\mathbf{F}_m\mathbf{P})$ will be invertible only if the columns of $\mathbf{F}_m$ are linearly independent. This condition will be violated as the number of users surpasses N/M and the decorrelator will not exist. However, as the observation window was increased, which would obviously be done in order to get good estimates of the fading processes, the existence of the decorrelator would be almost certain. A similar interpretation of the estimator would still apply, that is, matched filter/decorrelator/data removal.

The performance of this algorithm will be illustrated first for a multicarrier CDMA system with the same parameters as in [699]. A user with a data rate of 10 kHz, a carrier frequency of 900 MHz, a processing gain of 32 chips/bit, and an E_b/N_0 of 17 dB was simulated. With 15 users present, results for the average probability of bit error are shown in Figure 7.56(a) as a function of the vehicle speed, which was varied from 20 mph up to 200 mph. Single-carrier, two-carrier and four-carrier systems were considered. Results for the ideal Wiener solution are shown for comparison. It is seen that the performance is very close to ideal for vehicle speeds below about 80 mph, or a normalized Doppler frequency under 1 %.

In Figure 7.56(b), the identical system was simulated, this time with a fixed vehicle speed of 80 mph, and with the number of users varying between 1 and 30. Again, the results are seen to be close to ideal at this speed.

As mentioned previously, it is straightforward to extend this tracking algorithm to the frequency selective case. Without going into such details, in Figure 7.56(c), an identical system to that used to generate the results shown in Figure 7.56(a) for the multicarrier case was applied to a single-carrier system operating on a frequency selective channel with M resolvable paths. The average probability of bit error is shown as a function of the vehicle speed for 15 users. Again, the tracking algorithm gives results which are very close to ideal for vehicle speeds below about 80 mph, or a normalized Doppler frequency under 1%, In Figure 7.56(d), the vehicle speed was fixed at 80 mph, and the number of users was varied between 1 and 30, the results are again close to ideal at this speed.

Additional discussions on MMSE detectors for MC CDMA systems can be found in [699–711]

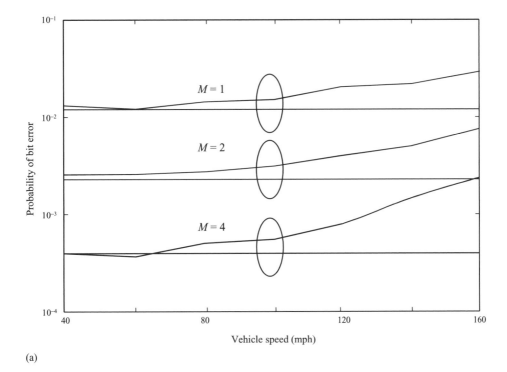

(a)

Figure 7.56 (a) Probability of error versus vehicle speed for multicarrier CDMA system with $E_b/N_0 = 17$ dB, composite processing gain of 32 chips/bit, 15 asynchronous users, bit rate of 10 000 bits/s, and carrier frequency of 900 MHz. M is the number of carriers used. Curves show the performance of the tracking algorithm, and solid straight lines show the Wiener solutions for 15 users. (b) Probability of error versus number of users for multicarrier CDMA system with $E_b/N_0 = 17$ dB, composite processing gain of 32 chips/bit, vehicle speed of 80 mph, bit rate of 10 000 bits/s, and carrier frequency of 900 MHz. M is the number of carriers used. Dashed lines show the performance of the tracking algorithm, and solid lines show the Wiener solutions. (c) Probability of error versus vehicle speed for single-carrier CDMA system with M resolvable paths. The system has 15 users, $E_b/N_0 = 17$ dB, composite processing gain of 32 chips/bit, bit rate of 10 000 bits/s, and carrier frequency of 900 MHz. Curves show the performance of the tracking algorithm, and solid straight lines show the Wiener solutions for 15 users. (d) Probability of error versus number of users for single-carrier system with M resolvable paths. System has $E_b/N_0 = 17$ dB, composite processing gain of 32 chips/bit, vehicle speed of 80 mph, bit rate of 10 000 bits/s, and carrier frequency of 900 MHz. Dashed lines show the performance of the tracking algorithm, and solid lines show the Wiener solutions.

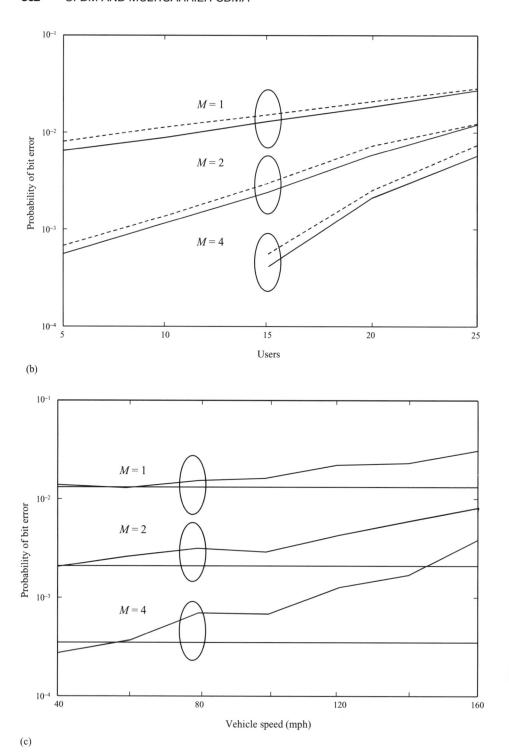

(b)

(c)

Figure 7.56 (*cont.*)

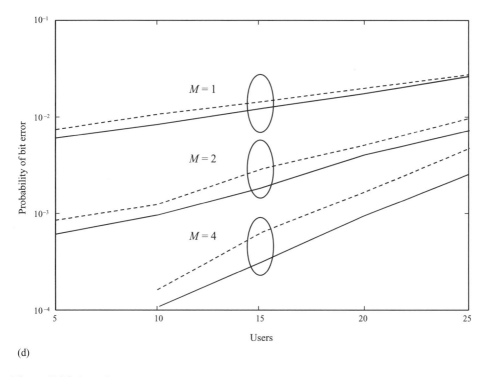

(d)

Figure 7.56 (*cont.*)

7.13 APPROXIMATION OF OPTIMUM MULTIUSER RECEIVER FOR SPACE–TIME CODED MULTICARRIER CDMA SYSTEMS

In this section we extend the discussion on multiuser detection for MC CDMA to the case when space–time coding is included. The presentation is based on [712–714]. The structure of the MC CDMA modulator of the kth user is illustrated in Figure 7.57. BPSK symbols of the kth user are first serial-to-parallel converted, by grouping every P symbols into a vector

$$\underline{s}^k \triangleq [s^k[1], \ldots, s^k[P]]^{\mathrm{T}} \tag{7.135}$$

In the next step, symbol $s^k[p]$ in $\underline{s}^k$ is spread by a spreading sequence $c^k[p]$, $p = 1, \ldots, P$. The sequences corresponding to all P binary phase shift keying (BPSK) symbols are represented by an N-vector as

$$\underline{c}^k \triangleq [c^k[1], \ldots, c^k[P]]^{\mathrm{T}} \tag{7.136}$$

and the spread signals are represented by a componentwise product of the above vectors

$$\underline{s}^k \underline{c}^k \triangleq [s^k[1]c^k[1]^{\mathrm{T}}, \ldots, s^k[P]c^k[P]^{\mathrm{T}}]^{\mathrm{T}} \tag{7.137}$$

In order to avoid the strong correlation among the G subcarriers occupied by a particular symbol, all N spread chips in Equation (7.137) are interleaved before they are transmitted

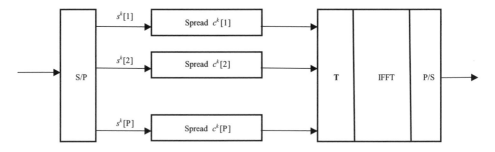

Figure 7.57 MC CDMA modulator structure of the kth user, with interleaver **T**.

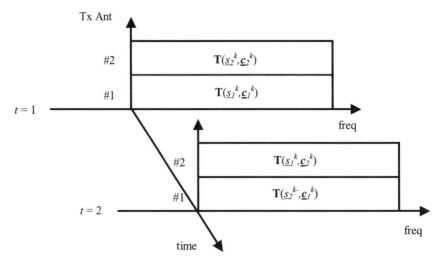

Figure 7.58 Transmitted signal structure of a particular STBC slot of the kth user in an STBC MC CDMA system.

from the N subcarriers. The interleaving function is denoted by **T** in Figure 7.57; for simplicity, the interleaver is assumed to be the same for all K users.

The Alamouti space–time block codes (STBC), discussed in Chapter 4, are employed to further increase the system capacity. In Chapter 4, the STBC was used to transmit scalar symbols. In this section, we extend it to the vector form and apply it in MC CDMA systems. In vector form, the simplest (2×2) STBC discussed in Chapter 4, as illustrated in Figure 7.58, takes two time slots to transmit two symbol vectors $\mathbf{s}_1^k$, $\mathbf{s}_2^k$ [cf. Equation (7.135)]. At the first time slot, signals $\mathbf{T}(\underline{\mathbf{s}}_1^k \underline{\mathbf{c}}_1^k)$ are transmitted from the *first* antenna on N subcarriers and $\mathbf{T}(\underline{\mathbf{s}}_2^k \underline{\mathbf{c}}_2^k)$ is transmitted from the *second* antenna. At the second time slot, $\mathbf{T}(-\underline{\mathbf{s}}_2^k \underline{\mathbf{c}}_1^k)$ is transmitted from the *first* antenna and $\mathbf{T}(\underline{\mathbf{s}}_1^k \underline{\mathbf{c}}_2^k)$ is transmitted from the *second* antenna. From Figure 7.58, we can see that different spreading sequences $\underline{\mathbf{c}}_1^k$ and $\underline{\mathbf{c}}_2^k$ are assigned to different transmitter antennas; this structure is shown to be an efficient way to resolve the so-called 'antenna-ambiguity' in blind algorithms, as will be discussed. The structure of the transmitter for the kth user is given in Figure 7.59.

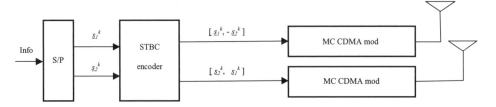

Figure 7.59 Transmitter structure of the kth user in an STBC MC CDMA system.

7.13.1 Frequency selective fading channels

The system with one receiver antenna is considered and the extension to multiple receiver antennas is straightforward. The time domain channel impulse response of the kth user between the jth transmitter antenna and the receiver antenna will be modeled as

$$h_j^k(\tau) = \sum_{l=0}^{L-1} \alpha_j^k(l)\delta\left(\tau - \frac{l}{\Delta_f}\right) \tag{7.138}$$

where $\delta(\cdot)$ is the Kronecker delta function; $L \overset{\Delta}{=} \lceil \tau_m \Delta_f + 1\rceil$, with τ_m being the maximum multipath spread of all K users (note that we have assumed the synchronous transmission of all k users) and Δ_f being the whole bandwidth of multicarrier systems; $\alpha_j^k(l)$ is the complex amplitude of the lth tap associated with the jth transmitter antenna of the kth user, whose relative delay is l/Δ_f.

For MC CDMA systems with proper cyclic extensions and sample timing, with tolerable leakage, the channel frequency response of the kth user at its jth transmitter antenna and at the nth subcarrier can be expressed as

$$H_j^k[n] \overset{\Delta}{=} H_j^k(n\Delta_f)$$

$$= \sum_{l=0}^{L-1} \alpha_j^k(l)\exp\left(-\frac{j2\pi\, nl}{N}\right) \tag{7.139}$$

$$= \mathbf{w}_f^H(n)\mathbf{h}_j^k$$

where

$$\mathbf{h}_j^k \overset{\Delta}{=} \left[\alpha_j^k(0), \alpha_j^k(1), \ldots, \alpha_j^k(L-1)\right]^T \tag{7.140}$$

contains the time response of all L taps; and

$$\mathbf{w}_f(n) \overset{\Delta}{=} \left[1, \exp\left(-\frac{j2\pi\, n}{N}\right), \ldots, \exp\left(-\frac{j2\pi\, n(L-1)}{N}\right)\right]^H \tag{7.141}$$

contains the corresponding discrete Fourier transform (DFT) coefficients.

7.13.2 Receiver signal model of STBC MC CDMA systems

The transmitted signals of all K users propagate through their respective frequency selective fading channels and finally reach the receiver antenna. We assume that the fading processes associated with different transmitter–receiver antenna pairs are uncorrelated. At the receiver, after matched filtering, the discrete Fourier transform (DFT) is applied to the received chip rate sampled discrete time signals. Consider all the DFT-ed signals in one signal frame, which spans M STBC slots, or equivalently, $2M$ time slots [recall that each STBC slot consists of two neighboring time slots (see Figure 7.58).] Using Equation (7.139) and assuming that the fading processes are time invariant within one signal frame, the received signal model can be represented as

$$
\mathbf{y}_i = \sum_{k=1}^{K} \underbrace{\begin{bmatrix} \mathbf{S}_{i,1}^k \mathbf{C}_1^k & \mathbf{S}_{i,2}^k \mathbf{C}_2^k \\ -\mathbf{S}_{i,2}^k \mathbf{C}_1^k & \mathbf{S}_{i,1}^k \mathbf{C}_2^k \end{bmatrix}}_{\mathbf{X}_i^k} \underbrace{\begin{bmatrix} \mathbf{TW_f} & 0 \\ 0 & \mathbf{TW_f} \end{bmatrix}}_{\mathbf{W}} \times \underbrace{\begin{bmatrix} \mathbf{h}_1^k \\ \mathbf{h}_2^k \end{bmatrix}}_{\mathbf{h}^k} + \mathbf{v}_i
$$

$$
= \sum_{k=1}^{K} \mathbf{X}_i^k \mathbf{W} \mathbf{h}^k + \mathbf{v}_i, \quad i = 0, \ldots, M-1 \tag{7.142}
$$

with

$$
\mathbf{C}_j^k \triangleq \mathrm{diag}\{\underline{\mathbf{c}}_j^k\}_{N \times N}
$$

$$
\underline{\mathbf{c}}_j^k \triangleq \left[\mathbf{c}_j^k[1], \ldots, \mathbf{c}_j^k[P]^T \right]^T, \quad j = 1, 2
$$

$$
\mathbf{S}_{i,j}^k \triangleq \mathrm{diag}\{\underline{\mathbf{s}}_{i,j}^k \otimes \mathbf{I_G}\}
$$

$$
= \mathrm{diag}\{\mathbf{s}_{i,j}^k[1]\mathbf{I_G}, \ldots, \mathbf{s}_{i,j}^k[P]\mathbf{I_G}\}_{N \times N}, \quad j = 1, 2
$$

$$
\mathbf{W_f} \triangleq [\mathbf{w_f}(0), \mathbf{w_f}(1), \ldots, \mathbf{w_f}(N-1)]_{N \times L}^H
$$

where $\otimes$ is the Kronecker matrix product; $\underline{\mathbf{s}}_{i,j}^k$, $j = 1, 2$ are the symbol vectors input to the STBC encoder at the ith STBC slot; $\underline{\mathbf{c}}_j^k$ is the spreading sequence assigned to the jth transmitter antenna of the kth user; $\mathbf{T}$ is an $(N \times N)$ permutation matrix, which acts as an interleaver mapping the N chips to their assigned subcarriers; $\mathbf{y}_i$ is the received signal during the ith STBC slot; $\mathbf{v}_i$ is circularly symmetric complex Gaussian ambient noise, with covariance matrix $\sigma^2 \mathbf{I}_{2N}$.

Note that Equation (7.142) can be used to describe both the slow-fading (when M is large) and the fast-fading (when M is small, e.g. $M = 1$) cases. In this section, the fading channels are assumed to be static during two neighboring time slots (i.e. one STBC slot), this is the only limitation of the signal model in Equation (7.142). Due to the orthogonality property of the STBC, i.e. $\mathbf{X}_i^{k^H} \mathbf{X}_i^k = 2\mathbf{I}_{2N}$, the orthogonality property of the OFDM multicarrier modulation, i.e. $\mathbf{W_f}^H \mathbf{W_f} = N \cdot \mathbf{I}_L$, the fact that the permutation matrix satisfies $\mathbf{T}^T \mathbf{T} = \mathbf{I}_N$ and the definitions given in Equation (7.142), we have

$$
\mathbf{W}^H \mathbf{X}_i^{k^H} \mathbf{X}_i^k \mathbf{W} = \begin{bmatrix} \mathbf{W_f}^H \mathbf{T}^T & 0 \\ 0 & \mathbf{W_f}^H \mathbf{T}^T \end{bmatrix} \begin{bmatrix} 2\mathbf{I}_N & 0 \\ 0 & 2\mathbf{I}_N \end{bmatrix} \times \begin{bmatrix} \mathbf{TW_f} & 0 \\ 0 & \mathbf{TW_f} \end{bmatrix}
$$

$$
= 2N \cdot \mathbf{I}_{2L} \tag{7.143}
$$

As we know from orthogonal design, discussed in Chapter 4, the structure of Equation (7.143) can be exploited to reduce the computational complexity of the optimal receiver for the STBC MC CDMA system. Note that the interleaver function $\mathbf{T}$ is the same for all K users; hence, the signal model Equation (7.142) can be written in an alternative form, which decouples the signal components corresponding to different symbols

$$\mathbf{y}_i[p] = \sum_{k=1}^{K} \mathbf{X}_i^k[p]\mathbf{W}[p]\mathbf{h}^k + \mathbf{v}_i[p], \quad i = 0, \ldots, M-1$$
$$p = 1, \ldots, P \tag{7.144}$$

where $\mathbf{X}_i^k[p]$ is a $(2G \times 2G)$ submatrix decimated from $\mathbf{X}_i^k$, which contains all the rows and columns related to the symbol $s_{i,1}^k[p]$ and $s_{i,2}^k[p]$; $\mathbf{y}_i[p]$, $\mathbf{W}[p]$ and $\mathbf{v}_i[p]$ are then the corresponding decimations from $\mathbf{y}_i$, $\mathbf{W}$, and $\mathbf{v}_i$. Equation (7.144) will be used later in deriving the conditional posterior distributions of the unknown symbols [see Equation (7.158)].

7.13.3 Blind approach

Since the channel state information is unknown to the receiver, there are two types of ambiguity inherent in the design of blind receivers for the STBC MC CDMA system: phase ambiguity and antenna ambiguity.

To resolve the phase ambiguity, differential encoding is employed before the STBC encoding. For each signal frame, a block of BPSK bits $\mathbf{b}^k = \{b^k[1], b^k[2], \ldots, b^k[2MP-1]\}$ is input to the differential encoder, and the output $\mathbf{d}^k = \{d^k[1], \ldots, d^k[2MP]\}$ is given by

$$\begin{cases} d^k[1] = 1 \\ d^k[n] = d^k[n-1]b^k[n-1], \quad n = 2, \ldots, 2MP \end{cases} \tag{7.145}$$

These differentially encoded bits $\{d^k[n]\}_n$ are the same set of bits $\{s_{i,j}^k[p]\}_{i,j,p}$ [defined by Equation (7.142)] input to the STBC encoder, where they are related as $s_{i,j}^k[p] = d^k[n]|_{n=(2i+j)P+p}$. Henceforth, the index n of $d^k[n]$ is understood as an implicit function of the index (i,j,p) of $s_{i,j}^k[p]$.

One possible approach to resolving the antenna ambiguity is to employ the differential space–time modulation as discusssed in Chapter 4; however, in that case, the signal constellation will be changed. Fortunately, in the STBC MC CDMA system, the antenna ambiguity can be resolved by using different spreading sequences on different transmitter antennas. Note that the usage of orthogonal sequences will result in a maximum 50 % system loading (defined as K/G) in the (2×2) STBC MC CDMA system. The problem with this is that the channel frequency selectivity destroys the orthogonality of the sequences. For this reason, in this section, random sequences are assumed and a sufficient number of these sequences, such that they impose no constraint on the system loading. When the system employs the outer channel code, it is possible to use the same spreading sequence at different antennas of the same user. In this case, the antenna ambiguity can be resolved by exploiting the coding structure. For example, for the system considered here, due to the antenna ambiguity, we have two possible code bit sequences at the output of the multiuser detector. We can send both of them to the channel decoder and count the number of bit corrections. The correct code bit sequence will have only a few bit errors, whereas the incorrect one will have many errors.

7.13.4 Bayesian optimal blind receiver

From here on the following matrix notation will be used

$$\mathbf{Y} \triangleq \{\mathbf{y}_i\}_{i=0}^{M-1}; \mathbf{H} \triangleq \{\mathbf{h}^k\}_{k=1}^{K}; \mathbf{D} \triangleq \{\mathbf{d}^k\}_{k=1}^{K}; \mathbf{B} \triangleq \{\mathbf{b}^k\}_{k=1}^{K}$$

The optimal blind receiver estimates the *a posteriori* probabilities (APP) of the multiuser data bits

$$P[b^k[n] = +1|\mathbf{Y}], \quad n = 1, 2, \ldots, MP - 1, \quad k = 1, \ldots, K \tag{7.146}$$

based on the received signals $\mathbf{Y}$, the signal structure Equation (7.142), the spreading sequences of all users, and the prior information of $\mathbf{B}$, without knowing the channel response $\mathbf{H}$ and the noise variance σ^2.

So, the unknown parameters have to be averaged out from the *a posteriori* function. The Bayesian solution to Equation (7.146) is given by

$$P[b^k[n] = +1|\mathbf{Y}]$$

$$= \sum_{\mathbf{B}:b^k[n]=+1} \int p[\mathbf{Y}|\mathbf{H}, \sigma^2, \mathbf{B}] p[\mathbf{H}] p[\sigma^2] p[\mathbf{B}] \, d\mathbf{H} \, d\sigma^2 \tag{7.147}$$

where $P[\mathbf{Y}|\mathbf{H}, \sigma^2, \mathbf{B}]$ is a Gaussian density function [see Equation (7.142)]; $p[\mathbf{H}]$, $p[\sigma^2]$ and $p[\mathbf{B}]$ are prior distributions of the independent and unknown quantities $\mathbf{H}$, σ^2 and $\mathbf{B}$ respectively. Clearly the computation in Equation (7.147) involves a very high-dimensional integral which is certainly infeasible for any practical implementations. Thus, the Gibbs sampler, a Monte Carlo method, is used to calculate the *a posteriori* probabilities of the unknown symbols.

7.13.5 Blind Bayesian Monte Carlo multiuser receiver approximation

In this section, we consider the problem of computing the *a posteriori* bit probabilities in Equation (7.146). The problem is solved under a Bayesian framework, by treating the unknown quantities as realizations of random variables with some prior distributions. The Gibbs sampler [715–719] is then employed to compute the Bayesian estimates.

7.13.6 Gibbs sampler

The Gibbs sampler [715–719] is a Markov chain Monte Carlo (MCMC) procedure for numerical Bayesian computation. Let $\theta = [\theta_1, \ldots, \theta_d]^T$ be a vector of unknown parameters. Let $\mathbf{Y}$ be the observed data. To generate random samples from the joint posterior distribution $p[\theta|\mathbf{Y}]$, given the samples at the $(j - l)$th iteration, $\theta^{(j-1)} = [\theta_1^{(j-1)}, \ldots, \theta_d^{(j-1)}]^T$ at the jth iteration, the Gibbs algorithm iterates as follows to obtain samples $\theta^{(j)} = [\theta_1^{(j)}, \ldots, \theta_d^{(j)}]^T$.

For $i = 1, \ldots, d$, draw $\theta_i^{(j)}$ from the conditional distribution

$$p[\theta_i|\theta_1^{(j)}, \ldots, \theta_{i-1}^{(j)}, \theta_{i+1}^{(j)}, \ldots, \theta_d^{(j)}, \mathbf{Y}]$$

It is known that under regularity conditions [716–719]:

1. The distribution of $\theta_i^{(j)}$ converges geometrically to $p[\theta|\mathbf{Y}]$, as $j \to \infty$;

2. $(1/J) \sum_{j=1}^{J} f(\theta^{(j)}) \xrightarrow{a.s.} \int f(\theta) p[\theta|\mathbf{Y}] \, d\theta$, as $J \to \infty$, for any integrable function f.

Hence, the marginal *a posteriori* distribution of any parameter θ_i can be computed easily from the samples drawn by the Gibbs sampler.

7.13.7 Prior distributions

For simplicity, we choose the sampling space, the set of unknown parameters sampled by the Gibbs sampler, to be $\{\mathbf{D}, \mathbf{H}, \sigma^2\}$, which are assumed to be independent of each other. Next, we specify their prior distributions $p[\mathbf{H}]$, $p[\sigma^2]$ and $p[\mathbf{D}]$.

1. For the unknown channel $\mathbf{h}^k$, a complex Gaussian prior distribution is assumed

$$p[\mathbf{h}^k] \sim \mathcal{N}_c(\mathbf{h}_{k,0}, \Sigma_{k0}) \qquad (7.148)$$

   Note that large value of Σ_{k0} corresponds to less informative prior distributions.

2. For the noise variance σ^2, an inverse chi squared prior distribution is assumed

$$p[\sigma^2] \sim \chi^{-2}(2\nu_0, \lambda_0) \qquad (7.149)$$

   a small value of $2\nu_0$ corresponds to the less informative prior distributions.

3. The data bit sequence $\mathbf{d}^k$ is a Markov chain, encoded from $\mathbf{b}^k$. Its prior distribution can be expressed as

$$
\begin{aligned}
p[\mathbf{d}^k] &= p(d^k[1])p(d^k[2]d^k[1]) \cdots p(d^k[2PM]d^k[2PM - 1]) \\
&= p(d^k[1])p(b^k[1] = d^k[2]d^k[1]) \\
&\qquad \cdots p(b^k[2PM - 1] = d^k[2PM]d^k[PM - 1]) \qquad (7.150) \\
&= \frac{1}{2} \prod_{n=2}^{PM} \frac{\exp(\rho^k[n-1]d^k[n-1]d^k[n])}{1 + \exp(\rho^k[n-1]d^k[n-1]d^k[n])}
\end{aligned}
$$

   where $\rho^k[n]$ denotes the *a priori* log likelihood ratio (LLR) of $b^k[n]$,

$$\rho^k[n] \triangleq \log \frac{P(b^k[n] = +1)}{P(b^k[n] = -1)} \qquad (7.151)$$

   Note that in Equation (7.150), we set $p(d^k[1]) = 1/2$ to account for the phase ambiguity in $d^k[1]$.

7.13.8 Conditional posterior distributions

The following conditional posterior distributions are required by the Bayesian multiuser detector. The derivations can be found in Appendix 7.1 [713].

1. The conditional distribution of the kth user's channel response $\mathbf{h}^k$ given σ^2, $\mathbf{D}$, $\mathbf{H}^k$, and $\mathbf{Y}$ is

$$p[\mathbf{h}^k|\mathbf{D}, \sigma^2, \mathbf{H}^k, \mathbf{Y}] \sim \mathcal{N}_c(\mathbf{h}_{k*}, \Sigma_{k*}) \qquad (7.152)$$

   where $\mathbf{H}^k \triangleq \mathbf{H} \backslash \mathbf{h}^k$ with

$$
\begin{aligned}
\Sigma_{k*}^{-1} &\triangleq \Sigma_{k0}^{-1} + \frac{1}{\sigma^2} \sum_{i=0}^{M-1} \mathbf{W}^H \mathbf{X}_i^{kH} \mathbf{X}_i^k \mathbf{W} \\
&= \Sigma_{k0}^{-1} + \frac{2MN}{\sigma^2} \mathbf{I}_{2L} \qquad (7.153)
\end{aligned}
$$

and

$$\mathbf{h}_{k*} \triangleq \Sigma_{k*} \left[\Sigma_{k0}^{-1} \mathbf{h}_{k0} + \frac{1}{\sigma^2} \sum_{i=0}^{M-1} \mathbf{W}^{k\mathrm{H}} \mathbf{X}_i^k \left(\mathbf{y}_i - \sum_{l \neq k} \mathbf{X}_i^l \mathbf{W}^l \mathbf{h}^l \right) \right] \tag{7.154}$$

Equation (7.153) follows from the orthogonality property in Equation (7.143).

2. The conditional distribution of the noise variance σ^2 given $\mathbf{H}$, $\mathbf{D}$, and $\mathbf{Y}$ is given by

$$p[\sigma^2 \mid \mathbf{H}, \mathbf{D}, \mathbf{Y}] \sim \chi^{-2} \left(2[\nu_0 + MN], \frac{\nu_0 \lambda_0 + s^2}{\nu_0 + 2MN} \right) \tag{7.155}$$

with

$$s^2 \triangleq \sum_{i=0}^{M-1} \left\| \mathbf{y}_i - \sum_{k=1}^{K} \mathbf{X}_i^k \mathbf{W}^k \mathbf{h}^k \right\|^2 \tag{7.156}$$

3. The conditional distribution of the data bit $d^k[n]$, given $\mathbf{H}, \sigma^2, \mathbf{D}_n^k$ and $\mathbf{Y}$ can be obtained from

$$\frac{P\left[d^k[n] = +1 \mid \mathbf{H}, \sigma^2, \mathbf{D}_n^k, \mathbf{Y} \right]}{P\left[d^k[n] = -1 \mid \mathbf{H}, \sigma^2, \mathbf{D}_n^k, \mathbf{Y} \right]}$$

$$= \exp \left[d^k[n+1]\rho^k[n] + d^k[n-1]\rho^k[n-1] - \frac{\Delta s^2}{\sigma^2} \right] \tag{7.157}$$

where $\mathbf{D}_n^k \triangleq \mathbf{D} \backslash d^k[n]$ and

$$\Delta s^2 \triangleq \left\| \mathbf{y}_i[p] - \sum_l \mathbf{X}_i^l \mathbf{W}[p]\mathbf{h}^l \right\|_{s_{i,j}^k[p]=+1}^2$$

$$- \left\| \mathbf{y}_i[p] - \sum_l \mathbf{X}_i^l \mathbf{W}[p]\mathbf{h}^l \right\|_{s_{i,j}^k[p]=-1}^2 \tag{7.158}$$

On the right-hand side of Equation (7.157), the first two items correspond to the prior information of the differentially encoded bits $\mathbf{D}$, through which the prior information of data bits $\mathbf{B}$ is incorporated.

7.13.9 Gibbs multiuser detection

Given the initial values of the unknown quantities $\{\mathbf{H}^{(0)}, \sigma^{2(0)}, \mathbf{D}^{(0)}\}$ drawn from their prior distributions (Equations 7.148–7.150), at the jth iteration, the Gibbs multiuser detector operates as follows.

1. For $k = 1, \ldots, K$, draw $\mathbf{h}^{k(j)}$ from $p[\mathbf{h}^k \mid \mathbf{H}^{k(j-1)}, \sigma^{2(j-1)}, \mathbf{D}^{(j-1)}, \mathbf{Y}]$, given by Equation (7.152), with $\mathbf{H}^{k(j-1)} \triangleq \{\mathbf{h}^{1(j)}, \ldots, \mathbf{h}^{(k-1)(j)}, \mathbf{h}^{(k+1)(j-1)}, \ldots, \mathbf{h}^{K(j-1)}\}$;

2. Draw $\sigma^{2(j)}$ from $p[\sigma^2 \mid \mathbf{H}^{(j)}, \mathbf{D}^{(j-1)}, \mathbf{Y}]$ given by Equation (7.155);

3. For $n = 1, \ldots, 2PM$, and for $k = 1, \ldots, K$, draw $d^{k(j)}[n]$ from $P[d^k[n] \mid \mathbf{H}^{(j)}, \sigma^{2(j)}, \mathbf{D}_n^{k(j-1)}, \mathbf{Y}]$ given by Equation (7.157), where $\mathbf{D}_n^{k(j-1)} \triangleq \{d^{1(j)}[1], \ldots, d^{1(j)}[2PM], \ldots, d^{k(j)}[n-1], d^{k(j-1)}[n+1], \ldots, d^{K(j-1)}[2PM]\}$.

To ensure convergence, the Gibbs iteration is usually carried out for $(J_0 + J)$ iterations. The initial J_0 iterations represent the burn-in period and only the samples from the last J iterations are used to calculate the Bayesian interference. In particular, the posterior distribution of the multiuser data bits $b^k[n]$ can be obtained by

$$P[b^k[n] = +1|\mathbf{Y}] \cong \frac{1}{J} \sum_{j=J_0+1}^{J_0+J} \delta_{kn}^{(j)}, \quad k = 1, \ldots, K; \quad n = 1, \ldots, 2PM - 1 \quad (7.159)$$

where $\delta_{kn}^{(j)}$ is an indicator such that

$$\delta_{kn}^{(j)} = \begin{cases} 1, & \text{if } d^{k(j)}[n]d^{k(j)}[n-1] = +1 \\ 0, & \text{if } d^{k(j)}[n]d^{k(j)}[n-1] = -1 \end{cases} \quad (7.160)$$

7.13.10 Sampling space of data

As we are interested in computing the posterior probabilities of the multiuser data bits, direct sampling can be done on the data bits **B**. Note that the conditional posterior distribution of $b^k[n]$, given $\mathbf{H}$, σ^2, $\mathbf{B}_n^k$ and $\mathbf{Y}$, involves a large number of received signals, i.e. $\{\mathbf{y}_i, \ldots, \mathbf{y}_{M-1}\}$. This long memory in the receiver signal processing will increase the computational complexity and decrease the convergence speed of the Gibbs procedure. To avoid these disadvantages, the Gibbs procedure described in this section samples the differentially encoded bits **D**, and outputs a sampling sequence $\{d^{k(j)}[n]\}$. It is shown in Equation (7.159) that the marginal posterior probability of $b^k[n]$ can be computed easily from the output samples $\{d^{k(j)}[n]\}$.

7.13.11 The orthogonality property

The dominant computations involved in the Gibbs sampler are Equations (7.153–7.154) By exploiting the orthogonality property, Equation (7.143), in STBC MC CDMA systems, the matrix Σ_{k*}^{-1} in Equation (7.153) is simply a constant matrix. In addition to this, with no matrix inversion involved in computing $\mathbf{h}_{k*}$, the numerical stability is also improved.

7.13.12 Blind turbo multiuser receiver

In this section, we consider employing iterative multiuser detection and decoding to improve the performance of the Bayesian multiuser receiver in a coded STBC MC CDMA system. Because it utilizes the *a priori* bit probabilities, and it produces the *a posteriori* bit probabilities, the Bayesian multiuser detector is well suited for iterative processing, which allows the multiuser detector to refine its processing based on the information from the decoding stage and vice versa. The kth user's transmitter structure is shown in Figure 7.60, with block of information bits $\{a^k[l]\}$ encoded using some channel code (e.g. block code, convolutional code or turbo code). A code bit interleaver is used to reduce the influence of error bursts at the input of the channel decoder. The interleaved code bits are then mapped to BPSK symbols $\{b^k[n]\}$. These BPSK symbols are differentially encoded to yield the symbol stream $\{d^k[n]\}$, which is then serial-to-parallel converted and reorganized in to a

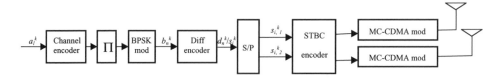

Figure 7.60 Transmitter structure of the *k*th user in an STBC MC CDMA system employing outer channel code, where Π represents an interleaver.

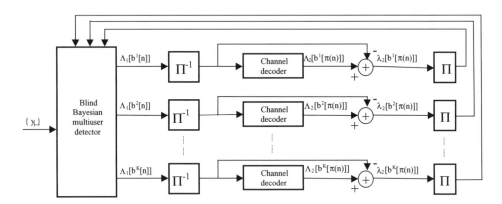

Figure 7.61 Turbo Bayesian multiuser receiver structure, where Π denotes an interleaver and Π^{-1} denotes the corresponding deinterleaver.

vector form as $\{\underline{s}_{i,j}^k\}$ to feed into the STBC encoder followed by the MC CDMA modulator, and finally transmitted from two antennas.

An iterative (turbo) receiver structure is shown in Figure 7.61. It consists of two stages: the Bayesian multiuser detector developed in the previous sections, followed by a soft-input soft-output channel decoder. The two stages are separated by a deinterleaver and an interleaver. Assume that $\{b^k[n]\}$ is mapped into $\{b^k[\pi(n)]\}$ after deinterleaving.

In the first stage, the blind Bayesian multiuser detector incorporates the *a priori* information $\{\lambda_2^p(b^k[n])\}$, which is computed by the channel decoder in the previous iteration. At the first iteration, it is assumed that all code bits are equally likely. At the output of the blind Bayesian multiuser detector, the *a posteriori* LLR is given by

$$\Lambda_1(b^k[n]) \triangleq \log \frac{P[b^k[n] = +1|\mathbf{Y}]}{P[b^k[n] = -1|\mathbf{Y}]} \tag{7.161}$$

According to the 'turbo principle' described in Chapter 2, the *a priori* information $\{\lambda_1^p(b^k[n])\}$ should be subtracted from the *a posteriori* LLR $\Lambda_1(b^k[n])$ to obtain the extrinsic information to deliver to the channel decoders. However, the posterior distribution delivered by the blind Bayesian multiuser detector is a quantized value instead of the true value, due to the finite number of samples. Therefore, to ensure numerical stability, the posterior LLR $\Lambda_1(b^k[n])$ is regarded as the approximated extrinsic information, deinterleaved and fed back to the channel decoder of the *k*th user.

The soft-input soft-output channel decoder, using the MAP decoding algorithm (Appendix 7.1), computes the *a posteriori* LLR of each code bit of the kth user

$$\Lambda_2(b^k[\pi(n)]) \triangleq \log \frac{P\big[b^k[n] = +1 \mid \{\Lambda_1^P(b^k[\pi(i)])\}_{i-1}^{PM-1}\big]}{P\big[b^k[n] = -1 \mid \{\Lambda_1^P(b^k[\pi(i)])\}_{i-1}^{PM-1}\big]}$$

$$= \lambda_2(b^k[\pi(n)]) + \Lambda_1^P(b^k[\pi(n)]) \qquad (7.162)$$

It is seen from Equation (7.162) that the output of the MAP decoder is the sum of the prior information $\Lambda_1^P(b^k[\pi(n)])$, and the *extrinsic* information $\lambda_2^P(b^k[\pi(n)])$ delivered by the channel decoder. After interleaving, the extrinsic information delivered by the channel decoder $\{\lambda_2^P(b^k[n])\}_{n=1}^{PM-1}$ is then fed back to the blind Bayesian multiuser detector as the refined prior information $\rho^k[n]$ [see Equation (7.151)] for the next iteration.

7.13.13 Decoder-assisted convergence assessment

Although it is desirable to have the Gibbs sampler reach convergence within the burn-in period (J_0 iterations), this may not always be the case. Hence, we need some mechanism to detect the convergence. In the coded system considered here, the blind multiuser detector is followed by a bank of channel decoders, we can assess convergence by monitoring the number of bit corrections made by the channel decoders [720]. The number of corrections is determined by comparing the signs of the code bit LLR at the input and output of the MAP channel decoder. If this number exceeds some predetermined threshold, then we decide convergence is not achieved, in which case, the Gibbs multiuser detector will be applied again to the same data block.

7.13.14 Performance example

In this section, we present some computer simulation results to illustrate the performance of the Bayesian multiuser receivers in the STBC MC CDMA system, where there are two transmitter antennas, one receiver antenna, N subcarriers and K users. The signal parameters are the same as in [713]. All spreading sequences used in simulations are randomly and independently generated for each transmitter antenna of each user. The frequency selective fading channels are assumed to be uncorrelated and have the same statistics for different transmitter–receiver antenna pairs. For simplicity, all L taps of a particular fading channel are assumed to be of equal power and normalized such that $\sum_{j=1}^{2} \|\mathbf{h}_j\|^2 = 1$, and have delays $\tau_i = (l/\Delta_f), l = 0, 1, \ldots, L - 1$; and all K users in the system are assumed to have equal transmission power. Such a system setup is also the worst case scenario from the interference mitigation point of view (Chapter 5). For STBC MC CDMA systems employing outer channel codes, a four-state, rate-1/2 convolutional code with generator (5,7) in octal notation is chosen for all users. For each block of received signals, $J_0 + J = 100$ samples are drawn by the Gibbs sampler, with the first $J_0 = 50$ samples discarded. As discussed in the previous section, at the end of the 100 Gibbs iterations, the convergence of the Gibbs sampler is tested. In very few cases, when the Gibbs sampler is not convergent, it is restarted for another round of 100 Gibbs iterations.

The performance is demonstrated in two forms: one (Figure 7.62) is in terms of the bit error rate (BER) and OFDM word error rate (WER) versus the number of users at a

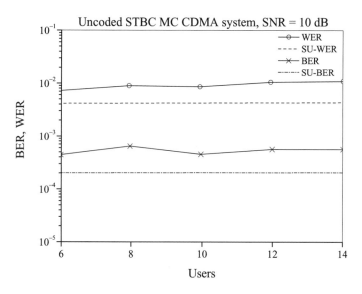

Figure 7.62 BER and OFDM WER of an STBC MC CDMA system in two-tap frequency selective fading channels, where $N = 16$, $G = 16$, $L = 2$, SNR = 10 dB.

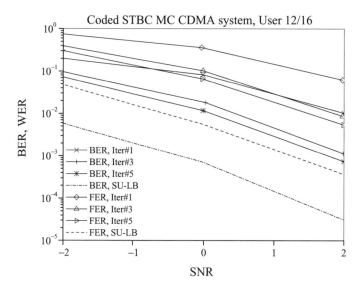

Figure 7.63 BER and OFDM WER of an STBC MC CDMA system employing outer convolutional channel code in five-tap frequency selective fading channels, where $N = 256$, $G = 16$, $L = 5$, $K = 12$.

particular signal to noise ratio (SNR), where SNR $= (1/\sigma^2)$ [cf. Equation (7.142)]; the other (Figure 7.63) is in terms of the BER/WBR versus SNR for the system with a particular number of users. Figure 7.62. demonstrates the expected effects of a multiuser detector where BER does not change significantly with the increase in the number of users. These curves are rather close to the single user (SU) case which is also expected.

APPENDIX 7.1

Derivation of Equation (7.152)

As explained earlier, $d^k[n]$ is also denoted as $s_{i,j}^k[p]$, and among all the received signals, only $\mathbf{y}_i[p]$ is related to $s_{i,j}^k[p]$. Hence, Δs^2 can be further simplified to

$$
\Delta s^2 \triangleq \left\| \mathbf{y}_i[p] + \sum_l \mathbf{X}_i^l[p]\mathbf{W}[p]\mathbf{h}^l \right\|^2_{s_{i,j}^k[p]=+1}
$$

$$
- \left\| \mathbf{y}_i[p] + \sum_l \mathbf{X}_i^l[p]\mathbf{W}[p]\mathbf{h}^l \right\|^2_{s_{i,j}^k[p]=-1}
$$

$$
p[\mathbf{h}^k|\mathbf{H}^k, \sigma^2, \mathbf{D}, \mathbf{Y}] \propto p[\mathbf{Y}|\mathbf{H}, \sigma^2, \mathbf{D}]p[\mathbf{h}^k]
$$

$$
\propto \exp\left\{ -\frac{1}{\sigma^2} \sum_{i=0}^{M-1} \left\| \mathbf{y}_i + \sum_{l=1}^{K} \mathbf{X}_i^l \mathbf{W}\mathbf{h}^l \right\|^2 \right\} \exp\left\{ -(\mathbf{h}^k - \mathbf{h}_{k0})^H \Sigma_{k0}^{-1}(\mathbf{h}^k - \mathbf{h}_{k0}) \right\}
$$

$$
\propto \exp\left\{ -\mathbf{h}^{k^H} \underbrace{\left(\Sigma_{k0}^{-1} + \frac{1}{\sigma^2} \sum_{i=0}^{M-1} \mathbf{W}^H \mathbf{X}_i^{k^H} \mathbf{X}_i^k \mathbf{W} \right)}_{\Sigma_{k*}^{-1}} \mathbf{h}^k \right.
$$

$$
\left. +2\mathcal{R} \left\{ \mathbf{h}^{k^H} \underbrace{\left[\Sigma_{k0}^{-1}\mathbf{h}_{k0} + \frac{1}{\sigma^2} \sum_{i=0}^{M-1} \mathbf{W}^H \mathbf{X}_i^{k^H} \left(\mathbf{y}_i - \sum_{l\neq k} \mathbf{X}_i^k \mathbf{W}\mathbf{h}^l \right) \right]}_{\Sigma_{k*}^{-1}\mathbf{h}_{k*}} \right\} \right\}
$$

$$
\propto \exp\left\{ -(\mathbf{h}^k - \mathbf{h}_{k*})^H \Sigma_{k*}^{-1}(\mathbf{h}^k - \mathbf{h}_{k*}) \right\} \sim \mathcal{N}_c(H_{k*}, \Sigma_{k*}) \tag{A.7.1}
$$

Derivation of Equation (7.155)

$$
p[\sigma^2|\mathbf{H}, \mathbf{D}, \mathbf{Y}] \propto p[\mathbf{Y}|\mathbf{H}, \sigma^2|\mathbf{D}]p[\sigma^2]
$$

$$
\propto \left(\frac{1}{\sigma^2}\right)^{2MN} \exp\left(-\frac{1}{\sigma^2} \underbrace{\sum_{i=0}^{M-1} \left\| \mathbf{y}_i + \sum_{k=1}^{K} \mathbf{X}_i^k \mathbf{W}\mathbf{h}^k \right\|^2}_{s^2} \right) \times \left(\frac{1}{\sigma^2}\right)^{\nu_0+1} \exp\left(-\frac{\nu_0\lambda_0}{\sigma^2} \right)
$$

$$
= \left(\frac{1}{\sigma^2}\right)^{\nu_0+2MN+1} \exp\left(-\frac{\nu_0\lambda_0 + s^2}{\sigma^2} \right)
$$

$$
\sim \chi^{-2}\left(2[\nu_0 + 2MN], \frac{\nu_0\lambda_0 + s^2}{\nu_0 + 2MN} \right) \tag{A.7.2}
$$

Derivation of Equation (7.157)

$$p\big[d^k[n] = +1 \mid \mathbf{H}, \sigma^2, \mathbf{D}_n^k, \mathbf{Y}\big] = \frac{p\big[d^k[n] = +1 \mid \mathbf{H}, \sigma^2, \mathbf{D}_n^k \mid \mathbf{Y}\big]}{p\big[\mathbf{H}, \sigma^2, \mathbf{D}_n^k \mid \mathbf{Y}\big]}$$

$$= p\big[\mathbf{Y} \mid d^k[n] = +1, \mathbf{D}_n^k, \mathbf{H}, \sigma^2\big] \frac{P\big[d^k[n] = +1 \mid \mathbf{D}_n^k\big] P[\mathbf{H}] P[\sigma^2]}{p\big[\mathbf{H}, \sigma^2, \mathbf{D}_n^k \mid \mathbf{Y}\big]}$$

$$\Rightarrow \frac{p\big[d^k[n] = +1 \mid \mathbf{H}, \sigma^2, \mathbf{D}_n^k, \mathbf{Y}\big]}{p\big[d^k[n] = -1 \mid \mathbf{H}, \sigma^2, \mathbf{D}_n^k, \mathbf{Y}\big]} = \frac{P\big[d^k[n] = +1, \mathbf{D}_n^k\big] p\big[\mathbf{Y} \mid d^k[n] = +1, \mathbf{D}_n^k, \mathbf{H}, \sigma^2\big]}{P\big[d^k[n] = -1, \mathbf{D}_n^k\big] p\big[\mathbf{Y} \mid d^k[n] = -1, \mathbf{D}_n^k, \mathbf{H}, \sigma^2\big]}$$

$$= \frac{P[d^k[n+1] \mid d^k[n] = +1]}{P[d^k[n+1] \mid d^k[n] = -1]} \times \frac{P[d^k[n] = +1 \mid d^k[n-1]]}{P[d^k[n] = -1 \mid d^k[n-1]]}$$

$$\times \exp\left\{ -\frac{1}{\sigma^2} \sum_{t=0}^{M-1} \underbrace{\left[\left\| \mathbf{y}_t + \sum_l \mathbf{X}_t^l \mathbf{W} \mathbf{h}^l \right\|_{d^k[n]=+1}^2 - \left\| \mathbf{y}_t + \sum_l \mathbf{X}_t^l (d^l) \mathbf{W} \mathbf{h}^l \right\|_{d^k[n]=-1}^2 \right]}_{\Delta s^2} \right\}$$

$$= \exp\left[\rho^k[n] d^k[n+1] + \rho^k[n-1] d^k[n-1] - \frac{\Delta s^2}{\sigma^2} \right]. \tag{A.7.3}$$

8

Ultra Wide Band Radio

In this chapter we discuss technology which is based on the spread spectrum concept, such as CDMA, as described in Chapter 5. The difference is that the pulse (called chip in Chapter 5) period used in this field is below 1 ns, resulting in a bandwidth of over 1 GHz, hence the name *Ultra Wide Band (UWB) Radio*. The second important characteristic is that the signal can be transmitted with no carrier. This is why very often the system is also referred to as *Impulse Radio* (IR). The above characteristics of the signal will require the modification of the signal format and detection concepts. In addition to these issues the chapter will also cover the basic characteristics of the UWB channel.

8.1 UWB MULTIPLE ACCESS IN A GAUSSIAN CHANNEL

A typical time-hopping format used in this case can be represented as [721–751]:

$$s_{\text{tr}}^{(k)}\left(t^{(k)}\right) = \sum_{j=-\infty}^{\infty} \omega_{\text{tr}}\left(t^{(k)} - jT_{\text{f}} - c_j^{(k)}T_{\text{c}} - \delta d_{[j/N_{\text{s}}]}^{(k)}\right) \tag{8.1}$$

where $t^{(k)}$ is the kth transmitter's clock time and T_{f} is the *pulse repetition time*. The transmitted pulse waveform ω_{tr} is referred to as a *monocycle*. To eliminate collisions due to multiple access, each user (indexed by k) is assigned a distinctive time shift pattern $\{c_j^{(k)}\}$ called a *time-hopping sequence*. This provides an additional time shift of $c_j^{(k)}T_{\text{c}}$ seconds to the jth monocycle in the pulse train, where T_{c} is the duration of addressable time delay bins. For a fixed T_{f}, the *symbol rate* R_{s} determines the number N_{s} of monocycles that are modulated by a given binary symbol as $R_{\text{s}} = (1/N_{\text{s}}T_{\text{f}})\text{s}^{-1}$. The modulation index δ is chosen to optimize performance. For performance prediction purposes, most of the time the data sequence $\{d_j^{(k)}\}_{j=-\infty}^{\infty}$ is modeled as a wide-sense stationary random process composed of equally likely symbols. For data, a pulse position modulation is used.

Advanced Wireless Communications. S. Glisic
© 2004 John Wiley & Sons, Ltd. ISBN: 0-470-86776-0

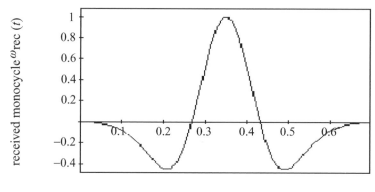

t in nanoseconds

Figure 8.1 A typical ideal received monocycle $\omega_{\text{rec}}(t)$ at the output of the antenna subsystem as a function of time in nanoseconds.

8.1.1 The multiple access channel

When K users are active in the multiple access system, the composite received signal at the output of the receiver's antenna is modeled as

$$r(t) = \sum_{k=1}^{K} A_k s_{\text{rec}}^{(k)}(t - \tau_k) + n(t) \tag{8.2}$$

The antenna/propagation system modifies the shape of the transmitted monocycle $\omega_{\text{tr}}(t)$ to $\omega_{\text{rec}}(t)$ at its output. An idealized received monocycle shape $\omega_{\text{rec}}(t)$ for a free-space channel model with no fading is shown in Figure 8.1.

8.1.2 Receiver

The optimum receiver for a single bit of a binary modulated impulse radio signal in additive white Gaussian noise (AWGN) is a correlation receiver defined as

$$decide \; d_0^{(1)} = 0 \quad if$$

$$\sum_{j=0}^{N_s-1} \overbrace{\int_{\tau_1+jT_f}^{\tau_1+(j+1)T_f} r(u,t)\, \upsilon\left(t - \tau_1 - jT_f - c_j^{(1)}T_c\right) dt}^{\text{pulse correlator output} \stackrel{\triangle}{=} \alpha_j(u)} > 0 \tag{8.3}$$

$$\underbrace{\phantom{\sum_{j=0}^{N_s-1} \int_{\tau_1+jT_f}^{\tau_1+(j+1)T_f} r(u,t)\, \upsilon\left(t - \tau_1 - jT_f - c_j^{(1)}T_c\right) dt}}_{\text{test statistic} \stackrel{\triangle}{=} \alpha(u)}$$

where $\upsilon(t) \stackrel{\triangle}{=} \omega_{\text{rec}}(t) - \omega_{\text{rec}}(t - \delta)$.

The *optimal detection* in a multiuser environment, with knowledge of all time-hopping sequences, leads to complex parallel receiver designs [722]. However, if the number of users is large and no such multiuser detector is feasible, then it is reasonable to approximate the combined effect of the other users dehopped interfering signals as a Gaussian random process [722]. Hence, the single-link reception algorithm (8.3) as shown in Figure 8.2 can be used for practical implementations. The test statistic in Algorithm (8.3) consists of summing

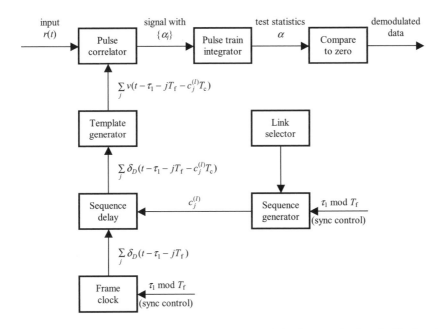

Figure 8.2 Receiver block diagram for the reception of the first user's signal. Clock pulses
are denoted by Dirac delta functions $\delta_D(\cdot)$ [751] © 2001, IEEE.

the N_s correlations α_j of the correlators template signal $\upsilon(t)$ at various time shifts with the
received signal $r(t)$.

For the monocycle waveform of Figure 8.1, the optimum choice of δ is 0.156 ns. By
choosing $\delta = 0.156$ ns and $T_f = 100$ ns, we achieve the results shown in Figure 8.3. For
the evaluation of the system efficiency, formulas from Chapter 4 are applicable with

$$D_r = K, \; \left(\frac{k_{02}}{k_{01}}\right) = K \text{ and } - g_{12} = \text{Additional Required Power (ARP)}$$

Both $K = $ total number of users and ARP are available in Figure 8.3.

8.2 THE UWB CHANNEL

8.2.1 Energy capture

In Section 8.1 a Gaussian channel was assumed. For a UWB signal, a high resolution of
multipath channel is expected. In this section we discuss some characteristics of a UWB
signal in such a channel.

8.2.2 The received signal model

In general, the received signal can be presented as

$$r(u, t) = s(u_s, t) + n(u_n, t) \tag{8.4}$$

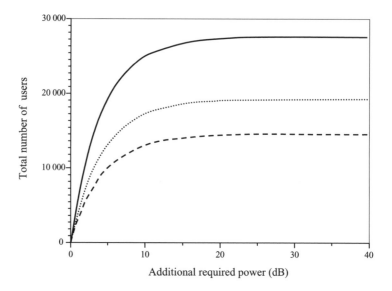

Figure 8.3 Total number of users versus additional required power (decibels) for the impulse radio example. Ideal power control is assumed at the receiver. Three different BER performance levels with the data rate set at 19.2 kb/s are considered.

where u characterizes the set of parameters defining the environment (position of the receiver in the room). The RAKE correlator structure is modeled as

$$\sum_{i=1}^{L_P} c_i \omega (t - \tau_i) \tag{8.5}$$

For experimental purposes, the pulse in Figure 8.1 that can be represented as $\omega_{\text{rec}}(t + 1.0) = \lfloor 1 - 4\pi(t/\tau_m)^2 \rfloor \exp\lfloor -2\pi(t/\tau_m)^2 \rfloor$ with $\tau_m = 0.78125$ is used. The *ML estimates* of the amplitude vector $\hat{\mathbf{c}}(\tilde{u})$ and delay vector $\hat{\tau}(\tilde{u})$ based on a specific observation $r(\tilde{u}, t)$ are the values $\mathbf{c}$ and τ which minimize the following mean squared error:

$$\mathrm{E}\left(\tilde{u}, L_p\right) = \int_0^T \left| r(\tilde{u}, t) - \sum_{i=1}^{L_p} c_i \omega (t - \tau_i) \right|^2 dt \tag{8.6}$$

The minimum value of the above mean squared error is denoted by $\mathrm{E}_{\min}(u, L_p)$. The *energy capture*, a function of L_p for each observation $r(\tilde{u}, t)$, is defined mathematically as

$$\mathrm{EC}\left(\tilde{u}, L_p\right) = 1 - \underbrace{\frac{\mathrm{E}_{\min}\left(\tilde{u}, L_p\right)}{E_{\text{tot}}\left(\tilde{u}\right)}}_{\overset{\Delta}{=}\quad \text{normalized} \quad \text{MSE}} \tag{8.7}$$

8.2.3   The UWB signal propagation experiment 1

A UWB signal propagation experiment performed in a typical modern office building [724] is described. The bandwidth of the signal used in this experiment is in

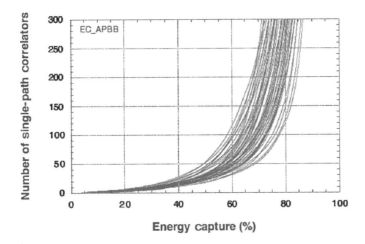

Figure 8.4 The required diversity level, L_p, in a UWB RAKE receiver as a function of percentage energy capture for each of the 49 received waveforms in an office representing a 'high SNR' environment (courtesy Moe Win, Massachusetts Institute of Technology) [724] © 1997, IEEE.

excess of 1 GHz, resulting in a differential path delay resolution of less than a nanosecond.

The transmitter is kept stationary in the central location of the building. Multipath profiles are measured using a digital sampling oscilloscope on one floor at various locations in 14 different rooms and hallways. In each office, multipath measurements are made at 49 different locations. They are arranged spatially in a level 7×7 square grid with six inch (15 cm) spacing, covering $3' \times 3'$ (90×90 cm).

Measurements from three different offices are used in the following discussions as typical examples of propagation environments. In these offices, the receiving antennas are located 6, 10 and 17 m away from the transmitter, representing typical UWB signal transmissions characterized as a 'high signal to noise ratio (SNR)' environment, 'low SNR' environment, and 'extremely low SNR' environment, respectively. The transmitter and receiving antenna are located in different rooms in these examples. Detailed results of this UWB signal propagation experiment can be found in [724, 733]. The results are shown in Figure 8.4 in the form of upper and lower bound curves.

8.2.4 UWB propagation experiment 2

The propagation experiment described in this section uses two vertically polarized diamond-dipole antennas [726], each 1.65 m above the floor and 1.05 m below the ceiling in an office/laboratory environment [727]. The equivalent received pulse at 1 m in free space can be estimated as the 'direct path' signal in an experiment in which there is no multipath signal, as shown in Figure 8.5.

A collection of results of recovered signal locations (delay and azimuth) is shown in Figure 8.6.

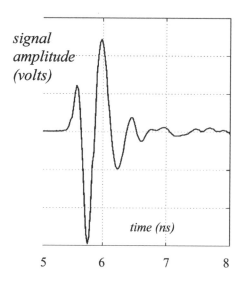

Figure 8.5 Transmitted pulse shape captured at 1 m separation from the transmit antenna.

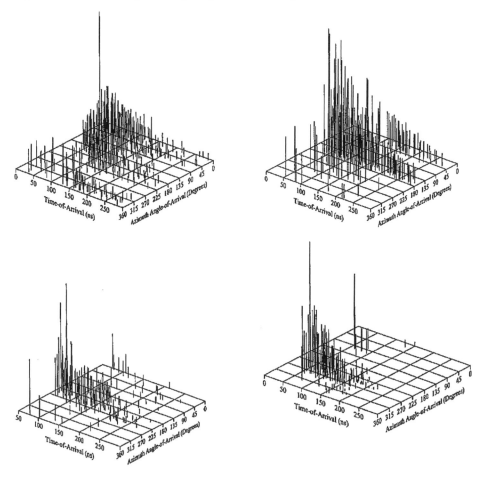

Figure 8.6 Recovered signal location and amplitude information [725] © 2002, IEEE.

8.2.5 Clustering models for the indoor multipath propagation channel

A number of models for the indoor multipath propagation channel [728–732] have reported a clustering of multipath components, in both time and angle. In the model presented in [731], the received signal amplitude β_{kl} is a Rayleigh distributed random variable with a mean square value that obeys a double exponential decay law, according to

$$\overline{\beta_{kl}^2} = \overline{\beta^2(0,0)}e^{-T_l/\Gamma}e^{-\tau_{kl}/\gamma} \qquad (8.8)$$

where $\overline{\beta^2(0,0)}$ describes the average power of the first arrival of the first cluster, T_l represents the arrival time of the lth cluster, and τ_{kl} is the arrival time of the kth arrival within the lth cluster, relative to T_l. The parameters Γ and γ determine the inter-cluster signal level rate of decay and the intra-cluster rate of decay, respectively. The parameter Γ is generally determined by the architecture of the building, while γ is determined by objects close to the receiving antenna, such as furniture. The results presented in [731] make the assumption that the channel impulse response as a function of time and azimuth angle is a separable function, or

$$h(t,\theta) = h(t)h(\theta) \qquad (8.9)$$

from which independent descriptions of the multipath time-of-arrival and angle-of-arrival are developed. This is justified by observing that the angular deviation of the signal arrivals within a cluster from the cluster mean does not increase as a function of time.

The cluster decay rate Γ and the ray decay rate γ can be interpreted for the environment in which the measurements were made. For the results, presented later in this section, at least one wall separates the transmitter and the receiver. Each cluster can be viewed as a path that exists between the transmitter and the receiver, along which signals propagate. This cluster path is generally a function of the architecture of the building itself. The component arrivals within a cluster vary because of secondary effects, e.g. reflections from the furniture or other objects. The primary source of degradation in the propagation through the features of the building is captured in the decay exponent Γ. Relative effects between paths in the same cluster do not always involve the penetration of additional obstructions or additional reflections, and therefore tend to contribute less to the decay of the component signals. Results for $p(\theta)$ generated from the data in Figure 8.6 are shown in Figure 8.7.

Interarrival times are hypothesized [731] to follow exponential rate laws, given by

$$p(T_l|T_{l-1}) = \Lambda e^{-\Lambda(T_l - T_{l-1})}$$
$$p(T_{kl}|T_{k-1,l}) = \lambda e^{-\lambda(T_l - T_{l-1})}$$

where Λ is the cluster arrival rate and λ is the ray arrival rate. Channel parameters are summarized in Table 8.1.

8.2.6 Path loss modeling

In this segment we are interested in a transceiver operating at approximately 2 GHz center frequency with a bandwidth in excess of 1.5 GHz, which translates to sub-nanosecond time resolution in the CIRs.

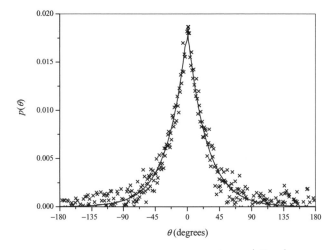

$$\textit{Laplacian distribution: } p\left(\theta\right) = \frac{1}{\sqrt{2}\,\sigma}\,e^{-\left|\sqrt{2}\theta\,/\,\sigma\right|}$$

(a)

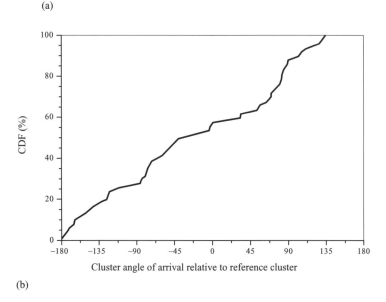

(b)

Figure 8.7(a) Ray arrival angles at $1°$ of resolution and a best fit Laplacian density with $\sigma = 38°$; (b) distribution of the cluster azimuth angle of arrival, relative to the reference cluster [725] © 2002, IEEE.

8.2.6.1 Measurement procedure

The measurement campaign is described in [734] and is conducted in a single-floor, hard-partition office building (fully furnished). The walls are constructed of drywall with vertical metal studs; there is a suspended ceiling ten feet (three metres) in height with carpeted

Table 8.1   Channel parameters

Parameter	UWB [725]	Spencer *et al.* [731]	Spencer *et al.* [731]	Saleh–Valenzuela [730]
Γ	27.9 ns	33.6 ns	78.0 ns	60 ns
γ	84.1 ns	28.6 ns	82.2 ns	20 ns
$1/\Lambda$	45.5 ns	16.8 ns	17.3 ns	300 ns
$1/\lambda$	2.3 ns	5.1 ns	6.6 ns	5 ns
σ	37°	25.5°	21.5°	. . .

concrete floor. Measurements are conducted with a stationary receiver and mobile transmitter, both transmit and receive antennas are five feet (1.5 metres) above the floor. For each measurement, a 300 ns time domain scan is recorded and the LOS distance from transmitter to receiver is recorded. A total of 906 profiles are included in the dataset with seven different receiver locations recorded over the course of several days. Except for a reference measurement made for each receiver location, all successive measurements are NLOS links, chosen randomly throughout the office layout, that penetrate anywhere from one to five walls. The remainder of the datapoints are taken in a similar fashion.

8.2.6.2   Path loss modeling

The average path loss for an arbitrary T-R separation is expressed using the power law as a function of distance. The *indoor* environment measurements show that at any given d, shadowing leads to signals with a path loss that is lognormally distributed about the mean [735, 736]. That is:

$$\text{PL}(d) = \text{PL}_0(d_0) + 10N \log \left(\frac{d}{d_0} \right) + X_\sigma \tag{8.10}$$

where N is the path loss exponent, X_σ is a zero mean lognormally distributed random variable with standard deviation σ(dB) and PL_0 is the free space path loss at reference distance, d_0. Some results are shown in Figure 8.8.

Assuming a simple RAKE with four correlators where each component is weighted equally, we can calculate the path loss versus distance using the peak CIR power plus RAKE gain, $\text{PL}_{\text{PEAK+RAKE}}$, for each CIR, as shown in Figure 8.8(c). The exponent N obtained from performing a least squares fit is 2.5, with a standard deviation of 4.04 dB. The results for delays are shown in Figure 8.9.

8.2.6.3   In-home channel

For the in-home channel, Equation (8.10) can also be used to model path loss. Some results are shown in Table 8.2 [737–740].

Table 8.3 presents the results for delay spread in the in-home channel [737].

Table 8.2   Statistical values of the path loss parameters

	LOS		NLOS	
	Mean	Standard deviation	Mean	Standard deviation
PL_0(dB)	47		51	
N	1.7	0.3	3.5	0.97
σ(dB)	1.6	0.5	2.7	0.98

Table 8.3   Percentage of power contained in profile, number of paths, mean excess delay and RMS delay spread for 5, 10, 15, 20 and 30 dB threshold level.

	50% NLOS				90% NLOS			
Threshold	% Power	L	τ_m (ns)	τ_{RMS} (ns)	% Power	L	τ_m (ns)	τ_{RMS} (ns)
5 dB	46.8	7	1.95	1.52	46.9	8	2.2	1.65
10 dB	89.2	27	7.1	5.77	86.5	31	8.1	6.7
15 dB	97.3	39	8.6	7.48	96	48	10.3	9.3
20 dB	99.4	48	9.87	8.14	99.5	69	12.2	11
30 dB	99.97	60	10.83	8.43	99.96	82	12.4	11.5

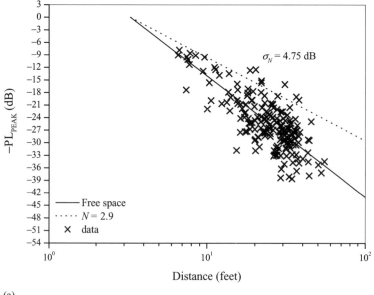

(a)

Figure 8.8   (a) Peak PL vs. distance; (b) total PL vs. distance; (c) peak PL + RAKE gain vs. distance [734] © 2002, IEEE.

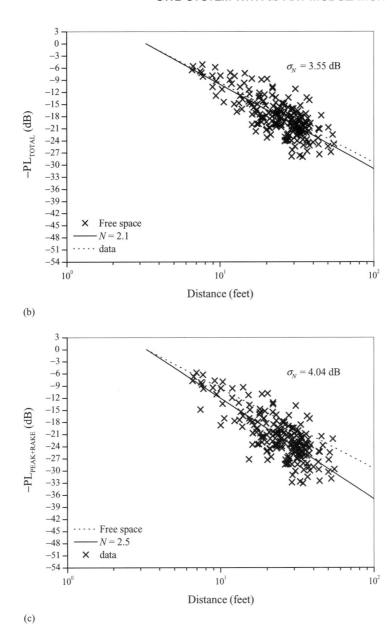

(b)

(c)

Figure 8.8 *(Cont.)*.

8.3 UWB SYSTEM WITH *M*-ARY MODULATION

8.3.1 Performance in a Gaussian channel

We will first assume that the transmitted pulse and the received signal are $p_{TX}(t) \stackrel{\Delta}{=} \int_{-\infty}^{t} p(\xi) d\xi$ and $p(t) + n(t)$ respectively (we ignore effects of propagation). The effect of the antenna system in the transmitted pulse is modeled as a differentiation operation.

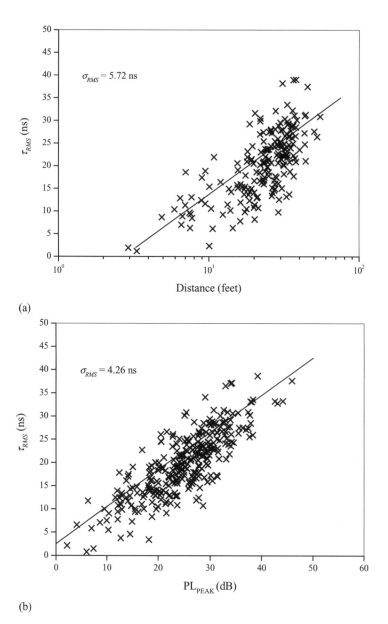

Figure 8.9 (a) RMS delay spread vs. distance; (b) RMS delay spread vs. path loss [734]
© 2002, IEEE.

The noise $n(t)$ is AWGN with two-sided power density $N_0/2$. The UWB pulse $p(t)$ has
duration T_p and energy $E_\mathrm{p} = \int_{-\infty}^{\infty} [p(t)]^2 dt$. The normalized signal correlation function of
$p(t)$ is

$$\gamma_\mathrm{p}(\tau) \stackrel{\Delta}{=} \frac{1}{E_\mathrm{p}} \int_{-\infty}^{\infty} p(t)\, p(t-\tau)\, dt > -1 \quad \forall \tau \tag{8.11}$$

Parameter $\gamma_{\min} \overset{\Delta}{=} \gamma_p(\tau_{\min})$ is defined as the minimum value of $\gamma_p(\tau)$, $\tau \in (0, T_p]$. The transmitted signal is a PPM signal, and each is composed of N_s time-shifted pulses

$$\psi_{TX}^{(j)}(t) = \sum_{k=0}^{N_s-1} p_{TX}\left(t - kT_f - a_j^k \tau_{\min}\right) \qquad (8.12)$$

$j = 1, 2, \ldots, M$. In the absence of noise, the received signals are composed of N_s time-shifted UWB pulses

$$\Psi_j(t) = \sum_{k=0}^{N_s-1} p\left(t - kT_f - a_j^k \tau_{\min}\right) \qquad (8.13)$$

Each $\Psi_j(t)$ represents the jth signal in an ensemble of M signals, each signal identified by the sequence of time shifts $a_j^k \tau_{\min} \in \{0, \tau_{\min}\}$ (this choice of time shifts allows us to produce M-ary PPM signals, which are equally correlated). The a_j^k is a 0, 1 pattern representing the jth cyclic shift of an m-sequence of length N_s. Since there are at most N_s cyclic shifts in an m-sequence, we require that $2 \leq M < N_s$.

The pulse duration satisfies $T_p + \tau_{\min} < T_f$, where T_f is the time shift value corresponding to the frame period. Each signal $\Psi_j(t)$ has duration $T_s \overset{\Delta}{=} N_s T_f$ and energy $E_\psi = N_s E_p$. The signals in Equation (8.133) have normalized correlation values

$$\beta_{ij} \overset{\Delta}{=} \frac{\displaystyle\int_{-\infty}^{\infty} \Psi_i(t)\,\Psi_j(t)\,\mathrm{d}t}{E_\psi} = \beta = \frac{1 + \gamma_{\min}}{2} \qquad (8.14)$$

for all $i \neq j$, i.e. they are equally correlated.

The optimum receiver is a bank of filters matched to the M signals $\Psi_j(t)$, $j = 1, 2, \ldots, M$. The receiver is assumed to be perfectly synchronized with the transmitter.

The union bound on the bit error probability using these equally correlated signals can be written [742]:

$$\mathrm{UBPb} = \frac{1}{M} \sum_{\substack{i=1 \\ i \neq j}}^{M} \sum_{j=1}^{M} Q\left(\sqrt{\frac{E_\psi}{N_0}(1 - \beta)}\right) = \frac{M}{2} \int_{\sqrt{\log_2(M)\mathrm{SNRb}}}^{\infty} \frac{\exp(-\xi^2/2)}{\sqrt{2\pi}}\,\mathrm{d}\xi \qquad (8.15)$$

where

$$\mathrm{SNRb} = \frac{1}{\log_2(M)} \frac{E_\psi}{N_0}(1 - \beta) \qquad (8.16)$$

is the received bit SNR and $Q(\xi)$ is the Gaussian tail function.

As an example for $p(t)$, we consider a UWB pulse that can be modeled by properly scaling the second derivative of a Gaussian function $\exp(-2\pi[t/t_n]^2)$. In this case, we have

$$p_{TX}(t) = t \exp\left(-2\pi\left[\frac{t}{t_n}\right]^2\right) \qquad (8.17)$$

$$p(t) = \left[1 - 4\pi\left[\frac{t}{t_n}\right]^2\right] \exp\left(-2\pi\left[\frac{t}{t_n}\right]^2\right) \qquad (8.18)$$

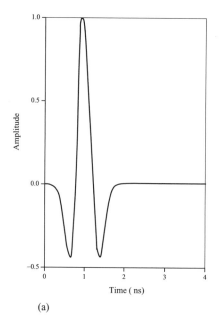
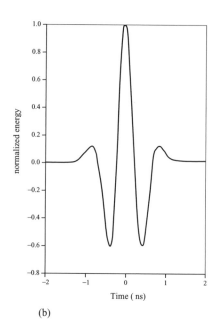

Figure 8.10 (a) The pulse $p\left(t - T_{\mathrm{p}}/2\right)$ as a function of time $0 \le t \le 4$ ns; (b) the signal autocorrelation $\gamma_{\mathrm{p}}(\tau)$ as a function of time shift $-2 \le \tau \le 2$ ns [741].

where the value $\tau_n = 0.7531$ ns was used to fit the model $p(t)$ to a measured waveform $p_m(t)$ from a particular experimental radio link [743]. This resulted in $T_{\mathrm{p}} \approx 2.0$ ns.

The normalized signal correlation function corresponding to $p(t)$ is calculated using Equation (8.11) to give:

$$\gamma_{\mathrm{p}}(t) = \left[1 - 4\pi\left[\frac{t}{t_n}\right]^2 + \frac{4\pi^2}{3}\left[\frac{t}{t_n}\right]^4\right]\exp\left(-\pi\left[\frac{t}{t_n}\right]^2\right) \qquad (8.19)$$

For this $\gamma_{\mathrm{p}}(t)$, we have $\tau_{\min} = 0.4073$ ns and $\gamma_{\min} = -0.6183$, so $\beta = 0.1909$ in Equation (8.14). Both $p(t - T_{\mathrm{p}}/2)$ and $\gamma_{\mathrm{p}}(\tau)$ are depicted in Figure 8.10. Figure 8.11 shows the spectrum of the impulse $p(t)$. The 3 dB bandwidth of the pulse is close to 1 GHz. The center frequency is around 1.1 GHz.

The specific values of N_{s} and T_{f} do not affect SNRb, as long as $M < N_{\mathrm{s}}$ and $T_{\mathrm{p}} + \tau_{\min} < T_{\mathrm{f}}$. Hence, we set arbitrarily $T_{\mathrm{f}} = 500$ ns and $N_{\mathrm{s}} > 1000$ [741]. The BER in AWGN can now be calculated using UBPb from Equation (8.15). Results for different values of M are shown in Figure 8.12. Values as large as $M = 128$ are easily obtained with the PPM signal design in Equation (8.13), allowing us to exploit the benefits of M-ary modulation without an excessive increase in the complexity of the receiver [744].

8.3.2 Performance in a dense multipath channel

In this section we discuss performance of M-ary UWB signals in a dense multipath channel with AWGN. The channel can be, for example, an indoor radio channel as discussed

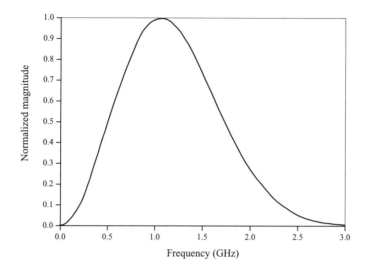

Figure 8.11 The magnitude of the spectrum of the pulse $p(t)$.

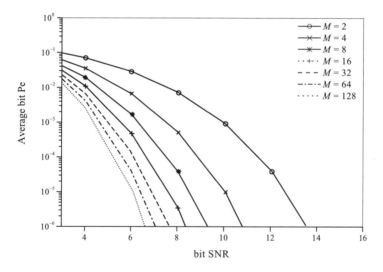

Figure 8.12 The UBPb in Equation (8.15). Curves for M are 2, 4, 8, 16, 32, 64, and 128
signals.

in Section 8.2. In the analysis, we will assume that the transmitter is placed at a certain
fixed location, and the receiver is placed at a variable location denoted u_0. The transmitted pulse is the same pulse $p_{TX}(t)$ as in the AWGN case, and the received UWB sig-
nal is $\sqrt{E_a}\tilde{p}(u_0, t) + n(t)$. The pulse $\sqrt{E_a}\tilde{p}(u_0, t)$ is a multipath spread version of $p(t)$
received at position u_0 with average duration $T_a \gg T_p$. The pulse has 'random' energy
$\tilde{E}(u_0) \stackrel{\Delta}{=} E_a\tilde{\alpha}^2(u_0)$, where E_a is the average energy and

$$\tilde{\alpha}(u_0) \stackrel{\Delta}{=} \int_{-\infty}^{\infty} [\tilde{p}(u_0, t)]^2 \, dt \tag{8.20}$$

is the normalized energy. The pulse has normalized signal correlation

$$\tilde{\gamma}(u_0, \tau) \overset{\Delta}{=} \frac{\int_{-\infty}^{\infty} \tilde{p}(u_0, t)\tilde{p}(u_0, t - \tau)\,dt}{\int_{-\infty}^{\infty} [\tilde{p}(u_0, t)]^2\,dt} \tag{8.21}$$

The transmitted signals are the same $\Psi_{\text{TX}}^{(j)}(t)$ as given in Equation (8.12). In the absence of noise, the received signals are composed of N_s time-shifted UWB pulses

$$\tilde{\Psi}_j(u_0, t) = \sum_{k=0}^{N_s-1} \sqrt{E_a}\,\tilde{p}\left(u_0, t - kT_f - \alpha_j^k \tau_{\min}\right) \tag{8.22}$$

for $j = 1, 2, \ldots, M$. The UWB PPM signal $\tilde{\Psi}_j(u_0, t)$ is a multipath spread version of $\Psi_j(t)$ received at position u_0. Assume that $\tilde{\Psi}_j(u_0, t)$ has fixed duration $T_s \approx N_s T_f$, provided that $T_a + \tau_{\min} < T_f$. The signals in Equation (8.22) have 'random' energy

$$\tilde{E}_\Psi(u_0) = \int_{-\infty}^{\infty} \left[\tilde{\Psi}_j(u_0, \xi)\right]^2\,d\xi = \bar{E}_\Psi \tilde{\alpha} u_0 \tag{8.23}$$

for $j = 1, 2, \ldots, M$, where $\bar{E}_\Psi = N_s E_a$ is the average energy. The signals in Equation (8.22) have normalized correlation values

$$\tilde{\beta}_{ij}(u_0) \overset{\Delta}{=} \frac{\int_{-\infty}^{\infty} \tilde{\Psi}_i(u_0, \xi)\,\tilde{\Psi}_j(u_0, \xi)\,d\xi}{\tilde{E}_\Psi(u_0)} = \tilde{\beta}(u_0) = \frac{1 + \tilde{\gamma}(u_0, \tau_{\min})}{2} \tag{8.24}$$

for all $i \neq j$, i.e. they are equally correlated. The multipath effects change with the particular position u_0, and therefore the M-ary set of received signals $\{\tilde{\Psi}_j(u_0, t)\}_{j=1}^{j=M}$ also changes with the particular position u_0.

8.3.3 Receiver and BER performance

Conditioned on a particular physical location u_0, the optimum receiver (matched filter) is a kind of perfect Rake receiver that is able to construct a reference signal $\tilde{\Psi}_j(u_0, T_s - t)$ that is perfectly matched to the signal received $\tilde{\Psi}_j(u_0, t)$ over the multipath conditions at that location u_0. We will assume that the receiver is perfectly synchronized with the transmitter. Performance analysis for the perfect Rake receiver can be calculated using standard techniques, conditioned on a particular physical location u_0, the union bound on the bit error probability using these equally correlated signals can be written

$$\text{UBPb}(u_0) = \frac{M}{2} \int_{\sqrt{\log_2(M)\text{SNRb}(u_0)}}^{\infty} \frac{\exp(-\xi^2/2)}{\sqrt{2\pi}}\,d\xi \tag{8.25}$$

where

$$\text{SNRb}(u_0) = \frac{1}{\log_2(M)} \frac{\tilde{E}_\Psi(u_0)}{N_0}(1 - \tilde{\beta}(u_0))$$

$$= \frac{1}{\log_2(M)} \frac{\bar{E}_\Psi \tilde{\alpha}^2(u_0)}{N_0}(1 - \tilde{\beta}(u_0)) \tag{8.26}$$

is the received bit SNR [745].

8.3.4 Time variations

The $\tilde{\beta}(u_0)$ value accounts for changes in the correlation properties of the received signals. These changes in $\tilde{\beta}(u_0)$ translate into changes in the Euclidean distance between signals. Therefore, the $(1 - \tilde{\beta}(u_0))$ value accounts for energy variations at the output of the perfect matched filter due to distortions in the shape of the signal correlation function caused by multipath. The $\tilde{\alpha}^2(u_0)$ value accounts for variations in the received signal energy due to fading caused by multipath. The average performance can be obtained by taking the expected value $\mathbf{E}_u(\cdot)$ over all values of u_0

$$\overline{\text{UBPb}} \left(\frac{\bar{E}_\Psi}{N_0} \right) = \mathbf{E}_u \left\{ \text{UBPb} (u) \right\} \tag{8.27}$$

where

$$\left(\frac{\bar{E}_\Psi}{N_0} \right) \stackrel{\Delta}{=} \mathbf{E}_u \left\{ \text{SNRb} (u) \right\} \tag{8.28}$$

is the average received bit SNR. This BER analysis provides a theoretical matched filter bound for the best performance attainable when the multipath channel is perfectly estimated.

Instead of using a channel model, the calculations based on the *received* waveforms can be used. This is possible since the expected value in Equation (8.27) is taken with respect to the quantities $\tilde{\alpha}^2(u_0)$ and $\tilde{\beta}(u_0)$. Histograms of these quantities can be calculated for a particular indoor channel environment and a first approximation can be obtained using the sample mean value

$$\overline{\text{UBPb}} \left(\frac{\bar{E}_\Psi}{N_0} \right) \approx \frac{1}{u_*} \sum_{u_0=1}^{u_*} \text{UBPb} (u_0) \tag{8.27a}$$

This calculation represents a rough approximation to the performance of UWB signals in the presence of dense multipath in a particular indoor radio environment. The histograms for $\tilde{\alpha}^2(u_0)$ and $\tilde{\beta}(u_0)$ can be derived from their definitions in Equations (8.20) and (8.24) using the ensemble of pulse responses

$$\{\tilde{p}(u_0, t)\}, \quad u_0 = 1, 2, \ldots, u_*$$

taken in a measurement experiment in the multipath channel of interest.

8.3.5 Performance example

The channel responses $\tilde{p}(u_0, t)$ were measured in eight different rooms and hallways in a typical office building described in Section 8.2 (see also [741]). In every room and hallway, 49 different locations are arranged spatially in a 7×7 square grid with six inch spacing. At every location u_0, the $T_a = 300$ ns-long pulses $\tilde{p}(u_0, t)$ are recorded, keeping the transmitter, the receiver and the environment stationary.

The UWB transmitter is placed at a fixed location inside the building. It consists of a step recovery diode-based pulser connected to a UWB omnidirectional antenna. The pulser produces a train of UWB 'Gaussian monocycles' $p_{TX}(t)$. The train of $p_{TX}(t)$ is transmitted as an excitation signal to the propagation channel. The train has a repetition rate of 500 ns with a tightly controlled average monocycle-to-monocycle interval. The clock driving the pulser has resolution in the order of picoseconds.

The $\tilde{p}(u_0, t)$ represents the convolution of $p_{\text{TX}}(t)$ with the channel impulse response at location u_0. The 500 ns repetition rate is long enough to make sure that pulse responses $\tilde{p}(u_0, t)$ corresponding to adjacent impulses $p_{\text{TX}}(t)$ do not overlap. The receiver consists of a UWB antenna and a low-noise amplifier. The output of this amplifier is captured using a high-speed digital sampling scope. The scope takes samples in windows of 50 ns at a sampling rate of 20.5 GHz. Noise in the measured $\tilde{p}(u_0, t)$ is reduced by averaging 32 consecutive received pulses measured at exactly the same location u_0. These samples are sent to a data storage and processing unit.

A total of $u_* = 392$ channel pulse responses $\tilde{p}(u_0, t)$ were measured. An equal number of normalized energy values $\tilde{\alpha}^2(u_0)$ and normalized correlation functions $\tilde{\gamma}(u_0, \tau)$ were calculated using Equations (8.20) and (8.21), respectively. Figure 8.13 shows histograms of $\tilde{\alpha}^2(u_0)$, $\tilde{\beta}(u_0)$ and the product $\tilde{\alpha}^2(u_0) \times (1 - \tilde{\beta}^2(u_0))$. The measured $\tilde{p}(u_0, t)$ have $T_a \approx$ 300 ns. The rest of the parameter values are the same as those used in the AWGN case, i.e. $\tau_{\min} = 0.4073$ ns, $T_f = 500$ ns and $N_s > 1000$. With these values, the conditions $M < N_s$ and $T_a + \tau_{\min} < T_f$ will be satisfied.

For the results from Figure 8.13, the average BER curves, Equation (8.27), are shown in Figure 8.14.

8.4   M-ARY PPM UWB MULTIPLE ACCESS

The signal format introduced in Section 8.3 is now used for the multiple access system [746]. The TH PPM signal conveying information exclusively in the time shifts is now represented as

$$x^{(v)}(t) = \sum_{k=0}^{\infty} \omega \left(t - kT_f - c_k^{(v)} T_c - \delta_{d_{[k/N_s]}^{(v)}}^k \right) \tag{8.29}$$

The superscript (v), $1 \leq v \leq N_u$, indicates user-dependent quantities, where N_u is the number of simultaneous active users. The index k is the number of time hops that the signal $x^{(v)}(t)$ has experienced, and also the number of pulses that have been transmitted. T_f is the frame (pulse repetition) time and equals the average time between pulse transmissions. The notation $[q]$ stands for the integer part of q. The $\{c_k^{(v)}\}$ is the pseudorandom time-hopping sequence assigned to user v. It is periodic with period N_p (i.e. $c_{k+lN_p}^{(v)} = c_k^{(v)}$, for all k, l integers) and each sequence element is an integer in the range $0 \leq c_k^{(v)} \leq N_h$. For a given time shift parameter T_c, the time-hopping code provides an additional time shift to the pulse in every frame, each time shift being a discrete time value $c_k^{(v)} T_c$, with $0 \leq c_k^{(v)} T_c \leq N_h T_c$. The time shift corresponding to the data modulation is

$$\delta_{d_{[k/N_s]}^{(v)}}^k \in \{\tau_1 = 0 < \tau_2 < \cdots < \tau_\eta\} \tag{8.30}$$

with $\eta \geq 2$ an integer. The data sequence $\{d_m^{(v)}\}$ of user v is an M-ary symbol stream, $1 \leq d_m^{(v)} \leq M$, that conveys information in some form. The system under study uses fast time-hopping, which means that there are $N_s > 1$ pulses transmitted per symbol. The data symbol changes only every N_s hops. Assuming that a new data symbol begins with pulse index $k = 0$, the data symbol index is $[k/N_s]$.

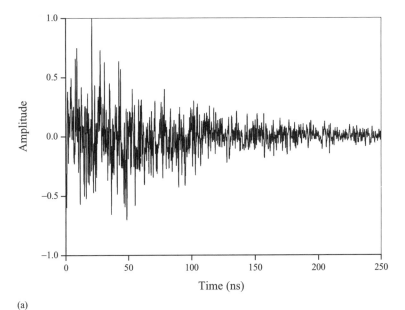

(a)

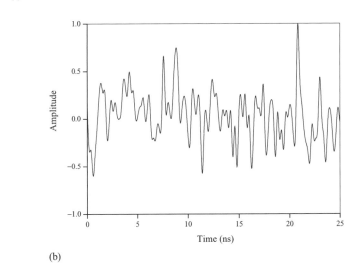

(b)

Figure 8.13 (a) Normalized correlation function $\tilde{\gamma}(u_0, t)$ of the pulse $\tilde{p}(u_0, t)$ in Figure 8.5; (b) a closer view of the correlation in (a). The spreading caused by multipath is notorious. The long tails in the correlation function are the effect of the pulse spreading [741]. The histogram of the normalized values of (c) the received energy $\tilde{\alpha}^2(u_0)$; (d) the correlation value $\tilde{\beta}(u_0)$; and (e) the product $\tilde{\alpha}^2(u_0) \times (1 - \tilde{\beta}^2(u_0))$. The ordinate represents appearance frequency, and the abscissa represents the value of the parameter. The size of the sample is $u_0 = 392$ [741] © 2001, IEEE.

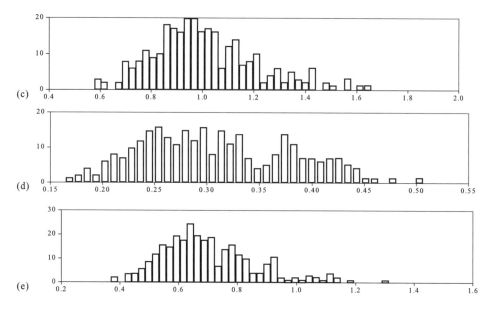

(c)

(d)

(e)

Figure 8.13 (*Cont.*).

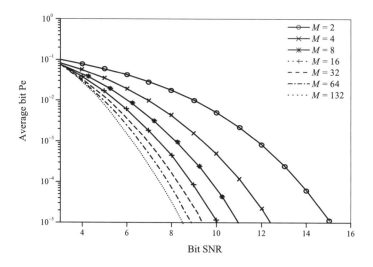

Figure 8.14 $\overline{\text{UBPb}}(\bar{E}_\Psi/N_0)$ in Equation (8.27a); $M = 2, 4, 8, 16, 32, 64$ and 132 signals.

We will use the following notation

$$H_m^{(v)}(t) \stackrel{\circ}{=} \sum_{k=mN_s}^{(m+1)N_s-1} T_c c_k^{(v)} p(t - kT_f)$$

$$p(t) = \begin{cases} 1, & \text{if } 0 \leq t \leq T_f \\ 0, & \text{otherwise} \end{cases} \tag{8.31}$$

and

$$S_i(t) \triangleq \sum_{k=0}^{N_s-1} \omega\left(t - kT_f - \delta_i^k\right) \tag{8.32}$$

for $i = 1, 2, \ldots, M$, then Equation (8.29) can be written

$$x^{(v)}(t) = \sum_{m=0}^{\infty} S_{d_m^{(v)}}\left(t - mN_sT_f - H_m^{(v)}(t)\right) \triangleq \sum_{m=0}^{\infty} X_{m,d_m^{(v)}}^{(v)}(t) \tag{8.29a}$$

The signal $S_i(t)$ in Equation (8.32) is the received signal corresponding to the transmitted signal $\int_{-\infty}^{t} S_i(\xi)d\xi = \sum_{k=0}^{N_s-1} \omega_{TX}(t - kT_f - \delta_i^k)$. With this notation, the signal correlation function is defined as

$$\begin{aligned}
R_{ij} &\triangleq \int_{-\infty}^{\infty} X_{m,i}^{(v)}(\xi)\, X_{m,i}^{(v)}(\xi)\, d\xi \\
&= \int_{-\infty}^{\infty} S_i(\xi)\, S_j d\xi = E_\omega \sum_{k=0}^{N_s-1} \gamma_\omega\left(\delta_i^k - \delta_j^k\right)
\end{aligned} \tag{8.33}$$

since the pulses are non-overlapping. The energy in the ith signal $X_{m,i}^{(v)}$ is $E_S = R_{ii} = N_s E_\omega$, and the normalized correlation value is

$$\alpha_{i,j} \triangleq \frac{R_{ij}}{E_S} = \frac{1}{N_s} \sum_{k=0}^{N_s-1} \gamma_\omega\left(\delta_i^k - \delta_j^k\right) \geq \gamma_{\min} \tag{8.34}$$

8.4.1 *M*-ary PPM signal sets

The PPM signal $S_i(t)$ in Equation (8.32) represents the ith signal in an ensemble of M information signals, each signal completely identified by the pulse shape $\omega(t)$ and the sequence of time shifts $\{\delta_i^k\}$, $k = 0, 1, 2, \ldots, N_s - 1$. The most interesting M-ary PPM sets from the practical point of view are orthogonal (OR), equally correlated (EC), and N-orthogonal (NO) signal sets.

In general, these signal designs have the property that the structure of the M-ary autocorrelation matrix is preserved for different $\omega(t)$. This is important because $\omega(t)$ is, in general, a non-standard pulse, and these signal designs reduce the dependence of the MA performance on the shape of $\omega(t)$. The time shift patterns defining each M-ary PPM signal set and their respective correlation properties are studied in detail in [747–749] and are summarized in Table 8.4. In the EC case, the a_i^k is a 0, 1 pattern representing the ith cyclic shift, $i = 1, 2, \ldots, M$, of an m-sequence [751] of length $N_s = 2^m - 1$, $m \geq 1$, and $N_s \geq M$.

The M-ary correlation receiver, shown in Figure 8.15, consists of M filters matched to the signals $\{X_{m,j}^{(1)}(t - \tau^{(1)})\}$, $j = 1, 2, \ldots M$, $t \in T_m$, followed by samplers and a decision circuit that selects the maximum among the decision variables

$$\int_{t \in T_m} r(t)\, X_{m,j}^{(1)}\left(t - \tau^{(1)}\right) dt, \quad j = 1, 2, \ldots M \tag{8.35}$$

Table 8.4   Time-shift patterns and normalized correlation values of the M-ary PPM signals under study. Orthogonal (OR), equally correlated (EC), N-orthogonal design 1 (NO1) and N-orthogonal design 2 (NO2) [746] © 2001, IEEE

Type of signal	Time shift pattern $\{\delta_i^k\}$ $i = 1, 2, \ldots, M$ $k = 0, 1, 2, \ldots, N_s - 1$	Normalized correlation coeficients
OR	$\delta_i^k = [(k + i - 1) \bmod M]\, T_{OR}$ $T_{OR} = 2T_\omega$	$\alpha_{ij}^{(OR)} = \begin{cases} 1, & i = j \\ 0, & i \neq j \end{cases}$
EC	$\delta_i^k = a_i^k \tau_2$ $\tau_2 \in (0, T_\omega]$ $a_i^k \in \{0, 1\}$	$\alpha_{ij}^{(EC)} = \begin{cases} 1, & i = j \\ \lambda, & i \neq j \end{cases}$ $\|\lambda\| < 1$
NO1	$\delta_i^k = \tau_1 + \left[(k + 2\tilde{I}) \bmod L\right] T_{NO1}$ $L \triangleq \left[\frac{M}{2}\right]$ $I = i - \left[\frac{i-1}{2}\right] 2$ $\tilde{I} \triangleq \left[\frac{i-1}{2}\right]$ $T_{NO1} \triangleq \tau_2 + T_{OR}$ $0 = \tau_1 < \tau_2 < T_\omega$	$\alpha_{ij}^{(NO1)} = \begin{cases} 1, & i = j \\ 0, & \left[\frac{i-1}{2}\right] \neq \left[\frac{j-1}{2}\right] \\ \beta_{ij}, & \left[\frac{i-1}{2}\right] = \left[\frac{j-1}{2}\right] \end{cases}$ $\beta_{ij} = \gamma_\omega (\tau_J - \tau_I)$ $J = j - \left[\frac{j-1}{2}\right] 2$
NO2	$\delta_i^k = a_i^k \tau_2 + \left[(k + 2\tilde{I}) \bmod L\right] T_{NO2}$ $T_{NO2} \triangleq \tau_2 + T_{OR}$ $0 = \tau_1 < \tau_2 < T_\omega$	$\alpha_{ij}^{(NO2)} = \begin{cases} 1, & i = j \\ 0, & \left[\frac{i-1}{2}\right] \neq \left[\frac{j-1}{2}\right] \\ \lambda, & \left[\frac{i-1}{2}\right] = \left[\frac{j-1}{2}\right] \end{cases}$

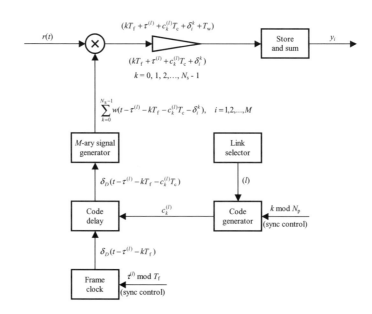

Figure 8.15 M-ary correlator receiver for the TH PPM signals.

8.4.2 Performance results

In this subsection, we illustrate the theoretical MA performance of this system for a specific $\omega(t)$ under perfect power control (i.e. $A^{(\nu)} = A^{(1)}$ for $\nu = 2, 3, \ldots, N_u$). The system parameters are the same as in [746]. The $\omega(t)$ considered here is the second derivative of a Gaussian function

$$\omega(t) = \left[1 - 4\pi\left[\frac{t}{t_n}\right]^2\right] \exp\left(-2\pi\left[\frac{t}{t_n}\right]^2\right) \tag{8.36}$$

where the value t_n is used to fit the model $\omega(t)$ to a measured waveform from a particular experimental radio link. The normalized signal correlation function corresponding to $\omega(t)$ in Equation (8.36) is

$$\gamma_\omega(\tau) = \left[1 - 4\pi\left[\frac{\tau}{t_n}\right]^2 + \frac{4\pi^2}{3}\left[\frac{\tau}{t_n}\right]^4\right] \exp\left(-\pi\left[\frac{\tau}{t_n}\right]^2\right) \tag{8.37}$$

In this case, T_ω and τ_{min} depend on t_n, and $\gamma_{min} = -0.6183$ for any t_n. Using $t_n = 0.4472$ ns, we get $T_\omega \approx 1.2$ ns and $\tau_{min} = 0.2149$ ns. Figure 8.16 depicts $\omega(t - T_\omega/2)$, $\gamma_\omega(\tau)$ and the spectrum of $\omega(t)$. The 3 dB bandwidth of $\omega(t)$ is in excess of 1 GHz.

Given $\omega(t)$, the signal design is complete when we specify N_s, T_f and δ_j^k in Table 8.4. For OR signals $T_{OR} = 2T_\omega$, for EC signals $\tau_1 = 0$ and $\tau_2 = \tau_{min}$, for NO1 signals $\tau_1 = 0$, $\tau_2 = \tau_{min}$ and $T_{NO1} \hat{=} \tau_{min} + 2T_\omega$, and for NO2 signals $\tau_1 = 0$, $\tau_2 = \tau_{min}$ and $T_{NO2} \hat{=} \tau_{min} + 2T_\omega$.

To choose a value for T_f, we need $0 < N_h T_c < (T_f/2) - 2(T_\omega + \tau_\eta)$. Also notice that $\tau_\eta = 16T_{OR}$ for $M = 16$ (the maximum value of M, in this example). By choosing $T_f = 100$ ns, we have that $0 < N_h T_c < 16$ ns.

To choose a value for N_s, notice that the M-ary PPM signal designs considered in [747] require that $N_s = 1/(R_s T_f) = \log_2(M)/(R_b T_f) = \log_2(M)N_s^{(2)}$. Hence, for a fixed T_f, the value of N_s is determined by R_b. However, in particular, the EC PPM signal design additionally requires $N_s \geq M$. Combining these two requirements on N_s we have that, in the EC PPM case, both R_b and N_s satisfy the relation $(\log_2(N_s)/N_s) \geq R_b T_f$. In this example, we use $R_b = 100$ kb/s, $N_s^{(2)} = 100$, $2 \leq M \leq 16$, and $N_s = \log_2(M)100$; hence, $(\log_2(N_s)/N_s) \geq R_b T_f$ holds, and both relations $N_s \geq M$ and $N_s = \log_2(M)/R_b T_f$ are satisfied.

For a single-link communications bit, $E_b/N_0 = 14.30$ dB, $\text{SNRb}_{OR}(1) = 14.30$ dB [747], and $\text{SNRb}_{TSK}(1) = 16.40$ dB. The BER results are given in Figure 8.17. If we define SNR(1) as the required signal to noise ratio for a certain BER with only one user in the network and SNR(N_u) as the required initial signal to noise ratio with one user which, after adding $N_u - 1$ additional users, will still give SNR(1), then we define the degradation factor DF as DF $= \text{SNR}(N_u)/\text{SNR}(1)$. In the next example we demonstrate some results for $N_u(\text{DF})$ when using the pulse $\omega(t)$ in Equation (8.36) with $t_n = 0.4472$ ns, $T_\omega = 1.2$ ns, $\tau_{min} = 0.2419$ ns, $\tau_1 = 0$, $\tau_2 = \tau_{min}$, and $T_f = 100$ ns. Figure 8.18 shows $N_u(\text{DF})$ for different values of DF using $R_b = 100$ kb/s, $N_s^{(2)} = 100$, $2 \leq M \leq 256$, and $N_s = \log_2(M)100$. Notice that $(\log_2(N_s)/N_s) \geq R_b T_f$ still holds for $M = 256$.

From $N_u(\text{DF})$ we can also find $R_b(\text{DF})$ for a particular value of N_u. Figure 8.19 shows $R_b(\text{DF})$ for different values of DF using $N_u = 1000$ active users.

The values of the upper bound on maximum capacity C_{sup} are given in Table 8.5. The upper bound on maximum data rate is shown in Figure 8.20.

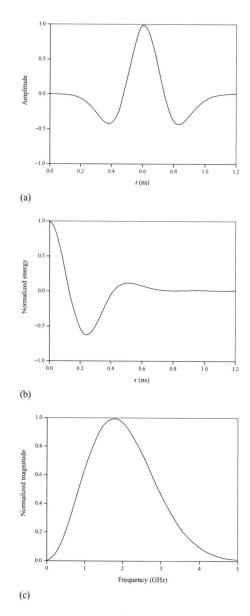

(a)

(b)

(c)

Figure 8.16 (a) Pulse $\omega(t - T_\omega/2)$ as a function of time t; (b) signal autocorrelation function $\gamma_\omega(\tau)$ as a function of time shift τ; (c) magnitude of the spectrum of the pulse $\omega(t)$ [746].

8.5 CODED UWB SCHEMES

We can consider the simple time-hopping spread spectrum from Section 8.4 as a coded system in which a simple repetition block code with rate $1/N_s$ is used. As outlined in Chapter 2 the efficiency of the repetition code is very low. Thus, by applying a near optimal code instead of the above simple repetition code, we should expect the system performance

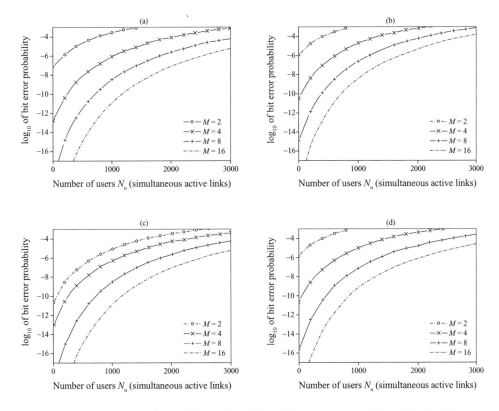

Figure 8.17 Base 10 logarithm of the probability of bit error as a function of N_u for different values of M, using $R_b = 100$ kb/s. (a) OR PPM signals, $\text{SNRb}_{\text{OR}}(1) = 14.30$ dB; (b) EC PPM signals, $\text{SNRb}_{\text{EC}}(1) = 13.39$ dB; (c) NO PPM signals, design 1, $\text{SNRb}_{\text{OR}}(1) = 14.30$ dB and $\text{SNRb}_{\text{TSK}}(1) = 16.40$ dB; (d) NO PPM signals, design 2, $\text{SNRb}_{\text{OR}}(1) = 14 : 30$ dB.

to improve significantly. In [752,753], a class of low-rate superorthogonal convolutional codes that have near optimal performance is introduced. In a superorthogonal code with constraint length K, the rate is equal to $1/2^{K-2}$. Since in the TH CDMA (UWB) system, N_s pulses are sent for each data bit, we must set $2^{K-2} = N_s$ or $K = 2 + \log_2 N_s$. The location of each pulse in each frame is determined by the user-dedicated pseudorandom sequence, along with the code symbol corresponding to that frame.

Decoding is performed using the Viterbi algorithm. The state diagram of this decoder consists of 2^{K-1} states. Two branches, corresponding to bit zero and bit one, exit from each state in the trellis diagram. To update the state metrics, it is first necessary to calculate the branch metrics, using the received signal $r(t)$. For this purpose, in each frame j, the quantity:

$$\alpha_j \triangleq \int_{\tau_1 + jT_f}^{\tau_1 + (j+1)T_f} r(t)\, v\!\left(t - \tau_1 - jT_f - c_j^{(1)} T_c\right) dt$$

(the pulse correlator output) is obtained. Because of the special form of the Hadamard–Walsh sequence that is used in the structure of superorthogonal codes, the branch metrics

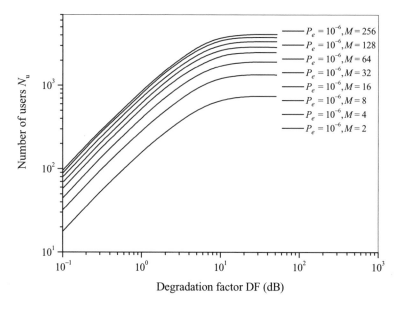

Figure 8.18 Number of simultaneous active links (users) $N_u(\mathrm{DF})$ for EC PPM signals for $2 \le M \le 256$ with $P_e(1) = \mathrm{UBP}_b^{(\mathrm{EC})}(1) \approx 10^{-6}$ and $R_b = 100$ kb/s.

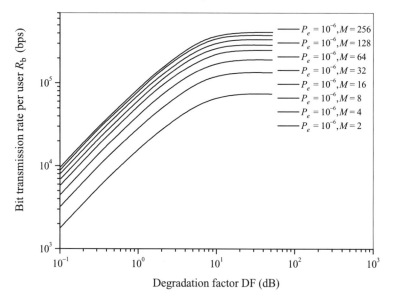

Figure 8.19 Data transmission rate per user $R_b(\mathrm{DF})$ for EC PPM signals for $2 \le M \le 256$ with $P_e(1) = \mathrm{UBP}_b^{(\mathrm{EC})}(1) \approx 10^{-6}$ and $N_u = 1000$ active users.

can be simply evaluated based on the outputs of pulse correlators $\alpha_j s$ [753]. This is also elaborated in Chapter 4 for space–time codes from orthogonal designs. The processing complexity of this decoder grows only linearly with K (or logarithmically with N_s); the required memory, however, grows exponentially with K (or equally linearly with N_s) [753].

Table 8.5 Values of C_{sup} in (Gb/s) calculated using three different pulse widths. Also included are the values of N_{sup} for $R_b = 100$ kb/s [746]

Set I of parameters	Set II of parameters	Set III of parameters
$t_n = 0.2877$ ns	$t_n = 0.4472$ ns	$t_n = 0.7531$ ns
$T_\omega = 0.75$ ns	$T_\omega = 1.2$ ns	$T_\omega = 2.0$ ns
$\tau_{\min} = 0.1556$ ns	$\tau_{\min} = 0.2419$ ns	$\tau_{\min} = 0.4073$ ns
$C_{\text{sup}}^{(\text{I})} = 3.6394$ (Gbps)	$C_{\text{sup}}^{(\text{II})} = 2.3412$ (Gbps)	$C_{\text{sup}}^{(\text{III})} = 1.3903$ (Gbps)
$N_{\text{sup}}^{(\text{I})} = 36394$ (users)	$N_{\text{sup}}^{(\text{II})} = 23412$ (users)	$N_{\text{sup}}^{(\text{III})} = 13903$ (users)

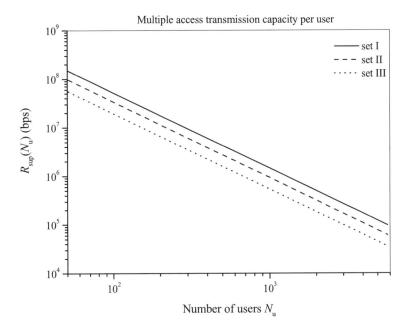

Figure 8.20 Upper bound on the bit transmission rate per user $R_{\text{sup}}(N_u)$ in b/s, calculated using the sets I, II and III of parameters in Table 8.5.

Since in time-hopping, spread spectrum applications, the value of K is relatively low (the typical value is in the range 3–12), the system can be considered to be practical.

8.5.1 Performance

The path generating function of the code for a superorthogonal code is computed as [753–754]:

$$T_{\text{SO}}(\gamma, \beta) = \frac{\beta W^{K+2}(1 - W)}{1 - W[1 + \beta(1 + W^{K-3} - 2W^{K-2})]} \tag{8.38}$$

in which $W = \gamma^{2^{K-3}}$ and K is the constraint length of the code. Expanding the above expression, we get a polynomial in γ and β. The coefficient and the powers of γ and β in each term of the polynomial indicate the number of paths and output–input path weights respectively. The free distance of this code is obtained from the first term of the expansion as $d_f = 2^{K-3}(K+2) = N_s(\log_2 N_s + 4)/2$.

8.5.2 The uncoded system as a coded system with repetition

In this case, the distance will be N_s. Comparing the free distances of these two schemes, it is clear that the coded scheme outperforms the uncoded scheme significantly.

An upper bound on the probability of error per bit for a memoryless channel is obtained using the union bound as follows:

$$P_b < \left. \frac{dT_{SO}(W, \beta)}{d\beta} \right|_{\beta=1} = \frac{W^{K+2}}{(1-2W)^2} \left(\frac{1-W}{1-W^{K-2}} \right)^2 \tag{8.39}$$

where $W = Z^{2^{K-3}}$. The parameter Z is calculated from the Bhattacharyya bound as

$$Z = \int_{-\infty}^{\infty} \sqrt{p_0(y) \, p_1(y)} \, dy \tag{8.40}$$

where $p_0(y)$ and $p_1(y)$ are pdfs of the pulse correlator output conditioned on the input symbol being zero and one, respectively.

A lower bound on the probability of error per bit is obtained by considering only the first term of the path generating function, Equation (8.38), The result is

$$P_b \geq P_{d_f} \tag{8.41}$$

where P_{d_f} is the probability of pairwise error in favor of an incorrect path that differs in d_f symbols from the correct path over the unmerged span in the trellis diagram.

For $\omega_{rec}(t + T_\omega/2) = \lfloor 1 - 4\pi(t/\tau_m)^2 \rfloor \exp(-2\pi(t/\tau_m)^2)$, $\tau_m = 0.2877$ and δ and T_f set to 0.156 and 100 ns respectively, the performance curves are given in Figures 8.21 and 8.22 for different R_s.

8.6 MULTIUSER DETECTION IN UWB RADIO

In this section we consider a system based on using orthogonal codes in a synchronous or quasi-synchronous context, coupled with TDD for transmissions through frequency selective channels. This is achieved by assigning to each user different orthogonal time-hopping sequences and designating two time slots for transmission: one for the uplink and one for the downlink. Binary PPM modulation from Section 8.5 is used with eight users ($N_c' = 8$) and two symbols per burst ($K = 2$). As per [755], frame duration $T_f = 100$ ns and there is a maximum delay spread equal to 100 ns.

Each symbol is repeated N_f frames. A time guard of duration T_c is used, hence, $N_g = 1$. The chip duration $T_c = T_f/(N_c' + N_g) = 11.11$ ns. The sampling rate is equal to t_c which leads to a channel of length $L_1 = 9$. The channel model from Section 8.2 is used. This model is based on clusters of rays. The received signal is composed of attenuated and delayed versions of the transmitted signal arriving in clusters. The times of arrival are modeled as a

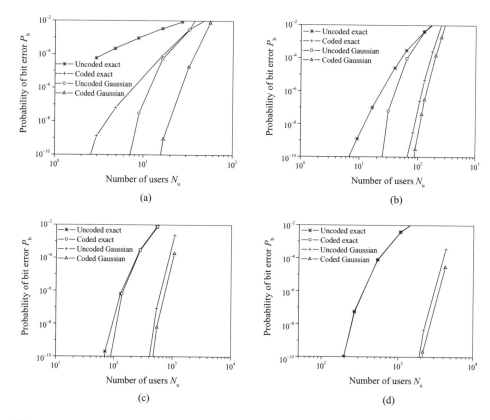

Figure 8.21 Probability of bit error as a function of number of users for synchronous un-coded and coded (upper bound) schemes in exact and Gaussian cases (a) at $R_s = 5$ Mb/s ($N_s = 2$); (b) at $R_s = 1.25$ Mb/s ($N_s = 8$); (c) at $R_s = 325$ kb/s ($N_s = 32$); (d) at $R_s = 78.1$ Mb/s ($N_s = 128$).

Poisson process and the amplitudes as Gaussian. For the simulations, the receivers will be assumed to be synchronized on the strongest path. In the PPM-IRMA system, the received signal is the second derivative of the Gaussian function $\sqrt{\tau^3/3}(2/\pi)^{1/4} \exp(-t^2/\tau^2)$ (normalized to have $r_\omega(0) = 1$); hence, we have $r_\omega(t) = \exp(-t^2/(2\tau^2))\lfloor 1 - 2(t/\tau)^2 + (t/\tau)^4/3 \rfloor$ where $r_\omega(t)$ is the correlation function of $\omega(t)$ and the parameter $\tau = 0.1225$ ns is adjusted to yield a pulse width equal to 0.7 ns.

For multiuser detection schemes (MF, zero forcing (ZF)-decorrelator and MMSE), based on the same logic as in Chapter 5, the results are shown in Figures 8.23 and 8.24. Details of the derivation of the detector transfer function are given in [756].

8.7 UWB WITH SPACE–TIME PROCESSING

8.7.1 Signal model

In this segment, the Generalized Gaussian Pulse (GGP) $\Omega(t)/E_0$ shown in Figure 8.25 will be used.

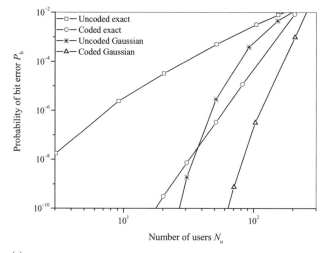

(a)

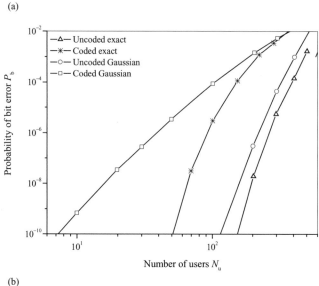

(b)

Figure 8.22 Probability of bit error as a function of number of users for asynchronous uncoded and coded (upper bound) schemes in exact and Gaussian cases (a) at $R_s = 5$ Mb/s ($N_s = 2$); (b) at $R_s = 2.5$ Mb/s ($N_s = 4$); (c) at $R_s = 1.25$ Mb/s ($N_s = 8$).

The self-steering array beamforming system for UWB impulse waveforms is shown in Figure 8.26.

The response of the ith sensor to the received wavefront can be expressed in terms of the voltage signal as

$$\upsilon_i(t, \phi) = \Omega(t - \tau_i(\phi))$$
$$= [E_0/(1 - \alpha)] (\exp\{-4\pi[(t - \tau_i(\phi))/\Delta T]^2\} \quad (8.42)$$
$$- \alpha \exp\{-4\pi[\alpha(t - \tau_i(\phi))/\Delta T]^2\})$$

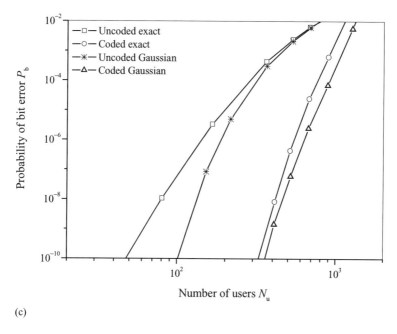

(c)

Figure 8.22 (*Cont.*).

The relative time delay $\tau_i(\phi)/\Delta T$ is a function of the angle of incidence ϕ and the distance d between adjacent sensors:

$$\tau_i\,(\phi)\,/\Delta T = (id/c\Delta T)\sin\phi = (i/2m)\,\rho\sin\phi \tag{8.43}$$

$$\rho = 2md/c\Delta T = L/c\Delta T = L\Delta f/c \tag{8.44}$$

In order to achieve electronic beamsteering for enhancing the quality of signal reception from a desired look direction, e.g. $\phi = \phi_0$ the variable delay circuit VDC_i applies to the incoming signal $\Upsilon_i(t)$ a time delay $\tilde{\tau}_i = (id/c)\sin\phi_0$. In analogy to Equation (8.44), the relative time delay $\tilde{\tau}_i/\Delta T$ can also be expressed in terms of ρ:

$$\tilde{\tau}_i/\Delta T = (id/c\Delta T)\sin\phi_0 = (i/2m)\,\rho\sin\phi_0 \tag{8.45}$$

Finally, the delayed signals $\Upsilon(t + \tilde{\tau}_i - \tau_i(\phi))$, $i = 0, \pm 1, \pm 2, \ldots, \pm m$ from the VDC_is are summed by SUM1, SUM2 and SUM3 to produce the beamformer's response function

$$\tilde{\Upsilon}(t, \phi) = \sum_{i=-m}^{m} \Upsilon(\tau + \tilde{\tau}_i - \tau_i\,(\phi)) \tag{8.46}$$

Some results are shown in Figures 8.27 to 8.31.

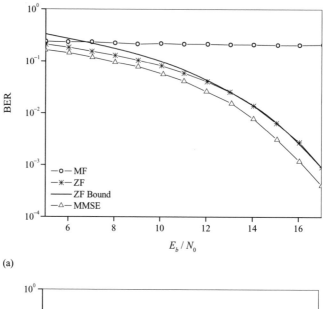

(a)

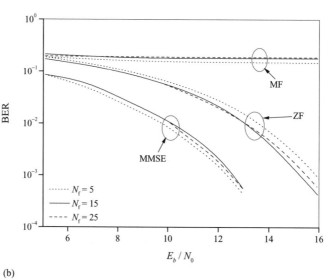

(b)

Figure 8.23   Comparing the different receivers (MF, ZF, MMSE) for the binary PPM-IRMA scheme with $K = 2$, $T_g = 1$ and eight users. (a) For $N_f = 1$; (b) for $N_f = 5$, $N_f = 15$ and $N_f = 25$.

8.7.2 The monopulse tracking system

The monopulse signal delivered to DAC by SUM4 is:

$$\Upsilon_\Delta(t, \phi) = \sum_{i=0}^{+m} \Upsilon(t + \tilde{\tau}_i - \tau_i(\phi)) - \sum_{i=-m}^{+0} \Upsilon(t + \tilde{\tau}_i - \tau_i(\phi)) \tag{8.47}$$

Angle-of-arrival estimation based on slope processing is illustrated in Figure 8.32. Basically, for different signs of ϕ, the control signal $\Upsilon_\Delta(t, \phi)$ has different slopes [see Figure 8.32(a)].

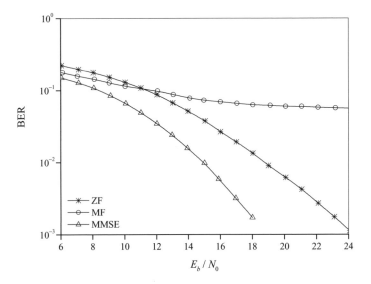

Figure 8.24 Comparing the different receivers (MF, ZF, MMSE) for the binary PPM-IRMA scheme for 100 Monte Carlo channel trials with the same parameters as in Figure 8.23(a).

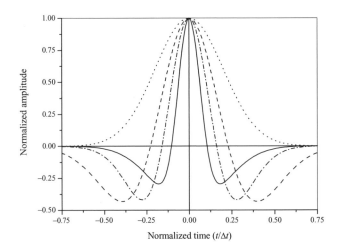

(a) $\Omega(t) = \frac{E_0}{1-\alpha} \left(\exp\left\{-4\pi[(t-t_0)/\Delta T]^2\right\} - \alpha \exp\left\{-4\pi[\alpha(t-t_0)/\Delta T]^2\right\} \right)$

Autocorrelation function $\Upsilon(t)/\Upsilon(0)$

Figure 8.25 (a) Normalized time variation of the generalized Gaussian pulse $\Omega(t)/E_0$; (b) autocorrelation function $\Upsilon(t)/\Upsilon(0)$; (c) energy density spectrum $\Psi(f) = |\Lambda(f)|^2$ for values of the scaling parameter $\alpha = 0$ (dotted line), $\alpha = 0.75$ (dashed line), $\alpha = 1.5$ (dashed-dotted line) and $\alpha = 3$ (solid line).

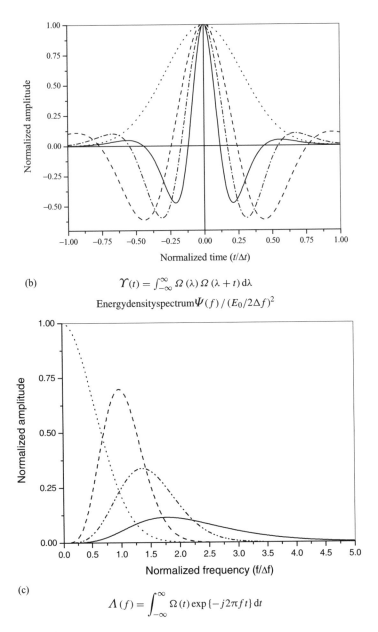

(b) $$\Upsilon(t) = \int_{-\infty}^{\infty} \Omega(\lambda)\,\Omega(\lambda+t)\,d\lambda$$

Energy density spectrum $\Psi(f)/(E_0/2\Delta f)^2$

(c) $$\Lambda(f) = \int_{-\infty}^{\infty} \Omega(t)\exp\{-j2\pi f t\}\,dt$$

Figure 8.25 (*Cont.*).

For larger ϕ, the control signal $\Upsilon_\Delta(t,\phi)$ is smaller. Based on this, the control mechanism is defined so that the delays of the VDC_is are adjusted by DAC according to the relationship

$$\tilde{\tau}_i > \tilde{\tau}_{i+1}, \quad \text{for} \quad -\pi/2 \le \phi < \phi_0$$

$$\tilde{\tau}_i < \tilde{\tau}_{i+1}, \quad \text{for} \quad \phi_0 < \phi \le \pi/2$$ (8.48)

More details can be found in [758].

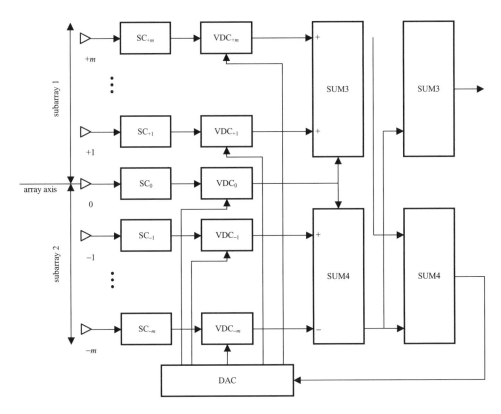

Figure 8.26 A self-steering array beamforming system for UWB impulse waveforms. The beamformer includes $M = 2m + 1$ sensors, sliding correlator (SC$_i$), variable delay circuit (VDC$_i$), delay adjustment computer (DAC) and summer circuits (SUM) [757].

8.8 BEAMFORMING FOR UWB RADIO

8.8.1 Circular array

The received signal is modeled as a Gaussian pulse [758–760]

$$\Omega(t) = (E/\Delta T) \exp\lfloor -\pi(t/\Delta T)^2 \rfloor \tag{8.49}$$

For the geometry shown in Figure 8.33 we have:

$$R_n = (r^2 + a^2 - 2ar \cos \Psi_n)^{1/2}$$

which, for $r \gg a$, can be approximated as

$$R_n = r - a \cos \Psi_n = r - a \sin \theta \cos (\phi - \phi_n)$$

with $\phi_n = 2\pi \left(\frac{n}{N} \right)$, $n = 1, 2, \ldots, N$

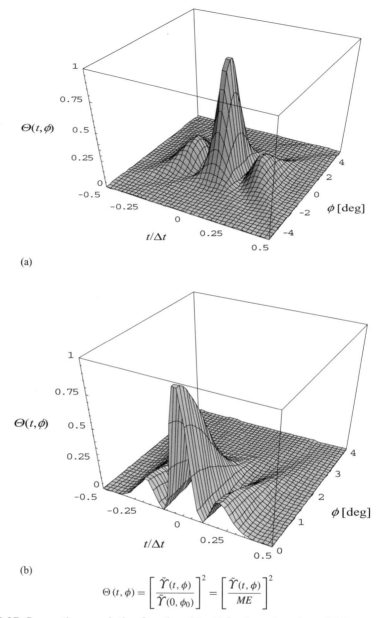

$$\Theta\left(t, \phi\right) = \left[\frac{\tilde{\Upsilon}_{(t, \phi)}}{\tilde{\Upsilon}_{(0, \phi_0)}}\right]^2 = \left[\frac{\tilde{\Upsilon}_{(t, \phi)}}{ME}\right]^2$$

Figure 8.27 Space–time resolution function $\Theta(t, \phi)$ for the value of spacial frequency bandwidth $\rho = 10$. (a) Scaling parameter $\alpha = 3$ and the angular range $-5^0 \le \phi \le +5^0$; (b) $0^0 \le \phi \le +3^0$.

We assume that the signal source is located in the far field at the point $P(r, \theta, \phi)$; $r \gg a$. So at the nth array element we have:

$$V_n(t) = \Omega(t + \tau_n) = (E/\Delta T) \exp\lfloor -\pi[(t + \tau_n)/\Delta T]^2 \rfloor \qquad (8.50)$$

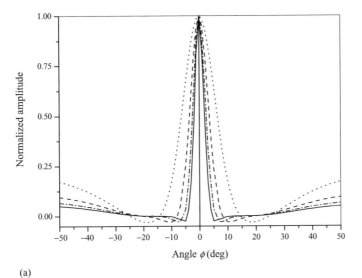

(a)

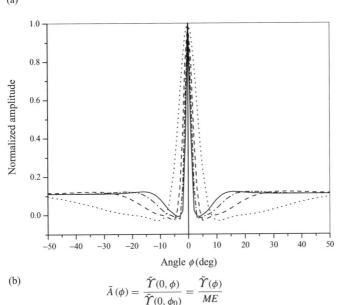

(b)

$$\tilde{A}(\phi) = \frac{\tilde{\Upsilon}(0,\phi)}{\tilde{\Upsilon}(0,\phi_0)} = \frac{\tilde{\Upsilon}(\phi)}{ME}$$

Figure 8.28 Peak amplitude pattern $\tilde{A}(\phi)$ for: (a) $\phi_0 = 0^0$, $\alpha = 3$, $d = c\Delta T/2$, and $M = 5$ (dotted line), 9 (dashed line), 13 (dashed-dotted line) and 17 (solid line); and (b) $M = 9$ and $d = c\Delta T/2$ (dotted line), $c\Delta T$ (dashed line), $3c\Delta T/2$ (dashed-dotted line) and $2c\Delta T$ (solid line).

with $\tau_n = (a/c)\sin\theta\cos(\phi - \phi_n)$. To steer the peak of the main beam of the array in the (θ_0, ϕ_0) direction, a delay

$$\alpha_n = -\tau_n(\theta_0, \phi_0) = -(a/c)\sin\theta_0\cos(\phi_0 - \phi_n) \tag{8.51}$$

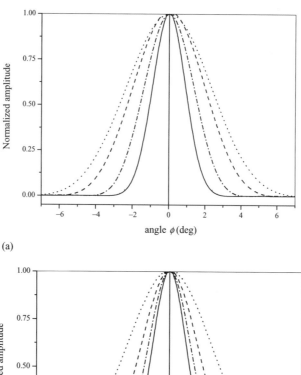

(a)

(b)

$$P(\phi) = [\tilde{A}(\phi)]^2 = \Theta(0, \phi)$$

Figure 8.29 Peak power pattern $P(\phi)$ for: (a) $\phi_0 = 0^0$ and $\rho = 10$, $\alpha = 0.5$ (dotted line), 0.75 (dashed line), 1.5 (dashed-dotted line) and 3 (solid line); and (b) $\alpha = 3$ and $\rho = 4$ (dotted line), 6 (dashed line), 8 (dashed-dotted line) and 10 (solid line).

must be applied to the voltage signal $V_n(t)$. So, the overall received signal is

$$V_T(t, \theta, \phi) = \sum_{n=1}^{N} \Omega(t + \tau_n + \alpha_n) \tag{8.52}$$

We will use notation

$$\rho_c = (a/c\Delta T) = 2\Delta f a/c, \quad \Delta f = 1/2\Delta T \tag{8.53}$$
$$\eta_s = \sin\theta \sin\phi - \sin\theta_0 \sin\phi_0$$

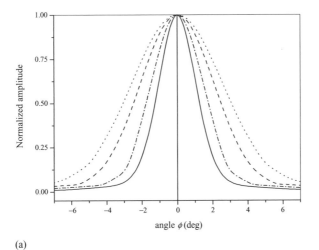

(a)

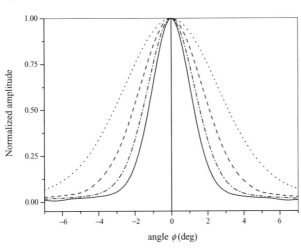

(b)

$$\tilde{U}(\phi) = \int_{-\infty}^{\infty} \left[\tilde{\Upsilon}(\tau\phi) \right]^2 dt$$

$$\tilde{W}(\phi) = \frac{\tilde{U}(\phi)}{\tilde{U}(\phi_0)}$$

Figure 8.30 Energy pattern $\tilde{W}(\phi)$ for: (a) $\phi_0 = 0^0$ and $\rho = 10$, $\alpha = 0.5$ (dotted line), 0.75 (dashed line), 1.5 (dashed-dotted line) and 3 (solid line); and (b) $\alpha = 3$ and $\rho = 4$ (dotted line), 6 (dashed line), 8 (dashed-dotted line) and 10 (solid line).

$$\eta_c = \sin\theta\cos\phi - \sin\theta_0\cos\phi_0$$

$$\eta_0 = [(\eta_c)^2 + (\eta_s)^2]^{1/2}$$

$$\cos\xi = \frac{\eta_c}{\eta_0}$$

$$\xi = \arctan\frac{\eta_s}{\eta_c}$$

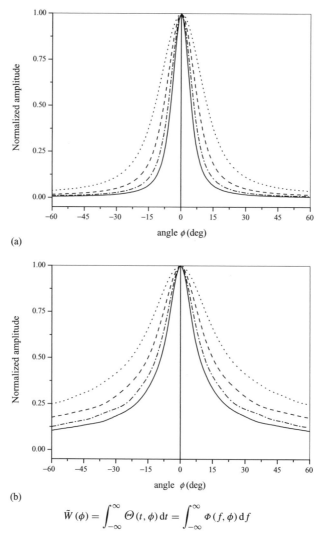

(a)

(b)

$$\tilde{W}(\phi) = \int_{-\infty}^{\infty} \Theta(t, \phi)\, dt = \int_{-\infty}^{\infty} \Phi(f, \phi)\, df$$

Figure 8.31 Peak power pattern $P(\phi)$ and energy pattern $\tilde{W}(\phi)$ for the case of an ideal Gaussian pulse with $\alpha = 0$. The plots are calculated for $\rho = 4$ (dotted line), 6 (dashed line), 8 (dashed-dotted line) and 10 (solid line).

Equations (8.52) and (8.54) together result in:

$$(\alpha_n + \tau_n)/\Delta T = \rho_c \eta_0 \cos(\xi - \phi_n) \tag{8.54}$$

and we have

$$V_T(t, \theta, \phi) = (E/\Delta T) \sum_{n=1}^{N} \exp\{-\pi[(t/\Delta T) + \rho_c \eta_0 \cos(\xi - \phi_n)]^2\} \tag{8.55}$$

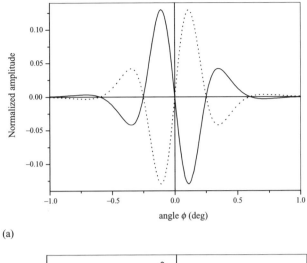

(a)

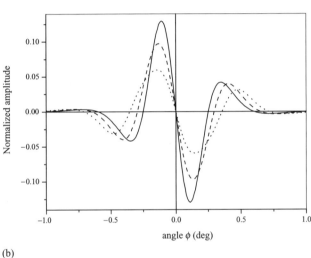

(b)

Figure 8.32 Normalized monopulse signal $\Upsilon_{\Delta}(t, \phi)/(m\Delta T E_0^2)$ plotted as a function of relative time $t/\Delta T$ for (a) $\alpha = 3$, $\rho = 10$, and the values of the angle of incidence $\phi = +2^0$ (solid line), and $\phi = -2^0$ (dotted line) and repeated for (b) $\phi = 2^0$ (solid line), $\phi = 3^0$ (dashed line), $\phi = 4^0$ (dotted line).

For a main beam in the vertical direction along the z-axis $\theta_0 = 0$ which yields $\alpha_n = \alpha_{nv} = 0$, $\eta_0 = \eta_{0v} = \sin\theta$, $\xi = \xi_v = \phi$ and we have:

$$V_v(t, \theta, \phi) = (E/\Delta T) \sum_{n=1}^{N} \exp\left\{-\pi[(t/\Delta T) + \rho_c \sin\theta \cos(\phi - \phi_n)]^2\right\} \quad (8.56)$$

A main beam can be formed in the principal horizontal plane ($z = 0$) by setting $\theta_0 = \pi/2$

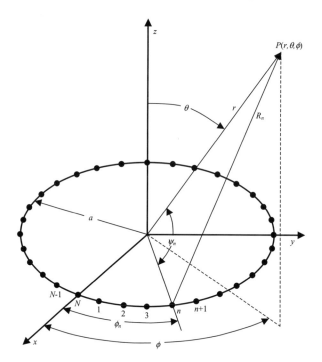

Figure 8.33 Geometry of an N-element circular array with radius a.

which gives

$$\alpha_n = \alpha_{nh} = -(a/c)\cos(\phi_0 - \phi_n)$$

$$\eta_0 = \eta_{0h} = 2\sin\left(\frac{\phi_0 - \phi_n}{2}\right) \tag{8.57}$$

$$\xi = \xi_h = \frac{\pi + \phi_0 + \phi_n}{2}$$

and we have

$$V_h(t, \phi) = (E/\Delta T) \times \sum_{n=1}^{N} \exp\left\{-\pi\left[(t/\Delta T) + 2\rho_c \sin\left(\frac{\phi - \phi_0}{2}\right)\right.\right.$$

$$\left.\left. \times \cos\left(\frac{\pi + \phi_0 + \phi - 2\phi_n}{2}\right)\right]^2\right\} \tag{8.58}$$

A graph of $V_v(t, \theta, \phi)/N(E/\Delta T)$ is shown in Figure 8.34
By using Figure 8.34, we define the peak amplitude pattern

$$A(\theta, \phi) = V_T(0, \theta, \phi)/N(E/\Delta T) = \frac{1}{N}\sum_{n=1}^{N}\exp\{-\pi[\rho_c\eta_0\cos(\xi - \phi_n)]^2\} \tag{8.59}$$

Similarly, we have, for the vertical and the horizontal peak amplitude patterns, the following

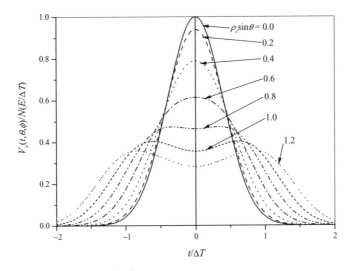

Figure 8.34 The time variation of the normalized voltage signal $V_V(t, \theta, \phi)$ for $\rho_c \sin \theta =$ 0, 0.2, 0.4, 0.6, 0.8, 1, and 1.2.

expressions

$$A_v(\theta, \phi) = V_v(0, \theta, \phi)/N(E/\Delta T) = \frac{1}{N} \sum_{n=1}^{N} \exp\{-\pi [\rho_c \sin \theta \cos (\phi - \phi_n)]^2\}$$

$$A_h(\phi) = \frac{1}{N} \sum_{n=1}^{N} \exp \left\{ -\pi \left[2\rho_c \sin \left(\frac{\phi - \phi_0}{2} \right) \cos \left(\frac{\pi + \phi_0 + \phi - 2\phi_n}{2} \right) \right]^2 \right\} \quad (8.60)$$

In a similar way, one can show that for sinusoidal waves with the time variation $\exp (j\omega t)$, the *Array factor*, AF, of an N-element circular array is given by [761, 762]:

$$AF(\theta, \phi) = \sum_{n=1}^{N} I_n \exp [j\rho_s \eta_0 \cos (\xi - \phi_n)] \quad (8.61)$$

where

$$\rho_s = 2\pi a/\lambda, \quad \lambda = \text{wavelength}, \quad (8.62)$$

I_n is the amplitude excitation of the array element n. For constant $I_n = I$ we have:

$$AF(\theta, \phi) = NI \sum_{n=1}^{N} J_{mN} (\rho_s \eta_0) \exp \left[jmN \left(\frac{\pi}{2} - \xi \right) \right] \quad (8.63)$$

where $J_x()$ is the Bessel function of the first kind. The vertical and the horizontal amplitude patterns are now given as:

$$AF_v(\theta, \phi) = NI \sum_{n=1}^{N} J_{mN} (\rho_s \sin \theta) \exp \left[jmN \left(\frac{\pi}{2} - \phi \right) \right] \quad (8.64)$$

$$AF_h(\phi) = NI \sum_{n=1}^{N} J_{mN} \left(2\rho_s \sin \frac{\phi}{2} \right) \exp \left[-\frac{jmN\phi}{2} \right]$$

For periodic sinusoidal signals, the square of the amplitude pattern, $AF(\theta, \phi)$, represents the average power pattern $P_{av}(\theta, \phi)$. We also define the vertical and horizontal power patterns as

$$P_{av,v}(\theta, \phi) = [AF_v(\theta, \phi)]^2$$
$$P_{av,h}(\theta, \phi) = [AF_h(\theta, \phi)]^2$$
(8.65)

The peak power pattern is defined as

$$P(\theta, \phi) = [A(\theta, \phi)]^2 = \frac{1}{N^2}\left[\sum_{n=1}^{N}\exp\{-\pi[\rho_c\eta_0\cos(\xi - \phi_n)]^2\}\right]^2$$
(8.66)

We also define the vertical and horizontal peak power patterns as

$$P_v(\theta, \phi) = [A_v(\theta, \phi)]^2$$
$$P_h(\phi) = [A_h(\phi)]^2$$
(8.67)

The average power or energy pattern is defined as

$$W(\theta, \phi) = \frac{\int\limits_{-\infty}^{\infty}[V_T(t, \theta, \phi)]^2\,dt}{\int\limits_{-\infty}^{\infty}[V_T(t, \theta_0, \phi_0)]^2\,dt}$$

$$= \frac{\int\limits_{-\infty}^{\infty}\left[\sum\limits_{n=1}^{N}\exp\{-\pi[(t/\Delta T) + \rho_c\eta_0\cos(\xi - \phi_n)]^2\}\right]^2\,dt}{N^2\int\limits_{-\infty}^{\infty}\exp\{-2\pi[(t/\Delta T)]^2\}dt}$$
(8.68)

As before, for the vertical and horizontal components, we have

$$W_v(\theta, \phi) = \frac{\int\limits_{-\infty}^{\infty}\left[\sum\limits_{n=1}^{N}\exp\left\{-\pi[(t/\Delta T) + \rho_c\sin\theta\cos(\phi - \phi_n)]^2\right\}\right]^2\,dt}{N^2\int\limits_{-\infty}^{\infty}\exp\left\{-2\pi[(t/\Delta T)]^2\right\}\,dt}$$
(8.69)

$$W_h(\theta, \phi)$$

$$= \frac{\int\limits_{-\infty}^{\infty}\left[\sum\limits_{n=1}^{N}\exp\left\{-\pi\left[(t/\Delta T) + 2\rho_c\sin\left(\frac{\phi - \phi_0}{2}\right)\cos\left(\frac{\phi + \phi_0 + \pi - 2\phi_n}{2}\right)\right]^2\right\}\right]^2\,dt}{N^2\int\limits_{-\infty}^{\infty}\exp\{-2\pi[(t/\Delta T)]^2\}dt}$$
(8.70)

Some numerical results are shown in Figures 8.35–8.44

A slope pattern $S(\theta, \phi)$, a *vertical slope pattern* $S_v(\theta, \phi)$ and a *horizontal slope pattern* $S_h(\phi)$ can be derived by using a linear regression algorithm to calculate the slope of the ramp that best fits the rising section of the Gaussian pulse $V_T(t, \theta, \phi)$, $V_v(t, \theta, \phi)$ and $V_h(t, \phi)$. A plot of the ramp slope versus angle results in a slope pattern.

More details on practical aspects of UWB antenna and receiver design can be found in [763–808].

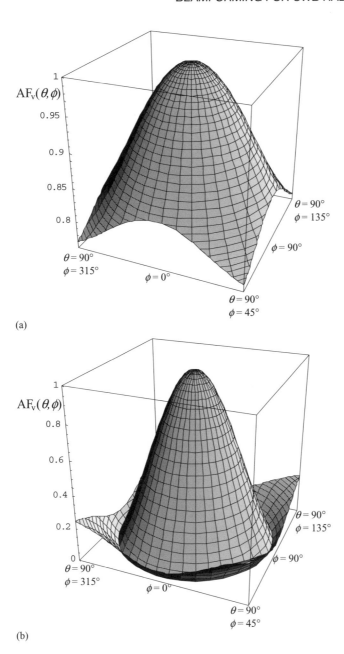

(a)

(b)

Figure 8.35 Normalized amplitude pattern $AF_v(\theta, \phi)$ given in Equation (8.65) for an infinitely extended periodic sinusoidal wave received by the circular array in Figure 8.33 with $N = 16$ elements and (a) $\rho_s = 1$; (b) $\rho_s = 3$; (c) $\rho_s = 6$; (d) $\rho_s = 12$.

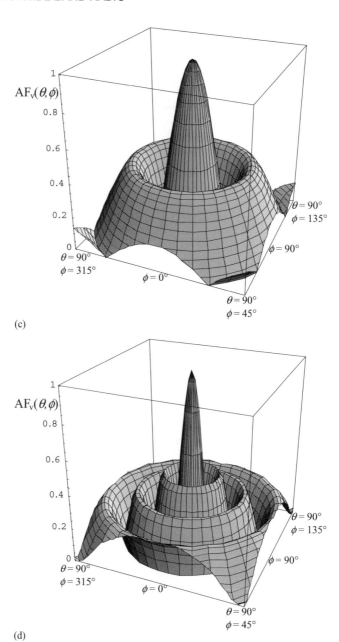

Figure 8.35 (*Cont.*).

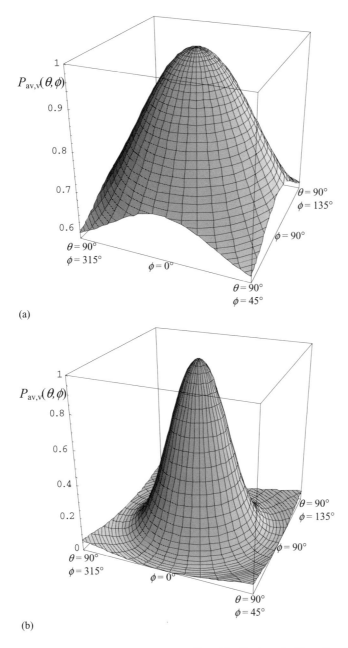

(a)

(b)

Figure 8.36 Normalized average power pattern $P_{av,v}(\theta, \phi)$ given in Equation (8.66) for an infinitely extended periodic sinusoidal wave received by the circular array in Figure 8.33 with $N = 16$ elements and (a) $\rho_s = 1$; (b) $\rho_s = 3$; (c) $\rho_s = 6$; (d) $\rho_s = 12$.

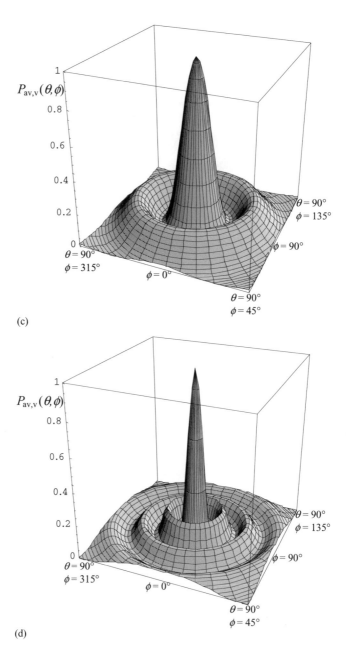

(c)

(d)

Figure 8.36 (*Cont.*).

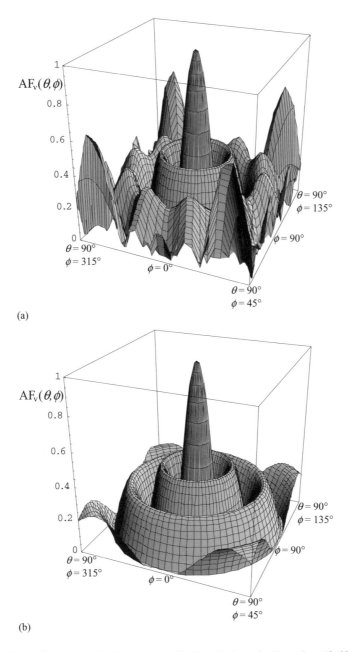

(a)

(b)

Figure 8.37 Normalized amplitude pattern $AF_v(\theta, \phi)$ given in Equation (8.65) for an infinitely extended periodic sinusoidal wave received by the circular array in Figure 8.33 with $\rho_s = 12$ and (a) $N = 10$ elements; (b) $N = 32$ elements.

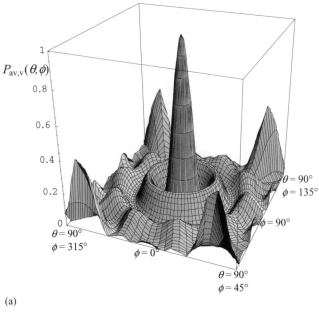

(a)

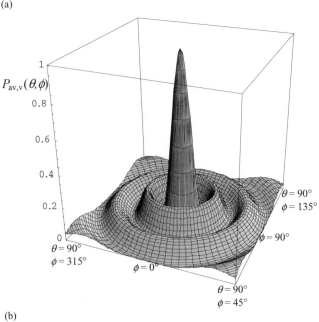

(b)

Figure 8.38 Normalized average power pattern $P_{av,v}(\theta, \phi)$ given in Equation (8.66) for an
infinitely, extended periodic sinusoidal wave received by the circular array in
Figure 8.33 with $\rho_s = 12$ and (a) $N = 10$ elements; (b) $N = 32$ elements.

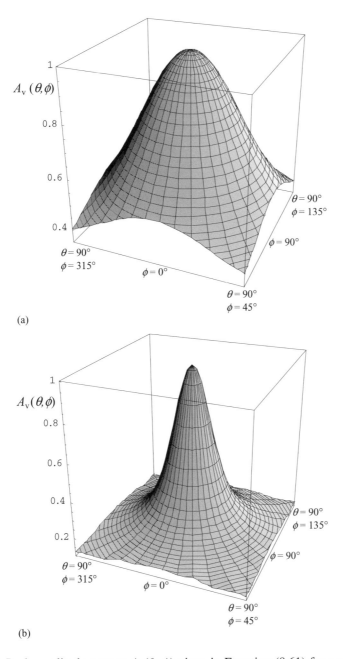

Figure 8.39 Peak amplitude pattern $A_v(\theta, \phi)$ given in Equation (8.61) for non-sinusoidal
Gaussian pulses received by the circular array in Figure 8.33 with $N = 16$
elements and (a) $\rho_s = 1$; (b) $\rho_s = 3$; (c) $\rho_s = 6$; (d) $\rho_s = 12$.

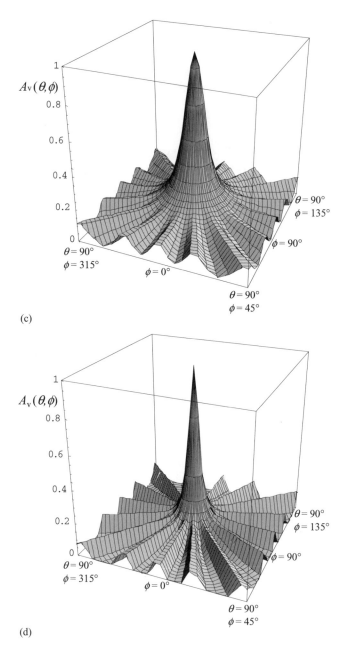

(c)

(d)

Figure 8.39 (*Cont.*).

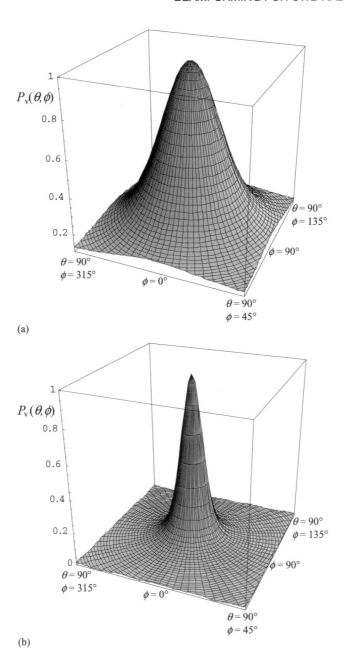

(a)

(b)

Figure 8.40 Peak power pattern $P_v(\theta, \phi)$ given in Equation (8.68) for non-sinusoidal Gaussian pulses received by the circular array in Figure 8.33 with $N = 16$ elements and (a) $\rho_s = 1$; (b) $\rho_s = 3$; (c) $\rho_s = 6$; (d) $\rho_s = 12$.

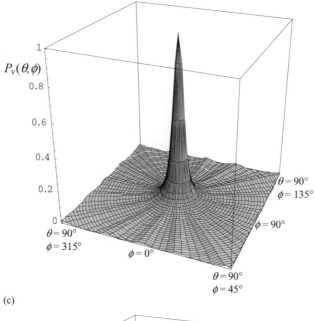

(c)

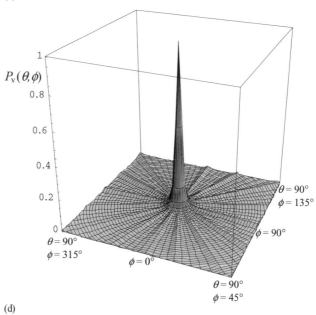

(d)

Figure 8.40 (*Cont.*).

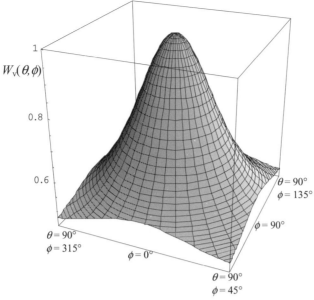

(a)

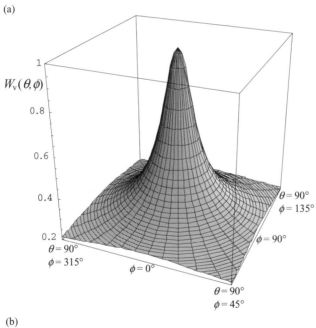

(b)

Figure 8.41 Energy pattern $W_v(\theta, \phi)$ given in Equation (8.70) for non-sinusoidal Gaussian pulses received by the circular array in Figure 8.33 with $N = 16$ elements and (a) $\rho_s = 1$; (b) $\rho_s = 3$; (c) $\rho_s = 6$; (d) $\rho_s = 12$.

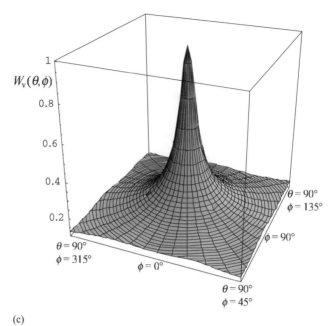

(c)

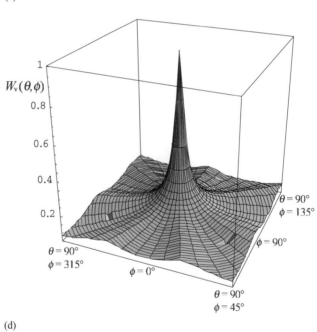

(d)

Figure 8.41 (*Cont.*).

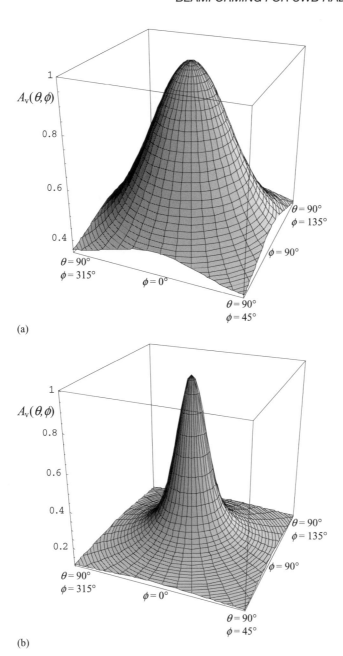

(a)

(b)

Figure 8.42 Peak amplitude pattern $A_v(\theta, \phi)$ for non-sinusoidal Gaussian pulses received by the circular array with a large number ($N \to \infty$) of elements and (a) $\rho_s = 1$; (b) $\rho_s = 3$; (c) $\rho_s = 6$; (d) $\rho_s = 10$.

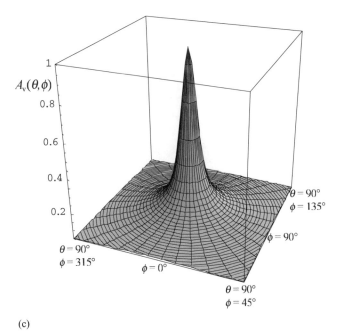

(c)

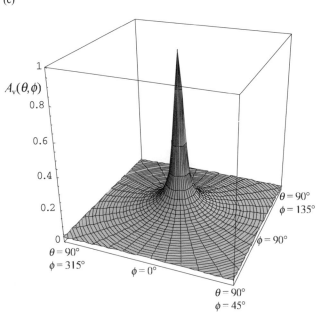

(d)

Figure 8.42 (*Cont.*).

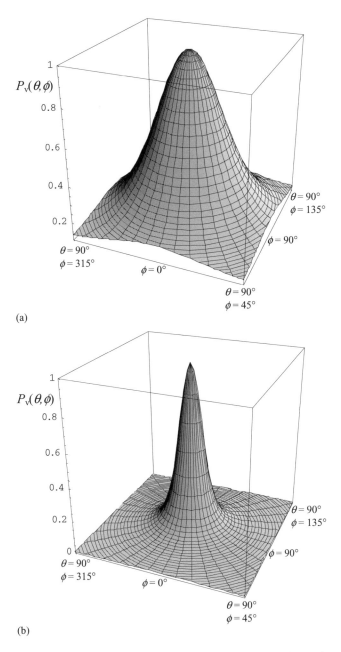

(a)

(b)

Figure 8.43 Peak power pattern $P_v(\theta, \phi)$ for non-sinusoidal Gaussian pulses received by the circular array with a large number ($N \to \infty$) of elements and (a) $\rho_s = 1$; (b) $\rho_s = 3$; (c) $\rho_s = 6$; (d) $\rho_s = 10$.

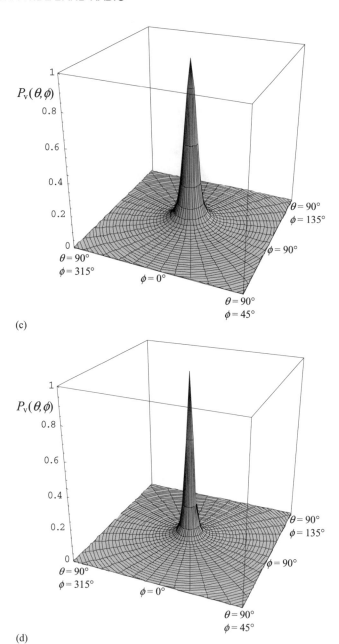

(c)

(d)

Figure 8.43 (*Cont.*).

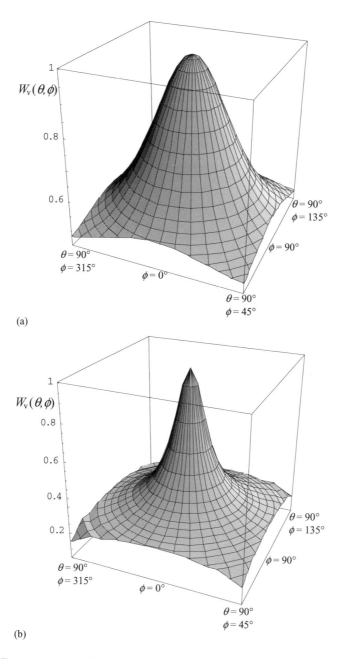

(a)

(b)

Figure 8.44 Energy pattern $W_v(\theta, \phi)$ for non-sinusoidal Gaussian pulses received by the circular array with a large number ($N \to \infty$) of elements and (a) $\rho_s = 1$; (b) $\rho_s = 3$; (c) $\rho_s = 6$; (d) $\rho_s = 10$.

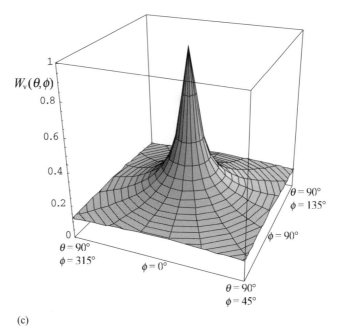

(c)

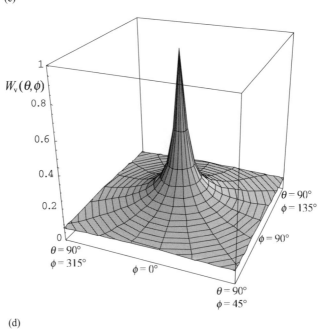

(d)

Figure 8.44 (*Cont.*).

III

Transceiver Integration

9

Antenna Array Signal Processing

9.1 SPACE–TIME RECEIVERS FOR CDMA COMMUNICATIONS

In this section we discuss the behavior of a MIMO system in a CDMA network.

9.1.1 Space–time signal model

We assume that only $M > 1$ antenna elements are used at the base station. The received continuous time $M \times 1$ signal vector can be expressed as

$$\mathbf{x}(t) = \sum_{q=1}^{Q} \mathbf{x}^q(t) + \mathbf{n}(t) \tag{9.1}$$

where Q is the number of users. The signal received from the user with index 'q' can be represented as

$$\mathbf{x}^q(t) = \sum_{n=-\infty}^{\infty} d^q(n)\mathbf{g}^q(t - nT_c) \tag{9.2}$$

with

$$\mathbf{g}^q(t) = \sum_{l=0}^{D-1} \alpha_l^q \mathbf{a}_l^q a(t - \tau_l^q) \tag{9.3}$$

where $a(t)$ is the chip pulse. D is the maximum number of paths for any of the users, each with complex amplitude and delay α_l^q and τ_l^q. $\mathbf{a}_l^q$ corresponds to the array response vector

Advanced Wireless Communications. S. Glisic
© 2004 John Wiley & Sons, Ltd. ISBN: 0-470-86776-0

for the *l*th path and

$$d^q(n) = s^q(k)\, c^q(k,\ n - kP)$$

$$\text{with} \quad k = \left[\frac{n}{P}\right]$$

(9.4)

$s^q(k)$ and $c^q(k)$ are the symbol and the code of the user q.

9.1.2 Assumptions

From here on we will assume:

- The inverse signal bandwidth is large compared to the travel time across the array. Therefore, the complex envelopes of the signals from a given path received by different antennas are identical except for phase and amplitude differences that depend on the path angle-of-arrival, array geometry and the element pattern. Thus, the use of multiple antennas at the receiver has merely converted a scalar channel $g^q(t)$ to a vector channel $\mathbf{g}^q(t)$.

- The channel responses $\{\mathbf{g}^q(t)\}$ are assumed to be unknown and completely arbitrary (with the exception of the pulse shape $a(t)$ in the blind RAKE receiver) and only knowledge of the spreading code for the user of interest is available.

- Linear (single user or multiuser) receivers are used.

- Blind methods exploiting solely the second order statistics of $\mathbf{x}(t)$ are used.

- The problem is to estimate the symbols $s^q(k)$ transmitted by a single or multiple users given a minimal amount of signal/channel information available at the receiver.

- Existing approaches to blind CDMA differ mostly on
  - the degree of *a priori* signal/channel information and the generality of the channel model: τ_l^q known or unknown, synchronized or unsynchronized users, delay spread smaller or larger than a chip period, a symbol period;
  - The receiver's structure: single-user or multiuser, linear or non-linear;
  - The estimation method of the receiver: use of finite alphabet information, second-order statistics or higher order statistics.

9.1.3 MIMO equivalent model

Another interesting way to represent the CDMA is to consider the stream of users' symbols $\{s^q(t)\}_{q=1...Q}$ as the signals driving a $Q \times M$ multiple inputs multiple outputs (MIMO) system.

Vector output $\mathbf{x}(t)$ can be represented as

$$\mathbf{x}(t) = \sum_{q=1}^{Q} \sum_{k=-\infty}^{\infty} s^q(t) \mathbf{h}^q(k, t - kT_s) + \mathbf{n}(t)$$

(9.5)

where $T_s = PT_c$ denotes the symbol period and $\mathbf{h}^q(k, t) = [h_1^q(k, t), \ldots, h_M^q(k, t)]^T$ is the $M \times 1$ impulse response of the channel from user q to the receive antenna array. Each symbol experiences a different channel in the case of aperiodic codes, hence the symbol index k in $\mathbf{h}^q(k, t)$. The channel response is given by

$$\mathbf{h}^q(k,\ t) = \sum_{j=0}^{P-1} c^q(k,\ j)\,\mathbf{g}^q(t-jT_c) \quad q = 1, \ldots, Q \tag{9.6}$$

In the particular case of periodic codes, the MIMO channel becomes *time-invariant*

$$\mathbf{x}(t) = \sum_{q=1}^{Q} \sum_{k=-\infty}^{\infty} s^q(t)\,\mathbf{h}^q(k,\ t-kT_s) + \mathbf{n}(t)$$

$$\mathbf{h}^q(t) = \sum_{j=0}^{P-1} c^q(j)\,\mathbf{g}^q(t-jT_c)$$

$$\forall k \quad q = 1, \ldots, Q \tag{9.7}$$

9.1.4 Single-user receiver with space–time RAKE

For the transmitted signature waveform of the qth user

$$p^q(k,\ t) = \sum_{j=0}^{P-1} c^q(k,\ j)\,a^q(t-jT_c) \tag{9.8}$$

the output of the coherent ST RAKE becomes

$$\hat{s}^q(k) = \mathrm{sign}\left\{ \sum_l \alpha_l^{q^*} \mathbf{a}_l^{q^*} \int_{kT_s+\tau_l}^{(k+1)T_s+\tau_l} \mathbf{x}(t)p^q(k,\ t-kT_s-\tau_l)\,\mathrm{d}t \right\} \tag{9.9}$$

9.1.5 Blind ST RAKE principle component method

The technique is based on the second-order statistics of both the precorrelation data $\mathbf{x}(t)$ and postcorrelation data $\mathbf{y}_l^q(k)$ where $\mathbf{y}_l^q(k)$ is defined for the lth path from the qth user by

$$\mathbf{y}_l^q(k) = \int_{kT_s+\tau_l}^{(k+1)T_s+\tau_l} \mathbf{x}(t)\,p^q(k,\ t-kT_s-\tau_l)\,\mathrm{d}t \tag{9.10}$$

Let $\mathbf{R}_x = E(\mathbf{x}(t)\mathbf{x}(t)^*)$ and $\mathbf{R}_x^q = E(\mathbf{y}(k)\mathbf{y}(k)^*)$.

$$\mathbf{R}_x = \varepsilon_l^q \left|\alpha_l^q\right|^2 \mathbf{a}_l^q \mathbf{a}_l^{q^*} + \mathbf{R}_{u,l}^q$$

$$\mathbf{R}_x^q = P\varepsilon_l^q \left|\alpha_l^q\right|^2 \mathbf{a}_l^q \mathbf{a}_l^{q^*} + \mathbf{R}_{u,l}^q \tag{9.11}$$

where ε_l^q denotes the signal power in the desired path and $\mathbf{R}_{u,l}^q$ denotes the correlation matrix of all other paths and users' signals.

The covariance of the interfering parts $\mathbf{R}_{u,l}^q$ remains the same before and after despreading. The path response $\alpha_l^q\,\mathbf{a}_l^{q^*}$ can be found up to a phase component as the dominant eigenvector of $\mathbf{R}_l^q - \mathbf{R}_x$. A problem with this approach is that the phase is lost. τ_l^q can be found by solving the problem above for a range of delays τ and selecting those offering the largest dominant eigenvalues of $\mathbf{R}_l^q - \mathbf{R}_x$. This is a computationally extensive and not necessarily robust approach.

9.1.6 Space–time equalizer

The coefficients of the chip rate N-long impulse response of the equalizer on the mth antenna will be denoted as

$$\left(f_m^q(0), \ldots, f_m^q(N-1)\right) \tag{9.12}$$

So, the equalizer output is

$$\hat{d}^q(n) = \sum_{m=1}^{M} \sum_{i=0}^{N-1} f_m^q(i)^* x_m((n-i)T_c) \tag{9.13}$$

Subsequent symbol estimates can be represented as

$$\hat{s}^q(k) = \sum_{j=0}^{P-1} c^q(k, j)\hat{d}^q(kP + j) \tag{9.14}$$

9.1.7 The MMSE chip equalizer

In this case the equalizer coefficients are defined by

$$\{f_m^q\} = \arg\min E|d^q(n) - \hat{d}^q|^2 \tag{9.15}$$

The solution is the Wiener filter (see Chapter 6):

$$\mathbf{R}_{N,x}\mathbf{f} = \mathbf{r}_N \tag{9.16}$$

where $\mathbf{R}_{N,x}$ is the TS correlation matrix of $\{\mathbf{x}(nT_c), \ldots, \mathbf{x}((n-N+1)T_c)\}$, $\mathbf{f}$ is the unknown ST equalizer vector of coefficients, and $\mathbf{r}_N$ is the unknown crosscorrelation (channel) vector.

9.1.8 Multiuser receivers

A total of MP samples are collected per symbol to create a vector

$$\mathbf{x}(k) = [\mathbf{x}^\mathrm{T}(kT_s), \mathbf{x}^\mathrm{T}(kT_s + T_c), \ldots, \mathbf{x}^\mathrm{T}(kT_s + (P-1)T_c)]^\mathrm{T} \tag{9.17}$$

The periodic codes are used, so that the time invariant MIMO model is used. It is assumed that $(L+1)T_s$ is the unrestricted maximum length of channels for the users. Hence, the degree of the vector channels is at most L. $L = 1$ is a typical value when considering practical CDMA data rates.

A severe asynchronism among the users can result in the case when $L = 2$. The received signal vector can be represented as

$$\mathbf{x}(k) = \sum_{q=1}^{Q} \mathbf{H}^q \mathbf{s}^q(k) + \mathbf{n}(k) \tag{9.18}$$

where $MP(L+1)$ dimensional matrix $\mathbf{H}^q$ is the channel matrix for user q. Its $(pM + m)$th row with $1 \le m \le M$ and $0 \le p \le P - 1$ is

$$\mathbf{H}^q(pM + m) = \left(h_m^q(pT_c), h_m^q(pT_c + T_s), \ldots, h_m^q(pT_c + LT_s)\right)$$
$$\mathbf{s}^q(k) \equiv (s^q(k), s^q(k-1), \ldots, s^q(k-L))^\mathrm{T} \tag{9.19}$$

Let the length of the linear MU (LMU) filter be N. Then we form

$$\mathbf{X}(k) = (\mathbf{x}^{\mathrm{T}}(k), \mathbf{x}^{\mathrm{T}}(k-1), \dots, \mathbf{x}^{\mathrm{T}}(k-N+1))^{\mathrm{T}} \tag{9.20}$$

In matrix form, Equation (9.18) can be now written as

$$\mathbf{X}(k) = \sum_{q=1}^{Q} \begin{bmatrix} \mathbf{H}^q & & \mathbf{0} \\ & \mathbf{H}^q & \\ & & \ddots & \\ \mathbf{0} & & \mathbf{H}^q \end{bmatrix} \begin{bmatrix} s^q(k) \\ s^q(k-1) \\ \vdots \end{bmatrix} + \begin{bmatrix} \mathbf{n}(k) \\ \mathbf{n}(k-1) \\ \vdots \end{bmatrix} \tag{9.21}$$

$$= \sum_{q=1}^{Q} \mathfrak{S}^q \mathbf{S}^q(k) + \mathbf{N}(k)$$

The global multiuser channel matrix $\mathfrak{S} \equiv (\mathfrak{S}^1, \dots, \mathfrak{S}^Q)$ has dimension MNP times $Q(L+N)$ where $MP > Q$. Based on the above, $\mathfrak{S}$ can be made tall (more rows than columns). Given the channel matrices $\mathfrak{S}$ for all users $u = 1, \dots, Q$, the linear multiuser detector finds a vector $\mathbf{w}$ with MNP entries that satisfies

$$\mathbf{w}^* \mathfrak{S}^q = (0, \dots, 1, \dots, 0)$$
$$\mathbf{w}^* \mathfrak{S}^u = (0, \dots, 0), \quad u \neq q \tag{9.22}$$

9.1.9 Subspace-based techniques

The signal subspace span($\mathfrak{S}$) is first obtained from the dominant eigenvectors of

$$\mathcal{R} = E(\mathbf{X}(k)\mathbf{X}^*(k))$$

Next, the coefficients of $\mathfrak{S}^q$, $q = 1, \dots Q$ are adjusted to match the signal subspace.

9.1.10 MMSE receiver

In the presence of noise, the best (in the MMSE sense) linear receiver with delay δ for user q, is given by

$$\mathbf{w} = \arg \min E |\mathbf{w}^* \mathbf{X}(k) - s^q(k - \delta)|^2 \tag{9.23}$$

The vector $\mathbf{w}$ satisfies the classical Wiener equation

$$\mathcal{R}\mathbf{w} = \mathbf{r}^q \tag{9.24}$$

where $\mathcal{R} = E(\mathbf{X}(k)\mathbf{X}^*(k))$ denotes the ST received covariance matrix and $\mathbf{r}^q = E(\mathbf{X}(k)s^{q*}(k-\delta))$ is the crosscorrelation vector. The goal is to identify $\mathbf{w}$ blindly without knowledge of $\mathbf{r}^q$. Assuming i.i.d symbols, we have

$$\mathbf{r}^q = \begin{bmatrix} \mathbf{H}^q & & \mathbf{0} \\ & \mathbf{H}^q & \\ & & \ddots & \\ \mathbf{0} & & \mathbf{H}^q \end{bmatrix} \begin{bmatrix} 0 \\ \vdots \\ 1 \\ \vdots \\ 0 \end{bmatrix} \tag{9.25}$$

$\mathbf{r}^q$ coincides with the $(\delta + 1)$th column of $\mathfrak{S}$. If the detector delay δ is chosen appropriately (namely $L \leq \delta \leq N - 1$ assuming $L \leq N - 1$ the vector $\mathbf{r}^q$ contains all the $(L + 1)MP$ channel coefficients for user q, as well as $(N - L - 1)MP$ zeros. For each antenna m, there exists a simple selection-permutation matrix $\mathbf{T}_m$ that selects $(L + 1)P$ channel coefficients in $\mathbf{r}^q$ associated with this antenna, together with $(N - L - 1)P$ zeros. This transformation puts the selected entries in a chronologically ordered vector so that

$$\mathbf{T}_m \mathbf{r}^q = \left(h_m^q(0), \ldots, h_m^q((LP + P - 1)T_c), 0, \ldots, 0 \right)^{\mathrm{T}} \tag{9.26}$$

Assume $\delta = L$ holds true for any admissible value of the delay. If the mth antenna known spreading sequence is $\{c^q(n)\}$

$$\mathbf{T}_m \mathbf{r}^q = \begin{bmatrix} c^q(0) & & 0 \\ \vdots & \ddots & \\ c^q(P-1) & & c^q(0) \\ & \ddots & \vdots \\ 0 & & c^q(P-1) \\ \vdots & & \\ 0 & & 0 \end{bmatrix} \begin{bmatrix} g_m^q(0) \\ g_m^q(T_c) \\ \vdots \\ g_m^q(LPT_c) \end{bmatrix} \equiv \mathbf{C}^q \mathbf{g}_m^q \tag{9.27}$$

$\mathbf{C}^q$ of dimension $PN(LP + 1)$ which can be made tall by selecting $N > L$ *is known to the receiver*. In contrast, $\mathbf{g}_m^q$ is unknown. $\mathbf{g}_m^q$ contains a certain number of zero entries. The knowledge of this number is not necessary, the knowledge of the actual channel length in chips is not required either. We introduce $\mathbf{U}^q$, an orthonormal basis for the orthogonal complement of $\mathbf{C}^q \cdot \mathbf{U}^q$ is of size $PN(P(N - L) - 1)$. So we have

$$\mathbf{U}^{q*}\mathbf{T}_m \mathcal{R}\mathbf{w} = \underbrace{\mathbf{U}^{q*}\mathbf{C}^q}_{0} \mathbf{g}_m^q = 0, \quad m = 0, \ldots, M \tag{9.28}$$

This gives a first set of $M(P(N - L) - 1)$ equations for $\mathbf{w}$.

Additive noise is white with variance σ^2. So, the correlation matrix of the received signal is

$$\mathcal{R} = \mathfrak{S}\mathfrak{S}^* + \sigma^2 \mathbf{I} \tag{9.29}$$

where $\mathbf{I}$ is the identity matrix of size MNP. The signal/noise subspace decomposition of $\mathcal{R}$ gives

$$\mathcal{R} = \mathbf{E}_s \Sigma_s \mathbf{E}_s^* + \sigma^2 \mathbf{E}_n \mathbf{E}_n^* \tag{9.30}$$

The columns of $\mathbf{E}_s$ (respectively $\mathbf{E}_n$) are given by the $Q(N + L)$ dominant eigenvectors (respectively the $MNP - Q(N + L)$ least dominant eigenvectors of $\mathcal{R}$. The desired MMSE detector always lies in the signal subspace defined by $\mathrm{span}(\mathfrak{S}) = \mathrm{span}(\mathbf{E}_s)$

$$(\mathfrak{S}\mathfrak{S}^* + \sigma^2 \mathbf{I})\mathbf{w} = \mathbf{r}^q$$
$$\mathbf{w} = (\mathbf{r}^q - \mathfrak{S}\mathfrak{S}^*\mathbf{w})/\sigma^2 \tag{9.31}$$

$\mathbf{r}^q$ is by construction in $\mathrm{span}(\mathfrak{S}^q)$, and hence in $\mathrm{span}(\mathfrak{S})$. $\mathbf{w}$ also belongs to $\mathrm{span}(\mathfrak{S})$ so we have

$$\mathbf{E}_n^* \mathbf{w} = 0 \tag{9.32}$$

This gives us an additional $MNP - Q(N + L)$ equations towards the determination of $\mathbf{w}$.

9.1.11 Algorithm

From Equations (9.28) and (9.32) $\mathbf{w}$ can be written as

$$\mathbf{Aw} = \begin{bmatrix} \mathbf{A}_1 \\ \mathbf{A}_2 \end{bmatrix} \mathbf{w} = 0 \tag{9.33}$$

$$\begin{aligned} \mathbf{A}_2 &= \mathbf{E}_n^* \\ \mathbf{A}_1 &= \boldsymbol{\mathcal{X}}^{q*}\mathbf{TR} \text{ with } \mathbf{T} = (\mathbf{T}_1^*, \ldots, \mathbf{T}_M^*)^* \end{aligned} \tag{9.34}$$

$$\boldsymbol{\mathcal{X}}^{q*} = \begin{bmatrix} \mathbf{U}^q & 0 & 0 \\ 0 & \mathbf{U}^q & \\ & & \ddots & \\ 0 & & & \mathbf{U}^q \end{bmatrix} \tag{9.35}$$

To determine $\mathbf{w}$ in a noisy situation, the algorithm optimizes the following simple quadratic cost function

$$J_{\text{MMSE}}(\mathbf{w}) = \mathbf{w}^* \left\{ \mathbf{A}_1^* \mathbf{A}_1 + \alpha \mathbf{A}_2^* \mathbf{A}_2 \right\} \mathbf{w} \tag{9.36}$$

where α is a tunable weight. Some additional constraint must be added in order to avoid trivial solutions. In the particular case of a unit-norm constraint, $\mathbf{w}$ is found as the minimum eigenvector of

$$\mathbf{A}_1^* \mathbf{A}_1 + \alpha \mathbf{A}_2^* \mathbf{A}_2 \tag{9.37}$$

Note that while $\boldsymbol{\mathcal{R}}$ is the estimated sample averaging, $\mathbf{T}$ and $\boldsymbol{\mathcal{X}}^q$ can be precomputed. Some results for blind MMSE and MOE (described in Chapter 5) are shown in Figures 9.1–9.3. More results on blind space–time signal estimation can be found in [809–826].

9.2 MUSIC AND ESPRIT DOA ESTIMATION

In this section we discuss the specific problem of estimating the direction of arrival (DOA) of a signal [827–867]. The problem is important for an efficient space–time receiver design, as well as for positioning, as will be discussed in Chapter 13. Within this section we will focus on MUSIC (MUltiple SIgnal Classifications) and ESPRIT (Estimation of Signal Parameters via Rotational Invariance Techniques) as presented in [827].

9.2.1 Signal parameters

We assume that all signals have the same known center frequency (ω_0). The ith signal can be written as $\tilde{s}_i(t) = u_i(t) \cos(\omega_0 t + v_i(t))$. $u_i(t)$ and $v_i(t)$ are slowly varying functions of

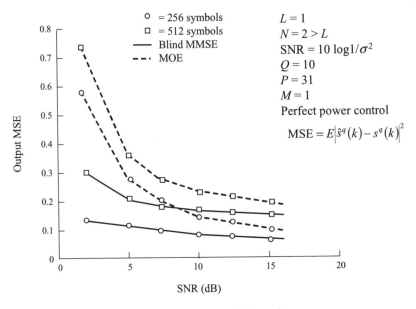

Figure 9.1 Performance results for MMSE and MOE receivers.

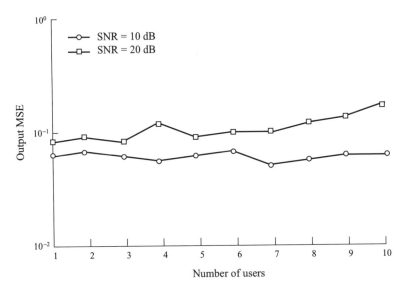

Figure 9.2 MSE as a function of user index for imperfect power control (user 1 has power 10 dB above user 10). The power (in dB) decreases linearly with user index. The SNR refers to the strongest user and 512 symbols are used in the estimation ($Q = 10$, $P = 9$, $M = 4$).

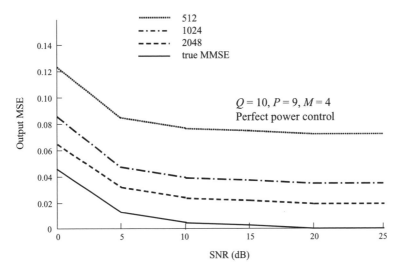

Figure 9.3 Performance results for different numbers of symbols used for estimation.

time. The signal can be represented as

$$\tilde{s}_i(t) = \mathrm{Re}\{s(t)\} \tag{9.38}$$

where $s(t) = u(t)e^{j(\omega_0 t + v(t))}$

The narrowband assumption implies $u(t) \approx u(t - \tau)$ and $v(t) \approx v(t - \tau)$ for all possible propagation delays τ. The effect of a time delay on the received waveforms is simply a phase shift, i.e. $s(t - \tau) \approx s(t)e^{-j\omega_0\tau}$.

$u_k(t)$ is the complex signal output of the kth sensor. For d point sources we have

$$\begin{aligned}
x_k(t) &= \sum_{i=1}^{d} a_k(\theta_i)s_i(t - \tau_k(\theta_i)) \\
&= \sum_{i=1}^{d} a_k(\theta_i)s_i(t)e^{-j\omega_0\tau_k(\theta_i)}
\end{aligned} \tag{9.39}$$

where $a_k(\theta_i)$ is the sensor element complex response (gain and phase). The sensor geometry is shown in Figure 9.4.

For m sensors, the overall received signal can be represented as

$$x(t) = \sum_{i=1}^{d} \mathbf{a}(\theta_i)s_i(t) \tag{9.40}$$

where

$$\mathbf{a}(\theta_i) = \left\lfloor a_l(\theta_i)e^{-j\omega_0\tau_l(\theta_i)}, \ldots, a_m(\theta_i)e^{-j\omega_0\tau_m(\theta_i)} \right\rfloor^{\mathrm{T}} \tag{9.41}$$

often termed the array response or array **a** steering vector for direction θ_i. By using vector notation the overall signal becomes

$$\mathbf{x}(t) = \mathbf{A}(\theta)\mathbf{s}(t) + \mathbf{n}(t) \tag{9.42}$$

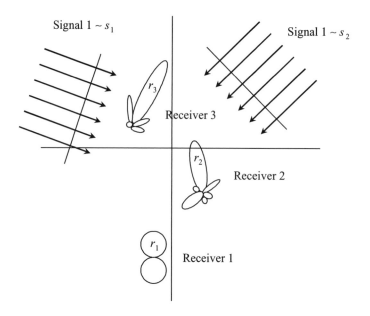

Figure 9.4 Passive sensor array geometry.

where $\mathbf{x}(t)$, $\mathbf{n}(t) \in \mathbb{C}^m$, $s(t) \in \mathbb{C}^d$, and $\mathbf{A}(\theta) \in \mathbb{C}^{m \times d}$, and it will be assumed that $m > d$ and

$$\mathbf{A}(\theta) = [\mathbf{a}(\theta_1), \ldots, \mathbf{a}(\theta_d)]$$

$$s(t) = [s_1(t), \ldots, s_d(t)]^{\mathrm{T}}$$

(9.43)

It is necessary to assume that the map from $\theta = \{\theta_1, \ldots, \theta_d\}$ to $\mathcal{R}\{\mathbf{A}(\theta)\}$, the subspace spanned by the columns of $\mathbf{A}(\theta)$, is one-to-one. This property can be ensured by proper array design.

9.2.2 MUSIC (MUltiple Signal Classification) algorithm

If $\mathbf{x}(t) = \mathbf{a}(\theta) s_\theta (t)$ is an appropriate data model (in the absence of noise) for a single signal, the data are confined to a *one-dimensional subspace* of $\mathbb{C}^m$ characterized by the vector $\mathbf{a}(\theta)$. For d signals, the observed data vector $\mathbf{x}(t) = \mathbf{A}(\theta)s(t)$ are constrained to the d-dimensional subspace of $\mathbb{C}^m$, termed the *signal subspace* ($\mathbb{S}_X$), that is spanned by the d vectors $\mathbf{a}(\theta_i)$, the columns of $\mathbf{A}(\theta)$. Let us define

$$\mathbf{R}_{SS} \overset{\text{def}}{=} \lim_{N \to \infty} \frac{1}{N} \sum_{t=1}^{N} s(t)s^*(t)$$

$$\mathbf{R}_{XX} \overset{\text{def}}{=} E\{xx^*\} = \mathbf{A}\mathbf{R}_{SS}\mathbf{A}^* + \sigma^2 \sum_{n}$$

(9.44)

The objective is to find a set of d linearly independent vectors that is contained in $\mathbb{S}_X = \mathcal{R}\{\mathbf{A}\}$, where *signal subspace* ($\mathbb{S}_X$) is spanned by the columns of $\mathbf{A}(\theta)$. The algorithm can be summarized as follows.

9.2.2.1 Summary of the MUSIC algorithm

1. Collect the data and estimate $\mathbf{R}_{XX} = E\{xx^*\} = \mathbf{A}\mathbf{R}_{SS}\mathbf{A}^* + \sigma^2\Sigma_n$ denoting the estimate $\hat{\mathbf{R}}_{XX}$.

2. Solve for the eigensystem; $\hat{\mathbf{R}}_{XX}\bar{\mathbf{E}} = \Sigma_n\bar{\mathbf{E}}\Lambda$, where $\Lambda = \mathrm{diag}\{\lambda_1, \ldots, \lambda_m\}$, $\lambda_1 \geq \ldots \geq \lambda_m$, and $\bar{\mathbf{E}} = [\mathbf{e}_1 | \ldots | \mathbf{e}_m]$ is the matrix of eigenvectors.

3. Estimate the number of sources $\hat{d}$ by identifying $\hat{d}$ significant eigenvalues.

4. Evaluate

$$P_M(\theta) = \frac{\mathbf{a}^*(\theta)\mathbf{a}(\theta)}{\mathbf{a}^*(\theta)\mathbf{E}_N\mathbf{E}_N^*\mathbf{a}(\theta)} \tag{9.45}$$

where $\mathbf{E}_N = \Sigma_n[\mathbf{e}_{d+1}| \ldots |\mathbf{e}_m]$

5. Find the d (largest) peaks of $P_M(\theta)$ to obtain estimates of the parameters. For this, knowledge of the array manifold is required.

9.2.3 ESPRIT array modeling

In this case we describe the array as being comprised of two subarrays, Z_X and Z_Y, identical in every respect although physically displaced (not rotated) from each other by a known displacement vector Δ of magnitude Δ as shown in Figure 9.5. The signals received at the

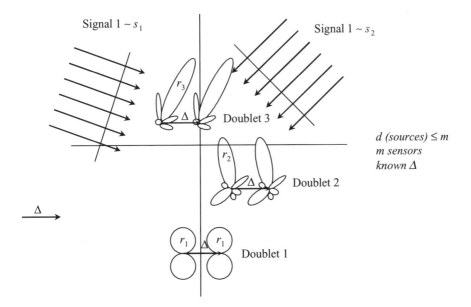

Figure 9.5 Array geometry for the ESPRIT algorithm.

*i*th doublet can be expressed as

$$x_i(t) = \sum_{k=1}^{d} s_k(t) a_i(\theta_k) + n_{xi}(t)$$

$$y_i(t) = \sum_{k=1}^{d} s_k(t) e^{j\omega_0 \Delta \sin \theta_k / c} a_i(\theta_k) + n_{yi}(t)$$

(9.46)

where θ_k is the direction of arrival of the *k*th source relative to the direction of the translation displacement vector $\mathbf{\Delta}$. The DOA estimates obtained are angles-of-arrival with respect to the direction of the vector $\mathbf{\Delta}$. Now we have

$$\mathbf{x}(t) = \mathbf{A}\mathbf{s}(t) + \mathbf{n}_x(t)$$

$$\mathbf{y}(t) = \mathbf{A}\mathbf{\Phi}\mathbf{s}(t) + \mathbf{n}_y(t)$$

(9.47)

where $\mathbf{s}(t)$ is the $d \times 1$ vector,

$$\mathbf{\Phi} = \text{diag}\{e^{j\gamma_1}, \ldots, e^{j\gamma_d}\}$$

$$\gamma_k = \omega_0 \Delta \sin \theta_k / c$$

$$\mathbf{z}(t) = \begin{bmatrix} \mathbf{x}(t) \\ \mathbf{y}(t) \end{bmatrix} = \bar{\mathbf{A}}\mathbf{s}(t) + \mathbf{n}_z(t)$$

$$\bar{\mathbf{A}} = \begin{bmatrix} \mathbf{A} \\ \mathbf{A}\mathbf{\Phi} \end{bmatrix}, \mathbf{n}_z(t) = \begin{bmatrix} \mathbf{n}_x(t) \\ \mathbf{n}_y(t) \end{bmatrix}$$

(9.48)

It is the structure of $\bar{\mathbf{A}}$ that is exploited to obtain estimates of the diagonal elements of $\mathbf{\Phi}$ *without having to know* $\mathbf{A}$.

One can see that the estimation problem posed is scale-invariant in the sense that absolute signal powers are not observable. For any non-singular diagonal matrix, $\mathbf{D}$, the data model, Equation (9.48), is invariant with respect to transformations

$$\mathbf{s}(t) \rightarrow \mathbf{D}^{-1}\mathbf{s}(t) \quad \text{and} \quad \bar{\mathbf{A}} \rightarrow \bar{\mathbf{A}}\mathbf{D}$$

(9.49)

Thus, estimates of the signals and the associated array manifold vectors derived herein are to be interpreted modulo an arbitrary scale factor, unless knowledge of the gain pattern of one of the sensors is available. The basic idea behind ESPRIT is to exploit the rotational invariance of the underlying signal subspace induced by the translational invariance of the sensor array. Simultaneous sampling of the output of the arrays leads to two sets of vectors, $\mathbf{E}_X$ and $\mathbf{E}_Y$, that span the same signal subspace (ideally, that spanned by the columns of $\mathbf{A}$). The ESPRIT algorithm is based on the following results for the case in which the underlying $2m$-dimensional signal subspace containing the entire array output is known. In the absence of noise, the signal subspace can be obtained as before by collecting a sufficient number of measurements and finding *any set of d linearly independent measurement vectors*. These vectors span the *d*-dimensional subspace of $\mathbb{C}^m$ spanned by $\bar{\mathbf{A}}$.

The signal subspace can also be obtained from knowledge of the covariance of the measurements

$$\mathbf{R}_{ZZ} = \bar{\mathbf{A}}\mathbf{R}_{SS}\bar{\mathbf{A}}^* + \sigma^2 \mathbf{\Sigma}_n$$

(9.50)

If $d \leq m$, the $2m-d$ smallest GEVs of $(\mathbf{R}_{ZZ}, \mathbf{\Sigma}_n)$ are equal to σ^2. The *d* GEVs corresponding to the *d* largest GEs are used to obtain $\mathbf{E}_s = \mathbf{\Sigma}_n[\mathbf{e}_1|\ldots|\mathbf{e}_d]$, where $\mathcal{R}\{\mathbf{E}_s\} = \mathcal{R}\{\bar{\mathbf{A}}\}$ is the signal subspace. Since $\mathcal{R}\{\mathbf{E}_s\} = \mathcal{R}\{\bar{\mathbf{A}}\}$, there must exist a unique (recall $d \leq m$), non-singular $\mathbf{T}$ such that

$$\mathbf{E}_s = \bar{\mathbf{A}}\mathbf{T} \qquad (9.51)$$

The invariance structure of the array implies $\mathbf{E}_s$ can be decomposed into $\mathbf{E}_X \in \mathbb{C}^{m \times d}$ and $\mathbf{E}_Y \in \mathbb{C}^{m \times d}$ (cf. Z_X and Z_Y subarrays) such that

$$\mathbf{E}_s = \begin{bmatrix} \mathbf{E}_X \\ \mathbf{E}_Y \end{bmatrix} = \begin{bmatrix} \mathbf{A}\mathbf{T} \\ \mathbf{A}\mathbf{\Phi}\,\mathbf{T} \end{bmatrix}$$
$$\mathcal{R}\{\mathbf{E}_X\} = \mathcal{R}\{\mathbf{E}_Y\} = \mathcal{R}\{\mathbf{A}\} \qquad (9.52)$$

Since $\mathbf{E}_X$ and $\mathbf{E}_Y$ share a common column space, the rank of

$$\mathbf{E}_{XY} \stackrel{\text{def}}{=} [\mathbf{E}_X \mid \mathbf{E}_Y] \qquad (9.53)$$

is d, which implies there exists a unique (recall $d \leq m$) rank d matrix $\mathbf{F} \in \mathbb{C}^{2d \times d}$ such that

$$0 = [\mathbf{E}_X | \mathbf{E}_Y]\,\mathbf{F} = \mathbf{E}_X\mathbf{F}_X + \mathbf{E}_Y\mathbf{F}_Y$$
$$= \mathbf{A}\mathbf{T}\mathbf{F}_X + \mathbf{A}\mathbf{\Phi}\,\mathbf{T}\mathbf{F}_Y \qquad (9.54)$$

$\mathbf{F}$ spans the null-space of $[\mathbf{E}_X \mid \mathbf{E}_Y]$. Defining

$$\mathbf{\Psi} \stackrel{\text{def}}{=} -\mathbf{F}_X[\mathbf{F}_Y]^{-1}$$
$$\mathbf{A}\mathbf{T}\mathbf{\Psi} = \mathbf{A}\mathbf{\Phi}\,\mathbf{T} \Rightarrow \mathbf{A}\mathbf{T}\mathbf{\Psi}\mathbf{T}^{-1} = \mathbf{A}\mathbf{\Phi} \qquad (9.55)$$

Assuming $\mathbf{A}$ to be full rank implies

$$\mathbf{T}\mathbf{\Psi}\mathbf{T}^{-1} = \mathbf{\Phi} \qquad (9.56)$$

The eigenvalues of $\mathbf{\Psi}$ must be equal to the diagonal elements of $\mathbf{\Phi}$.

The columns of $\mathbf{T}$ are the eigenvectors of $\mathbf{\Psi}$. The signal parameters are obtained as non-linear functions of the eigenvalues of the operator $\mathbf{\Psi}$ that maps (rotates) one set of vectors ($\mathbf{E}_X$) that span an m-dimensional signal subspace into another ($\mathbf{E}_Y$).

9.2.3.1 *Estimating the subspace rotation operator*

In practical situations where only a finite number of noisy elements are available, $\mathbf{E}_S$ is estimated from the covariance matrices of the measurements $\hat{\mathbf{R}}_{ZZ}$ or equivalently, from the data matrix $\mathbf{Z}$. The result is that $\mathcal{R}\{\mathbf{E}_s\}$ is only an estimate of S_Z, and with probability one, $\mathcal{R}\{\mathbf{E}_s\} \neq \mathcal{R}\{\mathbf{A}\}$. Furthermore $\mathcal{R}\{\mathbf{E}_X\} \neq \mathcal{R}\{\mathbf{E}_Y\}$. The objective of finding a $\mathbf{\Psi}$ such that $\mathbf{E}_X\mathbf{\Psi} = \mathbf{E}_Y$ is no longer achievable. A new criterion for obtaining a suitable estimate must be formulated. The most commonly employed criterion for this nature is the *least squares* (LS) criterion.

9.2.3.2 The standard LS criterion

When applied to the model $\mathbf{AX} = \mathbf{B}$ to obtain an estimate of $\mathbf{X}$, this assumes $\mathbf{A}$ is known and the error is to be attributed to $\mathbf{B}$. The LS solution is

$$\frac{\partial}{\partial \hat{\mathbf{X}}} |\mathbf{AX} - \mathbf{A}\hat{\mathbf{X}}|^2 = \frac{\partial}{\partial \hat{\mathbf{X}}} |\mathbf{B} - \mathbf{A}\hat{\mathbf{X}}|^2$$

$$\Rightarrow \hat{\mathbf{X}} = [\mathbf{A}^*\mathbf{A}]^{-1}\mathbf{A}^*\mathbf{B} \tag{9.57}$$

9.2.3.3 The total least squares criterion

The estimates $\mathbf{E}_X$ and $\mathbf{E}_Y$ are *equally* noisy. So the LS criterion is clearly inappropriate. A criterion that takes into account noise on both $\mathbf{A}$ and $\mathbf{B}$ is the *total least squares* (TLS) criterion. The TLS criterion can be stated as finding residual (error) matrices $\mathbf{R}_A$ and $\mathbf{R}_B$ of minimum Fröbenius norm, and $\hat{\mathbf{X}}$ such that

$$[\mathbf{A} + \mathbf{R}_A]\hat{\mathbf{X}} = \mathbf{B} + \mathbf{R}_B \tag{9.58}$$

This criterion is easily shown to be equivalent to replacing the zero matrix in Equation (9.45) by a matrix of errors, the Fröbenius norm of which is to be minimized (i.e. *total* least squared error). If the covariance of the errors, specifically the rows of $[\mathbf{R}_A \mid \mathbf{R}_B]$, is known to within a scale factor, the TLS estimate of $\mathbf{X}$ is *strongly consistent*. Appending a non-triviality constraint $\mathbf{F}^*\mathbf{F} = \mathbf{I}$ to eliminate the zero solution and applying standard Lagrange techniques leads to a solution for $\mathbf{F}$ given by the eigenvectors corresponding to the d smallest eigenvalues of $\mathbf{E}_{XY}{}^* \mathbf{E}_{XY}$. The eigenvalues of Ψ as defined above and calculated from the estimates $\mathbf{F}_X$ and $\mathbf{F}_Y$ are taken as estimates of the diagonal elements of Φ.

9.2.3.4 Summary of the TLS ESPRIT covariance algorithm

The TLS ESPRIT algorithm based on a covariance formulation can be summarized as follows.

1. Obtain an estimate of $\mathbf{R}_{ZZ}$, denoted $\hat{\mathbf{R}}_{ZZ}$, from the measurements $\mathbf{Z}$.

2. Compute the eigendecomposition of $\{\hat{\mathbf{R}}_{ZZ}, \Sigma_n\}$

$$\hat{\mathbf{R}}_{ZZ}\bar{\mathbf{E}} = \Sigma_n\bar{\mathbf{E}}\Lambda$$

   where $\Lambda = \text{diag}\{\lambda_1, \ldots, \lambda_{2m}\}$, $\lambda_1 \geq \cdots \geq \lambda_{2m}$, and $\bar{\mathbf{E}} = [\mathbf{e}_1|\ldots|\mathbf{e}_{2m}]$.

3. Estimate the number of sources $\hat{d}$.

4. Obtain, in signal subspace, estimate $\hat{\mathbb{S}}_Z = \mathcal{R}\{\mathbf{E}_s\}$, and decompose it to obtain $\mathbf{E}_X$ and $\mathbf{E}_Y$, where

$$\mathbf{E}_s \stackrel{\text{def}}{=} \Sigma_n \left[\mathbf{e}_1|\ldots|\mathbf{e}_{\hat{d}}\right] \Rightarrow \begin{bmatrix} \mathbf{E}_X \\ \mathbf{E}_Y \end{bmatrix}$$

5. Compute the eigendecomposition ($\lambda_1 > \cdots > \lambda_d$),

$$\mathbf{E}_{XY}{}^*\mathbf{E}_{XY} \stackrel{\text{def}}{=} \begin{bmatrix} \mathbf{E}_X^* \\ \mathbf{E}_Y^* \end{bmatrix} [\mathbf{E}_X \mid \mathbf{E}_Y] = \mathbf{E}\Lambda\mathbf{E}^*$$

and partition $\mathbf{E}$ into $\hat{d} \times \hat{d}$ submatrices

$$\mathbf{E} \stackrel{def}{=} \begin{bmatrix} \mathbf{E}_{11} & \mathbf{E}_{12} \\ \mathbf{E}_{21} & \mathbf{E}_{22} \end{bmatrix}$$

6. Calculate the eigenvalues of $\mathbf{\Psi} = -\mathbf{E}_{12}\,\mathbf{E}_{22}^{-1}$

$$\hat{\phi}_k = \lambda_k\left(-\mathbf{E}_{12}\mathbf{E}_{22}^{-1}\right), \quad \forall k = 1, \ldots, \hat{d}$$

7. Estimate $\hat{\theta}_k = f^{-1}(\hat{\phi}_k)$; e.g. for DOA estimation,

$$\hat{\theta}_k = \sin^{-1}\{c \arg(\hat{\phi}_k)/(\omega_0 \Delta)\}.$$

9.2.3.5 *Array calibration*

Using the TLS formulation of ESPRIT, the array manifold vectors associated with each signal (parameter) can be estimated (to within an arbitrary scale factor). The right eigenvectors of $\mathbf{\Psi}$ are given by $\mathbf{E}_\Psi = \mathbf{T}^{-1}$. This result can be used to obtain estimates of the array manifold vectors as

$$\mathbf{E}_s\mathbf{E}_\Psi = \bar{\mathbf{A}}\mathbf{T}\mathbf{T}^{-1} = \bar{\mathbf{A}}$$

No assumption concerning the source covariance is required.

9.2.3.6 *Limitations*

Although simple to compute, this estimate will not in general conform to the invariance structure of the array in the presence of noise. In low SNR scenarios, the deviation from the assumed structure $\bar{\mathbf{A}} = [\mathbf{A}^\mathrm{T} \mid (\mathbf{A}\mathbf{\Phi})^\mathrm{T}]^\mathrm{T}$ may be significant. In such situations, improved estimates of the array manifold vectors can be obtained by improving the formulation discussed in [827].

9.2.4 Performance illustration

A ten-element array is used with doublet spacing $\lambda/4$ and the five doublets randomly spaced on a line, resulting in an aperture of approximately 4λ [827]. Two sources are located at $24°$ and $28°$ (approximately 0.3 Rayleigh or 3 dB beamwidth separation), and are of unequal power, 20 dB and 15 dB, respectively. Sensor errors are introduced by zero mean normal random additive errors with sigmas of 0.1 dB in amplitude and $2°$ in phase (independent of angle). Sensor location errors (along the axis of the array) with sigmas 0.005 ($\lambda/2$) are included as well. The sources are 90 percent temporally correlated and 5000 trials are run.

The number of sources is assumed to be known in the implementation of both MUSIC and ESPRIT. The indicated failure rate for MUSIC of 37 percent is the percentage of trials in which the conventional MUSIC spectrum did not exhibit two peaks in the interval [$20°$, $32°$]. In ESPRIT, two parameter estimates are obtained every time. The sample means and sigmas of the ESPRIT estimates were $23.93° \pm 1.07°$ and $28.06° \pm 1.37°$. Those of the 3175 *successful* MUSIC trials were $24.35° \pm 0.28°$ and $27.48° \pm 0.38°$. The pdf of MUSIC and ESPRIT results is shown in Figure 9.6. The presence of a bias is evident even in the successful conventional MUSIC estimates. The ESPRIT estimates are unbiased, although

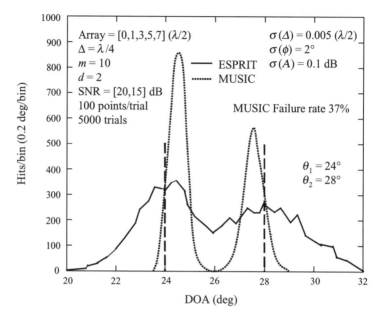

Figure 9.6 Pdf of MUSIC and ESPRIT results – random Ten-element linear array. Source correlation 90 %. Small array aperture ($\Delta = \lambda/4$) [827] © 1989, IEEE.

of large variance since less information concerning the array geometry is used. In comparing the estimate variances, there is no attempt to account for the 1825 trials in which MUSIC failed to provide two DOA estimates!

As the subarray separation increases, the ESPRIT parameter estimate variances approach those of MUSIC.

The same experiment was run for a subarray separation of 4λ. The resulting ESPRIT estimates were $24.003° \pm 0.062°$ and $28.002° \pm 0.089°$. The corresponding MUSIC estimates were $24.011° \pm 0.056°$ and $27.986° \pm 0.078°$. Due to the increased subarray spacing (array aperture), there were MUSIC failures.

Again, ESPRIT is unbiased. Now the sample parameter estimate sigmas are nearly equal to those obtained with MUSIC.

The primary computational advantage of ESPRIT is that it eliminates the search procedure inherent in all previous methods (ML, ME, MUSIC). ESPRIT produces signal parameter estimates directly in terms of (generalized) eigenvalues. As noted previously, this involves computations of the order d^3. On the other hand, MUSIC and the other high-resolution techniques, require a search over **a**, and it is this search that is computationally expensive.

The significant computational advantage of ESPRIT becomes even more pronounced in multidimensional parameter estimation where the computational load grows linearly with dimension in ESPRIT, while that of MUSIC grows exponentially. If r_l is the resolution (i.e. number of vectors) required in the calibration of **a** for the lth dimension in Θ, the computation required to search over L dimensions for d parameter vectors is proportional to $\prod_{l=1}^{L} r_l$. For $r_l = r$, the computational load is r^L.

9.3 JOINT ARRAY COMBINING AND MLSE RECEIVERS

An array combiner along with a maximum likelihood sequence estimator (MLSE) receiver is the basis for the derivation of a space–time processor presenting good properties in terms of cochannel and inter-symbol interference rejection [868–871]. The use of spatial diversity at the receiver front end together with a scalar MLSE implies a joint design of the spatial combiner and the impulse response for the sequence detector. We consider a system with R users, each transmitting simultaneously from a different location in space to a receiver consisting of an M-element array of arbitrary geometry. The transmitted baseband signal for the lth user is

$$s^{(l)}(t) = \sum_n d^{(l)}(n) p^{(l)}(t - nT) \tag{9.59}$$

where $d^{(l)}(n)$ are the symbols transmitted by the user l, $p^{(l)}(t)$ is the signature associated to the user and includes the effect of transmitter and receiver filters as well as the spreading signature. The signal reaches the array through a number $q^{(l)}$ of propagation paths, each one being characterized by a complex impulse response $c^{(lp)}(t)$. The $q^{(l)} \times 1$ vector of impinging signals is represented as

$$\mathbf{u}^{(l)}(t) = \sum_n d^{(l)}(n) \mathbf{f}^{(l)}(t - nT) \tag{9.60}$$

$\mathbf{f}^{(l)}(t)$ is the vector of path responses, so that its pth component is given by

$$f^{(lp)}(t) = p^{(l)}(t)^* \, c^{(lp)}(t) \tag{9.61}$$

The delay associated with each user is included in $c^{(lp)}(t)$. Every path received at the array is affected by a vector describing the propagation effect across the aperture, and depends on the angle-of-arrival of the path and the geometry of the array.

All these vectors can be arranged into an $M \times q^{(l)}$ matrix $\mathbf{A}^{(l)}$ (usually called the array manifold), thus resulting in a received signal

$$\mathbf{x}^{(l)}(t) = \sum_n d^{(l)}(n) \mathbf{A}^{(l)} \mathbf{f}^{(l)}(t - nT) + \mathbf{v}(t) \tag{9.62}$$

Matrix $\mathbf{A}^{(l)}$ may also include the effects of cable and RF receiver responses, I/Q imbalance, antenna elements' radiation patterns, element coupling, scattering from objects near the receiver, etc. The combined effects of multipath propagation and direction of arrival give rise to the combined space–time channel

$$\mathbf{g}^{(l)}(t) = \mathbf{A}^{(l)} \mathbf{f}^{(l)}(t) \tag{9.63}$$

The received signal can then be written as a sum of terms in which we can isolate the signal of the desired user 'i'

$$x(t) = \sum_n d^{(i)}(n) \mathbf{g}^{(i)}(t - nT) + \sum_{\substack{l=1 \\ l \neq i}}^{R} \sum_n \cdot d^{(l)}(n) \mathbf{g}^{(l)}(t - nT) + \mathbf{v}(t) \tag{9.64}$$

The noise vector process $\mathbf{v}(t)$ is independent of the transmitted symbols. Usually, single-user receivers consider interferent terms in the second summation as being noise, and rely either on the orthogonality of the signatures of different users or on spatial diversity, when using

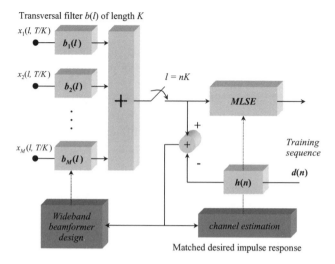

Figure 9.7 The MDIR receiver performing wideband array combining and symbol detection. A wideband spatial combiner of K samples per branch is represented.

array receivers, for good separation [872] in the space domain. Although this signal model is not strictly valid for non-linear modulations (as is the case in GSM), in practice a reasonable linear approximation is possible [873]. For this signal model, the joint detection optimum receiver has been developed in [874] for non-dispersive channels and in [875] for dispersive channels. A suboptimal approach considers only single-user multisensor detection and takes the interference as Gaussian white noise. This receiver is often called Vector MLSE [876]. The receiver block diagram is shown in Figure 9.7. The error signal is used to compute the combiner coefficients b and channel impulse response h using the MMSE criterion. Note that the scheme allows the removal of cochannel interference as well as late arrival paths of the desired user by the spatial combiner that cannot be accommodated into the length of $h(n)$. At the same time, some ISI is allowed (since $h(n)$ may be several samples long) and taken into account in the MLSE processor. This feature takes advantage of time diversity, and hence robustness to fading.

9.3.1 Joint combiner and channel response estimation

A training sequence of length N for the desired user is needed to jointly determine the parameters of the combiner and the taps of the equivalent channel $h(n)$. Assume that the space–time channel spans its response in a time interval of $(L + 1)T$ seconds. The operation performed by the array combiner and the sampler during the transmission of the training sequence may be expressed as (dropping the response of the first symbols):

$$\mathbf{Xb} = \begin{bmatrix} \mathbf{x}_1^{\mathrm{T}}(L) & \mathbf{x}_2^{\mathrm{T}}(L) & \cdots & \mathbf{x}_M^{\mathrm{T}}(L) \\ \mathbf{x}_1^{\mathrm{T}}(L+1) & \mathbf{x}_2^{\mathrm{T}}(L+1) & \cdots & \mathbf{x}_M^{\mathrm{T}}(L+1) \\ \vdots & \vdots & \ddots & \vdots \\ \mathbf{x}_1^{\mathrm{T}}(N-1) & \mathbf{x}_2^{\mathrm{T}}(N-1) & \cdots & \mathbf{x}_M^{\mathrm{T}}(N-1) \end{bmatrix} \begin{bmatrix} \mathbf{b}_1 \\ \mathbf{b}_2 \\ \vdots \\ \mathbf{b}_M \end{bmatrix} \quad \mathbf{X} \in \mathbb{C}^{(N-L)\times MK}, \mathbf{b} \in \mathbb{C}^{MK\times 1}$$

$$(9.65)$$

where $\mathbf{X}$ is the received signal matrix containing all users and noise (the subscripts inside $\mathbf{X}$ indicate the index of sensor) and M is the total number of sensors. The vectors in Equation (9.65) are defined as

$$\mathbf{x}_l^T(n) = [x_1(nT) \quad x_1(nT + T/K) \quad \cdots \quad x_1(nT + T(K-1)/K)]$$
$$\mathbf{b}_l^T = [b_{l,0} \, b_{l,1} \quad \cdots \quad b_{l,(k-1)}]. \tag{9.66}$$

The signal model may be conveniently written as

$$\mathbf{X} = \mathbf{D}\mathbf{G} + \mathbf{W} \tag{9.67}$$

where $\mathbf{W}$ contains the noise and undesired users and matrix $\mathbf{D}$ contains the symbols of the training sequence of the desired user

$$\mathbf{D} = \begin{bmatrix} d(L) & d(L-1) & \cdots & d(0) \\ d(L+1) & d(L) & \cdots & d(1) \\ \vdots & \vdots & \ddots & \vdots \\ d(N-1) & d(N-2) & \cdots & d(N-L-1) \end{bmatrix} ; \mathbf{D} \in \mathbb{C}^{(N-L)\times(L+1)} \tag{9.68}$$

$\mathbf{G}$ is the fractionally spaced space–time channel matrix for the user-of-interest

$$\mathbf{G} = \begin{bmatrix} g_{1,0} & g_{1,1} & \cdots & g_{1,(K-1)} & \cdots & g_{M,0} & g_{M,1} & \cdots & g_{M,(K-1)} \end{bmatrix}$$
$$\mathbf{G} \in \mathbb{C}^{(L+1)\times MK.} \tag{9.69}$$

Note that $g_{j,r}$ contains the channel impulse response between the user-of-interest and sensor j at the sampling phase $nT + rT/K$. It is implicitly assumed that the interferers are all frame-synchronized and their training sequences are sufficiently uncorrelated to allow separation on a time-reference basis.

The noise-plus-interference matrix $\mathbf{W}$ is approximated as spatially and temporally white. The joint design of the spatial combiner $\mathbf{b}$ and the impulse response $\mathbf{h}$ for the sequence detector is based on the minimization of the mean square error η

$$\eta = \|\mathbf{X}\mathbf{b} - \mathbf{D}\mathbf{h}\| \tag{9.70}$$

This optimization is aimed at keeping the multipath content of the signal so that it can be used later by the MLSE processor. A proper constraint has to be imposed in order to avoid the trivial solution. Several options are available [877]. In [878], the term $\mathbf{h}^H\mathbf{h}$ was fixed. Experimentally, the one proving more efficient is based on the control of the desired signal energy at the output of the spatial combiner, which can be formulated as

$$E = \mathbf{b}^H\mathbf{G}^H\mathbf{D}^H\mathbf{D}\mathbf{G}\mathbf{b} \tag{9.71}$$

We force this value to a non-zero constant α. A regular minimization procedure to the Lagrangian function

$$J = \eta - \lambda(\mathbf{b}^H\mathbf{G}^H\mathbf{D}^H\mathbf{D}\mathbf{G}\mathbf{b} - \alpha) \tag{9.72}$$

leads to the following two equations:

$$\mathbf{h} = (\mathbf{D}^H\mathbf{D})^{-1}\mathbf{D}^H\mathbf{X}\mathbf{b} \tag{9.73a}$$
$$\mathbf{X}^H(\mathbf{I} - \mathbf{D}(\mathbf{D}^H\mathbf{D})^{-1}\mathbf{D}^H)\mathbf{X}\mathbf{b} = \lambda \, \mathbf{G}^H\mathbf{D}^H\mathbf{D}\mathbf{G}\mathbf{b} \tag{9.73b}$$

If the training sequence and the noise-plus-interference are uncorrelated processes, then we can consider that, for sufficiently large

$$\mathbf{D}^H \mathbf{X} \cong \mathbf{D}^H \mathbf{D} \mathbf{G} \qquad (9.74)$$

we can rewrite Equation (9.73) as follows:

$$\mathbf{h} = \mathbf{G}\mathbf{b} \qquad (9.75a)$$

$$\mathbf{R}_w \mathbf{b} = \lambda \mathbf{G}^H \mathbf{D}^H \mathbf{D} \mathbf{G} \mathbf{b} \qquad (9.75b)$$

The noise-plus-interference matrix $\mathbf{R}_w$ is defined as

$$\mathbf{R}_w = \mathbf{X}^H \mathbf{X} - \mathbf{G}^H \mathbf{D}^H \mathbf{D} \mathbf{b} \qquad (9.76)$$

Equation (9.76) is an expected result: the impulse response seen by the MLSE block is the space–time channel filtered by the array combiner. The SINR at the output of the sampler is given by

$$\text{SINR} = \frac{\mathbf{b}^H \mathbf{G}^H \mathbf{D}^H \mathbf{D} \mathbf{G} \mathbf{b}}{\eta_{\text{opt}}} = \frac{\mathbf{b}^H \mathbf{G}^H \mathbf{D}^H \mathbf{D} \mathbf{G} \mathbf{b}}{\mathbf{b}^H \mathbf{R}_w \mathbf{b}} = \frac{1}{\lambda} \qquad (9.77)$$

The coefficients of the linear combiner are given by the generalized eigenvector of Equation (9.75b) associated with the minimum generalized eigenvalue. From Equation (9.77), it is concluded that the eigenvalues are positive. On the other hand, Equation (9.75a) reveals that the impulse response of the channel that is to be used in the MLSE block is the matched response of the linear combiner plus the sampler to the physical channel. Care has to be taken to guarantee that matrix $\mathbf{R}_w$ is full rank, otherwise the computation of the generalized EVD is not a well-conditioned problem. Inspecting Equation (9.76) when the number of rows in matrix $\mathbf{X}(N - L)$ is smaller than the number of columns MK (that is, the number of elements of $\mathbf{b}$), the product $\mathbf{X}^H \mathbf{X}$ is rank deficient and so is $\mathbf{G}^H \mathbf{D}^H \mathbf{D} \mathbf{G}$. Good performance has been observed when diagonal loading is used on matrix $\mathbf{R}_w$:

$$\mathbf{R}_w \leftarrow \mathbf{R}_w + \sigma^2 \mathbf{I} \qquad (9.78)$$

by giving a value to σ^2 just below the ground noise level. The degree of load is not critical. It should be noted that a large value would imply that the most important undesired signal is white noise and this would result in a distorted beamformer pattern. Equation (9.75) shows the procedure used to estimate the space–time signal channel $\mathbf{G}$. Of course, if the symbols of the training sequence are uncorrelated, then $\mathbf{D}^H \mathbf{D} \cong (N - L)\mathbf{I}$, and previous equations can be further simplified.

The ability to null interferers, as well as late arrivals of the signal of interest, is strongly dependent on two facts:

1. Consider the number of elements of the array and the spatial dispersion of the signals and interferers. It is widely known [879] that an M-element array is able to cancel up to $M - 1$ interferer paths.

   The direct consequence of this fact is that a high number of interfering signals in highly spatially dispersive channels cannot be fully canceled and this limits the performance of the proposed scheme since time-correlated noise (interference) $w(n)$ is induced at the output of the sampler. This noise might be dealt with at the MLSE block by using its covariance matrix in the computation of the metrics in the Viterbi lattice.

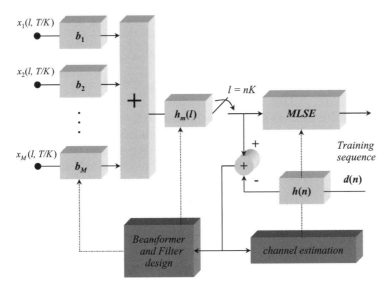

Figure 9.8 The spatial combiner, including a single linear filter, suitable for low dispersive channels.

2. The use of a fractionally spaced beamformer allows further interference rejection [880] provided that the incoming signal has some excess bandwidth.

9.3.2 Complexity reduction in wideband beamforming

Wideband spatial combining allows additional signal cancellation because it performs a different narrowband beamformer per delay. However, as the signal scenario presents low angular dispersion, it is reasonable to assume that all multipath rays with their associated time delays come from a narrow solid angle, and therefore each user can be considered a point source. In this case, there is no need to build a different narrowband combiner per delay, and all filters $\mathbf{b}_l$ present at the branches of the array are the same. The scheme for the receiver in this case is given in Figure 9.8.

Assume again that samples at a rate K/T are fed to the linear combiner. Then, let us write the output of the linear combiner plus sampler as

$$y(n) = \mathbf{x}^{\mathrm{T}}(n)\mathbf{b} = \left\lfloor x_1^{\mathrm{T}}(n) \quad x_2^{\mathrm{T}}(n) \quad \cdots \quad x_M^{\mathrm{T}}(n) \right\rfloor$$

$$\begin{bmatrix} \mathbf{b}_1 \\ \mathbf{b}_2 \\ \vdots \\ \mathbf{b}_M \end{bmatrix} = \mathrm{trace}(\mathbf{X}(n)^{\mathrm{T}}\mathbf{B}) \tag{9.79}$$

where

$$\mathbf{X}(n) = \mathrm{unvec}\,(\mathbf{x}(n)) \quad \in \mathbb{C}^{K \times M}$$
$$\mathbf{B} = \mathrm{unvec}\,(\mathbf{b}) \quad \in \mathbb{C}^{K \times M}$$

The beamformer plus matched filter is obtained as a rank-one approximation of the matrix **B**:

$$\mathbf{B} = \sum_{i=1}^{\min(K,M)} \sigma_i \mathbf{u_i} \mathbf{v}_i^H \cong \sigma_{\max} \mathbf{u_{max}} \mathbf{v}_{\max}^H \tag{9.80}$$

The left and right eigenvectors associated with the maximum eigenvalue provide the coefficients of the linear filter $\mathbf{h}_m$ and the conjugate beamformer coefficients **b**, respectively. The goodness of this approximation may be decided from the quotient between the maximum eigenvalue and the sum of singular values of **B**. The truncation in Equation (9.80) can be made on any order (not just keeping one term), therefore allowing simplification of the receiver according to the characteristics of the scenario.

9.3.3 Performance illustration

Since the MDIR receiver is a suboptimal single-user receiver, we will compare its performance to two practical single-user approaches usually proposed. Both consider interferers as Gaussian noise. The signal parameters are as in [868].

1. The multisensor vector MLSE (VMLSE) [876] is a popular, reasonably priced receiver which models the noise plus interference as spatially and temporally white.

2. The multisensor weighted vector MLSE (WVMLSE) receivers [881], are based on a modification of the vector MLSE metrics by using the inverse of the space correlation matrix of the interference. This matrix can be readily estimated as

$$\hat{\mathbf{R}}_w = (\mathbf{X} - \mathbf{DG})^H (\mathbf{X} - \mathbf{DG}) \tag{9.81}$$

This receiver is a fair competitor of MDIR since it performs some kind of spatial canceling, but at a higher complexity. No time modeling of the interference has been attempted since it would imply a larger number of states in the Viterbi algorithm [882].

9.3.4 Propagation channel model

Simulations based on a Gaussian stationary uncorrelated hypothesis for the channel are used, assuming independence between angular and Doppler spread. Details from measurements in the 1.8 GHz band are given in [883]. There, it is empirically shown that the azimuth spectrum follows a Laplacian law, along with Gaussian distribution for the directions of arrival (ϕ) around the mean angular position of the user. The angular spread (that is, the standard deviation of the Gaussian σ_ϕ) is taken to be $8°$. The number of rays impinging the array is fitted as a Poisson random variable of mean 25. An exponential law is found in [818] for the power delay spread. The delay associated with each impinging ray τ is taken as an exponential random variable of mean 1 microsecond (σ_τ). The amplitude associated with each propagation path (α) is a complex Gaussian random variable whose power decreases as the time delay and the angular direction of arrival with respect to the mobile position increase, according to the expression

$$E\{|\alpha|^2/\tau, \phi\} = \exp\left(-\frac{\tau}{6.88\sigma_\tau} + \frac{\phi^2}{3.81\sigma_\phi^2} - \sqrt{2}\frac{|\phi|}{\sigma_\phi}\right) \tag{9.82}$$

A classic Clarke's bath-shaped Doppler spectrum is obtained by assuming multiple reflections close around the mobile. The array works in a sectored area of 120°, with linearly and uniformly spaced elements at $d/\lambda = 0.5$. All plots shown in the simulations below are representations of the performance of the link level. The results can be used with convenient mapping to obtain the FER (frame erasure ratio) when considering channel coding, or other system level features (like power control, frequency-hopping, or discontinuous transmission).

9.3.5 Non-spread modulation

The air interface is TDMA-based, and training sequences are formed, as in the GSM standard [873], by 26 symbols located in the middle of a 116 traffic symbols frame. BPSK modulation with rolloff factor 0.2 has been used.

Cochannel users in other cells are the source of interference. It is assumed that they are frame-synchronous, with relative delays of up to two symbols. This implies base station synchronicity. The length of the estimated channel is $L + 1 = 4$. First, the three receivers (MDIR, VMLSE and WVMLSE) have been compared with respect to the number of interferent signals versus the frame instantaneous CIR, using four and eight sensors. Three and six interferers have been tested, all of equal mean power, which can be considered a pessimistic signal scenario when cells are sectored in 120°. The mean E_b/N_0 is 25 dB and the mobile speed is a vector of modulus 50 km/h and random direction. Training sequences are taken from the GSM standard [884]. Note that they are not completely orthogonal. The number of samples per symbol is one, which is the length of the filters at each branch of the array, and the right sampling time has been computed out from the channel estimated at four samples per symbol.

Plots showing the raw BER are given in Figure 9.9, where the estimated BER does not include the detected symbols of the training sequence. The results are compared to the single-sensor MLSE detector versus the instantaneous CIR in the slot. According to the figures, the CIR rejection of the MDIR is similar to the WVMLSE and its gain can be quantified in 10 dB at 0.1% BER when using four antennas with respect to VMLSE. Note that WVMLSE requires a higher computational cost.

In a second simulation, dependence on the mobile speed is illustrated. Figure 9.10 is equivalent to the graph in Figure 9.9(b), except for the mobile speed, which is set to 100 km/h. All receivers tend to exhibit some residual BER at high CIR due to the inaccuracy in the channel estimate, except for the VMLSE. MDIR seems to be the most affected.

In the third simulation, presented in Figure 9.11, the mean E_b/N_0 level has been changed for each receiver in values from 5 to 35 dB. As expected, the VMLSE shows the best performance when the interference levels are very low. WVMLSE should exhibit similar BER values in this condition, but the estimation error in matrix $\mathbf{R}_w$ is significant, due to the short length of the training sequence. On the contrary, when the E_b/N_0 is high, VMLSE exhibits some limiting performance with respect to interference cancellation, which is not found for MDIR and WVMLSE.

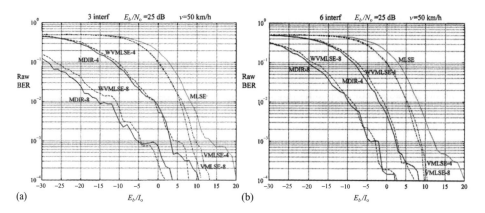

(a) E_b/I_o (b) E_b/I_o

Figure 9.9 Probability of error of the MDIR (solid), VMLSE (dashed-dotted), and WVMLSE (dashed) receivers versus instantaneous CIR for (a) three and (b) six interfering signals. The performance of the single-sensor MLSE is also displayed (dotted). Four and eight sensors have been tested [868] © 2000, IEEE.

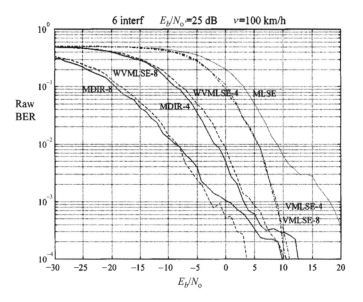

Figure 9.10 Probability of error of all receivers with increased mobile speed [868].

9.3.6 Spread modulation

In this case, signals are spread with equal length spreading codes for each user. Chip time is 0.9 μs. Except for the bandwidth, these conditions are similar to the TDD air interface for the third-generation mobile communications system [885]. The number of taps at each sensor of the array is a multiple of the spreading factor: one or two samples per chip are taken ($K = 4$ or $K = 8$). After the spatial combiner, a symbol-time sampler precedes the MLSE block. The codes chosen are orthogonal Walsh codes, and they have been taken

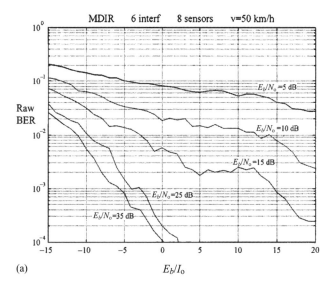

(a)

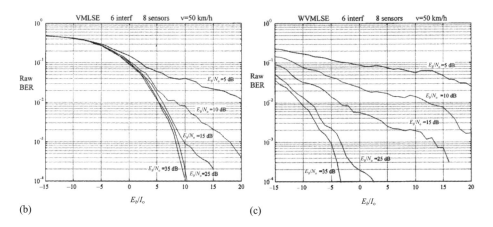

(b)

(c)

Figure 9.11 Probability of error of the (a) MDIR; (b) VMLSE; and (c) WVMLSE receivers versus instantaneous CIR for six interferers, a mobile speed of 50 km/h and different mean E_b/N_0 values. Eight sensors have been used in all cases [868].

randomly for each user in each Monte Carlo run. The channel model is the same as in previous simulations. Pulse shaping is raised cosine with rolloff factor of 0.2. Three users of spreading factor four have been used, two of them being considered as asynchronous interference users. The floor noise is 20 dB below the mean power of the desired user and may accommodate high spreading factor users in a multirate system. The broadband spatial combiner allows the design of an interference cancellation filter **b** in each array branch, which further reduces the interference level at the input of the MLSE block. The BER is displayed in Figure 9.12 along with the performance of the receiver with chip-time sampler

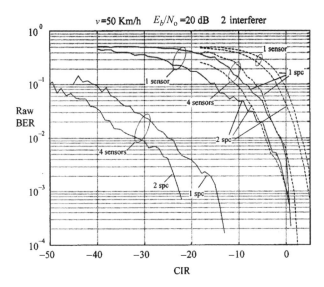

Figure 9.12 Probability of error of the MDIR (solid) and VMLSE (dashed) receivers versus instantaneous CIR in DS/SS channel access with spreading factor of four. Two interferer users are considered. One and four sensors have been used, each featuring a linear filter spanning one symbol time and one or two samples per chip (1 spc or 2 spc), so the number of coefficients is $K = 4$ or 8 [868] © 2000, IEEE.

after the spatial combiner, and using one or two samples per chip ($K = 1$ and $K = 2$). From the figure, it is apparent that the broadband combiner with symbol-time sampling allows a performance with one single sensor that is comparable to the performance of a four-sensor array using a chip-time signal as input to the MLSE. No significant gains are observed when four samples per symbol are used.

10

Adaptive Reconfigurable Software Radio

10.1 ENERGY-EFFICIENT ADAPTIVE RADIO

Within this section we discuss some practical solutions and results for the adaptation of receiver parameters in order to improve performance and reduce energy consumption. Depending on the channel state, different receiver parameters can be changed for these purposes. Table 10.1 presents a summary of these options. Figure 10.1 presents a generic block diagram of such a receiver. The discussion in this section is based on [886] © 1999, IEEE.

In the simulation environment, the sender and receiver entities are instrumental in tracking energy consumption. The sender entity keeps track of energy consumption due to the RF transmission and link layer computation, such as forward error correction (FEC) encoding. The receiver entity tracks energy consumption due to the RF receiver and computation tasks such as FEC decoding and equalization. In all of the simulation, the RF section is assumed to dissipate 0.1 W during sleep, 0.6 W during receive and 1.8 W during transmit mode. The data on dissipation are taken from a commercial (General Electric Company's Plessy) radio for wireless local area networks (WLANs). The details of the energy estimates for the computation tasks such as the FEC and equalization are described later in the section.

10.1.1 Frame length adaptation

When the BER due to random noise and interference changes, so does the frame error rate, which in turn determines the application level throughput, or goodput (excluding the header). Regardless of the transmission rate R_t, with large frame length the goodput of the link can drop to zero if every packet gets corrupted. The radio will still be transmitting (retransmissions) and therefore wasting battery energy. Reducing frame length can improve

Advanced Wireless Communications. S. Glisic
© 2004 John Wiley & Sons, Ltd. ISBN: 0-470-86776-0

Table 10.1   Adaptation of radio parameters to channel degradations

Channel degradations		Adaptive radio parameters	Adaptation mechanism
Control Interference	Low level	FEC/ARQ	Adaptive error
	Low level	Frame length	Adaptive frame length
	High level	Processing gain	Adaptive spread spectrum
	Bursty	Frame length	Adaptive frame length
Flat fading		Frame length	Adaptive frame length
Frequency selective fading		Channel impulse response	Adaptive equalizer

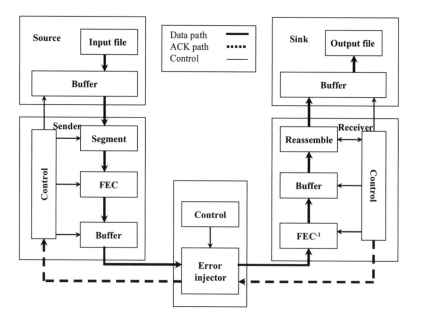

Figure 10.1 Simulation environment for adaptive radio.

the goodput and therefore energy efficiency by reducing the probability of frame errors and the need for excessive numbers of retransmissions. However, the relative overhead of the frame header also increases, thereby offsetting the improvement in goodput. Therefore, an optimum frame length exists that maximizes goodput for a given BER as shown in the measurements in Figure 10.2. These measurements have been obtained with a peer-to-peer wireless link using the 900 MHz WaveLAN (WLAN) radios under varying channel interference introduced using an AWGN noise generator. A link layer header of eight bytes has been used.

Conventionally, the relationship in Figure 10.2 is used to adapt the frame length to maximize the goodput for a given BER. However, if the frame length adaptation is applied for reducing energy consumption under a given constraint on goodput, the radio selects the frame length that maximizes the BER required to meet the goodput constraint, thereby minimizing the required transmit energy per bit E_b/N_0. The reduction in required E_b/N_0 is

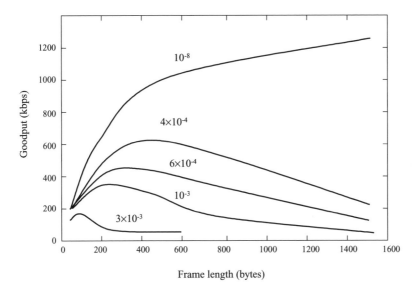

Figure 10.2 Goodput results.

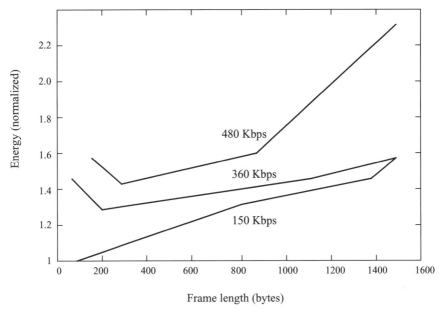

Figure 10.3 Energy versus frame length [886] © 1999, IEEE.

exploited to reduce the transmitter power to obtain energy saving. Note that in most radios the transmitter energy consumption dominates. This adaptation approach is illustrated in Figure 10.3, which is derived from the measurements in Figure 10.2 and shows the relative energy versus frame length for different goodput constraints.

For a given goodput constraint, the radio selects the frame length that minimizes energy according to the corresponding curve in Figure 10.3. The radio monitors the BER and adjusts the transmit power to hold the BER constant at the maximum allowable value. For example, for the best case in Figure 10.3, if the desired goodput is 150 kbit/s, which is 7.5 % of the transmission rate R_t in this experiment, then a 30 % reduction in battery energy consumption is achieved by reducing the frame size from the 1500 bytes Ethernet standard to 100 bytes.

The above discussion holds for the general case but the amount of energy saving will depend on the E_b/N_0 required for a particular modulation. The results in Figures 10.2 and 10.3 are valid for quadrature phase shift keying (QPSK modulation).

10.1.2 Frame length adaptation in flat fading channels

The simulations are now in a flat fading channel, where errors occur in bursts as a function of the relative motion between the transmitter and receiver. The burstiness of the channel is modeled with a two-state Discrete Time Markov Channel (DTMC) shown in Figure 10.4, with the following channel parameters [886]:

- $\rho = R/R_{rms}$ is the ratio of the Rayleigh fading envelop R to the local root mean square (RMS) level.

- The average number of level crossings in a positive direction per second is given by $N = (2\pi)^{1/2} f_m \rho\, e^{-\rho^2} N = (2\pi)^{1/2} f_m \rho\, e^{-\rho^2}$

- The average time T, for which the received signal is below a specified level R, is given by $T = (e^{\rho^2} - 1)/(f_m \rho (2\pi)^{1/2})$

- Assuming steady-state conditions, the probability that we will find a given channel in the Good or Bad condition is

$$\mu = \begin{bmatrix} \mu_0 \\ \mu_1 \end{bmatrix} = \begin{bmatrix} P(\text{Good}) \\ P(\text{Bad}) \end{bmatrix} = \begin{bmatrix} \dfrac{(1/N - T)}{1/N} \\ \dfrac{T}{1/N} \end{bmatrix} = \begin{bmatrix} e^{-\rho^2} \\ 1 - e^{-\rho^2} \end{bmatrix}$$

- $\text{BER} = P(\text{Good}) \cdot \text{BER}_G + P(\text{Bad}) \cdot \text{BEG}_B$.

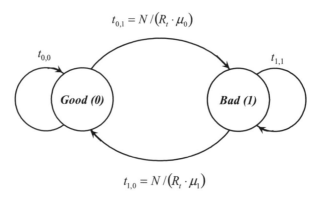

$$t_{0,1} = N/(R_t \cdot \mu_0)$$

$t_{0,0}$ $t_{1,1}$

Good (0)   **Bad (1)**

$$t_{1,0} = N/(R_t \cdot \mu_1)$$

Figure 10.4 Gilbert–Elliot model for flat fading channels.

- The Good state BER is a function of the signal to noise ratio (SNR) at the receiver (degraded by such things as path loss, local interferers, thermal noise, etc.).

- The Bad state BER is fixed at 0.5. The Good state BER is allowed to vary over a wide range, 10^{-2} to 10^{-8}.

- ρ is set to -20 dB throughout the simulation. This is a conservative assumption that the radio has a 20 dB fading margin.

- The second variable of interest is the speed of the mobile, which indicates the burstiness of the channel.

- Selective acknowledgment (SACK) has been chosen as the error control scheme with a link layer header of eight bytes.

- The carrier frequency is set at 900 MHz.

- The transmission rate R_t is set at 625 kbits/s.

Simulation results are shown in Figure 10.5 for frame lengths ranging from 50–1500 bytes. The results are based on the greedy data as the data source, with $L_n = (Lf_m)/R_t$ – the ratio of the frame length to the time interval between fades.

The achievable throughputs are shown in Figure 10.6. From these figures we can see that after the region of flat performance, the throughput decreases exponentially as L_n increases, and energy increases as L_n increases. The energy with $L = 50$ bytes becomes noticeably

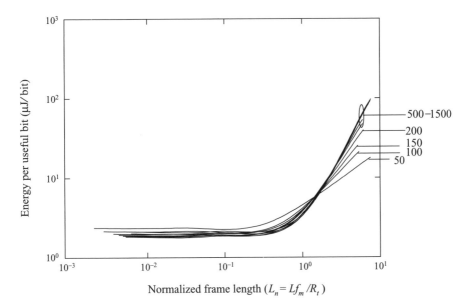

Figure 10.5 Energy per useful bit versus normalized frame length. Frame length size in bytes is annotated alongside the corresponding curves. The carrier frequency is 900 Hz and the transmission rate is 625 kbit/s. $\rho = -20$ dB, $\mathrm{BER_G} = 10^{-5}$, and $\mathrm{BER_B} = 0.5$.

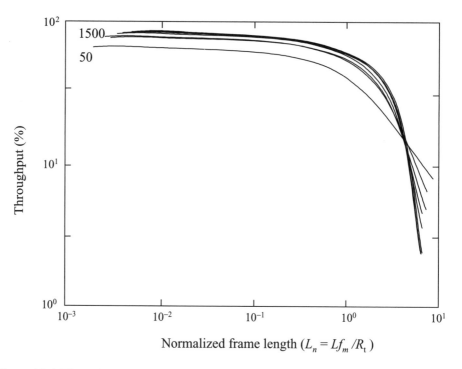

Figure 10.6 Throughput versus normalized frame length.

higher than the energy for $L > 50$ bytes, due to the increased header overhead in each frame. For $L > 50$ bytes, the energy and the throughput are relatively insensitive to the value of L and stay flat within a constant range for $L_n < 0.1$. Specifically, with $L_n < 0.1$, the energy stays between 2–2.5 μJ/bit and the throughput stays high, between 80–90 % of R_t. The radio maintains an energy-efficient link with a high application throughput by adapting the frame length within the following constraint in a flat fading channel

$$50 < L < 0.1\frac{R_t}{f_m} = 0.1\frac{\lambda R_t}{v} \tag{10.1}$$

where λ is the carrier wavelength.

10.1.3 The adaptation algorithm

The adaptation algorithm uses the frequency tracking loop in the radio receiver to estimate the speed v. Given the speed, the carrier frequency and the transmission rate, the upper bound on L is then computed using Equation (10.1). If, in addition to a throughput constraint, the delay is constrained, then the upper bound on L may be lower and can be obtained from the delay simulations as shown in Figure 10.7. The value of the lower bound in Equation (10.1) is determined by simulation and depends on the frame header size and the given throughput requirement (assumed to be eight bytes and 85 % of R_t, respectively, in the above case). Within the bounds on L for minimizing energy under flat fading conditions, the actual value of L is chosen based on the optimum for adapting to the channel interference, as illustrated

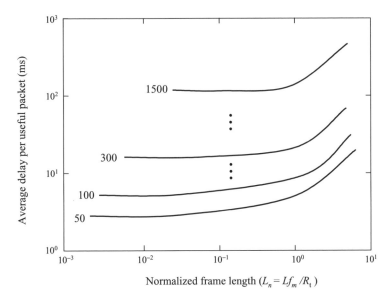

Figure 10.7 Average delay per useful packet versus normalized frame length.

in Figure 10.3 (energy vs. L). If the optimum L for adapting to channel interference is greater than the upper bound in Equation (10.1), then L is set to this upper bound.

The issue of frame length and data rate adaptation in a cellular network will be discussed in more detail in Chapter 15.

10.1.4 Energy-efficient adaptive error control

The above results in frame length adaptation are for uncoded transmissions based on SACK. Additional benefit is expected if FEC is introduced. The combination of FEC with automatic repeat request (ARQ), is known as hybrid ARQ.

Traditionally, the focus has been on satisfying a given throughput and/or delay requirement. As has already been pointed out, in portable multimedia devices, an additional requirement is to meet the throughput and delay requirements with minimum battery energy for a given media type. Simulations have been performed with RS block codes of rate 0.7 and the SACK retransmission protocol with a link layer header of eight bytes. Figure 10.8 shows the energy cost for implementing the RS coding on a StrongARM embedded microprocessor (http://www.develop.intel.com), which implements the link layer control in adaptive radio.

Results for the Gilbert–Elliot channel model from Figure 10.4 are presented in Figures 10.9–10.11 with the parameters shown in the figures. Using this data, the adaptive algorithm chooses an error control scheme that minimizes batteryenergy for transmission over the wireless link, while trading off QoS parameters of the link, such as throughput and delay over various channel conditions, traffic types and packet sizes. The simulations are performed for two packet lengths:

- 53 bytes ATM cell size packets;

- 1500 bytes Ethernet frame size IP packets.

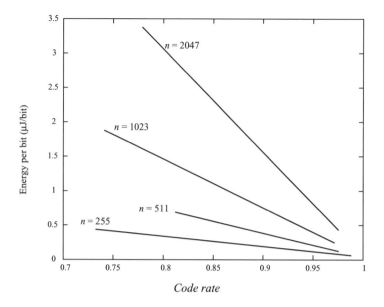

Figure 10.8 Energy cost of Reed–Solomon coding for differing block size (n) and code rates based on the StrongARM embedded processor [886] © 1999, IEEE.

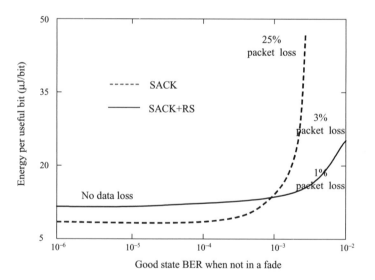

Figure 10.9 Speech transmission with ATM at 50 km/h. The carrier frequency is at 900 MHz, $R_t = 625$ kbit/s, $\rho = -20$ dB and $BER_B = 0.5$. The numbers next to the curves represent the packet loss [886] © 1999, IEEE.

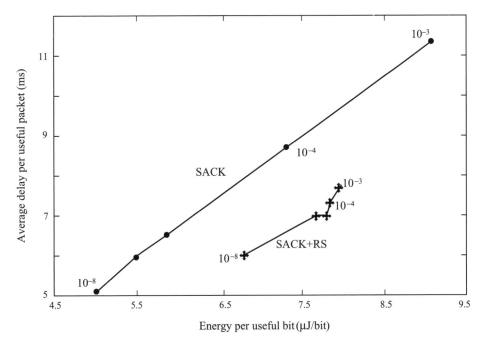

Figure 10.10 Data transmission with ATM at 50 km/h. The carrier frequency is at 900 MHz, $R_t = 625$ kbit/s, $\rho = -20$ dB and $BER_B = 0.5$[886] © 1999, IEEE.

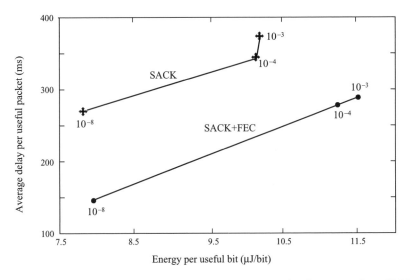

Figure 10.11 Data transmission with IP at 50 km/h. The carrier frequency is at 900 MHz, $R_t = 625$ kbit/s, $\rho = -20$ dB and $BER_B = 0.5$.

10.1.4.1 Error control for speech transmission

Figure 10.9 shows the effects of altering the Good state BER for real time data. Data is generated by the source at 32 kbits/s and forwarded to the sender with a delay of less than 50 ms. A delay greater than 50 ms is perceptually unacceptable for an interactive speech session and results in the packets being dropped. Figure 10.9 shows only small ATM packets, since larger packets do not meet the delay constraint. For a low Good state BER, no data is lost and both SACK and SACK with Reed–Solomon perform comparably. For energy-efficient speech transmission, the radio selects SACK error control for low BER (below 10^{-3}). For higher BER, the radio adapts the error control to include FEC.

10.1.4.2 Error control for data transmission

The plots in Figures 10.10 and 10.11 show the trend in energy consumption and delay as the channel quality increases from a Good state BER of 10^{-3}, as in packet cellular, to 10^{-8}, as in WLAN applications. In Figure 10.10, for small packets (ATM), hybrid FEC/ARQ consumes less battery energy for the poor channel case (e.g. 10^{-3} BER). As the channel improves, the energy for SACK without FEC becomes lower, but the delay can be higher (e.g. BER 10^{-4}). This is because the radio consumes the same amount of battery energy for the FEC, whereas the energy for SACK (without FEC) decreases because fewer retransmissions are required. In data transmission, a higher delay can be traded off for longer battery life. Therefore, for low BER (less than 10^{-3}) the radio selects SACK and for high BER $\geq 10^{-3}$ the hybrid FEC/SACK is selected.

10.1.4.3 IP packets

Figure 10.11 shows that FEC/ARQ reduces the average delay, but it also consumes more energy compared to SACK. If the delay constraint is high enough, the radio minimizes battery energy by selecting SACK. The reason that the hybrid FEC/ARQ does not help to reduce the battery energy is because the flat fading effects in the channel become more pronounced with larger packets. In the short packet case, the channel noise effects are more dominant than fading and FEC provides improvements. If the delay constraint is too tight, the radio selects the hybrid error control. In summary: for small data packets, the radio switches between SACK and FEC/ARQ based on the BER; for large packets, the radio adapts the error control based on the delay constraint.

10.1.5 Processing gain adaptation

In the presence of interference, adaptation of FEC and frame length can improve the system performance when the BER does not degrade appreciably beyond 10^{-3}. At this BER, the corresponding E_b/N_0 is 5–10 dB. In certain channel conditions, especially in shared bands, the interference level can be significantly higher, resulting in a signal to interference ratio (SIR) less than zero, implying a higher interference level than in the transmitted signal. Spread spectrum techniques, as discussed in Chapter 5, provide processing gain at the receiver, which can be used to provide protection in channels that exhibit negative SIRs. Transceiver IC, based on the architecture from Figure 10.12, is used to produce the

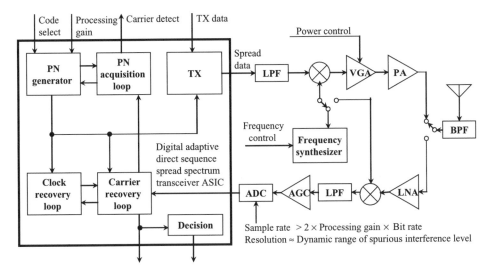

Figure 10.12 Adaptive processing gain direct sequence spread spectrum radio [886] © 1999, IEEE.

experimental results presented in this section. An adaptation interface for this radio has been built to enable the link layer to control the processing gain (PG) [886].

The radio has a fixed chip rate of 2 Mchips/s and the processing gain can be adapted to 12, 15 and 21 dB for date rates of 128, 64 and 16 kbit/s, respectively. The power dissipation of this radio in transmit, receive and sleep modes is 2.5, 0.6 and 0.2 W, respectively. This experiment was performed indoors, and the radio had no RAKE equalization. The performance improvements are only due to the processing gain.

The experimental results are shown in Figure 10.13, in which the measured application level throughput (goodput) and energy consumption are plotted against the SIR experienced in a channel with a narrowband interference. Curves for a processing gain of 12 dB are indicative of the performance one expects with a commercial radio based on IEEE Standard 802.11, such as WLAN. At this low processing gain, the throughput with commercial radios decreases as the SIR falls below −4 dB and the energy consumed per user bit sharply rises. By adapting the processing gain, the radio minimizes the energy by switching to a higher processing gain.

10.1.5.1 Receiver algorithm

The radio measures the SIR from the receive signal and sets the processing gain based on the thresholds in Table 10.2. When the SIR drops to −5 dB the radio adapts by changing the processing gain from 12 to 15 dB, thereby decreasing the energy consumed by 86 % as shown in Figure 10.14. In Figure 10.14, the achieved user throughput is 100 kbit/s for the lowest processing gain, and 9.6 kbit/s for the highest processing gain.

The tradeoff between data rate and processing gain is characteristic of spread spectrum, where the product of processing gain and data rate remains a constant and equals the occupied RF bandwidth.

To achieve a throughput of 1 Mbit/s required for most multimedia applications, and still provide a sufficiently high processing gain to adapt to a very low SIR value (e.g. −12 dB),

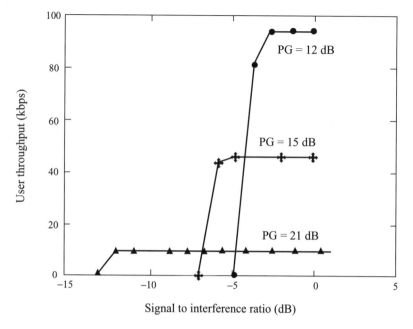

Figure 10.13 Energy consumption versus number of states in the MLSE estimated with a 3.3 V standard cell library in the 0.35 µm CMOS process. The operating data rate is 1 Mbit/s.

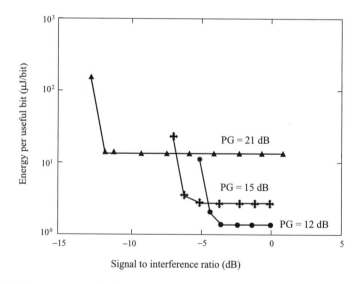

Figure 10.14 Energy per useful bit.

Table 10.2   SIR thresholds used for processing
gain adaptation

Threshold (dB)	Processing gain (dB)
SIR > −4	12
−4 > SIR > −6	15
−6 > SIR > −12	21

a significant amount of bandwidth (i.e. hundreds of megahertz) is required. Since such a wide bandwidth is not always available, a possible solution is to use non-contiguous bands and frequency hop (FH) across these bands. Current research is aimed at developing such a solution using a multiband radio that can achieve a tunable range from 25–2500 MHz. At a high data rate of 1 Mbit/s, the FH radio can experience ISI due to the frequency selectivity of the channel as the symbol duration becomes less than the delay spread of the channel. To combat that ISI, equalization in the physical layer can be employed in conjunction with the adaptive link layer control to improve the link performance.

10.1.6   Trellis-based processing/adaptive maximum likelihood sequence equalizer

At high data rates, the degradation due to ISI in a frequency selective channel cannot be mitigated adequately by adapting the link layer alone. Equalization at the physical layer is required for the radio to adapt to frequency selective fading. A tradeoff is possible between the energy consumed by equalization and the energy saved due to the corresponding increase in goodput.

In this section, a maximum likelihood sequence equalizer (MLSE) has been selected for the adaptive radio because it has two well known advantages over other equalizer algorithms. It also can be easily related to the other trellis-based algorithms presented so far in the book.

The energy consumption of the MLSE architecture shown in Figure 10.15 has been determined for a *low-power* 3.3 V 0.35 μm CMOS technology.

10.1.7   Hidden Markov channel model

DTMC models, such as the Gilbert–Elliot model, are a special case of the more general hidden Markov model (HMM) which is used in this section for performance evaluation. An HMM is a doubly stochastic process with an underlying stochastic process that is not observable but hidden. By simply observing the output sequence, the state that generated the output cannot be determined at any given time. The HMM has been used to obtain extremely accurate models for the mobile channel. The link performances over frequency selective channels are simulated using an HMM that is more accurate than the two-state Gilbert model.

A two-state HMM, shown in Figure 10.16 is used and is trained to match the error gap distribution in the error sequence, obtained through extensive simulation at the physical layer based on Jakes's model. In the error sequence, the position of an error event is marked

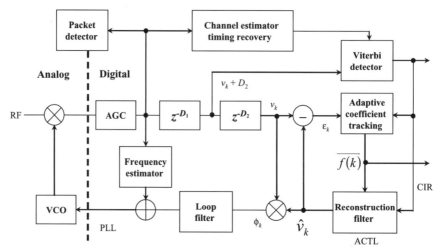

*delays D₁ and D₂ introduced by the packet detector and
trace-back operation in the Viterbi detector*

Figure 10.15 Maximum likelihood sequence equalizer modem architecture.

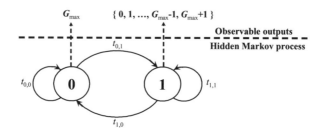

Figure 10.16 Two-state HMM channel model.

with a '1', while the position of an error-free event is marked with a '0'. An error gap
is defined as the distance (number of positions) between two consecutive 1s in the error
sequence. The output of the HMM consists of error gap values which can then be inverted
to obtain the actual error sequence.

Figure 10.17 shows the energy cost for the MLSE with different numbers of states M
in the Viterbi decoder. The results can be easily extrapolated to any type of trellis-based
algorithm.

- The carrier frequency is set at 2.4 GHz.

- The transmission rate is 1 Mbit/s.

- The typical urban (TU) channel model is used for the mobile frequency selective channel
  to generate the error sequence for the hidden Markov model.

- The greedy data source is used with an eight-byte link layer header.

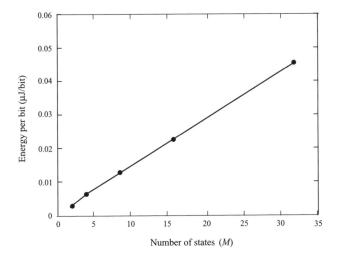

Figure 10.17 Energy consumption versus number of states in the MLSE estimated with a 3.3 V standard library on 0.35 μm in the CMOS process. The operating data rate is 1 Mbit/s.

10.1.8 Link layer performance with inadequate equalization

Adequate equalization occurs when the number of states M satisfies

$$\log_2(M) + 1 = P + L_{\text{mod}} - 1 \tag{10.2}$$

where P is the number of channel coefficients and L_{mod} is the memory of the modulation. In the simulation, Gaussian minimum shift keying (GMSK) is used as the modulation with a $BT = 0.3$ that results in $L_{\text{mod}} = 2$. As M increases, so does the energy consumption of the MLSE implementation. M is selected to be smaller than required by Equation (10.2).

For the TU channel, $P = 5$, but M is set inadequately to 2, i.e. a two-state MLSE. The objective of this simulation is to determine the tradeoff in the overall link energy, throughput and delay with a lower complexity equalizer that requires less energy consumption. Figure 10.18 shows the simulation results for frame length values of $L = 50, 500$ and 1500 bytes and speed values of $v = 5$ and 100 km/h. As expected, the performance degrades for the cases with larger frame lengths and higher speeds.

If the equalizer removes most of the ISI, then the channel appears to the receiver like an AWGN channel with some remaining level of burstiness due to mobility, similar to the behavior of a flat fading channel. To determine the effectiveness of the two-state equalizer, the results here are compared with the results for adapting frame length in a flat fading channel. As shown in Figure 10.18, the energy is minimal with a frame length of 50 bytes. The value of the energy consumption, throughput and delay are significantly worse than that obtained in a flat fading channel using the bounds on L as in Equation (10.1).

For the two-state equalizer with $L = 50$ bytes and $v = 5$ km/h, a user throughput of 35 % is achieved with delay of 70 ms, as compared to a user throughput of 73 % and a delay of 2.5 ms achieved with an adaptive frame length for the flat fading channel (Figures 10.6 and 10.7). The energy required is 5 μJ/bit, 100 % higher than that in Figure 10.5. We can conclude that using an equalizer that is smaller than required to satisfy Equation (10.2)

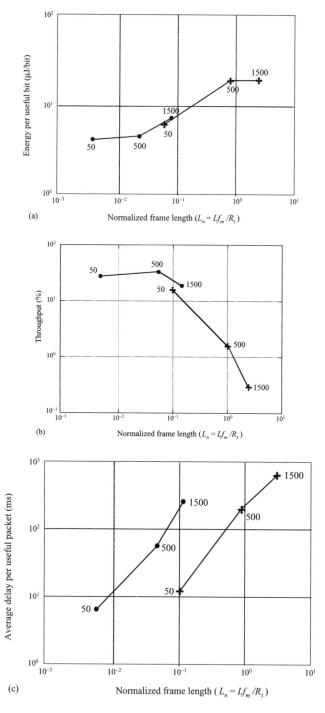

Figure 10.18 Energy, throughput and delay versus frame length for two-state MLSE for the mobile frequency selective TU channel. ✚ denotes $v = 5$ km/h, and ● denotes $v = 100$ km/h. The annotation to the data points indicates the frame length in bytes. SNR is 17 dB. The carrier frequency is 2.4 GHz. (a) Energy per useful bit versus normalized frame length; (b) throughput versus normalized frame length; (c) average delay per useful packet versus normalized frame length.

does not provide an efficient tradeoff between equalizer energy consumption and transmit energy.

10.1.9 Link layer performance with adequate equalization

In this case, M is chosen to satisfy Equation (10.2), i.e. sufficient equalization is provided for the TU channel. To meet Equation (10.2), an MLSE with 32 states ($M = 32$) is used. The results are shown in Figure 10.19.

Although the equalizer is more complex and consumes more energy, the overall system energy reduces since retransmissions are reduced. Compared to the case with $M = 2$ in Figure 10.18,

- the energy consumption is reduced by 100 %;

- the energy can be minimized for the high speed case to 2.5 µJ/bit with $L = 50$ bytes, and to 2.1 µJ/bit for the low speed case with $L = 500$ bytes.

The energy, as well as the throughput and delay, now match those obtained by applying the adaptive frame length in a flat fading channel (Figures 10.6 and 10.7). We can now conclude that in a mobile frequency selective channel, an energy-efficient link requires an equalizer with an adequate number of states, as determined by Equation (10.2).

For given system parameters, simulations can be performed to obtain results such as those in Figure 10.19 to determine optimum frame sizes that minimize energy.

10.1.9.1 *Implementation*

The radio monitors the CIR to determine the type of fading. A single peak in the CIR indicates flat fading. On the other hand, the occurrence of several peaks indicates frequency selective fading. The number of coefficients P in the CIR is used to determine the number of states required in the MLSE, using Equation (10.2) with $L_{mod} = 2$. The speed is estimated from the frequency tracking loop and is used to determine the optimum frame length L from simulation results as indicated above. The optimum frame length at different speeds can be intuitively explained by the limited ability of the equalizer to track the rapidly varying received amplitude in each of the CIR coefficients and the Doppler shift that causes an offset in the carrier frequency.

With a large frame size at high speed, the equalizer fails to maintain adequate tracking performance for the increased time duration of the given frame. At lower speed, the channel variation diminishes and therefore the equalizer can maintain tracking of the channel for longer frame sizes.

10.1.9.2 *Channel and frequency tracking performance*

The ability of the MLSE to equalize ISI degrades if it does not track the fast amplitude variation and a large Doppler shift in the channel. A conventional tracking subsystem has been implemented in the MLSE as shown in Figure 10.15, and consists of a phase-locked loop (PLL) and an adaptive channel tracking loop (ACTL). The PLL tracks the Doppler frequency and carrier offset in the down-converted RF signal. The ACTL uses the

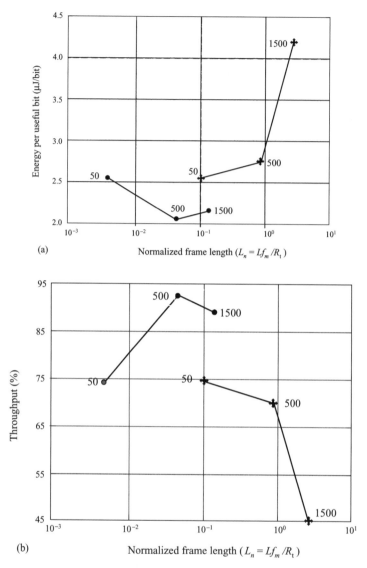

Figure 10.19 Energy, throughput and delay versus frame length for the 32-state MLSE for the mobile frequency selective TU channel. ● denotes v = 5 km/h and ✢ denotes v = 100 km/h. The annotation to the data points indicates the frame length in bytes. SNR is 17 dB. The carrier frequency is 2.4 GHz. (a) Energy per useful bit versus frame length; (b) throughput versus frame length; (c) average delay per useful packet versus frame length.

least mean squared (LMS) algorithm to track the time variations in the channel impulse response.

The tracking performance with the PLL and ACTL can be degraded due to the delays D_1 and D_2 introduced by the packet detector and traceback operation in the Viterbi detector

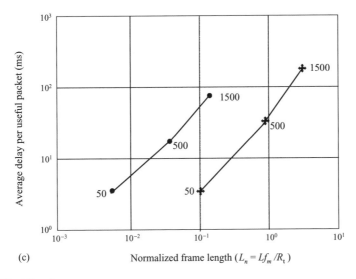

(c)

Figure 10.19 *(Cont.)*.

as discussed in Chapter 6.

10.1.9.3 Per-survivor processing (PSP)

To maintain performance of the tracking loops, PSP can be implemented in the MLSE. Figure 10.20 shows the performance of the MLSE with, and without, PSP operating at

- 1 Mbit/s over a TU channel;
- a delay spread of 5 μs at 100 km/h;
- a carrier frequency offset of 1200 Hz.

Without the PSP, the BER increases by nearly two orders of magnitude at an SNR of 20 dB.

10.1.9.4 System integration

The adaptive radio can be integrated in the system architecture of Figure 10.21. The architecture ties the adaptive link layer and physical layer in the radio with the rest of the network protocol stack. It is based on a multilevel QoS framework where the lower layers can adapt to channel variations without continually requiring renegotiation with the application layer. At the top of the protocol stack are reactive applications that specify their requirements to the QoS manager as a set of multiple values corresponding to a set of allowable operating points with different degrees of acceptability. The application adapts its behavior by reacting to events from a QoS manager indicating a change in the level of QoS being provided by the lower layers. Each QoS level is defined by the average sustained throughput, delay and packet loss rate.

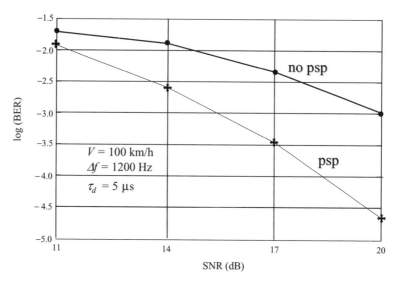

Figure 10.20 Performance of the 32-state MLSE at 100 km/h and with a 2.46 GHz RF
carrier. The packet length is 200 bytes. The transmission rate is 1 Mbit/s with
a delay spread (τ_d) of 5 μs, mobility of 100 km/h and a residual frequency
offset of 1200 Hz.

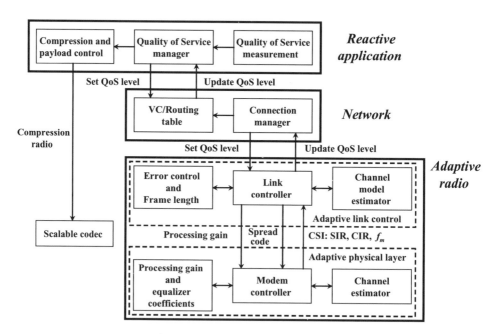

Figure 10.21 System architecture for maintaining QoS in mobile wireless multimedia networking.

10.1.9.5 Adaptive steps

When the channel estimator indicates a degradation in QoS parameters, the link controller first attempts to maintain the QoS level by adapting the error control and frame length control. If the interference levels are too high to be handled by the link layer, the processing gain is adapted by the modem controller. If none of the adaptations are sufficient to maintain the current QoS, the level of service quality for one or more applications is reduced, and an event indicating this is passed up the protocol stack. The application layer may respond by adjusting parameters such as the speech codec compression ratio to be compatible with the drop in QoS. The reverse sequence of events takes place when channel conditions improve. Similarly, the network layer itself may respond to events from the link layer indicating changes in current QoS level by, for example, performing a connection route optimization. These issues will be discussed later in Chapter 15.

The radio adaptation is based on the channel state information measured in the physical layer that includes:

- SIR;
- the channel impulse response;
- the Doppler frequency.

The SIR determines the interference level in the channel, while the CIR determines the type of fading in the channel. The CIR is measured by a 48-tap complex matched filter (MF) in the adaptive equalizer (Figure 10.15) that is matched to a training sequence inserted in the preamble of the packet.

The speed of the user is inferred from the Doppler frequency which is determined by the frequency tracking loop in Figure 10.15. The deviation from the average value of the phase error (ϕ_k) is proportional to the Doppler frequency. These three parameters allow the link controller to distinguish between a degradation due to fading or interference and, if due to fading, the speed of the mobile. This information enables the radio to appropriately adapt the frame length, error control, processing gain and equalization for the given channel condition.

10.1.9.6 Self-describing packets

The changes in the frame size, error control and processing gain used in the payload of each frame must be communicated by the sender node to enable the receiver to decode the packets. Communicating these parameters can result in a high signaling overhead and defeat many of the gains achieved by adaptive control. An alternative is to make each packet self-describing, such that no synchronization between the send and receive nodes is needed.

The physical layer header is encoded consistently from packet to packet and includes the information which tells the receiver how to decode the remainder of the packet. The physical layer header can, in this way, be decoded rapidly and consistently in the radio hardware while using a strong code, since it constitutes a relatively small amount of extremely important information. Even one bit error means the entire packet must be discarded; therefore, it should be heavily protected. Figure 10.22 shows the structure of the physical layer header used by the radio. After a preamble needed to lock onto the signal, the header bit map has single-bit flags that indicate which of the remaining header fields are actually present. In

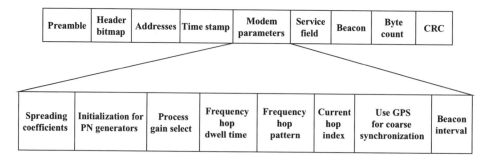

Figure 10.22 Packet format for maintaining QoS in mobile wireless multimedia networking.

other words, all other fields are optional so that the overhead is kept to a minimum. The remaining fields are used for:

1. The physical layer destination and origin addresses;

2. A time stamp field is inserted at the transmitter and used for synchronizing the MAC protocol;

3. The physical layer requires a number of parameters to successfully decode the incoming packet. These are specified in the modem field. For example, this field contains the processing gain to use for the body of the packet;

4. A service field which specifies the error control information for the receiving link layer to decode the rest of the packet;

5. A bit to indicate whether this is a beacon transmission or not (as from a base station);

6. A byte count and CRC for the header.

10.2   A SOFTWARE RADIO ARCHITECTURE FOR LINEAR MULTIUSER DETECTION

10.2.1   A unified architecture for linear multiuser detection and dynamic reconfigurability

From Chapter 5, in a CDMA multiuser system, the output of the matched filter for the kth user can be expressed as

$$y_k = A_k b_k + \sum_{j \neq k} A_j b_j \rho_{jk} + n_k \quad k = 1, \ldots, K \tag{10.3}$$

The MF outputs for the users in the system in vector form become

$$\mathbf{y} = \mathbf{R} \mathbf{A} \mathbf{b} + \mathbf{n} \tag{10.4}$$

10.2.1.1 *Linear multiuser schemes*

In general, a linear multiuser detector performs a linear transformation of the received vector $\mathbf{y}$ and makes the decision for user k as

$$\hat{b}_k = \text{sgn}(\mathbf{L}\mathbf{y})_k \tag{10.5}$$

where $\mathbf{L}$ is the appropriate transformation. From Chapter 5 we already know that for the MF, $\mathbf{L}$ takes the form

$$\mathbf{L}_{\text{MF}} = \mathbf{I} \tag{10.6}$$

where $\mathbf{I}$ is the identity matrix. For a decorrelator (DC),

$$\mathbf{L}_{\text{DC}} = \mathbf{R}^{-1} \tag{10.7}$$

For an approximate decorrelator (AD)

$$\mathbf{L}_{\text{AD}} = \mathbf{I} - \delta\mathbf{J} \tag{10.8}$$

where $\mathbf{J}$ is derived on the basis of the assumption that $\mathbf{R}$ is strongly diagonal as

$$(\mathbf{I} - \delta\mathbf{J})^{-1} = \mathbf{I} - \delta\mathbf{J} + \mathbf{o}(\delta)$$

For the MMSE detector we have

$$\mathbf{L}_{\text{MMSE}} = (\mathbf{R} + \sigma^2\mathbf{W}^{-1})^{-1} \tag{10.9}$$

where $\mathbf{W} = \mathbf{A}^T\mathbf{A}$.

10.2.1.2 *'Modified' filter $h_k(t)$*

The DC can also be considered as a matched filter with impulse response

$$h_k(t) = \sum_{j=1}^{K} \mathbf{R}^+_{kj} s_j(t) \tag{10.10}$$

where $\mathbf{R}^+_{kj}$ denotes $(\mathbf{R}^{-1})_{kj}$ The AD can be realized as

$$h_k(t) = s_k(t) - \sum_{j \neq k} \rho_{jk} s_j(t) \tag{10.11}$$

The MMSE detector as

$$h_k(t) = \sum_{j=1}^{K} (\mathbf{R} + \sigma^2\mathbf{W}^{-1})^+_{kj} s_j(t) \tag{10.12}$$

where $(\mathbf{R} + \sigma^2\mathbf{W}^{-1})^+_{kj}$ again denotes $[(\mathbf{R} + \sigma^2\mathbf{W}^{-1})^{-1}]_{kj}$. For the MF,

$$h_k(t) = s_k(t) \tag{10.13}$$

Any of the linear multiuser detectors described above can be realized by appropriately choosing the filter taps of the 'modified' MF, $h_k(t)$. This forms the basis of the software radio architecture for linear multiuser detection.

10.2.1.3 *Variable QoS*

When the measure of quality is the BER achieved by the user, the relative performance of these receivers (see Chapter 5) may be classified as

$$\text{BER}_{\text{MF}} \geq \text{BER}_{\text{AD}} \geq \text{BER}_{\text{DC}} \geq \text{BER}_{\text{MMSE}} \tag{10.14}$$

Thus, reconfiguring the detectors among the MF and the above structures allows the option of a variable QoS from moderate (for the MF) to very high (for the MMSE detector). As an example, a user may switch to an MF configuration when using voice traffic, while the MMSE mode may be preferred for data traffic. We can also account for different data rates by deriving appropriate single-user linear filters for the multirate schemes.

10.2.1.4 *Software radio architecture for linear multiuser detection*

Figure 10.23 shows a possible generic software radio architecture for linear multiuser detection. The architecture includes:

- channel processing (such as translation from IF to baseband);

- environment processing (e.g. estimation of signal and interference parameters and correlations);

- matched filtering and information bit-stream processing (e.g. FEC or convolutional decoding, soft decisions, etc.).

These functionalities are partitioned into two core technologies based on processing speed requirements. These two technologies are based on FPGA and DSP devices. The key idea behind the software architecture is to reconfigure the MF dynamically according to the desired QoS (corresponding to one of the appropriate linear detectors).

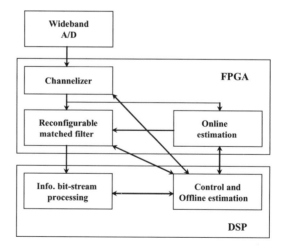

Figure 10.23 Functional architecture of a software radio for linear multiuser detection.

10.2.1.5 Logical partitioning of the architecture

Essentially, each user's radio could possibly have only one variable filter-tap receiver implemented using Xilinx FPGAs. All classes of the above receivers require two common generic operations:

- estimation of path delay of the users;

- the generation of PN sequences $\{s_k(t)\}$ of the users in the system.

The MF is probably the simplest in that, for any user k, it just uses the information from these two generic operations in determining the timing offset of the PN sequence $s_k(t)$ for the specific user.

For the AD, the complexity is slightly higher than that of the MF in that the 'modified' matched filter taps are adjusted according to the formulation given by Equation (10.10).

As seen in Figure 10.24, the additional functionality required here is the crosscorrelation values $\{\rho_{kj}\}$ and also the signature of sequences of all the users $\{s_k(t)\}$.

The functional operations required for the DC are additionally

- the computation of the inverse matrix of crosscorrelations $\mathbf{R}^{-1}$;

- the column vector corresponding to the kth user, i.e. $\mathbf{R}^+_{kj}$.

The MMSE receiver incurs additional complexity over the DC in that it also requires estimates of the received signal powers of the users in the system (matrix $\mathbf{W}$).

The issue of estimating the path delays or the received signal powers of the users in the system should be included.

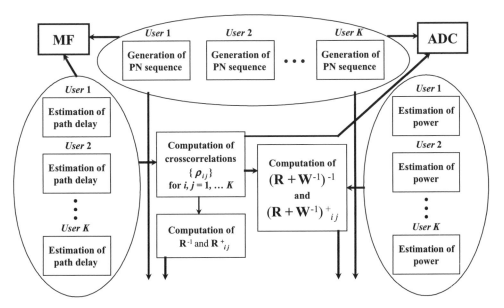

Figure 10.24 Logical partitioning of functionality in a software radio receiver for linear multiuser detection [887] © 1999, IEEE.

The software radio architecture does have the functionality required to do both the estimation operations, similar to that used in conventional radio designs.The software radio architecture should be versatile in that it can easily allow a variety of signal processing algorithms to be implemented for accomplishing the required estimation.

10.2.1.6 Testbed example

In the testbed (Figure 10.25) described in [887], the core of the hardware is based around an Aptix MP3 board with up to 12 FPGA components (up to 432 000 programmable gates) for sample level processing tasks with rates of up to 30.0 MHz. Information rate tasks (Viterbi decoding, deinterleaving, etc.) are performed by the VMEbus-based Pentak 4270 Quad 'C40 DSP Processor board. As a baseband front end, the testbed has two sets of dual (I and Q) A/D converters for input (Analog Devices AD 9762XR 12-bit, 41 MHz converters, and Pentak 6472 10-bit, 70 MHz converters) and one set of D/A converters (Analog Devices AD 9042ST 12-bit, 100 MHz).

Analog Devices converters are connected to the FPGA board through the custom adapters. Pentak converters are connected to the DSP via a multiband digital receiver (Pentak 4272) acting as a *channelizer* in Figure 10.23. The channelizer performs:

- frequency down-conversion;

- low pass filtering;

- decimation of the sampled baseband signal.

It is used for selection of the service bandwidth (i.e. the *tuning band*) from among those available in the sampled signal. The multiband digital receiver used as a channelizer has two

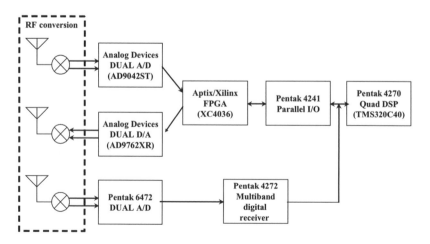

Figure 10.25 Testbed block diagram.

narrowband receivers with a dynamic range of 1 kHz – 1 MHz, and one wideband receiver with dynamic range 2 MHz – 35 MHz.

It is capable of supporting a wide range of output sample rates.

A Sun workstation accesses the VMEbus through a Bit3 Sun-Sbus to VMEbus adapter and is used as a primary data stream source as well as development host. Development tools are centered around Signal Processing Worksystem (SPW).

SPW is a computer-aided design (CAD) tool that allows for the simulation and design of the complex communication systems based on block diagrams. It has a rich library of common communication blocks as well as facilities for creation of custom-coded blocks. Once entered, the block diagram of the target system can be simulated by using SPW's signal flow simulator.

If these floating-point simulations are producing satisfactory results, the block diagram can be partitioned into DSP and hardware parts. The part of the design that is targeted for DSP implementation is prepared by SPW's code generation system (CGS) (or MultiProx in case of partitioning into multiple DSPs) and is directly downloadable into the Quad TMS320C40 floating-point DSP board.

SPW's hardware design system (HDS) is used to model the behavior of a fixed-point part of the design. Again, SPW's signal flow simulator is used to verify the fixed-point model functionality. After the design is verified, the corresponding hardware description language (HDL) code is automatically generated. This code is then synthesized (and/or simulated by the event-driven HDL simulator like Synopsys VSS) by an appropriate set of tools (e.g. Synopsys design compiler). The resulting design is further processed by the Xilinx XACT tool in order to generate the FPGA chip layout and routing, thereby producing the configuration bit stream. This configuration bit stream defines the combinatorial circuitry, flip-flops, interconnect structure and the I/O buffers inside a particular FPGA device.

Aptix tools are used to interconnect the FPGAs, connect FPGAs and DSPs through a parallel I/O board, and for routing of debugging signals to the control probes of the logic analyzer.

10.2.1.7 *Partitioning of the architecture*

The software radio architecture for linear multiuser detection is partitioned into two core technologies, namely FPGA and DSP devices. This partitioning is usually driven by the required functionality of the radio device and also the processing speed requirements.

The algorithmic complexity of the linear multiuser receivers increases with an increase in performance. Specifically, the complexity of the signal processing algorithms corresponding to the MF, AD, DC and MMSE receivers can be classified as

$$C_{\mathrm{MF}} < C_{\mathrm{AD}} < C_{\mathrm{DC}} < C_{\mathrm{MMSE}} \tag{10.15}$$

For a system with K users, processing gain N, and oversampling factor O_s, the floating-point complexity in terms of the number of multiply-accumulate (MAC) operations is

$$C_{\mathrm{MF}} = K(NO_s + 1)$$
$$C_{\mathrm{AD}} = K(NO_s + 1) + (K - 1)NO_s + (K - 1)\log_2 K$$
$$C_{\mathrm{DC}} = K(NO_s + 1) + 2K(K - 1)NO_s + (2/3)K^3$$
$$C_{\mathrm{MMSE}} = K(NO_s + 1) + 2K(K - 1)NO_s + 2K^2$$
$$+ (4/3)K^3 + C(\mathrm{amp}) \tag{10.16}$$

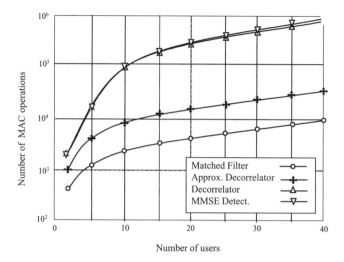

Figure 10.26 Processing speed constraints: MAC operations versus number of users.

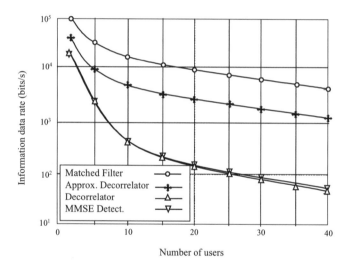

Figure 10.27 Achievable rates versus number of users (One 50 MIPS DSP).

C(amp) denotes the complexity due to amplitude estimation that is incurred in the MMSE receiver (which is not explicitly considered here).

Figure 10.26 presents the number of MAC operations. We can evaluate the number of users that can be supported as a function of achievable information data rates (for different receiver structures) where the active constraint is a limitation in the complexity or the processing speed of the DSP device. In Figure 10.27, a simple illustration is shown for a 50 MHz floating-point TMS320C40 DSP. In all DSP implementations, serious constraints on the achievable data rates are imposed when the number of users increases, especially when we operate with more complex receivers.

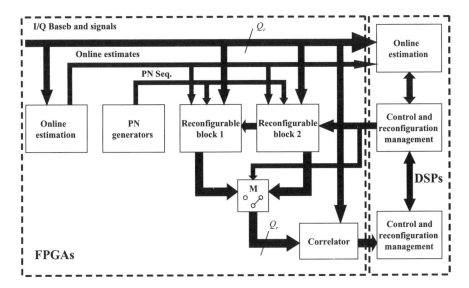

Figure 10.28 Block diagram of software radio implementation.

10.2.1.8 *The software radio architecture*

Reconfigurable linear multiuser detection is implemented by partitioning the resources between FPGAs and DSPs as shown in Figure 10.28. The FPGA segment of the architecture includes:

- the PN sequence generators;

- two reconfigurable blocks that determine the filter taps of the appropriate linear receivers;

- an online estimate module;

- the correlator.

The motivation for the particular selection of the constituents that comprise the FPGA segment is that the functionality provided by each constituent here can be easily handled by the processing speeds of the FPGA hardware. The two reconfigurable blocks contain the core of the sample level processing for each of the multiuser detectors and are actually implemented in separate FPGA components to facilitate on-the-fly reconfigurability.

FPGA

This segment enables us to reconfigure one of the blocks while the other is running and therefore, by oversizing the hardware, avoid loss of data during switchover to a different receiver structure. The rest of the sample level processing logic (PN sequence generators, control multiplexer and correlator) is implemented in a separate FPGA component since it does not require reprogrammability.

The online estimation block is used to perform timing estimates of the incoming signals, which are then used, along with the reference PN sequences, to compute the appropriate filter taps. There is also a provision for refining the online estimates by using more sophisticated offline algorithms. An FPGA component is used as an interface between the DSP segment and APTIX board.

DSP

The more algorithmically complex operations are partitioned into the DSP. This segment is implemented using a quad TMS320C40 DSP board. The operations include:

- offline estimation procedures;

- information bit-stream processing;

- control and reconfiguration management.

The offline estimation block includes:

- estimation of the received signals powers for the MMSE detector;

- computation of the inverse of the matrix of crosscorrelations (in the case of both the DC and the MMSE detectors);

- it can also refine estimates of the online estimation block in the FPGA segment.

The information bit-stream processing functions vary from error control techniques, such as FEC or convolutional decoding, to soft decision decoding.

The control and reconfiguration management block in the DSP segment of the architecture basically determines the variable QoS that can be achieved by dynamic reconfiguration of the receiver structures in the FPGA segment.

The reconfiguration of the blocks in the FPGA segment is directly controlled by the control and reconfiguration management block. The subsequent switching amongst receivers is achieved by a control multiplexer M (shown in Figure 10.28) that is also controlled by the control and reconfiguration management block. The DSP segment of the device acts to achieve the variable QoS requirements of the specific type of service. Implicit in the control and reconfiguration management block is also the capability to interact with higher layer protocols/stacks to facilitate the QoS demand of a specific data stream.

At the link layer, the actions of the DSP segment of the device also encompass environment processing such as sensing interference levels. The DSP devices have access to the configuration bit streams for each of the receiver structures and can download the particular configuration bit stream as and when required.

Table 10.3 shows the relative hardware complexity of the different linear multiuser detection schemes in terms of the required number of configurable logic blocks (CLBs).

Table 10.3   Complexity comparison of different detectors
(in number of CLBs)

Quantization	MF	AD	DC and MMSE
$Q_r = 4$ bits	25	$32 + 13K$	$37 + 3K(K - 1)$
$Q_r = 5$ bits	27	$40 + 14K$	$47 + (13/4)K(K - 1)$
$Q_r = 6$ bits	29	$57 + 15K$	$64 + (7/2)K(K - 1)$

A CLB, in the case of Xilinx 4000 series FPGA, comprises a pair of flip-flops and two independent (Boolean) logical four-input function generators.

The complexity for each detector increases with both increasing number of users K and increasing precision in quantization (Q_r denotes the number of bits used in quantization). The MF complexity remains invariant to the number of users in the system and depends only on the precision of quantization.

Even though the DC (and the MMSE) achieves better QoS, it comes at the expense of an exponential increase in complexity (with an increasing number of users) over the AD detector. This directly maps to an increase in processing power requirements of the FPGA segment of the software radio architecture. This again motivates switching to lower order receivers when the QoS requirements are moderate.

The complex operations required to be performed in the DSP segment of the architecture involve matrix inversion and offline estimation of amplitudes, etc.

10.2.2 Experimental results

First we consider floating-point implementations of the linear multiuser receivers. In all experiments, as described in [887], the transmitter powers of the users are controlled perfectly so that they are all received at the same power level. The floating-point results presented here agree very well with the analytical results on the performance of these receivers from Chapter 5. To show the flexibility of the software radio, consider two sets of experiments.

10.2.2.1 Example 1

Fix the number of users in the system at $K = 15$.

In Figure 10.29, at low SNRs, the range of QoS achievable is quite limited. At higher SNRs, the DC and MMSE provide a BER gain of up to three orders of magnitude compared to either the MF or the AD. The AD is again slightly better than the MF for the set of operating points considered here.

For the case when $N = 128$, it is seen that the dynamic range of QoS achievable is greater among the detectors at high SNRs. The AD provides up to an order of magnitude better performance than the MF.

10.2.2.2 Example 2: QoS achieved for different numbers of users

When the number of users in the system increases, the users can still operate at the same SNR, but can switch to a higher complexity detector to maintain the same QoS.

As an example, it is seen in Figure 10.30 that when $N = 64$, if users in a system desire a BER of $\approx 10^{-3}$, they can operate on

- an MF receiver (up to five users);

- an AD receiver (up to ten users);

- the DC receiver (up to 30 users)

without changing their received power level.

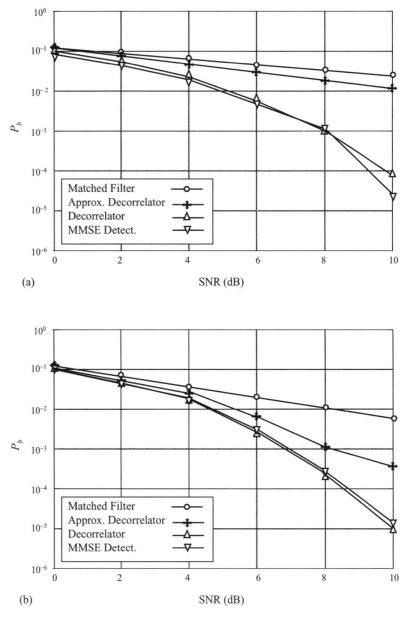

Figure 10.29 (a) Performance curves with 15 users and $N = 64$; (b) performance curves with 15 users and $N = 128$.

If some of the users require more stringent BERs, such as $\approx 10^{-5}$, they could operate entirely on a DC realization of the 'modified' MF.

When the processing gain is $N = 128$, it is seen that there is a greater range in achievable capacity by switching to the DC and MMSE receivers.

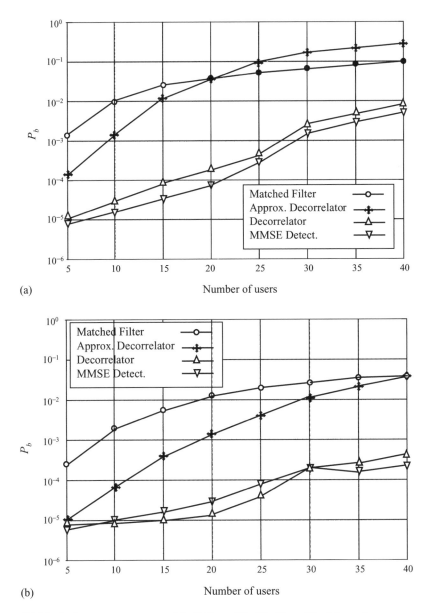

Figure 10.30 Performance curves with SNR = 10 dB, $N = 64$; (b) performance curves with SNR = 10 dB, $N = 128$.

10.2.3 The effects of quantization

Consider fixed-point operations using two-, three-, and four-bit quantization. The experimental results for the MF are shown in Figure 10.31, and for the approximate DC in Figure 10.32. All the detectors experience a degradation in performance, with the MF being the least sensitive and the DC being the most sensitive to quantization effects. The encouraging

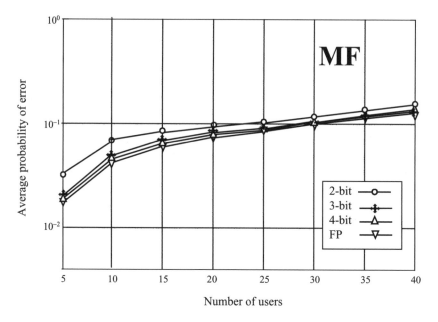

Figure 10.31 Performance curves for MF with SNR $= 5$ dB, $N = 64$, $O_s = 4$.

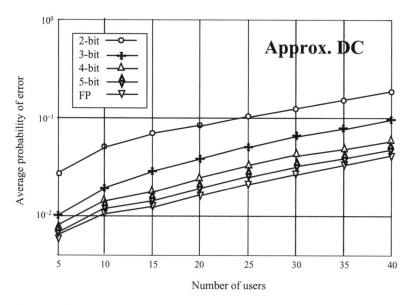

Figure 10.32 Performance curves for AD with SNR $= 5$ dB, $N = 64$, $O_s = 4$.

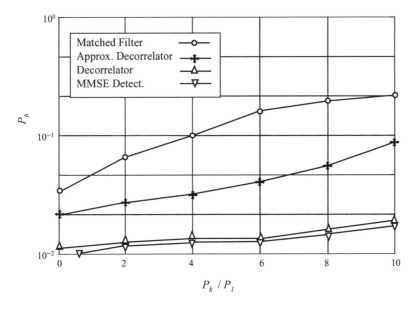

Figure 10.33 Performance curves in the presence of the near–far effect with SNR = 5 dB, $N = 64$, $O_s = 4$, $K = 10$, $Q_r = 5$.

note is that the use of even a six-bit precision quantizer seems to pull performance close to that of the floating-point reference results.

10.2.4 The effect on the 'near–far' resistance

Figure 10.33 shows the average BER achieved by the MF, AD and DC receivers for the case of four-bit quantization as a function of interference powers. The desired user's power is fixed, while the interfering powers are increased. The MF and AD are not near–far resistant, and show degradations in performance as the interfering powers increase. The AD shows a more graceful degradation in performance relative to the matched filter. The DC, which is theoretically near–far resistant, however, fails to maintain this property in the presence of quantization effects. However, the BER performance is still superior to the other two detectors. More details on practical solutions can be found in [886–949].

10.3 RECONFIGURABLE ASIC ARCHITECTURE

So far in this chapter we have discussed adaptive and reconfigurable schemes used for improving the system performance or reducing energy consumption. In this section, we extend this topic to include reconfiguration from one multiple access technology to another, mainly TDMA, OFDM and CDMA options. There are a number of different solutions to this problem [886–946] © 1999, IEEE. In this section we discuss the case where the reconfiguration is performed on the level of Application Specific Integrated Circuit (ASIC) implementation. In particular, we will have a closer look into an architecture that can be used to realize any one of several functional blocks needed for the physical layer implementation of data communication systems operating at symbol rates in excess of 125 Msymbols/s.

Multiple instances of a chip based on this architecture, each operating in a different mode, can be used to realize the entire physical layer of high-speed data communication systems based on different multiple access schemes. The presentation in this section is based on [950].

The architecture features the following modes (functions):

- real and complex FIR/IIR filtering;

- least mean square (LMS)-based adaptive filtering;

- discrete Fourier transforms (DFT); and

- direct digital frequency synthesis (DDFS) at up to 125 Msamples/s.

All of the modes are mapped onto regular data paths with minimal configuration logic and routing. Multiple chips operating in the same mode can be cascaded to allow for larger blocks.

Baseband signal processing requirements are measured in the units gigaoperations per second (GOPS), and are beyond the capabilities of present day DSP solutions (e.g. the TI TMS320 family [951] or the Motorola 56000 family). Designers of such systems must incur the high cost and extended development time associated with custom ASIC solutions, even at the prototype phase of their work.

Traditionally, the designers have had to settle for the usual ASIC versus DSP tradeoff. An ASIC provides low power and high performance, as shown in Table 10.4, but it is inflexible to change and takes considerable engineering time to build. The DSP approach provides design flexibility and short time to market, but it is power hungry and incapable of satisfying the increasing computational demands of high-speed communication systems. The norm has been to take the hybrid approach, which is to utilize ASIC blocks to handle computations that are beyond the capabilities of a DSP, and DSP code for flexibility and time to market. This approach was demonstrated in the previous sections of this chapter.

Table 10.4   ASIC, DSP, hybrid and flexible ASIC advantages and disadvantages

	Advantages		Disadvantages	
ASIC	Low power Small area	High performance	Inflexible Long TTM	Expensive High skill level
DSP	Flexible Quick TTM	Late changes possible Medium skill level	Large area High power	Low performance
Hybrid ASIC + DSP	Medium TTM	Some flexibility	Inflexible Long TTM Expensive	Large area High power Low performance
Flexible ASIC	Low power Flexible	Medium area High performance	High skill level	New paradigm

TTM – time to market

However, such a hybrid quite often leads to a system that processes the disadvantages of both the ASIC and the DSP.

One approach that overcomes this dilemma is to develop highly flexible VLSI data paths that are specifically designed and optimized for a single class of functions or tasks [950]. Such a class of circuits could be thought of as ASIC/FPGA hybrids, and they will be referred to as ASIC/FPGA. They would combine highly optimized processing units with programmable interconnects which can be changed in real time. However, unlike traditional FPGAs, the minimal functional block is not general purpose and the routing options are not global, rather, they are local and optimized for the set of applications envisioned for the architecture. Such a circuit is also fundamentally different from the minimally programmable arrays of custom computing elements used to solve very specific problems [952, 953]. The result is a highly optimized yet programmable VLSI circuit that can easily compete with ASICs in terms of performance, but at the same time provides a high degree of flexibility.

Another attempt to combine high speed with flexibility, generally known as *reconfigurable computing* (RC), has recently seen active development [952, 953]. Most reported RC implementations attempt to integrate an FPGA-type programmable block with a general purpose microprocessor [954] on the same die [955] or at the board level [954]. The programmable block is then configured to improve computational efficiency of the microprocessor. Thus, RC implementations can be thought of as a *DSP/FPGA hybrid*, while the example architecture is an *ASIC/FPGA hybrid*.

The first approach has the advantage of being maximally flexible, but it cannot achieve the very high speeds required for the target application of the example architecture, since it offers no speed advantages over a regular FPGA.

Other efforts are directed at making FPGAs faster at the expense of decreased flexibility, and could some day make the ASIC/FPGA tradeoff unnecessary [956]. A design has been reported that uses a high-speed reconfigurable ASIC data path that is quite similar to the one discussed in this section but is limited to different versions of the same algorithm [956].

Finally, a major research effort has resulted in a chip that attempts a three-way hybrid (DSP/FPGA/ASIC), but stops short of implementing sufficient computational power for the applications envisioned for the example architecture [957].

At this time, no reported design has the combination of high speed, low power and flexibility of the example architecture. This example is a highly versatile VLSI architecture targeted at the high-speed data communications market. The architecture is highly regular and operates at data rates of up to 125 Msamples/s. Most importantly, it can be reconfigured in different modes in order to realize many of the functional blocks required in the transmitters and receivers of high-speed data communication systems. All modes utilize the same I/O pins and have been mapped onto a common, highly regular, data path with minimal control and configuration circuitry. A summary of the modes supported by the example architecture is provided in Table 10.5.

10.3.1 Motivation and present art

Figures 10.34 and 10.35 illustrate how multiple instances of a chip, based on the example architecture, can be used to realize different blocks within an OFDM transmitter and a QAM receiver. Each of the shaded blocks in these figures represents a single instance of the chip

Table 10.5 Modes of operation

Mode	Parameters	ID
Real FIR	Max. number of taps $= 64$	RFIR
Dual real FIR	Max. number of taps $= 32$	R2FIR
Complex FIR	Max. number of taps $= 16$	CFIR
Real IIR	Max. number of taps $= 64$	RIIR
Adaptive FIR	Max. number of taps $= 8$	AFIR
Digital frequency synthesis	fmax. $= 60$ MHz; fmin $= 4$ kHz; Df $= 2$ Hz; SFDR $= 72$ dB	DDFS
Discrete Fourier transformation	Max. block size $= 32$ Min. block size $= 4$	GFFT

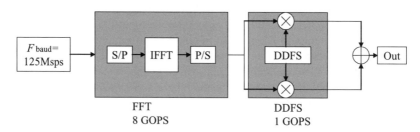

Figure 10.34 Block diagram of an OFDM transmitter.

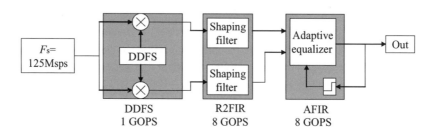

Figure 10.35 Block diagram of a QAM receiver.

under a different configuration. A chip based on the example architecture is assigned the acronym *RADComm*, for *R*econfigurable *A*SIC for *D*ata *Comm*unications.

The computational requirements are annotated in gigaoperations per second (GOPS, where an operation is defined as a single real multiplication) for a system operating at the maximum data rate.

10.3.2 Alternative implementations

The need for a high-performance flexible architecture becomes obvious when the limitations of the classical FPGA and DSP solutions are considered. The RC designs discussed above cannot implement all the desired features at the required data rates, and thus will not be used in this comparison. Let us compare the example architecture to the standard FPGA and DSP solutions using three basic metrics, namely, required sources, power consumption and ease of implementation.

The example architecture is synthezised onto a 0.25 mm/2.5 V CMOS technology, and is compared to the present state-of-the-art FPGA and DSP solutions available from leading manufacturers [950].

The FPGA field is represented by the largest and most advanced chip from the Xilinx Virtex series, XVC1000. The XVC1000 [958] features over one million gates partitioned into 8464 logic blocks and is manufactured in 0.22 mm/2.5 V technology [959]. DSPs are represented by the top-of-the-line processor from TI, TMS320C6201B. This processor operates at up to 233 MHz and is based on a 0.18 mm/1.8 V technology [960].

The example architecture is very flexible and easily scalable. For the purposes of illustration, we will assume an implementation with 64 computational elements (defined subsequently) operating on 16-bit data and capable of realizing the features in Table 10.5 at a rate of 125 Ms/s. Based on results obtained from the synthesis of the VHDL code, this implementation requires 70 kgates, occupies an area of just 4 mm^2, and consumes approximately 1.6 W power. We will make comparisons on two representative modes of operation: a 16-tap complex FIR (CFIR) and a 16-point DFT (GFFT) [961]. The example architecture is ideally suited for implementing a complex FIR filter and is less suitable for computing the DFT. It will be shown that the example architecture is superior to the FPGA and the DSP solutions for both of these modes.

10.3.3 Example architecture versus an FPGA

The CFIR mode utilizes 100 % of the computational resources in the example architecture. Over 90 % of the computational core is taken up by fast 16×16 bit multipliers, making the size and performance of a multiplier a critical parameter. The fastest 16×16 multiplier that can be implemented in an XCV1000 operates at 59 MHz and requires 96 logic blocks [962]. Thus, the desired data rates simply cannot be achieved using an FPGA (pipelining will not be considered since it would also increase RADComm's performance).

Let us decrease the requirements in order to compare power consumption and implementation complexity.

A fully utilized XCV1000 can just implement 64 multipliers, leaving no room for additional logic. Thus, two FPGAs would be needed to realize a 16-tap complex filter operating at 59 MHz. This greatly complicates the design for two reasons:

1. A single chip solution becomes a board level problem.

2. Complicated partitioning of logic elements is needed, with special attention paid to delays introduced by inter-chip connections.

A first estimate based on the number of logic blocks required to implement the needed functionality, gives a power dissipation of *6.4 W when operating at 59 Ms/s (13.5 W at 125 Ms/s)*. The development time required for an FPGA design of this scope is considerably longer than the time required to configure and use a chip based on the example architecture. Moreover, a chip based on this architecture can be switched to a different mode in just a few clock cycles, whereas an FPGA of this size takes a long time to reconfigure (over 750 kcycles, or 15 ms).

Table 10.6   RADComm, FPGA and DSP for complex filtering [950]

	Simple to use	Single chip	Fast reconfiguration
RADComm	•	•	•
FPGA			
DSP	•		•

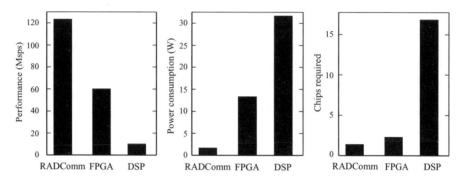

Figure 10.36   RADComm versus FPGA and DSP for complex filtering [950] © 2000, IEEE.

10.3.4   DSP against the example architecture

This comparison is not as straightforward as the comparison to an FPGA. The TMS320C6201B operates at a high frequency of 233 MHz but can only perform two multiplications per cycle and incurs delay in memory access. TI benchmarks indicate that the DSP requires $2N$ cycles per symbol for an N-tap complex FIR filter [961]. Thus, the TMS320C6201B could only process 7.3 Ms/s for a 16-tap complex filter. Again, the desired data rates are considerably above those achievable. Indeed, it would take 17 of these processors to provide the needed computational resources. One TMS320C6201B consumes over 1.9 W when operating at 233 MHz [905]. Thus, a total of 32 W would be used for filtering at the desired data rate. Clearly, a board full of these very expensive and power-hungry chips is not an acceptable solution. The advantages of the example architecture operating in the CFI mode over the existing solutions are summarized in Figure 10.36 and Table 10.6.

10.3.5   Computation of a complex 16-point DFT – the Goertzel FFT mode

The DFT can be computed using a very efficient radix-4 FFT algorithm. This algorithm does not map well onto the example computational core and is not used in this architecture. No such constraints exist for an FPGA or a DSP, and these solutions have a potential advantage over the example architecture. The DFT operating at the desired data rate uses 1386 logic blocks (16 %) in an XCV1000, and consumes 3.3 W power. The TMS320C6201B DSP requires ten cycles per symbol to compute 16-point FFTs, and could only process the input at up to 23 Ms/s, requiring five processors to achieve the desired data rate and consuming

Table 10.7   RADComm, FPGA and DSP for DFT computation [950]

	Simple to use	Single chip	Fast reconfiguration
RADComm	●	●	●
FPGA		●	
DSP	●		●

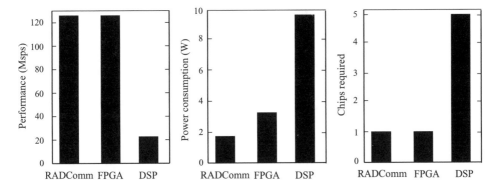

Figure 10.37 RADComm versus FPGA and DSP for DFT computation.

9.5 W power. Despite using a less efficient algorithm, the example architecture operating in the GFFT mode has significant advantages over the existing solutions, as summarized in Figure 10.37 and Table 10.7.

The main challenge in designing such a highly versatile architecture, capable of supporting high symbol rates (over 125 Msymbols/s), is to identify suitable realizations of each function (mode of operation) such that all functions could be mapped onto a computational fabric with a minimal amount of routing and control logic. Indeed, chips exits that implement one or two of the features in Table 10.5 (e.g. [963]), but no existing architecture can implement all of them at high sampling rates. Identification of such an architecture together with the selection of the algorithms to implement each function will be presented hereafter.

At first sight, the architectures' modes of operation do not reveal a great deal of similarity between them. For example, the conventional architecture for DFT [964] is quite different from adaptive filters using the LMS algorithm [965], which in turn have little in common with architectures used for the implementation of direct digital frequency synthesizers [966–968]. The main reason for these differences is the optimization of each architecture for the implementation of a single function. The first task is iteratively to consider various implementations for each mode to identify those that have the greatest computational overlap with all other modes. Once the implementations have been selected, we can define *a single computational element that forms the least common denominator for all functions* listed in Table 10.5. This rather simple computational unit, hereafter referred to as a *tap*, is shown in Figure 10.38. Having identified the tap, a computational core is created by grouping a number of taps in an array, as shown in Figure 10.39. The core consists of four parallel rows (labeled RR, II, IR, RI) with N ($N = 16$ for this implementation) taps each, and fixed

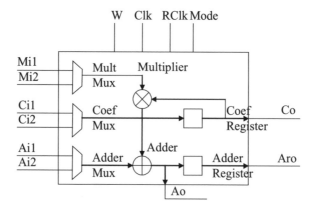

Figure 10.38 Computational unit (TAP) [950] © 2000, IEEE.

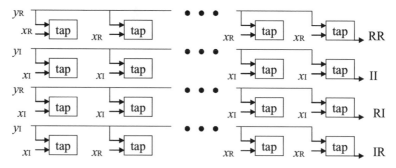

Figure 10.39 Computational core.

inter-tap routing. Reconfiguration of the chip is accomplished by proper utilization of the three multiplexers within each tap (Figure 10.38).

Each tap is pipelined at its output, and the critical path in any one of the operating modes is never greater than the delay through a multiplier plus two adders, a register and a multiplexer. When synthesized in 0.25 mm CMOS technology with a 16-bit databus, the critical path delay is under 7 ns, well below what is needed for operation at 125 MHz. Almost all of the fixed routing is local – taps connect only to their four immediate neighbors. This allows for a very compact layout, low routing area overhead and reduced capacitive loading. This, in turn, translates into higher clock frequencies.

10.3.6 Fixed coefficient filters

The transfer function of an FIR filter

$$y(n) = \sum_{k=0}^{N-1} x(n-k)C_k$$

can be implemented in either direct or transposed forms [964]. The transposed form implementation, shown in Figure 10.40, allows the critical path to be reduced to a single

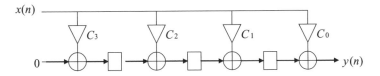

Figure 10.40 Transposed form FIR ($N = 4$).

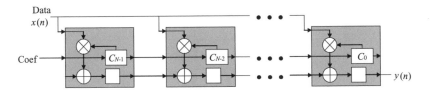

Figure 10.41 Mapping of the transposed form FIR.

'multiply/add,' and is a natural choice for the high-speed operations envisioned for this architecture. The transposed form FIR filter structure can be mapped onto the example core directly, as shown in Figure 10.41. For all fixed coefficient filters (Table 10.5), the coefficients are shifted in serially from an external source during the chip startup sequence. If a filter requires more taps than are available in a single chip, an unlimited number of chips can be cascaded.

10.3.7 Real FIR/correlator

The architecture can be configured as a single 64-tap structure (RFIR) by routing the output of each row to the input of the row below (e.g. output of row RR to input of row II, Figure 10.39). If a lower filter order is acceptable, a single chip based on the example architecture can be configured to process two data streams (in-phase and quadrature) simultaneously (R2FIR), with 32 taps each. In this mode, one filter is realized using rows RR and II, while the second is realized using rows RI and IR. A single 16-tap complex FIR filter can be implemented by using the chip in the CFIR mode.

Since filtering is a linear operation, the filter operating on complex values can be decomposed into four parallel real-valued filters, with the final output obtained by combining the four outputs

$$xc = (x_R c_R - x_I c_I) + i(x_R c_I + x_I c_R)$$

where x is the input and c is the coefficient vector, see Figure 10.39). A pair of adders is used to combine the outputs of rows (RR, II) and (RI, IR).

10.3.8 Real IIR/correlator

The example architecture can also be configured to operate as a single 64-tap, real-valued infinite impulse response filter (RIIR). This configuration is almost identical to that of the RFIR mode described above. To realize an IIR filter, we simply feedback the output of the FIR filter to an additional adder at the input, as shown in Figure 10.42.

Figure 10.42 Realization of a real IIR filter.

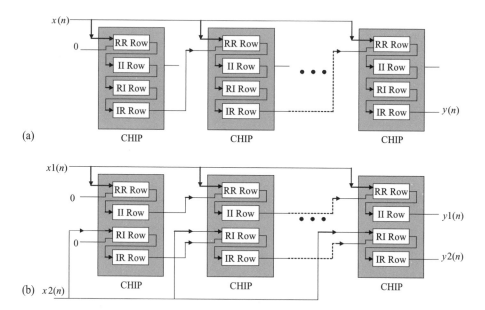

Figure 10.43 Cascading for (a) RFIR; and (b) R2FIR. Unused I/Os are not shown.

10.3.9 Cascading fixed coefficient filters

An unlimited number of chips based on the example architecture can be cascaded to achieve filters with a larger number of taps. The cascading is accomplished by routing the outputs of each chip to the X0 inputs of the next chip, as shown in Figure 10.43 for the RFIR and R2FIR modes (cascading routing for the CFIR and IIR modes is very similar). All of the chips load the coefficients from the same bus, requiring only one external source (ROM) for the entire filter. Each chip uses its sequential position in the cascade to determine the starting and ending times for shifting in the coefficients. The unused inputs on all chips are set to '0.'

10.3.10 Adaptive filtering

The adaptive filtering mode can operate on both real and complex data. For complex operations, the filter coefficient adaptation is based on the popular LMS algorithm [890], summarized in Figure 10.44. The adaptive filter consists of two distinct parts – the filtering circuit, Equation (10.17), and the coefficient update circuit, Equation (10.18). In the example architecture, this functionality is implemented by allocating half of the taps to the

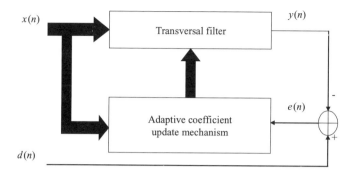

Figure 10.44 The LMS adaptive algorithm.

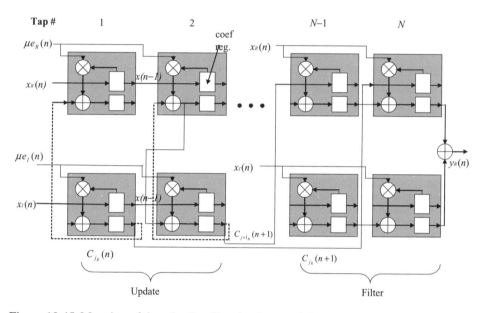

Figure 10.45 Mapping of the adaptive filter (real part only).

update circuit, and the rest to the filter circuit. From Equation (10.18) we observe that $\mu e(n)$ is a term common to all of the new coefficients.

$$y(n) = \sum_{j=1}^{N} C_j x(n - j) \tag{10.17}$$

$$C_j(n + 1) = C_j(n) + \mu x^*(n - j)e(n) \tag{10.18}$$

For this reason, in Figure 10.45, $\mu e(n)$ is fed to all the update taps. The $x^*(n - j)$ values are obtained by shifting the values on $x(n)$ into the coefficient registers. At time n, the jth tap will have $x(n - j)$ stored in the coefficient register. The complex conjugate is obtained by multiplying $\mu e(n)$ by -1 in the RI row. The new coefficient value, $C_j(n)$, is obtained by feeding the output of the II row to the adder input of the RR row (dashed line). The filtering part of the circuit is identical to the CFIR mode discussed above.

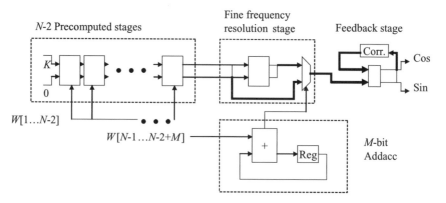

Figure 10.46 DDFS using modified CORDIC.

10.3.11 Direct digital frequency synthesis

There exist a number of different methods for digital generation of a sinusoid of variable frequency. Most DDFS designs used today store precomputed samples of a sinusoid in a ROM lookup table and output these samples at different rates [895]. This method requires a large ROM to achieve acceptable spectral purity and does not map well on any flexible computational core. An alternative method for generation of a sinusoid is based on trigonometric definitions and properties of the *sine* and *cosine functions*. This method, known as coordinate rotation (CORDIC) [966–968] (see Figure 10.46), requires very few constant coefficients and is more suitable for implementation on a flexible ASIC architecture such as the one used in this example.

10.3.12 CORDIC algorithm [968]

Signal $e^{j\omega t}$ at sampling instants $t = nT_s$ can be represented as $e^{j\omega(n-1)T_s}e^{j\omega T_s}$. So, to synthesize a signal of frequency ω with sampling rate $1/T_s$, each previous sample should be multiplied by $e^{j\omega T_s}$ (rotated for phase, the ωT_s). In this case, the sine and cosine of an angle are calculated using a cascade of N 'subrotation' stages. The kth stage rotates the input complex number, considered as a two-element vector (x_k, y_k), by $\pm \delta/2^k (\delta = \pi/2)$ radians depending on the kth bit of the phase control word (W) [see Equation (10.19)]. By changing W, we can rotate an initial vector by an angle in the range $[0 \cdots \pi - \delta/2^{N+1}]$ in increments of $\delta/2^N$ radians.

$$(x_k, y_k) = \begin{cases} x_{k-1} - T_k y_{k-1}, \ y_{k-1} + T_k x_{k-1} & \text{when } W[k] = 1 \\ x_{k-1} - T_k y_{k-1}, \ y_{k-1} - T_k x_{k-1} & \text{when } W[k] = 0 \end{cases} \tag{10.19}$$

where $T_k = \tan(\frac{\delta}{2^k})$.

The standard CORDIC algorithm suffers from two major problems, namely, low frequency resolution and high power consumption. The particular realization of the DDFS in the example architecture introduces two modifications to the conventional CORDIC algorithm that circumvent both of these problems. Modifications involve the fine frequency resolution and the feedback stages shown in Figure 10.46, which depicts the top-level block

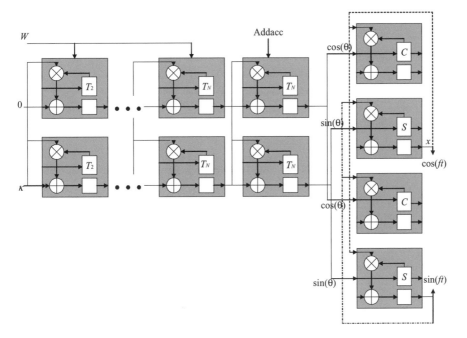

Figure 10.47 Mapping of the DDFS architecture.

diagram of the modified CORDIC. A detailed discussion of these enhancements to the standard CORDIC algorithm and their effect on the DDFS performance are given in [966]. The algorithm is easily decomposed into a sum of products formulation, and is ideally suited for implementation on the example core.

The detailed mapping of the precomputed stages and the feedback section on the proposed computational core is shown in Figure 10.47. Only a small accumulator is needed in addition to the example computational core to implement the CORDIC algorithm.

A DDFS implemented using the example core can generate 125 Ms/s quadrature sinusoids in the frequency range of 4 kHz to 40 MHz in 2-Hz steps while maintaining SFDR above 72 dB [966].

10.3.13 Discrete Fourier transform

The discrete Fourier transform (DFT) can be represented as

$$X[k] = \sum_{n=0}^{N-1} x(n) W_N^{kn} \tag{10.20}$$

where $W_N = e^{-j(2\pi/N)}$.

Two approaches can be used: direct evaluation, and the Cooley–Tukey FFT algorithm [969]. Most dedicated DFT processors use the FFT algorithm because of its computational efficiency for large block sizes. Two major problems prevent the use of the FFT algorithm in the example architecture; FFT requires complex global routing, and the routing must change significantly to process different DFT sizes.

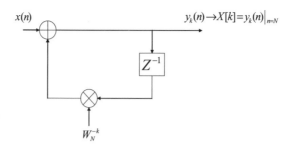

Figure 10.48 Goertzel algorithm for computing $X[k]$.

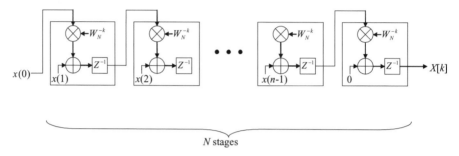

Figure 10.49 Unrolling the Goetzel recursive stage.

10.3.14 Goertzel algorithm

The example architecture implements a form of direct DFT evaluation known as the Goertzel algorithm [969]. The Goertzel algorithm calculates each DFT point using a simple recursive circuit, shown in Figure 10.48. The major advantages of this algorithm are 'in-place' computation, no external memory requirements, a highly regular structure, and easy cascadability that allows multiple chips to process a single large DFT block. If multiplication by W_N^{-k} is viewed as a rotation by $2\pi k/N$ radians, the Goertzel algorithm is very similar to the CORDIC algorithm discussed in the previous section. The similarity can be extended to the set of CORDIC precompute stages by 'unrolling' the recursive loop. N recursive cycles in Figure 10.48 can also be computed using N multiplication stages, with the output of the kth stage being the input of the $(k+1)$th stage, as shown in Figure 10.49. The main difference between the two algorithms is that the same coefficient is used for all the stages of the DFT computation.

The DFT algorithm operates on blocks of data at a time and therefore requires proper scheduling of resources to process the serial input data. Since the algorithm is first-order recursive, only the value $y_k(n-1)$ is needed to compute $y_k(n)$. Thus, only the nth stage is used at time n to compute $y_k(n)$. This property is exploited by scheduling the data and coefficients on each stage so that $y_k(n)$, $k = 0 \ldots N - 1$ can be computed simultaneously. The following schedule allows full utilization of the core: the coefficient in the kth stage at time n is given by W_N^{n-k}. An example of this schedule for a four-point FFT using four columns of taps is given in Table 10.8, where W refers to the coefficient and A refers to the input to the adder (see Figure 10.49). The input data stream is given by

Table 10.8 Scheduling for the Goertzel algorithm

Time	Tap number / (W, A)				Out
n	0	1	2	3	X
0	$W^0, x(1)$	$W^3, v(2)$	$W^2, v(3)$	$W^1, 0$	
1	$W^1, x(1)$	$W^0, x(2)$	$W^3, v(3)$	$W^2, 0$	$V[1]$
2	$W^2, x(1)$	$W^1, x(2)$	$W^0, x(3)$	$W^3, 0$	$V[2]$
3	$W^3, x(1)$	$W^2, x(2)$	$W^1, x(3)$	$W^0, 0$	$V[3]$
0	$W^0, y(1)$	$W^3, x(2)$	$W^2, x(3)$	$W^1, 0$	$X[0]$
1	$W^1, y(1)$	$W^0, y(2)$	$W^3, x(3)$	$W^2, 0$	$X[1]$
2		$W^1, y(2)$	$W^0, y(3)$	$W^3, 0$	$X[2]$
3			$W^1, y(3)$	$W^0, 0$	$X[3]$
0				$W^1, 0$	$Y[0]$

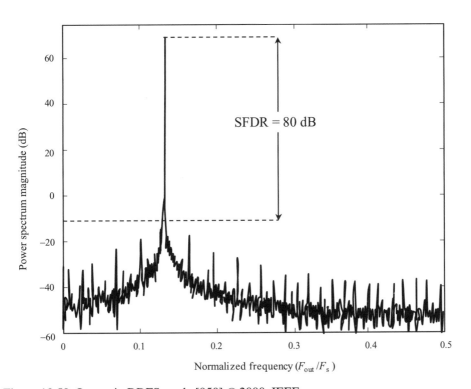

Figure 10.50 Output in DDFS mode [950] © 2000, IEEE.

$[v(0), \ldots v(3), x(0), \ldots, x(3), y(0), \ldots, y(3)]$. The highlighted part of the table shows the processing of $x(n)$.

This scheduling can be easily implemented on the example core. Coefficient values are simply read from the ROM in sequential order and shifted in. The adder inputs (A) are stored in the adder registers of rows RR and RI. The correct load/hold pattern for these registers is achieved using a simple N-bit shift register. A single '1' circulates in the shift register. Whenever the kth bit is '1,' the adder registers in rows RR and RI of the kth stage load the current value of the input, $x(n)$; if the bit is '0,' the value is unchanged. This shift register, and the two registers to hold values $x(0)$ and $x(1)$ (see Figure 10.49), make up all the overhead of this algorithm. The algorithm implicitly allows for taking DFTs of blocks smaller than the number of stages. Assuming that the coefficient values are stored sequentially in a ROM, to compute DFTs of size L, $(L < N)$, the ROM address is incremented in steps of $m = N/L$ (m is an integer).

Using this method, the desired value, $X[k]$, is generated by the first L stages. Processing of these values by the remaining $N - L$ stages does not change them. This can be verified by observing that each stage multiplies the input by W_N^{-k}, and $N - L$ stages result in a multiplication by $W_N^{-k(N-L)}$. Using the definition of W_L, we obtain

$$W_L^{-k(N-L)} = e^{i2\pi k((N-L)/L)} = e^{-i2\pi k} e^{i2\pi k(N/L)}$$
$$= e^{-i2\pi k} e^{i2\pi km} = e^{i2\pi(m-k)} = 1^{m-k} = 1$$

Figure 10.50 shows the PSD of a sinusoid generated by a chip configured in the DDFS mode [950].

11

Examples of Software Radio Architectures

In this chapter we present some examples of software radio architectures, including different military and TDMA- and WCDMA-based civil mobile communication systems.

11.1 A LOW-POWER DSP CORE-BASED SOFTWARE RADIO ARCHITECTURE

In this section we discuss briefly an approach to developing a low-power digital signal processor (DSP) subsystem architecture for advanced software radio platforms. The architecture is intended to support next generation wideband spread spectrum military waveforms. The platform should provide the reconfigurability between the systems listed in Table 11.1 [970–972].

The signal forms to be included in the system are given in Table 11.2 and the power needed for such a system is presented in Figure 11.1. The system level block diagram is shown in Figure 11.2, while options considered for the RF section are shown in Figure 11.3.

The required power dissipation of ADC conversion versus the required resolution is shown in Figure 11.4. It is possible to reduce ADC power by interleaving the sample-and-hold and quantizer circuits. Interleaved ADCs repeat functional blocks of the serial ADC (Figure 11.5(a)) to distribute functions over parallel sample-and-hold circuits (Figure 11.5(b), left), or over parallel quantizers (Figure 11.5(b), right). The interleaved circuits can be integrated on monolithic chips. In many cases, greater performance is achieved at lower total power due to the lower circuit frequencies per parallel path. This is achieved at the expense of a larger chip area. Parallel circuits may not be the best low-power option for passband ADCs in which the highest IF frequency is much greater than the signal bandwidth. In this case, a single sample-and-hold with parallel quantization yields lower power.

Advanced Wireless Communications. S. Glisic
© 2004 John Wiley & Sons, Ltd. ISBN: 0-470-86776-0

Table 11.1   Representative DoD and commercial systems

Waveform family	Frequency range	Channel bandwidth	Modulation
FH HF	2 to 20 MHz	3 kHz	AM/FM
SINCGARS	30 to 88 MHz	25 kHz	FH (100 b/s)
Have Quick	225 to 400 MHz	25 kHz	FH
EPLRS	420 to 450 MHz	5 MHz	DSSS/FH
JTIDS	960 to 1310 MHz	6 MHz	DSSS/FH
GPS	1.5 GHz	10 MHz	DSSS
Speakeasy	All of the above		
SATCOM	7 GHz	variable	variable
IS-95	800 MHz, 1.9 GHz	1.25 MHz	DSSS/CDMA
IS-54/136	800 MHz	30 kHz	TDMA
ITU 3G	1.8 to 2.1 GHz	5, 10, 15 MHz	DSSS/CDMA

Table 11.2   Waveforms. Adapted from [973]

Wave-form	Frequency range	Bandwidth	Chip rate	FH bandwidth	FH Rate	Modulation format
1	30 to 88 MHz 225 to 400 MHz 0.8 to 2.45 GHz	Variable to 10 MHz	0.32 to 16 Mcps			Quasi-bandlimited MSK
2	77 to 88 MHz 225 to 400 MHz 1.8 to 2.0 GHz	Variable to 20 MHz	N/A	>= 100 MHz	100hps	Wavelet (feature-less, good side lobe suppr-ession)
3	20 to 2000 MHz	Variable to 12 MHz	20 Mcps	200MHz	400 hps	MSK
4	6 to 2000 MHz	Variable to 26 MHz	N/A	N/A	N/A	transform domain DSPN
5	10 to 2000 MHz	1.6 MHz	680 Kcps I and O	Maximum 60 MHz or 20 % carrier	1200 hps	non-LPI: MSK LPI: Filtered DS

Another approach to low power is the oversampling delta–sigma (also called sigma–delta) converter [976] shown in Figure 11.6. The sample-and-hold circuit oversamples the input analog signal by a factor of N. This allows the use of a simpler, lower power converter such as a threshold comparator (a 1 V bit converter). In addition, the quality of the antialiasing analog filter (measured as the ratio of the −3 dB bandwidth to the −23 dB bandwidth) need

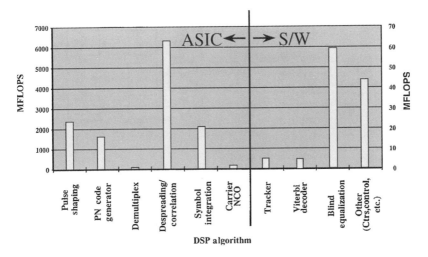

Figure 11.1 Software radio receiver DSSS processing capacity estimates for 10 Mchips/s, 40 Msamples/s (ADC), 9600 bits/s, 5 μs delay spread; total: 12.78 GFLOPS; ASIC: 12.67 GFLOPS, S/W: 0.11 GFLOPS.

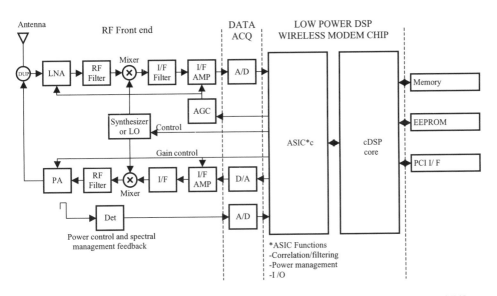

Figure 11.2 Example system-level software radio transceiver block diagram [973, 974].

not be as high as the Nyquist ADC. The cascade of the decimator and digital filter with the antialiasing filter yields a product filter, each stage of which need not be as effective as the single antialiasing filter of the Nyquist ADC.

The delta–sigma and Nyquist converters differ in dissipated power, number of components and pace of development of the underlying technology. Since digital technology traditionally advances faster than RF and analog technology in terms of integration and power reduction, delta–sigma converters have a technology advantage. Delta–sigma converters

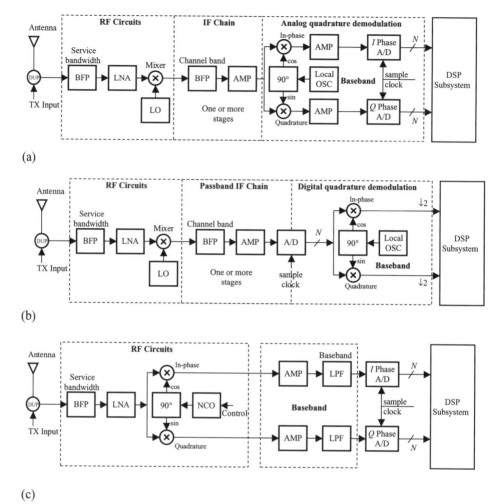

(a)

(b)

(c)

Figure 11.3 Common RF receiver configurations for software radios. (a) I,Q superheterodyne with passband alternatives; (b) passband superheterodyne; and (c) direct conversion (zero IF or homodyne).

with 2–4 bits of resolution are now deployed in commercial low-power, spread spectrum wireless terminals.

As already indicated in Chapter 10, increased dynamic range is required for multiuser interference mitigation, near–far performance and jammer suppression. In analyzing options for the low-power applications, a reasonable aggressive compromise of a few years ago envisioned 8–12 bits of dynamic range, a 20–30 MHz sampling rate, and 150–300 mW of power dissipation. In the meantime, the sampling rate has advanced significantly with the same accuracy and dissipated power.

As an example, a member of the Texas Instruments (TI) ASIC family [977], the TSC6000, has a 0.18 μm geometry, and operates at 1.8 V. Its density is 30 000 gates/mm^2, with an

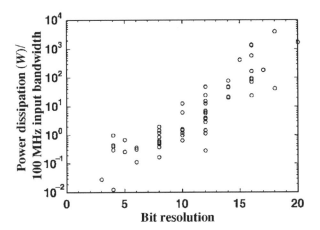

Figure 11.4 ADC power dissipation versus resolution [975] © 1999, IEEE.

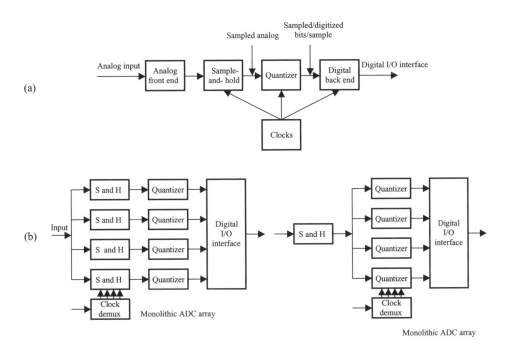

Figure 11.5 ADC functional block diagrams. (a) ADC functional diagram; and (b) inter-leaved ADCIS.

average power dissipation of 0.025 μW/MHz/gate. In comparison to the 5 V devices, the gate density has increased by a factor of twelve. The average power per gate has decreased by a factor of 40. This reduction is considerably more than the difference resulting from the square of the supply voltage.

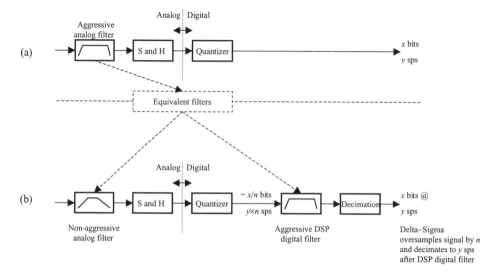

Figure 11.6 Nyquist (traditional) versus delta–sigma ADC. (a) Traditional ADC; and (b) delta–sigma ADC.

11.1.1 CMOS power dissipation

The power consumption of CMOS circuits includes switching power, short-circuit power, static power and leakage power. Switching power, the dominant component, is essentially proportional to the square of the supply voltage, and is linearly proportional to the load capacitance, the frequency and the percentage of time the circuit is active. Short-circuit power depends on the current owing during switching transients. An increase in transition speed reduces this effect, since short-circuit current spikes are present only when multiple circuits are in transition at the same time. Static power is dissipated only when bias currents are present in analog circuits. Leakage power is static power that is unintentionally dissipated by the current in the device during the 'off' state. This can become a significant factor in low-voltage designs using low-threshold voltages [978]. The low-power ASIC designs for core-based DSP must minimize the effects of the entire range of sources of CMOS power dissipation.

11.1.2 Low-power design techniques

A number of techniques can be used to reduce power. Lowering the supply voltage dramatically reduces power, but also degrades performance. Reducing the threshold voltage can then enhance performance. But this results in greater leakage of current. Parallel circuits yield higher performance at the expense of greater complexity and chip area. Spurious transitions occur from switching a circuit several times during the same clock cycle because of multiple input changes or differences in signal path length. Power dissipated in spurious transitions can be reduced by latches and by balancing logic paths. Switching can also be postponed by reordering inputs to introduce the most frequently changing inputs later in the logic path.

Placement and routing can minimize the product of inter-connection capacitance times, switching activity through the localization of high-activity networks. Static random access memory (SRAM) designs that only activate a small portion of the memory array and that use latch-style sense amplifiers will essentially eliminate static current.

Power supply switching transistors, that turn off internal power to circuits not in use, reduce standby power. For example, high-threshold transistors can turn off power to circuits using low-threshold transistors. Power management techniques, such as 'sleep' and 'standby' modes, minimize power during times when only a portion of the circuit is needed. Other portions can then be selectively turned on as necessary, and active circuits can be selected in software or hardware. As an example, clock gating can turn off the clock to inactive circuits, based on a sleep bit set by software. Alternatively, clock gating could be based on an instruction decode that clocks only those circuits required for the instruction.

11.1.3 Configurable digital processors

The processing capacity needed for wideband, high-bit rate, and spread spectrum waveforms exceeds the capabilities of programmable DSP technology available at this moment.

The flexibility needed for software radio implementations can be achieved using configurable digital ASICs with a microprocessor or DSP that is tailored for software radio applications as discussed in Chapter 10.

In a 'ideal' software radio, practically all of the transceiver's functions would be implemented in a general purpose processor. Transmitter functions ranging from the source encoder to the up-conversion of the baseband signal to the final carrier frequency would be performed by this processor.

Likewise, the converse functions in the receiver would also be accomplished by the processor, including carrier phase recovery and symbol or pseudonoise (PN) code timing recovery in a spread spectrum application.

In principle, this would allow the same hardware platform to support *any* physical layer imaginable, as well as the higher layers of the protocol stack.

This ideal radio would only be limited by the capabilities of the analog components (e.g. ADCs, DACs, power amplifiers, low-noise amplifiers, antialiasing filters and antenna subsystems), and by the capacity of the processor.

11.1.4 The DSP core and ASIC approach

The flexibility goal of the software radio can be approximated by:

- moving the ADC and DAC as close to the antenna as possible;

- implementing functions with very high processing demands in ASICs that can be run-time configured to support a wide range of signal structures;

- maximizing the number of functions performed by the DSP.

Current DSP, ASIC and semiconductor technology is rapidly evolving to support DSP core macro cells via an ASIC library [977].

The DSP core is a microprocessor-like programmable DSP that has been designed for efficient gate count, die area and performance.

The DSP core can be augmented with a customized ASIC on a single chip. For software radio applications, the ASIC includes high-throughput modulation functions, such

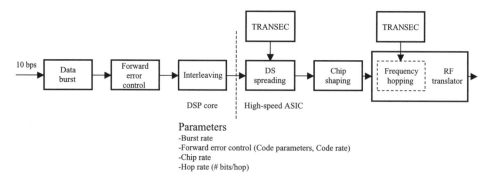

Figure 11.7 DSSS transmitter block diagram.

as correlators, and high-speed filters. It also includes power management and input/output functions.

The combination of the DSP core and other ASIC functions can be closely tailored to the application. This offers greater potential throughput, lower gate count, smaller die area and lower power consumption than other semicustom approaches, such as field-programmable gate arrays (FPGAs).

To some, it seems inappropriate to use the term 'flexible' when describing an ASIC, because the logic of an *application-specific* chip cannot be modified once the part has been fabricated. However, the generic functions used in software radio are common to a broad spectrum of wireless waveforms. Thus, one may be able to design an ASIC that accommodates many waveforms, as discussed in Chapter 10. A waveform may be characterized by a parameter set defining modulation type, symbol rate, chip rate, pulse shape, constellation, etc. If implemented in an ASIC with programmable parameters, the ASIC can support a variety of waveforms.

11.1.4.1 Example

Consider the pulse-shaping function of the modulated block in Figure 11.7. Suppose this function is implemented in an ASIC using a finite impulse response filter. If the filter coefficients are stored in RAM, and if this RAM is accessible to the DSP, the DSP can change the pulse shape by changing the contents of the RAM.

A reasonable ASIC design would also allow the number of taps in the filter to be programmable. As long as the number of taps required does not exceed the maximum number available on the ASIC, an arbitrary pulse shape can be supported.

11.1.4.2 Example

Similar analysis applies to the receiver ASIC of Figure 11.8. Timing recovery, transmission security (TRANSEC), and a multifingered correlation receiver that is programmable, probably consumes more chip area and power than its waveform-specific counterpart. It is therefore clear that significant effort will be expended in the design of such ASICs. The implementation of parameterized ASICs might not be as efficient with respect to gate count and power consumption as that of a point design. Thus, waveform flexibility requires

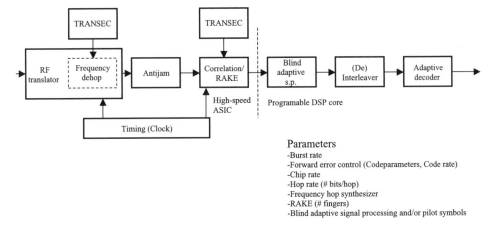

Figure 11.8 DSSS receiver algorithm block diagram [973, 974].

tradeoffs that include waveform types supported, gate count, die area, power consumption and cost.

11.1.5 Waveform supportability

For handheld radios, size, power, and weight constraints generally limit the amount of processing capacity available for hosting the functions in software. This implies that the computationally intensive functions must be implemented in ASICs. A partitioning of DSSS/CDMA waveforms onto ASIC and DSP components is presented in Figures 11.7 and 11.8. The most crucial parameters in determining waveform supportability by the DSP subsystem include:

- ADC sampling rate;

- dynamic range;

- processing capacity requirement for the translation of digital IF signals to baseband;

- processing capacity requirements of modulation and demodulation algorithms;

- processing capacity requirements of error coding and decoding algorithms (especially, e.g. Viterbi decoders);

- processing capacity of synchronization algorithms, especially the demanding burst mode and satellite applications.

   Partitioning a function for ASICs generally emphasizes increased throughput, but can also emphasize lower power consumption.

   The precision of arithmetic operations can be customized through the analysis of dynamic range requirements. Lower precision arithmetic can greatly reduce power consumption in the DSSS correlator, for example, if a multiuser receiver is not used.

   In addition to the physical layer considerations, protocols can limit the flexibility of the software radio. In particular, it may be difficult or impossible to support burst protocols

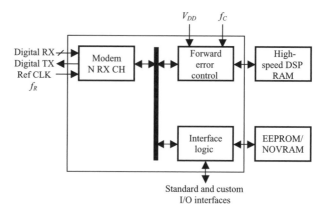

Figure 11.9 Multipurpose modem chip [973, 974].

that place hard requirements on carrier-phase and symbol (or PN code) timing recovery. Such protocols may place a higher processing demand on the system than can be provided in a DSP architecture. Satellite signal tracking has similar requirements. These timing constraints may be met using dedicated ASIC tracking circuits. Thus, a software radio may support the desired range of waveforms and protocols if RF conversion and digital ASICs with the requisite flexibility are teamed with an appropriate DSP.

11.1.6 A DSP core-based software radio ASIC

Based on the above discussion, we now present an example design of the system defined by Table 11.1.

The design consists of a single ASIC that includes

- wireless modem functions;

- advanced power management;

- a Texas Instruments TMS320C6xx programmable DSP core [977, 978]; and

- input/output interface logic.

This ASIC is presented in Figure 11.9 and may be called the multipurpose modem chip (MMC). The spread spectrum modem logic within the MMC may be programmed by the DSP core for a wide range of functions, including:

- narrowband jammer suppression;

- digital-to-digital (D/D) conversion;

- demodulation;

- modulation;

- transmit power management;

- military transmission security (TRANSEC) features;

- power management.

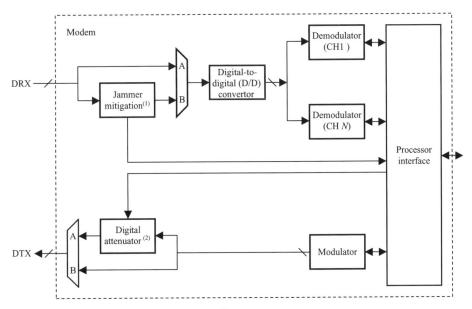

(1) Jammer mitigation includes space–time adaptive antenna processing, an adaptive transversal filter or a combination
(2) Digital attenuator required for precise transmit power management

Figure 11.10 Front end processing [973, 974].

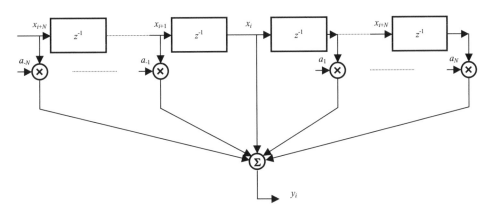

Figure 11.11 Adaptive transversal filter.

11.1.6.1 *Jammer suppression*

Figure 11.10 illustrates the main functions of the digital front end of the ASIC, including the placement of the jammer suppression block. Interference mitigation algorithms for this part of the ASIC [979] are chosen to address the characteristics of expected jammers and interference. The adaptive transversal filter (ATF) shown in Figure 11.11, for example, excises narrowband noise and continuous-wave (CW) jammers. The processing for this filter is described in Chapter 12.

A typical two-sided ATF design for the MMC calls for twelve bits of precision, 33 taps, and the Widrow–Hoff least mean square (LMS) algorithm (see Chapter 12) to determine the weight values.

The tap weights are maintained in 16 read-only registers within the ASIC, and are accessible to the DSP core. Since the weights are accessible to the DSP core, it computes the ATF transfer function, jammer frequency and bandwidth in parallel while the ATF is operating. As shown by the multiplexer (MUX) block after the ATF in Figure 11.10, the filter can be bypassed when there is no need for jammer mitigation. Under these circumstances, the low-power ASIC design allows the ATF reference clock to be gated off, thereby conserving power.

11.1.6.2 Digital-to-digital conversion

The digital-to-digital converter reduces the number of bits of precision from the jammer mitigation block to the minimum required to demodulate the signal. This substantially reduces the overall gate count and power consumption required for the computationally intensive front end filtering and correlation functions. Existing DSSS receivers have used lower precision since the dynamic range can be recovered through integration in despreading.

11.1.6.3 The general demodulator

The number of independent demodulator channels is determined by the application. In terrestrial systems, at least four channels are needed to implement a traditional RAKE receiver because of the nature of the channel. Reception of global positioning satellite (GPS) signals requires as few as four, and as many as 12, independent channels.

Depending on the number of active channels required, power management provisions in the design allow the DSP to disable the reference clock to unused channels to conserve power.

A more detailed description of the programmable general demodulator is shown in Figure 11.12. Conventional digital modulation formats supported by the general demodulator in the MMC design include

- M-ary phase shift keying (M-PSK);

- offset quaternary phase shift keying (OQPSK);

- M-ary pulse amplitude modulation (M-PAM); and

- M-ary quadrature amplitude modulation (M-QAM).

Spectral shaping may be used with any of these waveforms. Trellis-coded modulation (TCM) may be generated and demodulated in the DSP core. Further design tradeoffs include the possibility of a hardware Viterbi decoder in the ASIC. Minimum shift keying (MSK) and Gaussian MSK may also be demodulated primarily using the DSP core.

The first operation performed by each demodulator block of the general demodulator is final quadrature down-conversion of the passband signal to baseband. Doppler and reference oscillator drift are compensated by a numerically controlled oscillator (NCO) having a mean phase and frequency precision of 2^{-16} degrees and 2^{-32} Hz, respectively. The phase and

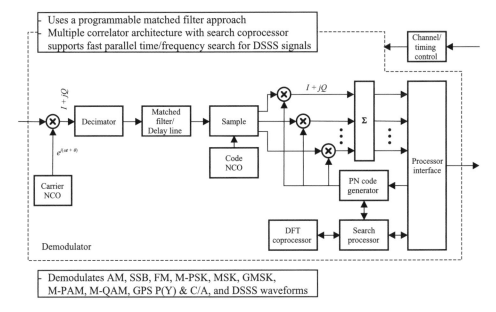

Figure 11.12 General demodulator architecture [970] © 1999, IEEE.

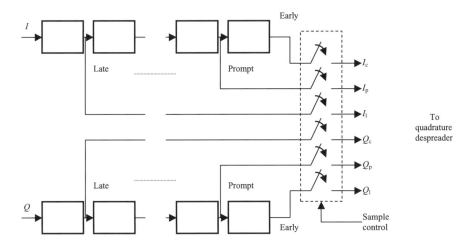

Figure 11.13 Delay line and sample functions.

frequency of the NCO are set by the DSP via a memory-mapped I/O (MMIO) interface. The complex baseband samples are then decimated so that matched filter processing uses the lowest possible frequency. The decimator includes a complex finite impulse response (FIR) filter followed by a decimator. The filter coefficients and decimation rate are controlled by the DSP core.

Following decimation, the complex signal is applied to a matched filter, the coefficients of which are specified by the DSP via the MMIO. The output of the matched filter is sent to the delay line and sample functions, which are shown in greater detail in Figure 11.13.

The PN chip separation associated with each output in the delay line is a function of the final sample frequency at the output of the decimator. In non-spread applications, the delay line can be used to recover symbol timing.

The output of the code NCO determines when the switches in Figure 11.13 are closed. The outputs of the delay line are then mixed with the PN code generator output and accumulated during one symbol period or less. At least three code mixers are needed to track the PN code phase using a delay-locked loop (DLL). As many as 32 code mixers could be used in an alternative ASIC design to reduce PN code acquisition time.

The demodulator design allows the DSP to specify precisely when the accumulation period begins, as well as how long the period will be. This flexibility allows the DSP to track the phase of the data edges in satellite communications systems in which propagation path distances are changing rapidly. It can also track Doppler that measurably lengthens and shortens the symbol period, particularly at low data rates. In non-spread spectrum applications, the PN code is inhibited, allowing the same demodulator structure to accommodate other waveforms. After accumulation, the I and Q samples are stored in ASIC RAM with an interrupt to the DSP. The DSP reads the I and Q samples via the MMIO interface to complete the demodulation process in software.

The code mixer and accumulator (or code correlator) design is such that the DSP can disable the reference clock to unused correlators in order to conserve power. To simplify the logic and conserve gates, code correlators are turned off in groups of four or eight. The demodulator contains a search processor and discrete Fourier transform (DFT) coprocessor that accelerate the PN code-timing acquisition process. The search processor forms a time–frequency array of data by computing the DFT of the I and Q samples from each correlator. A sequential probability ratio test is then performed on the time–frequency data to determine the location of the signal rapidly in both time and frequency, independently of the DSP core. To initiate a search, the DSP specifies the number of PN chips to be searched and the sample integration period, the inverse of which yields the desired DFT frequency coverage. Then, the search processor signals the DSP with a flag indicating whether the signal was found, and with the PN chip offset and frequency bin of the signal if appropriate. As with other parts of the chip, the search processor may be powered down by the DSP when not in use.

11.1.6.4 The general modulator

The general modulator function of the MMC is illustrated in Figure 11.14. It is designed to generate the following baseband signals: amplitude modulation (AM), single sideband (SSB), frequency modulation (FM), M-PSK, MSK, GMSK, M-PAM, and M-QAM. Others may be programmed using the MMC and DSP core. The modulation process begins when the DSP writes the coded bits to be transmitted into an ASIC data buffer via the MMIO interface. The vector mapper arranges these bits into an index, which selects the appropriate constellation point from the constellation lookup table. These constellation points have been previously initialized by the DSP core. The I and Q samples associated with the selected constellation point are mixed with the applicable PN code sequence. They are then interpolated to the transmit clock frequency, exciting two pulse-shaping filters. These filter coefficients are also initialized by the DSP. A programmable delay can be inserted into the quadrature signal path as needed to generate offset QPSK, MSK or GMSK.

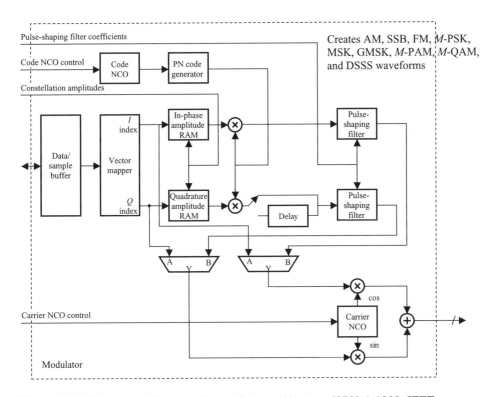

Figure 11.14 Programmable general modulator architecture [970] © 1999, IEEE.

The outputs of the pulse-shaping filters are then fed through a multiplexer to a quadrature up-converter consisting of the carrier NCO and two mixers. The quadrature up-converter operates like the quadrature down-converter in the demodulator. If the sample buffer contains complex samples of an analog waveform, the digital waveform logic is bypassed, feeding the vector mapper output directly to the quadrature up-converter. The DSP core programs the frequency and phase of the quadrature up-converter and the PN code. This allows the radio to perform Doppler precorrection and to correct reference clock drift.

11.1.6.5 *Precision programmable attenuator*

The attenuator of Figure 11.10 scales the modulator output amplitude using a fixed-point multiplier, the value of which is set by the DSP core. This assures precise power management when required.

The numerical precision of the attenuator depends on the dynamic range over which the output power must be controlled. If the dynamic range is large, the precision of the attenuator output could exceed that of the modulator, distorting the waveform. The attenuator block may be bypassed, sending the output of the modulator directly to the DAC.

11.1.6.6 *Complexity and power consumption*

The receiver complexity drives gate count, which is nearly linear in the number of parallel receiver channels. In addition, the single unit operation (SUO) configurations include

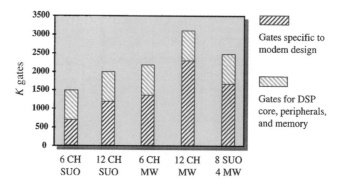

Figure 11.15 Total MMC gate count [970] © 1999, IEEE.

the complex DSSS waveform and a GPS receiver. Multiple-waveform (MW) configurations accommodate multiple conventional waveforms. Device configurations comparing gate count therefore vary the number of receive channels and the mix of SUO and MW waveforms, as shown in Figure 11.15. The corresponding gate counts show the fixed gates required for the DSP core and the variable gates required for the ASIC portions of the MMC design.

The gate count of the DSP core is based on the C6xxx series high-performance DSP chip [980]. This core has an instruction cache and internal data memory with minimal local memory and DMA interfaces. The matched filter and increased numerical precision of the correlators increase the gate count of the MW channels. Demodulators have 32 correlators each. For the SUO waveform, the correlators are based on an existing design consisting of separate accumulators for I and Q. This consists of a two-stage accumulator with seven-bit precision in the first stage, followed by a 23-bit accumulator. To save power, the 23-bit accumulator operates at a lower rate, using carriers from the first stage accumulator. The precision of the MW first correlator stage was increased to 12 bits.

Figure 11.16 shows the estimated power for the demodulator implemented in 1 V technology using a 40 MHz clock. All channels and all 32 correlators per channel are active in the acquisition mode. In the track mode, four channels are used, and only four of the correlators are used per channel.

The six-channel SUO device operates communications and GPS as separate modes, picking one or the other. In each case, all six channels are used for acquisition and four channels are used for tracking. The dissipated power is therefore the same for either the GPS or the communication mode. The 12-channel MW device may operate the GPS and communication modes simultaneously. The eight SUO channel device could use eight channels for GPS acquisition and four for tracking. At the same time, the four MW channels could be used for communications. In this case, four channels are used for both acquisition and tracking. When all 12 channels are active for acquisition and eight are active for tracking, the device dissipates the power shown in the figure. For the 12-channel MW device, simultaneous operation is also shown, with six channels each for acquisition and four each for tracking. The 12-channel MW demodulator dissipates over 700 mW in the acquisition mode. Track mode power shown in the figure is reasonably low for all of the

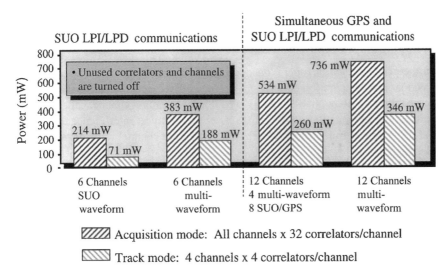

Figure 11.16 1 V demodulator power consumption [970] © 1999, IEEE.

configurations since the unused channels and correlators are inactive, consuming essentially zero power. These power estimates are based on scaling the power per gate from 1.8 to 1.0 V. There would be some additional benefit to reducing feature size from 0.18 to 0.13 or 0.12 μm.

11.1.6.7 *Design methodology*

The design of low-power DSP-based ASICs for high-performance applications requires appropriately structured tradeoffs. Advanced wideband waveforms, such as those contemplated for DARPA's SUO program, present large processing demands. Satisfying these demands at low power requires attention to power dissipation throughout in the design of ASICs like the MMC. The design methods used to develop this architecture are shown in Figure 11.17. In each design, power consumption must be minimized per gate, function and subsystem, in the context of overall power management strategies such as clock control, sleep and off modes.

More information on military software radio can be found in [980–1004].

11.2 A SOFTWARE RADIO ARCHITECTURE WITH SMART ANTENNAS

The architecture presented in Section 11.1 is motivated mainly by military requirements that include wideband spread spectrum waveforms, as well as legacy military waveforms. The approaches used to develop this architecture are also relevant to commercial applications. However, the specific device designs might not be cost effective for commercial applications. Table 11.3 compares military and commercial communication goals. Military requirements for security force hardware commitment to PN codes and channel modulations with features such as low probability of intercept and detection that may be inappropriate to commercial applications. Jammer suppression may also be less appropriate for the

Table 11.3 Comparison of military and commercial communication goals

Military	Commercial
Peer-to-peer (no base station) operation mode	Base station/mobile operation/control
Base station/mobile operation/control	Fixed network infrastructure
Low probability of intercept	Maximum channel capacity
Low probability of detection	Multiuser interference mitigation
Special PN code requirements	Pilot channel for synchronization
Jammer suppression	

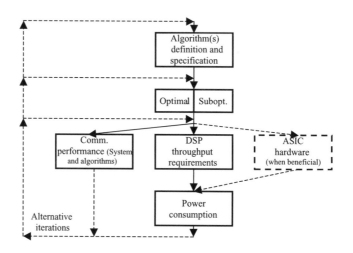

Figure 11.17 Low-power ASIC development methodology [970] © 1999, IEEE.

commercial sector. In addition, Section 11.1 has not addressed emerging multicarrier and wavelet-based communication waveforms.

The architecture can support such emerging waveforms as long as the necessary ASIC circuits are provided. Fast Fourier transform (FFT) circuits may be necessary for multi-carrier waveforms, as disussed in Chapter 10. Wavelet waveforms may require filter bank hardware.

Within this section we will present software radio solutions for applications in wireless networks with beamforming capabilities at the receiver, which allow two or more transmitters to share the same channel to communicate with the base station. The presentation in this section is based on Chapter 9 of this book and [1005].

An antenna array consisting of M elements is considered at the receiver. Adaptive beamforming capabilities of antenna arrays are used to maintain a constant gain for the signal along the direction of interest, and to adjust the nulls so as to reject the cochannel interference. In this way, the interference is minimized and the CIR for the signal of interest is maximized. This concept is illustrated in Figures 11.18–11.19.

Let $s_j(k)$ for $j = 1, 2, \ldots, J$ denote the jth transmitted signal. At most N multipath signals from each user arrive in the base station with different delays (τ_n). The received

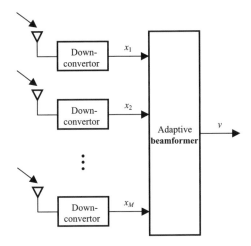

Figure 11.18 A diversity combining system.

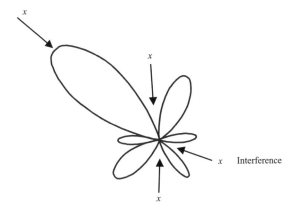

Figure 11.19 Sample array pattern.

signal vector is

$$\mathbf{x}(k) = \sum_{j=1}^{J} \sqrt{P_j G_j} \sum_{n=1}^{N} \alpha_j^n \mathbf{a}(\theta_j^n) s_j(t - \tau_n) + n(k) \tag{11.1}$$

where θ_j^n is the arrival direction of the nth multipath signal from the jth user, P_j is the power of the jth transmitter, G_j is the link gain between the jth transmitter and the base station, α_j^n is the nth path fading coefficient, and $\mathbf{a}(\theta_j^n)$ is the array response to the multipath signal arriving from direction θ_j^n with

$$\mathbf{a}(\theta_j^n) = \left[a_1(\theta_j^n), a_2(\theta_j^n), \ldots, a_M(\theta_j^n) \right]^{\mathrm{T}} \tag{11.2}$$

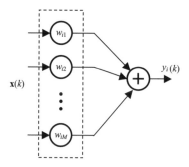

Figure 11.20 Space diversity combiner.

11.2.1 Space diversity combining

A weighted sum of the outputs of the array elements is generated by a beamformer, shown in Figure 11.20, in the following way:

$$y_i(k) = \mathbf{w}_i^H \mathbf{x}(k) \tag{11.3}$$

where $\mathbf{w}_i^T = [w_{i1}, \ldots, w_{iM}]$ is the weight and $\mathbf{x}^T(k) = [x_1(k), \ldots, x_M(k)]$ is the received signal vector sampled at the output of the down-converters.

11.2.1.1 Minimum variance distortionless response (MVDR)

One can choose the weight vector $\mathbf{w}_i$ to steer a beam toward the direction of the signal of interest θ_i and adjust the nulls to reject the interference. This is done by attempting to maintain a distortionless response in the direction of interest and placing the nulls in the directions of other cochannel interferers. The average output power is given by

$$E_i = E\lfloor y_i(k)y_i^H(k)\rfloor = \mathbf{w}_i^H E\lfloor \mathbf{x}(k)\mathbf{x}^H(k)\rfloor \mathbf{w}_i = \mathbf{w}_i^H \boldsymbol{\varphi}_i \mathbf{w}_i \tag{11.4}$$

where

$$\boldsymbol{\varphi}_i = P_i G_i \alpha_i^1 \mathbf{a}_i\left(\theta_i^1\right)\mathbf{a}_i^H\left(\theta_i^1\right) + \sum_{j=1, j\neq i}^{j} P_j G_j \sum_{n=2}^{N} \alpha_j^n \mathbf{a}_j\left(\theta_j^n\right)\mathbf{a}_j^H\left(\theta_j^n\right) + \frac{N_0}{2}I \tag{11.5}$$

The correlation matrix due to the interference terms is

$$\boldsymbol{\varphi}_N = \sum_{j=1, j\neq i}^{j} P_j G_j \sum_{n=2}^{N} \alpha_j^n \mathbf{a}_j\left(\theta_j^n\right)\mathbf{a}_j^H\left(\theta_j^n\right) + \frac{N_0}{2}\mathbf{I} \tag{11.6}$$

If the array response in the direction of the desired user is known, the beamformer tries to minimize the output power E_i subject to maintaining a distortionless response in the direction of interest such that $\mathbf{w}_i^H \mathbf{a}_i(\theta_i^1) = 1$ For this reason, this adaptive beamformer is called an MVDR beamformer. From Equations (11.4) and (11.5), with the constraint that $\mathbf{w}_i^H \mathbf{a}_i(\theta_i^1) = 1$, the received signal power plus the interference power as a function of $\mathbf{w}_i$ is

$$E_i = P_i G_i + \sum_{j=1, j\neq i}^{j} P_j G_j \sum_{n=2}^{N} \alpha_j^n \mathbf{w}_i \mathbf{a}_j\left(\theta_j^n\right)\mathbf{a}_j^H\left(\theta_j^n\right)\mathbf{w}_i + \frac{N_0}{2}\mathbf{w}_i^H \mathbf{w}_i. \tag{11.7}$$

Let I_i denote the total interference plus noise power given by

$$I_i = \sum_{j=1, j \neq i}^{j} P_j G_j \sum_{n=2}^{N} \alpha_j^n \mathbf{w}_i \mathbf{a}_j \left(\theta_j^n\right) \mathbf{a}_j^H \left(\theta_j^n\right) \mathbf{w}_i + \frac{N_0}{2} \mathbf{w}_i^H \mathbf{w}_i \tag{11.8}$$

In Equation (11.7), $P_i G_i$ is the received power from the signal of interest, while I_i given by Equation (11.8) is the contribution to the output power E_i from the interference and noise.

The optimum weight vector minimizes the interference I_i while maintaining a unity gain in the direction of interest by imposing $\mathbf{w}_i^H \mathbf{a}_i(\theta_i^1) = 1$. The solution of this problem is

$$\hat{\mathbf{w}}_i = \frac{\boldsymbol{\varphi}^{-1} \mathbf{a}_i \left(\theta_i^1\right)}{\mathbf{a}_i^H \left(\theta_i^1\right) \boldsymbol{\varphi}^{-1} \mathbf{a}_i \left(\theta_i^1\right)} \tag{11.9}$$

Since the desired signal (arriving along θ_i) will not be affected by the beamforming process and only the interference is rejected, the CIR is maximized for the signal in the direction of interest, i.e.

$$\Gamma_{i,\max} = \frac{P_i G_i}{\sum\limits_{j=1, j \neq i}^{J} P_i G_i \sum_{n=2}^{N} \alpha_j^n \hat{\mathbf{w}}_i^H \mathbf{a}_j \left(\theta_j^n\right) \mathbf{a}_j^H \left(\theta_j^n\right) \hat{\mathbf{w}}_i + \frac{N_0}{2} \hat{\mathbf{w}}_i^H \hat{\mathbf{w}}_i} \tag{11.10}$$

11.2.1.2 *Minimum mean square error (MMSE)*

If the array response is not known, one may employ a training sequence and minimize the difference between the training sequence and the output of the beamformer in the mean square sense

$$\hat{\mathbf{w}}_i = \arg\min_{w_i} E\left[\left| d_i - \mathbf{w}_i^H \mathbf{a}_i \left(\theta_i^1\right) \right|^2 \right] \tag{11.11}$$

The solution to this problem is given by

$$\hat{\mathbf{w}}_i = \boldsymbol{\varphi}_i^{-1} \mathbf{p}_i \tag{11.12}$$

$\boldsymbol{\varphi}_i$ is defined as before and $\mathbf{p}_i$ is the crosscorrelation between the received vector and the training sequence $\mathbf{p}_i = E[\mathbf{x}_i d_i^*]$. The maximum CIR in this case is given by

$$\Gamma_{i,\max} = \frac{P_i G_i \hat{\mathbf{w}}_i^H \mathbf{a}_i \left(\theta_i^1\right) \mathbf{a}_i^H \left(\theta_i^1\right) \mathbf{w}_i}{\sum\limits_{j=1, j \neq i}^{J} P_i G_i \sum_{n=2}^{N} \alpha_j^n \hat{\mathbf{w}}_i^H \mathbf{a}_j \left(\theta_j^n\right) \mathbf{a}_j^H \left(\theta_j^n\right) \hat{\mathbf{w}}_i + \frac{N_0}{2} \hat{\mathbf{w}}_i^H \hat{\mathbf{w}}_i} \tag{11.13}$$

11.2.2 Space–time diversity combining

In the space diversity combining system of Figure 11.20, the interference and multipath signals are rejected by placing nulls at the directions of those signals. In a broadband linear combiner, illustrated in Figure 11.21, the desired signal and its multipath are combined at the combiner output to estimate the desired signal.

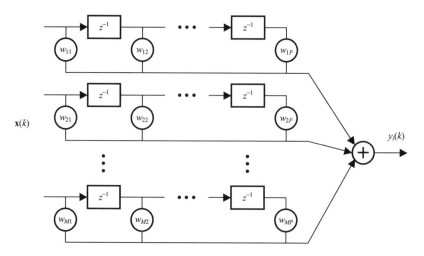

Figure 11.21 Space–time diversity combiner.

11.2.2.1 MMSE Space–time combining

The output of a broadband combiner can be expressed as

$$y(k) = \sum_{m=1}^{M} \mathbf{w}^H(m)\mathbf{x}(m) \tag{11.14}$$

where $\mathbf{w}^T(m) = [w_{m1}, \ldots, w_{mP}]$ is the weighting vector and $\mathbf{x}^T(m) = [x_{m1}, \ldots, x_{mP}]$ is the received signal vector at the mth element equalizer.

Let $\mathbf{W} = [\mathbf{w}^T(1), \ldots, \mathbf{w}^T(P)^T]$ and $\mathbf{X} = [\mathbf{x}^T(1), \ldots, \mathbf{x}^T(P)^T]$. The output of the beamformer can be written as

$$y(k) = \mathbf{W}^H \mathbf{X} \tag{11.15}$$

The objective is to minimize the mean square error between the output of the combiner and the desired signal d, i.e.

$$\hat{\mathbf{W}} = \arg \min_{W} \left| d - \mathbf{W}^H \mathbf{X} \right|^2 \tag{11.15a}$$

The optimal beamformer coefficients are similar to space diversity MMSE:

$$\hat{\mathbf{W}} = E\{\mathbf{X}\mathbf{X}^H\}^{-1} E\{\mathbf{X}d^*\} \tag{11.16}$$

Many adaptive methods that update weight vectors according to the incoming data have been developed, such as recursive least squares (RLS) [1006] and minimum mean square error (MMSE) [1007]. Most of these algorithms are discussed in Chapter 5 and Chapter 9.

Blind methods may be used in order to save the bandwidth allocated to the training sequence. For single-input single-output (SISO) systems, numerous blind identification algorithms [1008–1011] and blind equalization algorithms [1012–1021] have been proposed that exploit the higher order statistics of channel output. Most of these algorithms are discussed in Chapters 5 and 6. Among these algorithms, the Godard algorithm (GA) [1014], also known as the constant modulus algorithm (CMA) [1019–1020], is one of the best and simplest adaptive blind equalization algorithms. These algorithms are discussed in Chapter 5.

Single-input multiple-output (SIMO) systems may be viewed as fractionally spaced sampled communication systems which receive the distorted versions of one input signal. Fractionally spaced CMA adaptive blind equalizers, under symbol timing offsets, are considered in [1022]. Multiple-input multiple-output (MIMO) transmission systems are studied in [1023] and [1024], where the MIMO channel impulse response is known. When the channel parameters of the MIMO systems are unknown, blind identification and equalization techniques must be used to separate and capture signals. The capture properties of the CMA algorithm used in MIMO systems with constant modulus input signals are studied in [1020], [1025], and [1026]. The blind identification of MIMO systems using second-order statistics or higher order statistics are also studied in [1027–1028]. Some subspace algorithms with fast convergence rates are proposed in [1029]. Most of these algorithms are covered in Chapter 4.

11.2.3 Space–time diversity in CDMA systems

As discussed in Chapter 5 and Chapter 7, in CDMA all users are sharing the same channel and each user has a different pseudonoise (PN) sequence. The received signal due to J cochannel users at the receiver is given by

$$\mathbf{x}(k) = \sum_{j=1}^{J} P_j G_j \sum_{n_b} s_{n_b j} \mathbf{p}_j(t - n_b T) + \mathbf{n}(k) \tag{11.17}$$

where $\mathbf{x}(k)$ is the received signal at the array, s_{n_b} is the transmitted sequence, and

$$\mathbf{p}_j(k) = \sum_n \mathbf{a}_j(\theta_n)\alpha_j^n c_j\left(t - \tau_j^n\right) \tag{11.18}$$

τ_j^n is the nth path delay associated with the jth user, and c_j is the jth user PN sequence. In a space diversity combiner, the signals from the main path are despread and combined at the beamformer output, as shown in Figure 11.22.

Code filters (CFs) are matched to the desired user code and the delay is matched to the nth path delay, i.e.

$$\mathbf{z}_j^n(k) = \int_T \mathbf{x}(k)c_j(t - kT - \tau_j^n)\,dt \tag{11.19}$$

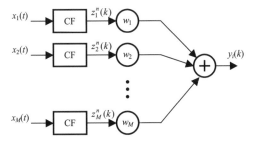

Figure 11.22 Space–time diversity combiner in a CDMA system.

and

$$y_j^n(k) = \mathbf{w}^H \mathbf{z}_j^n(k) \tag{11.20}$$

It has been shown [1030] that the optimal beamformer for this architecture is given by

$$\hat{\mathbf{w}} = \frac{\boldsymbol{\varphi}_{in}^{-1} \mathbf{a}(\theta_j^n)}{\mathbf{a}(\theta_j^n)^H \boldsymbol{\varphi}_{in}^{-1} \mathbf{a}(\theta_j^n)} \tag{11.21}$$

φ and φ_{in} are defined as before, and

$$\varphi_{in} = \frac{\rho}{\rho - 1}\left(\varphi - \frac{1}{p}\varphi_0\right) \tag{11.22}$$

ρ is the processing gain and

$$\varphi_0 = \varphi_{in} + P\mathbf{a}(\theta_j^n)\mathbf{a}(\theta_j^n)^H \tag{11.23}$$

In a multiple-antenna system, one can design a beamformer for each resolvable path, and then combine the outputs with a standard RAKE receiver [1031], i.e.

$$\hat{\mathbf{z}}_j(k) = \sum_{n=1}^{N} \mathbf{z}_j^n(k)\alpha_j^n \tag{11.24}$$

where N is the maximum number of paths in each link.

11.2.4 Software radio architecture with smart antenna

The software implementation of this concept is shown in Figure 11.23. Each antenna element has its own down-converter and ADC. But the subsequent beamforming and demodulation are implemented in software and are shared among all of the elements.

Figure 11.24 illustrates the software architecture of the beamformer for each channel. The received signal from each antenna element is passed through the same software block as depicted in Figure 11.23 to generate I and Q signals. These signals are combined using the 'combiner' block shown in Figure 11.24. The received signal from the user would be the output of the demodulator in Figure 11.24. Using K similar blocks for each channel, one allows up to K users to share the same channel. Since each user generates its carrier locally, there are separate carrier tracking and down-converting blocks per user. It was mentioned that one can assign K users to each channel by using K beams for each channel. In this manner, a system with L physical channels serves at most KL users. Such a software radio with a smart antenna is scalable. Increasing the number of beams and users is a software process within the constraints of the hardware costs. The proportion of increase in computational resources required for the smart antenna depends on the cell site architecture.

With baseband DSP, one converts the smart antenna channels back to analog IF to provide only L channels in the conventional cell site as the maximum capacity of each cell. In principle, with an IF software radio architecture, the smart antenna processing could be part of the larger pooled DSP resources, and one could implement KL or KL/α service channels. Hereafter, a software radio architecture is assumed in which one synthesizes up to KL servers in the integrated DSP architecture of the software radio cell site.

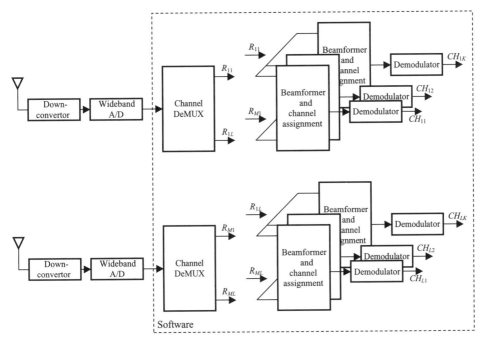

Figure 11.23 Functional block diagram of the software radio for a base station with smart antenna.

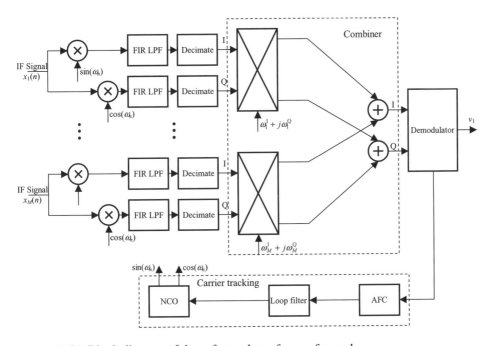

Figure 11.24 Block diagram of the software beamformer for each user.

Table 11.4 Computational complexity of various beamforming algorithms [1005]
© 1999, IEEE

Algorithm	Multiplications	Divisions	Additions
LMS	$2Q + 1$	0	$2Q$
RLS	$2Q^2 + 7Q + 5$	$Q^2 + 4Q + 3$	$2Q^2 + 6Q + 4$
FTF	$7Q + 12$	4	$6Q + 3$
LSL	$10Q + 3$	$6Q + 2$	$8Q + 2$

The beamforming process can be implemented on DSPs. Usually, adaptive algorithms are used to update the weight vectors. The optimum beamformer weight vector presented above requires knowledge of second-order statistics. These statistics are usually not known, but with the assumption of ergodicity, they can be estimated. Statistics may also change over time (due to moving interferences). To resolve these problems, weight vectors are typically determined adaptively. One may employ training sequences to update the weight vectors. The training sequence is known at the receiver.

Let $d_i, i = 1, \ldots, Q$ denote the training sequence. The received training sequence at the receiver is denoted by $X_i, i = 1, \ldots, Q$. The direct matrix inversion (DMI) is the straightforward method for calculating the weight vectors. DMI minimizes the difference between the training sequence and the output of the beamformer in a mean square sense

$$\hat{\mathbf{w}}_i = \arg\min_{w_i} \sum_{q=1}^{Q} \left| d_i - \mathbf{w}_i^H X_i(q) \right|^2 \tag{11.25}$$

The solution to this problem is given by

$$\hat{\mathbf{w}}_i = \boldsymbol{\varphi}_i^{-1} \mathbf{p}_i \tag{11.26}$$

$\boldsymbol{\varphi}_i$ is the estimated correlation matrix and $\mathbf{p}_i$ is the crosscorrelation between the receiver vector and the training sequence given by

$$\boldsymbol{\varphi}_i = \frac{1}{Q} \sum_{q=1}^{Q} \mathbf{X}_i(q) \mathbf{X}_i^H(q) \tag{11.27}$$

$$\mathbf{p}_i = \frac{1}{Q} \sum_{q=1}^{Q} \mathbf{X}_i(q) d_i \tag{11.28}$$

The complexity of the DMI method is $O(Q^3)$, where Q is the length of the training sequence. Adaptive algorithms that update the weight vector taps reduce the complexity as shown in Table 11.4. In the recursive estimation procedure, the weights are adjusted iteratively. Such an approach eliminates the estimation of the correlation matrix or crosscorrelation vector.

11.2.5 Traffic and handoff improvement with smart antennas

This section analyzes the network with an adaptive array under the traffic policies proposed in [1032]. In this network model it is assumed that when a new call (handoff or originating

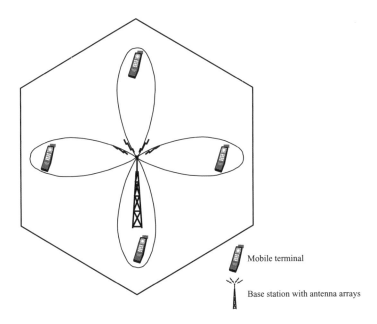

Figure 11.25 Capturing four mobile users over the same channel successfully by an adaptive antenna array at the base station.

call) arrives, the adaptive array points one beam towards that user, and assigns one channel out of the L channels to that user. Each channel c_i for $i = 1, 2, \ldots, L$ may be assigned to one of the K users by K separate beams, using the K beamformers in parallel for each channel, as shown in Figure 11.25. If the first beam of all channels is occupied, the new call is assigned to another beam. If there are no multipaths in the system, with K beamformers for each channel and L RF frequency channels, at most KL users can be accepted into the system.

On the other hand, if there are multipath signals with α effective paths per user, then the number of antennas has to increase to null those multipaths. In this case, the effective number of usable channels would be KL/α. The flowchart of the channel assignment with smart antennas is illustrated in Figure 11.26.

11.2.5.1 Example

In this example the system parameters are the same as in [1005]. Assume that $(i - 1)$ cochannel transmitters successfully share the same channel. For acceptable link quality, the newly arrived ith transmitter shares that channel if $\Gamma_i \geq \gamma$, where Γ_i is given by Equation (11.10) and γ is a system parameter which is dictated by the governing standard. For instance, in the IS-54 standard, γ is 14 dB, while in AMPS, it is 18 dB. The probability of establishing an acceptable link, i.e. $P(\Gamma_i \geq \gamma)$ is estimated by Monte Carlo simulation. In the network model, the probability density function for the angular position of the transmitters in the system is assumed to be $f_\theta(\theta_j)$ for $1 \leq j \leq J$. The Monte Carlo simulations of the success probabilities $p_{2|1,M}$ and $p_{3|2,M}$ for two-beam and three-beam adaptive arrays, respectively (with M antenna elements), are shown in Figures 11.27 and 11.28.

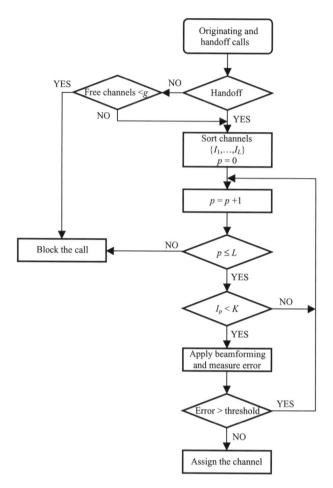

Figure 11.26 Flow chart of the channel assignment algorithm in a system with smart antennas.

A call admission control (CAC) is required to avoid degradations in system performance. If $\Gamma_i < \gamma$ or $i > K$, that user is not accepted into the system. The probability of the event $\Gamma_i \geq \gamma$ in the system, would yield the success probability $p_{i|i-1,M}$ that the ith transmitter can share the same channel, given that i transmitters are already using that channel. Clearly, the success probability $p_{i|i-1,M}$ depends on M. Since there are L distinct channels in the system and each channel may be reused up to K times, define the *probability of successful reception of the* $(n-1)$*th user into the system, given that there are already* n *users in the system* as

$$q_{N_{t-1}|N_t}(N_{t-1} = n + 1 | N_t = n) \qquad (11.29)$$

where N_t is the number of users in the system at time t before a new call (user) arrives into the system. Parameter t is the time index which increases by 1 at each epoch corresponding to a new handoff or originating call.

Let $\alpha = \lambda + \gamma$, $\mu = \eta + \nu$, $a = \alpha/\mu$, $b = \gamma/\mu$, $c = \lambda/\mu$, where γ and η are the arrival rate and service rate for originating calls, and λ and ν are the arrival rate and service rate

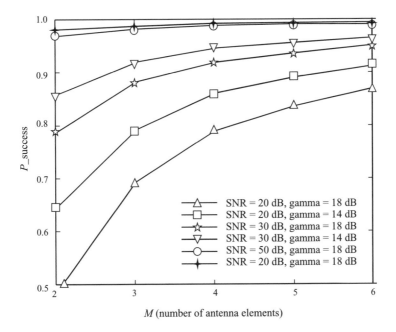

Figure 11.27 Success probability $p_{2|1M}$ for a two-beam adaptive array for different values of M (number of antenna elements), SNR and threshold γ.

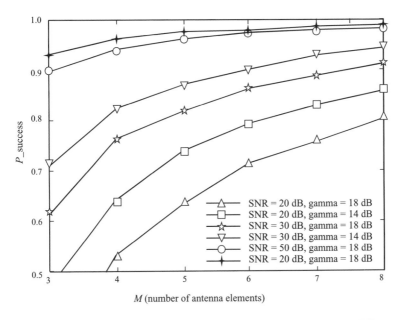

Figure 11.28 Success probability $p_{3|2M}$ for a three-beam adaptive array for different values of M (number of antenna elements), SNR and threshold γ.

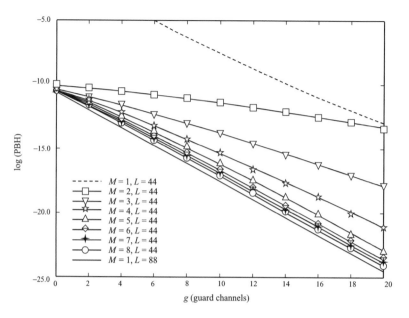

Figure 11.29 Blocking probabilities of handoff calls (B_H) for different numbers of antenna elements M, with SNR = 20, λ = 18 dB, a = 40; c = 8, β = 4, and a two-beam adaptive array system.

for handoff calls, respectively. The path loss exponent β is assumed to be four. Simulation results for the probability of blocking the handoff or original call are shown in Figures 11.29–11.30.

11.2.6 Power control and beamforming in CDMA networks

The transmitted power updating is defined as

$$P_i^{n+1} = \frac{\lambda}{G_{ii}} \left(\sum_{j \neq 1} G_{ji} P_j^n + N_i \right) = \frac{\lambda}{G_{ii}} I_i^n \tag{11.30}$$

where P_i^n is the transmitted power at the nth iteration and G_{ji} is the link gain between the ith mobile and the jth base station. N_i is the thermal noise and I_i^n the interference at the ith receiver. The right-hand side of this equation is a function of the interference at the ith receiver. In order to update the power, the interference I_i is measured at each receiver, and the power is updated by multiplying a constant λ/G_{ii}. The flowchart of the algorithm is shown in Figure 11.31.

Each receiver measures the interference power by averaging power at the output of the receiver during a window of length W.

$$E_i = \sum_{i=1}^{W} x_i^2 \tag{11.31}$$

It subtracts the received power due to the desired transmitter power from the total received

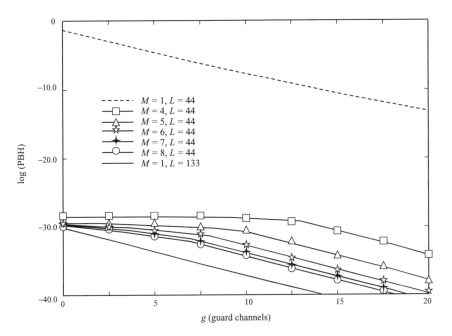

Figure 11.30 Blocking probabilities of handoff calls (B_{H}) for different numbers of antenna elements M; with SNR $= 20$, $\lambda = 18$ dB, $a = 40$; $c = 8$, $\beta = 4$, and a three-beam adaptive array system.

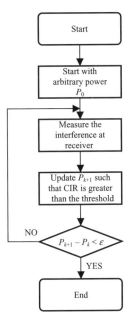

Figure 11.31 Power control algorithm.

power

$$I_i^n = E_i - G_{ii}P_i \tag{11.32}$$

The complexity of power estimation is $4W + 1$ multiplications and $W + 1$ additions. The transmitted power can be updated readily up to a few hundred times per second. Joint power control and beamforming algorithms are proposed in [1033,1034]. In these algorithms, the beamforming weight vectors and power allocations are updated jointly by a distributed algorithm. The nth iteration of the algorithm is as follows [1034]. The minimum error is calculated by

$$E_i = \sum_k \left| d(k) - w_i^{\mathrm{H}} x_i(k) \right|^2 \tag{11.33}$$

The power is updated by

$$P_i^{n+1} = P_i^n \frac{\lambda}{\Gamma_i} = \lambda P_i^n \frac{E_i}{1 - E_i} \tag{11.34}$$

The weight vectors are updated by the beamforming algorithm only during the training phase, and are constant between the training intervals. The complexity of the beamforming algorithms is as discussed in the previous sections. The complexity of the first step of the power control algorithm is $4W + 1$ multiplications, one addition for the calculation of $E_{\min}$ and one division and addition for updating the power. This algorithm is depicted in Figure 11.32.

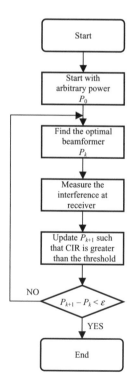

Figure 11.32 The joint power control and beamforming algorithm.

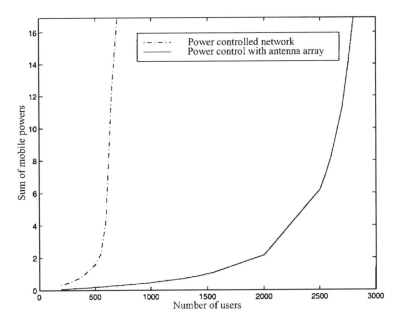

Figure 11.33 The total mobile powers versus the number of users.

11.2.6.1 *Example*

The above algorithm converges to the jointly optimal beamforming weight vector and power allocation such that the transmitted power is minimized [1035]. In order to show the performance of the algorithm, a network of mobiles and base stations is simulated where mobiles are randomly distributed in the network. The total transmitted power as a function of the total number of users in the network is shown in Figure 11.33. As the number of users approaches the capacity of the network, the total power is increased significantly. As illustrated in this figure, with smart antennas with four elements at the base station, for the same number of users, one can reduce the transmitted power. It is also possible to increase the maximum number of users in the network while the SNR is above a threshold.

Additional details on adaptive antenna implementation in software radio can be found in [1036–1040].

11.3 SOFTWARE REALIZATION OF A GSM BASE STATION

In this section we discuss an example of software realization of a TDMA-based system. The performance of each software module is evaluated using both a % CPU metric and a processor-independent metric based on SPEC benchmarks defined with a specific number of logical circuits (operations). SPEC stands for system performance evalution cooperative (for details see www.specbench.org/spec/ and Tables 11.18–11.19 at the end of this section). The results can be used to dimension systems, e.g. to estimate the number of software-based GSM channels that can be supported by a given processor configuration, and to predict the impact of future processor enhancements on BTS capacity.

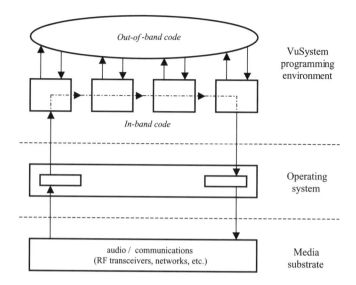

Figure 11.34 VuSystem programming environment.

Figure 11.35 Block diagram of a GSM sender BTS (downlink).

The virtual system programming environment is shown in Figure 11.34. The system is partitioned into an *in-band* axis, that supports the flow of temporally sensitive information, and an *out-of-band* axis that supports the event-driven program components, including the user interface and the configuration and control of the in-band processing pipeline. In-band processing and out-of-band processing are best handled in separate partitions, instead of together, because choices of language and architecture can then be made for each partition separately (e.g. C++ for in-band and Tcl scripting language [1041] for out-of-band processing).

11.3.1 Estimation of computational requirements

The system block diagram is shown in Figure 11.35. We will now provide brief descriptions of the system components and computational requirements for their software realization [1042].

11.3.1.1 *GSM speech coding*

The GSM full-rate speech coding algorithm is based on regular pulse excitation with long-term prediction (RPE–LTP) [1043]. The analog voice is first digitized in an ADC to 8000 samples/s, and then uniformly coded at 13 bits each. Then, the encoder divides the speech into a short-term predictable part, a long-term predictable part, and the remaining residual

Table 11.5   GSM speech coding performance [1042] © 1999, IEEE

Speech coder (RPE–LTP 06.10)	% CPU	Performance (metric/user-channel)			
		SPECint		SPECfp	
		92	95	92	95
SUN Ultra M170	8.1	20	0.45	28	0.73
DEC 3000/800	10	14	—	19	—
Pentium 166	10	20	0.48	—	0.34
Pentium Pro 200	5.9	19	0.48	17	0.40

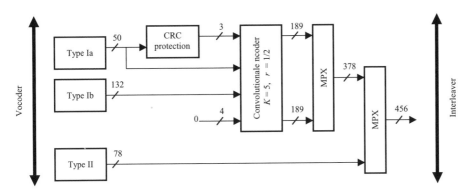

Figure 11.36 Channel encoder.

pulse. Finally, it encodes the residual pulse and parameters describing the two predictors. The speech coding algorithm produces a speech block of 260 bits every 20 ms. These 260 bits are classified into three classes (Ia, Ib and II), according to their importance. The required computational resources for realization of the speech coder are given in Table 11.5.

11.3.1.2 Channel coding

Class Ia bits are first protected by three CRC bits for error detection. The Class Ib bits are then added to this result. A convolutional code with rate $r = 1/2$ and constraint length $K = 5$ is then applied to this complete Class I sequence. The resulting 378 bits are used in conjunction with the 78 unprotected Class II bits, to form a complete coded speech frame of 456 bits (see Figure 11.36).

The performance of this channel coding algorithm is summarized in Table 11.6.

11.3.1.3 Interleaving

The aim of interleaving is discussed in Chapter 2. GSM coding blocks are interleaved on eight bursts. The 456 bits of one block are split into eight groups of 57 bits. Each group of 57 bits is then carried in a different burst. The performance of the software implementation is shown in Table 11.7.

Table 11.6   Channel coding performance [1042] © 1999, IEEE

		Performance (metric/user-channel)			
		SPECint		SPECfp	
Channel coder	% CPU	92	95	92	95
SUN Ultra M170	0.20	0.50	0.011	0.70	0.018
DEC 3000/800	0.45	0.73	—	0.85	—
Pentium 166	0.38	0.76	0.018	—	0.013
Pentium Pro 200	0.24	0.76	0.019	0.68	0.016

Table 11.7   Interleaving performance [1042]

		Performance (metric/user-channel)			
		SPECint		SPECfp	
Interleaver	% CPU	92	95	92	95
SUN Ultra M170	0.04	0.11	0.002	0.15	0.004
DEC 3000/800	0.12	0.17	—	0.23	—
Pentium 166	0.15	0.30	0.007	—	0.005
Pentium Pro 200	0.045	0.14	0.004	0.13	0.003

Table 11.8   Ciphering and deciphering performance [1042]

		Performance (metric/user-channel)			
		SPECint		SPECfp	
De/Cipher	% CPU	92	95	92	95
SUN Ultra M170	0.08	0.19	0.004	0.26	0.007
DEC 3000/800	0.16	0.22	—	0.30	—
Pentium 166	0.09	0.17	0.004	—	0.005
Pentium Pro 200	0.05	0.16	0.004	0.14	0.003

11.3.1.4 Ciphering

Ciphering is achieved by performing an XOR (exclusive OR) operation between a pseudo-random bit sequence and the 114 bits of each burst. The deciphering operation is identical to ciphering. The pseudorandom sequence is derived from the burst number and a session key that itself is determined through signaling when a call is established. The algorithm used to generate the pseudorandom sequence is not fully included in the public GSM specification. The performance of the algorithm is shown in Table 11.8.

Table 11.9   GMSK modulator performance

		Performance (metric/user-channel)			
		SPECint		SPECfp	
GMSK modulator	% CPU	92	95	92	95
SUN Ultra M170	2.1	5.3	0.12	7.4	0.19
DEC 3000/800	4.5	6.2	—	8.4	—
Pentium 166	3.7	7.3	0.18	—	0.12
Pentium Pro 200	1.7	5.4	0.14	4.8	0.11

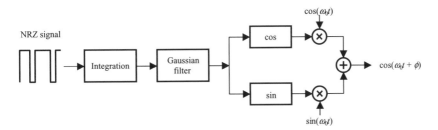

Figure 11.37 GMSK modulation block diagram.

11.3.1.5 *Modulation*

GSM uses Gaussian modulation shift keying (GMSK) [1044] with modulation index $h = 0.5$, BT (filter bandwidth times bit period) equal to 0.3, and a modulation rate of 271 kbaud. There are many ways to implement a GMSK modulator. A block diagram of such a modulator is shown in Figure 11.37. The Gaussian filter is achieved by directly using the phase-shaping response of the Gaussian filter $\phi(t)$. Basically, $\phi(t)$ consists of a $\pi/2$ step function, smoothed in order to have a more narrow spectrum than if the step were steeper. The algorithm precomputes the function $\phi(t)$, and then obtains the sum over all input bits k_i using

$$\phi(t) = \phi_0 + \sum_{i=0}^{147} k_i \phi(t - iT) \quad (\phi_0 \text{ may take any value})$$

Theoretically, a bit k_i influences the output indefinitely. However, in practice, this influence becomes negligible outside a $3T$ period. So, the algorithm can precompute pieces of output of bit sequences of length three, and find the whole output by assembling pieces together.

Table 11.9 reports the performance of the modulator.

11.3.1.6 *Channel partitioning*

The number of frequency channels used by a GSM BTS depends mainly on the density of the network and the type of cells used (micro cells, urban, rural or road cells). It can be one in a low-density area, and two to four in high-density areas. Typically, adjacent channels are not

Table 11.10   Polyphase transform performance (200 kHz spacing between channels)

Polyphase transform	Performance (% CPU) number of frequency channels			
	1	2	3	4
SUN Ultra M170	20	60	180	620
DEC 3000/800	25	80	220	670
Pentium 166	40	110	340	1200
Pentium Pro 200	25	70	21	720

Figure 11.38 Block diagram of a GSM receiver BTS (uplink).

used within the same cell to avoid spectral recovering, and a GSM BTS uses every second channel (i.e. 200 kHz spacing) to avoid high-level interference. In the GSM terminology, a logical user channel corresponds to a speech channel (defined by its frequency and its time slot number), whereas a physical channel stands for an FDMA channel (which contains eight logical channels). Multichannel receivers for digitized data can be synthesized using a polyphase transform algorithm [1045]. Basically, this algorithm performs the following tasks:

1. Frequency translation (each center frequency to baseband).

2. Bandwidth reduction of the translated spectrum to match the signal bandwidth.

3. Resampling of the output to match the reduced channel bandwidth.

Table 11.10 presents the percentage of CPU per platform required to partition N GSM frequency channels ($N \in [1, 8]$) at a sampling rate equal to 2.5 times the maximal frequency and with a 200 kHz spacing scheme. The filter used in the implementation has 255 coefficients. The corresponding SPEC performances are shown in Table 11.10. The high CPU requirement is due to the high input data rate and the need to compute imaginary and real components of output samples for each frequency channel (requiring multiple floating-point operations).

11.3.1.7 Demodulation and equalization

A typical implementation of a GSM receiver is shown in Figure 11.38. The GSM specifications do not impose a particular demodulation algorithm. However, they impose minimal performance criteria, and the algorithm used is expected to cope with two multipaths of equal power received at intervals of up to 16 μs (i.e. more than four symbols). With such a level of inter-symbol interference, simple demodulation techniques are ineffective, and an equalizer is required. If an equalizer based on the Viterbi algorithm discussed in Chapter 6

Table 11.11 Deinterleaving performance

Deinterleaver	% CPU	SPECint		SPECfp	
		92	95	92	95
SUN Ultra M170	0.04	0.11	0.003	0.16	0.004
DEC 3000/800	0.17	0.24	—	0.32	—
Pentium 166	0.16	0.32	0.008	—	0.005
Pentium Pro 200	0.04	0.13	0.003	0.12	0.003

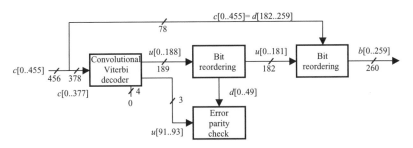

Figure 11.39 GSM channel decoding.

is used, this module is expected to be processor intensive. A typical solution for a GSM BTS equalizer can be implemented using a full TMS 320/C40 DSP, i.e. consuming 50 MIPS.

11.3.1.8 Deciphering

Deciphering performs the same operation as the ciphering module, hence its performance is summarized by the results in Table 11.8.

11.3.1.9 Deinterleaving

Deinterleaving performs the inverse operation to the interleaving. Every 40 ms, eight 114 bit bursts are merged into a 456 bit buffer. The performance of this module is shown in Table 11.11.

11.3.1.10 Channel decoding

Channel decoding involves the retrieval of the original compressed speech data from the (possibly corrupted) received flow. Figure 11.39 shows the different steps of the channel decoding algorithm. The deinterleaving algorithm generates a 456 bit data buffer from eight GSM bursts. Some of these bits (the 378 bits corresponding to the Class I data bits) are fed into the convolutional decoder, which tries to reconstitute the 189 bits corresponding to the original sequence. After a bit-reordering step, the 50 bits that have the highest priority

Table 11.12   Channel decoding performance

Channel decoder	% CPU	Performance (metric/user-channel)			
		SPECint		SPECfp	
		92	95	92	95
SUN Ultra M170	1.7	4.2	0.09	5.9	0.15
DEC 3000/800	2.2	3.1	—	4.2	—
Pentium 166	2.1	4.2	0.10	—	0.07
Pentium Pro 200	1.5	4.8	0.12	4.2	0.10

(Class Ia) are checked using an error control algorithm. If there is no error, a final bit reordering is performed. These steps are further described in the following paragraphs.

Convolutional decoding   Convolutional decoding can be performed using the Viterbi algorithm. The encoder memory is limited to K bits; a Viterbi decoder in steady-state operation takes only 2^{K-1} paths. Its complexity increases exponentially with the constraint length K, which is equal to five for the GSM convolutional code. The implementation discussed here is a modified version of the Viterbi decoder 16 developed by Karn for the NASA standard code ($K = 7, r = 1/2$) [1046].

Reordering algorithm   The 189 bits generated by the Viterbi decoder ($u(0), u(1), \ldots, u(187), u(188)$) consist of 181 information bits of Class I ($d(0), d(1), \ldots, d(180)$), three CRC bits and four tail bits. The relation between u bits and d bits is defined as

$$u(k) = d(2k) \quad \text{and} \quad u(184 - k) = d(2k + 1), k \in [0, 90]$$

Error-detecting codes   The polynomial operation is applied to the CRC-protected data. If the remainder differs with the received CRC bits, an error is detected, and the audio frame is ignored and eventually discarded.

Reordering algorithm   The last step of the channel decoding regroups the 260 bits of the speech block from the position that corresponds to the order of decreasing importance ($d(0), d(1), \ldots, d(259)$) to the position that matches their classification ($b(0), b(1), \ldots, b(258), b(259)$) as specified in [1043]. This operation (which requires no effort when implemented in hardware) can be efficiently implemented in software using lookup tables. The performance of the whole channel decoding algorithm is summarized in Table 11.12.

11.3.1.11   GSM speech decoding

The decoder reconstructs the speech by passing the residual pulse first through the long-term prediction filter, and then through the short-term predictor (see Figure 11.40). The performance of this module is shown in Table 11.13.

Table 11.13   GSM 06.10 decoding performance

Speech decoder (RPE–LTP 06.10)	% CPU	Performance (metric/user-channel)			
		SPECint		SPECfp	
		92	95	92	95
SUN Ultra M170	4.4	11	0.24	15	0.40
DEC 3000/800	5.2	7.2	—	9.8	—
Pentium 166	3.9	7.7	0.19	—	0.13
Pentium Pro 200	2.1	6.7	0.17	5.9	0.14

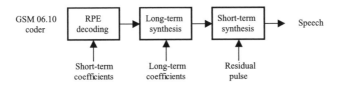

Figure 11.40 Block diagram of the GSM speech decoder.

11.3.1.12 System integration

Table 11.14 summarizes the performance of the modules required to encode and decode one GSM logical (TDM) channel, exclusive of the demodulator (including equalizer) and partitioner modules. The minimum and maximum SPEC numbers, obtained with the four different platforms are given. Table 11.15 (based on Table 11.14) shows the computational requirements for the encoding and decoding of N GSM physical FDMA channels ($N \in [1, 8]$), where each physical channel supports eight logical user channels. Here, again, the entries are exclusive of the partitioner and demodulator/ equalizer modules.

Furthermore, it is assumed that, for each group of up to four physical channels, one of the derived logical channels is dedicated to signaling, and the remaining logical channels require speech processing. This table suggests that a single Sun UltraSparc M170 or Pentium Pro 200 workstation can just about support the encoding/decoding requirements of one physical GSM channel. Figure 11.41 shows the corresponding percentage of CPU required per GSM module on the Pentium Pro 200 platform; a total of 93 % of CPU is used.

Table 11.16 presents the SPEC-normalized performance of the partitioning algorithm. The results suggest that our current Sun UltraSparc M170 platform would be sufficiently powerful to perform the partitioning of two physical FDMA channels (i.e. 16 logical channels). Table 11.17 provides a summary of computational demands, including the partitioning algorithm, and exclusive of the demodulator/equalizer. All together, the above tables suggest that a *minicell* base station supporting two physical channels, i.e. up to 15 logical channels, could be assembled using: one processor to perform the partitioning of incoming samples; one processor to multiplex the outgoing sample streams; two processors to perform the encoding/decoding; and some number of additional processors to realize the equalization/demodulation stages.

Even after allowing a considerable margin for management and overhead, it will be possible to realize a complete two-channel system using a cluster of 8–15 workstations, or

Table 11.14   Aggregate GSM module performance [1042] © 1999, IEEE

| GSM modules (for one GSM logical channel) | Performance (metric/user-channel) | | | | | | | |
| | SPECint92 | | SPECfp92 | | SPECint95 | | SPECfp95 | |
	min	max	min	max	min	max	min	max
Speech coder	14	20	17	28	0.45	0.48	0.34	0.73
Channel coder	0.50	0.76	0.68	0.85	0.011	0.019	0.013	0.018
Interleaver	0.11	0.30	0.13	0.23	0.002	0.007	0.003	0.005
Cipher	0.16	0.22	0.14	0.30	0.004	0.004	0.003	0.007
Modulator	5.3	7.3	4.8	8.4	0.12	0.18	0.11	0.19
Decipher	0.16	0.22	0.14	0.30	0.004	0.004	0.003	0.007
Deinterleaver	0.11	0.32	0.12	0.32	0.003	0.008	0.003	0.005
Channel decoder	3.1	4.8	4.2	5.9	0.09	0.12	0.07	0.15
Speech decoder	6.7	11	5.9	15	0.17	0.24	0.13	0.40
Total	30	45	33	59	0.85	1.1	0.68	1.51

Table 11.15   GSM BTS performance (without partitioning and equalization) [1042]

| BTS number of frequency channels | number of log.ch data/total | SPECint92 | | SPECfp92 | | SPECint95 | | SPECfp95 | |
		min	max	min	max	min	max	min	max
1	7/8	14	20	17	28	0.45	0.48	0.34	0.73
2	15/16	0.50	0.76	0.68	0.85	0.011	0.019	0.013	0.018
4	31/32	0.11	0.30	0.13	0.23	0.002	0.007	0.003	0.005
8	62/64	0.16	0.22	0.14	0.30	0.004	0.004	0.003	0.007

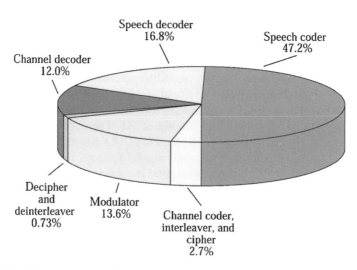

Figure 11.41 CPU requirement for eight logical channels on a Pentium Pro 200 [1042].

Table 11.16 Partitioning performance

BTS number of frequency channels	SPECint92		SPECfp92		SPECint95		SPECfp95	
	min	max	min	max	min	max	min	max
1	35	80	47	70	1.1	2.0	1.4	2.0
2	110	220	150	201	3.3	5.6	3.7	5.6
4	300	670	410	630	10	17	11	17
8	930	2400	1300	2300	34	57	40	58

Table 11.17   GSM BTS performance (including partitioning and without equalization)

BTS number of frequency channels	Number of data/total	SPECint92		SPECfp92		SPECint95		SPECfp95	
		min	max	min	max	min	max	min	max
1	7/8	275	440	307	540	7.9	10.8	6.9	14
2	15/16	590	940	680	1150	17.3	23.6	14.7	29.6
4	31/32	1260	2110	1470	2520	37	52	33	65
8	62/64	2850	5280	3410	6080	88	127	84	154

Table 11.18   SPEC92 benchmarks

Platform	Clk (MHz) ext/in	Cache ext+I/D	SPECint 92	SPECfp 92	Information date	Source obtained
SUN Ultra M170	83/167	512+16/16	252	351	Nov 95	SunIntro
DEC 3000/800	40/200	2M+8/8	138.4	187.6	May 94	c.sun.hw
Intel XXpress Pentium	66/166	1M+8/8	197.5	—	Jan 96	www.intel
Intel Alder Pentium Pro	200	256+8/8	318.4	283.2	Jan 96	www.intel

possibly, a single multiprocessor server. For convenience, details about SPEC benchmarks are contained in Tables 11.18–11.19.

11.4 SOFTWARE REALIZATION OF WCDMA (FDD) DOWNLINK IN BASE STATION SYSTEMS

In this section we discuss the computational requirements for software realization of the downlink channel for the WCDMA (FDD) system. The system radio interface protocol architecture is shown in Figure 11.42.

Table 11.19   SPEC95 benchmarks

Platform	Clk (MHz) ext/in	Cache ext+I/D	SPECint 95	SPECfp 95	Information date	Source obtained
SUN Ultra M170	83/167	512+16/16	5.56	9.05	Mar 96	SunIntro
DEC 3000/800	40/200	2M+8/8	—	—	—	—
Intel XXpress Pentium	66/166	1M+8/8	4.76	3.37	Jan 96	www.intel
Intel Alder Pentium Pro	200	256+8/8	8.09	6.75	Jan 96	www.intel

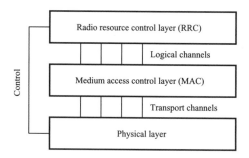

Figure 11.42 Radio interface protocol architecture.

The different downlink transport channels are

1. Speech channel.

2. 64 kbps data channel.

3. Simultaneous speech and 64 kbps channels.

4. 144 kbps channel.

5. Simultaneous speech and 144 kbps channels.

6. 348 kbps channel.

7. Simultaneous speech and 348 kbps channels.

8. 2 Mbps channel

9. ISDN channel.

10. Modem/fax channel.

11. Common packet channels such as FACH and DSCH.

12. Control channels such as pilot channel, SCH and indicator channels.

All these transport channels are mapped onto physical channels by performing the physical layer operations. All the physical layer operations on all of these channels are implemented in C language on a general purpose processor for estimating the computational requirements of downlink physical channels. The implementation procedure for all channels is identical, except for the difference in block lengths, transmission time intervals, SF, etc. Implementation details of the simultaneous speech and 64 kbps channel are given here for illustration. The procedure is identical for all other channels. Complete details of all channels are given in [1047].

An adaptive multirate (AMR) speech codec delivers speech at a rate of 12.2 kbps. It outputs 244 or 39 or 0 bits once every 20 ms (TTI) depending upon speech activity. Each output is divided into three sub blocks and each subblock is individually coded. The 244 bit output is split into three subblocks of 81, 103, 60 bits; the 39 bit output into 39, 0, 0 bit subblocks; and all 0 sized output into 0 bit subblocks. After fragmenting each output into three subblocks, 12-bit CRC coding is applied to the first subblock. Rate 1/3 convolutional coding (CC) is applied to the CRC coded first subblock and to the uncoded second sub block, while CC of rate 1/2 is applied to the uncoded third subblock. Eight tail bits are added to each subblock before convolutional coding. Tail bits are set as 0 and are used to clear the convolutional coder after encoding each block. For the data channel of 64 kbps rate, data blocks of fixed size of 336 bits (including MAC overhead of 16 bits) arrive from the MAC layer every 20 ms (TTI).

The number of blocks for each TTI can be any one of 0, 1, 2, 4. Each block is 16-bit CRC coded and all blocks are serially concatenated into a code block. The concatenated block is turbo coded. In order to convey control information, a dedicated control channel (DCCH) at 3.4 kbps rate is multiplexed and transmitted along with the speech and data channel.

MAC delivers a 148-bit block of DCCH for every 40 ms duration. This block is 16-bit CRC coded and then convolutional coded (rate 1/3). As the number of bits of the speech and data channel varies for every TTI, bits are to be rate matched (punctured or repeated) so that they can be mapped onto DPCH slots.

11.4.1 Example

For the 64 kbps channel with four transport blocks per TTI, the concatenated block size after CRC coding and turbo coding is 4236 bits. For the speech channel, assuming full speech activity (with 81, 103, 60 bit input blocks), the number of bits per TTI (20 ms here) after convolutional coding is 303, 333, 136.

For DCCH, the number of bits for each TTI (40 ms here) after CRC and convolutional coding is 516. As the TTI of DCCH is 40ms and the TTI of speech and data channels is 20 ms, two successive frames of speech channel are concatenated so that they can be multiplexed with DCCH. For the same reason, two successive frames of 64 kbps data channel are also concatenated. A *total* of 10 532 bits, containing 8472 bits of 64 kbps data channel (4236 turbo coded bits for each 20 ms), 1544 bits of speech channel (722 convolutional coded bits for 20 ms) and 516 convolutional coded bits of DCCH should be transmitted for every 40 ms duration.

38 400 chip slots on I channel and 38 400 chip slots on Q channel are available for every 10 ms on DPCH after spreading. A total of 307 200 ($2 \times 38400 \times 4$) slots are available for every 40 ms. A spreading factor of 32 is preferred in mapping the 10 532 bits of all transport channels onto downlink DPCH. For SF of 32, 8400 bit positions are available for transport channel data [1048]. A total of $10532 - 8400 = 2132$ bits of all data channels are being punctured (rate matched). After rate matching, bits of each transport channel are interleaved (1st interleaving) and segmented into radio frames of 10 ms time duration each.

The first 10 ms radio frames of all transport channels are multiplexed, this is called transport channel (TrCH) multiplexing and is repeated for all other radio frames. Second interleaving is performed on each 10 ms frame. Bits on each 10 ms radio frame are multiplexed with physical layer control bits and mapped onto each slot of DPCH. Each pair of consecutive bits on DPCH is serial to parallel converted, mapped onto I and Q channels and spread by using the same OVSF code. The chips on I and Q channels are combined into a complex sequence and then scrambled. In the same manner, all other downlink transport channels are mapped onto downlink physical channels.

In order to estimate the computational requirements for each physical channel, all the physical layer operations for each physical channel are implemented in software and are executed on a PC for a specific number (N) of frames. The % CPU required is taken as performance metric and is divided by the time during which those N frames have to be actually transmitted. In this classification, the computational power required by each downlink channel is expressed in terms of % CPU required on an Intel Pentium III 450 MHz processor. In order to express the computational power in a processor independent way, similar to the previous section, SPEC benchmarks are chosen which express each of the commercially available processors' computational power in SPEC ratings [976]. SPEC 95 ratings are expressed with two sub components, 'SPECint 95', focusing on integer/non-floating-point computationally intense activity, and 'SPECfp 95', focusing on floating-point computationally intense activity.

Table 11.20 gives the computational requirements of all the downlink channels in terms of the % CPU metric and in SPEC 95 ratings. With 60 MHz of allocated bandwidth for

Table 11.20   Downlink channels performance metric

| | Performance metric | | |
| | | SPEC ratings | |
Physical channel	%CPU	SPECint95	SPECfp95
PCCPCH	2.178	0.41	0.30
SCCPCH	76.96	14.39	10.54
PDSCH	38.00	7.15	5.21
Modem/fax channel	82.35	15.40	11.28
ISDN channel	84.05	15.72	11.51
Speech channel	54.90	10.27	7.52
64 kbps channel	84.90	15.88	11.63
144 kbps channel	109.83	20.54	15.05
384 kbps channel	158.42	29.62	21.7
Speech and 64 kbps channel	107.55	20.11	14.73
Speech and 144 kbps channel	153.70	28.74	21.06
Speech and 384 kbps channel	243.59	45.55	33.37
2 Mbps channel	1452.13	271.55	198.94
Sync. channel	7.364	1.38	1.01
AICH or AP-AICH or CD/CA-AICH	130.20	24.35	17.84
CPICH	52.68	9.85	7.22
CSICH	4.96	0.93	0.68
PICK	47.92	8.96	6.57

Table 11.21 Number of channels supported on each RF carrier for different systems

RF carrier	Type of channel	No. of channels supported by each system		
		I	II	III
1	64 kbps	32	32	32
2	Speech and 64 kbps	32	32	32
3	144 kbps	16	16	16
4	Speech and 144 kbps	16	16	16
5, 6	2 Mbps	2	2	2
	384 kbps ch.	4	4	4
7	Control and indicator	32	32	32
	Speech	40	96	—
	64 kbps	5	—	12
	Speech and 64 kbps	5		12
	384 kbps	2	1	1
8	Speech	128	128	128
9, 10	Speech	—	16	—
	ISDN	16	16	16
	Modem/fax	16,16	16,16	24,16
	Speech and 384 kbps	6	5	5
11	SCCPCH	32	32	32
12	PDSCH	16	16	16

downlink WCDMA, 12 RF carriers, each of 5 MHz, are available for downlink WCDMA. The number of physical channels that can be transmitted on each RF carrier depends on the SF used. For a base station assuming a frequency reuse factor of 1 (ideal case), and without any sectorization, the number of downlink physical channels that are mapped onto the available 12 RF carriers is shown in Table 11.21.

Three systems are considered, each with a different number of different types of channel. System I corresponds to an equal number of voice and data channels. System II corresponds to 60 % voice channels and 40 % data channels. System III corresponds to 40 % voice channels and 60 % data channels.

The computational requirement of each system gives an estimate of the computational requirements for the downlink in practical base stations. Simultaneous speech and data channels are regarded as data channels. By maintaining orthogonality in assigning OVSF codes to physical channels, different downlink channels for the assumed three systems are mapped onto RF carriers as shown in Table 11.21.

From the total computational power requirements for each system from Table 11.23, and from the SPEC ratings of the processor from Table 11.22, it is observed that the number of Pentium III processors at 1 GHz required to realize each system is around 150, quite a high value.

One would require a number of more powerful processors than have been used in this study to realize the downlink of WCDMA base stations in software.

Table 11.22   SPEC 95 ratings [1049]

Platform	Clock	SPECint95	SPECfp95
Intel Pentium III	450 MHz	18.7	13.7
Intel Pentium III	1 GHz	46.8	32.2

Table 11.23   Downlink channels performance metric in SPEC ratings

	SPEC ratings					
	System I		System II		System III	
Channel type	int 95	fp 95	int 95	fp 95	int 95	fp 95
PCCPCH	0.4	0.3	0.4	0.3	0.4	0.3
SCCPCH	460	337	460	337	460	337
PDSCH	114	83.3	114	83.3	114	83.3
Modem/fax	492	360	492	360	616	451
ISDN	251	184	251	184	251	184
Speech	1725	1263	2464	1804	1314	962
64 kbps	587	430	508	372	698	511
144 kbps	328	240	328	240	328	240
384 kbps	177	130	148	108	148	108
Speech and 64 kbps	744	545	643	471	884	648
Speech and 144 kbps	459	336	459	336	459	336
Speech and 384 kbps	273	200	227	166	227	166
2 Mbps	543	397	543	397	543	397
Sync.	1.4	1.0	1.4	1.0	1.4	1.0
Indicator	438	321	438	312	438	312
CPICH	9.8	7.2	9.8	7.2	9.8	7.2
CSICH	5.6	4.0	5.6	4.0	5.6	4.0
PICH	53.7	39.4	53.7	39.4	53.7	39.4
TOTAL	6662	4878	7146	5222	6551	4787

One could conjecture that the computational requirements of uplink physical channels for 3G WCDMA base stations in software would be similarly high.

11.5 DESIGNING A DS-CDMA INDOOR SYSTEM OVER FPGA PLATFORMS

In this section we discuss designing a DS-CDMA [1050–1058] indoor system which works at the 2.4 GHz frequency band and has a star architecture. All the terminals interact with their own base station that has the transmission, reception and control capabilities, as well as user control, channel assignment, code distribution, power control, etc.

Each base station can bear up to 64 different channels of 32 kbps simultaneously assigned to a maximum of 16 different users.

The following user terminals are distinguished:

- a voice terminal, using DPCM (differential pulse code modulation) at 32 kbps, similar to that used in DECT (digital European cordless telephone);

- a data terminal at 9.6 kbps and delays below 30 ms; and

- a video terminal, using standard H261 which is capable of adapting its rate up to 2 Mbps depending on the image quality desired.

The CDMA radio link will use traffic channels (TCH) to carry data, voice and video. These channels will be complemented with associated control channels (ACCH) for power control, channel information, status of communication, etc. Other signaling channels used are:

- the pilot channel (PICH), to allow the correct synchronism of the mobile to the network;

- the broadcast paging channel (BPCH), for paging, access to the network or handover; and

- the random access channel (RACH), for mobile traffic channel requests.

The parts of this system implemented using an FPGA approach, range from the physical channels to intermediate frequency. This does not include channel coding, puncturing, interleaving, etc. of information coming from upper layers. The service that the FPGA-based layer provides is a simple raw service of transport of data over the radio channel.

11.5.1   Downlink structure

The downlink (see Figure 11.43) is mainly based on an orthogonal multiplexing of information and control channels using pseudonoise sequences. The sequences used to multiplex channels are Walsh or Hadamard sequences, and those used to isolate adjacent cells are Gold sequences. Each user has a maximum of four QPSK-CDMA channels, each of them at 64 kbits/s, separated by different Walsh sequences at 1024 Mchips/s. This provides a maximum of 256 kbits/s per user, which is the limit assumed in the system. In transmission, two separated antennas are used to transmit twice the same signal with a determined delay between them longer than the CDMA chip resolution. This kind of space–time coding is discussed in Chapter 4. In the receiver side, a RAKE structure can take advantage of this transmission diversity. To simplify the receiver, a pre-RAKE [1051] structure is included in the transmitter (base station) for each user, this is adjusted using the channel estimation values obtained by the mobile terminal.

11.5.2   Uplink structure

In this case, a simple QPSK scheme is used for transmission, where the maximum speed for both the in-phase and quadrature channels is 128 kb/s. This data rate is spread using Gold sequences up to 4096 kchips/s, obtaining a processing gain of 32 when transmitting

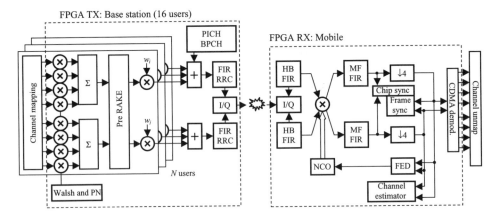

Figure 11.43 System model.

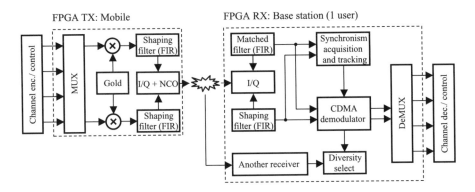

Figure 11.44 The FPGA sections of the uplink transmitter and receiver schemes.

at the maximum speed (256 kb/s total). In Figure 11.44, the FPGA sections of the uplink transmitter and receiver schemes can be seen.

11.5.3 Physical implementation

Figures 11.43 and 11.44 give a general view of the blocks implemented using the FPGA approach for the up- and downlinks of the base station and mobile terminal respectively. In each transmitter the information bits are given to the FPGA, which generates a QPSK modulation centred at 8192 kHz and sampled at 32 768 kHz. Inversely, in the receiver, the bits are obtained after synchronization, frequency adjustment, channel estimation, etc. As can be observed, some blocks in the downlink will be similar to those in the uplink. But when each block is optimized to reduce the resources required to implement it, reusing the block is not possible.

What certainly can be reused is the way a function is implemented in every case. The concrete implementation can be adjusted in a more or less automatic way.

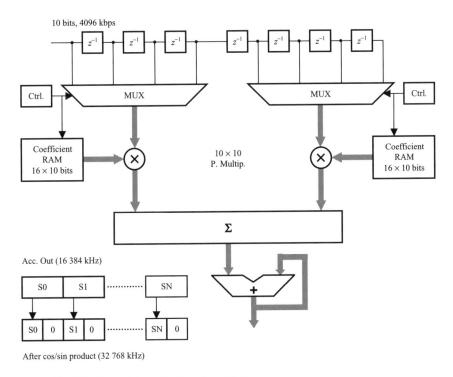

Figure 11.45 FIR filter with multiplexed multipliers.

Consider, for instance, two different filtering stages of the downlink. First, in transmission, the FIR filter is used to generate a pulse with root raised cosine shape. To this filter only one of every eight input samples is non-zero (4096 kHz input zero padded and filtered to get 32 768 kHz output). The property can be used to reduce the number of multipliers and, consequently, reduce the amount of logic. Another filtering stage would be the half band filter used in the downlink receiver before decimating by 2 (HB FIR). The samples entering this filter are non-zero and simplifications can only be done considering the decimation process and through the selection of an adequate filter with, for instance, zero-valued coefficient in every odd sample. Of course, these simplifications are not only useful for FPGA-based implementations, but also for any other kind of implementation.

On the basis of the previously suggested simplifications, the FPGA implementation can have different shapes, where considerations about number of bits, number of coefficients, etc. are very important. The first filtering process mentioned can be implemented efficiently as shown in Figure 11.45, where only two multipliers with their inputs multiplexed are used to compute the output. In the case of the second filtering process, the transposed FIR structure shown in Figure 11.46 fits well to our purposes (and, in general in most FPGA FIR implementations). Here, the multipliers are implemented as distributed arithmetic multipliers taking advantage to the fact that the coefficients of the filter are constants. In both filtering cases, FPGA internal RAM and fast internal adders are the keys that allow an efficient implementation of the blocks. Other examples in the transmission/reception chains can be found. After general simplifications, efficiently mapping the block onto FPGA consists mainly of identifying how the algorithm can be translated to the internal structures. The

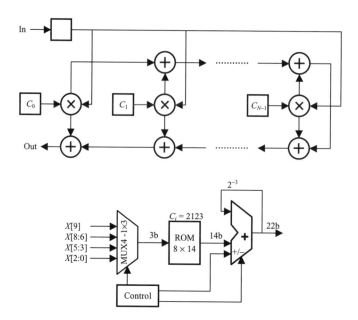

Figure 11.46 FIR transposed form and block distributed arithmetic multiplier example.

use of internal RAM/ROM and fast adders provides a good mechanism to implement typical signal processing tasks, but sometimes requires algorithm reshaping or even a special control.

11.5.4 Resource utilization

The system roughly described above, although representing a simplification with respect to commercial products, is an example that incorporates all the functions that imply a higher signal processing demand (intermediate frequency processing and synchronization). The blocks that perform these tasks must use resources optimally in order both to reduce the amount of resources required, and also to reduce the power consumption of the devices. This has been the main goal of the design. The information provided here on resource allocation is expressed in terms of logic elements (LEs). One LE is defined as the composition of one four-input lookup table (LUT), equivalent to a RAM of one bit wide and 16 addresses deep, and one flip-flop. It should be noted that figures given are approximated because they are extracted from CAD tool reports, where several blocks are mixed and/or other support functions are considered, like the µprocessor access interface. The hardware platform used to check the validity of the system is not tuned for this specific design. It is a predesigned platform with enough flexibility to accommodate different applications. This platform is called SHaRe (Reconfigurable Hardware System) [1052], and provides up to eight user programmable FPGAs interconnected in a flexible way with access to additional resources like RAM, ROM, FIFOs and programmable clocks. Each one of the FPGAs can be (re)programmed individually at any time by simply performing write cycles to a memory address over the host bus (VME bus). The main processor over the bus controlling the

Table 11.24   LEs required to implement the mobile terminal [1058] © 2001, IEEE

Block description	LEs
Terminal TX spreading and root raised cosine filters	400
Terminal TX frequency adjust NCO and IF translation	620
Terminal RX I/Q down-conversion with half band filter	1060
Terminal RX frequency adjust (NCO + FED)	830
Terminal RX matched filters	1450
Terminal RX chip, bit and frame synchronism	1520
Terminal RX channel estimation and CDMA demodulation	500
Total mobile terminal	6380

Table 11.25   LEs required to implement the base station [1058]

Block description	LEs
Base station TX 16 users spreading, pre-RAKE and PICK + BPCH	1140
Base station TX 16 users I/Q shaping filter and IF translation	1090
Base station RX 1 user I/Q down-conversion and matched filters	4390*
Base station RX 1[a] user chip synchronism and tracking	1180*
Base station RX 1[a] user CDMA demodulation	500*
Total base station, 16 users	99350

* Multiply by 16 to compute the total base station LE requirements

whole system is a Sparc CPU running Solaris. This host processor provides connection to an IP network, thus obtaining a higher functionality of the system. The LE utilization for the mobile terminal and base station is summarized in Tables 11.24 and 11.25 . The values concern only the specific part named. The figures show that the complete terminal system would fit into a state-of-the-art FPGA with about 7000 LEs. This number of LEs can be found in relatively small FPGAs like Xilinx Virtex XCV400 [1053]. The base station part for a single user would need less than 9000 LEs, which are also available in that FPGA.

The increase in resource utilization for more users (16 in this case) is not linear because most of the transmitter is reused, as can be observed in Figure 11.43, where IF stages are not repeated for each user. In any case, about 100 000 LEs should be employed to construct this part of the base station.

The actual implementation presented is not based on the Xilinx Virtex family but on the Xilinx 4K family [1054,1055]. The basic elements of both families are quite similar but Virtex includes more RAM bits, more and faster interconnect logic, more input/output standards, dedicated functions (e.g. multipliers), etc., as expected for a more modern family. As stated before, the platform used, called SHaRe, has eight user-programmable FPGAs with a count of up to 50 176 LEs (with eight XC4085 FPGAs on the board), 2 Mbytes of SRAM distributed along the devices, up to 512 Kbytes of FIFOs and up to 512 Kbytes of ROM. One can see that *the mobile terminal fits over a single board.*

Also, *the base station for one user receiver and 16 transmitters fits over a single board, but every new base station receiver requires a new board.*

With this distribution chosen, the maximum number of resources required over a single board is about 9000 LEs, this can be obtained by completing a SHaRe board with the smallest devices it can bear to achieve a total of 9216 LEs.

Although the numbers fit, it is not possible to adjust the design to that level and some room must be allowed to install accessories, control and even leave some resources free to ease the process of mapping of the blocks over the devices. All the boards involved in the design are joined over a VMEbus to allow correct configuration and management of the application. In general, the blocks have been implemented exchanging area for speed. Reducing area at the cost of an increment of the frequency clock does not save on power consumption.

Power consumption is minimized only when the mapping of a function over FPGA resources is optimally carried out the power demanded by the FPGA devices depends highly on the physical properties of them, but the more advanced the technology used, the less power is required.

The flexibility provided by FPGAs over ASICs is paid for as an increase in power consumption when the FPGA option is selected. On the other hand, the amount of LEs stated here is relatively low compared to the availability of them in commercial FPGAs, so a different structure with higher sampling rates can be considered by simply increasing area. Note that in the digital implementation of a radio receiver, only the very first filtering stages after sampling have high rates. As soon as the channel of interest has been selected, the sampling rate is reduced and/or transformed to be adequate for the required demodulation.

In conclusion we can say that there are several ways to implement a reconfigurable radio terminal, but one that offers a wide application flexibility, ranging from highly computationally intensive tasks to algorithmically complex but relatively low-speed tasks, and at the same time the flexibility to modify its behavior, is the one based on FPGAs. Their increasing capacity and speed, together with the progressive reduction of power consumption, make them a good candidate to occupy relevant positions in future radio terminals designed under the Software Radio line of thought. A DS-CDMA system taking advantage of the structural properties of FPGAs has been described in this section to illustrate their adaptation to that kind of application, and to explore the pros and cons when designing specific parts. As in many digital systems, there appears a tradeoff between speed and logic resources used: the higher the speed required, the higher the number of resources required to implement the functions.

The commercial FPGAs offer the range of resources required to build a system with these features. Designing with FPGA predefined structures tends to modify the way the different blocks are implemented to use resources efficiently. This, in general, makes an impact on the complexity of the solution, because typical signal processing structures cannot be translated directly. Finding the right mechanism will improve the final design in terms of die and power consumption.

All of this has an interesting application when considering restrictions related to fully digital radio terminals, because a better adaptation to FPGA structure will allow implementation of more computationally intensive tasks. Because of this, it is important to identify clearly the blocks which require a more accurate tuning, and get a set of possible solutions that may fit in many different applications. Also, it is important to investigate which are the more interesting architectures for FPGA arrays plus complements (memory, I/O, programmable clocks, etc.) to extend its flexibility and reusability. To find the right use of FPGA devices in a wide range of structures and simulations, a UMTS physical layer implementation based on FPGAs is a good testbed that will also serve to analyze which are good solutions to get a reconfigurable fully digital terminal. Of course AD and DA technology has a lot to say about that, but it is clear that optimal implementation of digital processing blocks (whatever algorithm is used) is a cornerstone.

IV

Wireless Communication Networks

12

Network Overlay in 4G

The coexistence between WCDMA (or UWB) signals and narrowband signals like ATDMA, may be further improved by using additional schemes for interference suppression in wideband systems. In this section we present schemes for interference suppression in CDMA wireless networks. Modification of these schemes for UWB systems is straightforward. Depending on the interfering signal, these schemes adaptively change not only their parameters but their structures too. For the relevant type of interfering signal, suppression is also possible for the wideband interference occupying the same frequency band as the CDMA signal. The main applications are CDMA and UWB overlay type wireless networks colocated in the same frequency band with high bit rate ATDMA [1059–1062] or the multimedia CDMA network, where a high bit rate signal due to lower processing gain must use higher power in order to provide the required quality of transmission. The focus is on suppressing high constellation, high bit rate, QPSK or QAM signals, which are likely to be used in advanced TDMA systems.

12.1 ADAPTIVE SELF-RECONFIGURABLE INTERFERENCE SUPPRESSION SCHEMES FOR CDMA WIRELESS NETWORKS

In general, narrowband interference canceler structures for DSSS systems have been studied extensively [1063–1076]. In the 1980s, a number of papers was published using single-sided (linear prediction) and two-sided (linear interpolator) transversal filters to suppress a significant portion of the interference. The optimum algorithm, in the sense of minimum mean square error, results in the Wiener filter [1063]. Practical implementation of such a filter requires signal correlation matrix inversion. In the case of non-stationary interference, this matrix must be updated to follow the changes in the interfering signal parameters and the overall computation complexity becomes excessive. The problem becomes even more serious for UWB systems. On the other hand, simpler methods, like LMS algorithms, will

Advanced Wireless Communications. S. Glisic
© 2004 John Wiley & Sons, Ltd. ISBN: 0-470-86776-0

demonstrate worse performance and convergence problems. In these papers, different types of interfering signal model are used: a simple CW format [1066], an impulsive signal [1067, 1072], an autoregressive model [1069, 1070], frequency-hopping [1077], and frequency sweeping [1078] signals.

At the beginning of the 1990s, non-linear methods were introduced to further improve the system performance [1071, 1072]. Additional improvements to these non-linear schemes were published later [1079–1081]. It has also been shown that an adjustable center weight in the filter (instead of the fixed one) can also offer some improvement [1075]. The linearly constrained scheme, which takes care that the useful signal degradation is minimized in the interference suppression process, is described in [1076].

In addition to these applications, a broadband CDMA overlay type of communication network has become an attractive solution to further increase capacity of the existing cellular radio networks [1082, 1083]. In this concept, narrowband and low-level CDMA signals use the same frequency band. Prior to the standard signal processing, narrowband signals are additionally suppressed in the CDMA receiver using adaptive interference suppression schemes (most of the time, the LMS algorithm). Due to modulation, the interfering signal is non-stationary, and special care is taken to improve the system performance in the transient state, when the suppression scheme is readjusting the filter coefficients due to the change of the interfering signal parameters.

The traditional schemes based on adaptive transversal filtering (TF) are efficient only as long as the interfering signal bandwidth occupies less than 20 % of the CDMA signal bandwidth. The schemes based on transversal self-reconfigurable filters (TSRF) suppress efficiently wideband interference and still preserve simplicity.

The main applications of these schemes in 4G are:

1. CDMA overlay type networks collocated in the same frequency band with the ATDMA system using standard modulation formats (PSK, QPSK, QAM, etc.).

2. The multimedia CDMA network, where a high bit rate signal due to lower processing gain must use higher power in order to provide the required quality of transmission.

12.1.1 Signal model

The useful DSSS signal is assumed to have the following form:

$$s(t) = m_1(t) \cdot e_1(t) \cos \omega_c t + m_2(t) \cdot e_2(t) \sin \omega_c t \qquad (12.1)$$

where e_1 and e_2 are PN sequences with chip interval Δ, m_1 and m_2 are two independent information streams with bit interval $T_b = Q\Delta$ and ω_c is the carrier frequency. Parameter Q is here referred to as the system processing gain

12.1.2 Interference model

In the analysis we will be using the following forms of the interfering signal

$$
\begin{aligned}
\text{ASK} &\rightarrow j_a(t) = A \left\{ \frac{p_1(t) + 1}{2} \right\} \cos(\omega_1 t + \theta) \\
\text{PSK} &\rightarrow j_P(t) = A \cdot p_1(t) \cos(\omega_1 t + \theta) \\
\text{FSK} &\rightarrow j_f(t) = A \cos \varphi(t)
\end{aligned}
\qquad (12.2)
$$

where

$$\varphi(t) = \int\limits_0^t \omega_1 \left\{ \frac{1 + p_1(t)}{2} \right\} dt + \int\limits_0^t \omega_2 \left\{ \frac{1 - p_1(t)}{2} \right\} dt + \theta \qquad (12.3)$$

$p_1(t) = \pm 1$ is the binary information in the interfering signal with bit interval T_h and ω_1, θ and A are the interfering signal frequency, initial phase and amplitude, respectively. In the case of FSK interference, the signal frequency may be either ω_1 or ω_2 depending on the data being transmitted.

12.1.3 Receiver structures

The system block diagram is shown in Figure 12.1. After frequency down-conversion and low pass filtering, the signal samples in the kth sampling moment can be represented in complex form as

$$x(k) = j(k) + s(k) + n(k) \qquad (12.4)$$

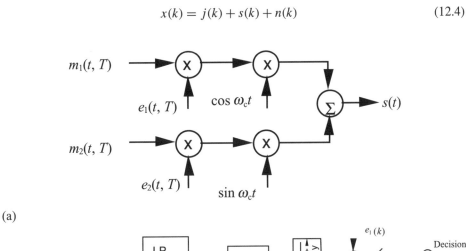

(a)

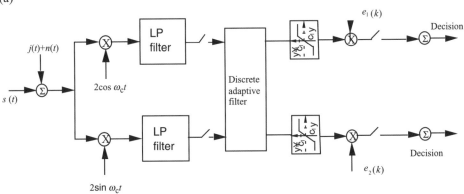

(b)

Figure 12.1 System block diagram. (a) Transmitter; (b) receiver.

Signal $j(k)$ for ASK, PSK and FSK interference is designated as $j_a(k)$, $j_p(k)$ and $j_f(k)$, respectively, and can be represented as

$$j_a(k) = A\frac{p_1(k)+1}{2}e^{-j(\delta\omega_1 \cdot \Delta \cdot k + \theta)}; \quad \delta\omega_1 = \omega_1 - \omega_c, \quad t_k = k\Delta$$

$$j_p(k) = Ap_1(k) \cdot e^{-j(\delta\omega_1 \cdot \Delta \cdot k + \theta)} \tag{12.5}$$

$$j_f(k) = Ae^{-j\varphi(k,l)}$$

where

$$\varphi(k) = \int_0^{k\Delta} \delta\omega_1 \cdot \frac{p_1(t)+1}{2} \, dt + \int_0^{k\Delta} \delta\omega_2 \cdot \frac{1-p_1(t)}{2} \, dt + \theta \tag{12.6}$$

$$\delta\omega_2 = \omega_2 - \omega_c$$

Signal $s(k)$ has the following form

$$s(k) = s_1(k) + js_2(k) = [s_1(k), s_2(k)] \tag{12.7}$$

where

$$s_1(k) = \frac{1}{\sqrt{2}}m_1(k)e_1(k)$$

$$s_2(k) = \frac{1}{\sqrt{2}}m_2(k)e_2(k) \tag{12.8}$$

$n(k)$ in Equation (12.4) represents noise samples having Gaussian distribution with zero mean and variance σ^2. For the normalized signal amplitude, the signal to noise ratio is defined as SNR $= 1/2\sigma^2$, and the interference (jammer) to signal ratio as JSR $= (A^2/2)/(1/2) = A^2$.

12.1.4 Filter model

For these applications, two forms of the filter are used, either the two-sided filter shown in Figure 12.2, or the one-sided filter (forward prediction error filter FPEF). The latter can be represented as a half of the two-sided filter, framed by dotted lines in Figure 12.2. For our purposes we will also need a backward prediction error filter (BPEF), which is represented

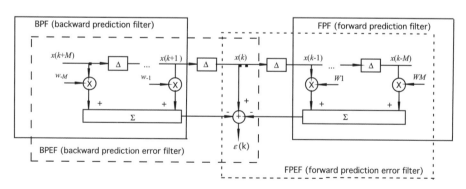

Figure 12.2 Complex adaptive two-sided filter block diagram.

as another half of the two-sided filter, framed by dashed lines in Figure 12.2. It can be shown that, in the case of the CW interfering signal, the optimum (Wiener) values of each pair of filter coefficients, symmetrical with respect to the central point, are complex conjugates, i.e. [1063, 1075]:

$$w_{-i}(k) = w_i^*(k) \quad i = 1, 2, \ldots, M \tag{12.9}$$

where '*' stands for the complex conjugate and $2M$ is the filter length. In addition to this, we also have

$$w_{i+1}(k) = w_i(k) \cdot w_1(k) \quad \text{for} \quad i = 1, \ldots, M - 1 \tag{12.9a}$$

and

$$w_{i-1}(k) = w_i(k) \cdot w_{-1}(k) \quad \text{for} \quad i = -1, \ldots, -M + 1$$

By using a standard methodology [1063,1078,1084], one can show that, in the case of a PSK interference, each time the phase of the interference is changed, the suppression gain for 2SF and FPEF filters will change.

The suppression gain G is defined as the filter input interference, plus the noise power to output interference, plus the noise power ratio. Interference suppression gain reduction is caused in the periods when filter coefficients are changed (updated) in order to adjust to the new values of the interfering signal parameters. We will use parameters k_c and k_r to designate the sampling index when the filter coefficients start to change and are recovered, respectively. So, the readjustment period is $k_r - k_c$ sampling intervals long. The degradation time is approximately $2(k_r - k_c) = 2M$ chips for all three types of the filter – 2SF, FPEF and BPEF. The rate of these degradations is characterized by parameter $1/T_h$ (interference signal bit rate). The positions of readjustment periods on the time scale for these filters are different, due to the propagation delay of the signal changes in the filter. So, in order to eliminate any impact of interfering signal phase change on suppression gain, we should use only segments of the 2SF filter in such a way that for $k < k_c$, FPEF is used, and for $k \geq k_c$, BPEF, with initial coefficient obtained from Equation (12.9). Because the 2SF performs better than any 1SF, the two-sided filter can be used in the period when its coefficients are recovered ($k > k_r$). In this way, the effective filter will have different configurations in different time segments, hence the name self-reconfigurable interference suppression schemes.

We will now present several schemes which use different ways to detect the moments when the configuration should be changed. The change of configuration will be referred to as reconfiguration.

12.1.4.1 U structure

As the first step, the 2SF from Figure 12.2 is represented in the form shown in Figure 12.3. This form will be referred to as a *U structure*. In the figure, BPF represents the backward predictor filter that includes all elements contributing to the summation on the left-hand side of Figure 12.2, and FPF represents the forward predictor filter that includes all elements contributing to the summation on the right-hand side of Figure 12.2. Based on Figure 12.3, we create variable $c_m(k)$, which is used to decide about the moments when changes in configuration should take place. Those moments will be referred to as reconfiguration

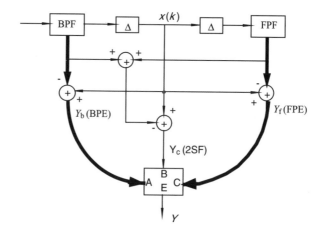

Figure 12.3 U structure; segmentation of 2SF configuration.

times. The variable $c_m(k)$ is defined as:

$$c_m(k) = \min \{| \operatorname{Re}[y_f(k)]| + | \operatorname{Im}[y_f(k)]| ,$$
$$| \operatorname{Re}[y_b(k)]| + | \operatorname{Im}[y_b(k)]| , \qquad (12.10)$$
$$| \operatorname{Re}[y_c(k)]| + | \operatorname{Im}[y_c(k)]|\}$$

where, by definition,

$$y_b(k) = x(k) - \sum_{i=-1}^{-M} x(k-i)w_i(k)$$

$$y_f(k) = x(k) - \sum_{i=1}^{M} x(k-i)w_i(k) \qquad (12.11)$$

and

$$y_c(k) = x(k) - \sum_{\substack{i=-M \\ i \neq 0}}^{M} x(k-i)w_i(k)$$

From Equation (12.10), one can see that there are three different configurations the filter can choose from. If

$$c_m(k) = |\operatorname{Re}[y_b(k)]| + |\operatorname{Im}[y_b(k)]| \qquad (12.12)$$

which means that BPF is active, we will use the notation UCB (U structure, Configuration with Backward predictor being active). We will consider two options:

1. UCB$\bar{\text{C}}$ – coefficients are not copied ($\bar{\text{C}}$) from the active to the non-active side of the filter.

2. UCBC – coefficients are copied (C) from the active to the non-active side of the filter.

For the UCBC̄ option, we have:

$$y(k) = y_b(k)$$
$$e(k) = \mu y(k)/(\eta_{kom} \cdot 0.8)^2$$
$$\eta_{kom} = \overline{|Re[x(k)]| + |Im[x(k)]|}$$

$$w_i(k+1) = w_i(k) + e(k) \cdot x^*(k-i), \quad i = -M, \ldots, -1 \qquad (12.13)$$

Constant 0.8 is derived from the condition to normalize the algorithm to the input signal power. In UCBC, Equation (12.13) applies, together with:

$$w_i(k+1, l) = w^*_{-i}(k+1, l), \quad i = 1, 2, \ldots, M \qquad (12.14)$$

If

$$c_m(k) = |Re[y_f(k)]| + |Im[y_f(k)]| \qquad (12.15)$$

we will use the notation UCF (U structure, Configuration with Forward predictor being active).

For the UCFC̄ option, we have:

$$y(k) = y_f(k)$$

$$w_i(k+1) = w_i(k) + e(k) \cdot x^*(k-i), \quad i = 1, 2, \ldots, M \qquad (12.16)$$

For the UCFC option, in addition to Equation (12.16), the following relation holds:

$$w_{-i}(k+1, l) = w^*_i(k+1, l), \quad i = 1, 2, \ldots, M \qquad (12.16a)$$

If

$$c_m(k) = |Re[y_c(k)]| + |Im[y_c(k)]|$$

the complete (C) two-sided filter is used and the notation will be UCC.

For the UCCC̄ option, the corresponding system of equations is:

$$y(k) = y_c(k)$$

$$w_i(k+1) = w_i(k) + e(k) \cdot x^*(k-i), \quad i = -M, \ldots, -1, 1, \ldots, M \qquad (12.17)$$

For the UCCC case, an additional averaging of the filter coefficients is performed in accordance with the following set of equations:

$$w'_i(k+1) = w_i(k) + e(k) \cdot x^*(k-i); \quad i = -M, \ldots, -1, 1, \ldots, M$$
$$w_{-i}(k+1) = \frac{1}{2}[w'^*_i(k+1) + w'_{-i}(k+1)]; \quad i = 1, \ldots, M \qquad (12.18)$$
$$w_i(k+1) = \frac{1}{2}[w'_i(k+1) + w'^*_{-i}(k+1)]; \quad i = 1, \ldots, M$$

As long as $2M \leq T_h/\Delta$, this structure will completely remove degradations in suppression gain G. In other word, G will be more or less constant in time, regardless of sharp transitions in interfering signal parameters. This is due to the fact that filter recovering time $2(k_r - k_c)$ is approximately $2M$. If $2M > T_h/\Delta$, gaps in G, although less deep, will reappear. This can be further improved by using limiters at the output of the U structure. The reconfiguration command is sensitive to noise. In the case of PSK interference, significantly better performance is achieved if the so-called Ψ *structure*, described in the next section, is used.

12.1.4.2 Ψ structure

In this case, an additional one-tap forward prediction filter is used to detect the moment when the interfering signal is changing the phase. The basic block diagram of the circuit shown in Figure 12.4 has the form of letter Ψ, hence the name Ψ structure.

The 'detector filter' output can be represented as

$$
\begin{aligned}
y_d(k) &= x(k) - x(k-1)w_d(k) \\
e_d(k) &= \frac{\mu_d \cdot y_d(k)}{(\eta_{kom}0.8)^2} \\
w_d(k+1) &= w_d(k) + e_d(k)x^*(k-1)
\end{aligned}
\tag{12.19}
$$

In addition to this, we introduce a control parameter (counter) $b_{kom}(k, l)$ which initially takes value zero. The circuit operates in three different modes. If

$$
\begin{aligned}
|\text{Re}[y_d(k)]| + |\text{Im}[y_d(k)]| &\leq |\text{Re}[x(k)]| + |\text{Im}[x(k)]| \\
b_{kom}(k) &= 0
\end{aligned}
\tag{12.20}
$$

the configuration will be referred to as ΨCF$\bar{\text{C}}$ & ΨCFC. In this case, FPF is active so that

$$
y(k) = y_f(k) \quad \text{(in Figure 12.4 } Q \equiv D) \tag{12.21}
$$

The LMS algorithm is defined by Equation (12.16). For Ψ CFC, the filter coefficients are copied from the active to the passive side in accordance with Equation (12.16a).

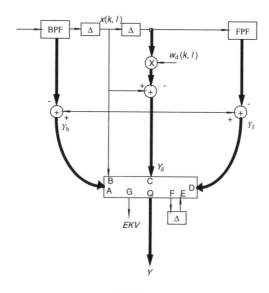

Figure 12.4 Ψ structure; segmentation of 2SF configuration.

If

$$b_{\text{kom}}(k) = 0 \text{ and}$$

$$|\text{Re}[y_{\text{d}}(k)]| + |\text{Im}[y_{\text{d}}(k)]| > |\text{Re}[x(k)]| + |\text{Im}[x(k)]| \qquad (12.22)$$

we have the $\Psi\text{CB}\bar{\text{C}}$ & ΨCBC configuration. Parameter $b_{\text{kom}}(k+1)$ becomes

$$b_{\text{kom}}(k+1) = M - 1 \qquad (12.23)$$

and

$$y(k) = y_{\text{b}}(k) \quad (\text{in Figure} 12.4 \quad Q \equiv A) \qquad (12.24)$$

For the ΨCBC system, the filter coefficients are copied in accordance with Equation (12.14). The system stays in this mode for the next $(M - 1)$ sampling intervals. This is controlled by decreasing parameter b_{kom}, as $b_{\text{kom}}(k+1) = b_{\text{kom}}(k) - 1$ as long as $b_{\text{kom}}(k) > 0$. As for the U structure, the Ψ structure also requires $2M \leq T_{\text{h}}/\Delta$ for good performance. This structure is even more sensitive if this condition is not met. The reconfiguration command is less sensitive to noise and the Ψ structure performs better only in the case of PSK interference when the additional, one-tap, filter detects the changes in the input signal with the highest reliability

12.1.4.3 ξ structure

This structure is obtained as a solution for the previous limitations defined by the request $2M \leq T_{\text{h}}/\Delta$. As a price, the computational complexity will be increased. The structure consists of a number $(2M)$ of one-tap filters and is intended for PSK interference suppression. Every one-tap filter acts independently and the resulting output is the average value of the individual outputs. Every one-tap filter uses a signal sample from a different time slot to approximate the instantaneous value of the interfering signal. In addition to the regular LMS algorithm for course adjustments, a new set of filter coefficients $\{w_i^{(s)}(k) = \pm 1\}$ is used. The responses of each of the $2M$ one-tap filters and corresponding coefficients $w_i^{(s)}$ are obtained by minimizing the following relation:

$$y(k - i) = \min_{w_i^{(s)}(k)} \left\{ \begin{array}{l} \left|\text{Re}\big[x(k) - w_i^{(s)}(k)x(k - i)w_i(k)\big]\right| \\ +\left|\text{Im}\big[x(k) - w_i^{(s)}(k)x(k - i)w_i(k)\big]\right| \end{array} \right\} \qquad (12.25)$$

$$i = -M, \ldots, M, \quad i \neq 0$$

Minimization of this equation will require higher computational complexity because a larger number of combinations must be searched for the minimum. The overall error signal is then formed as

$$y(k) = \frac{1}{2M} \sum_{\substack{i=-M \\ i \neq 0}}^{M} y(k - i) \qquad (12.26)$$

In this case, the LMS algorithm is defined as

$$w_i(k + 1) = w_i(k) + e(k)\Big[x(k - i) \cdot w_i^{(s)}(k)\Big]^* \qquad (12.27)$$

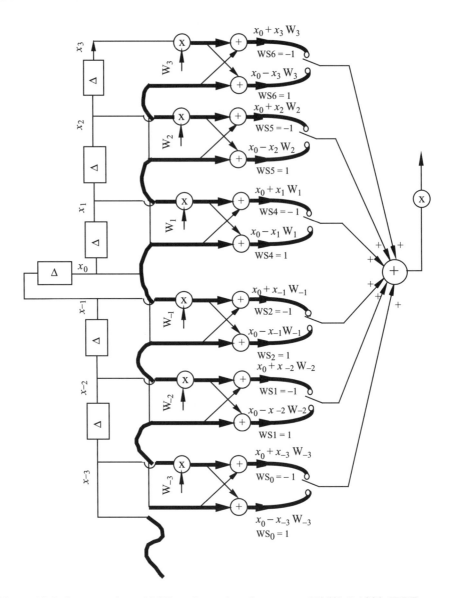

Figure 12.5 Segmentation of 2SF configuration; ξ structure [1059] © 1999, IEEE.

Switching between ± 1 is symbolically represented in Figure 12.5 by letter ξ, hence the name. This filter provides the best suppression of PSK signals, including wideband interference of the same bit rate as the chip rate of the DSSS signal. In other words, the interfering signal may be another CDMA signal. An important application is the UMTS system, where a higher bit rate (2 Mbit/s) signal should be transmitted, together with a number of voice channels. Due to low processing gain, such a signal must be much stronger than voice signals (2 Mbits/10 kbits = 200 times). Also, due to the limited capacity, only one such signal could be expected per cell. In the case of ASK interference, coefficients $w_i^{(s)}(k, l)$ are chosen in accordance with

$$w_i^{(s)}(k) = \begin{cases} 0, & \sum(x(k-i)) < \eta_{kom} \\ 1, & \text{elsewhere} \end{cases} \tag{12.28}$$

where

$$\sum(x(k-i)) = |\text{Re}\{x(k-i)\}| + |\text{Im}\{x(k-i)\}| \tag{12.29}$$

The output of each one-tap filter is obtained as

$$y(k-i) = x(k) - x(k-i)w_i(k)w_i^{(s)}(k) \tag{12.30}$$

The overall output signal is given as

$$y(k) = \begin{cases} \dfrac{1}{M_k}\sum y(k-i), & M_k \neq 0 \wedge \sum(x(k-i)) \geq \eta_{kom} \\ x(k), & \sum(x(k-i)) < n_{kom} \\ 0, & M_k = 0 \wedge \sum(x(k-i)) \geq \eta_{kom} \end{cases} \tag{12.31}$$

where

$$M_k = \sum_{\substack{i=-M \\ i \neq 0}}^{M} \left|w_i^{(s)}(k)\right| \tag{12.32}$$

One can see that, for binary zero of the interfering ASK signal, the input useful signal is forwarded directly to the output.

This structure will be referred to as the ξ_A structure, to indicate that it is intended for suppressing ASK-type interference. For the same reason, the structure described by the set of equations (12.25–12.27) will be referred to as the ξ_p structure. By combining the previous two structures we get the ξ_{Ap} structure, that will be equally good for suppressing either ASK or PSK interference. In this case, we have $\{w_i^{(s)}(k) = 0, -1, 1\}$, if

$$\sum(x(k-i)) < \eta_{kom}, \; w_i^{(s)}(k) = 0$$

and $y(k-i)$ is given by Equation (12.30). Otherwise, the $w_i^{(s)}(k)$ are chosen from the set $\{w_i^{(s)}(k) = -1, 1\}$ by minimizing the following expression:

$$y(k-i) = \min_{w_i^{(s)}(k)} \left\{ \begin{array}{l} \left|\text{Re}(x(k) - x(k-i)w_i(k)w_i^{(s)}(k))\right| \\ + \left|\text{Im}(x(k) - x(k-i)w_i(k)w_i^{(s)}(k))\right| \end{array} \right\} \tag{12.33}$$

The overall output is given again by Equation (12.31).

12.1.4.4 H algorithms

In order to reduce the number of incorrect reconfiguration commands in U structures, we introduce the so-called *H (hysteresis) algorithm*. In this case, a set of weighting coefficients is introduced for different sections of the filter. This will depend on whether or not a given section was active in the previous sampling interval. In this way, we introduce a delay (hysteresis) into the reconfiguration command and reduce the impact of noise on the filter performance. The reconfiguration decision variable $c_m(k)$, given by Equation (12.10), can be represented as

$$c_m(k) = \min\{c_f(k), c_b(k), c_c(k)\} \tag{12.34}$$

where $c_f(k)$, $c_b(k)$ and $c_c(k)$ are represented by the first, second and third line of Equation (12.10), respectively. Each $c_i(k)$ is now replaced by $h_i C_i(k)$ with $i \in \{f, b, c\}$, which defines the priority (or complete removal) of each coefficient. In addition to that, the coefficient of the active segment of the structure is multiplied by factor $h_{ai} < 1$, $i \in \{f, b, c\}$, which defines hysteresis for the small error variations of FPF and BPF filters.

12.1.4.5 Limiter

Further improvements to the system characteristics can be obtained by using a limiter at the filter output. The limiter cut off level is adjusted to the relative signal level. For these purposes, we have to measure the signal level first in such a way that the impact of residual interference pulses is minimized. This is achieved by using a three-stage, adaptive, delta-type detector whose operations are defined by the following set of equations. The output of the first stage is defined as

$$D_0(k + 1) = D_0(k)(1 - d_0) + d_0 C_y(k) \tag{12.35}$$

where d_0 is the delta step, $D_0(0)$ an arbitrary constant and $C_y(k)$ the input signal level defined as

$$C_y(k) = |\text{Re}\{y(k)\}| + |\text{Im}\{y(k)\}| \tag{12.36}$$

The output of the second and the third stages is defined as:

$$D_i(k + 1) = D_i(k) \pm d_i D_{i-1}(k + 1); \quad i = 1, 2 \tag{12.37}$$

where '+' is used if $D_i(k) < C_y(k)$ and '−' if $D_i(k) \geq C_y(k)$ and d_i is the delta step of stage i. Now, the limiter output y' is defined as:

$$\text{Re}\{y'\} = \begin{cases} C_1 \text{sgn}\,(\text{Re}\{y\}), & C_1 D_2\,(k+1) < |\text{Re}\{y\}| \\ \text{Re}\{y\}, & C_1 D_2\,(k+1) \geq |\text{Re}\{y\}| \end{cases}$$

$$\text{Im}\{y'\} = \begin{cases} C_1 \text{sgn}\,(\text{Im}\{y\}), & C_1 D_2\,(k+1) < |\text{Im}\{y\}| \\ \text{Im}\{y\}, & C_1 D_2\,(k+1) \geq |\text{Im}\{y\}| \end{cases} \tag{12.38}$$

One should be aware that d_i should be low in order to reduce the impact of residual, impulse-like, interference. At the same time, this parameter should be large enough to follow (track) the input signal dynamics.

12.1.4.6 Performance

The filter suppression gain can be represented as

$$G(k) = \frac{E\{\text{Re}\{[j(k, l) + n(k, l)][j(k, l) + n(k, l)]*\}\}}{E\{\text{Re}\{[y(k, l) - s(k, l)][y(k, l) - s(k, l)]*\}\}} \tag{12.39}$$

where $E\{\ \}$ stands for averaging with respect to noise. The BER for the nth bit of the in-phase (index $\alpha = 1$) and the quadrature phase (index $\alpha = 2$) channel is given as

$$P_\alpha(n) = \frac{1}{2} erfc[\sqrt{\frac{SNR_\alpha(n)}{2}}], \quad n = 1, 2, \ldots, M_b, \quad \alpha = 1, 2 \tag{12.40}$$

and M_b is the bit ensemble size (measured in number of information bits) used for averaging the result. Parameter $SNR_\alpha(n)$ is the output signal to noise ratio of the nth bit . This can be further represented as

$$SNR_\alpha(n) = SNR'_\alpha(k) \cdot Q \tag{12.41}$$

where

$$SNR'_\alpha(k) = \frac{\left[\dfrac{1}{K}\displaystyle\sum_{k=1}^{K} DS_\alpha(k)\right]^2}{\left|\dfrac{1}{Q}\displaystyle\sum_{k=nQ}^{(n+1)Q} DS_\alpha^2(k) - \left[\dfrac{1}{K}\displaystyle\sum_{k=1}^{K} DS_\alpha(k)\right]^2\right|} \tag{12.42}$$

and K is the ensemble length measured in chips. Parameters $DS_\alpha(k, l)$ are given as

$$DS_1(k) = Re\left\{Y'(k)\right\} Re\left\{s(k)\right\}$$
$$DS_2(k) = Im\left\{Y'(k)\right\} Im\left\{s(k)\right\} \tag{12.43}$$

The average BER becomes

$$P_\alpha = \frac{1}{M_b} \sum_{n=1}^{M_b} P(n) \tag{12.44}$$

The final averaging gives

$$P_e = \sum_{\alpha=1}^{2} p(\alpha) P_\alpha \tag{12.45}$$

that will, most of the time, be calculated as

$$P_e = 0.5 \left(P_1 + P_2\right) \tag{12.46}$$

As an illustration, bit error rate (BER) P_e versus the normalized symbol interval T_h/Δ is presented in Figure 12.6. The interference is a PSK signal and results are obtained with the following set of parameters: $(\omega_1 - \omega_c) \Delta = 20°$, $2M = 10$, signal to noise ratio $SNR = -2$ dB, interference to signal ratio $JSR = 20$ dB and $Q = 70$. A high SNR is chosen in order to demonstrate the high range of improvements with respect to interference. The common practice in this field is to present these results for $\sigma_n^2 = 0$ or $SNR \rightarrow \infty$.

The curve 'a' represents the BER for a standard 2SF filter. A significant difference in performance between this filter and all other self-reconfigurable structures is self-evident. The set of four curves labeled b_1, b_2, b_3 and b_4 is obtained for the U structure, U structure with limiter, U structure with hysteresis and U structure with hysteresis and limiter, respectively. Curves b_1 and b_2 start to rise after the point where $T_h/\Delta > 2M$, due to the

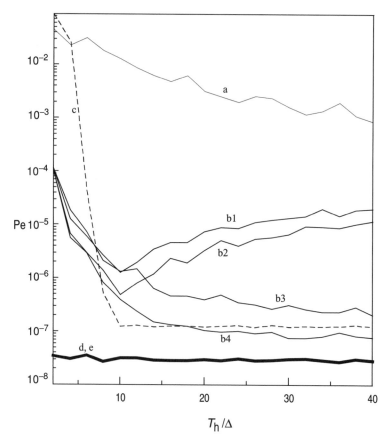

Figure 12.6 Bit error rate versus the normalized symbol interval of the interfering signal: $(\omega_1 - \omega_0)\Delta = 20°, 2M = 10$, SNR $= -2$ dB, JSR $= 20$ dB, $Q = 70$, PSK interference. (a) Standard 2SF filter; (b1) U structure; (b2) U structure with limiter; (b3) U structure with hysteresis; (b4) U structure with hysteresis and limiter; (c) Ψ structure; (d) ξ_p structure; (e) ξ_{AP} structure [1059] © 1999, IEEE.

propagation effect of incorrect detection of the PSK signal phase change. This is compensated by introducing hysteresis, as shown in curves b_3 and b_4. The hysteresis coefficients should be obtained experimentally for each type of interference. In Figure 12.6, for curves b_3 and b_4, it was found by simulation that the optimum values are $h_i = 1(i \in \{f, b, c\})$ and $h_{ai} \cong 0.5(i \in \{f, b, c\})$. The BER for the Ψ structure is represented by curve c. This structure requires $T_h/\Delta > 2M$ for good operation. Finally, the BERs for ξ_p and ξ_{Ap} structures are represented by curves d and e respectively. The superior performance of these structures is self-evident. For a PSK interfering signal, after remodulation, the problem is equivalent to suppressing only a CW signal. If such a signal is suppressed by using the Wiener optimal solutions for the filter coefficients [1063], the performance is almost the same as for the ξ structure suppressing complete PSK interference without any side information.

12.2 MULTILAYER LMS INTERFERENCE SUPPRESSION ALGORITHMS FOR CDMA WIRELESS NETWORKS

In this section the general theory of adaptive self-reconfigurable interference suppression schemes is applied to several specific practical problems, mainly the suppression of m-level (m-ASK, m-PSK and m-QAM) signals. This is a practical situation when a CDMA network is overlaid with standard microwave or ATDMA systems. Another example is a multirate CDMA network where a limited number of high bit rate CDMA signals are allowed to use a much higher power level, due to lower processing gain. The algorithm is well suited for a modular software radio concept, which is believed to be more and more accepted for future wireless communications.

Further modifications of the schemes necessary for these applications are described, and numerous results are presented to illustrate performance improvements. A general interpretation of these techniques based on the so-called multilayer LMS algorithm is introduced and discussed. The algorithm is based on estimating fast changing interfering signal parameters by using parallel structures, which are fast, but complex. At the same time, estimation of slow varying signal parameters over a large range is accomplished by using an LMS algorithm that is simple but slower. In this way, suppression of the interference occupying the same bandwidth as the CDMA signal is possible with reasonable implementation complexity.

12.2.1 Multilayer LMS algorithms (ML LMS)

All algorithms derived in Section 12.1 were obtained intuitively step by step, and each modification was motivated either by performance improvement or circuit simplification. In order to provide a unified interpretation of these schemes, we introduce a so-called multilayer LMS concept, which is described below. The general form of the overall received signal after frequency down-conversion is

$$x(t) = s(t) + i(t) + n(t) \tag{12.47}$$

where $n(t)$ is Gaussian noise. The complex signal component is given by Equation (12.1):

$$s(t) = m(t)e^{j\theta}$$
$$m(t) = m_1(t)e_1(t) - jm_2(t)e_2(t) \tag{12.48}$$

where $e_1(t)$ and $e_2(t)$ are two PN sequences with chip interval Δ, $m_1(t)$ and $m_2(t)$ are two independent information streams with bit interval $T_b = Q\Delta$, and θ is the slowly varying signal phase, uniformly distributed in the range $\theta \in [0, 2\pi]$. For the analysis to follow, the interfering signal will be represented in the general form as

$$i(t) = \sqrt{J}a(t)e^{j\varphi} \tag{12.49}$$

where J is the peak signal power, $a(t) = a_1(t) - ja_2(t)$ is complex symbol of period T_h, $\varphi = \Omega t + \varphi_0$, Ω is the offset between the local reference and interference carrier frequencies and φ_0 is the slowly varying phase. In general, $a(t)$ can be represented also in exponential

Table 12.1   Interfering m-ary signal formats

m-PSK	m-ASK	m-QAM
$\|a(t)\| = 1$	$\|a(t)\| = \left\|\frac{2k-m-1}{m-1}\right\|$	$\|a(t)\| = \sqrt{\left(\frac{2k_1-\sqrt{m}-1}{\sqrt{m}-1}\right)^2 + \left(\frac{2k_2-\sqrt{m}-1}{\sqrt{m}-1}\right)^2}$
$\phi(t) = \frac{k2\pi}{m}$	$\phi(t) = \frac{\pi}{2}\left[1 - sign\left(\frac{2k-m-1}{m-1}\right)\right]$	$\phi = \arctan\left((2k_2 - \sqrt{m} - 1)/(2k_1 - \sqrt{m} - 1)\right)$
$k \in (0, 1, \ldots, m-1)$	$k \in (0, 1, \ldots, m)$	$k_1, k_2 \in \left(1, 2, \ldots, \sqrt{m}\right),$
	$m = 2^n$ $n = 1,2,3, \ldots$	$m = 4^n,$ $n = 1,2,3, \ldots$

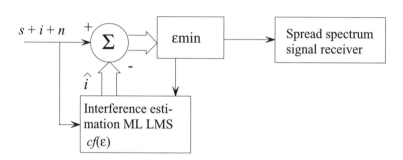

Figure 12.7 A generalized interference suppression scheme.

form as

$$a(t) = |a(t)|e^{j\phi(t)} \tag{12.50}$$

For the purpose of this analysis we will be particularly interested in the types of interference shown in Table 12.1.

A generalized interference suppression scheme is shown in Figure 12.7. As the first step, the interfering signal is estimated and then subtracted from the overall input signal.

A maximum likelihood approach for interference estimation would require generation of all possible versions of the interfering signal; i.e. all possible combinations of different values of parameters J, $|a(t)|$, $\phi(t)$ and φ. The interfering signal estimation, $\hat{i}$, would be the solution to $\arg \max_p \lambda(p)$, where $\lambda(p)$ is a properly defined likelihood function and p is the set of all different combinations of parameters.

With reference to Equation (12.47), most of the time, prior to despreading, the noise power is higher than the signal power and $s(t) + n(t)$ is predominantly Gaussian. For this case we have

$$\hat{i} = \arg \max_p \lambda(p) = \arg \max_p \int x(t)\tilde{i}(p, t)\, \mathrm{d}t \tag{12.51}$$

where $\tilde{i}(\,)$ is a possible candidate (trial) of the interfering signal and

$$p \in \{J, |a(t)|, \phi(t), \varphi\} \tag{12.52}$$

Instead of maximizing $\lambda(p)$, some other suitable form of cost function $cf(\varepsilon)$ can be minimized, where

$$\varepsilon = s + \left(i - \hat{i}\right) + n \tag{12.53}$$

and time subscripts are dropped for simplicity. This would provide the best estimation in only one observation interval, but at the same time, would also require a large number of parallel operations. Independent of the type of the optimization function, these algorithms will be referred to as parallel algorithms (PAs). Another option is the LMS algorithm, which is rather simple to implement but is not able to follow closely enough fast changes of signal parameters. A possible solution is a combination of the PA and LMS algorithms in such a way that the PA algorithm (parallel structure) is used to estimate a relatively small number of different values of fast changing parameters, and the LMS algorithm is used to estimate slow varying parameters which may vary over a very large range. This combination will be called the parallel algorithm aided LMS algorithm, or PAA LMS. A further modification would be to estimate some of the parameters by using simpler demodulation techniques, eliminate these parameters from the signal and then estimate the remaining signal parameters by using the PAA LMS algorithm. These techniques will be referred to as decision-directed PAA LMS, or DD PAA LMS, algorithms. For the purpose of this analysis we will represent the interfering signal $i(t)$, given by Equation (12.49), as:

$$i(t) = \sqrt{J} i_1(t) i_2(t) i_3(t) \tag{12.54}$$

where

$$\begin{aligned} i_1(t) &= |a(t)| \\ i_2(t) &= e^{j\phi(t)} \\ i_3(t) &= e^{j\varphi(t)} \end{aligned} \tag{12.55}$$

From Equation (12.50) we also have

$$i_{12}(t) = \sqrt{J} i_1(t) i_2(t) = \sqrt{J} |a(t)| e^{j\phi(t)} = \sqrt{J} a(t) \tag{12.56}$$

At a given sampling instant $(k + i)\mathbf{\Delta}$, for the lth noise ensemble member, these components can be represented as

$$\begin{aligned} i_{1i}(k, l) &= |a_i(k, l)| \\ i_{2i}(k, l) &= e^{j\phi_i(k,l)} \\ i_{3i}(k, l) &= e^{j\varphi_i(k,l)} \end{aligned} \tag{12.57}$$

A noise ensemble member is a realization of the noise signal. Now, by using Equation (12.54), the overall interfering signal can be represented as

$$\begin{aligned} i_i(k, l) &= \sqrt{J} |a_i(k, l)| e^{j[\phi_i(k,l)+\varphi_i(k+i.l)]} \\ &= \sqrt{J} \cdot i_{1i}(k, l) \cdot i_{2i}(k, l) \cdot i_{3i}(k, l) \end{aligned} \tag{12.58}$$

If the standard transversal filter, shown in Figure 12.2, was used for the interference suppression, then the filter coefficients updating process would be defined as

$$\begin{aligned} w_i(k + 1, l) &= w_i(k, l) + \mu x_i^*(k, l) e(k, l) \\ &= w_i(k, l) + \mu x_i^*(k, l) [x_0(k, l) - \hat{x}_0(k, l)] \end{aligned} \tag{12.59}$$

where $i \in (-M, M)$, for a two-sided filter (2SF), and $i \in (1, 2M)$, for a single-sided filter (1SF), is the filter tap index. Parameter k is the sampling index and $\hat{x}_0$ is the signal estimate which depends on the filter coefficients. The general form of the 2SF and the 1SF (forward prediction error filter) is given in Figure 12.2. When the interfering signal is predominant, $x(t) = s(t) + i(t) + n(t) \Rightarrow x(t) \cong i(t)$ and Equation (12.59) can be represented as

$$w_i(k + 1, l) \cong w_i(k, l) + \mu \, i^*_i(k, l)\left[i_0(k, l) - \hat{i}_0(k, l)\right]$$

$$= w_i(k, l) + \mu \, i_i^*(k, l) i_0(k, l)\left[1 - \hat{i}_0(k, l)/i_0(k, l)\right]$$

$$= w_i(k, l) + \mu \, J |a_0(k, l)|^2 e^{-j\Delta\varphi_i(k,l)}\left[1 - \hat{i}_0(k, l)/i_0(k, l)\right]\frac{|a_i(k, l)|}{|a_0(k, l)|} \cdot e^{-j\Delta\phi_i(k,l)}$$

$$= w_i(k, l) + \mu \cdot J \cdot |a_0(k, l)|^2 \tilde{w}_i(k, l) \cdot e(k, l) \cdot \left(\tilde{w}_i^{(s)}\right)^*$$

where

$$\tilde{w}_i(k, l) = e^{-j\Delta\varphi_i(k,l)} \tag{12.60}$$

is a slow varying term that can be tracked by the LMS algorithm and

$$\tilde{w}_i^{(s)}(k, l) = \frac{|a_i(k, l)|}{|a_0(k, l)|} e^{j\Delta\phi_i(k,l)} \tag{12.61}$$

is a fast varying term due to modulation and must be estimated in a parallel way. For binary PSK, $\tilde{w}_i^{(s)}(k, l) = \pm 1$, which leads to the so-called ξ filter structure shown in Figure 12.5. The switching coefficients $\tilde{w}_i^{(s)}(k, l) = \pm 1$ will connect each tap of the filter to the output with a $+$ or $-$ sign, depending on the result of the cost function minimization.

For multilevel modulation, these coefficients and the filter structure will be further elaborated. In the previous equation we are also using the following definitions

$$\Delta\varphi_i(k, l) = \varphi_i(k, l) - \varphi_0(k, l)$$
$$\Delta\phi_i(k, l) = \phi_i(k, l) - \phi_0(k, l) \tag{12.62}$$

Using Equation (12.58) in (12.60) gives

$$w_i(k + 1, l) \cong w_i(k, l) + \mu\left[i_{10}(k, l)i_{20}(k, l)i_{30}(k, l)i^*_{1i}(k, l)i^*_{2i}(k, l)i^*_{3i}(k, l)\right] \cdot$$

$$\times J\left[1 - \frac{\hat{i}_{10}(k, l)\hat{i}_{20}(k, l)\hat{i}_{30}(k, l)}{i_{10}(k, l)i_{20}(k, l)i_{30}(k, l)}\right] \tag{12.63}$$

where the estimations $\hat{i}_{n0}(n = 1, 2, 3)$ are represented as $\hat{i}_{n0} = i_{n0} - \Delta i_{n0}$. If we introduce notation for the relative estimation errors, $e_{n0} = \Delta i_{n0}/i_{n0}$, then we have $\hat{i}_{n0}/i_{n0} = 1 - e_{n0}$ and Equation (12.63) becomes

$$w_i(k + 1, l) \cong w_i(k, l) + \mu \, i^*_{1i}(k, l)i^*_{2i}(k, l)i^*_{3i}(k, l)i_{10}(k, l)i_{20}(k, l)i_{30}(k, l) \cdot$$

$$\times J\left[1 - (1 - e_{10})(1 - e_{20})(1 - e_{30})\right] \tag{12.64}$$

This equation will be used in the next section as a generic equation to define several LMS algorithms for interference suppression of the general form given by Equation (12.49).

12.2.2 Filter modeling

12.2.2.1 Normalization

As the first step we would like to minimize the impact of the interfering signal format on the LMS algorithm efficiency. For example, in the case of an amplitude modulated interfering signal, $|\tilde{w}_i^{(s)}| \neq 1$ and so the algorithm for updating filter coefficients w_i has to be modified. The modification should provide an update process for w_i that will not depend on the instantaneous power of the interfering signal. So, the filter coefficients update process will be defined as

$$
\begin{aligned}
w_i\,(k+1,l) &= w_i\,(k,l) + \frac{\mu \cdot e\,(k,l) \cdot x_i^*\,(k,l) \cdot w_i^{(s)}\,(k,l)}{J \cdot |a_0(k,l)|^2} \\[2mm]
&= w_i\,(k,l) + \frac{\mu \cdot e\,(k,l) \cdot x_i^*\,(k,l) \cdot w_i^{(s)}\,(k,l)}{x_i\,(k,l)\,x_i^*\,(k,l)\,w_i^{(s)}\,(k,l)\,w_i^{(s)*}\,(k,l)} \\[2mm]
&= w_i\,(k,l) + \frac{\mu \cdot e\,(k,l)}{x_i\,(k,l)\,w_i^{(s)}\,(k,l)}
\end{aligned}
\tag{12.65}
$$

where $w_i^{(s)}$ are estimates of the coefficients $\tilde{w}_i^{(s)}$. In Appendix 12.1, we demonstrate that coefficients $w_i(k,l)$ do not depend upon modulation. So, the slow changes of the interference carrier will be tracked and suppressed by an LMS algorithm implemented by a transversal filter with coefficients $w_i(i \in (1, 2M))$ for 1SF, and $i \in (-M, M)$ for 2SF. The fast changes due to modulation are left in the 'switching coefficients', $\tilde{w}_i^{(s)}\,(k,l)$, defined by Equation (12.61), and these coefficients must be estimated by using faster but more complex schemes. In what follows we discuss different options to estimate these coefficients.

12.2.2.2 Compactness of the LMS algorithm

In the case of a constant envelope, signal suppression is more efficient if a global estimation error is used for filter coefficient updates. The global error is the difference between the input signal and sum of weighted outputs of all filter taps. For amplitude modulated signals, the update process is based upon using the partial errors between the input signal and the weighted output of the given tap whose coefficient is being updated. This is more efficient. The contribution of the global and the partial error in the update process of coefficient w_i is defined through the compactness factor z given in

$$
E_i(k,l) = z \cdot e(k,l) + (1-z)e_i(k,l)
\tag{12.66}
$$

This error is now used for updating the coefficient weight with index i. In Equation (12.66), $e(k,l)$ is the average (global) error defined by Equation (12.69), $e_i(k,l)$ is the tap error defined as

$$e_i(k, l) = x_0 - y_i(k, l) \cdot w_i^{(s)}(k, l)$$
$$= x_0 - x_i(k, l)w_i(k, l)w_i^{(s)}(k, l) \tag{12.67}$$

and z is the compactness factor.

12.2.2.3 Multilevel ξ structures

The basic characteristic of the ξ structure is that coefficients $w_i^{(s)}$ are differential, i.e. they represent the ratio x_o/x_i.

ξ m-PSK filter
In this case, the signal constellation consists of m points in the complex plane. The filter is similar to the ξ filter presented in Section 12.1, where, based on Equation (12.61), coefficients $w_i^{(s)}$ now can take values from the set

$$w_i^{(s)} = \left\{ e^{j\frac{2\pi}{m}k}; \qquad k \in (1, 2, \ldots, m) \right\} \tag{12.68}$$

The filter output signal (global error) is defined as

$$e(k, l) = y(k, l) = x_0(k, l) - \frac{1}{2M}\sum_i y_i(k, l)w_i^{(s)*}(k, l) \tag{12.69}$$

where

$$i \in \begin{cases} -M \ldots -1, 1, \ldots, M; & \text{for } 2\text{SF} \\ 1, \ldots, 2M; & \text{for } 1\text{SF} \end{cases} \tag{12.70}$$

$$y_i(k, l) = x_i(k, l)\, w_i(k, l)$$

Parameters $w_i^{(s)}(k, l)$ are estimates of $\tilde{w}_i^{(s)}(k, l)$, defined by Equation (12.61), and these estimates are obtained as

$$w_i^{(s)}(k, l) = e^{i\,\varphi_{yi}\,(k,l)} \tag{12.71}$$

$$\varphi_{yi}(k, l) = \perp_m (\arg x_{m0}(k, l) - \arg y_i(k, l)) \tag{12.72}$$

where $\perp_m(\ )$ stands for the projection of the argument to the closest one of m possible phases of the m-PSK signal.

ξ m-ASK filter
For m-ASK signal suppression, a standard ξ structure, will be used to estimate the phase of coefficients $w_i^{(s)}$, together with an additional one-tap filter for the estimation of the interfering signal power. A block diagram of the filter is shown in Figure 12.8. The one-tap filter for signal power estimation is defined by the following set of equations:

$$w^{(p)}(k + 1, l) = w^{(p)}(k, l) + \mu \cdot e_{01}(k, l)/|WS(k, l)|^2 \tag{12.73}$$

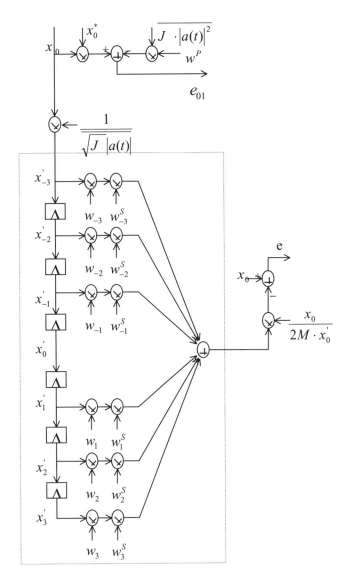

Figure 12.8 ξ m-ASK filter block diagram.

where $|WS|^2 = \overline{J|a(t)|^2}$ is the estimation of the instantaneous signal power, and e_{01} is obtained as

$$e_{01}(k, l) = x_0(k, l)x_0^*(k, l) - |WS(k, l)|^2 \qquad (12.74)$$

After estimation of $|WS(k, l)|$, the rest of the operation is the suppression of a constant amplitude binary phase modulated signal (see Table 12.1). The operation of the main filter with $2M$ taps is characterized by the set of equations

$$y(k,l) = e(k,l) = x_0(k,l) - \frac{1}{2M} \cdot \frac{x_0(k,l)}{x_0'} \sum_{\substack{i=-M \\ i \neq 0}}^{M} x_i'(k,l) \cdot w_i(k,l) \cdot w_i^{(s)}(k,l)$$

$$e_i(k,l) = x_0(k,l) - \frac{x_0(k,l)}{x_0'} \cdot x_i'(k,l) \cdot w_i(k,l) \cdot w_i^{(s)}(k,l) \tag{12.75}$$

$$\min_{ws_i} \left| e_i e_i^* \right| \Rightarrow w_i^{(s)} \qquad \text{(Equation (12.68) } for \ m = 2)$$

$$w_i(k+1,l) = w_i(k,l) + \frac{\mu \cdot E_i(k,l)}{x_i'(k,l) \cdot w_i^{(s)}(k,l) \cdot |WS(k,l)|}$$

$$E_i(k,l) = z \cdot e(k,l) + (1-z) \cdot e_i(k,l)$$

In this notation, $x'(k,l)$ represents the signal sample after normalization, $x'(k,l) = x(k,l)/|WS(k,l)|$. The filter works best with $z = 0$.

ξ m-QAM filter
A differential structure for an ξ m-QAM filter is also possible. The number of $w_i^{(s)}$ coefficients in this case is m^2 (see Table 12.1). Due to the large number of coefficients $w_i^{(s)}$ used in this configuration, we will introduce a new 'simpler' structure described in section 12.2.2.4.

12.2.2.4 Multilevel G (global) structures

In this case there is only one global $w^{(s)}$ coefficient, which is supposed to represent the overall modulation of the interfering signal. A one-tap filter estimates this coefficient, which is used to remove modulation from the interfering signal, so that afterwards the rest of the LMS algorithm adaptively removes the unmodulated signal carrier.

G m-PSK and G m-ASK filters
These filters have two components: demodulation filter, DF, and main filter, MF. Operation of the DF filter can be described with the following set of equation:

$$e_0(k,l) = x_0(k,l) - x_1'(k,l) \cdot w_0(k,l) \cdot WS^*(k,l)$$
$$w_0(k+1,l) = w_0(k,l) + \mu e_0(k,l)/(x_0' \cdot WS(k,l))$$
$$x_0'(k,l) = x_0(k,l)/WS(k,l) \tag{12.76}$$
$$WS(k,l) = \sqrt{J}a(k,l) = \sqrt{J}\,|a(k,l)|\,e^{j\phi(k,l)}$$

The WS coefficient is a specific case of the general coefficient $w^{(s)}$, and is obtained by minimizing the following cost function

$$cf(e) = \arg \min_{WS} |e_0(k,l) \cdot e_0^*(k,l)|$$

for all possible values of a and ϕ specified in Table 12.1 for m-PSK and m-ASK signals. The DF filter will remove modulation and the MF will suppress the remaining cw (continuous wave) part of the interference by using signal samples x_i'. Operation of the MF filter is described by the set of equations:

$$y(k, l) = e(k, l) = x_0(k, l) - \frac{1}{2M} \sum_{i=1}^{2M} x_i'(k, l) \cdot w_i(k, l) \cdot WS(k, l)$$

$$w_i(k + 1, l) = w_i(k, l) + \frac{\mu \cdot E_i(k, l)}{x_i'(k, l) \cdot WS(k, l)} \tag{12.77}$$

$$e_i(k, l) = x_0(k, l) - x_i'(k, l) \cdot WS(k, l) \cdot w_i(k, l)$$

$$x_i'(k, l) = x_i'(k, l), \quad i = 1, \ldots, 2M$$

where $E_i(k, l)$ is defined by Equation (12.75).

G m-QAM filter

A block diagram of this filter is shown in Figure 12.9. It consists of three segments, power weight (PW) filter PWF, demodulation filter DF and main filter MF. The PW filter is described with the set of equations:

$$e_{01}(k, l) = x(k, l) \cdot x^*(k, l) - WS(k, l) \cdot WS^*(k, l) w^{(p)}(k, l)$$

$$w^{(p)}(k + 1, l) = w^{(p)}(k, l) + \mu e_{01}(k, l) / |WS(k, l)|^2 \tag{12.78}$$

The operation of the DF filter is defined by the following set of equations:

$$e_0(k, l) = x_0(k, l) - x_1' \cdot w_0(k, l) \cdot WS(k, l)$$

$$w_0(k + 1, l) = w_0(k, l) + \mu e_0(k, l) / x_1' \cdot WS(k, l) \tag{12.79}$$

Finally, the MF operates as follows:

$$x_0'(k, l) = x_0(k, l) / WS(k, l)$$

$$y(k, l) = e(k, l) = x_0(k, l) - \frac{1}{2M} \sum_{i=1}^{2M} x_i'(k, l) \cdot w_i(k, l) \cdot WS(k, l)$$

$$w_i(k + 1, l) = w_i(k, l) + \frac{\mu \cdot E_i(k, l)}{x_i'(k, l) \cdot WS(k, l)} \tag{12.80}$$

$$e_i(k, l) = x_0(k, l) - x_i'(k, l) \cdot WS(k, l) \cdot w_i(k, l)$$

$$x_i'(k + 1, l) = x_{i-1}'(k, l), \quad i = 1, \ldots, 2M$$

and $E_i(k, l)$ is given again by Equation (12.75). Parameter $WS(k, l)$ is the solution to

$$\arg \min_{ws} cf(e) = \arg \min_{ws}(|e_{01}(k, l)| + |e_0(k, l) \cdot e_0^*(k, l)|) \tag{12.81}$$

for all possible values of a and ϕ specified in Table 12.1 for the m-QAM signal. Equations (12.58–12.62) define the G m-QAM-PWF filter. If in Equation (12.78), instead of PW, proportionality constant k_p is used, then we have $PPW = k_p \overline{xx^*}$ and a filter for suppressing m-QAM signals that is based on using the average power of the input signal is obtained (G m-QAM-PPW filter). Parameter k_p is different for different types of modulation. In the case when the structure is composed of both DF and MF filters ($e_{01} = 0$), we have the basic G m-QAM filter.

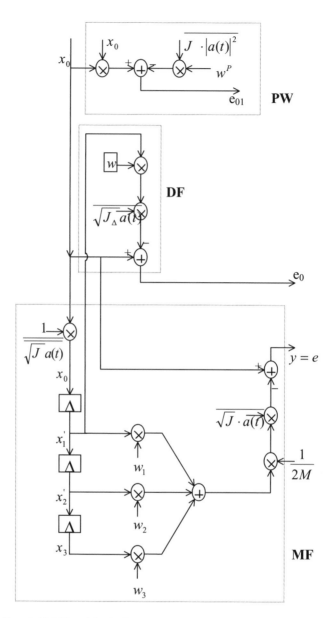

Figure 12.9 G m-QAM filter block diagram.

In accordance with the classification described earlier, the ξ m-PSK filter uses the PAA LMS algorithm. The PA algorithm is used to estimate parameter $\phi(t)$ by choosing proper values for coefficients $w_i^{(s)}$, whereas all other slowly varying signal parameters (j and φ) are estimated by using the LMS algorithm. The ξ m-ASK and ξ m-QAM filters use PAA. They can also use the DD segment of the algorithm to simplify the implementation. The G m-ASK, G m-PSK and G m-QAM algorithms use the DD approach.

12.2.3 Performance analysis

12.2.3.1 Error probability

We will use the following notation:

- the information symbol period, $T_b = Q\Delta$;

- the interfering symbol period, T_h;

- the noise ensemble length, $K\Delta$ (the period used for time averaging of the results with respect to information and the interfering signal);

- noise ensemble size, B (number of noise samples used in ensemble averaging of the results with respect to Gaussian noise);

- N, the number of information bits in period $K\Delta (N = K/Q)$.

By using the above notation, the filter suppression gain can be represented as

$$G(k) = E_l\{G(k, l)\} = \frac{\displaystyle\sum_{l=1}^{B} \text{Re}\{[i(k, l) + n(k, l)][i(k, l) + n(k, l)]^*\}}{\displaystyle\sum_{l=1}^{B} \text{Re}\{[y(k, l) - s(k, l)][y(k, l) - s(k, l)]^*\}} \tag{12.82}$$

where $E_l()$ stands for ensemble averaging and $y(k, l)$ is the filter output signal defined by equation (12.69–12.80). According to Equation (12.47), the numerator in Equation (12.82) represents the power of the overall input interference $i(k, l) + n(k, l)$. Also, according to Equation (12.69–12.80) the denominator in Equation (12.82) represents the power of the overall residual interference. The BER for the nth bit of the in-phase (index $\alpha = 1$) and quadrature phase (index $\alpha = 2$) channel is given as:

$$P(n \cdot Q, l) = \frac{1}{2}erfc\left[\sqrt{\frac{\text{SNR}_\alpha(nQ, l)}{2}}\right], \quad n = 1, 2, \ldots N, \quad \alpha = 1, 2 \tag{12.83}$$

The parameter $\text{SNR}_\alpha (nQ, l)$ is the SNR of the nth bit for the lth ensemble member. This can be further represented as

$$\text{SNR}_\alpha(nQ, l) = \text{SNR}'_\alpha(k, l) \cdot Q \tag{12.84}$$

where

$$\text{SNR}'_\alpha(k, l) = \frac{\left[\dfrac{1}{K}\displaystyle\sum_{k=1}^{K} DS_\alpha(k, l)\right]^2}{\left|\dfrac{1}{Q}\displaystyle\sum_{k=nQ}^{(n+1)Q} DS_\alpha^2(k, l) - \left[\dfrac{1}{K}\displaystyle\sum_{k=1}^{K} DS_\alpha(k, l)\right]^2\right|} \tag{12.85}$$

Parameters $DS_\alpha(k, l)$ are given as

$$DS_1(k, l) = \text{Re}\{y(k, l)\}\text{Re}\{s(k, l)\}$$
$$DS_2(k, l) = \text{Im}\{y(k, l)\}\text{Im}\{s(k, l)\}$$
(12.86)

The average BER becomes

$$P_\alpha = \frac{1}{NB} \sum_{l=1}^{B} \sum_{n=1}^{N} P(nQ, l)$$
(12.87)

Final averaging gives

$$P = \sum_{\alpha=1}^{2} p(\alpha) P_\alpha$$
(12.88)

where $p(\alpha)$ is a prior distribution of binary one and binary zero.
Most of the time this results in

$$P = 0.5 (P_1 + P_2)$$
(12.89)

12.2.3.2 Convergence

For algorithm convergence analysis we will start with the representation of the interfering signal given by Equation (12.54)

$$i(t) = i_1(t)i_2(t)i_3(t)$$
(12.90)

All algorithms presented in this section are based on the idea of removing the modulation from the signal and then using the LMS algorithm to track and suppress the remaining CW signal $i'(t)$. This, in general, can be represented as

$$i'(t) = \frac{i_1(t)i_2(t)}{\hat{i}_1(t)\hat{i}_2(t)}i_3(t) = (1 - e_1(t))(1 - e_2(t))i_3(t)$$
(12.91)

where e_1 and e_2 are the complex estimation errors for the signal components i_1 and i_2. For an approximate analysis of the algorithm convergence, the standard results should apply where the equivalent signal power is modified by the factor

$$\sigma_e^2 = \overline{(1 + e_1)(1 + e_2)(1 + e_1)^*(1 + e_2)^*} = \overline{|1 + e_1|^2 |1 + e_2|^2}$$
$$= \overline{\left(1 + e_{1r}^2 + e_{1i}^2\right)\left(1 + e_{2r}^2 + e_{2i}^2\right)} = \left(1 + \sigma_{e1}^2\right)\left(1 + \sigma_{e2}^2\right)$$
(12.92)

In general, this would modify the condition for adaptation rate factor μ, and adaptation time constant τ, as follows. From the condition

$$1/\text{tr}\mathbf{R} > \mu > 0$$
(12.93)

where $\mathbf{R}$ is the signal correlation matrix [1085], and from Equation (12.91–12.93), we can see that parameters $\text{tr}\mathbf{R}$ and μ should be modified by the factors

$$\lambda'_{max} \Rightarrow \sigma_e^2 \lambda_{max}$$
$$\text{tr}'\mathbf{R} \Rightarrow \sigma_e^2 \text{tr}\mathbf{R}$$
$$\mu' \Rightarrow \mu/\sigma_e^2$$
(12.94)

where $\lambda_{\max}$ is the maximum eigenvalue of matrix $\mathbf{R}$. From an expression given in [1085],

$$\tau = \frac{1}{2\mu\,\lambda_{\max}} \tag{12.95}$$

we can see that parameter τ will not be changed if parameter μ is modified accordingly.

12.2.3.3 Performance example

For the numerical analysis we use expressions for the filter gain G and BER as given by Equations (12.82) and (12.88) with $K = 100\,000$. In addition to this, we also present results of extensive Monte Carlo simulations. Within this section we use the following notation:

- m-PSK filter; a filter designed to suppress an m-PSK signal;

- $TF = AB.C$; transversal filter designed to suppress a QAM signal if $A = 1$ or an ASK signal if $A = 2$, with 2^B levels for QAM and 2^{B+1} for ASK. The signal levels are detected either based on the information about the average signal power level ($C = 1$); the instantaneous power level ($C = 2$); or without power estimation ($C = 0$);

- Parameter z; if $w^{(s)}$ coefficients are based on errors on the corresponding filter cells, then $z = 0$. If these coefficients are based on the overall filter error, then $z = 1$.

Figure 12.10 presents BER curves versus the filter length. For m-PSK interfering signals, the BER gets worse when m is increased, as expected. The same trend occurs with m-QAM interfering signals. Curve g, obtained from traditional transversal filtering is by far the worst curve from all sets of results. From the figure, one can see that for $2M > 20$, the slope of the curves is low, and $2M = 20$ seems to be a reasonable selection as a tradeoff between the system performance and filter complexity. Figures 12.11–12.14 represent the BER curves for different variations of the parameters.

APPENDIX 12.1

For illustration purposes, when the interfering signal is predominant i.e. $x(t) = s(t) + i(t) + n(t) \cong i(t)$, Equation (12.65) becomes

$$w_i\,(k+1, l) \cong w_i\,(k, l) + \mu\,e^{-j\Delta\varphi(i,l)}[1 - \hat{i}_0(k, l)/i_0(k, l)]$$

If $\hat{i}_0$ is given as

$$\hat{i}_0 = \sum_m x_m\,(k, l) \cdot w_m\,(k, l) \cdot w_m^{(s)}\,(k, l)$$

$$\cong \sum_m i_m\,(k, l) \cdot w_m\,(k, l) \cdot w_m^{(s)}\,(k, l)$$

then one can show that

$$1 - \hat{i}_0\,(k, l)\,/\,i_0\,(k, l) \text{ becomes } 1 - \sum_m w_m(k, l)e^{j\Delta\varphi_m(k,l)}$$

which does not depend on the modulation.

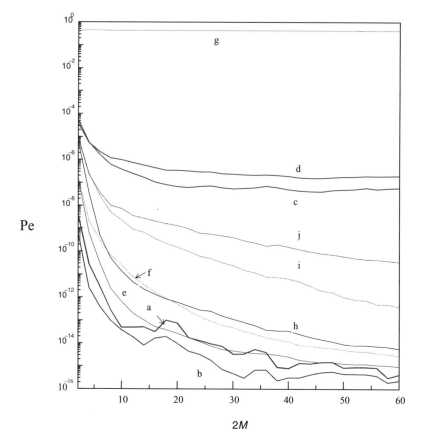

Figure 12.10 BER versus the filter length: $\mu = 2 \times 10^{-2}$, $Q = 100$, SNR $= 0$ dB, $T_h = 1\Delta$. (a) BPSK interfering signal, ξ_p structure, JSR $= 25$ dB; (b) 8PSK interfering signal, 8PSK filter, JSR $= 25$ dB; (c) 32PSK interfering signal, 32PSK filter, JSR $= 25$ dB; (d) 128PSK interfering signal, 128PSK filter, JSR $= 25$ dB; (e) TF $= 14.1$; (f) TF $= 14.2$; (g) TF $= 14$, 16 QAM interfering signal, JSR $= 46$ dB; (h)TF $= 16.1$; (i)TF $= 16.2$; (j) TF $= 16$, 64-QAM interfering signal, JSR $= 46$ dB.

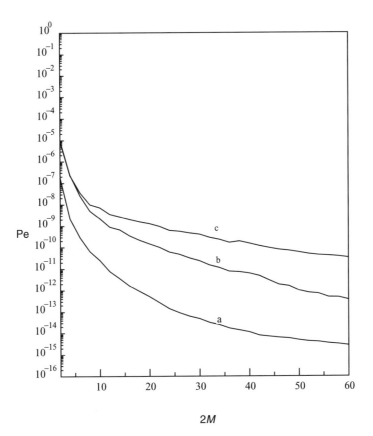

Figure 12.11 BER versus the filter length: $\mu = 2 \times 10^{-2}$, $Q = 100$, SNR $= 0$ dB, $T_h = 1\Delta$, JSR $= 46$ dB, 64-QAM interfering signal. (a) TF $= 16.1$; (b) TF $= 16.2$; (c) TF $= 16$.

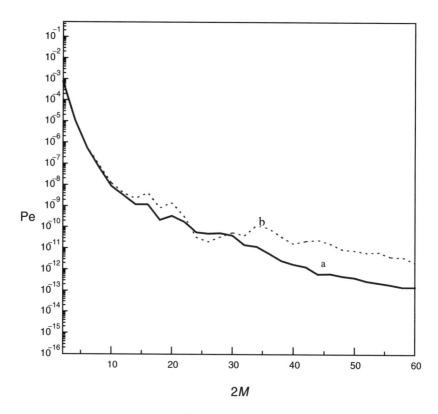

Figure 12.12 BER versus the filter length: $\mu = 2 \times 10^{-2}$, $Q = 100$, SNR = 0 dB, $T_h = 1\Delta$, JSR = 40 dB 16-ASK interference signal, TF = 23.1, for (a) $z = 0$; (b) $z = 1$.

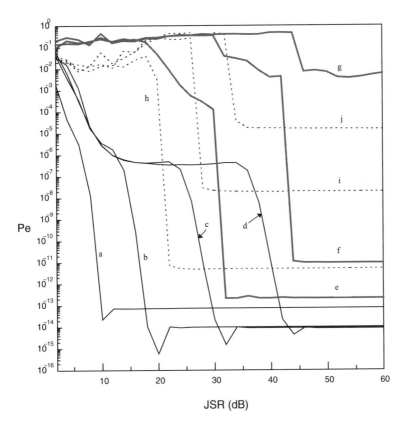

Figure 12.13 BER versus JSR: $\mu = 2 \times 10^{-2}$, $Q = 100$, SNR $= 0$ dB, $T_h = 1\Delta$, $2M =$ 20. (a)BPSK interfering signal, ξ_p structure; (b)8PSK interfering signal, 8PSK filter; (c)32PSK interfering signal, 32PSK filter; (d)128PSK interfering signal, 128PSK filter; (e)16-QAM interfering signal, TF $= 14.2$; (f)64-QAM interfering signal, TF $= 16.2$; (g)256-QAM interfering signal; (h)4-ASK interfering signal, TF $= 21.1$; (i)8-ASK interfering signal, TF $= 22.1$; (j)16-ASK interfering signal, TF $= 23.1$.

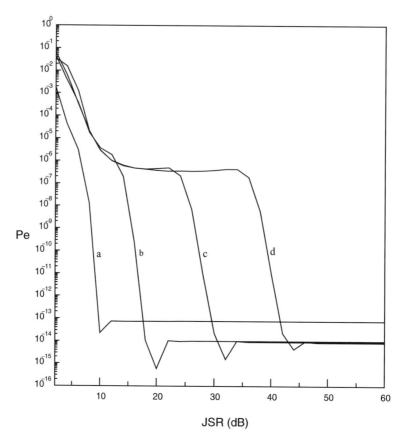

Figure 12.14 BER versus JSR. $\mu = 2 \times 10^{-2}$, $Q = 100$, SNR = 0 dB, $T_h = 1\Delta$, $2M = 20$. (a)PSK interfering signal, ξ_p structure; (b)8PSK interfering signal, 8PSK filter; (c)32PSK interfering signal, 32PSK filter; (d)128PSK interfering signal, 128PSK filter.

13

User Location in 4G Networks

13.1 BASIC LOCATION TECHNOLOGIES

Radio-based technology typically uses base stations, satellites, or other devices emitting radio signals to the mobile receiver to determine the position of its user. Signals can also be emitted from the mobile device to the base. Commonly studied techniques are *angle of arrival* (AOA) positioning, *time of arrival* (TOA) positioning, and *time difference of arrival* (TDOA) positioning.

The AOA system determines the mobile phone position based on triangulation, as shown in Figure 13.1. It is also called direction of arrival positioning in some literature. The intersection of two directional lines of bearing defines a unique position, each formed by a radial from a base station to the mobile phone in a 2D space. This technique requires a minimum of two stations (or one pair) to determine a position. If available, more than one pair can be used in practice. Because directional antennas or antenna arrays are required, it is difficult to realize AOA at the mobile phone. Signal processing for angle of arrival estimation is discussed in Section 9.2.

The time of arrival (TOA) system determines the mobile phone position based on the intersection of the distance (or range) circles (Figure 13.2 shows a 2D example). Since the propagation time of the radio wave is directly proportional to its traversed range, multiplying the speed of light by the time obtains the range from the mobile phone to the communicating base station. Two range measurements provide an ambiguous fix, and three measurements determine a unique position. The same principle is used by GPS, where the circle becomes a sphere in space and a fourth measurement is required to solve the receiver–clock bias for a 3D solution. The bias is caused by the unsynchronized clocks between the receiver and the satellite.

Advanced Wireless Communications. S. Glisic
© 2004 John Wiley & Sons, Ltd. ISBN: 0-470-86776-0

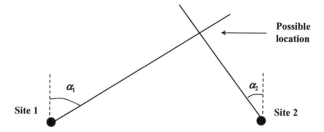

Figure 13.1 Angle of arrival (AOA) based positioning.

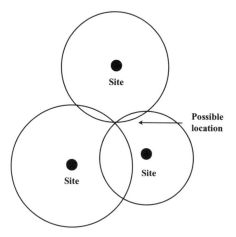

Figure 13.2 Time of arrival (TOA) or distance-based positioning.

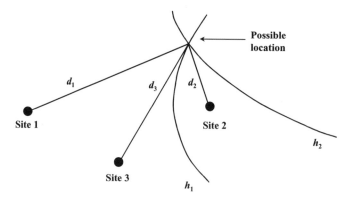

Figure 13.3 Time difference of arrival (TDOA) based positioning.

The time difference of arrival (TDOA) system determines the mobile phone position based on trilateration, as shown in Figure 13.3. This system uses time difference measurements rather than absolute time measurements as used in TOA. It is often referred to as the *hyperbolic system* because the time difference is converted to a constant distance difference to two base stations (as foci) to define a hyperbolic curve. The intersection of two hyperbolas determines the position. Therefore, it utilizes two pairs of base stations (at least three for the 2D case shown in Figure 13.3) for positioning. The accuracy of the system is a function of the relative base station geometric locations. For terrestrial-based systems, it also requires either precisely synchronized clocks for all transmitters and receivers, or a means to measure these time differences. Otherwise, a 1 μs timing error could lead to a 300 m position error. The time of arrival estimation (frame synchronization) is discussed in Section 13.2. The network synchronization is discussed in Section 13.3. A number of references covering this issue are included at the end of the book [1086–1128].

13.2 FRAME SYNCHRONIZATION

In a digital communication system a rapid initial frame synchronization is required. The good solution assumes: rapid initial frame synchronization acquisition, rapid detection of timing anomalies and frame synchronization recovery, reliability of the lock indication, simplicity of the clock synchronization algorithm and minimal insertion of redundancy in the data bit stream.

13.2.1 Problem definition

We will start with an initial discussion based on [1128]. In a system, D data bits are separated by a known marker sequence m, which is M bits long. Assume that a perfect bit clock is available and that the channel is essentially a binary symmetric channel (BSC) with error rate P_e, i.e. each bit transmitted over the channel is received erroneously with probability P_e, independently of all prior channel transmissions.

13.2.2 Algorithms

For the received bit stream $r_1 r_2 \ldots$, the receiver forms the M-tuple $r(t)$

$$r(t) = r_t r_{t+1} \ldots r_{t+M-1}$$

Subsequently, the number of positions in which two vectors $\mathbf{x}$ and $\mathbf{y}$ disagree will be denoted by $H(\mathbf{x}, \mathbf{y})$, i.e. the Hamming distance between $\mathbf{x}$ and $\mathbf{y}$. One possible frame synchronization acquisition algorithm is *the BSC frame synchronization acquisition algorithm:*

1. $t \leftarrow 1$.

2. If $H(r(t), m) \leq h$, go to 4.

3. $t \leftarrow t + 1$, go to 2.

4. Output 'Marker detected at time t'.

The threshold h is called the *error tolerance* of the frame synchronization algorithm. The performance of this algorithm depends on m, M, D, P_e, h, and on the statistics of the data bits. Performance curves for such an algorithm are given in Figure 13.4.

The real synchronization system must also deal with the unlikely event of a false synchronization acquisition, or a slip of the bit synchronization after synchronization acquisition.

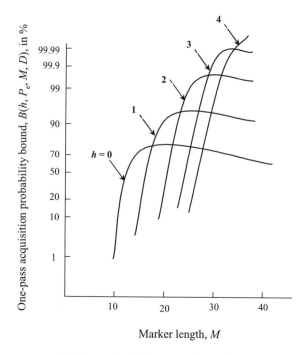

Figure 13.4 One-pass acquisition probability bound as a function of M and h for $P_e = 10^{-2}, D = 10^3$.

One such frame synchronization verification (or *lock*) algorithm, based on initial marker detection at time t_0, is *the BSC frame synchronization verification algorithm* defined as

1. $\tau \leftarrow t_0, i \leftarrow 0, k \leftarrow 0$.
2. $\tau \leftarrow \tau + M + D, i \leftarrow i + 1$.
3. If $H(r(t), m) \le h'$, then $k \leftarrow k + 1$.
4. If $k \ge K$, then
   i. Say 'frame sync verified at τ,'
   ii. $i \leftarrow 0, k \leftarrow 0$.
5. If $i < J$, go to 2.
6. Reinitiate frame synchronization acquisition algorithm.

For communication over many channels the binary sequence $\{b_t\}$ consisting of zeros and ones is modulated on a pulse train to create the signal

$$s(t') = \sum_t a_t p(t' - tT) \tag{13.1}$$

where T is the bit time, $p(t')$ is a pulse of duration T, and

$$a_t = (-1)^{b_t} \tag{13.2}$$

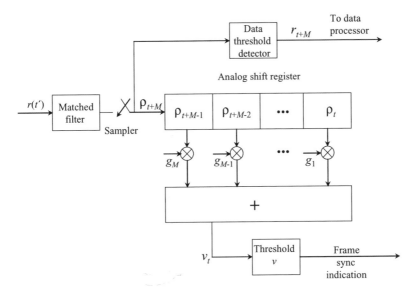

Figure 13.5 A coherent frame synchronization acquisition system.

Assuming that $s(t')$ is immersed in additive white Gaussian noise to create the received signal $r(t')$, a typical block diagram for frame synchronization acquisition based on $r(t')$ is shown in Figure 13.5.

All discrete time devices in this system are driven by the output of the bit synchronizer. The *coherent frame synchronization acquisition algorithm* is defined as

1. $t \leftarrow 1$.

2. If $v_t > V$, go to 4.

3. $t \leftarrow t + 1$, go to 2.

4. Output 'r_t is the first bit in a marker.'

where v_t is the output of an analog correlator matched to the marker waveform

$$v_t = \sum_{j=1}^{M} g_j \rho_{t-j-1} \qquad (13.3)$$

with

$$g_i = (-1)^{m_i}$$

$$\rho_t = \int_{tT}^{(t+1)T} r(t') p(t' - tT) dt' \qquad (13.4)$$

13.2.3 Marker design

An acceptable marker design is one which achieves the probability of a data M-tuple looking like the marker greater than the probability of an M-tuple containing both data and marker bits looking like the marker.

$$P_A(h, n, h_0) \le P_A(h, 0, 0) \quad \text{for} \quad 0 \le n < M \tag{13.5}$$

$$P_{\text{FAD}} = P_A(h, 0, 0) = \left(\frac{1}{2}\right)^M \sum_{k=0}^{h} \binom{M}{k}$$

This probability can be expressed as

$$P_A(h, n, h_0) = \sum_{j=0}^{\min(n,h)} \left[\sum_{l=\max(0,j-h_0)}^{\min(j,n-h_0)} \binom{n-h_0}{l} \right.$$
$$\left. \bullet \binom{h_0}{j-l} (1 - P_e)^n \left(\frac{P_e}{1-P_e}\right)^{h_0-j+2l} \right]$$
$$\left[\left(\frac{1}{2}\right)^{M-n} \sum_{k=0}^{\min(h-j,M-n)} \binom{M-n}{k} \right] \tag{13.6}$$

13.2.3.1 *Example*

Choose the marker bits to complete the design with parameters $M = 36$, $h = 3$, and $P_e = 0.01$. Figure 13.6 shows plots of $P_A(3, n, h_0)$ as a function of n for various h_0. Noting that $P_A(3, 0, 0)$ is approximately 0.114×10^{-6}, the condition is satisfied by any marker which

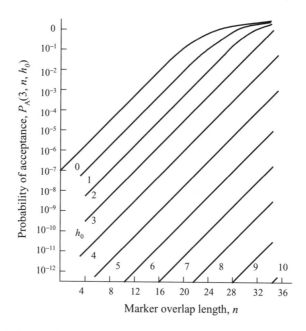

Figure 13.6 Probability of acceptance $P_A(3, n, h_0)$ for a 36-bit marker transmitted over a BSC with $P_e = 0.01$, where h_0 is the contribution of the received marker bits to the Hamming distance test measurement at time t [1128] © 1980, IEEE.

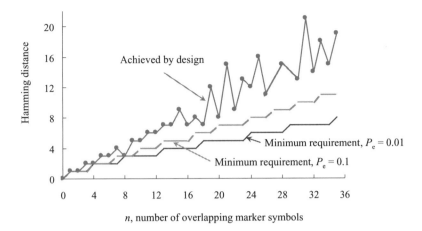

Figure 13.7 Acceptable marker design requirement for $M = 36$, $P_e = 0.01$, and $h = 3$, along with the achieved values of h_0 for the sequence: 000000011110010111010011001101010101 [1128] © 1980, IEEE.

has $h_0 \geq 1$ for $1 \leq n \leq 3$, $h_0 \geq 2$ for $4 \leq n \leq 7$, etc. The complete requirements for an acceptable design at both $P_e = 0.01$ and $P_e = 0.1$, along with one possible solution, are shown in Figure 13.7. As expected, the design requirements become more severe as the BSC's probability of error P_e increases. Even at the relatively extreme value of $P_e = 0.1$, the design in Figure 13.7 is acceptable.

Results for binary marker designs for lengths up to 34 bits are given in Table 13.1.

For a specific application, the frame synchronization problem will be dependent on the signal format. More details can be found in [1129–1141].

13.3 NETWORK SYNCHRONIZATION

Network synchronization is important for simplifying positioning and cellular network operation. The presentation in this section is based on [1142]. The classification of the possible solutions is given in Figure 13.8.

13.3.1 Plesiochronous and mutually synchronized networks

In a plesiochronous network each node contains its own precise clock and there are no control signals coordinating the operation of these clocks. Initially, the clocks are set such that the time difference between them is zero (or at least one tries to get as close to zero as possible). This calibration can be done centrally before shipping the clocks to their final locations, or it can be done by a 'traveling clock'. An example of a plesiochronous network employing cesium clocks is the TRI-TAC timing system which has an updating period of 24 h. The Global Positioning System (GPS) also employs this technique.

Principles of master–slave and mutually synchronized networks are given in Figures 13.9 and 13.10 respectively. More detailed block diagrams are shown in Figures 13.11–13.14 for a two-node network, and Figures 13.15–13.16 for an N-node network.

Table 13.1   Binary marker designs of Maury and Styles for the BSC and coherent channels, and of Turyn for the differentially coherent channel. All markers listed in octal with the left-most bit in each marker being 1

Length	Maury and Styles	Turyn
7	130 → 001011000	130
8	270	—
9	560	—
10	1560	—
11	2670	2670
12	6540	—
13	16540	17465
14	34640	37145
15	73120	76326
16	165620	167026
17	363240	317522
18	746500	764563
19	1746240	1707355
20	3557040	3734264
21	7351300	7721306
22	17155200	17743312
23	36563200	23153702
23	76571440	77615233
25	174556100	163402511
26	372323100	—
27	765514600	664421074
28	1727454600	1551042170
29	3657146400	3322104360
30	7657146400	7754532171
31	—	16307725110
32	—	37562164456
33	—	77047134545
34	—	177032263125

Different methods for delay compensation are presented in Figure 13.17, and block diagrams of the synchronizers using these methods are shown in Figures 13.18–13.20. Next we will discuss mathematical models for these synchronizers.

13.3.2 Mathematical model of a synchronous network

The instantaneous radian frequency function $\Phi(t)$ of a oscillator can be modeled in the form

$$\dot{\Phi}(t) = \omega_0 + \sum_{k=0}^{M-1} \frac{L(k)}{k!} t^k + \dot{\xi}(t) \tag{13.7}$$

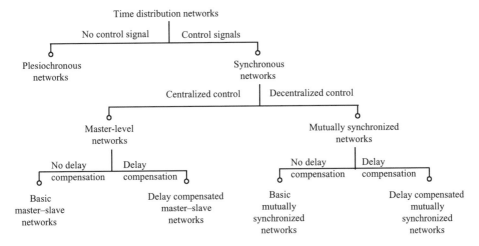

Figure 13.8 Classification of time and frequency distribution networks [1142] © 1985, IEEE.

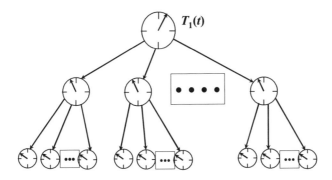

Figure 13.9 A hierarchical master–slave network.

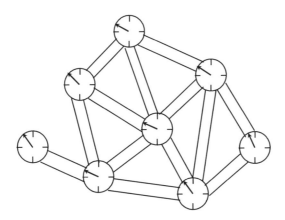

Figure 13.10 A partially connected mutually synchronized *N*-node network.

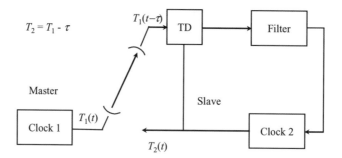

Figure 13.11 Block diagram of basic two-node master–slave network. TD: time difference (or PD: phase detector).

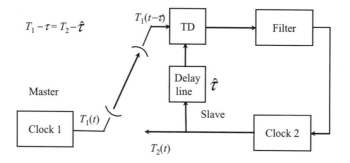

Figure 13.12 Delay line compensated master–slave two-node network.

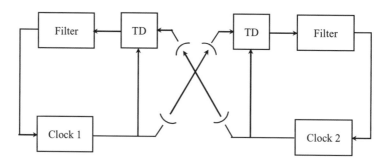

Figure 13.13 Basic mutual synchronization in two-node network.

where ω_0 is a constant denoting the nominal value of the free-running frequency of the oscillator, $L(0)$ is a zero mean random variable representing the initial frequency error (departure). This error arises from the uncertainty which exists in the initial (setability) of the free-running frequencies of the oscillators. The $L(k)$s $(k = 1, \ldots, M - 1)$, specify a set of time-independent random variables modeling the kth-order frequency drifts. $\xi(t)$ is a stationary zero mean random process characterizing the short-term oscillator instabilities.

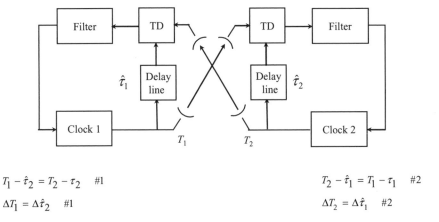

$$T_1 - \hat{\tau}_2 = T_2 - \tau_2 \quad \#1 \qquad\qquad\qquad T_2 - \hat{\tau}_1 = T_1 - \tau_1 \quad \#2$$

$$\Delta T_1 = \Delta \hat{\tau}_2 \quad \#1 \qquad\qquad\qquad\qquad \Delta T_2 = \Delta \hat{\tau}_1 \quad \#2$$

Figure 13.14 Delay line compensated mutually synchronized two-node network.

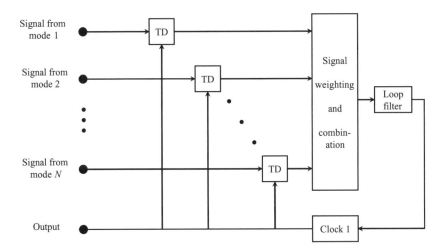

Figure 13.15 Phase averaging at node i of an N-node network with no delay compensation.

By integrating Equation (13.7) we have

$$\Phi(t) = \Phi(0) + \omega_0 t + \sum_{k=1}^{M} \frac{L(k-1)}{k!} t^k + [\xi(t) - \xi(0)] \qquad (13.8)$$

for $M \geq 1$. The 'time process' of the clock is obtained by dividing the oscillator phase by the nominal free-running frequency of the oscillator ω_0. Hence, the time process $T(t)$ can be written as:

$$T(t) = T(0) + t + \sum_{k=1}^{M} \frac{q(k)}{k!} t^k + \Psi(t) \qquad (13.9)$$

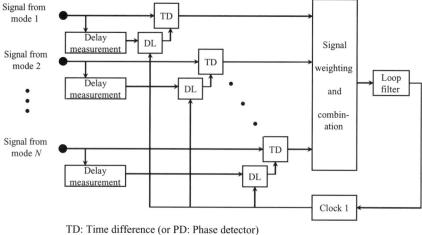

TD: Time difference (or PD: Phase detector)
DL: Delay line

Figure 13.16 Nodal processing at node *i* of an *N*-node network with delay compensation.

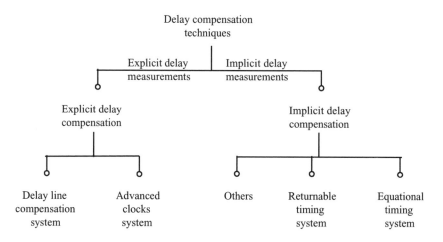

Figure 13.17 Different methods of delay compensation in a synchronous network.

where $q(k) = L(k-1)/\omega_0 (k = 1, \ldots, M)$ are a set of random variables, modeling the
$(k-1)$th-order time drifts and $\Psi(t) = \{\xi(t) - \xi(0)\}/\omega_0$ is, in general, a non-stationary
stochastic process characterizing the short-term clock instabilities.

13.3.3 Mathematical model of a plesiochronous network

In plesiochronous networks, the clocks are located at different geographical positions with
no interconnections between them. A clock located at node *i* generates a time process $T_i(t)$
which is independent of all time processes generated by other clocks in the network. $T_i(t)$

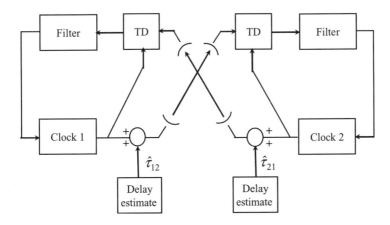

Figure 13.18 A two-node mutually synchronized network employing the advanced clock method.

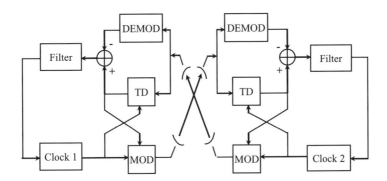

Figure 13.19 Equational timing system in a mutually synchronized two-node network.

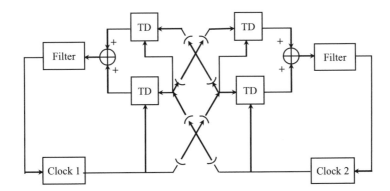

Figure 13.20 Returnable timing system in a mutually synchronized two-node network.

is of the form

$$T_i(t) = T_i(0) + t + \sum_{k=1}^{m} \frac{q_i(k)}{k!} t^k + \Psi_i(t) \tag{13.10}$$

$i = 1, 2, \ldots, N$, where N is the number of nodes in the network. The derivative of the time process $T_i(t)$ is given as:

$$\dot{T}_i(t) = 1 + \sum_{k=1}^{M} \frac{q_i(k)}{(k-1)!} t^{k-1} + \dot{\Psi}_i(t) \quad i = 1, 2, \ldots, N \tag{13.11}$$

whereas the instantaneous frequency Ω_i (t) is defined as

$$\Omega_i(t) = \frac{\dot{\Phi}(t)}{w_0} = 1 + \sum_{k=1}^{M} \frac{q_i(k)}{(k-1)!} t^{k-1} + \dot{\Psi}_i(t) \tag{13.12}$$

The system of equations describing the operation of plesiochronous networks is given by

$$\dot{T}(t) = \Omega(t) \tag{13.13}$$

where

$$\dot{T}(t) = [\dot{T}_1(t), \ \dot{T}_2(t), \ldots, \dot{T}_N(t)]^{\mathrm{T}}$$
$$\Omega(t) = [\Omega_1(t), \ \Omega_2(t), \ldots, \Omega_N(t)]^{\mathrm{T}} \tag{13.14}$$

where T indicates the transpose operation on the vectors.

13.3.4 Mathematical model of synchronous networks with no delay compensation

For a two-node network we have

$$\dot{T}_1(t) = \Omega_1(t) + B_1 F_1(p)$$
$$\cdot [g_{12}[T_2(t - \tau_{12}) - T_1(t)] + N_{12}(t)]$$
$$\dot{T}_2(t) = \Omega_2(t) + B_2 F_2(p) \tag{13.15}$$
$$\cdot [g_{21}[T_1(t - \tau_{21}) - T_2(t)] + N_{21}(t)]$$

where B_i is the loop gain at node i, $F_i(p)$ is the loop filter at node i, $g_{ij}(\cdot)$ is the characteristic of the time (phase) detector at node j receiving time signal of node i, τ_{ij} is the delay encountered by the signal traveling from node i to node j and N_{ij} (t) is the equivalent thermal noise in that channel. For an N-node net we have:

$$\dot{T}_i(t) = \Omega_i(t) + B_i F_i(p)$$
$$\cdot \sum_{j=1}^{N} a_{ij} \{ g_{ij}[T_j(t - \tau_{ij}) - T_i(t)] + N_{ij}(t) \} \tag{13.16}$$

for $i = 1, 2, \ldots, N$, where a_{ij} are normalized weighting coefficients with

$$\sum_{j=1}^{N} a_{ij} = 1 \quad \text{and} \quad a_{ii} = 0, \tag{13.17}$$

for all $i = 1, 2, \ldots, N$.

13.3.5 Mathematical model of synchronous networks with delay compensation

13.3.5.1 Delay line compensation

For a two-node network, we have:

$$
\begin{aligned}
\dot{T}_1(t) &= \Omega_1(t) + B_1 F_1(p) \\
&\quad \cdot \{g_{12}[T_2(t - \tau_{12}) - T_1(t - \hat{\tau}_{12})] + N_{12}(t)\} \\
\dot{T}_2(t) &= \Omega_2(t) + B_2 F_2(p) \\
&\quad \cdot \{g_{21}[T_1(t - \tau_{21}) - T_2(t - \hat{\tau}_{21})] + N_{21}(t)\}
\end{aligned}
\tag{13.18}
$$

For the N-node network this becomes

$$
\begin{aligned}
\dot{T}_i(t) &= \Omega_i(t) + B_i F_i(p) \\
&\quad \cdot \sum_{j=1}^{N} a_{ij}\{g_{ij}[T_j(t - \tau_{ij}) - T_i(t - \hat{\tau}_{ij})] + N_{ij}(t)\}
\end{aligned}
\tag{13.19}
$$

where $i = 1, 2, \ldots, N$.

13.3.5.2 Advanced clock method

The mathematical model for a two-node network using this compensation method can be written by inspection as follows:

$$
\begin{aligned}
\dot{T}_1(t) &= \Omega_1(t) + B_1 F_1(p) \\
&\quad \cdot \{g_{12}[T_2(t - \tau_{12}) - \hat{\tau}_{12} - T_1(t)] + N_{12}(t)\} \\
\dot{T}_2(t) &= \Omega_2(t) + B_2 F_2(p) \\
&\quad \cdot \{g_{21}[T_1(t - \tau_{21}) + \hat{\tau}_{21} - T_2(t)] + N_{21}(t)\}
\end{aligned}
\tag{13.20}
$$

The generalization of this model to N-node networks is very similar.

13.3.5.3 Equational timing system (ETS)

In this section we have

$$
\begin{aligned}
\dot{T}_1(t) &= \Omega_1(t) + F_1(p) \{B_1 g_{12}[T_2(t - \tau_{12}) - T_1(t)] + N_{12}(t) \\
&\quad + \tilde{B}_1 \tilde{g}_{12}[T_2(t - d_{12}) - T_1(t - \tau_{21} - d_{12}) + \tilde{N}_{12}(t)]\} \\
\dot{T}_2(t) &= \Omega_2(t) + F_2(p) \{B_2 g_{21}[T_1(t - \tau_{21}) - T_2(t)] + N_{21}(t) \\
&\quad + \tilde{B}_2 \tilde{g}_{21}[T_1(t - d_{21}) - T_2(t - \tau_{12} - d_{21}) + \tilde{N}_{21}(t)]\}
\end{aligned}
\tag{13.21}
$$

and for N-node networks

$$
\begin{aligned}
\dot{T}_i(t) &= \Omega_i(t) + F_i(p) \Bigg\{ B_i \sum_{j=1}^{N} a_{ij}[g_{ij}[T_j(t - \tau_{ij}) - T_i(t)] + N_{ij}(t)] \\
&\quad + \tilde{B}_i \sum_{j=1}^{N} b_{ij}[\tilde{g}_{ij}[T_j(t - d_{ij}) - T_i(t - \tau_{ji} - d_{ij}) + \tilde{N}_{ij}(t)]] \Bigg\}
\end{aligned}
\tag{13.22}
$$

13.3.5.4 *Returnable timing system (RTS)*

For this system Equation (13.21) becomes

$$
\begin{aligned}
\dot{T}_1(t) &= \Omega_1(t) + F_1(p)\,\{B_1 g_{12}[T_2(t - \tau_{12}) - T_1(t)] + N_{12}(t) \\
&\quad + \tilde{B}_1 \tilde{g}_{12}[T_2(t - \tau_{12}) - T_1(t - \tau_{21} - \hat{\tau}_{12}) + \tilde{N}_{12}(t)]\} \\
\dot{T}_2(t) &= \Omega_2(t) + F_2(p)\,\{B_2 g_{21}[T_1(t - \tau_{21}) - T_2(t)] + N_{21}(t) \\
&\quad + \tilde{B}_2 \tilde{g}_{21}[T_1(t - \tau_{21}) - T_2(t - \tau_{12} - \hat{\tau}_{21}) + \tilde{N}_{21}(t)]\}
\end{aligned}
\tag{13.23}
$$

13.3.6 Generalized mathematical model of synchronous networks

In general, previous equations can be presented as

$$
\dot{T}(t) = \Omega(t) + F(p)[\,G + \tilde{G} + N + \tilde{N}\,] \cdot J
\tag{13.24}
$$

where

$$
\begin{aligned}
T(t) &= [T_1(t), T_2(t), \ldots, T_N(t)]^{\mathrm{T}} \\
\Omega(t) &= [\Omega_1(t), \Omega_2(t), \ldots, \Omega_N(t)]^{\mathrm{T}} \\
F(p) &= \mathrm{diag}\,[F_1(p), F_2(p), \ldots, F_N(p)]_{N \times N} \\
G &= [B_i a_{ij} g_{ij}\,[T_j(t - \tau_{ij}) - T_i(t - \eta_{ij}) + \mu_{ij}]]_{N \times N} \\
\tilde{G} &= [\tilde{B}_i a_{ij} g_{ij}\,[T_j(t - \sigma_{ij}) - T_i(t - \delta_{ij})]]_{N \times N} \\
N &= [B_i N_{ij}(t)]_{N \times N} \\
\tilde{N} &= [\tilde{B}_i \tilde{N}_{ij}(t)]_{N \times N} \\
J &= [1, 1, \ldots, 1]^{\mathrm{T}}
\end{aligned}
\tag{13.25}
$$

$$
\sum_{j=1}^{N} a_{ij} = \sum_{j=1}^{N} b_{ij} = 1, \quad \forall i
$$

and $a_{ii} = b_{ii} = 0, \quad \forall i = 1, \ldots, N.$

13.3.7 Linearized mathematical model of synchronous networks

If function $g_{ij}(\cdot)$ and $\tilde{g}_{ij}(\cdot)$ are assumed linear, we have

$$
\begin{aligned}
\dot{T}_i(t) &= \Omega_i(t) + F_i(p) \Bigg\{ B_i \sum_{j=1}^{N} a_{ij}[[T_j(t - \tau_{ij}) - T_i(t - \eta_{ij})] + \mu_{ij} + N_{ij}(t)] \\
&\quad + \tilde{B}_i \sum_{j=1}^{N} b_{ij}[[T_j(t - \sigma_{ij}) - T_i(t - \delta_{ij}) + \tilde{N}_{ij}(t)]] \Bigg\}
\end{aligned}
\tag{13.26}
$$

13.3.8 Network performance measures: steady-state network behavior

The *time error process* is defined as

$$
T_{ij}(t) \equiv T_i(t) - T_j(t), \quad i, j = 1, \ldots, N; \quad i \neq j.
\tag{13.27}
$$

The phase error between the nodes is defined as

$$\Phi_{ij}(t) \equiv \Phi_i(t) - \Phi_j(t), \quad i, j = 1, \ldots, N; \quad i \neq j. \tag{13.28}$$

The *time interval process*, a time interval of length h as observed on the time scale at node i, is given by

$$\Delta T_i(t; h) = T_i(t + h) - T_i(t), \quad i = 1, \ldots, N. \tag{13.29}$$

The phase version becomes

$$\Delta \Phi_i(t; h) = \Phi_i(t + h) - \Phi_i(t), \quad i = 1, \ldots, N. \tag{13.30}$$

The *time interval error process* is defined as

$$\Delta T_{ij}(t; h) \equiv [T_i(t + h) - T_i(t)] - \lfloor T_j(t + h) - T_j(t) \rfloor$$
$$\Delta T_{ij}(t; h) = \Delta T_i(t; h) - \Delta T_j(t; h) \tag{13.31}$$
$$\Delta T_{ij}(t; h) = T_{ij}(t + h) - T_{ij}(t)$$

The phase version can be expressed as

$$\Delta \Phi_{ij}(t; h) \equiv \Phi_{ij}(t + h) - \Phi_{ij}(t) \tag{13.32}$$

13.3.9 Network stability: sufficient stability conditions

By taking the Laplace transform of the linearized mathematical model we have

$$\mathbf{Q}(s) \cdot \mathbf{T}(s) = \mathbf{P}(s) \tag{13.33}$$

where $\mathbf{Q}(s)$ is an $N \times N$ matrix and $\mathbf{P}(s)$ and $\mathbf{T}(s)$ are n-vectors. $\mathbf{T}(s)$ is the vector of the Laplace transforms of the nodal time scale, i.e. it is the desired system output. For stability conditions $\mathbf{P}(s)$ should be a vector with no poles in $R^+ = \{s : \text{Re}\,(s) \geq 0, \, s \neq 0\}$. The equation can be solved for $\mathbf{T}(s)$ if $\mathbf{Q}^{-1}(s)$ exists. The derivation of the sufficient stability conditions are based on the existence of $\mathbf{Q}^{-1}(s)$, in particular on the following theorem.

13.3.9.1 *Theorem*

Let $\mathbf{Q}(s) = [q_{ik}]_{N \times N}$ be a complex matrix which is diagonally dominant in R^+, i.e.

$$|q_{ii}| > \sum_{\substack{j=1 \\ j \neq i}}^{N} |q_{ij}|, \quad i = 1, 2, \ldots, N \tag{13.34}$$

Then the roots of $\det[\mathbf{Q}(s)]$ have negative real parts with the possible exception of roots at the region. Sufficient conditions for the network stability are summarized in Table 13.2.

13.3.10 Steady-state behavior of synchronous networks

In determining steady-state behavior, it is always assumed that the network is stable and that there is no clock instability or channel noise present. In the steady state, the parameters of interest are the steady-state frequency of the clocks and the time error between them. It is

Table 13.2   Sufficient conditions for the stability of synchronous networks

Network type	No loop filter	$F_k(s) = \dfrac{1}{1 + \rho_k s}$
Master–slave networks	Always stable	
Mutually synchronized with no delay compensation	Always stable	$B_k \cdot \rho_k < 1/2, \quad \forall k$
Mutually synchronized with delay line compensation	$B_k \cdot \hat{\tau}_{kl} < 1/2, \quad \forall k^a$	$B_k \cdot [\hat{\tau}_{kl} + \rho_k] < 1/2, \quad \forall k^a$
Mutually synchronized with ETS or RTS^c	$B_k \cdot \delta_{kl} < \sqrt{2} - 1, \quad \forall k^b$	$B_k \cdot [\hat{\tau}_{kl} + \rho_k] < 1/2, \quad \forall k^b$

$^a \tau_{kl} = \max\limits_{j \neq k} \hat{\tau}_{kj}$

$^b \delta_{kl} = \max\limits_{j \neq k} \delta_j$

$^c \delta_{kj} = \hat{\tau}_{kj} + \tau_{jk}, \quad$ for RTS; $\delta_{kj} = d_{kj} + \tau_{jk}, \quad$ for ETS

assumed that there are no channel noise or loop filters in the network, although the results will hold for networks with loop filters if the filters satisfy the condition

$$\lim_{s \to 0} F_i(s) = 1, \quad \forall i \tag{13.35}$$

If the network is assumed to be stable with its steady-state frequency (this is in fact the normalized frequency) denoted by Ω_s, then

$$\lim_{t \to \infty} \dot{T}(t) = \Omega_s, \quad \forall i \tag{13.36}$$

The results are summarized in Tables 13.3–13.5.

13.3.11   Influence of clock phase noise

The mathematical model of the clock indicates that the time process of each clock depends on the short-term clock instabilities $\Psi(t)$. Although, in general, the process $T(t)$ is a non-stationary stochastic process, $\Psi(t)$ can be assumed to be stationary. Since the time process $T(t)$ is a non-stationary random process, its behavior cannot be characterized in terms of power spectral density. The nth increment ($n \geq M$) of $T(t)$ is a stationary process and can be used as a measure of performance of the clock.

13.3.12   Structure function

The nth increment of the time error process for $n > 1$, is defined recursively by

$$\Delta^n T_{ij}(t; h) = \Delta^{n-1}[\Delta T_{ij}(t; h)] \tag{13.37}$$

The process $T_{ij}(t)$ is called a process with stationary nth increment if $\Delta^n T_{ij}(t; h)$ is a stationary process. The nth *structure function* of a process $T_{ij}(t)$ with stationary nth increment

Table 13.3   Steady-state frequency of a fully connected mutually synchronized network of N nodes

Network configuration	Steady-state frequency Ω_s
No delay compensation	$$\dfrac{\displaystyle\sum_{i=1}^{N}\Omega_i/B_i}{\displaystyle\sum_{i=1}^{N}1/B_i+\dfrac{1}{N-1}\sum_{i=1}^{N}\sum_{\substack{j=1\\j\neq i}}^{N}\tau_{ij}}$$
Delay-line compensation	$$\dfrac{\displaystyle\sum_{i=1}^{N}\Omega_i/B_i}{\displaystyle\sum_{i=1}^{N}1/B_i+\dfrac{1}{N-1}\sum_{i=1}^{N}\sum_{\substack{j=1\\j\neq i}}^{N}\Delta\tau_{ij}}$$
Equational timing system	$$\dfrac{\displaystyle\sum_{i=1}^{N}\Omega_i/B_i}{\displaystyle\sum_{i=1}^{N}1/B_i}$$
Returnable timing system	$$\dfrac{\displaystyle\sum_{i=1}^{N}\Omega_i/B_i}{\displaystyle\sum_{i=1}^{N}1/B_i+\dfrac{1}{N-1}\sum_{i=1}^{N}\sum_{\substack{j=1\\j\neq i}}^{N}\Delta\tau_{ij}}$$

Table 13.4   Steady-state time error between master clock and clock i in the hierarchical master–slave network of N nodes

Network configuration	Steady-state time error T_{1i}
No delay compensation	$\displaystyle\sum_{j\in M_i}\left(\Omega_{\text{mas}}-\Omega_j\right)/B_j+\Omega_{\text{mas}}\sum_{\substack{j\in M_j\\k\to j}}\tau_{jk}$
Delay-line compensation	$\displaystyle\sum_{j\in M_i}\left(\Omega_{\text{mas}}-\Omega_j\right)/B_j+\Omega_{\text{mas}}\sum_{\substack{j\in M_j\\k\to j}}\Delta\tau_{jk}$
Equational timing system	$\displaystyle\sum_{j\in M_i}\left(\Omega_{\text{mas}}-\Omega_j\right)/2B_j+\dfrac{\Omega_{\text{mas}}}{2}\sum_{\substack{j\in M_j\\k\to j}}\left(\tau_{jk}-\tau_{kj}\right)$
Returnable timing system	$\displaystyle\sum_{j\in M_i}\left(\Omega_{\text{mas}}-\Omega_j\right)/2B_j+\dfrac{\Omega_{\text{mas}}}{2}\sum_{\substack{j\in M_j\\k\to j}}\left[\Delta\tau_{jk}+\left(\tau_{jk}-\tau_{kj}\right)\right]$

Ω_{mas}: Frequency of master clock, $\Omega_s=\Omega_{\text{mas}}$

Table 13.5   Steady-state time error between the nodes of a fully connected mutually synchronized network of N nodes

Network configuration	Steady-state time error T_{ri}
No delay compensation	$\dfrac{N-1}{N}\left[\dfrac{\Omega_r - \Omega_i}{B}\right] - \dfrac{1}{N}\Omega_s\left[\displaystyle\sum_{\substack{j=1\\j\neq r}}^{N}\tau_{rj} - \sum_{\substack{j=1\\j\neq i}}^{N}\tau_{ij}\right]$
Delay-line compensation	$\dfrac{N-1}{N}\left[\dfrac{\Omega_r - \Omega_i}{B}\right] - \dfrac{1}{N}\Omega_s\left[\displaystyle\sum_{\substack{j=1\\j\neq r}}^{N}\Delta\tau_{rj} - \sum_{\substack{j=1\\j\neq i}}^{N}\Delta\tau_{ij}\right]$
Equational timing system	$\dfrac{N-1}{N}\left[\dfrac{\Omega_r - \Omega_i}{B}\right] - \dfrac{1}{N}\Omega_s\left[\displaystyle\sum_{\substack{j=1\\j\neq r}}^{N}(\tau_{rj} - \tau_{jr}) - \sum_{\substack{j=1\\j\neq i}}^{N}(\tau_{ij} - \tau_{ji})\right]$
Returnable timing system	$\dfrac{N-1}{N}\left[\dfrac{\Omega_r - \Omega_i}{B}\right] - \dfrac{1}{N}\Omega_s\left[\displaystyle\sum_{\substack{j=1\\j\neq r}}^{N}[\Delta\tau_{rj} + (\tau_{rj} - \tau_{jr})]\right.$ $\left. - \displaystyle\sum_{\substack{j=1\\j\neq r}}^{N}[\Delta\tau_{ij} + (\tau_{ij} - \tau_{ji})]\right]$

is defined by

$$D^n_{T_{ij}}(h) \equiv E[\Delta^n T_{ij}(t; h)]^2 \qquad (13.38)$$

which is basically the variance of the nth increment process. This relates to the power spectral density as

$$D^n_{T_{ij}}(h) = \int_{-\infty}^{\infty} s_{\Delta^n T_{ij}}(\omega)\,\frac{d\omega}{2\pi} \qquad (13.39)$$

One can define the nth increment, and hence the nth structure function, of the time processes observed at the network nodes by

$$D^n_{T_i}(h) = E\left[\Delta^n T_i(t; h)\right]^2 = \int_{-\infty}^{\infty} s_{\Delta^n T_i}(\omega)\,\frac{d\omega}{2\pi} \qquad (13.40)$$

$D^n_{T_i}(h)$ and $D^n_{T_{ij}}(h)$ have been calculated for plesiochronous, master–slave, and mutually synchronized networks of clocks and the results are tabulated in Tables 13.6 and 13.7 and presented in Figures 13.21–13.24.

Table 13.6 The nth structure function of the time processes generated at network nodes

Network type	nth structure function
Plesiochronous	$$D_{T_i}^n(h) = 2^{2n} \int_{-\infty}^{\infty} \sin^{2n}\left(\frac{\omega h}{2}\right) \frac{s_{\Psi_i}(\omega)}{2} \frac{d\omega}{2\pi}$$
Master–slave	$$D_{T_i}^n(h) = 2^{2n} \int_{-\infty}^{\infty} \sin^{2n}\left(\frac{\omega h}{2}\right) |G_{ms}(\omega)|^2 \frac{s_{\Psi_i}(\omega)}{\omega^2} \frac{d\omega}{2\pi}$$ $$|G_{ms}(\omega)|^2 = |H(\omega)|^{2(L-1)} + \sum_{k=0}^{L-1} |H(\omega)|^{2K} \cdot |1 - H(\omega)|^2$$
Mutually synchronized	$$D_{T_i}^n(h) = 2^{2n} \int_{-\infty}^{\infty} \sin^{2n}\left(\frac{\omega h}{2}\right) |G_{mu}(\omega)|^2 \frac{s_{\Psi_i}(\omega)}{\omega^2} \frac{d\omega}{2\pi}$$ $$|G_{mu}(\omega)|^2 = \frac{\left|(N-1) - (N-2)H(\omega)e^{-j\omega\tau}\right|^2 + (N-1)|H(\omega)|^2}{\left|(N-1) + H(\omega)e^{-j\omega\tau}\right|^2 \left|1 - H(\omega)e^{-j\omega\tau}\right|^2} |1 - H(\omega)|^2$$ where $H(s) = \dfrac{BF(s)}{s + BF(s)}$ with $F(s) = \dfrac{1}{1 + Ts}$

Table 13.7 The nth structure function of the time error processes generated between network nodes

Network type	nth structure function
Plesiochronous	$$D_{T_{ij}}^n(h) = 2^{2n} \int_{-\infty}^{\infty} \sin^{2n}\left(\frac{\omega h}{2}\right) \frac{s_{\Psi_{ij}}(\omega)}{\omega^2} \frac{d\omega}{2\pi}$$
Master–slave	$$D_{T_{ij}}^n(h) = 2^{2n} \int_{-\infty}^{\infty} \sin^{2n}\left(\frac{\omega h}{2}\right) |H_{ms}(\omega)|^2 \frac{s_{\Psi_{ij}}(\omega)}{\omega^2} \frac{d\omega}{2\pi}$$ $$|H_{ms}(\omega)|^2 = \frac{1}{2} \left\{ 1 + |H(\omega)|^{2(L-1)} + \sum_{k=0}^{L-1} |H(\omega)|^{2K} \cdot |1 - H(\omega)|^2 \right\} - \mathrm{Re}\left\{ \left[H^*(\omega)e^{j\omega\tau} \right]^{L-1} \right\}$$
Mutually synchronized	$$D_{T_{ij}}^n(h) = 2^{2n} \int_{-\infty}^{\infty} \sin^{2n}\left(\frac{\omega h}{2}\right) |H_{mu}(\omega)|^2 \frac{s_{\Psi_{ij}}(\omega)}{\omega^2} \frac{d\omega}{2\pi}$$ $$|H_{mu}(\omega)|^2 = H_1(\omega)[H_2(\omega) + H_3(\omega)]$$ where $$H_1(\omega) = \frac{|1 - H(\omega)|^2}{\left|(N-1) + H(\omega)e^{-j\omega\tau}\right|^2 \left|1 - H(\omega)e^{-j\omega\tau}\right|^2}$$ $$H_2(\omega) = \left|(N-1) - (N-2)H(\omega)e^{-j\omega\tau}\right|^2$$ $$H_3(\omega) = (2N-3)|H(\omega)|^2 - 2(N-1)\,\mathrm{Re}\left\{H(\omega)e^{-j\omega\tau}\right\}$$

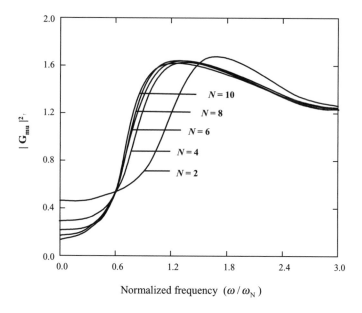

Figure 13.21 Frequency response of the nodal filtering function for the mutually synchronized network.

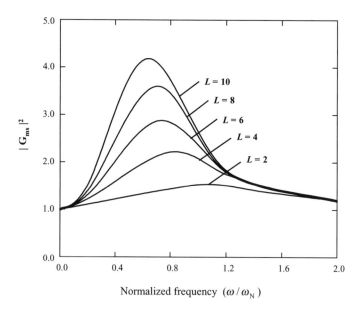

Figure 13.22 Frequency response of the nodal filtering function for the master–slave synchronized network.

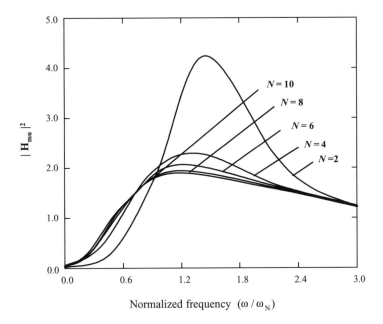

Figure 13.23 Frequency response of the time error filter function for the mutually synchronized network.

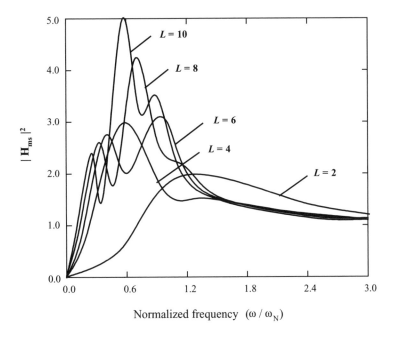

Figure 13.24 Frequency response of the time error filter function for the master–slave synchronized network.

13.3.13 Synchronization behavior in the presence of channel noise

Assuming a first-order PLL, the solution of the Fokker–Planck equation gives the phase error pdfs shown in Figures 13.25 and 13.26.

Additional details on network synchronization can be found in [1143–1153].

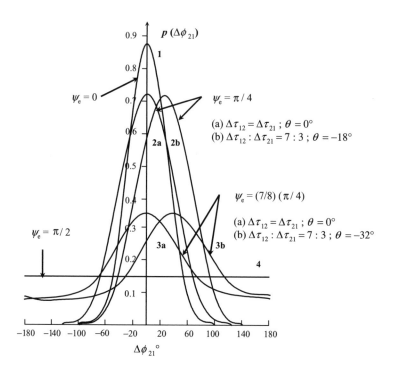

$$\psi_e = \overline{\omega}\left(\Delta\tau_{21}+\Delta\tau_{12}\right)/2$$

$$\theta_e = \overline{\omega}\left(\Delta\tau_{21}-\Delta\tau_{12}\right)/2$$

$$\rho = 5 \text{ (SNR)}$$

Figure 13.25 The effect of residual delays on the probability density function $p(\Delta\phi_{21})$ of the phase error in a two-node mutually synchronized network [1142].

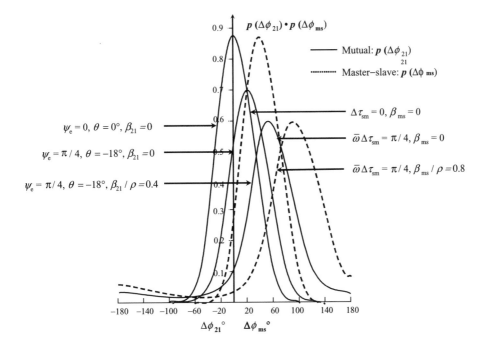

$$\psi_e = \overline{\omega}\left(\Delta\tau_{21}+\Delta\tau_{12}\right)/2$$

$$\theta_e = \overline{\omega}\left(\Delta\tau_{21}-\Delta\tau_{12}\right)/2$$

$$\rho = 5 \text{ (SNR)}$$

Figure 13.26 Comparison of the combined effects of the residual delays and detuning on the probability density functions of the phase error in mutual and master–slave synchronized networks [1142].

14

Channel Modeling and Measurements for 4G

14.1 MACROCELLULAR ENVIRONMENTS (1.8 GHZ)

In this section we present analysis of the joint statistical properties of azimuth spread, delay spread and shadowing fading in macrocellular environments. The analysis is based on data reported from a measurement campaign in typical urban (TU), bad urban (BU), and suburban (SU) [1154] areas. In the experiment, a BS equipped with an eight-element uniform linear antenna array and an MS with an omnidirectional dipole antenna are used. The MS is equipped with a differential global positioning system (GPS) and an accurate position encoder so its location is accurately known by combining the information from these two devices. MS displacements of less than one centimeter can, therefore, be detected. The system is designed for transmission from the MS to the BS. Simultaneous channel sounding is performed on all eight branches, which makes it possible to estimate the azimuth of the impinging waves at the BS. The sounding signal is a maximum length linear shift register sequence of length 127 chips, clocked at a chip rate of 4.096 Mbps. This chip rate has been initially used in WCDMA proposals in Europe. The testbed operates at a carrier frequency of 1.8 GHz. Additional information regarding the stand-alone testbed can be found in [1154, 1155 and 1158]. A summary of macrocellular measurement environments is given in Table 14.1.

The channel's azimuth delay spread function at the BS is modeled as

$$h(\phi, \tau) = \sum_{l=1}^{L} \alpha_l \delta(\phi - \phi_l, \tau - \tau_l) \qquad (14.1)$$

where the parameters α_l, τ_l, and ϕ_l are the complex amplitude, delay, and incidence azimuth of the lth impinging wave at the BS. In general, $h(\phi, \tau)$ is considered to be a time-variant

Advanced Wireless Communications. S. Glisic
© 2004 John Wiley & Sons, Ltd. ISBN: 0-470-86776-0

Table 14.1 Summary of macrocellular measurement environments

Class	BS antenna height	Description of environment
TU Typical urban	10 m and 32 m	The city of Aarhus, Denmark. Uniform density of buildings ranging from 4–6 floors. Irregular street layout. Measurements carried out along six different routes with an average length of 2 km. No line-of-sight between MS and BS. MS–BS distance varies from 0.2 km to 1.1 km.
TU Typical urban	21 m	Stockholm city, Sweden (Area #1). Heavily built-up area with a uniform density of buildings, ranging from 4–6 floors. Ground is slightly rolling. No line-of-sight between MS and BS. MS–BS distance varies from 0.2 km to 1.1 km.
BU Bad urban	21 m	Stockholm city, Sweden (Area #2). Mixture of open flat areas (river) and densely built-up zones. Ground is slightly rolling. No line-of-sight between MS and BS. MS–BS distance varies from 0.9 km to 1.6 km.
SU Suburban	12 m	The city of Gistrup, Denmark. Medium-sized village with family houses of one–two floors and small gardens with trees and bushes. Typical Danish residential area. The terrain around the village is rolling with some minor hills. No line-of-sight between MS and BS. MS–BS distance varies from 0.3 km to 2.0 km.

function, since the constellation of the impinging waves is likely to change as the MS moves along a certain route. The local average power azimuth delay spectrum is given as

$$P(\phi, \tau) = E\left\{\sum_{l=1}^{L} |\alpha_l|^2 \delta(\phi - \phi_l, \tau - \tau_l)\right\} \tag{14.2}$$

From Equation (14.2), the local power azimuth spectrum (PAS) and the local power delay spectrum (PDS) are given as

$$P_A(\phi) = \int P(\phi, \tau) \, d\tau \tag{14.3}$$

$$P_D(\tau) = \int P(\phi, \tau) \, d\phi \tag{14.4}$$

The radio channel's local azimuth spread (AS) σ_A and the local delay spread (DS) σ_D are defined as the root second central moments of the corresponding variables. The values of the local AS and DS are likely to vary as the MS moves within a certain environment. Hence, we can characterize σ_A and σ_D as being random variables, with the joint pdf $f(\sigma_A, \sigma_D)$.

Their individual pdfs are

$$f_A(\sigma_A) = \int f(\sigma_A, \sigma_D) \, d\sigma_D \tag{14.5}$$

$$f_D(\sigma_D) = \int f(\sigma_A, \sigma_D) \, d\sigma_A \tag{14.6}$$

The function $f(\sigma_A, \sigma_D)$ can be interpreted as the global joint pdf of the local AS and DS. If the expectation in Equation (14.2) is computed over the radio channel's fast fading component, we can, furthermore, apply the approximation

$$\int \int P(\phi, \tau) \, d\phi \, d\tau = h_{\text{channel}} \cong h_{\text{loss}}(d) h_s \tag{14.7}$$

where h_{channel} is the radio channel's integral path loss, $h_{\text{loss}}(d)$ is the deterministic long-term distance- dependent path loss, while h_s is the channel's shadow fading component, which is typically modeled with a log-normal distributed random variable [1157, 1158]. The global pdf of h_s is denoted $f_s(h_s)$. The global degree of shadow fading is described by the root second central moment of the random shadow fading component expressed in decibels, i.e.

$$\sigma_s = \text{Std} \{10 \, \log_{10}(h_s)\} \tag{14.8}$$

where Std{ } denotes standard deviation. Empirical results for cumulative distribution functions (cdfs) for σ_A and σ_D are given in Figure 14.1 and Figure 14.2 respectively. The log-normal fit for σ_A results is given as

$$\sigma_A = 10^{\varepsilon_A X + \mu_A} \tag{14.9}$$

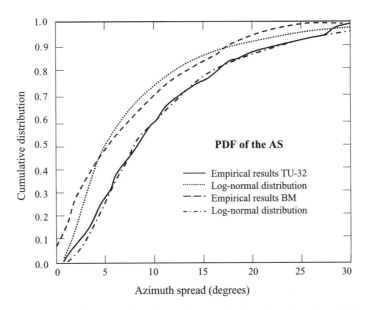

Figure 14.1 Examples of the empirical cumulative distribution function (cdf) of AS obtained in different environments. The cdf of a log-normal distribution is fitted to the empirical results for comparison [1154] © 2002, IEEE.

Table 14.2   Summary of the first and second central moments of the AS, DS and shadow fading in the different environments [1154]

Class	σ_s	$E\{\sigma_A\}$	μ_A	ε_A	$E\{\sigma_D\}$	μ_D	ε_D
TU-32	7.3 dB	8^0	0.74	0.47	0.8 μs	−6.20	0.31
TU-21	8.5 dB	8^0	0.77	0.37	0.9 μs	−6.13	0.28
TU-20	7.9 dB	13^0	0.95	0.44	1.2 μs	−6.08	0.35
BU	10.0 dB	7^0	0.54	0.60	1.7 μs	−5.99	0.46
SU	6.1 dB	8^0	0.84	0.31	0.5 μs	−6.40	0.22

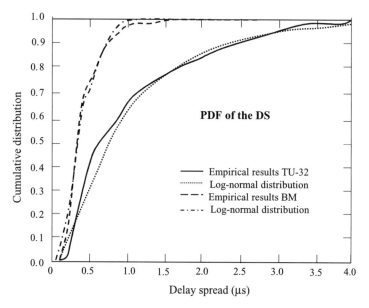

Figure 14.2   Examples of empirical cdfs of the DS obtained in different environments. The cdf of a log-normal distribution is fitted to the empirical results for comparison [1154] © 2002, IEEE.

where X is a zero mean Gaussian distributed random variable with unit variance, $\mu_A = E\{\log_{10}(\sigma_A)\}$ is the global logarithmic mean of the local AS, and $\varepsilon_A = \mathrm{Std}\{\log_{10}(\sigma_A)\}$ is the logarithmic standard deviation of the AS.

Similarly,

$$\sigma_D = 10^{\varepsilon_D Y + \mu_D} \tag{14.10}$$

where Y is a zero mean Gaussian distributed random variable with unit variance, $\mu_D = E\{\log_{10}(\sigma_D)\}$ is the global logarithmic mean of the local DS, and $\varepsilon_D = \mathrm{Std}\{\log_{10}(\sigma_D)\}$ is the logarithmic standard deviation of the DS. A summary of the results for these parameters is given in Table 14.2.

14.1.1 PDF of shadow fading

The shadow fading component is extracted from the measurement data under the assumption that the deterministic distance path loss can be expressed in decibels as

$$10 \log_{10}(h_{\text{loss}}(d)) = A + B \log_{10}(d) \tag{14.11}$$

where d is the distance between the BS and MS expressed in kilometers. As an example, the Okumura–Hata model [1159] is based on a similar structure. From Equations (14.11) and (14.7) we have

$$10 \log_{10}(h_{\text{channel}}(d)) = A + B \log_{10}(d) + 10 \log_{10}(h_{\text{s}}) \tag{14.12}$$

Assuming that $E\{10 \log_{10}(h_{\text{s}})\} = 0.0$ dB, the parameters A and B can be obtained as the least squares estimates from a large number of measurements. Subsequently, the shadow fading component can be isolated from each measurement segment by rearranging Equation (14.12).

From the measurements it is found that a log-normal distribution function provides a good match to the empirical pdf of the shadow fading component. This observation is in coherence with numerous other studies, see [1157, 1160 and 1161] among others. The shadow fading standard deviation is found to be in the range $\sigma_{\text{s}} = [6\text{–}10]$ dB depending on the environment class, with the largest standard deviation observed in the BU, and the smallest in SU environments. These findings are in accordance with previous findings in the open literature as well. Hence, the random variable describing the shadow fading component can be expressed as

$$h_{\text{s}} = 10^{\sigma_{\text{s}} Z / 10} \tag{14.13}$$

where Z is a zero mean Gaussian random variable with unit variance. The experimental results and analytical approximation for the spatial autocorrelation function are given in Figures 14.3 and 14.4.

Now,

$$\rho_{\text{A}}(d) = \exp(-d/d_{\text{A}}) \tag{14.14}$$

A double exponential decaying function is matched to the empirical results

$$\rho_{\text{A}}(d) = k \exp\left(-\frac{d}{d_{\text{A},1}}\right) + (1 - k) \exp\left(-\frac{d}{d_{\text{A},2}}\right); \quad k \in [0, 1] \tag{14.15}$$

The same results for DS are given in Figures 14.5 and Equations (14.16) and (14.17)

$$\text{BU environment: } \rho_{\text{D}}(d) = \exp(-d/d_{\text{D}}) \tag{14.16}$$

$$\text{SU environment: } \rho_{\text{D}}(d) = k \exp(-d/d_{\text{D},1}) + (1 - k) \exp(-d/d_{\text{D},2}) \tag{14.17}$$

Finally, the results for the decorrelation distance are summarized in Table 14.3.

The mutual interdependence between the different components of the fading is characterized by their crosscorrelation functions. In general, the crosscorrelation coefficient between

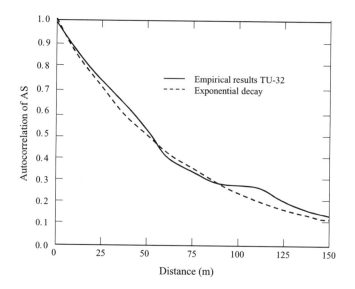

Figure 14.3 Empirical spatial autocorrelation function of AS in TU-32. An exponential decaying function is matched to the empirical results with $d_A = 70$ m.

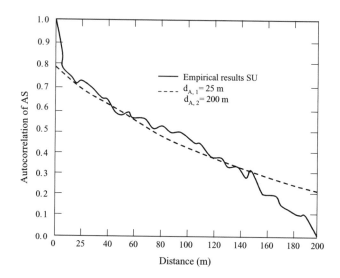

Figure 14.4 Empirical spatial autocorrelation function of AS in SU.

a and b is computed according to

$$\rho\langle a, b \rangle = \frac{\displaystyle\sum_{i=1}^{N} (a(i) - \bar{a})(b(i) - \bar{b})}{\sqrt{\displaystyle\sum_{i=1}^{N} (a(i) - \bar{a})^2 \sum_{i'=1}^{N} (b(i') - \bar{b})^2}} \tag{14.18}$$

Table 14.3   Spatial decorrelation distance for AS, DS and shadow fading in different environments expressed in meters. The two numbers presented for SU, correspond to the short and long decorrelation coefficients [1154]

Class	d_A	d_D	d_s
TU-32	50	40	45
TU-21	50	50	25
TU-20	75	50	55
BU	65	95	120
SU	25/200 $(k = 0.2)$	15/150 $(k = 0.3)$	30/200 $(k = 0.4)$

Table 14.4   Empirical crosscorrelation coefficients for the different environment classes for the AS, DS and shadow fading, expressed in both linear and logarithmic forms [1154]

Class	$\rho\langle\sigma_A,\sigma_D\rangle$	$\rho\langle\sigma_A,\sigma_D\rangle$ log	$\rho\langle\sigma_A,h_s\rangle$	$\rho\langle\sigma_A,h_s\rangle$ log	$\rho\langle\sigma_D,h_s\rangle$	$\rho\langle\sigma_D,h_s\rangle$ log
TU-32	0.39	0.44	−0.51	−0.7	−0.38	−0.4
TU-21	0.34	0.36	−0.47	−0.54	−0.34	−0.5
TU-20	0.6	0.6	−0.65	−0.72	−0.44	−0.48
BU	0.69	0.67	−0.44	−0.53	−0.55	−0.69
SU	0.46	0.46	−0.47	−0.5	−0.38	−0.46

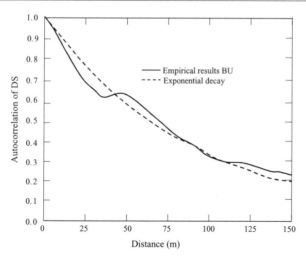

Figure 14.5   Empirical spatial autocorrelation function of DS in the BU environment. An exponential decaying function is matched to the empirical results.

where $\bar{a}$ and $\bar{b}$ are the sample means of the sets $\{a(i)\}$ and $\{b(i)\}$ with set size N. The results for AS, DS and shadow fading, based on measurements, are summarized in Table 14.4.

Additional details on the topic can be found in [1154–1182].

14.2   URBAN SPATIAL RADIO CHANNELS IN MACRO/MICROCELL (2.154 GHZ)

The discussion in this section is based on experimental results collected with a wideband channel sounder using a planar antenna array [1183]. The signal center frequency was 2154 MHz and the measurement bandwidth 100 MHz. A periodic PN sequence, 255 chips long, was used. The chip rate was 30 MHz and the sampling rate 120 MHz, giving an oversampling factor 4. The correlation technique is used for the determination of the impulse response. Hence, the delay range is 255/30 MHz = 8.5 μs, with a resolution of 1/30 MHz = 33 ns. The transmit antenna at the MS was a vertically polarized omnidirectional disconne antenna. The vertical 3 dB beamwidth was 87° and the transmit power 40 dBm. Approximately 80 different transmitter positions were investigated.

The receiving BS was located at one of three different sites: below, at, and above the rooftop level (RX 1–RX 3, see, Figure 14.6). A 16-element physical array with dual polarized $\lambda/2$-spaced patch antennas was combined with a synthetic aperture technique to build a virtual two-dimensional (2D) antenna structure. The patches were linearly polarized at 0° (horizontal direction) and 90° (vertical direction). With these 16×62 elements, the direction of arrival (DOA) of incoming waves, both in azimuth (horizontal angle), and elevation (vertical angle), could be resolved by using the super-resolution Unitary ESPRIT algorithm [1184–1186]. Note that the number of antenna elements limits the number of identifiable waves, but not the angular resolution of the method. Together with a delay resolution of 33 ns, the radio channel can be characterized in all three dimensions separately for the two polarizations. Array signal processing – including estimation of the DOAs and a comparison of ESPRIT with other algorithms–was discussed in Chapter 13.

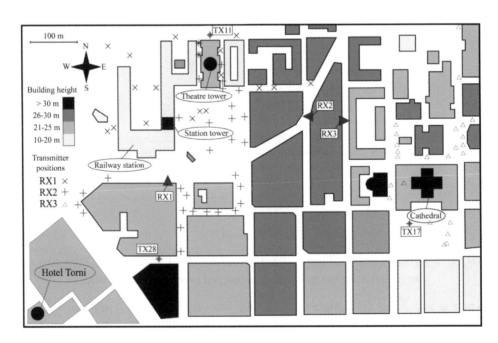

Figure 14.6 The measurement area with all three RX sites; TX positions of the sample plots are marked [1183] © 2002, IEEE.

One prerequisite for the applicability of the synthetic aperture technique is that the radio channel is static during the whole data collection period. To avoid problems, the whole procedure was done at night with minimum traffic conditions.

14.2.1   Description of environment

A typical urban environment is shown in Figure 14.6 [1183] with three receiver locations (RX 1–3) marked by triangles pointing broadside in the direction of the array. Figure 14.6 also provides information about all the corresponding TX positions. The location RX 1 (height $h_{RX} = 10$ m) is a typical microcell site below the rooftop height of the surrounding buildings, and measurements are performed with 20 different TX positions. RX 2 (height $h_{RX} = 27$ m) is at the rooftop level, and 32 TX positions are investigated. RX 3 (height $h_{RX} = 21$ m) is a typical macrocell BS position above rooftop height, and 27 TX positions are measured.

14.2.2   Results

The measurement results show that it is possible to identify many single (particular, different) multipath components, impinging at the receiver from different directions. But these components are not randomly distributed in the spatial and temporal domain, they naturally group into clusters. These clusters can be associated with objects in the environment due to the high angular and temporal resolution of evaluation. Sometimes even individual waves within a cluster can be associated with scattering objects. The identification of such clusters is facilitated by inspection of the maps of the environment.

A cluster is defined as a group of waves whose delay, azimuth and elevation at the receiver are very similar, while being notably different from other waves in at least one dimension. Additionally, all waves inside a cluster must stem from the same propagation mechanism. The definition of clusters always involves a certain amount of arbitrariness. Even for mathematically 'exact' definitions, arbitrary parameters (e.g. thresholds or numbers of components) must be defined. Clustering by human inspection, supported by maps of the environment, seems to give the best results.

The received power is calculated within each cluster (cluster power) by means of Unitary ESPRIT and a following beamforming algorithm. The results are plotted in the azimuth–elevation, azimuth–delay, and elevation–delay planes

According to the obvious propagation mechanism, each cluster is assigned to one of three different classes.

Class 1. Street-guided propagation: Waves arrive at the receiver from the street level after traveling through street canyons.

Class 2. Direct propagation–over the rooftop: The waves arrive at the BS from the rooftop level by diffraction at the edges of roofs, either directly or after reflection from buildings surrounding the MS. The azimuth mostly points to the direction of the transmitter, with some spread in azimuth and delay.

Class 3. Reflection from high-rise objects–over the rooftop: The elevation angles are near the horizon, pointing at or above the rooftop. The waves undergo a reflection at an object rising above the average building height before reaching the BS. The azimuth shows the direction of the reflecting building, the delay is typically larger than for Class 1 or Class 2.

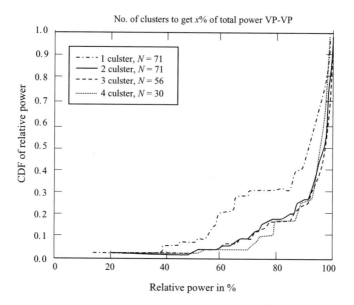

Figure 14.7 CDF of the relative power for one to four clusters, all TX positions, transmitter and receiver vertically polarized (VP-VP), N is the number of samples [1183] © 2002, IEEE.

The sum of the powers of all clusters belonging to the same class is called class power. In some cases, the propagation history is a mixture of different classes, e.g. street guidance followed by diffraction at rooftops. Such clusters are allocated to the class of the final path to the BS.

Some statistical evaluations are now presented to illustrate and quantify the clusterization. First we look at the number of clusters that is required to get a specific percentage of the total power; secondly, the powers of the clusters versus the delay; thirdly, the crosspolarization discrimination (XPD) versus the delay; fourthly, the relative class powers; and finally, the distribution of the number of clusters. Such questions are important for designing algorithms for adaptive antennas, e.g. should one capture, in uplink, the power of one or more clusters; or, in downlink, how does one distribute the available transmit power, and to which directions?

The first example illustrates the percentage contribution to the total power of the strongest cluster. If we refer to the 10 % level in Figures 14.7–14.8, we can see that in 90 % of all cases, the power in (in Figure 14.7) and 40 % (in Figure 14.8) the strongest cluster is at least 55 % of the total power. The same results for clusters separated by class are given in Figures 14.9 and 14.10.

These results are summarized in Table 14.5.

For the evaluation of delays we define the vector **P** containing the powers of the clusters and vector τ of corresponding mean delays. A particular cluster i has mean delay τ_i, and power P_i. The relation between the delays τ and the powers **P** is modeled as exponential

$$P_n \propto P(\tau_n) = ae^{-\tau_n/b} \tag{14.19}$$

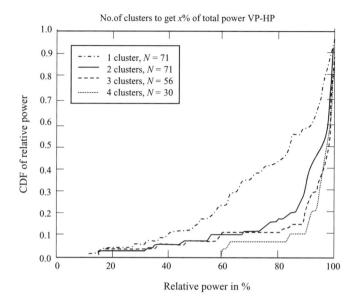

Figure 14.8 CDF of the relative power for one to four clusters, all TX positions, transmitter vertically and receiver horizontally polarized (VP-HP). N is the number of samples.

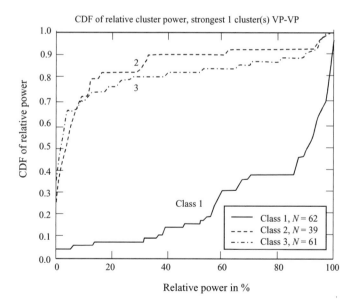

Figure 14.9 CDF of the relative power for the strongest cluster separated by class, for all TX positions. N is the number of samples.

Table 14.5   Average relative power in dB (relation of cluster power to total received power P_{tot}) of the strongest cluster of RX 1, RX 2, and RX 3, and the ratio in %

RX	VP-VP dB (% of P_{tot})	VP-HP dB (% of P_{tot})
1	−0.9 (81%)	−1.2 (76%)
2	−0.4 (91%)	−0.6 (87%)
3	−1.3 (74%)	−1.8 (66%)

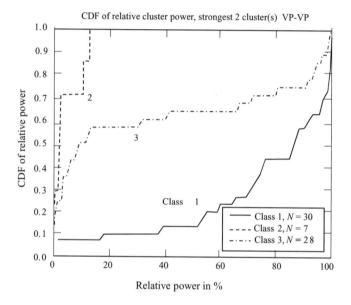

Figure 14.10 CDF of the relative power for the two strongest clusters separated by class, all TX positions. N is the number of samples.

Experimental data are fit into the model in Equation (14.19) by using the least squares (LS) estimation. The logarithmic estimation error $\mathbf{v}$ is defined as

$$\mathbf{v} = 10 \log \mathbf{P} - 10 \log s(\theta) \tag{14.20}$$

and its standard deviation σ_v as

$$\sigma_v = \sqrt{\text{var}\{\mathbf{v}\}} \tag{14.21}$$

Some results are shown in Figures 14.11 and 14.12. The summary of the results for parameters a and b is shown in Tables 14.6 and 14.7.

In Equation (14.21) σ_v was defined as the standard deviation of the logarithmic estimation error in dB. This estimation error was found to be log-normally distributed and up to a delay

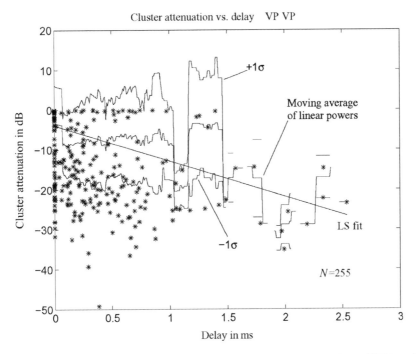

Figure 14.11 Relation of the cluster power to the total power versus delay, all TX positions and copolarization. N is the number of clusters [1183] © 2002, IEEE.

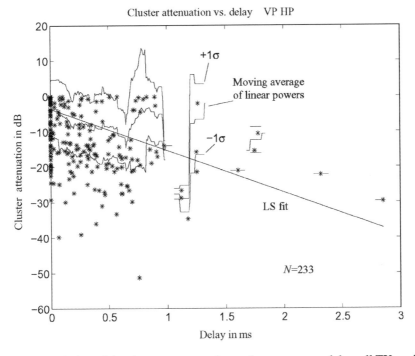

Figure 14.12 Relation of the cluster power to the total power versus delay, all TX positions and cross polarization. N is the number of clusters [1183].

Table 14.6   The model parameters a and b for both received polarizations (VP and HP) averaged over all available clusters. The transmitter was VP

	VP-VP	VP-HP
a	−3.9 dB	−3.6 dB
b	8.9 dB/μs	11.8 dB/μs

$P_i = ae^{-\tau_i/b}$

Table 14.7   Average delay of the strongest cluster of RX 1, RX 2 and RX 3

RX	Average delay VP-VP μs	Average delay VP-HP μs
1	0.068	0.071
2	0.38	0.28
3	0.11	0.048

Table 14.8   Averaged class powers of RX 1, RX 2 and RX 3

RX	Class	Horizontal power as % of total power	Vertical power as % of total power
1	1	96.5	95.7
	2	2.4	3.8
	3	1.1	0.4
2	1	93.5	97.2
	2	4.0	2.7
	3	2.5	0.1
3	1	46.7	78.0
	2	37.2	12.8
	3	16.0	9.2

of about 1 μs, σ_v is independent of the delay τ. The value of σ_v is 9.0 dB and 10.0 dB (co- and crosspolarization), respectively, averaged over the first microsecond.

The XPD is defined as the ratio of the received power of the copolarized component to the power of the crosspolarized component, evaluated for each cluster. Some results are shown in Figure 14.13.

The average powers for different classes of cluster are shown in Table 14.8.

Additional data on the topic can be found in [1183–1194].

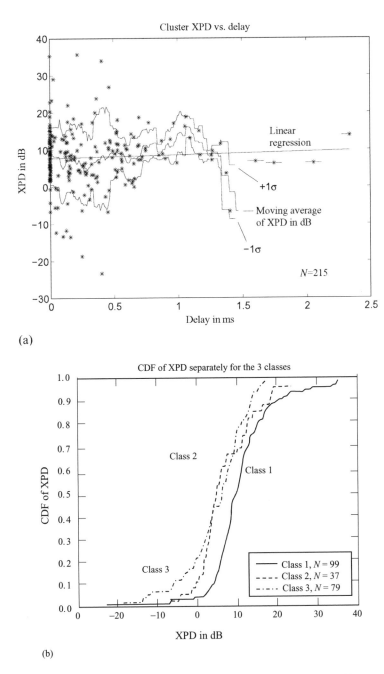

(a)

(b)

Figure 14.13 (a) XPD versus delay, all TX positions; (b) CDF of the XPD separately for the three classes. N is the number of clusters [1183] © 2002, IEEE.

14.3 MIMO CHANNELS IN MICROCELL AND PICOCELL ENVIRONMENTS (1.71/2.05 GHZ)

The model presented in this section is based upon data collected in both picocell and microcell environments [1195]. The stochastic model has also been used to investigate the capacity of MIMO radio channels, considering two different power allocation strategies, water filling and uniform, and two different antenna topologies, 4×4 and 2×4. It will be demonstrated that the space diversity used at both ends of the MIMO radio link is an efficient technique in picocell environments, achieving capacities within 14 b/s/Hz and 16 b/s/Hz in 80 % of the cases for a 4×4 antenna configuration implementing water filling at an SNR of 20 dB.

The basic parameters of the measurements setup are shown in Figure 14.14. The following notation is used subsequently: $d_{\text{MS–BS}}$ stands for distance between MS and BS, h_{BS} for the height of BS above the ground floor, and AS for azimuth spread [1195].

The vector of received signals at BS can be represented as $\mathbf{y}(t) = [y_1(t), y_2(t), \ldots y_M(t)]^T$, where $y_m(t)$ is the signal at the mth antenna port and $[\cdot]^T$ denotes transposition. Similarly, the signals at the MS are $\mathbf{s}(t) = [s_1(t), s_2(t), \ldots s_N(t)]^T$. The NB MIMO radio channel $\mathbf{H} \in X^{M \times N}$ which describes the connection between the MS

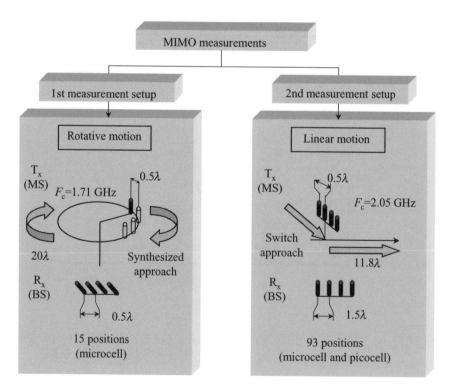

Figure 14.14 Functional sketch of the MIMO model [1195] © 2002, IEEE.

and the BS can be expressed as

$$
\mathbf{H} = \begin{bmatrix}
\alpha_{11} & \alpha_{12} & \cdots & \alpha_{1N} \\
\alpha_{21} & \alpha_{22} & \cdots & \alpha_{2N} \\
\vdots & \vdots & \ddots & \vdots \\
\alpha_{M1} & \alpha_{M2} & \cdots & \alpha_{MN}
\end{bmatrix}
\tag{14.22}
$$

where α_{mn} is the complex transmission coefficient from the antenna at the MS to the antenna at the BS. For simplicity, it is assumed that α_{mn} is complex Gaussian distributed with identical average power. However, this latest assumption can be easily relaxed. Thus, the relation between the vectors $\mathbf{y}(t)$ and $\mathbf{s}(t)$ can be expressed as

$$
\mathbf{y}(t) = \mathbf{H}(t)\,\mathbf{s}(t)
\tag{14.23}
$$

Subsequently, we will use the following correlations:

$$
\rho_{m_1 m_2}^{\mathrm{BS}} = \langle \alpha_{m_1 n}, \alpha_{m_2 n} \rangle
\tag{14.24a}
$$

$$
\rho_{n_1 n_2}^{\mathrm{MS}} = \langle \alpha_{m n_1}, \alpha_{m n_2} \rangle
\tag{14.24b}
$$

$$
\mathbf{R}_{\mathrm{BS}} = \begin{bmatrix}
\rho_{11}^{\mathrm{BS}} & \rho_{12}^{\mathrm{BS}} & \cdots & \rho_{1M}^{\mathrm{BS}} \\
\rho_{21}^{\mathrm{BS}} & \rho_{22}^{\mathrm{BS}} & \cdots & \rho_{2M}^{\mathrm{BS}} \\
\vdots & \vdots & \ddots & \vdots \\
\rho_{M1}^{\mathrm{BS}} & \rho_{M2}^{\mathrm{BS}} & \cdots & \rho_{MM}^{\mathrm{BS}}
\end{bmatrix}_{M \times M}
\tag{14.24c}
$$

$$
\mathbf{R}_{\mathrm{MS}} = \begin{bmatrix}
\rho_{11}^{\mathrm{MS}} & \rho_{12}^{\mathrm{MS}} & \cdots & \rho_{1N}^{\mathrm{MS}} \\
\rho_{21}^{\mathrm{MS}} & \rho_{22}^{\mathrm{MS}} & \cdots & \rho_{2N}^{\mathrm{MS}} \\
\vdots & \vdots & \ddots & \vdots \\
\rho_{N1}^{\mathrm{MS}} & \rho_{N2}^{\mathrm{MS}} & \cdots & \rho_{NN}^{\mathrm{MS}}
\end{bmatrix}_{N \times N}
\tag{14.24d}
$$

The correlation coefficient between two arbitrary transmission coefficients connecting two different sets of antennas is expressed as

$$
\rho_{n_2 m_2}^{n_1 m_1} = \langle \alpha_{m_1 n_1}, \alpha_{m_2 n_2} \rangle
\tag{14.25}
$$

which is equivalent to

$$
\rho_{n_2 m_2}^{n_1 m_1} = \rho_{n_1 n_2}^{\mathrm{MS}} \rho_{m_1 m_2}^{\mathrm{BS}}
\tag{14.26}
$$

provided that Equations (14.24a) and (14.24b) are independent of n and m, respectively. In other words, this means that the spatial correlation matrix of the MIMO radio channel is the Kronecker product of the spatial correlation matrix at the MS and the BS and is given by

$$
\mathbf{R}_{\mathrm{MIMO}} = \mathbf{R}_{\mathrm{MS}} \otimes \mathbf{R}_{\mathrm{BS}}
\tag{14.27}
$$

where $\otimes$ represents the Kronecker product. This has also been confirmed in [1196].

14.3.1 Simulation of channel coefficients

Correlated channel coefficients, α_{mn}, are generated from zero mean complex independent identically distributed (i.i.d.) random variables a_{mn} shaped by the desired Doppler spectrum such that

$$\mathbf{A} = \mathbf{Ca} \tag{14.28}$$

where $\mathbf{A}_{MN\times1} = [\alpha_{11}, \alpha_{21}, \ldots, \alpha_{M1}, \alpha_{12}, \ldots, \alpha_{MN}]^{\mathrm{T}}$ and $\mathbf{a}_{MN\times1} = [a_1, a_2, \ldots, a_{MN}]^{\mathrm{T}}$. The symmetrical mapping matrix $\mathbf{C}$ results from the standard Cholesky factorization of the matrix $\mathbf{R}_{\mathrm{MIMO}} = \mathbf{CC}^{\mathrm{T}}$, provided that $\mathbf{R}_{\mathrm{MIMO}}$ is non-singular [1197].

Subsequently, the generation of the simulated MIMO channel matrix $\tilde{\mathbf{H}}$ can be deduced from the vector $\mathbf{A}$. Note that the correlation matrices and the Doppler spectrum cannot be chosen independently, as they are connected through the PAS at the MS [1198].

14.3.2 Measurement setups

The Tx is at the MS and the stationary Rx is located at the BS. The two setups from Figure 14.14 provide measurement results with different correlation properties of the MIMO channel for small antenna spacings of the order of 0.5λ or 1.5λ . The BS consists of four parallel Rx channels. The sounding signal is an MSK-modulated linear shift register sequence of length 127 chips, clocked at a chip rate of 4.096 Mcps. At the Rx, the channel sounding is performed within a window of 14.6 μs, with a sampling resolution of 122 ns (1/2 chip period) to obtain an estimate of the complex impulse response (IR). The NB information is subsequently extracted by averaging the complex delayed signal components. A more thorough description of the stand-alone testbed (i.e. Rx and Tx) is documented in [1176, 1199]. The description of the measurement environments is summarized in Table 14.9.

A total of 107 paths are investigated within these seven environments. The first measurement setup is used to investigate 15 paths in a microcell environment, i.e. environment A in Table 14.9. The MS is positioned in different locations inside a building, while the BS is mounted on a crane and elevated above roof top level (i.e. 9 m) to provide direct LOS to the building. The antenna is located 300 m away from the building. The second setup is used to investigate 92 paths for both microcell and picocell environments, i.e. environments B and C to G, respectively, as shown in Table 14.9. The distance between the BS and the MS is 31 to 36 m for microcell B, with the BS located outside.

14.3.3 Validation of the stochastic MIMO channel model assumptions

The validity of the underlying assumptions has been verified for a 4×4 MIMO configuration. These assumptions are that: (1) the spatial correlation at the BS, Equation (14.24a), and the MS, Equation (14.24b), is independent of n and m, respectively; and (2) the spatial correlation matrix of the MIMO radio channel is the Kronecker product of the spatial correlation matrices at the BS and the MS, Equation (14.27).

To verfy assumption (1), the standard deviation (std) of each measured spatial correlation coefficient $|\rho_{m_1m_2}^{\mathrm{BS}}|$ and $|\rho_{n_1n_2}^{\mathrm{MS}}|$ is computed over the N and M reference antennas, respectively, for each environment. The std at the BS is expressed as

$$\mathrm{std}\_\rho_{m_1m_2}^{\mathrm{BS}} = \mathrm{std}\left(\{|\rho_{nm_2}^{nm_1}|\}\right), \quad \forall n \in [1\ldots4] \tag{14.29}$$

Table 14.9 Summary and description of the different environments [1195] © 2002, IEEE

Cell type	Environment	MS locations	Measurement setup	Description
Microcell	A	15	1st	The indoor environment consists of small offices with windows metallically shielded – 300 m between MS and BS
Microcell	B	13	2nd	The indoor environment consists of small offices – 31 to 36 m between MS and BS
Picocell	C	21	2nd	The indoor environment is the same as in A
Picocell	D	12	2nd	Reception hall – large open area
Picocell	E	18	2nd	Modern open office with windows metallically shielded
Picocell	F	16	2nd	The indoor environment is the same as in B
Picocell	G	12	2nd	Airport – very large indoor open area

and at the MS

$$\text{std\_}\rho_{n_1 n_2}^{\text{MS}} = \text{std}\left(\left\{\left|\rho_{n_1 m}^{n_2 m}\right|\right\}\right), \quad \forall\, m \in [1\ldots 4] \tag{14.30}$$

Figure 14.15 presents the empirical cumulative distribution function (cdf) of $\text{std\_}\rho_{m_1 m_2}^{\text{BS}}$ (a) and $\text{std\_}\rho_{n_1 n_2}^{\text{MS}}$ (b) computed over the 92 paths considered with the second measurement setup for the six different correlation coefficients, i.e. the upper triangular coefficient of the correlation matrix, when a 4×4 MIMO configuration is used. To validate the statistical significance of the empirical results, the empirical cdf is compared with a cdf obtained from simulations performed under similar conditions. The matching of the two cdfs demonstrates that assumption (1) is fulfilled, as explained hereafter. For each of the 92×6 different measured correlation coefficients, two correlated, Rayleigh-distributed signals of length 1000λ are generated. These 1000λ-long vectors are truncated into 11.8λ-long runs over which the correlation coefficient is computed once again. Hence, a wider new set of correlation values is collected, exhibiting a standard deviation $\text{std}_{11.8\lambda}$. This operation is repeated 92×6 times. A simulated cdf of $\text{std}_{11.8\lambda}$ is then obtained under similar conditions as for the measured cdf.

14.3.4 Input parameters to the validation of the MIMO model

The input parameters used in the validation stage are illustrated by the shaded areas of Figure 14.16. They are the average spatial complex correlation matrices $\mathbf{R}_{\text{BS}}$ and $\mathbf{R}_{\text{MS}}$, and the associated average Doppler spectrum.

The measured spatial complex correlation matrices are the results of an average over the reference antennas n and m with respect to which the matrices are computed.

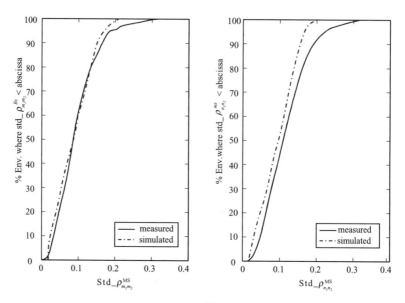

Figure 14.15 Example of the cdf of (a) $\text{std}\_\rho^{\text{BS}}_{m_1m_2}$ and (b) $\text{std}\_\rho^{\text{MS}}_{n_1n_2}$. The cdf is performed over all the measured environments and for all seven correlation coefficients.

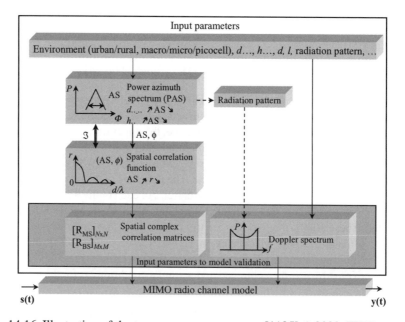

Figure 14.16 Illustration of the two measurement setups [1195] © 2002, IEEE.

The averaged measured Doppler spectrum is obtained by averaging over all the *MN* channel coefficients. It is defined at the MS, since the BS is fixed. This limitation is due to the measurement setup implementation, but is not inherent to the model. If both MS and BS were moving, the Doppler spectrum of the channels would have been defined as the convolution of separate Doppler spectra, defined either at the MS or at the BS, considering, respectively, the BS or the MS as fixed. The corresponding complex coefficients of the vector **a** in Equation (14.28) have their amplitudes shaped by the average measured Doppler spectrum and assigned a random phase uniformly distributed over $[0, 2\pi]$ such that MN independent and identically distributed variables are generated. Two examples of typical paths are presented by Equations (14.31) and (14.32) [1195] .

Example 1: Picocell decorrelated

$$
\mathbf{R}_{BS} =
\begin{bmatrix}
1 & -0.45 + 0.53i & 0.37 - 0.22i & 0.19 + 0.21i \\
-0.45 - 0.53i & 1 & -0.35 - 0.02i & 0.02 + 0.27i \\
0.37 + 0.22i & -0.35 + 0.02i & 1 & -0.10 + 0.54i \\
0.19 - 0.21i & 0.02 + 0.27i & -0.10 - 0.54i & 1
\end{bmatrix}
$$

$$
\mathbf{R}_{MS} =
\begin{bmatrix}
1 & -0.13 - 0.62i & -0.49 + 0.23i & 0.15 + 0.28i \\
-0.13 + 0.62i & 1 & -0.13 - 0.52i & -0.38 + 0.12i \\
-0.49 - 0.23i & -0.13 + 0.52i & 1 & 0.02 - 0.61i \\
0.15 - 0.28i & -0.38 - 0.12i & 0.02 + 0.61i & 1
\end{bmatrix}
\tag{14.31}
$$

Example 2: Microcell correlated

$$
\mathbf{R}_{BS} =
\begin{bmatrix}
1 & -0.61 + 0.77i & 0.14 - 0.94i & 0.24 + 0.89i \\
-0.61 - 0.77i & 1 & -0.85 + 0.50i & 0.57 - 0.78i \\
0.14 + 0.94i & -0.85 - 0.50i & 1 & -0.91 + 0.40i \\
0.24 - 0.89i & 0.57 + 0.78i & -0.91 - 0.40i & 1
\end{bmatrix}
$$

$$
\mathbf{R}_{MS} =
\begin{bmatrix}
1 & -0.12 - 0.18i & -0.08 + 0.05i & -0.02 - 0.13i \\
-0.12 + 0.18i & 1 & -0.17 - 0.16i & 0.11 + 0.04i \\
0.08 - 0.05i & -0.17 + 0.16i & 1 & -0.17 - 0.16i \\
-0.02 + 0.13i & 0.11 - 0.04i & -0.17 + 0.16i & 1
\end{bmatrix}
\tag{14.32}
$$

In both Examples 1 and 2, $|\mathbf{R}_{MS}|$ is decorrelated. This is expected, since the MS is surrounded by scatterers. On the other hand, $\mathbf{R}_{BS}$ presents two different behaviors. In Example 1, the spatial correlation coefficients remain low, as expected in the case of an indoor termination. On the other hand, the spatial correlation coefficients at the BS are highly correlated in Example 2, with a mean absolute value of the coefficient of 0.96. The high correlation is explained by the fact that regarding this specific example, the BS is identified to be located above any surrounding scatterer. Therefore, it experiences a low azimuth spread (AS), which causes its antenna array elements to be highly correlated. An illustration of the averaged measured Doppler spectrum of Example 1 is presented in Figure 14.17. The spectrum is

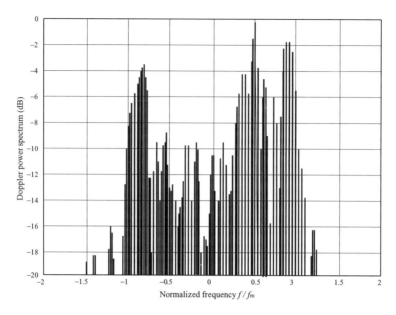

Figure 14.17 Averaged measured Doppler power spectrum for Example 1 (picocell decorrelated) [1195] © 2002, IEEE.

normalized in frequency to its maximum Doppler shift f_m and in power to its maximum value.

14.3.5 The eigenanalysis method

The eigenvalue decomposition (EVD) of the instantaneous correlation matrix $\mathbf{R} = \mathbf{HH}^H$ (not to be confused with $\mathbf{R}_{MIMO}$), where $[\cdot]^H$ represents Hermitian transposition, can serve as a benchmark of the validation process. The channel matrix $\mathbf{H}$ may offer K parallel subchannels with different mean gains, with $K = \text{Rank}(\mathbf{R}) \le \min(M, N)$, where the functions Rank$(\cdot)$ and min$(\cdot)$ return the rank of the matrix and the minimum value of the arguments, respectively, [1180]. The kth eigenvalue can be interpreted as the power gain of the kth subchannel [1180]. In the following, λ_k represents the eigenvalues. In order to assess the qualitative accuracy of the model, the comparison between measured and simulated eigenvalues is made for an antenna configuration where the largest number of eigenvalues is achievable within the limitation of the measurement setup antenna topology. This is the case for a 4×4 scenario, since at most four eigenvalues can be expected. In the following, the eigenvalues are normalized to the mean power of the single Tx and a single Rx channel coefficient $(1/MN)\sum_{m=1}^{M}\sum_{n=1}^{N}|\alpha_{mn}|^2$.

14.3.5.1 Validation procedure

For each of the 107 paths, the input parameters are fed into the proposed stochastic MIMO model and a Monte Carlo simulation consisting of 100 iterations is performed to generate the elements of the simulated matrix $\tilde{\mathbf{H}}$. $\tilde{\mathbf{H}}$ is a three-dimensional (3D) matrix ($M \times N \times L$), where L is the number of samples equivalent to the time domain definition in Equation (14.23). For each iteration, the seed of the random generator which defines the phase of the complex coefficient of the vector **a** is different.

At iteration q, $\tilde{\mathbf{H}}_{4\times4\times L}$ counts as many samples L as in the measured $\mathbf{H}_{4\times4\times L}$ collected during one antenna array run, that is to say 20λ or 11.8λ, depending on the measurement setup used. The EVD of $\tilde{\mathbf{H}}_{4\times4\times L}\tilde{\mathbf{H}}_{4\times4\times L}^{H}$ is then performed for each sample l in order to identify the corresponding simulated eigenvalues denoted by the vector,

$$\lambda_{\text{sim}\_kq\ L\times1} = [\lambda_{\text{sim}\_kq\ 1}, \ldots, \lambda_{\text{sim}\_kq\ l}]^{\text{T}}, \ k = 1, \ldots, K.$$

From these eigenvalues, k vectors $\lfloor\lambda_{\text{sim}\_k}\rfloor_{1\times QL}$ containing the 100 iterations of the simulated eigenvalues λ_{sim_kq} are deduced, so that $\lambda_{\text{sim}\_k} = \{\lambda_{\text{sim}\_kq}\}$, where $\{\cdot\}$ represents a set of variables. For the measured data, the eigenvalues were deduced so that $\lambda_{\text{meas}\_k\ L\times1} = [\lambda_{\text{meas}\_k\ 1}, \ldots, \lambda_{\text{meas}\_k\ l}]^{\text{T}}$. Some results are presented in Figure 14.18.

14.3.5.2 Definition of the power allocation schemes

In the situation where the channel is known at both Tx and Rx, and is used to compute the optimum weight, the power gain in the kth subchannel is given by the kth eigenvalue, i.e. the signal to noise ratio (SNR) for the kth subchannel equals

$$\gamma_k = \lambda_k \frac{P_k}{\sigma_N^2} \tag{14.33}$$

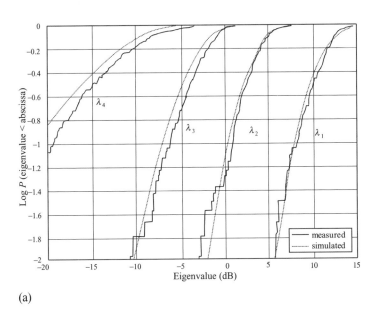

(a)

Figure 14.18 (a) Local validation. Cdf of λ_{meas_k} and λ_{sim_k} from Example 1 (picocell decorrelated); (b) local validation. Cdf of λ_{meas_k} and λ_{sim_k} for each of the 107 paths from Example 2 (microcell correlated); (c) global validation. Cdf of $|\Delta_{\text{error}\_k}|$ over the 107 paths.

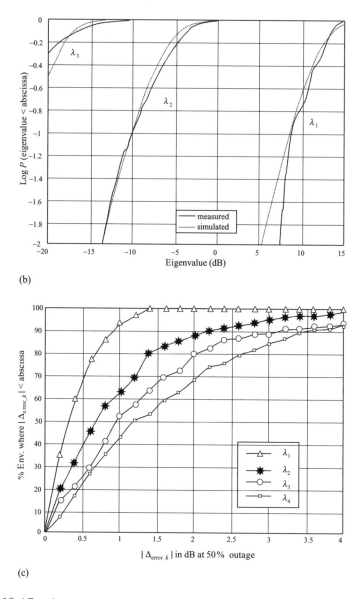

(b)

(c)

Figure 14.18 (*Cont.*).

where P_k is the power assigned to the kth subchannel, λ_k is the kth eigenvalue and σ_N^2 is the noise power. For simplicity, it is assumed that $\sigma_N^2 = 1$. According to Shannon, the maximum capacity normalized with respect to the bandwidth (given in terms of b/s/Hz spectral efficiency) of parallel subchannels equals [1200]

$$C = \sum_{k=1}^{K} \log_2(1 + \gamma_k) \tag{14.34}$$

$$= \sum_{k=1}^{K} \log_2\left(1 + \lambda_k \frac{P_k}{\sigma_N^2}\right) \tag{14.35}$$

where the mean SNR is defined as

$$\text{SNR} = \frac{E[P_{\text{Rx}}]}{\sigma_N^2} = \frac{E[P_{\text{Tx}}]}{\sigma_N^2} \tag{14.36}$$

Given the set of eigenvalues $\{\lambda_k\}$, the power P_k allocated to each subchannel k is determined to maximize the capacity by using Gallagher's water filling theorem [1180] such that each subchannel is filled up to a common level D, i.e.

$$\frac{1}{\lambda_1} + P_1 = \cdots = \frac{1}{\lambda_K} + P_k = \cdots D \tag{14.37}$$

with a constraint on the total Tx power such that

$$\sum_{k=1}^{K} P_k = P_{\text{Tx}} \tag{14.38}$$

where P_{Tx} is the total transmitted power. This means that the subchannel with the highest gain is allocated the largest amount of power. In the case where $1/\lambda_k > D$, then $P_k = 0$.

When the uniform power allocation scheme is employed, the power P_k is adjusted according to

$$P_1 = \cdots = P_K \tag{14.39}$$

Thus, in the situation where the channel is unknown, the uniform distribution of the power is applicable over the antennas [1180] so that the power should be equally distributed between the N elements of the array at the Tx, i.e.

$$P_n = \frac{P_{\text{Tx}}}{N}, \quad \forall n = 1 \ldots N \tag{14.40}$$

Some results are given in Figure 14.19. In Figure 14.19(a), C_k is the capacity of the kth subchannel of Figure 14.18(a).

14.4 OUTDOOR MOBILE CHANNEL (5.3 GHZ)

In this section we discuss the mobile channel at 5.3 GHz. The discussion is based on measurement results collected at six different sites [1201]. *Site A* is an example of a dense urban environment, the transmitting antenna is about 45 m above ground level, representing a case with the BS antenna over the rooftops. *Site B* is a dense urban residential environment. Here, the transmitting antenna is placed on a mast with a height of 4 m, which is a typical case with the BS antenna lower than the rooftops. The measurement routes for this site are shown in Figure 14.20(a). The receiving antenna mobile station is at a height of 2.5 m on top of a car for both of the sites mentioned above. *Site C* is located in a typical city center. The goal was to place the transmitter at some elevation relative to the ground, but still keep it below the rooftops. The transmitting antenna is placed at a height of 12 m, and the receiving antenna is on top of a trolly with a height of 2 m above ground level. In Site C, the rotation measurements are taken using a directive horn antenna. The 3 dB beamwidth of the horn antenna is 30° in the H-plane and 37° in the E-plane, and the peak sidelobe level was 26 dB. The specific environment for rotation measurements is shown in Figure 14.20(b). *Site D* represents a semiurban/semirural residential area. The three-story buildings are the tallest ones around, and the transmitting antenna is placed over the rooftops at a height of 12 m

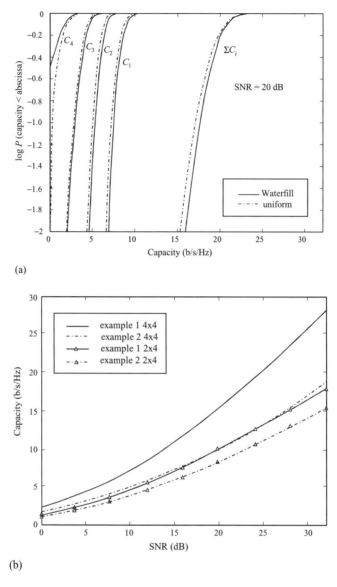

Figure 14.19 (a) Cdf of the capacity per subchannel C_k and its total results $\sum C_i$ for Example 1 (picocell decorrelated). A 4×4 antenna topology is presented here [1195]; (b) capacity (10 % level) versus SNR for Example 1 (picocell decorrelated) and Example 2 (microcell correlated). The water filling power allocation scheme; (c) cdf over all the 79 picocell paths of the total capacity deduced from (b) at SNR = 20 dB for 4×4 and 2×4 antenna configurations [1195]; (d) cdf over all the 28 microcell paths of the total capacity deduced from (b) at SNR = 20 dB for 4×4 and 2×4 antenna configurations [1195] © 2002, IEEE.

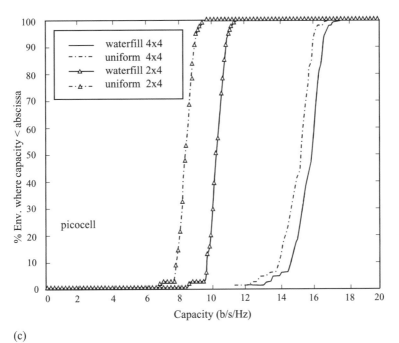

(c)

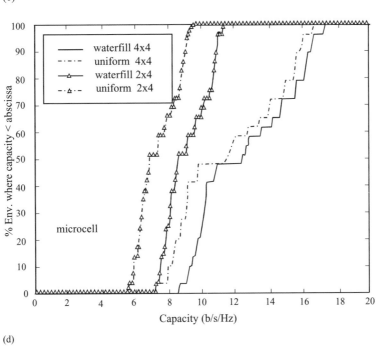

(d)

Figure 14.19 (*Cont.*).

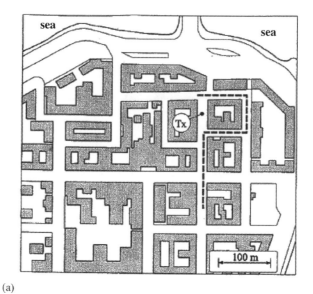

Figure 14.20 (a) Measurement routes for Site B with Tx height of 4m; (b) Rotation measurements in an urban environment, Site C [1201] © 2002, IEEE.

from ground level. *Site E* was selected to represent the rural case. The transmitting antenna is placed on top of a 5 m mast at the hilltop so that the antenna is about 55 m above the surroundings. *Site F* represents a typical semiurban/urban case. The transmitter is placed on top of a 5 m mast. The receiving antenna is always on top of a car at a height of 2.5 m. The routes are measured using the wideband channel sounder.

System parameters are summarized in Table 14.10.

Table 14.10   System configuration for mobile measurements in urban (U),
suburban (S) and rural (R) areas

Receiver	Direct sampling / 5.3 GHz
Transmitter power	30 dBm
Chip frequency	30 MHz
Delay range	4.233 µs
Doppler range	124 Hz (U), 62 Hz (S, R)
Measurement rate	248 sets/s (U), 124 sets/s (S, R)
Sampling frequency	120 Ms/s
IRs / wavelength	5 (U), 4.2 (S, R)
Receiver velocity	2.80 m/s (U), 1.67 m/s (S, R)
Antennas and polarization	Omnidirectional antenna with 1 dBi gain; vertical polarization

Table 14.11   Path loss models for urban environments

Urban models	Tx height: 4 m			Tx height: 12 m			Tx height: 45 m		
	n	b (dB)	std (dB)	n	b (dB)	std (dB)	n	b (dB)	std (dB)
LOS	1.4	58.6	3.7	2.5	35.8	2.9	3.5	16.7	4.6
NLOS	2.8	50.6	4.4	4.5	20.0	1.7	5.8	−16.9	2.8

Table 14.12   Path loss models for suburban and rural environments

Models	Rural Tx height: 55 m			Suburban LOS: (Tx height: 5 m), NLOS: (Tx height 12 m)		
	n	b (dB)	std (dB)	n	b (dB)	std (dB)
LOS	3.3	21.8	3.7	2.5	38.0	4.9
NLOS	5.9	−27. 8	1.9	3.4	25.6	2.8

14.4.1 Path loss models

Similarly to Equation (14.11), we have

$$\text{PL (dB)} = b + 10n \log_{10} d \tag{14.11a}$$

where $d_0 = 1$ m, n is the attenuation exponent, b is the intercept point in the semilog coordinate, and d_m is the distance from the receiver to the transmitter. The measurement distances are about 30–300 m in this case. Some results are given in Tables 14.11 and 14.12 and Figure 14.21.

The measurement results for delays are summarized in Table 14.13.

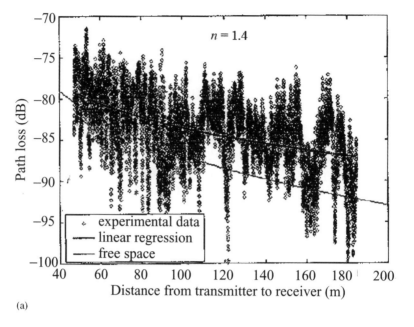

(a)

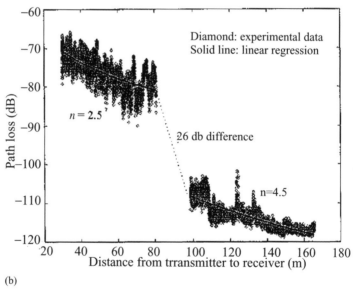

(b)

Figure 14.21 (a) Path loss for urban LOS with Tx height of 4 m; (b) mobile terminal turning around a corner with Tx height of 12 m in an urban environment [1201] © 2002, IEEE.

14.4.2 Window length for averaging fast fading components at 5 GHz

Multipath propagation causes fast fading in mobile communications. Thus, an important consideration in experimental data processing is how to average out the fast fading components and still preserve the slow fading characteristics. In [1202], it was suggested that a

Table 14.13 Measured values for mean excess delay and rms delay spread

( ) Tx height in meters			Urban		Suburban		Rural	
Mean excess delay (ns)		LOS	38	(4)				
			42	(12)	36	(5)	29	(55)
			102	(45)				
		NLOS	70	(4)	68	(12)		
Rms delay spread (ns)	Mean	LOS	44	(4)				
			43	(12)˙	25	(5)	22	(55)
			88	(45)				
		NLOS	44	(4)	66	(12)		
	Median	LOS	25	(4)				
			31	(12)	13	(5)	15	(55)
			86	(45)				
		NLOS	37	(4)	63	(12)		
	CDF <90%	LOS	93	(4)				
			64	(12)	57	(5)	44	(55)
			120	(45)				
		NLOS	63	(4)	105	(12)		

suitable window length for data taken from macrocells is 4λ. However, examination of data taken from microcells showed that the local mean could suffer quite large variations over short distances, and in [50], 5λ (about 1.7 m) was considered a more appropriate window length for microcells from the experimental data at 900 MHz. In this experiment, the least squares method with wide and narrowband received power is used to give the linear regression curves. Let us take the regression curves as the reference values, and then change the window length to 5, 10, 20, and 40λ for averaging the fading signals. Figure 14.22(a) shows the wideband received power for urban LOS with a transmitter height of 12 m. If we now take the linear regression values as the average received power, the standard deviations (std) are 2.47, 2.25, 1.93, and 1.62 dB, corresponding to the window lengths of 5, 10, 20 and 40λ, respectively. It is seen that the fast fading components are averaged out if the window length is in the range from 20λ to 40λ, namely, 1–2 m. The same conclusion can also be obtained for averaging narrowband fast fading components, and the corresponding result can be found in Figure 14.22(b). So, based on [1203] and the experience of processing much measured data at 5 GHz, it seems that the practical window length for averaging out fast fading components is 1–2 m in micro- and picocells at 900 MHz–5 GHz frequency bands.

14.4.3 Spatial and frequency correlations

Spatial and frequency correlation study is useful for the design of antenna diversity to reduce the multipath fading. Because the correlation behavior is a small-scale effect, a wide-sense stationary uncorrelated scattering (WSSUS) situation should be assumed. To meet this condition, about 40λ is used as the window length to give the average correlation function. The formulas for calculating spatial and frequency correlation functions can be

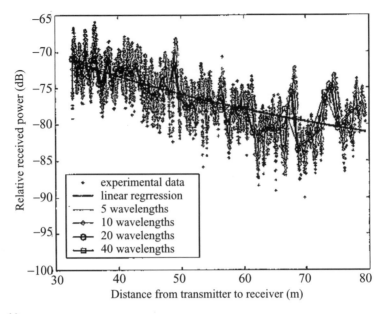

(a)

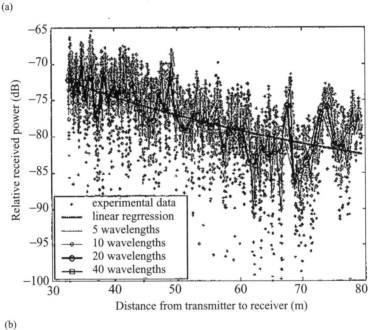

(b)

Figure 14.22 Window length for averaging out fast fading components at 5.3 GHz. (a) Wideband, (b) narrowband [1201] © 2002, IEEE.

found in [1204, 1205]. In this section, envelope correlation is considered for narrowband signals. However, recent research [1206] has shown that spatial correlation characteristics do not largely depend on frequency bandwidth up to approximately 20 % of the carrier frequency ($B/f_c = 0.2$, where B is the bandwidth of a transmitted signal and f_c is the carrier frequency). Therefore, the narrowband model is sufficient for computing the spatial correlation characteristics within $B \leq 0.2 f_c$.

Figure 14.23 shows the spatial and frequency envelope correlation functions for LOS outdoor environments at different transmitter heights. It is seen that the correlation distances are strongly dependent on the transmitter heights. The correlation distances with the envelope correlation coefficient of 0.7 are between 1 and 11λ (about 0.06–0.62 m). The respective correlation bandwidths are between 1.2 and 11.5 MHz. In LOS cases, due to the direct wave superimposed by only weak scattered waves, the coherence is high and the correlation length is large.

14.4.4 Path number distribution

The multipath number distribution was regarded as Poisson's and modified Poisson's in [1207], and modified Poisson's distribution has been shown to have good agreement with the experimental results in some cases. However, the modified Poisson's distribution does not have an explicit expression, just a process. Therefore, it is not convenient for practical use. In [1201], another simple and useful path number distribution was suggested, by considering the path number variation of radio waves in land mobile communications a Markov process at finite state space, and it was shown to have good agreement with the experimental results. The path number distributions given by Poisson and Gao can be expressed as

$$P(N) = \frac{\eta^{N_T - N}}{(N_T - N)!} e^{-\eta} \tag{14.41}$$

$$P(N) = C_{N_T}^N \frac{\eta^{N_T - N}}{(1 + \eta)^{N_T}} \tag{14.42}$$

where N is variable and C means combination. N_T is the maximum number of paths that the mobile can receive. The parameters η and N_T can be fitted by the experimental data. For Poisson's probability density function (PDF), the mean path number is $\langle N \rangle = \eta$. For Gao's PDF, the mean value is $\langle N \rangle = N_T/(1 + \eta)$.

The empirical path number distributions for the outdoor measurements are fitted by using Equations (14.41) and (14.42), respectively. The path numbers are obtained from measured data by counting the peaks of the power delay profiles. The best fit is obtained by minimizing the following standard deviation

$$\text{std} = \sqrt{\frac{1}{N_T} \sum_{i=1}^{N_T} \left(p^i - p_e^i \right)^2} \tag{14.43}$$

where p_e^i is the experimental probability corresponding to path number i, and p^i is the fitted probability using Equations (14.41) and (14.42). The fitted parameters are available in Table 14.14. If the dynamic range is cut at different levels, for example, $-25, -20$ and -15 dB, the fitting parameters in Equations (14.41) and (14.42) will be changed, but the path number distributions still follow Poisson's and Gao's distributions.

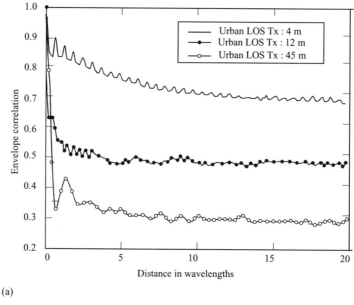

(a)

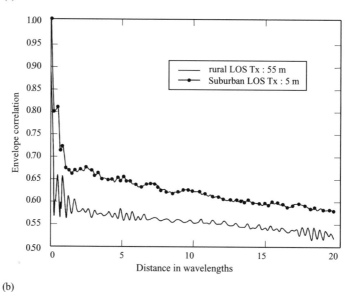

(b)

Figure 14.23 (a) Spatial correlations in LOS outdoor environments for urban cases with three transmitter heights. (b) Spatial correlations in LOS outdoor environments for rural and suburban cases with two transmitter heights. (c) Frequency correlations in LOS outdoor environments for urban cases with three transmitter heights. (d) Frequency correlations in LOS outdoor environments for rural and suburban cases with two transmitter heights.

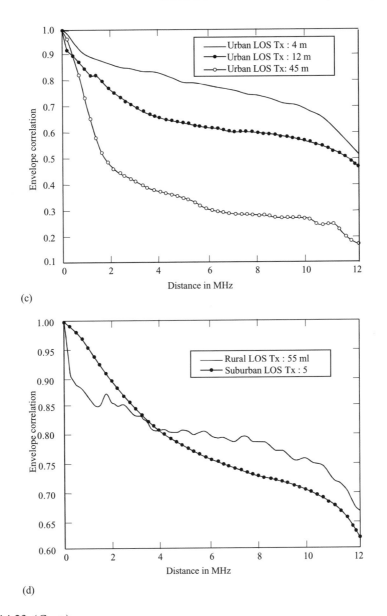

Figure 14.23 (*Cont.*).

14.4.5 Rotation measurements in an urban environment

The rotation measurements at points P_1 and P_2 were performed at Site C and are shown in Figure 14.20(b). The transmitter height was 12 m and the receiver was on a rotating stand at a height of 1.6 m, close to the receiver is a large open square. In the experiments, large excess delays up to 1.2 µs and rms delay spread of about 0.42 µs are found (Figure 14.24).

Table 14.14   Path number distributions for outdoor environments

| | | Urban | | | | Suburban | | Rural |
| | | Tx 4 m | | Tx 12 m | Tx 45 m | Tx 12 m | | Tx 55 m |
		LOS	NLOS	LOS	LOS	LOS	NLOS	LOS
η	Poisson	2.8	4.2	3.3	6.0	1.2	4.5	1.8
	Gao	4.7	4.5	3.5	2.7	9.0	3.3	4.0
	N_T	16	21	14	22	13	20	9
$\langle N \rangle$	Poisson	2.8	4.2	3.3	6.0	1.2	4.5	1.8
	Gao	2.8	3.8	3.2	6.0	1.3	4.7	1.8
	Experiment	3.4	4.2	3.5	6.2	2.4	5.0	1.7

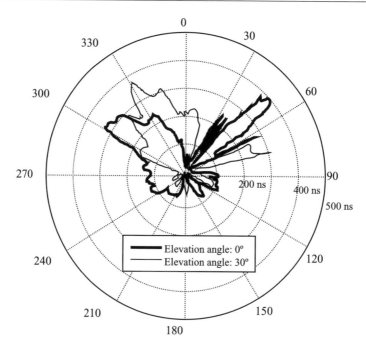

Figure 14.24 RMS delay spread with different rotation angles in the azimuth plane.

The power angular profiles (PAPs) $P_r(\phi)$ of the measurements were calculated by using the maximal ratio combining algorithm in the delay domain [1205]

$$P_r(\phi) = \alpha_{\text{cal}} \int_{\tau_{\text{min}}}^{\tau_{\text{max}}} |h(\tau, \phi)|^2 \, d\tau \tag{14.44}$$

where α_{cal} is a factor which is obtained from the calibration measurement with a cable and an attenuator, τ_{min} and τ_{max} are the delays of the first and last detectable IR components, and ϕ is the angle of arrival of the waves in the azimuth plane. In the rotation measurements, the dynamic range is cut at -26 dB relative to the strongest path. Some results for angular profiles of relative received power are shown in Figure 14.25.

The number of paths as a function of azimuth angle is shown in Figure 14.26.

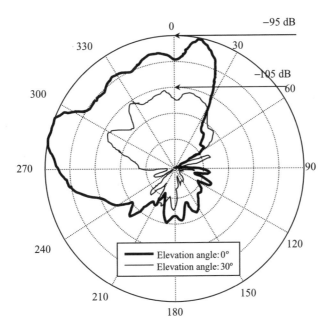

Figure 14.25 Angular profiles of relative received power [1201] © 2002, IEEE.

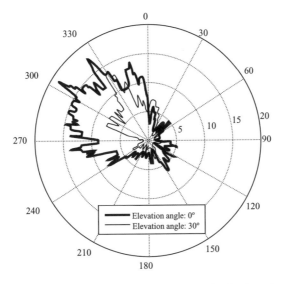

Figure 14.26 Numbers of paths with different rotation angles in the azimuth plane [1201].

14.5 MICROCELL CHANNEL (8.45 GHZ)

In this section spatio-temporal channel characterization in a suburban non line-of-sight microcellular environment is discussed. Figure 14.27 shows a map of the environment under consideration [1208]. This is a residential area with predominantly wooden houses of 8 m average height and is considered to be a typical suburban microcellular environment

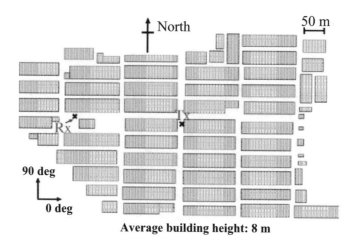

Figure 14.27 The microcellular environment [1208] © 2002, IEEE.

Table 14.15   Electrical parameters used in the ray-tracing simulation

	ε_r	$\sigma\,[\text{s/m}]$
Concrete [1158]	5.5	0.023
Foliage [1169]	1.2	0.0003
Ground [1171]	15.0	1.3
Metal	—	∞

of size 600×600 m^2. The traffic was very light and the environment was considered to be static throughout the experiments. A non line-of-sight (NLOS) transmitter and receiver shown in Figure 14.27 are considered. The distance between the transmitter and the receiver is 219 m. As a transmitter antenna, simulating the mobile station (MS), a vertically polarized omnidirectional half wave sleeve dipole is set at a height of 2.7 m. At the receiving point, corresponding to the base station (BS), the azimuth delay profile is measured. This is done by rotating a vertically polarized parabolic antenna which has azimuth and elevation beamwidths of about $4°$, at an azimuth step of $4°$ and is set at a height of 4.4 m. At each azimuth step, the delay profile is measured by using a delay profile measurement system described in [1208, 1209]. At a center frequency of 8.45 GHz and with a chip rate of 50 Mcps, a seven stage M sequence with a dynamic range of 42 dB is transmitted. A correlation receiver is used to produce the delay profile.

The electric parameters used in the ray-tracing simulation are presented in Table 14.15. The wooden houses are modeled by concrete, as the surfaces of the houses are covered by concrete-like paint. To compare with the experimental results, the directivity of the antenna and the autocorrelation of the pseudorandom noise (PN) sequence were convolved with the result of the ray-tracing simulation. Therefore, the path gain includes the antenna gain.

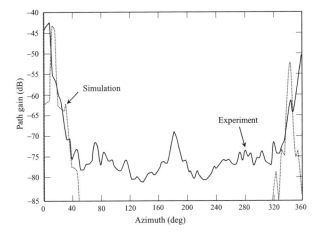

Figure 14.28 Azimuth profile. Solid line: experiment; dotted line: simulation [1208] © 2002, IEEE.

14.5.1 Azimuth profile

Azimuth profiles are obtained by summing up the power of azimuth delay profiles with respect to the delay time. Figure 14.28 shows the azimuth profiles obtained from the experiment as the solid line, and from the simulation as the dotted line. From Figure 14.28, the forward arrival waves within the range from $-40°$ to $44°$ are in agreement for both data sets with respect to the level and the shape of the profile. The experimental profile has the floor level of about 30 dB below the peak, but this level is low enough so that its effect on the transmission property is negligibly small.

14.5.2 Delay profile for the forward arrival waves

For the forward arrival waves ($-40°$ to $-44°$), the delay profiles are obtained by summing up the azimuth delay profiles with respect to the azimuth. The experimental result is shown in Figure 14.29(a), and the simulation result is shown in Figure 14.29(b). The experimental result exhibits an exponential decay. The results of least squares fitting are shown as the dashed line in Figure 14.29(a). The function is expressed as

$$P(\tau) = -0.038\tau - 28.6 \qquad (14.45)$$

where P is the path gain in dB, and τ is the delay time in ns.

The ray-tracing simulation can predict accurately the first two peaks in the delay profile. However, the exponential decay of the profile cannot be accurately predicted. The problem here seems to be due to incomplete modeling of the effect of random scattering in the ray-tracing simulation. If, however, the gradient of this exponential function can be determined by some independent means, the ray-tracing results can be extrapolated to predict accurately the delay profile.

The cumulative distribution function (CDF) of the fluctuation component of the experimental data from its exponential fit is shown in Figure 14.30(a). This fluctuation component can be very accurately approximated by a log-normal distribution with a standard deviation of 5.3 dB. In the ray-tracing simulation, waves with a long delay time cannot be predicted. Instead, their statistical properties are used for the extrapolation.

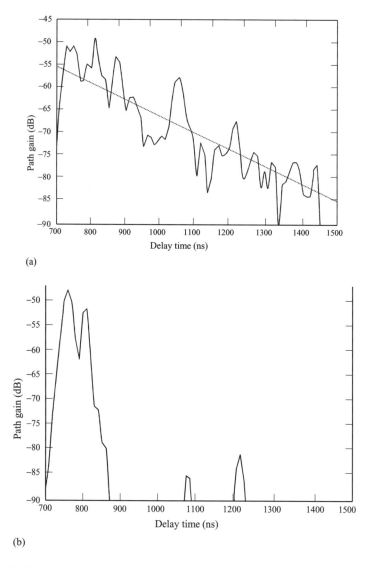

Figure 14.29 Delay profile of forward arrival waves: (a) experiment; (b) simulation.

The autocorrelation of this fluctuation component is presented in Figure 14.30(b). The correlation decreases monotonously, with a correlation distance and a time at a correlation coefficient of 0.5 of about 3 m and 10 ns, respectively. Considering the resolution of the delay profile, which is equal to the chip duration 20 ns, the fluctuation is modeled as uncorrelated, just as in the wide-sense stationary uncorrelated scattering (WSSUS) model.

14.5.3 Short-term azimuth spread (AS) for forward arrival waves

Figure 14.28 showed the azimuth profile averaged over the delay time. Here, we focus on the short-term azimuth profiles, which are obtained every 10 ns. The short-term AS for the

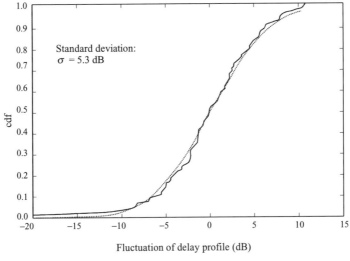

(a)

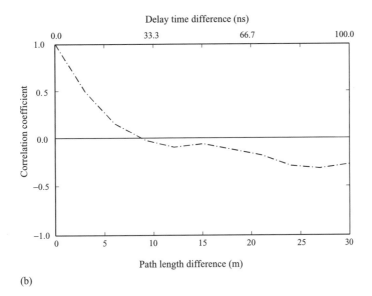

(b)

Figure 14.30 (a) CDF of the fluctuation component of the experimental delay profile from its exponential fit for forward arrival waves: experiment; (b) autocorrelation of the fluctuation component of the experimental delay profile from its exponential fit for forward arrival waves: experiment.

forward arrival waves, σ_φ, is defined as

$$\sigma_\varphi(\tau) = \sqrt{\langle \varphi^2(\tau) \rangle - \langle \varphi(\tau) \rangle^2} \qquad (14.46)$$

where $\langle \cdot \rangle$ is the average of the φ weighted by the power for a fixed delay time, τ. At each delay time, the threshold level is set to be 30 dB below the peak in the profile in order to calculate the short-term AS in Equation (14.46).

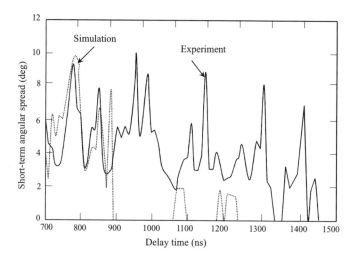

Figure 14.31 Short-term AS for forward arrival waves.

Figure 14.31 shows the resultant short-term angular spread. The experimental result and the result of the simulation agree well within the range 740–860 ns in which the ray-tracing simulation is known to predict the delay profile accurately (as shown in Figure 14.29(b)). The experimental results are observed to exhibit the same properties for a larger delay time. Therefore, this behavior can be used for an extrapolation of the angular profile beyond the range in which the simulation can predict the delay profile.

Figure 14.31 indicates that the variation of the short-term AS can be modeled as a stationary process. To characterize this stationary process, the CDF of the short-term AS is shown in Figure 14.32. The solid line indicates the experimental result and the dotted line indicates the simulation result. It is noted that the simulation result is obtained within a range of delay times from 700 ns to 880 ns. Although the distribution functions of the experiment and the simulation look slightly different, their average and their standard deviation are both in agreement. A Gaussian distribution has been used as an approximation in the figure, although the most appropriate distribution function to use is still under consideration.

Figure 14.33 shows the autocorrelation function of the short-term AS. The solid line indicates the experimental result and the dotted line indicates the simulation result. In a similar way to Figure 14.30(b), the correlation decreases monotonously. The correlation distance at a correlation coefficient of 0.5 is about 2 m for the experimental data and about 5 m for the simulation. The correlation distance is smaller using the experimental data, since random scattering is not taken into account in the simulation. As in Figure 14.30(b), considering a chip duration of 20 ns, which corresponds to a distance of 6 m, the short-term angular spread is also modeled as uncorrelated. It is noted that the correlation distances for Figures 14.30(b) and 14.33 are comparable and that, therefore, these two fluctuations seem to be related to each other.

Since the fluctuation component from the exponential function of the delay profile and the short-term azimuth profile for the forward arrival waves have similar correlation lengths, this may suggest some relationship between them.

A Nakagami–Rice fading model for the short-term azimuth profile, as shown in Figure 14.34 is used. This is composed of a stable strong signal component plus a

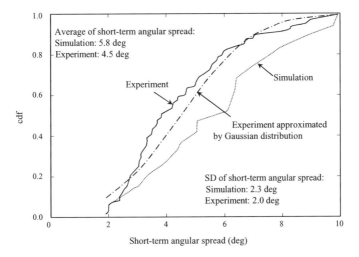

Figure 14.32 CDF of the short-term AS for forward arrival waves. Solid line: experiment; dotted line: simulation [1208] © 2002, IEEE.

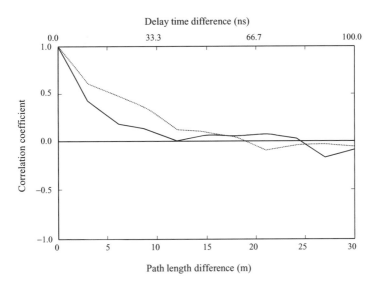

Figure 14.33 Autocorrelation of the short-term AS for forward arrival waves. Solid line: experiment; dotted line: simulation.

weak scattered signal component. The scattered signal component is assumed to be stationary. Under these assumptions, the power increases and the AS decreases when the stable signal component is large, i.e. there is a negative correlation between these parameters.

To evaluate this model, the crosscorrelation is calculated between the fluctuation component of the experimental delay profile from its exponential fit and the short-term azimuth

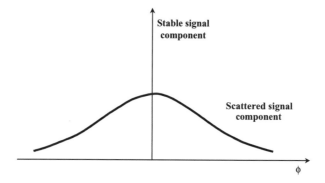

Figure 14.34 Nakagami–Rice fading model for the short-term azimuth profile.

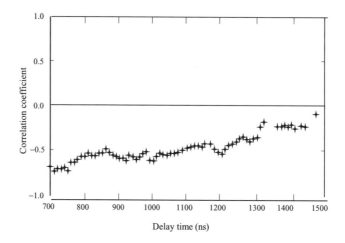

Figure 14.35 Crosscorrelation between fluctuation component of the experimental delay profile from its exponential fit and the short-term azimuth profile for forward arrival waves: experiment [1208].

profile for the forward arrival waves, which is defined as

$$\rho_{P\Delta\varphi} = \frac{\langle(P(\tau) - \langle P(\tau)\rangle)(\Delta\varphi(\tau) - \langle\Delta\varphi(\tau)\rangle)\rangle}{\sqrt{\langle(P(\tau) - \langle P(\tau)\rangle)^2\rangle\langle(\Delta\varphi(\tau) - \langle\Delta\varphi(\tau)\rangle)^2\rangle}} \tag{14.47}$$

where $\langle\cdot\rangle$ is the sample average within the given time window, $P(\tau)$ is the fluctuation component from the exponential function at a delay time, τ, in dB, and $\Delta\varphi(\tau)$ is the short-term angular spread at a delay time, τ, in degrees. Figure 14.35 shows the crosscorrelation with a time window of 200 ns for the experimental data. It is clear from Figure 14.35 that the fluctuation component of the delay profile and the short-term AS are negatively correlated. This result suggests that the proposed Nakagami–Rice fading model can be used successfully to model the short-term azimuth profile.

14.6 WIRELESS MIMO LAN ENVIRONMENTS (5.2 GHZ)

The presentation in this section is based on the results of a measurement campaign [1210] in two courtyards in the 5.2 GHz band assigned for wireless LANs (e.g. HYPERLAN (see www.etsi.org), or IEEE 802.11a). These standards specify wireless communication between computers, which is a compelling application for MIMO systems. For measurement, a channel sounder with a bandwidth of 120 MHz connected via a fast RF switch to a uniform linear receiver antenna array is used. This array consists of $N_R = 8$ antenna elements ($\pm 60^0$ element beam width), plus two dummy elements at each end of the array. All these components together constitute a single-directional channel sounder. A virtual array at the transmitter consists of a monopole antenna mounted on an X–Y-positioning device with stepping motors.

The experiment starts by positioning the transmit antenna at a certain position. At the receiver, the RF switch is connected to the first antenna element of the array, so that the transfer function (measured at 192 frequency samples) from the first transmit to the first receive element of the array is measured. Then, the switch is connected to the next receive antenna element, and the next transfer function is measured. The measurement of all the transfer functions is repeated 256 times, in order to assess the time variance of the channel (see below). Then, the transmit antenna is moved to the next position, and the procedure is repeated. $N_T = 16$ transmit antenna positions are situated on a cross (i.e. 8 positions on each axis of the cross) and bursts of complex channel transfer functions are recorded. Any virtual array requires that the channel remains static during the measurement period. One complete measurement run (2×8 antenna positions at TX times eight spatial samples at RX times 192 frequency samples and 256 temporal samples gives $16 \times 8 \times 192 \times 256 = 6\,291\,456$ complex samples) takes about five minutes.

14.6.1 Data evaluation

Starting from the four-dimensional transfer function (time, frequency, position of RX antenna, position of TX antenna), the Doppler-variant transfer function is computed first, by Fourier transforming the 256 temporal samples (with Hanning windowing). Next, all components that do not exhibit zero Doppler shift (Doppler filtering) are elliminated. Those components correspond, for example, to MPCs scattered by leaves moving in the wind. The eliminated components carry on the order of 1 % of the total energy. The three-dimensional (static) transfer function obtained in that way is then evaluated by Unitary ESPRIT [1186] to estimate the delays τ_i. Unitary ESPRIT is an improved version of the classical ESPRIT algorithm discussed in Chapter 13. They both estimate the signal subspace for extraction of the parameters of (spatial or frequency) harmonics in additive noise. One important step in ESPRIT is the estimation of the model order. Different methods have been proposed in the literature for that task. The relative power decrease between neighboring eigenvalues with additional correction by visual inspection of the *scree graph* showing the eigenvalues is an option used for generating the results presented subsequently.

After estimation of the parameters τ_i, we can determine the corresponding 'steering' matrix $\mathbf{A}_\tau$. Subsequent beamforming with its Moore–Penrose pseudoinverse [1186, 1211–1214] $\mathbf{A}_\tau^+$ gives the vector of delay weights for all $\mathbf{x}_R$, $\mathbf{x}_T$

$$\mathbf{h}_\tau(\mathbf{x}_T, \mathbf{x}_R) = \mathbf{A}_\tau^+ \mathbf{T}_f(\mathbf{x}_T, \mathbf{x}_R) \tag{14.48}$$

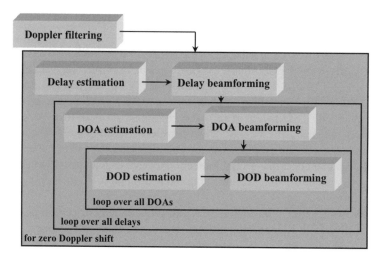

Figure 14.36 Sequential estimation of the parametric channel response in the different domains: alternating estimation and beamforming [1210] © 2002, IEEE.

where $\mathbf{T}_f$ is the vector of transfer coefficients at the 192 frequency subbands sounded. This gives us now the transfer coefficients from all positions $\mathbf{x}_T$ to all positions $\mathbf{x}_R$ separately for each delay τ_i. Thus, one dimension, namely the frequency, has been replaced by the parameterized version of its dual, the delays.

For the estimation of the directions of arrival (DOA) in each of the two-dimensional transfer functions, ESPRIT estimation and beamforming by the pseudoinverse are used

$$\mathbf{h}_{\varphi R}(\tau_i, \mathbf{x}_T) = \mathbf{A}_{\varphi R}^+ \mathbf{h}_{xR}(\tau_i, \mathbf{x}_T) \tag{14.49}$$

Finally, for the direction of departure (DOD) we have

$$\mathbf{h}_{\varphi T}(\tau_i, \varphi_{R,i,j}) = \mathbf{A}_{\varphi T}^+ \mathbf{h}_{xT}(\tau_i, \varphi_{R,i,j}) \tag{14.50}$$

Figure 14.36 illustrates these steps.

The procedure gives us the number and parameters of the MPCs, i.e. the number and values of delays, which DOA can be observed at these delays, and which DOD corresponds to each DOA at a specific delay. Furthermore, we also obtain the powers of the MPCs. One important point in the application of the sequential estimation procedure is the sequence in which the evaluation is performed. Roughly speaking, the number of MPCs that can be estimated is the number of samples we have at our disposal.

14.6.2 Capacity computation

In a fading channel, the capacity is a random variable, depending on the local (or instantaneous) channel realization. In order to determine the cumulative distribution function (cdf) of the capacity, and thus the outage capacity, we would have to perform a large number of measurements, either with slightly displaced arrays, or with a temporally varying scatterer arrangement. Since each single measurement requires a huge effort, such a procedure is highly undesirable.

To improve this situation, an evaluation technique that requires only a single measurement of the channel is used. This technique relies on the fact that we can generate different realizations of the transfer function by changing the phases of the multipath components. It is a well established fact in mobile radio that the phases are uniformly distributed random variables, whose different realizations occur as either transmitter, receiver or scatterers move [1180]. We can thus generate different realizations of the transfer function from the mth transmit to the kth receive antenna as

$$h_{k,m}(f) = \sum_i a_i \exp\left(-j\frac{2\pi}{\lambda}d\left[k\sin\left(\phi_{\mathrm{R},i}\right) + m\sin\left(\phi_{\mathrm{T},i}\right)\right]\right)$$
$$\times \exp\left(-j2\pi f \tau_i\right)\exp\left(j\alpha_i\right) \tag{14.51}$$

where α_i is a uniformly distributed random phase, which can take on different values for the different MPCs numbered i. Note, however, that α_i stays unchanged as we consider different antenna elements k and m. To simplify discussion, we, for now, consider only the flat fading case, i.e. $\tau_i = 0$. We can thus generate different realizations of the channel matrix $\mathbf{H}$

$$\mathbf{H} = \begin{bmatrix} h_{11} & h_{12} & \cdots & h_{1N_{\mathrm{T}}} \\ h_{21} & h_{22} & \cdots & h_{2N_{\mathrm{T}}} \\ \cdots & \cdots & \cdots & \cdots \\ h_{N_{\mathrm{R}}1} & h_{N_{\mathrm{R}}2} & \cdots & h_{N_{\mathrm{R}}N_{\mathrm{T}}} \end{bmatrix} \tag{14.52}$$

by the following two steps.

1. From a single measurement, i.e. a single snapshot of the channel matrix, determine the DOAs and DODs of the MPCs as described earlier in the section.

2. Compute synthetically the impulse responses at the positions of the antenna elements, and at different frequencies. Create different realizations of one ensemble by adding random phase factors (uniformly distributed between 0 and 2π) to each MPC. For each channel realization, we can compute the capacity (from Section 4.12)

$$C = \log_2 \det\left(\mathbf{I} + \frac{\rho}{N_{\mathrm{T}}}\mathbf{H}^{\mathrm{H}}\mathbf{H}\right) \tag{14.53}$$

where ρ denotes the SNR. $\mathbf{I}$ is the identity matrix and superscript H means Hermitian transposition. For the frequency selective case, we have to evaluate the capacity by integrating over all frequencies

$$C = \int \log_2 \det\left(\mathbf{I} + \frac{\rho}{N_{\mathrm{T}}}\mathbf{H}^{\mathrm{H}}(f)\mathbf{H}(f)\right) df \tag{14.54}$$

Here, $\mathbf{H}(f)$ is the frequency-dependent transfer matrix. The integration range is the bandwidth of interest.

14.6.3 Measurement environments

As an example, the following scenarios are evaluated with the procedure described above [1210]:

- *Scenario I: A courtyard with dimensions 26 m × 27 m, open on one side.* The RX array broadside points into the center of the yard, the transmitter is located on the positioning

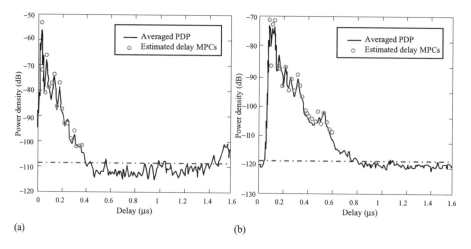

(a)   (b)

Figure 14.37 Power delay profiles (lines) in (a) the LOS Scenario I and (b) in the obstructed
LOS Scenario IV. Superimposed circles are the identified MPCs that are
further used to compute the simulated capacities.

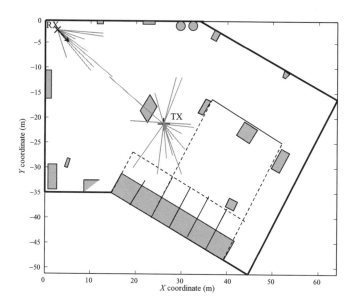

Figure 14.38 Geometry of the environment of Scenarios II to IV (backyard) in plan view.
Superimposed are the extracted DOAs and DODs for Scenario III [1210]
© 2002, IEEE.

device 8m away in LOS. The power delay profile (PDP) in this scenario is given in
Figure 14.37 (a).

- *Scenario II: Closed backyard of size 34 m × 40 m with inclined rectangular extension.*
  The RX array is situated in one rectangular corner with the array broadside of the linear array pointing under 45° inclination directly to the middle of the yard. The LOS

Table 14.16 B_c at 17 GHz

	Coherence bandwidth (MHz)	
Place	Mean	Standard deviation
Hall	24.85	12.35
Floors	14.44	9.85
Building	22.86	10.24
Total	20.72	11.56

connection between TX and RX measures 28 m. Many metallic objects are distributed irregularly along the building walls (power transformers, air conditioning fans, etc.). This environment looks very much like the backyard of a factory (Figure 14.38).

• *Scenario III: Same closed backyard as in II but with artificially obstructed LOS path.* It is expected that the metallic objects generate serious multipath and higher order scattering that can only be observed within the dynamic range of the device if the LOS path is obstructed.

• *Scenario IV: Same as Scenario III but with different TX position and LOS obstructed.* The TX is situated nearer to the walls. Figure 14.37 (b) gives the measured power delay profile. The PDPs in Scenarios II and III look similar, besides the LOS component that occurs dominantly for Scenario II. More details about the scenarios can be found in [1215].

Some of the measurement results for these scenarios are presented in Figure 14.39.

14.7 INDOOR WLAN CHANNEL (17 GHZ)

In this section we discuss the indoor radio propagation channel at 17 GHz. The presentation is based on results reported in [1216]. Wideband parameters, such as coherence bandwidth or rms delay spread, and coverage are analyzed for the design of an OFDM-based broadband WLAN. The method used to obtain the channel parameters is based on using a simulator described in [1216]. This simulator is a site-specific propagation model based on three-dimensional (3D) ray-tracing techniques, which has been specifically developed for simulating radio coverage and channel performance in enclosed spaces such as buildings, and for urban microcell and picocell calculations. The simulator requires the input of the geometric structure and the electromagnetic properties of the propagation environment, and is based on a full 3D implementation of geometric optics and the uniform theory of diffraction (GO/UTD). Examples of the measurement environments are given in Figure 14.40.

The results for coherence bandwidth $B_c = 1/\alpha\tau_{rms}$ are given in Table 14.16 and Figure 14.41.

A further requirement related to the correct and efficient channel estimation process by the receiver is the selection of a number of subcarriers in OFDM satisfying the condition

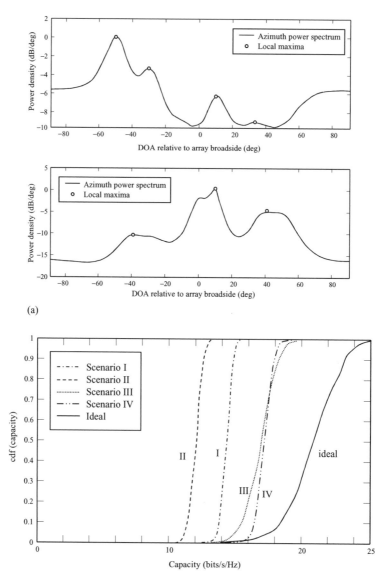

Figure 14.39 (a) Azimuth power spectra at the transmitter (upper plot) and receiver (lower plot) for Scenario III (obstructed LOS). Spectra computed with MVM (minimum variance method, Capon's beamformer). Angles refer to array broadside, so that (due to the array position) +8 and −53 degrees correspond [1210]; (b) cdfs of the MIMO channel capacity encountered in Scenarios I–IV, and the cdf for an ideal channel. The SNR is 20 dB, and 4 × 4 antenna elements were used; (c) outage capacity at the 10 % level in Scenarios I (LOS) and IV (LOS) over the number of antenna element pairs; (d) capacity of a 4 × 4 antenna arrangement in Scenario I at different bandwidths and SNR = 20 dB (capacity is normalized to unit bandwidth); (e) capacity distribution for the narrowband case (dashed) and 100 MHz bandwidth (solid) and 10 dB SNR in Scenario I for array sizes $N_T = N_R = 1, 2, 4, 8$.

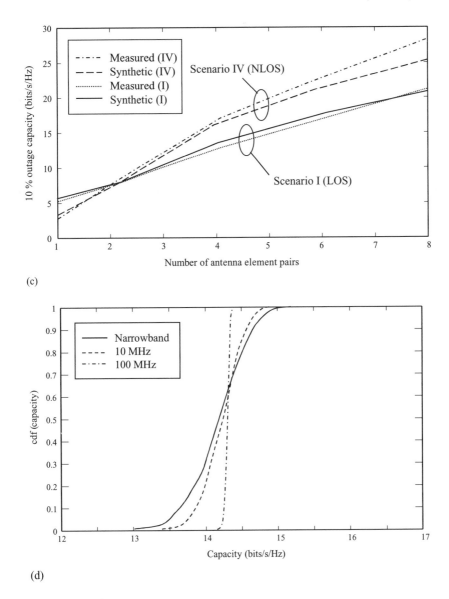

(c)

(d)

Figure 14.39 (*Cont.*).

of being separated between approximately $B_c/5$ and $B_c/10$. Results for delay spread are shown in Figure 14.42 and Tables 14.17–14.20.

Results for the path loss exponent are given in Figure 14.43 and Tables 14.21 and 14.22.

For channel modeling purposes, the mean power of the received signal will be represented as

$$
\begin{aligned}
P_{\mathrm{RX}}|_{\mathrm{dB}} = {} & P_{\mathrm{TX}}|_{\mathrm{dB}} + G_{\mathrm{TX}}|_{\mathrm{dB}} + G_{\mathrm{RX}}|_{\mathrm{dB}} - L_{\mathrm{fs}}|_{\mathrm{dB}} \\
& + 10 \cdot \log \left(\int_0^\infty \mathrm{PDP}\,(t) \, . \, \mathrm{d}t \right)
\end{aligned} \tag{14.55}
$$

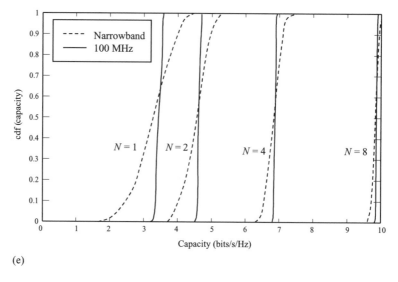

(e)

Figure 14.39 (*Cont.*).

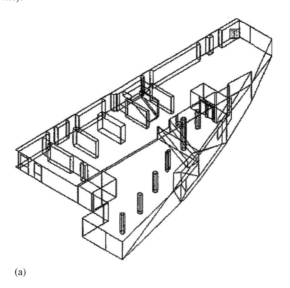

(a)

Figure 14.40 (a) Hall (49 m × 26 m) [1216]; (b) Floors −2 and −3 below top T (34 m × 20 m) [1216]; (c) office building (72 m × 38 m), 3D representation [1216].

where P_{TX} is the mean power at the transmitting antenna input, G_{TX} is the transmitting antenna gain, while G_{RX} is the receiving antenna gain. L_{fs} is free space propagation loss, given by

$$L_{fs}|_{dB} = 32.45 \text{ dB} + 20 \cdot \log_{10}(d_{Km} + f_{MHz})$$

and PDP(*t*) is the modeled power delay profile. Once the PDF is modeled, to obtain the discrete channel impulse response, h_i, we only have to add a random phase to the square

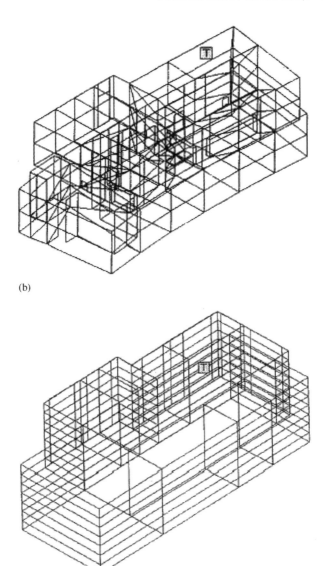

(b)

(c)

Figure 14.40 (*Cont.*).

root of each delay bin amplitude, as follows:

$$h_i = \sqrt{p_i}e^{j\phi_i} \quad \phi_i \quad r.\upsilon. \quad \text{unif}[0, 2\pi] \tag{14.56}$$

where h_i is the ith bin of the modeled channel impulse response and p_i, the module of the ith bin of the modeled power delay profile.

It can be assumed that phases of different components of the same channel impulse response are uncorrelated at the frequency of interest (17 GHz), because their relative range is higher than a wavelength, even for high resolution models [1217].

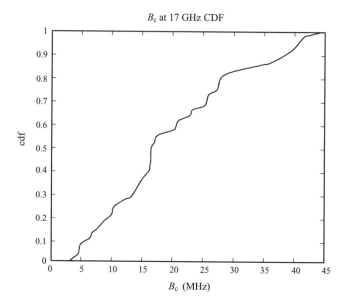

Figure 14.41 B_c CDF at 17 GHz [1216].

Table 14.17 RMS delay spread CDF for Figure 14.42(a) [1216] © 2002, IEEE

CDF value	RDS value (ns)
0.2	12.1
0.4	14.3
0.6	17.5
0.8	34.3
1	58.3

Table 14.18 Maximum delay CDF, 30 dB criterion (Figure 14.42(b)) [1216]

CDF value	T_{max} value (ns)
0.2	62
0.4	76
0.6	101
0.8	122
1	197

Table 14.19   Maximum delay CDF, 20
dB criterion (Figure 14.42(c))
[1216] © 2002, IEEE

CDF value	T_{max} value (ns)
0.2	51
0.4	56
0.6	69
0.8	94
1	156

Table 14.20   Alpha CDF, $B_c = 1/\alpha\tau_{rms}$
(Figure 14.42(d))

CDF value	Alpha value
0.2	2.17
0.4	2.67
0.6	3.75
0.8	4.44
1	5.78

Table 14.21   Mean values of 'n'

Type of path	LOS	OLOS	NLOS
'n' mean value	1.68	2.14	2.61

Table 14.22   Fading statistic over
distance, LOS case

Radius (m)	K factor
4	17
5	10
6	9
7	8
8	6
9	5
10	1

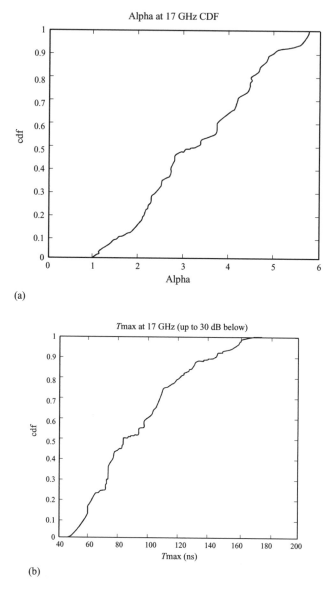

Figure 14.42 (a) RMS delay spread CDF ($B_c = 1/\alpha\tau_{rms}$); (b) maximum delay CDF, 30 dB criterion; (c) maximum delay CDF, 20 dB criterion; (d) alpha CDF.

As the total bandwidth assigned to the communication is 50 MHz, a selection of 10 ns for the bin size must be made. Using 99 % of the total power criterion for the maximum duration of the PDF, the former bin size selection leads to a total of nine taps for the LOS case and seventeen for the NLOS case.

The statistical variability of the bin amplitudes has been modeled following different probability density functions. Taking into account the fact that the area of service of future applications (SOHO–small office home office) has small ranges, the variability has been analyzed considering a medium scale, that is, the environment is divided in the LOS area

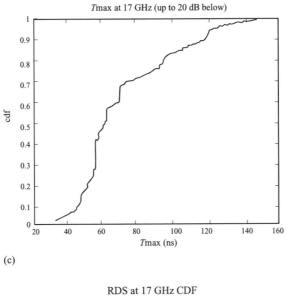

(c)

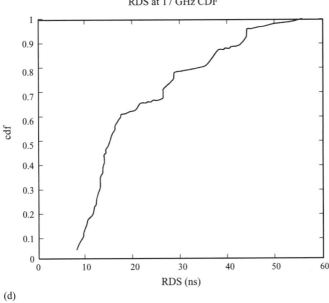

(d)

Figure 14.42 (*Cont.*).

and the NLOS one. In the LOS case, a Frechet PDF [1218] is chosen for the first bin and exponential PDFs for the rest.

A continuous random variable X has a Frechet distribution if its PDF has the form

$$f(x; \sigma; \lambda) = \frac{\lambda}{\sigma} \left(\frac{\sigma}{x}\right)^{\lambda+1} \exp\left\{-\left(\frac{\sigma}{x}\right)^{\lambda}\right\}; \qquad (14.57)$$

$$x \geq 0; \quad \sigma, \lambda > 0.$$

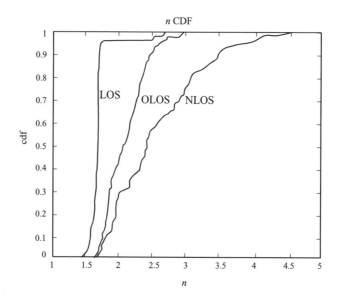

Figure 14.43 CDF of path loss exponent '*n*'.

A Frechet variable X has the CDF

$$F(x;\sigma;\lambda) = \exp\left\{-\left(\frac{\sigma}{x}\right)^{\lambda}\right\}$$

(14.58)

This model has a scale structure, with σ a scale parameter and λ a shape parameter. A continuous random variable X has an exponential distribution if its PDF has the form

$$f(x;\mu) = \frac{1}{\sigma} \exp\left\{-\left(\frac{x-\mu}{\sigma}\right)\right\}$$

(14.59)

$$x \geq 0; \quad \mu, \sigma > 0$$

This PDF has a location–scale structure, with a location parameter, μ, and a scale, σ. The CDF of the exponential variable X is

$$F(x;\mu) = 1 - \exp\left\{-\left(\frac{x-\mu}{\sigma}\right)\right\}$$

(14.60)

These PDFs were considered the most suitable after a fitting process.

The NLOS case needs a combination of exponential and Weibull PDFs for the first bin and exponential PDFs for the others. A continuous random variable X has a Weibull distribution if its PDF has the form

$$f(x;\sigma;\lambda) = \frac{\lambda}{\sigma} \left(\frac{x}{\sigma}\right)^{\lambda-1} \exp\left\{-\left(\frac{x}{\sigma}\right)^{\lambda}\right\}$$

(14.61)

$$x \geq 0; \quad \sigma, \lambda > 0$$

While the CDF is

$$F(x;\sigma;\lambda) = 1 - \exp\left\{-\left(\frac{x}{\sigma}\right)^{\lambda}\right\}$$

(14.62)

Table 14.23   Wind-flex channel model PDFs, LOS case [1216]

WIND-FLEX LOS Channel model pdfs

Bin 1	Frechet $(\sigma = 2.66 \times 10^{-8}, \lambda = 7)$	**Bin 4**	Exponential $(\sigma = 1.45 \times 10^{-7})$	**Bin 7**	Exponential $(\sigma = 0.41 \times 10^{-7})$
Bin 2	Exponential $(\sigma = 5.44 \times 10^{-7})$	**Bin 5**	Exponential $(\sigma = 1.03 \times 10^{-7})$	**Bin 8**	Exponential $(\sigma = 0.27 \times 10^{-7})$
Bin 3	Exponential $(\sigma = 2.51 \times 10^{-7})$	**Bin 6**	Exponential $(\sigma = 0.79 \times 10^{-7})$	**Bin 9**	Exponential $(\sigma = 0.71 \times 10^{-7})$

Table 14.24   Wind-flex channel model PDFs, NLOS case [1216]

WIND-FLEX NLOS Channel model pdfs

Bin 1	0.5^*[Exponential $(\sigma = 4.378 \times 10^{-6})$ + Weibull$(\sigma = 4.207)$ $\times 10^{-7}, \lambda = 5$	**Bin 7**	Exponential $(\sigma = 1.88 \times 10^{-5})$	**Bin 13**	Exponential $(\sigma = 9.21 \times 10^{-4})$
Bin 2	Exponential $(\sigma = 3.04 \times 10^{-6})$	**Bin 8**	Exponential $(\sigma = 2.51 \times 10^{-5})$	**Bin 14**	Exponential $(\sigma = 1.27 \times 10^{-5})$
Bin 3	Exponential $(\sigma = 2.47 \times 10^{-6})$	**Bin 9**	Exponential $(\sigma = 5.69 \times 10^{-5})$	**Bin 15**	Exponential $(\sigma = 2.76 \times 10^{-4})$
Bin 4	Exponential $(\sigma = 2.14 \times 10^{-6})$	**Bin 10**	Exponential $(\sigma = 1.53 \times 10^{-5})$	**Bin 16**	Exponential $(\sigma = 6.71 \times 10^{-4})$
Bin 5	Exponential $(\sigma = 1.1 \times 10^{-6})$	**Bin 11**	Exponential $(\sigma = 3.29 \times 10^{-5})$	**Bin 17**	Exponential $(\sigma = 6.42 \times 10^{-4})$
Bin 6	Exponential $(\sigma = 3.71 \times 10^{-5})$	**Bin 12**	Exponential $(\sigma = 2.67 \times 10^{-5})$		

This model has a scale structure, that is, σ is a scale parameter, while λ is a shape parameter. Tables 14.23 and 14.24 show the probability density functions employed for LOS and NLOS channel models [1216]. For both tables, the units of parameter σ are Hz (s^{-1}), while λ has no units. These units have no physical correlation but make the last term of Equation (14.55) non-dimensional, as it represents a factor scale between the free space behavior and the real one. The mean value of the probability density functions is so high due to the ulterior integral over the time (in seconds) required, and the PDF duration (tens of nanoseconds). As expected, the mean value of the first bin is the highest, since it includes the direct ray (LOS case).

Additional details on the topic can be found in [1217–1229].

14.8 INDOOR WLAN CHANNEL (60 GHZ)

Based on the results reported in [1230], in this section we present spatial and temporal characteristics of 60 GHz indoor channels. In the experiment, a mechanically steered directional antenna is used to resolve multipath components. An automated system is used to

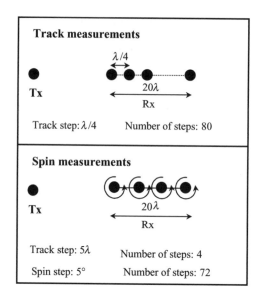

Figure 14.44 Track and spin measurement procedure [1230] © 2002, IEEE.

position the receiver antenna precisely along a linear track and then rotate the antenna in the azimuthal direction as illustrated in Figure 14.44. Precision of the track and spin positions is less than 1 mm and 1°, respectively. When a highly directional antenna is used, the system provides high spatial resolution to resolve multipath components with different angles of arrival (AOAs). The sliding correlator technique was used to further resolve multipath components with the same AOA by their times of arrival (TOAs). The spread spectrum signal has an RF bandwidth of 200 MHz, which provides a time resolution of approximately 10 ns.

For this measurement campaign, an open-ended waveguide with 6.7 dB gain is used as the transmitter antenna, and a horn antenna with 29 dB gain is used as the receiver antenna. These antennas are chosen to emulate typical antenna systems that have been proposed for millimeter-wave indoor applications. In these applications, a sector antenna is used at the transmitter and a highly directional antenna is used at the receiver. Both antennas are vertically polarized and mounted on adjustable tripods about 1.6 m above the ground. The theoretical half power beamwidths (HPBW) are 90° in azimuth and 125° in elevation for the open-ended waveguide, and 7° in azimuth and 5.6° in elevation for the horn antenna. Some measurement results for specific environments and locations are shown in Figure 14.45(a)–(c).

14.8.1 Definition of the statistical parameters

14.8.1.1 Path loss and received signal power

The free space *path loss* at a reference distance of d_0 is given by

$$\overline{PL}_{\text{fs}}(d_0) = 20 \log \left(\frac{4\pi d_0}{\lambda} \right) \tag{14.63}$$

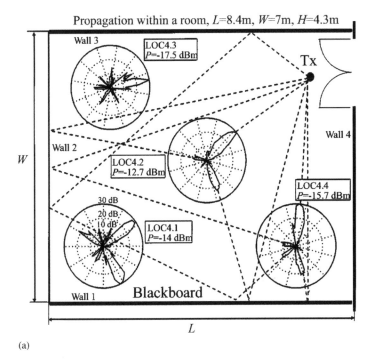

(a)

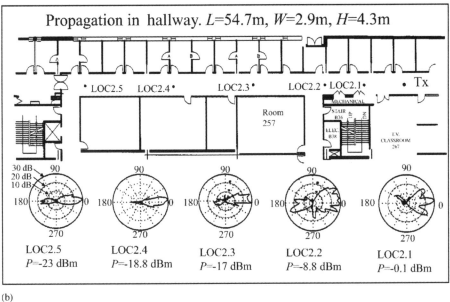

(b)

Figure 14.45 (a) AOA measurements for propagation within a room (location 4), relative power levels are shown on the polar plots, and peak multipath power (P) is given in the text boxes. Rays are shown only for locations 4.2 and 4.4 in the figure, although a similar procedure can be performed for all the locations; (b) AOA measurements for propagation along a hallway (location 2), relative power levels are given in the polar plots, and peak multipath power (P) is given in the text at the bottom of the figure; (c) AOA measurements for propagation into rooms (locations 5 and 6), relative power levels are given in the polar plots, and peak multipath power (P) is given in the text boxes [1230] © 2002, IEEE.

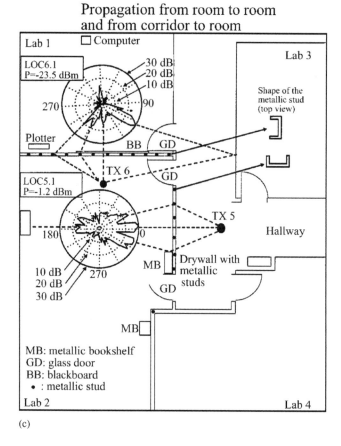

Propagation from room to room and from corridor to room

(c)

Figure 14.45 (*Cont.*).

where λ is the wavelength. Path loss over distance d can be described by the path loss exponent model as follows:

$$\overline{PL}(d)[\text{dB}] = PL_{\text{fs}}(d_0)(\text{dB}) + 10n \log_{10}\left(\frac{d}{d_0}\right) \qquad (14.64)$$

where $\overline{PL}(d)$ is the average path loss value at a TR separation of d, and n is the *path loss exponent* that characterizes how fast the path loss increases with the increase of TR separation. The path loss values represent the signal power loss from the transmitter antenna to the receiver antenna. These path loss values do not depend on the antenna gains or the transmitted power levels. For any given transmitted power, the received signal power can be calculated as

$$P_r(\text{dBm}) = P_t(\text{dBm}) + G_t(\text{dB}) + G_r(\text{dB}) - \overline{PL}(d)(\text{dB}) \qquad (14.65)$$

where G_t and G_r are transmitter and receiver gains, respectively. In this measurement campaign, the transmitted power level was 25 dBm, the transmitter antenna gain was 6.7 dB, and the receiver antenna gain was 29 dB.

14.8.1.2 TOA parameters

TOA parameters characterize the time dispersion of a multipath channel. The calculated TOA parameters include mean excess delay ($\bar{\tau}$), rms delay spread (σ_τ), and also timing jitter ($\delta(x)$) and standard deviation ($\Delta(x)$), in a small local area. Parameters $\bar{\tau}$ and σ_τ are given as [1231]:

$$\bar{\tau} = \frac{\sum_{i=1}^{N} P_i \tau_i}{\sum_{i=1}^{N} P_i} \tag{14.66}$$

$$\sigma_\tau = \sqrt{\overline{\tau^2} - (\bar{\tau})^2} \tag{14.67}$$

$$\overline{\tau^2} = \frac{\sum_{i=1}^{N} P_i \tau_i^2}{\sum_{i=1}^{N} P_i} \tag{14.68}$$

where P_i and τ_i are the power and delay of the ith multipath component of a PDF, respectively, and N is the total number of multipath components. Timing jitter is calculated as the difference between the maximum and minimum measured values in a local area. Timing jitter $\delta(x)$ and standard deviation $\Delta(x)$ are defined as

$$\delta(x) = \max_{i=1}^{M}\{x_i\} - \min_{i=1}^{M}\{x_i\} \tag{14.69}$$

$$\Delta(x) = \sqrt{\overline{x^2} - (\bar{x})^2} \tag{14.70}$$

$$\bar{x} = \frac{1}{M} \sum_{i=1}^{M} x_i \tag{14.71}$$

$$\overline{x^2} = \frac{1}{M} \sum_{i=1}^{M} x_i^2 \tag{14.72}$$

where x_i is the measured value for parameter $x(\bar{\tau}$ or $\sigma_\tau)$ in the ith measurement position of the spatial sampling, and M is the total number of spatial samples in the local area. For example, for the track measurements, M was chosen to be 80.

Mean excess delay and rms delay spread are the statistical measures of the time dispersion of the channel. Timing jitter and standard deviation of $\bar{\tau}$ and σ_τ show the variation of these parameters over the small local area.

These TOA parameters directly affect the performance of high-speed wireless systems. For instance, the mean excess delay can be used to estimate the search range of RAKE receivers and the rms delay spread can be used to determine the maximum transmission data rate in the channel without equalization. The timing jitter and standard deviation parameters can be used to determine the update rate for a RAKE receiver or an equalizer.

Table 14.25   Spin measurements: transmitter–receiver separations (TR) in m, time dispersion parameters ($\bar{\tau}$ and σ_τ) in ns, angular dispersion parameters (Λ and γ) are dimensionless, maximum fading angle (θ_{max}) and AOA of maximum multipath (max AOA) in degrees, ratio of maximum multipath power to average power (Peak/avg) in dB and maximum multipath power (P_{max}) in dBm [1230]

Site information	#	TR	$\bar{\tau}$	σ_τ	Λ	γ	θ_{max}	max AOA	$\dfrac{Peak}{avg}$	P_{max}	Comments
	1.1	5	80.0	14.7	0.46	0.83	−80.7	−4.0	12.3	−14.9	
	1.2	10	52.0	18.8	0.44	0.74	−86.6	4.0	12.0	−18.2	
LOS, hallway	1.3	20	85.9	40.1	0.56	0.28	−61.9	8.0	14.5	−28.8	
Durham Hall	1.4	30	116.6	38.7	0.42	0.22	−66.4	5.0	14.7	−28.3	open area
	1.5	40	84.9	60.0	0.69	0.25	4.3	5.0	13.9	−38.2	
	1.6	50	52.1	26.1	0.66	0.26	8.2	10.0	13.3	−38.2	
	1.7	60	53.2	30.3	0.78	0.36	4.0	2.0	13.2	−40.8	
	2.1	5	51.0	20.7	0.48	0.88	−73.5	5.0	12.5	−13	
	2.2	10	62.1	29.4	0.66	0.79	−72.3	21.0	11.4	−21.7	intersection
LOS, hallway	2.3	20	90.7	14.6	0.36	0.43	−73.8	4.0	12.9	−29.8	
Whittemore	2.4	30	41.2	12.3	0.41	0.15	−64.8	10.0	13.8	−31.7	
	2.5	40	83.7	53.8	0.72	0.19	5.0	1.0	13.2	−36.0	
LOS, room	3.1	4.2	42.6	16.2	0.86	0.64	−79.2	0.0	12.5	−11.8	corner
Durham Hall	3.2	3.3	47.7	17.5	0.81	0.70	−79.1	5.0	13.1	−12.1	center
	4.1	7.1	46.6	13.0	0.84	0.55	−88.0	−60.0	12.3	−26.8	corner
LOS, room	4.2	3.8	64.3	13.3	0.62	0.74	−89.6	−1.0	13.1	−25.6	center
Whittemore	4.3	5.2	66.3	17.7	0.73	0.84	−35.2	49.0	14.0	−30.4	corner, ⊥ to Tx
	4.4	4.2	77.8	13.3	0.78	0.72	−38.2	−49.0	14.2	−28.6	corner, ⊥ to Tx
	5.1	2.4	49.1	21.4	0.81	0.13	−76.3	0.0	12.0	−6.0	LOS
Hallway to	5.2	2.4	41.6	18.1	0.74	0.44	−89.6	5.0	10.3	−14.1	through wall
room	5.3	2.4	95.8	14.6	0.63	0.40	−88.1	0.0	12.1	−5.6	LOS
	5.4	2.4	80.3	16.0	0.68	0.27	72.3	5.0	11.9	−8.9	through glass
Room to room	6.1	3	42.7	16.6	0.80	0.40	−25.3	52.0	11.5	−36.4	through wall
LOS, outdoor	7.1	1.9	41.3	17.4	0.12	0.97	−81.2	2.0	13.9	−15.0	Tx pattern
parking lot	7.2	1.9	56.6	16.1	0.49	0.94	−66.7	20.0	8.5	−29.9	Rx pattern
LOS, outdoor	8.1	2	24.4	7.7	0.26	0.76	−66.3	3.0	13.9	−10.1	near Durham Hall

14.8.1.3   AOA parameters

AOA parameters characterize the directional distribution of multipath power. The recorded AOA parameters include angular spread Λ, angular constriction γ, maximum fading angle θ_{max}, and maximum AOA direction. Angular parameters Λ, γ and θ_{max} are defined based on the Fourier transform of the angular distribution of multipath power, $p(\theta)$ [1232]:

$$\Lambda = \sqrt{1 - \frac{||F_1||^2}{||F_2||^2}} \tag{14.73}$$

$$\gamma = \frac{||F_0 F_2 - F_1^2||}{||F_0||^2 - ||F_1||^2} \tag{14.74}$$

Table 14.26   Track measurement results: TR separations (TR) in m, time dispersion parameters ($\bar{\tau}$ and σ_τ) in ns, variations of time dispersion parameters ($\delta\bar{\tau}$, $\Delta\bar{\tau}$, $\delta\sigma_\tau$ and $\Delta\sigma_\tau$) in ns and average received power (P_{rx}) in dBm [1230]

Site information	LOC #	TR	$\bar{\tau}$	σ_τ	$\delta\bar{\tau}$	$\Delta\bar{\tau}$	$\delta\sigma_\tau$	$\Delta\sigma_\tau$	P_{rx}	Comments
	1.1	5	1.20	6.95	6.33	1.91	1.20	0.29	−13.7	
	1.2	10	6.16	5.88	5.06	1.20	6.16	1.73	−20.3	
LOS, hallway	1.3	20	32.61	47.25	32.89	8.43	32.61	9.02	−36.6	
Durham Hall	1.4	30	15.50	31.15	10.16	3.43	15.50	5.69	−31.2	open area
	1.5	40	27.60	37.04	25.89	8.81	27.60	9.76	−40.5	
	1.6	50	46.42	28.17	36.70	8.10	46.42	10.73	−42.8	
	1.7	60	6.38	22.57	5.99	1.82	6.38	1.57	−41.5	
	2.1	5	2.22	6.24	7.52	2.38	2.22	0.73	−16.7	
LOS, hallway	2.2	10	2.78	6.48	8.24	2.61	2.78	0.82	−24.4	intersection
Whittemore	2.3	20	2.3	4.56	7.81	2.55	2.30	0.55	−32.86	
	2.4	30	22.02	33.87	13.17	4.60	22.02	6.30	−34.7	
LOS, room	2.5	40	77.3	45.07	105.04	34.41	77.30	25.86	−36.3	
Durham Hall	3.1	4.2	0.74	4.85	6.20	1.88	0.74	0.20	−12.1	corner
	3.2	3.3	0.92	4.95	5.97	1.87	0.92	0.23	−12.9	center
LOS, room	4.1	7.1	2.74	4.72	11.16	3.08	2.47	0.36	−29.7	corner
Whittemore	4.2	3.8	2.4	4.98	11.11	3.17	2.40	0.47	−24.2	center
	4.3	5.2	12.88	31.10	26.36	6.86	12.88	2.95	−56.2	corner, ⊥ to Tx
	4.4	4.2	21.3	33.94	31.5	7.4	21.3	5.43	−57.9	corner, ⊥ to Tx
	5.1	2.4	0.83	5.50	2.41	0.69	0.83	0.32	−5.5	LOS
	5.2	2.4	2.46	7.41	2.61	0.84	2.46	0.94	−14.3	through wall
Hallway to	5.3	2.4	0.71	5.36	1.30	0.41	0.71	0.25	−6.7	LOS
room	5.4	2.4	1.16	5.19	1.85	0.61	1.16	0.36	−9.1	through glass
Room to room	6.1	3	10.67	14.72	23.07	6.62	10.67	1.30	−12.8	LOS
	6.2	3	14.82	21.78	34.30	8.57	14.82	3.37	−48.3	through wall
LOS, outdoor	8.1	2	7.63	24.59	10.24	2.66	7.63	1.75	−2.4	near Durham Hall

$$\theta_{max} = \frac{1}{2}\text{phase}\left\{F_0 F_2 - F_1^2\right\} \tag{14.75}$$

where

$$F_n = \int_0^{2\pi} p(\theta)\exp(jn\theta)\,d\theta \tag{14.76}$$

F_n is the nth Fourier transform of $p(\theta)$. As shown in [1232], angular spread, angular constriction and maximum fading angle are three key parameters to characterize the small-scale fading behavior of the channel. These new parameters can be used for diversity techniques, fading rate estimation, and other space–time techniques. The maximum AOA provides the direction of the multipath component with the maximum power. It can be used in system installation to minimize the path loss. The results of measurements for the parameters defined by Equations (14.63–14.76) are given in Tables 14.25–14.27 and Figure 14.46.

More details on the topic can be found in [1232–1243].

Table 14.27   Measured penetration losses and results from literature

Material	Penetration loss (dB)	Reference
Composite wall with studs not in the path	8.8	[1230]
Composite wall with studs in the path	35.5	[1230]
Glass door	2.5	[1230]
Concrete wall one week after concreting	73.6	[1233]
Concrete wall two weeks after concreting	68.4	[1233]
Concrete wall five weeks after concreting	46.5	[1233]
Concrete wall 14 months after concreting	28.1	[1233]
Plasterboard wall	5.4 to 8.1	[1234]
Partition of glass wool with plywood surfaces	9.2 to 10.1	[1234]
Partition of cloth-covered plywood	3.9 to 8.7	[1234]
Granite with width of 3 cm	>30	[1235]
Glass	1.7 to 4.5	[1235]
Metalized glass	>30	[1235]
Wooden panels	6.2 to 8.6	[1235]
Brick with width of 11 cm	17	[1235]
Limestone with width of 3 cm	>30	[1235]
Concrete	>30	[1235]

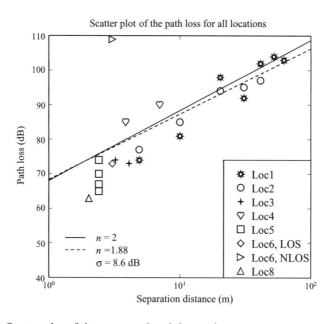

Figure 14.46 Scatter plot of the measured path loss values.

14.9 UWB CHANNEL MODEL

UWB channel parameters were discussed to some extent in Chapter 8. In this section we will present some additional results with the focus on channel modeling, mainly based on [1244].

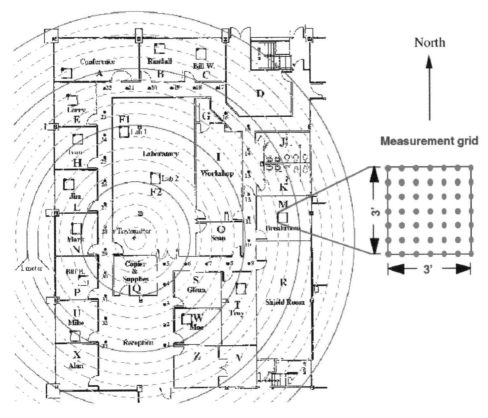

Figure 14.47 The floor plan of a typical modern office building where the propagation measurement experiment was performed. The concentric circles are centered on the transmit antenna and are spaced at 1 m intervals [1244].

The measurements environment is presented in Figure 14.47 and the signal format used in these experiments in Figure 14.48.

The repetition rate of the pulses is 2×10^6 pulses per second, implying that multipath spreads up to 500 ns could have been observed unambiguously. Multipath profiles with a duration of 300 ns were measured. Sample results are shown in Figure 14.49. Multipath profiles were measured at various locations in 14 rooms and hallways on one floor of the building presented in Figure 14.47. Each of the rooms is labeled alphanumerically. Walls around offices are framed with metal studs and covered with plasterboard. The wall around the laboratory is made from acoustically silenced heavy cement block. There are steel core support pillars throughout the building, notably along the outside wall and two within the laboratory itself. The shield room's walls and door are metallic. The transmitter is kept stationary in the central location of the building near a computer server in a laboratory denoted by F. The transmit antenna is located 165 cm from the floor and 105 cm from the ceiling.

In each receiver location, impulse response measurements were made at 49 measurement points, arranged in a fixed-height, 7×7 square grid with 15 cm spacing, covering

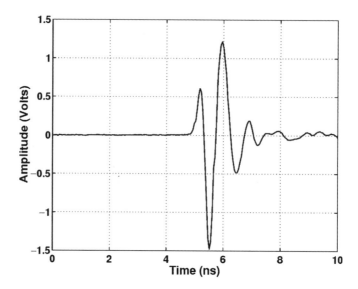

Figure 14.48 The transmitted pulse measured by the receiving antenna located 1 m away from the transmitting antenna with the same height.

90 cm × 90 cm. A total of 741 different impulse responses were recorded. One side of the grid is always parallel to the north wall of the room. The receiving antenna is located 120 cm from the floor and 150 cm from the ceiling.

Profiles measured in offices U, W and M are shown in Figure 14.49. The approximate distances between the transmitter and the locations of these measurements are 10, 8.5 and 13.5 m, respectively. Figure 14.49 also shows that the response to the first probing pulse has decayed almost completely in roughly 200 ns, and has disappeared before the response to the next pulse arrives at the antenna. The multipath profiles recorded in the offices W and M have a substantially lower noise floor than those recorded in office U. This can be explained, with the help of Figure 14.47, by observing that office U is situated at the edge of the building with a large glass window, and is subject to more external interference (e.g. from radio stations, television stations, cellular and paging towers), while offices W and M are situated roughly in the middle of the building. In general, an increased noise floor was observed for all the measurements made in offices located at the edges of the building with large glass windows.

The large-scale fading characterizes the changes in the received signal when the receiver position varies over a significant fraction of the transmitter–receiver (T–R) distance and/or the environment around the receiver changes. This situation typically occurs when the receiver is moved from one room to another room in a building. The small-scale effects, on the other hand, are manifested in the changes of the PDP caused by small changes of the receiver position, while the environment around the receiver does not change significantly. This occurs, for instance, when the receiver is moved over the measurement grid within a room in a building.

In the following, we refer to the PDP measured at one of the 14(rooms) × 49 locations as *local* PDP, while we denote the PDP averaged over the 49 locations within one room as the *small-scale averaged* PDP (SSA-PDP). This spatial averaging (mostly) removes

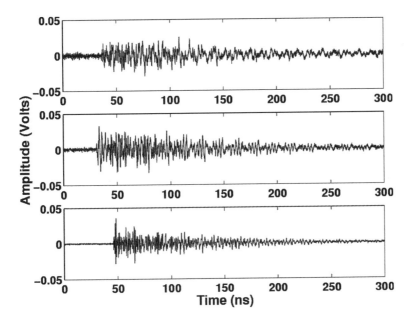

Figure 14.49 Average multipath measurements of 32 sequentially measured multipath profiles where the receiver is located at exactly the same locations in offices U (upper trace), W (middle trace) and M (lower trace). The measurement grids are 10, 8.5, and 13.5 m away from the transmitter, respectively [1244].

the effect of small-scale fading. The small-scale statistics are derived by considering the deviations of the 49 local PDPs from the respective SSA-PDP. The large-scale fading may be investigated by considering the variation of the SSA-PDPs over the different rooms. We also make a distinction between the 'local' parameters, which refer to the small-scale effects, and the 'global' parameters, extracted from the SSA-PDPs. For clarity, all the symbols and parameters are listed in Table 14.28.

14.9.1 The large-scale statistics

All SSA-PDPs exhibit an exponential decay as a function of the excess delay. Since we perform a delay axis translation, the direct path always falls in the first bin in all the PDPs. It also turns out that the direct path is always the strongest path in the 14 SSA-PDPs, even if the LOS is obstructed. The energy of the subsequent MPCs decays exponentially with delay starting from the second bin. This is illustrated by the fit (linearly on a decibel scale) in Figure 14.50 using the SSA-PDP of a typical high signal to noise ratio (SNR) room. Let $\overline{G}_k \hat{=} A_{\text{Spa}}\{G_k\}$ be the locally averaged energy gain, where the $A_{\text{Spa}}\{\cdot\}$ denotes the spatial average over the 49 locations of the measurement grid. The average energy of the second MPC may be expressed as a fraction r of the average energy of the direct path, i.e. $r = \overline{G}_2/\overline{G}_1$. We refer to r as the *power ratio*. As we will show later, the SSA-PDP is completely characterized by $\overline{G}_1$, the power ratio r, and the decay constant ε (or equivalently, by the total average received energy $\overline{G}_{\text{tot}}$, r, and ε). The number of resolved MPCs is given by the number of the MPCs that exceed a threshold and thus, given the threshold, it depends on

Table 14.28   Symbols and parameters [1244]

	Delay axis
τ	Excess delay
τ_{Ref}	Absolute propagation delay
$\tau_k = (k - 1)\Delta\tau$	kth Delay Bin
$\Delta\tau = 2$ ns	Bin width
N_{bins}	Number of bins
	Energy gains
$\bar{G}_{\text{tot}}$	Total average energy gain
$\bar{G}_k$	Average energy gain of the kth delay bin
G_k	Energy gain of the kth delay bin
$\bar{g}(\tau)$	Average received energy at excess delay τ
	Exponential time decay
ε	Decay constant
$r = G_2/G_1$	Power ratio

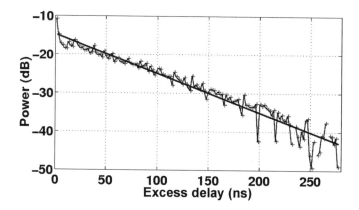

Figure 14.50   The average power delay profile versus the excess delay in a semi-logarithmic scale for a typical high SNR room. The wavy line is the measured profile, the straight line is the exponential decay obtained by a best fit procedure.

the shape of the SSA-PDP, characterized by the parameters $\bar{G}_1, r$, and ε. Best fit procedures are used to extract the εs and the rs from the SSA-PDP of each room.

The power ratio r and the decay constant ε vary from location to location, and should be treated as stochastic variables. As only 14 values for ε and r were available, it was not possible to extract the *shape* of their distribution from the measurement data. Instead, a model was assumed *a priori* and the *parameters* of this distribution were fitted. Previous narrowband studies showed that the decay constants are well modeled as log-normal variables [1159]. It was found that the log-normal distribution, denoted by $\varepsilon \sim \mathcal{LN}(\mu_{\varepsilon_{\text{dB}}}; \sigma_{\varepsilon_{\text{dB}}})$, with $\mu_{\varepsilon_{\text{dB}}} = 16.1$ and $\sigma_{\varepsilon_{\text{dB}}} = 1.27$ gives the best agreement with the empirical distribution. The histograms of the experimental decay constants and the theoretically fitted distribution are shown in Figure 14.51. Applying the same procedure to characterize the power ratios r, it was found that they are also log-normally distributed, i.e. $r \sim \mathcal{LN}(\mu_{r_{\text{dB}}}; \sigma_{r_{\text{dB}}})$,

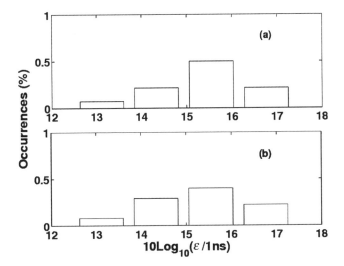

Figure 14.51 Histograms of (a) the experimental decay constants ε and (b) the theoretically fitted distribution. The decay constants are expressed in logarithmic units by referring to nanoseconds.

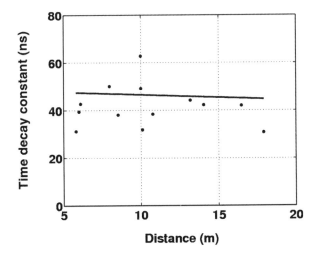

Figure 14.52 Scatter plot of the decay constants ε versus the T–R distance. The solid line is the regression fit whose slope is -0.22 ns/m.

with $\mu_{r_{\mathrm{dB}}} = -4$ and $\sigma_{r_{\mathrm{dB}}} = 3$, respectively.

The possible correlation of the decay constant with the T–R separation was also investigated, by applying a linear regression to the εs versus the distance. As Figure 14.52 shows, the regression fit of the decay constants ε decreases with the increasing distance so slightly that we can conclude that it is *de facto* independent.

By integrating the SSA-PDP of each room over all delay bins, the total average energy $\bar{G}_{\mathrm{tot}}$ within each room is obtained. Then its dependence on the T–R separation is analyzed. As suggested by the scatter plot of Figure 14.53, a breakpoint model, commonly referred

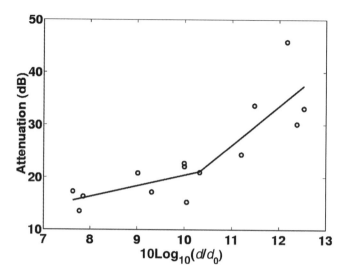

Figure 14.53 Scatter plot of the large-scale attenuation versus the logarithm of the distance. The solid line represents the best fit with the path loss model of Equation (14.77).

to as a *dual slope* model, can be adopted for path loss PL as a function of the distance. The regression lines are shown in Figure 14.53 and the parameters extracted by performing a best fit of the empirical attenuation are

$$
PL = \begin{cases} 20.4 \log_{10}(d/d_0), & d \leq 11\,\mathrm{m} \\ -56 + 74 \log_{10}(d/d_0), & d > 11\,\mathrm{m} \end{cases} \tag{14.77}
$$

where PL is expressed in decibels, $d_0 = 1$ m is the reference distance, and d is the T–R separation distance in meters. Because of the shadowing phenomenon, the $\bar{G}_{tot}$ varies statistically around the value given by Equation (14.77). A common model for shadowing is the log-normal distribution [1245, 1246]. By assuming such a model, it was found that $\bar{G}_{tot}$ is log-normally distributed about Equation (14.77) with a standard deviation of the associated normal random variable equal to 4.3.

14.9.2 The small-scale statistics

The differences between the PDPs at the different points of the measurement grid are caused by small-scale fading. In 'narrowband' models, it is usually assumed that the magnitude of the first (quasi-LOS) multipath component follows Rician or Nakagami statistics and the later components are assumed to have Rayleigh statistics [1247]. However, in UWB propagation, each resolved MPC is due to a small number of scatterers, and the amplitude distribution in *each* delay bin differs markedly from the Rayleigh distribution. In fact, the presented analysis showed that the best fit distribution of the small-scale magnitude statistics is the Nakagami distribution [1248], corresponding to a Gamma distribution of the energy gains. This distribution has been used to model the magnitude statistics in mobile radio when the conditions of the central limit theorem are not fulfilled [1249].

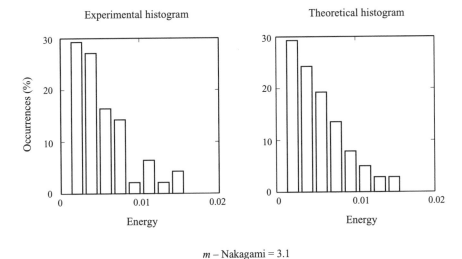

m – Nakagami = 3.1

Figure 14.54 Histogram of the received energy in the 34th bin of a typical high SNR room, compared with the theoretical Gamma distribution, whose mean Ω_{34} and m_{34} were extracted from the experimental PDF. The energies on the horizontal axes are expressed in arbitrary units [1244] © 2002, IEEE.

The small-scale statistics are characterized by fitting the received normalized energies $\{G_k^{(i)}\}$ in *each* bin at the 49 locations of the measurement grid to a distribution. The variations over the measurement grid are treated as stochastic. The result shows that the statistics of the energy gain vary with delays. Let us denote by $\Gamma(\Omega; m)$ the Gamma distribution with parameters Ω and m. The $\Gamma(\Omega; m)$ gives a good fit of the empirical distribution of the energy gains. The accuracy of the fit has been quantified in terms of the relative mean squared error, which varies between 0.0105 (for the highest SNR) to 0.1137 (for the lowest SNR). A comparison between experimental and theoretical histograms for one exemplary bin in a typical high SNR is shown in Figure 14.54.

The parameters of the Gamma distribution vary from bin to bin: $\Gamma(\Omega; m)$ denotes the Gamma distribution that fits the energy gains of the local PDPs in the kth bin within each room. The Ω_k are given as $\Omega_k = \bar{G}_k$, i.e. the magnitude of the SSA-PDP in the kth bin. The m_k are related to the variance of the energy gain of the kth bin. Figure 14.55 shows the scatter plot of the m_k, as a function of excess delay for all the bins (except the LOS components). It can be seen from Figure 14.55 that the m_k values range between 1 and 6 (rarely 0.5), decreasing with the increasing excess delay. This implies that MPCs arriving with large excess delays are more diffused than the first arriving components, which agrees with intuition.

The m_k parameters of the Gamma distributions themselves are random variables distributed according to a truncated Gaussian distribution, denoted by $m \sim \mathcal{T}_N(\mu_m; \sigma_m^2)$, i.e. their distribution looks like a Gaussian for $m \geq 0.5$ and zero elsewhere

$$f_m(x) = \begin{cases} K_m e^{-((x-\mu_m)^2/2\sigma_m^2)}, & \text{if } x \geq 0.5 \\ 0, & \text{otherwise} \end{cases} \quad (14.78)$$

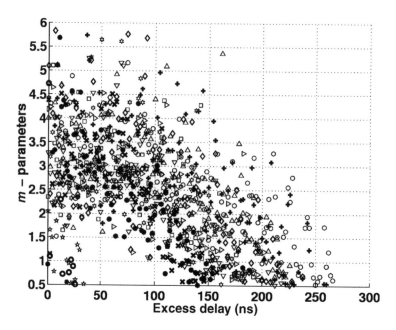

Figure 14.55 Scatter plot of the *m*-Nakagami of the best fit distribution versus excess delay for all the bins except the LOS components. Different markers correspond to measurements in different rooms[1244] © 2002, IEEE.

where the normalization constant K_m is chosen so that the integral over the $f_m(x)$ is unity. Figure 14.56 shows the mean and variance of such Gaussian distributions that fit the m_k as a function of the excess delay, along with the respective regression lines. The regression lines are given by

$$\mu_m(\tau_k) = 3.5 - \frac{\tau_k}{73} \qquad (14.79)$$

$$\sigma_m^2(\tau_k) = 1.84 - \frac{\tau_k}{160} \qquad (14.80)$$

where τ_k is in nanoseconds.

14.9.3 Correlation of MPCs among different delay bins

We next evaluate the correlation between the energy gain of the MPCs arriving in the same room at different excess delays as

$$\rho_{k,k+m} = \frac{A_{\text{Spa}}\{(G_k - \overline{G_k})(G_{k+m} - \overline{G_{k+m}})\}}{\sqrt{A_{\text{Spa}}\{(G_k - \overline{G_k})^2\}A_{\text{Spa}}\{(G_{k+m} - \overline{G_{k+m}})^2\}}} \qquad (14.81)$$

The analysis shows that the correlation coefficients remain below 0.2 for almost all rooms and delay bins and are, thus, negligible for all practical purposes.

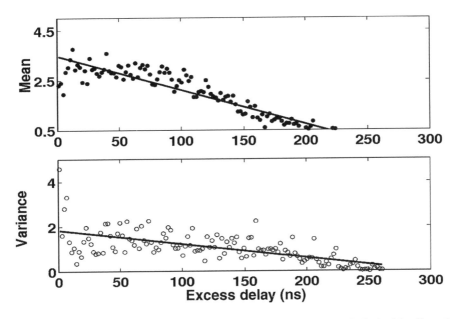

Figure 14.56 Scatter plot of the mean values (dots) and the variance (circles) of the Gaussian distributions that fit the experimental distribution of m values at each excess delay. The solid lines represent the linear regression for these parameters, respectively [1244] © 2002, IEEE.

14.9.4 The statistical model

The received signal is a sum of the replicas (echoes) of the transmitted signal, being related to the reflecting, scattering and/or deflecting objects via which the signal propagates. Each of the echoes is related to a single such object. In a narrowband system, the echoes at the receiver are only attenuated, phase-shifted and delayed, but undistorted, so that the received signal may be modeled as a linear combination of N_{path} delayed basic waveforms $w(t)$

$$r(t) = \sum_{i=1}^{N_{path}} c_i w(t - \tau_i) + n(t) \tag{14.82}$$

where $n(t)$ is the observation noise. In UWB systems, the frequency selectivity of the reflection, scattering and/or diffraction coefficients of the objects via which the signal propagates, can lead to a distortion of the transmitted pulses. Furthermore, the distortion and, thus, the shape of the arriving echoes, varies from echo to echo. The received signal is given as

$$r(t) = \sum_{i=1}^{N_{path}} c_i \tilde{w}_i(t - \tau_i) + n(t) \tag{14.83}$$

If the pulse distortion was greater than the width of the delay bins (2 ns), one would observe a significant correlation between adjacent bins. The fact that the correlation coefficient remains very low for all analyzed sets of the data implies that the distortion of a pulse due to a single echo is not significant, so that in the following, Equation (14.82) can be used. The

SSA-PDP of the channel may be expressed as

$$\bar{g}(\tau) = \sum_{k=1}^{N_{\text{bins}}} \bar{G}_k \delta(\tau - t_k) \tag{14.84}$$

where the function $\bar{g}(\tau)$ can be interpreted as the average energy received at a certain receiver position and a delay τ, normalized to the total energy received at one meter distance, and N_{bins} is the total number of bins in the observation window. Assuming an exponential decay starting from the second bin, we have

$$\bar{g}(\tau) = \bar{G}_1 \delta(\tau - \tau_1) + \sum_{k=2}^{N_{\text{bins}}} \bar{G}_2 \exp\left[-(\tau_k - \tau_2)/\varepsilon\right] \delta(\tau - t_k) \tag{14.85}$$

where ε is the decay constant of the SSA-PDP. The total average energy received over the observation interval T is:

$$\bar{G}_{tot} = \int_0^T \bar{g}(\tau)\,d\tau = \bar{G}_1 + \sum_{k=2}^{N_{\text{bins}}} \bar{G}_2 \exp\left[-(\tau_k - \tau_2)/\varepsilon\right] \tag{14.86}$$

Summing the geometric series, gives

$$\bar{G}_{\text{tot}} = \bar{G}_1[1 + rF(\varepsilon)] \tag{14.87}$$

where $r = \bar{G}_2/\bar{G}_1$ is the power ratio, and

$$F(\varepsilon) = \frac{1 - \exp\left[-(N_{\text{bins}} - 1)\Delta\tau/\varepsilon\right]}{1 - \exp\left(-\Delta\tau/\varepsilon\right)} \approx \frac{1}{1 - \exp\left(-\Delta\tau/\varepsilon\right)} \tag{14.88}$$

The total normalized average energy is log-normally distributed, due to the shadowing, around the mean value given from the path loss model in Equation (14.77):

$$\bar{G}_{\text{tot}} \sim \mathcal{LN}(-\text{PL};\ 4.3) \tag{14.89}$$

From Equation (14.87), we have, for the average energy gains,

$$\bar{G}_k = \begin{cases} \dfrac{\bar{G}_{\text{tot}}}{1 + rF(\varepsilon)}, & \text{for } k = 1 \\[3mm] \dfrac{\bar{G}_{\text{tot}}}{1 + rF(\varepsilon)} r e^{-(\tau_k - \tau_2)/\varepsilon}, & \text{for } k = 2, \ldots, N_{\text{bins}} \end{cases} \tag{14.90}$$

and Equation (14.84) becomes

$$\bar{g}(\tau) = \frac{\bar{G}_{\text{tot}}}{1 + rF(\varepsilon)} \left\{ \delta(\tau - \tau_1) + \sum_{k=2}^{N_{\text{bins}}} \left[r e^{-(\tau_k - \tau_2)/\varepsilon}\right] \delta(\tau - t_k) \right\} \tag{14.91}$$

14.9.5 Simulation steps

In the model, the local PDF is fully characterized by the pairs $\{G_k,\ \tau_k\}$, where $\tau_k = (k - 1)\Delta\tau$, with $\Delta\tau = 2$ ns. The G_k are generated by a superposition of large- and small-scale statistics.

The process starts by generating the total mean energy $\overline{G_{\text{tot}}}$ at a certain distance according to Expression (14.89). Next, the decay constant ε and the power ratio r are generated as

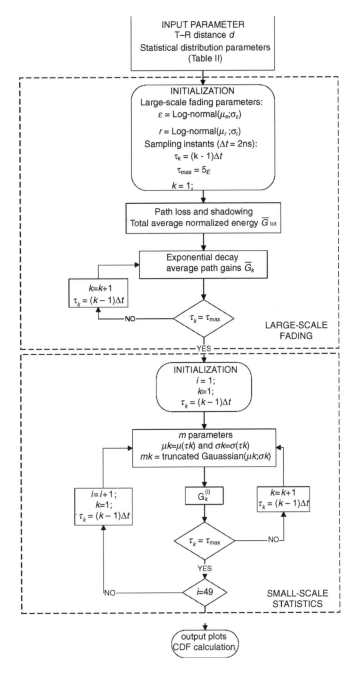

Figure 14.57 The flowchart of the simulation procedure.

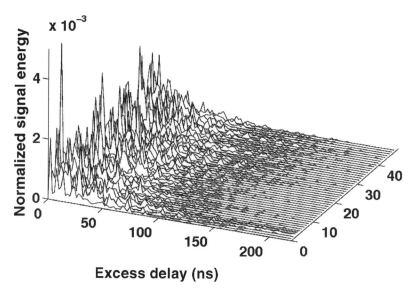

Figure 14.58 The measured 49 local PDPs for an exemplary room [1244] © 2002, IEEE.

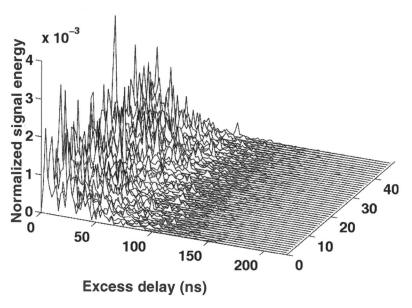

Figure 14.59 Simulated 49 local PDPs for an exemplary room [1244].

log-normal distributed random numbers

$$\varepsilon \sim \mathcal{L_N}(16.1; \ 1.27) \tag{14.92}$$

$$r \sim \mathcal{L_N}(-4; \ 3) \tag{14.93}$$

The width of the observation window is set to be $T = 5\varepsilon$. Thus, the SSA-PDP is completely specified according to Equation (14.91). Finally, the local PDPs are generated by computing the normalized energy gains $G_k^{(i)}$ of every bin k and every location i as Gamma distributed

Table 14.29   Statistical models and parameters

Global parameters $\Rightarrow \bar{G}_{tot}$ and $\bar{G}_k$	
Path loss	$PL = \begin{cases} 20.4 \log_{10}(d/d_0), & d \leq 11 \text{ m} \\ -56 + 74 \log_{10}(d/d_0), & d > 11 \text{ m} \end{cases}$
Shadowing	$\bar{G}_{tot} \sim \mathcal{L_N}(-PL; \, 4.3)$
Decay constant	$\varepsilon \sim \mathcal{L_N}(16.1; \, 1.27)$
Power ratio	$r \sim \mathcal{L_N}(-4; \, 3)$
	Local parameters $\Rightarrow G_k$
Energy gains	$G_k \sim \Gamma(\bar{G}_k; \, m_k)$
	$m_k \sim \mathcal{T_N}\big(\mu_m(\tau_k); \, \sigma_m^2(\tau_k)\big)$
m values	$\mu_m(\tau_k) = 3.5 - \dfrac{\tau_k}{73}$
	$\sigma_m^2(\tau_k) = 1.84 - \dfrac{\tau_k}{160}$

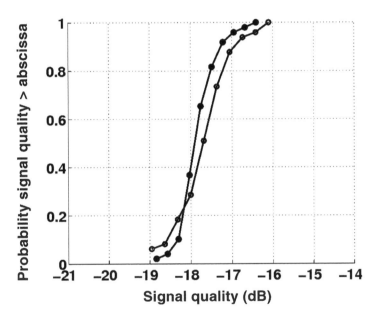

Figure 14.60 Comparison between the measured and simulated CDF of the signal quality received in an exemplary room. The circles represent the experimental data, and the dots represent the simulation results.

independent variables. The Gamma distributions have averages given by Equation (14.90), and the m_ks are generated as independent truncated Gaussian random variables

$$m_k \sim T_\mathcal{N}\big(\mu_m(\tau_k); \, \sigma_m^2(\tau_k)\big) \qquad (14.94)$$

with $\mu_m(\tau_k)$ and $\sigma_m^2(\tau_k)$ given by Equations (14.79) and (14.80). These steps are summarized in Figure 14.57 and Table 14.29.

Some results are shown in Figures 14.58–60.

15

Adaptive 4G Networks

15.1 ADAPTIVE MAC LAYER

In this section we discuss a centralized quasi-asynchronous DS-CDMA packet radio network (PRN) with adaptive bit rate and optimal packet size in low mobility environments. The channel load sensing protocol (CLSP) is used to control the packet access on the uplink so that contention is avoided and throughput is maximized. Due to the high uncertainty of radio channels and imperfect power control (PC), the performance of CLSP/DS-CDMA may suffer from notable degradations. The PC problem is essential to the capacity and coverage of interference-limited DS-CDMA systems, in particular on the uplink. It often requires a fast closed loop PC along with an open loop PC for initial power setting. In a PRN, the closed loop PC appears to be impractical due to the connectionless nature of datagram transmissions. So, the near–far effect represents a major factor affecting the design and performance of PC. This section introduces a system with bit rate adaptation, based on a location dependent rate and power (LDRP) resource allocation. The protocol compensates for PC inaccuracy and improves spectral efficiency of the system. The optimal packet size and data rates are derived to enhance robustness of radio transmissions over correlated fading to reduce the inter-cell interference and improve energy efficiency over the protocol overhead. The modeling includes the impacts of radio propagation attenuation, correlated fading, spatial user distribution, user mobility, and traffic load. Based on such modeling, optimal adaptive mechanisms are developed and performance characteristics are derived.

An ALOHA/DS-CDMA PRN [1250–1255] is a simple and practical system for providing a wide range of communication services for wireless data users in low mobility environments. It is used for both military and commercial applications, such as tactical networks, satellite communications, ad hoc networks, wireless LANs, packet-access domains of cellular networks, etc. The pure ALOHA systems, however, provide a poor throughput delay performance. Throughout the years, a significant amount of research effort has been invested in studying the media access control (MAC) for improving those characteristics,

Advanced Wireless Communications. S. Glisic
© 2004 John Wiley & Sons, Ltd ISBN: 0-470-86776-0

resulting in numerous MAC schemes. The centralized quasi-asynchronous (unslotted or spread-slotted) CLSP/DS-CDMA, described in [1250–1253], is a simple and effective system that outperforms the conventional ALOHA/DS-CDMA. It also overcomes the hidden-terminal problem of the distributed carrier sense multiple access, CSMA/DS-CDMA. In a CLSP/DS-CDMA system, a hub is responsible for sensing the channel load, which represents the received multiple access interference (MAI). If overload is about to occur, it starts using interference cancellation techniques, described in Chapter 5, to remove the interference or to force users to refrain from transmission with feedback control. Thus, in ideal conditions, the contention of the multiaccess DS-CDMA channel can be avoided, resulting in optimal resource utilization and throughput. The performance of non-adaptive CLSP systems, under the perfect PC assumption, has been investigated extensively [1250–1253]. It was found that, due to the high uncertainty of radio channels and imperfect PC, the system suffers from notable performance degradations.

The PC problem is essential to the capacity, coverage and performance of interference-limited DS-CDMA systems, in particular on the uplink [1254–1258]. It is desired to have all signals received at the hub with exact targets of signal to interference ratio (SIR) for the required radio performance. To solve the PC problem properly, it often requires a fast closed loop PC along with an open loop PC for initial power setting. Such solutions are used, for example, in mobile cellular systems [1254–1258]. In a PRN, the closed loop PC over a short transmission time interval (TTI) of a radio packet appears to be impractical. The accuracy of open loop PC, on the other hand, is inversely proportional to the dynamic range of the signal variation. This variation is often large and uncertain in radio networks [1254–1255]. Therefore, in order to guarantee the performance requirements, the design of the non-adaptive PRN tends to be based on worst-case scenarios of power consumption and resource utilization. Recent research results have demonstrated that adaptive mechanisms can be used at all layers of protocol stacks to accommodate the dynamics of the wireless channel [1258, 1259], leading to crosslayer optimization. This section discusses the possibility of bit rate adaptation and packet-size optimization to solve the PC problem and to enhance the performance of the CLSP/DS-CDMA PRN in low mobility environments.

The concept of trading off transport formats, including bit rate, packet length in bits and TTI in milliseconds, with radio performance is discussed in [1254–1258], [1259–1267]. There are numerous papers investigating adaptive rate and packet size in different contexts, such as adaptive modulation and coding, adaptive spreading factor, adaptive multicodes, etc. [1258–1261]. In this section we focus explicitly on PC problems and inter-cell interference reduction in energy-efficient design of MAC protocols. The system keeps the transmit power of mobile terminals at a constant level (within a small range of a preset target) throughout the cell. In the meantime, it adapts the user bit rate to the distance between the terminal and the hub using a specified LDRP RAP. The abbreviation stands for location dependent rate and power resource allocation protocol. The closer the user is to the hub, the higher the bit rate required to transmit with, and vice versa. As we will see later, this bit rate adaptation substantially reduces the PC headroom and stabilizes the transmit power of mobile terminals. It reduces the peak transmit power (and power consumption) of the mobile unit. It increases throughput of datagram packet transmissions, due to reduced inter-cell interference, and enhances system capacity and coverage.

To discuss the packet length adaptation, we should be aware that the open loop PC, in general, is not efficient against the impact of fading, because using an additional power rise to eliminate the fading impact means a significant increase in the interference. The diversity techniques, using a RAKE receiver operating in frequency selective fading channels, for example, can reduce the impact of deep fades [1268–1269]. In low mobility environments of the PRN, the Doppler bandwidth is small. Therefore, the fading process is highly correlated causing burst error states of the channel. The burst error states expand over tens or hundreds of milliseconds. So, in this case, the forward error correcting (FEC) channel coding is not efficient [1256, 1263–1265]. It is also well known that the smaller the packet size (length in bits or TTI in milliseconds), the higher the average probability of successful packet transmission over fading channels. At the same time, the shorter the packet, the heavier the protocol overhead is. Thus, there is an optimal packet size for radio transmission over fading channels so that the actual goodput, defined as the effective data rate successfully transmitted, excluding the protocol overhead, is maximized. The goodput is used to measure the energy efficiency of mobile terminals as well as the spectral efficiency of the PRN. This section presents such an optimal packet size for the PRN to improve robustness of packet transmissions over correlated fading channels, while keeping the variations of transmit power of mobile terminals low. The correlation between the fade/interfade duration statistics and the packet size is also discussed.

So, location dependent bit rate adaptation and the packet size optimization together, not only help to resolve PC problems, but also improve efficiency of resource utilization for the PRN. They stabilize the transmit power of mobile terminals throughout the cell allowing the terminals to operate with much lower maximum power requirements, reduce inter-cell interference, and therefore give the possibility for introducing new systems with minimal environmental and health risks. Furthermore, under certain conditions, such adaptive schemes improve the performance of the PRN without the need for excessive and power consuming signal processing. This section also provides a comprehensive analytical tool for investigating the simultaneous impacts of radio propagation attenuation, correlated fading, spatial user distribution, user mobility, and MAI statistics on the PRN performance.

15.1.1 Signal variations and the power control problem

In a centralized PRN, the received signal power varies depending on the distance, the shadowing and the multipath fading between the mobile terminal and the hub. From Chapter 14, it can be expressed as

$$P_{rx} = P_{tx} d^{-\alpha} 10^{(\zeta/10)} \phi \qquad (15.1)$$

where P_{tx} and P_{rx} are the transmitted and the received power of the signal at the hub respectively, d is the distance between the terminal and the hub; α is the path loss exponent (in the range of 2 to 5); ζ is the attenuation in dB due to shadowing (assumed to be a Gaussian random variable with zero mean and standard deviation 8dB); and ϕ represents the impact of correlated fading in low mobility environments.

The received energy per bit per interference density, for the given bit rate R is

$$(E_b/I_0)_R = (P_{rx}/I)(W/R) \qquad (15.2)$$

where I is the received wideband interference power including the background noise, and W is the signal chip rate. The ratio (P_{rx}/I) is the SIR, and (W/R) is the processing gain. From Equations (15.1) and (15.2) we have

$$(E_b/I_0)_R = (P_{tx}/I)W(d^{-\alpha}/R)10^{(\zeta/10)} \phi \qquad (15.3)$$

Due the randomness of signal variations, $(E_b/I_0)_R$ is a random variable as well. The PC is used for keeping $(E_b/I_0)_R$ above a specified target, γ_R, for the required BER/PER performance, yet as close to γ_R as possible to minimize MAI for capacity enhancement. The probability that the link quality requirement is maintained is $P\{(E_b/I_0)_R \geq \gamma_R\}$.

15.1.2 Spectral efficiency and effective load factor of the multirate DS-CDMA PRN

In the case of N simultaneous packet transmissions with different bit rates, (E_b/I_0) of user j with bit rate R_j is given by

$$y_j = (E_b/I_0)_j = \frac{P_{\text{rx},j}}{I_{\text{total}} - P_{\text{rx},j}} \frac{W}{R_j} \tag{15.4}$$

where $P_{\text{rx},j}$ is the received signal power from user j, I_{total} is the total received wideband power at the hub. Parameter I_{total} may include own-cell interference, I_{own}, other-cell interference, I_{other}, and background noise, N_0. Solving Equation (15.52) for $P_{\text{rx},j}$ gives

$$P_{\text{rx},j} = u_j I_{\text{total}}, \quad u_j = \left[1 + \frac{W}{(E_b/I_0)_j R_j}\right]^{-1} \tag{15.5}$$

The variable u_j in Equation (15.5) represents the load factor of user $J's$ transmission. Thus, the normalized load of the multiaccess channel is

$$\sum_j u_j = \sum_j P_{\text{rx},j}/I_{\text{total}} = I_{\text{own}}/I_{\text{total}} \tag{15.6}$$

In the equilibrium condition with the maximum tolerable cell interference, $I_{\max}$, we have

$$\lim_{I_{\text{total}} \to I_{\max}|N} E\left[\sum_{j|N} u_j\right] = \lim_{I_{\text{total}} \to I_{\max}|N} E\left[I_{\text{own}}/I_{\text{total}}\right] \tag{15.7}$$

Providing that $I_{\text{total}} = I_{\text{own}} + I_{\text{other}} + N_0$, $E[N_0/I_{\text{total}}|I_{\text{total}} \to I_{\max}] = \eta$, $E[I_{\text{other}}/I_{\text{own}}] = \iota$, and the link quality requirement is maintained, by using Equation (15.4–15.7) the upper limit on Equation (15.6), which is an effective load representing the spectral efficiency, can be approximated with the ratio $(1 - \eta)/(1 + \iota)$. From Equation (15.4–15.7) the average cell capacity for a class of users with the same bit rate R and target E_b/I_0 ratio γ_R, defined as the maximum number of tolerable simultaneous transmissions, C_R, can be approximated by

$$C_R = \left\lfloor \frac{1 - \eta}{(1 + \iota)\varpi_R} \right\rfloor \tag{15.8}$$

where $\lfloor x \rfloor$ denotes the largest integer that does not exceed the real number x, ϖ_R is the effective load factor of a packet transmission within the user class, which can be calculated with the formula of u_j in Equation (15.5) given that $R_j = R$ and $(E_b/I_0)_j = \gamma_R$. The contention or the system outage state occurs in the DS-CDMA channel if

$$\sum_{j|N} u_j > (1 - \eta)/(1 + \iota) \tag{15.9}$$

The objective of MAC schemes in PRN is to prevent the contention and to optimize the system performance in terms of energy and spectral efficiency.

15.1.3 CLSP/DS-CDMA packet access and traffic model

In the quasi-asynchronous CLSP/DS-CDMA, the users communicate via the hub using different code sequences for packet transmissions with the same link quality requirement. The user data is coded and segmented into transport blocks. Then, a header that contains system-specific addresses, transmission control information and error control fields is added to each transport block to form a radio packet. The packet header is assumed to have a fixed length of H bits. The transmission of radio packets over the air toward the hub is controlled by the CLSP, which attempts to prevent the condition in Inequality (15.9). For instance, in the non-adaptive system, it keeps the number of simultaneous transmissions under the average system capacity given by Equation (15.8). The hub is responsible for sensing the channel load or the number of simultaneous transmissions. It broadcasts control information periodically in a downlink control channel. The control signal is either a soft 'inhibit' signal when the channel load is reaching a certain threshold, or a 'transmission free' signal otherwise. The terminal that has a packet to send will listen to the control channel and decide whether to transmit or to refrain from transmission in a non-persistent fashion. The above discussion assumes identical cell structures, zero propagation delays, perfect sensing of the number of ongoing transmissions, and perfect radio reception on the downlink.

The traffic model is based on the following assumptions. The system population is infinite. The scheduling of packet transmissions at the mobile terminal, including the re-transmission of unsuccessful packets, is randomized sufficiently so that the offered traffic of each user is the same and the overall number of packet arrivals at the hub is generated according to the Poisson process with rate λ.

15.1.4 Bit rate adaptation

As discussed above, in PRNs the PC often relies on an initial power setting at the beginning of each packet transmission. The transmit power is then kept unchanged during the short packet interval. The power setting in the open loop PC is based on an estimate of the path loss at the mobile terminal by using a downlink beacon signal. The dynamic range of signal variations largely affects the accuracy of the PC, and therefore the performance of interference-limited DS-CDMA systems. The bit rate adaptation compensates the dominant component of the path loss, the near–far effect. It adapts the bit rate R to the distance d, by keeping factor $(d^{-\alpha}/R)$ in Equation (15.3) approximately constant throughout the cell, resulting in approximately constant P_{tx}.

In particular, we assume that the system supports $(M + 1)$ different bit rates, denoted by R_m with $m \in \mathbf{M}, \mathbf{M} = \{0, 1, \ldots, M\}$. Let R_0 be the basic bit rate, $R_0 = 32$ kbps, specified for the cell coverage taking into account the power restriction of the mobile terminal on the uplink, and $R_m = 2^m R_0$. Let d_0 be the normalized radius of the cell coverage, $d_0 \equiv 1$. The distance d between the terminal and the hub therefore takes a value in the (0, 1] interval. In the near–far resistant bit rate adaptation, the mobile location resolution is specified with a division of the cell area into $(M + 1)$ rings centered to the hub. The normalized radius of the ring boundary, denoted by d_m, is given by $d_m = 2^{-m/\alpha}$ for all $m \in \mathbf{M}$. Thus, $d_m^{-\alpha}/R_m \equiv 1/R_0$ is constant regardless of the distance. The bit rate adaptation mechanism can be formally

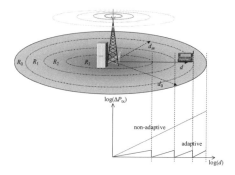

Figure 15.1 Rate-location-power resolution for near–far resistant bit rate adaptation (ΔP_{tx} is the normalized average of transmit power due to the near–far effect).

expressed as

$$\text{If} \quad d_{m+1} < d \leq d_m \quad \text{then} \quad R = R_m \quad \text{for all} \quad m \in \mathbf{M}, \quad \text{and} \quad d_{M+1} = 0. \quad (15.10)$$

The packet transmission of mobile terminals from an outer ring with a smaller index m will use a lower bit rate R_m. The location awareness of mobile terminals is feasible, especially in low mobility environments. The fineness of the rate-location-power resolution increases with larger M and smaller α. This is illustrated in Figure 15.1.

The near–far resistant bit rate adaptation, stabilizes the transmit power of mobile terminals, thus improving the PC accuracy, reducing the inter-cell interference, and enhancing the spectral efficiency and system performance. As an illustration, let us consider the following simplified scenario. The system consists of identical hexagonal cells using omnidirectional antennas. The cell of interest is in the middle. The transmit power of mobile terminals, P_{tx}, is kept constant. It is assumed that the active users located outside the typical $5d_0$ radius circle area around the hub do not cause any interference to the cell. Let the spatial user distribution (SUD) density function be $f(d, \theta)$. The mean of $I_{\text{other}}/I_{\text{own}}$ has an upper bound that can be represented as

$$E\left[I_{\text{other}}/I_{\text{own}}\right] \leq \frac{\displaystyle\int_0^{2\pi} \int_{d_0}^{5d_0} P_{\text{tx}} d^{-\alpha} f(d, \theta) \, \mathrm{d}d \, \mathrm{d}\theta}{\displaystyle\int_0^{2\pi} \int_{0+}^{d_0} P_{\text{tx}} d^{-\alpha} f(d, \theta) \, \mathrm{d}d \, \mathrm{d}\theta}$$

In the case where the SUD is a two-dimensional uniform distribution and $\alpha = 3$, the right-hand side of the above expression gives the numerical result of about 5 %. This is remarkably small compared to the 40 % − 65 % of the same factor in the non-adaptive system with perfect PC as shown in [1254]. In general, the near–far resistant bit rate adaptation allows the transmit power of mobile terminals to be kept within a small range of a preset target throughout the cell. The impact of fading and the fade-margin setting, as well as the correlation between fade/interfade duration and packet size are investigated in the next section.

Table 15.1   Fade statistics of fading channels with diversity (based on the analytical results reported in [1268]) © 1988, IEEE

F	Non-diversity		SC		EGC		MRC	
(dB)	LCR$/f_d$	AFD$*f_d$	LCR$/f_d$	AFD$*f_d$	LCR$/f_d$	AFD$*f_d$	LCR$/f_d$	AFD$*f_d$
20	0.2482	0.0401						
15	0.4319	0.0721						
10	0.7172	0.1327	0.1365	0.0663	0.0938	0.0657	0.0717	0.0652
5			0.5571	0.1319	0.4076	0.1277	0.3249	0.1250
0			1.1658	0.3427	1.0279	0.3066	0.9221	0.2866

15.1.5 The correlated fading model and optimal packet size

The application of a PRN is primarily targeted for low mobility environments. The fading process is therefore highly correlated. Let us consider two important fade statistics: the level crossing rate (LCR) and the average fade duration (AFD), which characterize the rate of occurrence and the average length of burst errors in fading channels, respectively [1258, 1267, 1268–1273] . These are functions of the ratio $\rho = R_s/R_{rms}$, where R_s is the specified signal level and R_{rms} is the local root mean square amplitude of the fading envelope, and the maximal Doppler frequency $f_d = f_c/c$, where f_c is the carrier frequency, v is the velocity of the mobile user, and c is the speed of light [1257]. The fade margin F, in decibels, is defined as $F = -20\log(\rho)$ that represents the power rise in the transmit power target in order to compensate the impact of fading. Thus, studying LCR and AFD allows us to relate the rate of change of the received signal to the target signal level and the user velocity. This is necessary for the design of optimal packet radio transmission over burst error correlated fading channels.

The statistics of LCR and AFD in correlated fading channels with diversity have been studied, for example, in [1268–1269]. Table 15.1, based on [1268], illustrates the behavior of LCR and AFD for different fade margins and diversity techniques with an assumption of perfect two-branch combining in multipath Rayleigh fading channels.

It includes results for non-diversity, selection combining (SC), equal-gain combining (EGC), and maximal ratio combining (MRC) cases. There is a significant gain in the fade margin, and thus in the power efficiency as well, with the diversity cases for the same range of fade statistics or similar fading conditions. It is therefore natural to use diversity techniques to combat deep fades in interference-limited DS-CDMA systems. In the following, we will assume that the MRC is used in the system. It is desirous to have as small a fade margin as possible for the required PER performance. The probability of correct packet transmission over fading channels, in general, depends on the number of fades occuring during TTI, duration of burst errors, and capability of FEC coding. Let us define the following notation: t_f is the fade duration, i.e. the period of time the received signal r spends below the threshold level R_s, having PDF $g(t_f)$ and mean t_{f_avrg}; t_{if} is the interfade duration, i.e. the period of time between two successive fades, having PDF $h(t_{if})$ and mean t_{if_avrg}; t_{fia} is the fade inter-arrival time interval, i.e. the time interval between the time instants of two successive fades: $t_{fia} = t_f + t_{if}$, having PDF $s(t_{fia})$ and mean t_{fia_avrg}.

The mean value of the above variables relates to LCR and AFD as follows: $t_{fia\_avrg} = 1/$LCR; $t_{f\_avrg} = (1/\text{LCR})P\{r \le R_s\} = \text{AFD}$; and $t_{if_avrg} = (1/\text{LCR})P\{r > R_s\} = 1/\text{LCR} - $ AFD [1257, 1267]. Thus, these are functions of ρ and f_d as well. For $f_c = 2$ GHz and v less than 10 miles per hour, f_d is obtained in the range of 0–30 Hz. From the values of AFD$*f_d$ in Table 15.1, t_{f_avrg} can be expected to be of the order of tens of milliseconds. Therefore, a faded error block probably contains hundreds of bits, depending on bit rate and packet duration. In order to correct them with FEC coding, it may need an impracticably low coding rate or large interleaving depth. The FEC coding is therefore not yet efficient in burst error correlated fading channels [1256]. Let us denote T as the packet duration, or TTI, in milliseconds. The probability of successful packet transmission over fading P_{sf}, as a function of T, ρ, and f_d, can be expressed by [1262]:

$$P_{sf}(T, \rho, f_d) = \int_0^\infty \int_T^\infty (t_{if} - T)(t_{if} + t_f)^{-1} h(t_{if})\, g(t_f)\, dt_{if}\, dt_f \qquad (15.11)$$

The closed-form expressions for the PDF of the fade/interfade duration are not available for a general case, but are, rather, assessed on an individual basis. However, in the related literature so far, the Markov channel model is applied extensively [1254–1265, 1267, 1270, 1273–1276]. In particular, in the case of correlated fading channels with $t_{f\_avrg}/t_{if\_avrg} \ll 1$, it is reasonably accurate to assume that t_{if} and t_f are identically distributed and independent random variables having exponential PDF [1270–1271, 1274–1276]. The condition $t_{f\_avrg}/t_{if\_avrg} \ll 1$ is required for reliable data communications over burst error fading channels. In the system model based on the numerical results for the MRC case in Table 15.1, the ratio t_{f_avrg}/t_{if_avrg} ranges from 0.5 % to 36 % for the fade margin of 10 dB and 0 dB, respectively. It is 4 % if the fade margin is 5 dB, which is reasonable for the above condition as well as for the transmit power rise needed. It is below a 6 dB transmit power rise corresponding to $\rho = 0.5$, or 3 dB for the average SIR thereof. In the following, we set the fade margin to 5 dB.

It is obvious that the smaller the T, the higher the P_{sf}, and at the same time, the heavier the protocol overhead. For given constraints of the required PER, the user velocity v, the bit rate R, the fade margin F, and the length of the packet header H, one can find an optimal packet duration T so that the normalized goodput G_{sf}

$$G_{sf}(T, \rho, f_d) = P_{sf}(T, \rho, f_d)(TR - H)/TR \qquad (15.12)$$

is maximized. The goodput is defined above as the effective data rate successfully transmitted excluding the protocol overhead. The product TR in Equation (15.12) gives the value of the packet length L in bits, $L = TR$. One can further exploit the potential of adaptive packet size based on Equations (15.11) and (15.12). For example, T and/or L can be adapted to ρ and/or f_d along with adaptive R. These details are not included due to limited space. In the mobility equilibrium, we have

$$P_{sf}(T) = \int_{f_d} P_{sf}(T, \rho, f_d) p(f_d)\, df_d \qquad (15.13)$$

where $p(f_d)$ is the PDF of f_d resulting from the distribution of the user velocity in the

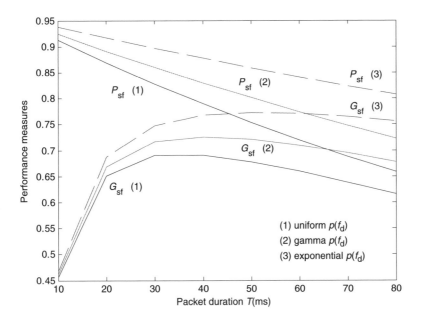

Figure 15.2 Optimal packet size for packet transmissions over correlated fading channels.

system equilibrium. Let us assume that f_d takes a discrete value in $\{1, 2, \ldots, f_{d-\max}\}$. Then the right-hand side of Equation (15.13) becomes

$$P_{sf}(T) = \sum_{f_d} P_{sf}(T, f_d) P(f_d) \qquad (15.14)$$

Figure 15.2 shows the equilibrium P_{sf} and G_{sf} characteristics versus T with $R = R_0 = 32$ kbps, $H = 160$ bits, $\rho = 10^{-5/20}$, and $f_{d-\max} = 30$ Hz for three different mobility scenarios. The most dynamic one is characterized with a uniform $p(f_d)$ over the limit range of f_d, for which the user velocity v is distributed evenly within 0–4.5 m/s. The least dynamic one is characterized with an exponential $p(f_d)$ with a mean of f_d about 7 Hz, for which the user velocity is distributed exponentially around a mean of 1m/s. For the one in between, we use a generic gamma PDF for $p(f_d)$ because of its flexibility and richness in modeling [1277]:

$$p(f_d) = \frac{1}{b^a \Gamma(a)} f_d^{a-1} e^{-f_d/b}$$

where a is the shape parameter, b is the scale parameter, and $\Gamma(x)$ is the gamma function. Thus, we can adjust the values of parameters a and b to obtain suitable PDFs for a number of mobility scenarios and expansions of the Doppler frequency range. This PDF family also takes the exponential PDF form as its special member when the shape parameter a is set to 1. For the moderate mobility scenario of Figure 15.2, we choose a gamma PDF with $a = 2$ and $b = 6$ for instance. The optimal packet size is obtained at T_0 around 30 ms, which is also the maximum TTI if PER is not to exceed 15 %. Based on that, we set the optimal packet size for the most dynamic scenario as $T_0 = 20$ ms and $L_0 = T_0 R_0 = 640$ bits; for the least dynamic one, $T_0 = 40$ ms and $L_0 = T_0 R_0 = 1280$ bits; and for the moderate one, $T_0 = 32$ ms and $L_0 = T_0 R_0 = 1024$ bits.

15.1.6 Performance

Let T_0 be the normalized unit of time, $T_0 \equiv 1$. Let us define the following notation. Λ is the offered system traffic, i.e. the average number of packet arrivals per normalized unit of time. Given that the packets arrive at the hub according to the Poisson process with rate λ and $T_0 \equiv 1$, we have $\Lambda = \lambda T_0 = \lambda$. The offered system traffic in terms of data rate is kept the same for both the non-adaptive and the adaptive systems, and takes the value ΛL_0. S is the system throughput (the average number of successful packet transmissions per normalized unit of time for a given offered traffic Λ). D is the average packet delay (the average time interval from the instant a given packet is generated to the instant the packet is transmitted successfully). G is the system goodput (the effective average data rate successfully transmitted excluding protocol overhead). For instance, in the system with a fixed packet length of L bits, including a packet header of H bits, the system goodput is $G = S(L - H)$.

15.1.6.1 Performance characteristics of the non-adaptive system

In this case we have $R = R_0$, $T = T_0$ and $L = L_0$. The system capacity, $C = C_0$, is obtained from Equation (15.8). The formal notation of the queuing system model for this system is: exponentially distributed inter-arrival time; deterministic service time as the packet duration T; finite number of servers equal to the system capacity C; no waiting room and infinite population. Let us define n as the system state (the number of ongoing packet transmissions in the system) and p_n as the steady-state probability of the system given by:

$$p_n = \frac{\beta^n}{n!} \left(\sum_{i=0}^{C} \frac{\beta^i}{i!} \right)^{-1} , \quad 0 \leq n \leq C \tag{15.15}$$

where $\beta = \lambda T$ [1278]. The equilibrium probability of successful packet transmission, P_{succ}, depends on the following three factors. The first one is the equilibrium probability that the packet is not blocked by the CLSP, which is given by

$$P_{nb} = 1 - p_C \tag{15.16}$$

The second one is the equilibrium probability that the packet is not corrupted by the system outage, which can be expressed as $P_{nso} = E[$the number of transmissions not hit by the system outage$] / E[$the number of transmissions in the system$]$. This probability can be expressed as

$$P_{nso} = \sum_{n=0}^{C} n p_n P_{ok}(n) / \sum_{n=0}^{C} n p_n \tag{15.17}$$

where $P_{ok}(n)$ is the probability that n simultaneous transmissions are not corrupted by the system outage, i.e. the condition in Inequality (15.9) does not occur. The impact of imperfect PC, in the long run, is often characterized with a log-normal error of the average E_b/I_0 ratio with a standard deviation of σ in decibels [1254–1255]. Therefore, $P_{ok}(n)$ can be expressed as

$$P_{ok}(n) = 1 - Q \left(\frac{C - E[Z|n]}{\sqrt{\text{Var}[Z|n]}} \right) \tag{15.18}$$

where $Q(x)$ is the standard Gaussian integral, Z is the normalized MAI caused by n users, $E[Z|n] = n \exp((\varepsilon\sigma)^2/2)$; $\mathrm{Var}[Z|n] = n \exp(2(\varepsilon\sigma)^2)$; $\varepsilon = \ln(10)/10$ [1254]. The third factor is the equilibrium probability that the packet is not faded, which is given by Equation (15.14) with $T = T_0$. Therefore, we have

$$P_{\text{succ}} = P_{\text{nb}} P_{\text{nso}} P_{\text{sf}} \tag{15.19}$$

The average system throughput for the offered system traffic Λ is given by:

$$S = \Lambda P_{\text{succ}} \tag{15.20}$$

The ratio $S/\Lambda \equiv P_{\text{succ}}$ is called the normalized throughput.

The normalized average packet delay is decomposed into two components. The first one is the normalized average waiting time for accessing the channel, D_{w}. This can be calculated by the normalized average delay time, which follows Little's result at the arrival-side [1278], minus the normalized serving time T

$$D_{\text{w}} = (\lambda T/E[n] - 1)T/T_0 \tag{15.21}$$

where $E[n]$ is the average number of packets being served in the system. The second one is the normalized average resident time from the instant the packet enters the system to the instant the packet leaves the system successfully, D_{r}. This is obtained by using Little's result at the departure-side:

$$D_{\text{r}} = E[n]/S \tag{15.22}$$

where S is the average system throughput. The normalized average packet delay is therefore obtained as

$$D = D_{\text{w}} + D_{\text{r}} \tag{15.23}$$

The system goodput is given by

$$G = S(L - H) \tag{15.24}$$

The normalized goodput is given by the ratio $G/(\Lambda L_0) \equiv P_{\text{succ}}(L - H)/L$.

15.1.6.2 *Performance characteristics of the adaptive system*

This system uses bit rate adaptation while having the same cell coverage and offered system traffic in terms of data rate as those of the non-adaptive system. Either the packet length L, or the packet duration T, can be kept the same as in the non-adaptive system, i.e. L_0 or T_0, which is optimized. In the first option, with $L = RT = L_0$, the packet transmission with higher bit rate will have shorter TTI. That compensates the increase of MAI on a time scale within the T_0 interval, resulting in the same normalized power consumption per packet for all packet transmissions. In the latter option, with $T = L/R = T_0$, the packet transmission with higher bit rate will have longer packet length and smaller offered traffic intensity. That compensates the increase of MAI on a time scale over the T_0 interval, resulting in less protocol overhead but longer packet delay compared to the first option. Let us denote T_m as the packet duration corresponding to bit rate R_m. Thus, we have $T_m = 2^{-m} T_0$ for the first option, and $T_m = T_0$ for the latter one.

The queuing system model for the adaptive system is the stochastic model of multirate loss networks [1279]. The offered system traffic is composed of portions corresponding to different bit rates, which are generated by active users from different ring areas of the cell due to the adaptation mechanism given by Expression (15.10). Let us define λ_m as the rate of packet arrivals from ring m using bit rate R_m. The value of λ_m depends on SUD, and since the offered system traffic in terms of data rate is $\Lambda L_0 = \lambda R_0 T_0$, the same as that of the non-adaptive system, λ_m is given by

$$\lambda_m = \lambda \frac{R_0 T_0}{R_m T_m} \int_{d_{m+1}}^{d_m} \int_0^{2\pi} f(d, \theta) \, dd \, d\theta \tag{15.25}$$

where $f(d, \theta)$ is the PDF of SUD.

Let us define $\mathbf{n}$ as the variable vector of the system states, $\mathbf{n} = [n_m, m \in \mathbf{M}]$, where n_m is the number of ongoing packet transmissions using bit rate R_m; and ω as the constraint vector of the effective load factors, $\omega = [\omega_m, m \in \mathbf{M}]$, where ω_m is defined as ω_R in Equation (15.8) with $R = R_m$.

The CLSP controls the packet access on the uplink so that $\mathbf{n}$ is kept in the set of allowed system states defined as $\Omega = \{\mathbf{n}, \mathbf{n}\omega \le (1 - \eta)/(1 + \iota)\}$ based on Inequality (15.9). Let us define $p(\mathbf{n})$ as the steady-state probability of the system, $\mathbf{n} \in \Omega$, with the product form solution [1279]:

$$p(\mathbf{n}) = \frac{1}{G_0} \prod_m \frac{\beta_m^{n_m}}{n_m!}, \quad G_0 = \sum_{\mathbf{n} \in \Omega} \prod_m \frac{\beta_m^{n_m}}{n_m!} \tag{15.26}$$

where $\beta_m = \lambda_m T_m$. For a large set of Ω, the computational complexity with the above formulas is prohibitively high.

Let us define s as the effective load state of $\mathbf{n}$ packet transmissions, $\mathbf{n} \in \Omega$, given by the product $s = \mathbf{n}\omega$. It forms a set Ψ of possible effective load states corresponding to the set Ω of system states. The steady-state probability of the system, in terms of effective load states, can be calculated based on the so-called stochastic knapsack-packing approximation [1279] as follows.

$$p(s) = \frac{q(s)}{\sum_{s \in \Psi} q(s)} \tag{15.27}$$

with $q(s)$ given in recursive form as:

$$q(s) = \frac{1}{s} \sum_m \varpi_m \beta_m q(s - \omega_m), s \in \Psi^+, q(0) = 1, q(-) = 0$$

Based on the results of the non-adaptive system above, the equilibrium probability of successful packet transmission with bit rate R_m, denoted by P_{succ_m}, depends on the following three factors. The first one is the equilibrium probability that the packet is not blocked by the CLSP. This is given by

$$P_{\text{nb}\_m} = 1 - \sum_{s \in \Psi : s > 1 - \eta - \varpi_m} p(s) \tag{15.28}$$

The second is the equilibrium probability that the packet is not corrupted by the system outage. This can be expressed as $P_{\text{nso}} = E[$the number of transmissions not hit by the system outage]$/E$[the number of transmissions in the system]

$$P_{\text{nso}} = \sum_{\mathbf{n}\in\Omega} \mathbf{n}p(\mathbf{n})P_{\text{ok}}(\mathbf{n}) / \sum_{\mathbf{n}\in\Omega} \mathbf{n}p(\mathbf{n}) \tag{15.29}$$

where $P_{\text{ok}}(\mathbf{n})$ is the probability that $\mathbf{n}$ simultaneous transmissions are not corrupted by the system outage, i.e. the condition in Inequality (15.9) does not occur. Let us denote the standard deviation of the log-normal error of the average E_b/I_0 ratio corresponding to bit rate R_m, with σ_m in decibels. It can be expected that σ_m of the adaptive system is smaller than σ of the non-adaptive counterpart because the PC in the adaptive system is more stable and accurate as discussed above. To simplify the computation, we assume that $\sigma_m = \varphi$ for all m. Based on Equation (15.18), $P_{\text{ok}}(\mathbf{n})$ can be expressed by

$$P_{\text{ok}}(\mathbf{n}) = 1 - Q\left(\frac{\dfrac{1-\eta}{1+\iota} - E\left[Z\,|\mathbf{n}\right]}{\sqrt{\text{Var}\left[Z\,|\mathbf{n}\right]}}\right) \tag{15.30}$$

where Z is the normalized MAI caused by $\mathbf{n}$ users,

$$E\left[Z|\mathbf{n}\right] = \sum_m n_m \varpi_m \exp((\varepsilon\sigma_m)^2/2)$$

$$= \exp((\varepsilon\varphi)^2/2)\sum_m n_m \varpi_m = s\exp((\varepsilon\varphi)^2/2)$$

$$\text{Var}\left[Z|\mathbf{n}\right] = \sum_m n_m \varpi_m \exp(2(\varepsilon\sigma_m)^2) = s\exp(2(\varepsilon\varphi)^2)$$

Thus, $P_{\text{ok}}(\mathbf{n})$ can now be replaced with $P_{\text{ok}}(s)$ and Equation (15.29) is equivalent to

$$P_{\text{nso}} = \sum_{s\in\Psi} sp(s)P_{\text{ok}}(s) / \sum_{s\in\Psi} sp(s)$$

The third factor is the equilibrium probability that the packet is not faded. This is given by Equation (15.14) with $T = T_m$, i.e. $P_{\text{sf}}(T_m)$. Therefore, we have

$$P_{\text{succ}\_m} = P_{\text{nb}\_m}P_{\text{nso}}P_{\text{sf}}(T_m) \tag{15.31}$$

The average system throughput is given by:

$$S = \sum_m \lambda_m T_0 P_{\text{succ}\_m} \tag{15.32}$$

The normalized system throughput is given by the ratio $S/\sum_m \lambda_m T_0$. The normalized average packet delay is given by Equation (15.23) with the components

$$D_{\text{w}} = \left(\sum_m \lambda_m T_m \varpi_m / \sum_{s\in\Psi} sp(s) - 1\right)\left(\sum_m \lambda_m T_m / \sum_m \lambda_m T_0\right) \tag{15.33}$$

$$D_{\text{r}} = \sum_{s\in\Psi} sp(s) / \sum_m \lambda_m T_0 P_{\text{succ}\_m}\varpi_m \tag{15.34}$$

The system goodput is given by

$$G = \sum_m \lambda_m T_0 P_{\text{succ}\_m}(R_m T_m - H) \tag{15.35}$$

and the normalized goodput is $G/(\Lambda L_0)$.

15.1.6.3 Performance examples

Table 15.2 summarizes the system parameters. The adaptive system supports four different bit rates: 32 kbps, 64 kbps, 128 kbps and 256 kbps. The target E_b/I_0 is set to 5 dB, which includes an increase of 2.5 dB due to the fade margin of 5 dB.

The value of the above analytical method is in the modeling of the system dynamics due to effects of numerous simultaneous factors. These factors include imperfect PC, propagation attenuations, SUD, user mobility, and offered traffic intensity over burst error correlated fading channels. The optimal packet size, obtained from the optimization presented in the above section is used for both the non-adaptive and the adaptive systems. That is, $T_0 = 20$ ms and $L_0 = 640$ bits for the dynamic mobility scenario characterized by a uniform $p(f_d)$; or $T_0 = 40$ ms and $L_0 = 1280$ bits for the less dynamic one with an exponential $p(f_d)$. To investigate the effects of SUD, two simple functions for $f(d, \theta)$ are used, the one-dimensional uniform PDF and the two-dimensional uniform PDF. The first one is often used to model indoor office environments, where the users are located along the corridors.

Table 15.2   Summary of the system parameters

Name	Definition	Values
f_c	The carrier frequency	2 GHz
W	The DS-CDMA chip rate	4.096 Mcps
η	The ratio of the background noise power normalized to the maximum tolerable received interference	-10 dB
α	The path loss law exponent	3 (4)
F	The fade margin	5 dB
R_0	The primary bit rate	32 kbps
T_0	The optimal packet duration	20 ms (40 ms)
L_0	The optimal packet length	640 bits (1280 bits)
$M+1$	The number of supported bit rates	4
γ_m	The target E_b/I_0 ratio of packet transmission with bit rate R_m	5 dB
σ	The standard deviation of log-normal error of the average E_b/I_0 for the non-adaptive system	3 dB
φ	The standard deviation of log-normal error of the average E_b/I_0 for the adaptive system	2 dB
f_{d_max}	The upper bound of maximal Doppler frequency	30 Hz
H	The length of the packet header plus trailer	160 bits

The latter one is for more open areas. The offered system traffic Λ varies and takes values in $\{4, 8, 16, \ldots, 60\}$ packet arrivals per normalized unit of time T_0. The simulation results are based on Omnet $++$ [1277], a multipurpose discrete-event simulator. The channel sampling period is set to 0.1 ms. The simulation is run over 100 000 packet arrivals for each rate of the offered system traffic.

Figures 15.3–15.7 present the performance of both the non-adaptive and the adaptive systems in different system scenarios under the effects of different system parameters. Each figure consists of two subfigures: (a) the normalized system goodput and (b) the throughput-delay tradeoffs. The adaptive system outperforms the non-adaptive counterpart in all cases.

Figures 15.3–15.4 show the impact of user mobility on the performance and how to choose a practical option of the optimal packet size for the adaptive system. For these figures, we use a single-cell scenario with the two-dimensional uniform $f(d, \theta)$ and the path loss law exponent $\alpha = 3$. The optimal packet size, as mentioned above, includes the optimal packet length in bits and the optimal packet duration in milliseconds. The less dynamic mobility scenario in Figure 15.4 has larger optimal packet size and better goodput performance than the more dynamic one in Figure 15.3, due to fading. This is also shown in Figure 15.2. There are two options for the optimal packet size, with the optimal packet length in bits, or the optimal TTI in milliseconds for the adaptive system. In the first option, keeping the packet length L constant at the optimized L_0 results in a constant $(L - H)/L$ ratio. Thus, there is no further gain in the goodput, but instead, a significant gain in the packet delay, due to the shorter TTI of packet transmissions with higher bit rates. However, if the optimal packet length is not large enough, the packet transmission with a certain high bit rate may require an impractically short TTI. In the latter one, keeping the packet duration T constant at the optimized T_0 results in a better $(RT - H)/RT$ ratio for larger R. Thus, there is a gain in the goodput, but not in the packet delay. If the optimal packet duration is large enough, the gain in the goodput is not significant and the packet length in bits may become impractically large for a certain high bit rate. Based on this consideration and the results shown in Figures 15.3–15.4, $T = T_0 = 20$ ms for the dynamic mobility scenario, and $L = L_0 = 1280$ bits for the less dynamic one, are used.

Figures 15.5–15.6 present the impacts of SUD and path loss exponent on the performance characteristics. For these figures, the less dynamic mobility scenario is used with $L = L_0 = 1280$ bits and a single cell. The adaptive system is sensitive to the SUD and the path loss exponent, which affect the patterns of the user location resolution and the offered system traffic as formulated in Expression (15.10) and Equation (15.25). The more users put to inner rings with one-dimensional uniform $f(d, \theta)$ in Figure 15.5, and/or $\alpha = 4$ in Figure 15.6, the more the bit rate is boosted and the greater the reduction in TTI of packet transmissions, resulting in better goodput and throughput delay performance for the adaptive system. The differences are not of significance though. One should also keep in mind that the larger path loss law exponent and the non-uniform SUD cause larger headroom of the transmit power, especially in the non-adaptive system.

The increase in goodput, i.e. energy efficiency, offered by the proposed bit rate adaptation and packet length optimization, can well be over 100 %, as shown in Figure 15.7 for a multicell system scenario with identical cells and two-dimensional uniform SUD. The inter-cell interference in the adaptive system (with the factor $\iota = E[I_{\text{other}}/I_{\text{own}}]$ set to the maximum of 5 %) is much smaller than in the non-adaptive counterpart (with ι set to the minimum of 40 %) as presented above. This, added to the advantages of the adaptive system as presented above with a single-cell scenario, significantly enhances system performance and cell deployment.

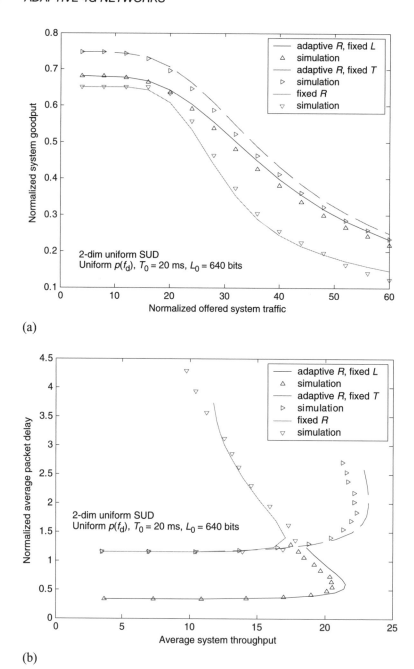

(a)

(b)

Figure 15.3 Single-cell system performance in dynamic mobility scenario. (a) Normalized system goodput; (b) throughput – delay tradeoffs [1280] © 2002, IEEE.

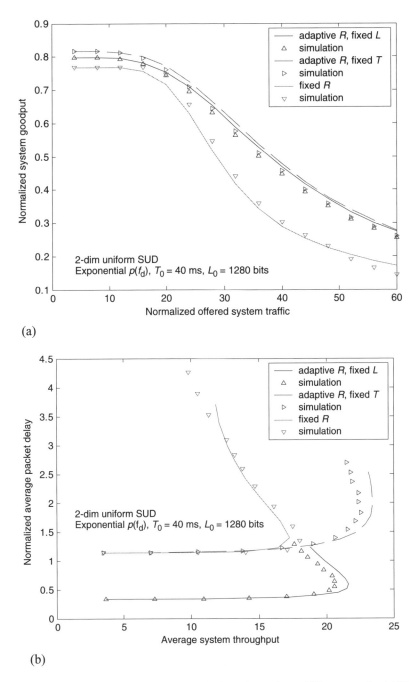

(a)

(b)

Figure 15.4 Single-cell system performance in less dynamic mobility scenario. (a) Normalized system goodput; (b) throughput – delay tradeoffs [1280] © 2002, IEEE.

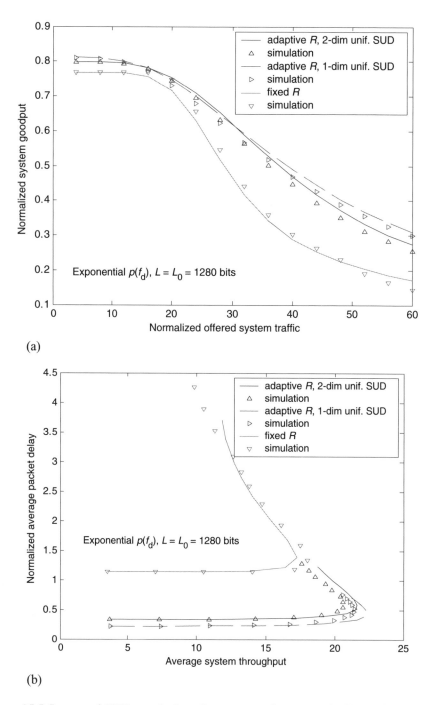

(a)

(b)

Figure 15.5 Impact of SUD on single-cell system performance. (a) Normalized system goodput; (b) throughput – delay tradeoffs [1280] © 2002, IEEE.

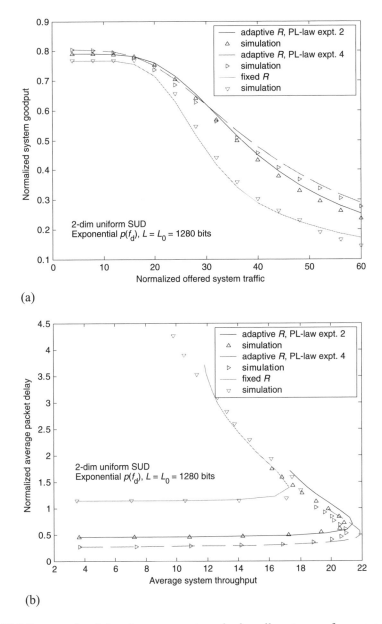

Figure 15.6 Impact of path loss law exponent on single-cell system performance. (a) Normalized system goodput; (b) throughput – delay tradeoffs.

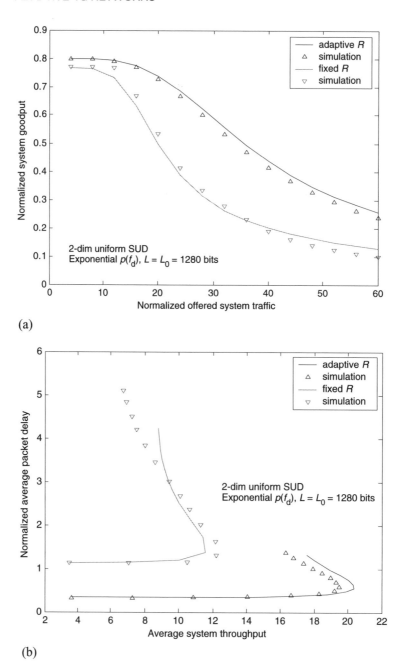

Figure 15.7 Multicell system performance. (a) Normalized system goodput; (b) throughput – delay tradeoffs.

15.2 MINIMUM ENERGY PEER-TO-PEER MOBILE WIRELESS NETWORKS

This section describes a distributed network protocol optimized for achieving the minimum consumption energy for randomly deployed ad hoc networks. A position-based algorithm, based on [1281], is presented, which is supposed to set up and maintain a minimum energy network between users that are randomly deployed over an area and are allowed to move with random velocities. These mobile users are referred to as 'nodes' over the two-dimensional plane. The network protocol reconfigures the links dynamically as nodes move around, and its operation does not depend on the number of nodes in the system. Each mobile node is assumed to have a portable set with transmission, reception and processing capabilities. In addition, each has a low-power global positioning system (GPS) receiver on board, which provides position information within at least 5 m of accuracy. The principles of GPS system operation are discussed in Chapter 14. The recent low-power implementation of a GPS receiver [1282] makes its presence a viable option in minimum energy network design.

15.2.1 Network layer requirements

In peer-to-peer communications, each node is both an information source and an information sink. This means that each node wishes to both send messages to, and receive messages from, any other node. An important requirement of such communications is strong connectivity of the network. A network graph is said to be 'strongly connected' if there exists a path from any node to any other node in the graph [1283]. A peer-to-peer communications protocol must guarantee strong connectivity.

For mobile networks, since the position of each node changes over time, the protocol must be able to update its links dynamically in order to maintain strong connectivity. A network protocol that achieves this is said to be 'self-reconfiguring.' A major focus of this section is the design of a self-reconfiguring network protocol that consumes the least amount of energy possible.

In order to simplify the discussion, we take one of the nodes to be the information sink for all nodes in the network. We call this node the 'master site.' The master site can be thought of as the headquarters located at the edge of the digital battlefield, the supervisory station in a multisensor network, or the base station in a cellular phone system. All of these scenarios are special cases of peer-to-peer communications networks.

Each node knows its own instantaneous position via GPS, but not the position of any other node in the network, and its aim is to send its messages to the master site whenever necessary.

A protocol that solves the minimum energy problem with a single master site simultaneously solves the general peer-to-peer communications problem, because each node can independently be taken as a master site, and the optimal topologies can be superimposed. We take advantage of this simplification and concentrate on the problem with a single master site without loss of generality.

15.2.2 The power consumption model

In the network configuration algorithm, this section concentrates only on path loss that is distance-dependent. The algorithm does not depend on the particular value of the path

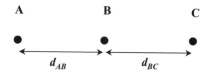

Figure 15.8 Three collinear nodes A, B and C.

loss exponent $n(n > 2$ for outdoor propagation models; see Chapter 14) and thus offers the flexibility to be applied in various propagation environments. So, since the transmit power falls as $1/d^n$, $n \geq 2$, as given by the path loss model, relaying information between nodes may result in lower power transmission than communicating over large distances.

For illustration, consider three nodes A, B and C on a line, as in Figure 15.8. Assume that all three nodes use identical transmitters and receivers. Node A wants to send a message to C. Let t denote the predetection threshold (in mW) at each receiver. In other words, the minimum power that a transmitter must radiate in order to allow detection at distance d meters away is td^n, where n is the exponent in the path loss model. *Assuming that node A knows the positions of B and C*, it has two options: it can transmit the signal directly to C, which requires a power consumption of td^n_{AC} at node A, or it can relay the message through node B and have it retransmit it with the minimum power needed for B to reach C. In this second case, the total transmit power consumption is $td^n_{AB} + td^n_{BC}$. In the case of three collinear nodes, it is easily seen that relaying the message through the middle node always comes at a lower total transmit power consumption than transmitting directly.

When the three nodes are allowed to lie on a two-dimensional plane, which is denoted by $\Re^2$, the option that costs less total power becomes a function of where the receive node is positioned. In the next section, we find the positions for the receive node, where relaying will always consume less total power than transmitting directly.

There is another source of power consumption that must be considered in addition to path loss. In the previous example, when node A relays through B, node B has to devote part of its receiver to receive and store node A's message. This additional power will be referred to as the receiver power at the relay node, and will be denoted by c. Each relay induces an additional receiver power to be consumed at the relay node. For the previous example, the total power consumption, including transmit and receiver power consumption in the transmission, is thus $td^n_{AB} + td^n_{BC} + c$ when node B is used as a relay.

The main goal of the section is to arrive at an algorithm that requires only local computation for updates, and requires as little global information as possible. A protocol requiring only local information is extremely advantageous for networks with mobile nodes, since delays associated with disseminating global information would be intolerable. From the perspective of power consumption, a distributed protocol running almost exclusively on local information requires transmission only over small distances. This in turn conserves the total power required for transmitting that information. A third advantage of the use of only local information is that it reduces the interference levels dramatically, since a user's communication with only nodes in its immediate surroundings causes little interference to nodes further away.

15.2.3 Minimum power networks

In order to investigate the implications of local information on power-efficient transmission, we consider three nodes in $\Re^2$, denoted by i, r and j. Node i is a node that wishes to transmit information to node j. Node i is referred to as the 'transmit node' and node j the 'receive node.' Node i considers the third node, r, to be used as a relay for transmission from i to j. Node r is called the 'relay node.' The aim is to transmit information from i to j with minimum total power incurred by i, j and r. By varying the position of j, we investigate under which conditions it consumes less power to relay through r. Below, the position of j is denoted by (x, y).

The following definitions will be used:

Relay region: The relay region $R_{i \to r}$ of the transmit–relay node pair (i, r) is defined to be

$$R_{i \to r} \equiv \left\{ (x, y) | P_{i \to r \to (x,y)} < P_{i \to (x,y)} \right\} \tag{15.36}$$

where $P_{i \to r \to (x,y)}$ denotes the power required to transmit information from node i to (x, y) through the relay node R, whereas $P_{i \to (x,y)}$ denotes the power required to transmit information from i to (x, y) directly.

Deployment region: Any bounded set in $\Re^2$ that has the position of the nodes in $\aleph$ as a subset is said to be a deployment region for the node set $\aleph$. The deployment region is introduced because, in practice, there is a finite area beyond which no nodes should be looking for neighbors with which to communicate. The boundaries of deployment regions can also be taken as known and impenetrable obstacles to communication. Then, the nodes near the edge can use this fact not to search unnecessarily beyond the deployment region.

Enclosure and Neighbor: The enclosure of a transmit node i is defined as the non-empty solution ε_i to the set of equations defined as

$$\varepsilon_i = \bigcap_{k \in N(i)} R^c_{i \to k} \cap D_\aleph \tag{15.37}$$

and

$$N(i) = \{ n \in \aleph \, | (x_n, y_n) \in \varepsilon_i, n \neq i \} \tag{15.38}$$

In Equations (15.37–15.38) R^c represents the complemnt of set R, and $D_\aleph$ denotes the deployment region for the node set $\aleph$. Each element of $N(i)$ is said to be a 'neighbor' of i and $N(i)$ is called the 'neighbor set' of i. A node i is said to be enclosed if it has communication links to each of its neighbors and no other node.

Enclosure graph: The enclosure graph of a set of nodes $\aleph$ is the graph whose vertex set is $\aleph$ and whose edge set is

$$\bigcup_{i \in \aleph} \bigcup_{k \in N(i)} l_{i \to k} \tag{15.39}$$

where $l_{i \to k}$ is the directed communications link from i to k.

Minimum power topology: A graph on the stationary node set $\aleph$ is said to be a minimum power topology on $\aleph$ if:

1. Every node has a directed path to the master site;

2. The graph consumes the least total power over all possible graphs on $\aleph$ for which 1) holds.

15.2.4 Distributed network routing protocol

The main idea in this protocol is that a node does not need to consider all the nodes in the network to find the global minimum power path to the master site. By using a very localized search, it can eliminate any nodes in its relay region from consideration and pick only those few links in its immediate neighborhood to be the only potential candidates. The protocol effectively operates in two phases: first, a local search executed by each node to find the enclosure graph, and second, a minimum cost search from the master site to every node. The cost metric is the total power required for a node to reach the master site along a directed path.

15.2.4.1 Search for enclosure (Phase 1)

In order for a protocol to find the enclosure graph, each node must find its enclosure and its neighbor set. Since computing enclosure requires knowledge of the positions of nearby nodes, each node broadcasts its position to its search region. The search region is defined as the region where a node's transmitted signal (and hence its position) can be correctly detected by any node in the region. *When searching for neighbors, a node must keep track of whether a node found is in the relay region of previous nodes found in the search.* The relay graph defined below is, in effect, a data structure which stores this information.

Relay graph of a node   Let A denote the set of all nodes that transmit node i has found thus far in its search. Let j and k be two nodes in A. Whenever $k \in R(j)$, we form a directional edge from j to k and denote it by $e_{j \to k}$. The relay graph of a transmit node i is defined to be the directed graph whose vertex set is a and whose edge set is

$$\bigcup_{j \in A} \bigcup_{k \in R(j)} e_{j \to k} \tag{15.40}$$

The relay graph of i is denoted by $G(i)$. $e_{j \to k}$ represents a relation between j and k based on their positions. It indicates that k lies in the relay region $R_{i \to j}$. It does not represent a communication link between j and k.

To find $N(i)$, namely the neighbor set of i, each node in the algorithm starts a search by sending out a beacon search signal that includes the position information of that node. Since every node executes the same algorithm, we will focus on a particular node, referred to as the transmit node. The transmit node also listens for signals from nearby nodes. When it receives and decodes these signals, it finds out the position of the nearby nodes and calculates the relay region for them. As we described in the discussion preceding the definitions of enclosure and the relay graph, the transmit node must keep only those nodes that do not lie in the relay regions of previously found nodes. Therefore, each time new nodes are found, the transmit node must update its relay graph.

The nodes that have been found thus far in the neighbor search fall into two categories: if a node found (call it node k) falls in the relay region of some other found node (call it j), then we mark k as 'dead', otherwise the node is marked 'alive'. The set of alive nodes when the search is over constitutes the set of neighbors for transmit node i. When the search terminates, the transmit node is enclosed, and the nodes that enclose the transmit node are not in the relay region of any node found, as required by the definition of a neighbor.

15.2.4.2 Cost distribution (Phase 2)

After the enclosure graph has been found in Phase 1, the distributed Bellman–Ford minimum cost algorithm [1262] on the enclosure graph is applied, using power consumption as the cost metric. In Phase 2, each node broadcasts its cost to its neighbors. The *cost* of a node i is defined as the minimum power necessary for i to establish a path to the master site. Each node calculates the minimum cost it can attain given the cost of its neighbors. Let $n \in N(i)$. When i receives the information Cost(n), it computes

$$C_{i,n} = \text{Cost}(n) + P_{\text{transmit}}(i, n) + P_{\text{receiver}}(n) \tag{15.41}$$

where $P_{\text{transmit}}(i, n)$ is the power required to transmit from i to n, and $P_{\text{receiver}}(n)$ is the additional receiver power that i's connetion to n would induce at n. $P_{\text{receiver}}(n)$ is either known to i, if for instance every user carries an identical receiver, or can be transmitted to i as a separate piece of information along with Cost(n). Then, node i computes

$$\text{Cost}(i) = \min_{n \in N(i)} C_{i,n} \tag{15.42}$$

and picks the link corresponding to the minimum cost neighbor. This computation is repeated, and the minimum cost neighbor is updated each time. The data transmission from i to the master site can start on the minimum cost neighbor link, which is the global minimum power link.

15.2.4.3 Computation of the relay region

In the following example, we illustrate the relay region of a single node, assuming the two-ray propagation model for terrestrial communications, which implies a $1/d^4$ transmit power rolloff . The close-in reference distance is taken as 1 m. The carrier frequency is 1 GHz, and the transmission bandwidth 10 kHz. We assume omnidirectional antennas with 0 dB gain, -160 dBm/Hz thermal noise, 10 dB noise figure in the receiver, and a predetection signal to noise ratio (SNR) of 10 dB. Using the Friis free space formula gives -67.5 dBm as the minimum transmit power required for detection at 1 m. We take this to be roughly -70 dBm for our simulations. This can be treated as an effective predetection threshold to be used with the $1/d^4$ rolloff formula to compute the minimum required transmit power for any distance.

We assume the following model for receiver power at any relay node: a fixed receiver power of 80 mW is consumed at each node, with a 20 mW increase for each additional node from which transmission is received. This model can be easily modified according to the actual receiver design [1284–1285], see also Chapters 10 and 11.

With the previous assumptions, the relay region is obtained by solving the following two equations simultaneously:

$$d_{ij}^4 \geq d_{ir}^4 + d_{rj}^4 + c/t \tag{15.43}$$

and

$$d_{ij}^2 = d_{ir} + d_{rj}^2 - 2d_{ir}d_{rj}\cos\theta \tag{15.44}$$

where θ is the angle between position vectors $\mathbf{r}_{r \to i}$ and $\mathbf{r}_{r \to j}$. In Inequality (15.43), c denotes the additional receiver power cost of 20 mW for relays, and t the predetection threshold of

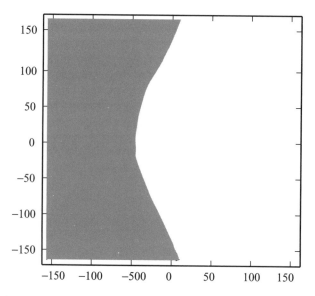

Figure 15.9 Relay region for 80 m inter-nodal distance [1281] © 1999, IEEE.

10^{-7} mW. Figure 15.9 displays the relay region in the case where the relay node is at (0,0), and the transmit node is at (80,0). The relay region has been shaded. The units are meters.

15.2.4.4 Stationary network simulation

We now present the simulation results for a stationary network with nodes deployed over a square region of 1 km on each side. The (x, y) coordinates of the nodes are generated as independent, identically distributed (i.i.d.) uniform random variables over this region. Since the nodes are stationary, once each node is enclosed and obtains a valid cost, the network remains in the minimum power topology.

The transmit and receiver powers for providing point-to-point connections are as described above. In this simulation, we investigate how the total power consumption of the minimum power topology varies with the number of nodes. Figure 15.10 illustrates this relationship. As the number of nodes grows larger, the average power decreases toward its asymptote of 300 mW receiver power/node. The plot has been normalized to the receiver power.

15.2.5 Distributed mobile networks

The protocol presented so far has been for stationary networks. However, due to the localized nature of its search algorithm, it proves to be an effective energy-conserving protocol for the mobile case as well.

Synchronization in a mobile network can be achieved by use of the absolute time information provided by GPS up to 100 ns resolution [1286]. The network synchronization is discussed in Section 14.3. In a synchronous network, each node wakes up regularly to 'listen' for change and goes back to the sleep mode to conserve power. The time between

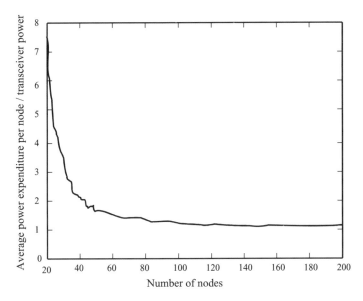

Figure 15.10 Average power expenditure per node [1281] © 1999, IEEE.

successive wakeups is referred to as the *cycle period* of the network. If the cycle period is too long, the power costs to the master site can change significantly from one wakeup to the next. In this case, the network cannot track the correct costs. If the cycle period is too short, then the network consumes unnecessary energy to compute costs that change only slowly. The choice of the cycle period for energy-efficient operation of a wireless network must address this tradeoff. In our simulation, we assume that the cycle period has been chosen to meet these two constraints.

After wakeup, each node executes *Phase 1* of the protocol, as described in the previous section. When a node completes *Phase 2*, it either starts data transmission on the optimal link, or it goes to the sleep mode to conserve power.

The protocol is self-reconfiguring since strong connectivity is ensured within each cycle period, and the minimum power links are dynamically updated. It can be seen that this protocol is also fault tolerant. A network protocol is 'fault tolerant' if it is self-reconfiguring when nodes leave or new nodes join the network. Under such a scenario, each node employing the protocol would compute its new enclosure and find the minimum power topology.

15.2.5.1 Mobile network simulation

In this case, the initial positions of 100 nodes are generated as i.i.d. uniform random variables over a square field, 1 km on each side. The velocity in each coordinate direction is uniformly distributed on the interval $(-v_{max}, v_{max})$. The velocity is the vector sum of the velocities in each coordinate direction. Parameter v_{max} is varied to observe how the energy consumption changes.

The choice of the *SetSearchRegion* function in the search algorithm, which is optimized to perform the minimum energy neighbor search, is a separate topic. In this simulation,

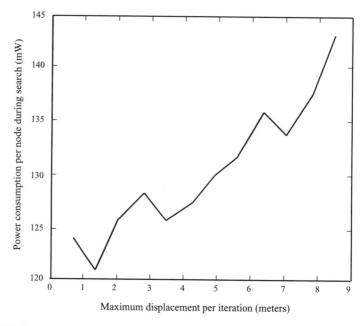

Figure 15.11 Power consumption per node during search period.

omnidirectional antennas and an heuristic strategy for the choice of the search radius are assumed. The results indicate that even with an heuristic strategy, the energy consumption is very low.

Let T be the cycle period of the network. Assume that node i is enclosed in the nth iteration, and let e_n be the distance of i to its furthest neighbor in the nth iteration. In the next iteration, if i sets its search radius to

$$r_{n+1} = e_n + 2\sqrt{2}v_{max}T \tag{15.45}$$

then its neighbors in the nth iteration must fall within this radius. Because the cycle period is small enough to allow positions to vary only slightly from one iteration to the next, in most cases the node will have its previous neighbors in its new enclosure as well. Nodes employing this strategy are enclosed within one iteration of the search algorithm presented earlier in this section.

From a system perspective, the measure of mobility is not the velocities, but rather the displacements of nodes in a cycle period of the network. The maximum displacement of a node in a cycle period is $\sqrt{2}v_{max}T$ from the previous analysis.

Figure 15.11 displays the search period power level per node averaged over 10 000 iterations and averaged over all the nodes. The horizontal axis on this graph is the maximum displacement in meters. Since the average distance between nodes is about 100 m in this particular simulation, it was estimated that the network cannot track correct costs for maximum displacements greater than 8 m, so that power consumption over only this range was graphed.

Figure 15.12 displays the search period power consumption per meter of maximum displacement. The graph indicates that the power consumption per node scales better than

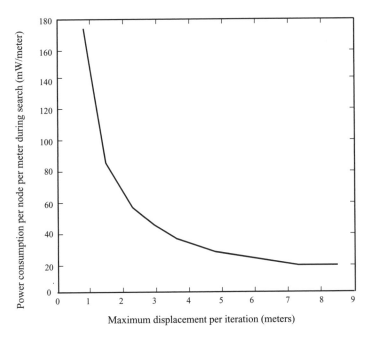

Figure 15.12 Power consumption per node per meter of maximum displacement during a search.

linearly with maximum displacement for the range of displacements for which the network can track the correct costs.

15.3 LEAST RESISTANCE ROUTING IN WIRELESS NETWORKS

In the previous section we have already seen that packet radio networks can provide wireless communication and data distribution among mobile terminals, but adaptive protocols are required in order for them to do so. Two important characteristics of a communication link in a mobile packet radio network are its unreliability and its variability. The links in such a network are unreliable because of fading, interference, noise, and perhaps failure of the transmitting or receiving radios. They are variable because of the mobility of the radios, the dynamic nature of the propagation medium and the interference. The interference can be generated externally (e.g. another system operating in the same frequency band or intentional jamming), or it can be generated within the network (e.g. multiple access interference).

If a network is to provide reliable service over communication links with these characteristics, the link and network protocols must adapt to changes in the network. In order to adapt effectively, these protocols must be provided with information about the current status of certain elements of the network, such as the communication channels and radio buffers. The mobile networks considered in this section are multiple hop networks with distributed control, including distributed forwarding and routing protocols. The presentation is based on [1287]. Since there is no central controller or other entity that can provide status information for all of the network elements, the required information must be derived by the radios themselves. As in the previous section, the exchange of information is among neighboring radios (i.e. local exchange rather than global).

15.3.1 Least resistance routing (LRR)

An important feature of lrr is its use of link quality information as a basis for route selection. The *link resistance,* which is determined for each link in the network, provides a measure of the noise and interference in the communication channel and the congestion in the receiving radio. The link resistance is a quantitative measure of the receiving radio's ability to demodulate, decode, store and forward a packet that is transmitted to it on that link.

The *metric* in LRR specifies how a radio is to calculate the resistance for each of its incoming links. The probability that a transmission from radio A to radio B is successful depends on both the condition of the communication channel from A to B and the ability of B to store and forward the packet. In this section we restrict attention to metrics of the form

$$\text{LR}(A, B) = \alpha I(A, B) + \beta W(B) \tag{15.46}$$

where $\text{LR}(A, B)$ is the resistance of the link from radio A to radio B. The term $I(A, B)$ represents the *resistance of the communication channel from A to B* and the term $W(B)$ represents the *resistance of radio B*. The coefficients α and β are selected to give the desired relative weightings to these two components of the link resistance,

The resistance of the communication channel accounts for fading, propagation loss and other features that are unique to transmissions from A to B. The resistance of radio B characterizes the conditions at radio B that equally affect each of the transmissions to radio B. Included in these conditions are the number of packets in the radio's buffer, the amount of traffic near the radio, interference that equally affects each transmission to radio B, and the expected delay in forwarding packets. If all packets transmitted to radio B are received at approximately the same power level, any radio frequency (RF) interference has approximately the same effect on each transmission to radio B, in which case the interference could be accounted for in the term $W(B)$ only. In practice, however, the packets may be transmitted at different power levels, and the propagation losses for the different communication channels to radio B usually differ greatly. Because a given source of RF interference may have substantially different effects on transmissions to radio B from different radios, it is better to include the effects of interference in the resistance of the communication channel rather than in the resistance of the receiving radio.

15.3.2 Multimedia least resistance routing (MLRR)

In MLRR, each link in the network is assigned a link resistance for each of the types of traffic that it may need to handle. The link resistance can be thought of as a vector with a component for each different type of traffic. A different metric can be employed for each component in order to provide a mechanism for accounting for different service requirements for different message types.

In order to simplify the presentation this section will focus on two types of multimedia packet, type-D packets and type-V packets. The intent is that type-D packets have the characteristics associated with data traffic (e.g. computer file transfers), while type-V packets have the characteristics associated with voice or video traffic. Type-D packets must be delivered to their destinations with no errors or erasures; however, a moderate delay is permitted. On the other hand, type-V packets are required to be delivered with much less delay than type-D packets, but the type-V packets may have a small number of frame erasures and still be considered acceptable to the destination. In describing the components of the link resistance, the subscripts d and v are used to distinguish between the link resistance for the two types of packet. The two components of the link resistance for MLRR are

$$LR_d(A, B) = \alpha_d I_d(A, B) + \beta_d W_d(B) \tag{15.47}$$

and

$$LR_v(A, B) = \alpha_v I_v(A, B) + \beta_v W_v(B) \tag{15.48}$$

In the example used to illustrate MLRR in this section, the importance of the channel resistance relative to the receiver resistance is determined for each type of traffic from the following specifications. A type-D packet is accepted by its destination if all the received words in the packet are decoded correctly. A type-V packet is accepted by its destination if λ or fewer of the received words in the packet do not decode correctly.

The selection of the coefficients to provide the proper weighting of the channel resistance and the radio resistance will be aided by the simulation results, in which the general trends will be observed. Because type-D packets are more sensitive to errors than to delay, α_d is set to a relatively large value compared to β_d. Type-V packets are more sensitive to delay than to errors, however, so β_v is set to a relatively large value compared to α_v.

If a received word in a type-V does not decode correctly, that word is erased. If the receiving radio is not the final destination for the packet, the packet is forwarded with the erasure inserted in place of the missing word. Because of the requirements on type-V packets, any radio along the route discards any type-V packet that has accumulated more than λ word erasures. The choice for the value of λ, the maximum number of word erasures that are permitted, depends primarily on the speech or video compression method.

The *packet rejection probability* is defined as the probability that a packet with no previous word erasures is discarded as a result of a single transmission. For frequency-hopping (FH) systems this probability depends on the code, the modulation and demodulation, the number of other simultaneous FH transmissions, the number of words per packet and the signal to noise ratio on the channel.

The curves in Figure 15.13 illustrate the sensitivity of the packet reception probability to the value of λ. In this graph, the packet rejection probability is shown as a function of E_b/N_0 for six different values of λ, where E_b is the energy per information bit and N_0 is the one-sided spectral density for the thermal noise. For convenience, we refer to E_b/N_0 as the signal to noise ratio in the text of this section. For the results given in Figure 15.13, binary orthogonal modulation and optimum non-coherent demodulation are employed, there are three interfering FH transmissions, a (32, 22) extended Reed–Solomon code is used, and there are 15 code words per packet.

In general, as discussed in Chapters 2 and 3, one may want to use different modulation techniques or different combinations of modulation and error control coding for different types of multimedia traffic. If the modulation or coding is different for different types of packet, the metrics for the channel resistance may differ also. For the numerical examples

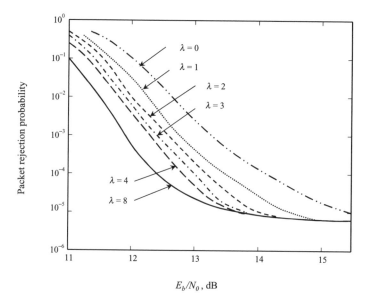

Figure 15.13 Packet rejection probability for six different values of λ.

presented in this section, the coding and modulation are the same for both types of packet, and I_v and I_d, the metrics used for the channel resistance, are identical. The metric for each type of message is the errors and erasures (EE) metric, and the coefficients α_d and α_v are adjusted to account for the differences in the requirements for the two types of packet.

The resistance components for the receiver (i.e. W_d and W_v) are used to represent the delay that a packet is expected to experience before it is forwarded by a radio that has decoded the packet successfully. The number of packets waiting in the receiving radio's buffer is a simple metric for estimating the expected forwarding delay. The number of type-D packets at a radio is denoted by N_d and the number of type-V packets is N_v. The receiver resistance components are then defined as

$$W_d(B) = N_d + \omega_d N_v \quad \text{and} \quad W_v(B) = N_v + \omega_v N_d.$$

15.3.3 Network performance examples: LRR versus MLRR

For the numerical examples included in this section, the (32, 22) extended Reed–Solomon code is employed with errors-and-erasures decoding. There are 15 code words per packet, and the code words are fully interleaved. Ten test symbols are included in each dwell interval. A type-V packet is accepted if no more than three words have been erased (i.e. $\lambda = 3$).

For both the MLRR and LRR routing protocols, the forwarding protocol gives type-V packets a higher priority than type-D packets. A radio that is preparing to transmit a packet first checks for a type-V packet in its buffer, and transmits such a packet if possible. If there are no type-V packets in the buffer, or if none of the type-V packets can be forwarded because of the retransmission policy, then the radio checks for a type-D packet to transmit.

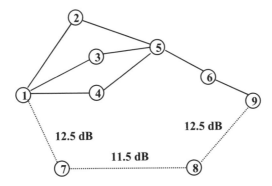

Figure 15.14 Nine-node network.

The service requirements of a packet also affect the number of times a radio may retransmit that packet. For the results presented here, type-V packets are permitted up to two forwarding attempts at a given radio, while type-D packets are permitted six attempts at most. To prevent outdated packets from congesting the network, packets expire and are discarded after a certain number of packet intervals since they were generated. For the performance curves that follow, this number is 500 for type-D packets and 50 for type-V packets.

Various combinations of parameters for the MLRR resistance metrics have been examined, and it was found that the following two metrics perform well over a wide range of network conditions. For the results presented in this section, the two metrics employed for MLRR are $LR_d = 2t + e$ and $LR_v = t + 0.5e + N_v + 0.7N_d$. The metric employed for LRR is the EE metric, $LR = 2t + e$. A word that has no more than e erasures and t errors is decoded correctly if $2t + e$ does not exceed $n - k$.

15.3.3.1 Example 1: Nine-node network

An example of a network with nine radios is illustrated in Figure 15.14 [1287]. The signal to noise ratios for the channels represented by solid lines is 15 dB, which results in a very low error probability on the channel. The signal to noise ratios for the channels shown with dashed lines are indicated by the labels on the channels. Both type-D and type-V packets are generated at radio 1 and routed to radio 9. In a given packet interval, the probability that a marked type-D packet is generated is denoted by q_d, and the probability that a marked type-V packet is generated is denoted by q_d. A radio can generate no more than one packet of each type within a packet interval. Additionally, the interfering type-D packets are generated at radios 5 and 6 and sent to radios 9 and 1, respectively, and their generation probability is fixed at 0.05. The inclusion of these interfering packet results in increased congestion in the upper three routes of the network shown in Figure 15.14.

This simple network topology is selected to illustrate the advantages of using multimedia routing. The two types of packet generated at radio 1 must be routed to radio 9, but there is not a single route that is preferred for both types of packet. The upper route has large delays because of the congestion caused by the extra packets generated at radios 5 and 6. On the other hand, the lower route has poorer quality channels that give a bit error rate in the range of 1–5 %. Although type-V packets can be accepted with some word

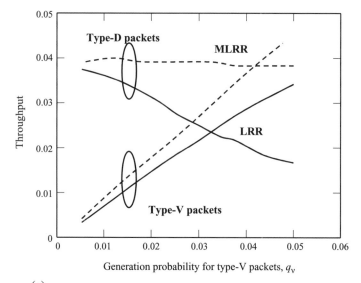

(a)

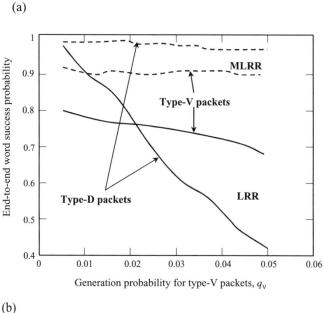

(b)

Figure 15.15 (a) Throughput for $q_d = 0.04$; (b) word success probability for $q_d = 0.04$; (c) Delay of type-V packets, $q_d = 0.04$.

erasures, type-D packets cannot. For a given channel, this difference in service requirements leads to a larger expected number of retransmissions for type-D packets than for type-V packets.

Two scenarios are presented that illustrate the value of selecting a route for each packet type rather than using a single 'best' route for all packet types. For the first scenario, the

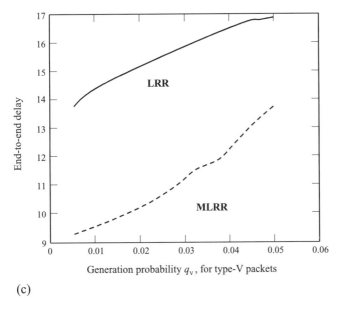

(c)

Figure 15.15 (*Cont.*).

generation probability for type-D packets is fixed at $q_d = 0.04$, and the network performance is evaluated as the generation probability for type-V packets is increased. This scenario is of particular interest, because it has been reported that some existing radio networks suffer from severely degraded service in the delivery of data traffic when voice traffic increases significantly in the network. The results of the simulation for this scenario are shown in Figure 15.15. The results for MLRR are shown as dashed curves, and the results for LRR are shown as solid curves.

In the second scenario the situation is reversed: the generation probability for type-V packets is held constant at $q_v = 0.04$, and the generation probability for type-D packets is increased. The results in Figure 15.16 for this scenario are similar to those discussed above for the first scenario. In particular, the MLRR protocol maintains a high level of throughput and a large end-to-end success probability for type-V packets, even as the generation probability for type-D packets is increased. However, for the LRR protocol, the throughput and success probability drop as the type-D traffic increases in the network. The MLRR protocol gives better performance than the LRR protocol for all situations depicted in Figure 15.16.

15.3.3.2 *Example 2: Twelve-node network*

In this case, the MLRR and LRR protocols are used in the 12-node network illustrated in Figure 15.17 [1287], in which the number of possible routes and the amount of traffic are larger than in the nine-node network of *Example 1*. The 12 radios are connected by communication channels for which the signal to noise ratios are as labeled in Figure 15.17. Packets of both types have the origin–destination pairs $(5, 6), (6, 5), (7, 8)$ and $(8, 7)$. Marked

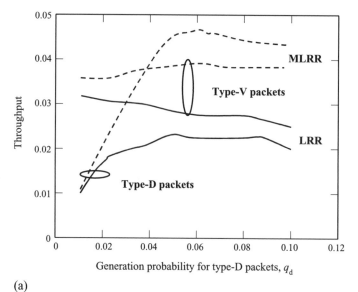

(a)

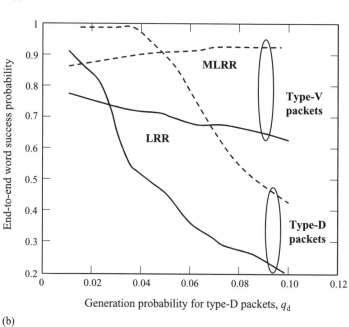

(b)

Figure 15.16 (a) Throughput for $q_v = 0.04$; (b) word success probability for $q_v = 0.04$; (c) delay of type-V packets, $q_v = 0.04$.

type-D packets have generation probability $q_d = 0.04$, and the generation probability for marked type-V packets is denoted by q_v. Radios 1–4 each generate additional interfering type-D packets with origin-destination pairs (1, 4), (4, 1), (2, 3) and (3, 2). The generation probability for these interfering packets is fixed at 0.05.

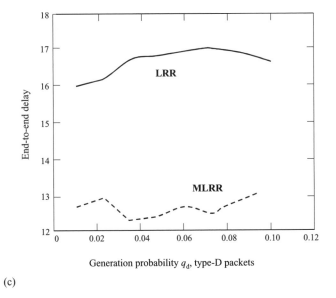

(c)

Figure 15.16 (*Cont.*).

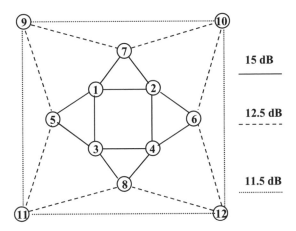

Figure 15.17 Network of Example 2.

The throughput for each packet type is shown in Figure 15.18(a), and the end-to-end word success probability is shown in Figure 15.18(b). As q_v increases, the throughput for type-D packets declines rapidly for the LRR protocol, but a higher throughput is maintained by MLRR. Furthermore, the word success probability for type-D packets is uniformly larger for MLRR than for LRR. For type-V packets, the MLRR protocol gives higher throughput, as shown in Figure 15.18(a), and larger word success probabilities, as shown in Figure 15.18(b).

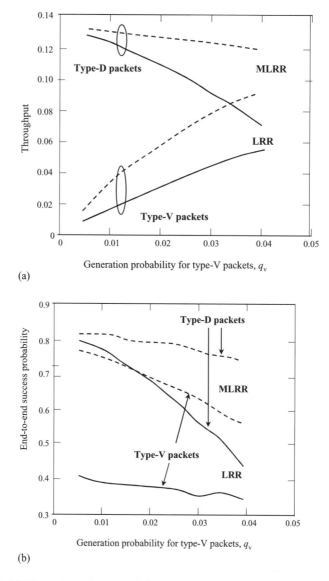

Figure 15.18 (a) Throughput for $q_d = 0.04$; (b) word success probability for $q_d = 0.04$.

15.3.4 Sensitivity to the number of allowable word erasures

Because the performance of MLRR depends on λ, the number of word erasures that are permitted in a type-V packet, it is important to examine the sensitivity of the throughput and word success probability to the value of this parameter. The LRR protocol does not attempt to distinguish between type-D and type-V packets, so the performance of LRR does not depend significantly on the value of λ. To illustrate the dependence of MLRR on the choice of λ, results are presented for the nine-node network of *Example 1*.

The generation probability for type-D packets is fixed at $q_d = 0.04$, and the network performance is evaluated as a function of the generation probability for type-V packets. The results for the nine-node network of *Example 1* are shown in Figure 15.19. We see that the performance of MLRR for type-D packets is approximately independent of the value of λ, but the performance of MLRR for type-V packets improves as the number of allowable word erasures is increased. If $\lambda = 0$, the performance for type-V packets is still significantly better for MLRR than for LRR, but the performance advantage obtained from MLRR is even greater if $\lambda = 1$ (i.e. a type-V packet is accepted if no more than one of its words is erased). A small performance increase is obtained as the number of allowable word erasures is increased from $\lambda = 1$ to $\lambda = 2$, but additional increases in λ produce no significant benefit for this network. If the number of relays required for the type-V packets is much larger than in the networks we consider in this section, there may be a significant difference in performance if a larger number of word erasures can be permitted.

More information on the topic can be found in [1288–1291].

15.4 POWER OPTIMAL ROUTING IN WIRELESS NETWORKS FOR GUARANTEED TCP LAYER QoS

15.4.1 Constant end-to-end error rate

In this section we consider the case when a packet is transmitted from a source to its destination along multiple hops, where there is some probability of error per hop that depends on the distance between the hops and the transmit power. In Section 15.2, the assumption was that if the received power (signal to noise ratio) $P_{recv} \geq \gamma$, where γ is some constant, then the packet is successfully received; otherwise the packet is lost. The symbol error rate (SER) is a monotonically decreasing function of P_{recv}, therefore, for each hop, P_{recv} can be made large enough that the SER for the hop satisfies

$$\text{SER} \leq \text{SER}(\gamma) \tag{15.49}$$

Since any transmission will use the minimum amount of power required to meet the necessary error rate, this means that $\text{SER} = \text{SER}(\gamma)$. Assuming that errors per hop are independent, this implies that the error seen by the transport control protocol layer (end-to-end error rate) SER_{e2e} is given by

$$\text{SER}_{e2e} = 1 - (1 - \text{SER}(\gamma))^N \tag{15.50}$$

where N is the number of hops along the path. SER_{e2e} is monotonically increasing with N.

Instead of using a routing scheme and power cost metric that allows the end-to-end error rate to increase with hop count, we examine the effect of constraining the end-to-end error rate to be a constant. This criterion will be referred to as TCP layer power optimal routing.

15.4.2 Optimization problem

The presentation in this section is based on [1292]. Let $\mathbf{X} = (X_0, X_1, \ldots, X_N)$ be an *N-hop* path from node X_0 to X_N that passes through nodes $X_1, X_2, \ldots, X_{N-1}$ in that order. Let P_i be the power allocated for the hop (X_{i-1}, X_i), and let SER_i be the corresponding symbol

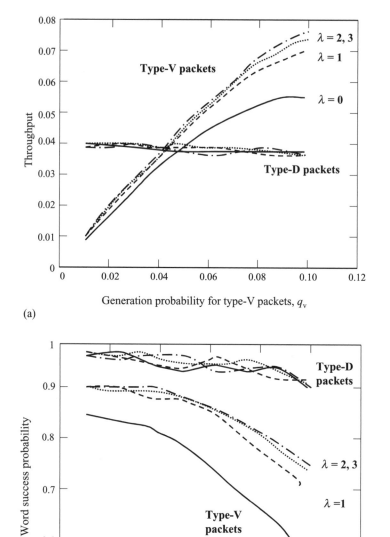

Figure 15.19 (a) Throughput for MLRR with $q_d = 0.04$; (b) end-to-end word success probability for MLRR with $q_d = 0.04$; (c) end-to-end delay for type-V packets with $q_d = 0.04$.

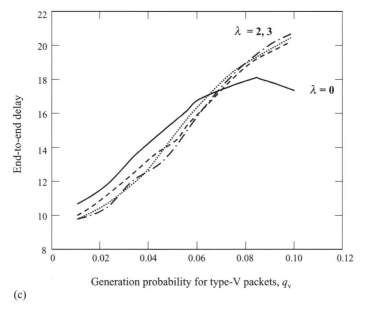

(c)

Figure 15.19 (*Cont.*).

error rate for the hop ($i = 1, 2, \ldots, N$). Assuming independent errors per hop, the TCP layer QoS (the end-to-end SER) is given by

$$SER_{e2e} = 1 - \prod_{i=1}^{N} (1 - SER_i) \tag{15.51}$$

We define the power cost function as follows:

$$\mathcal{PC}^*(\varepsilon; \mathbf{X}) = \min_{P_1, \ldots, P_N} \sum_{i=1}^{N} P_i \quad \text{given} \quad SER_{e2e} \leq \varepsilon \tag{15.52}$$

We will assume that we are interested in the case when the quantity ε is much smaller than one, i.e. when there is a very low probability of error.

A quick analysis shows that the optimization problem in Equation (15.52) leads to equations that do not have simple closed form solutions in terms of the power per hop and the power cost, even for the simplest models for the SER. We formulate our approximate power cost metric by the following equations:

$$\mathcal{PC}(\varepsilon; \mathbf{X}) = \min_{P_1, \ldots, P_N} \sum_{i=1}^{N} P_i \quad \text{given} \quad \sum_{i=1}^{N} SER_i \leq \varepsilon \tag{15.53}$$

We can justify this approximation by the following result: assuming $\varepsilon \geq 0$ and $\varepsilon + 4\varepsilon^2 < 1/\sqrt{2}$, and given the power cost metrics $\mathcal{PC}^*$ and $\mathcal{PC}$ defined by Equations (15.52) and (15.53) respectively, we have:

$$\mathcal{PC}(\varepsilon + 4\varepsilon^2; \mathbf{X}) \leq \mathcal{PC}^*(\varepsilon; \mathbf{X}) \leq \mathcal{PC}(\varepsilon; \mathbf{X}) \tag{15.54}$$

The proof of Inequality (15.54) only relies on the power cost metric being a minimization problem where the error rate constraint is met with equality at the solution. In particular, it does not depend on the expression for SER$_i$, or the fact that the metric being minimized is the sum of the power per hop.

For the rest of this section, we use $\mathcal{PC}$ as the power cost metric because it will lead to analytical expressions for the power cost of a path. While this is an approximation, Inequality (15.54) provides a way of bounding the error $E(\varepsilon; \mathbf{X}) = \mathcal{PC}(\varepsilon; \mathbf{X}) - \mathcal{PC}^*(\varepsilon; \mathbf{X})$ since

$$E(\varepsilon; \mathbf{X}) \leq \mathcal{PC}(\varepsilon; \mathbf{X}) - \mathcal{PC}(\varepsilon + 4\varepsilon^2; \mathbf{X}) \tag{15.55}$$

When we compute the power cost with specific models for SER, we will use this expression to show that the approximation error is $\mathcal{O}(\varepsilon)$, and thus small compared to the power cost of the path.

Error correction mechanisms (both ARQ and FEC) can be easily included in this framework with minor changes, as discussed in [1293].

15.4.3 Error rate models

15.4.3.1 *Time-invariant attenuation*

The first model is a deterministic power model where we assume that the receive power of a link is attenuated by a time-invariant quantity. This attenuation coefficient can be given a physical interpretation by setting it to d^α/a, where d is the distance between the transmitter and receiver, and $\alpha > 2$ and a are constants. We assume that the expression for the SER of a link is given by:

$$\text{SER}_i = be^{-P_i/a_i} \tag{15.56}$$

where P_i is the transmit power and a_i is the time-invariant attenuation coefficient of the link. Equation (15.56) is sufficiently parameterized to be able to provide a bound for the probability of error for most digital modulations that are detected optimally in the presence of additive Gaussian noise [1294].

Under the assumptions of Inequality (15.54) and the link SER assumption given by Equation (15.56), the optimal power cost $\mathcal{PC}(\varepsilon; \mathbf{X})$ for path $\mathbf{X} = (X_0, X_1, \ldots, X_N)$ is obtained when

$$
\begin{aligned}
&1. \quad \text{SER}_j = \varepsilon \frac{a_j}{\sum_{i=1}^{N} a_i} \\
\\
&2. \quad P_j = a_j \left(\log(b/\varepsilon) + \log \left(\sum_{i=1}^{N} a_i/a_j \right) \right) \\
\\
&3. \quad \mathcal{E}(\varepsilon; \mathbf{X}) \leq \log(1 + 4\varepsilon) \sum_{j=1}^{N} a_j
\end{aligned}
\tag{15.57}
$$

where a_j, SER$_j$, and P_j are the attenuation coefficient, link error rate, and link power allocation, respectively, for link (X_{j-1}, X_j).

15.4.3.2 *Large- and small-scale fading*

In a wireless mobile scenario the received power is affected by many more factors than the mere distance, and it is common to represent the received power as a doubly stochastic random variable, with long-term and short-term variations (see Chapter 14). The transmit power is attenuated by two factors: $G_L(t)$ caused by large-scale fading, and $G_S(t)$ caused by small-scale fading. To account for the effects of large- and small-scale fading, the time-varying SER of a link is given by the random process

$$SER_i = be^{-a\Omega_{\tau_i}} \tag{15.58}$$

where a and b are constants, and Ω_{τ_i} is the received power, whose statistics are functions of position and time. As is standard, we assume a log-normal distribution for the large-scale fading coefficient. For the small-scale fading, several distributions have been introduced [1275] and using Equation (15.58), the corresponding average SER for a given large-scale fading parameter is the characteristic function of the small-scale fading density. For example, for the Nakagami m-distribution, the expected SER_i for a link is given by [1275]:

$$SER_i = b(1 + P_i/a_i)^{-m} \tag{15.59}$$

where a_i is the contribution from the slow-varying large-scale fading coefficient and P_i is the average transmit power.

The Nakagami m-distribution captures the intermediate ground between strong line-of-sight and non-line-of-sight systems. Note, in fact, that Rayleigh fading is a special case of Nakagami fading when $m = 1$, while the deterministic case is obtained as $m \to \infty$.

One can generalize the previous power cost expressions and power optimal routing to these scenarios.

Under the assumptions of Inequality (15.54) and the link SER assumption given by Equation (15.59), the optimal power cost $\mathcal{PC}(\varepsilon; \mathbf{X})$ for path $\mathbf{X} = (X_0, X_1, \ldots, X_N)$ is obtained when

$$1. \quad SER_j = \varepsilon \frac{a_j^{m/(m+1)}}{\sum_{i=1}^N a_i^{m/(m+1)}}$$

$$2. \quad P_j = a_j \left((b/\varepsilon)^{1/m} \left(\sum_{i=1}^N (a_i/a_j)^{m/(m+1)} \right)^{1/m} - 1 \right) \tag{15.60}$$

$$3. \quad \mathcal{E}(\varepsilon; \mathbf{X}) \le \frac{4\varepsilon}{m} (b/\varepsilon)^{1/m} \left(\sum_{j=1}^N a_i^{m/(m+1)} \right)^{(m+1)/m}$$

where a_j, SER_j, and P_j are the large-scale attenuation coefficient, link error rate, and link power allocation, respectively, for link (X_{j-1}, X_j).

15.4.4 **Properties of power optimal paths**

In this section we discuss some of the consequences of adopting a TSP layer QoS (end-to-end SER) constraint for power optimization. In particular, we compare these results to the concepts similar to those discussed in Section 15.2, which assume that the amount of power required for a link (X_{i-1}, X_i) is given by

$$P_i = \log(b/\varepsilon)\frac{d_i^\alpha}{a} \tag{15.61}$$

where a and $\alpha > 2$ are constants, and $d_i = |X_{i-1} - X_i|$ is the distance between points X_{i-1} and X_i. This model assumes that the link SER is constant $(= \varepsilon)$, and the attenuation coefficient a_i is given by d_i^α/a.

Under this model, the power cost of using a link does not depend on the path under consideration, unlike the link power allocation in Equations (15.57) and (15.60). The power cost using this model, which we denote $\mathcal{KC}(\varepsilon; \mathbf{X})$, is given by

$$\mathcal{KC}(\varepsilon; \mathbf{X}) = \log(b/\varepsilon) \sum_{i=1}^{N} \frac{d_i^\alpha}{a} = \log(b/\varepsilon) \sum_{i=1}^{N} \alpha_i \tag{15.62}$$

where $d_i = |X_{i-1} - X_i|$ is the distance between points X_{i-1} and X_i. Note that since ε only appears in the factor in front of this particular power cost metric, the best path to route a packet according to $\mathcal{KC}$ will not depend on ε. [1295].

We compare the properties of power optimal paths obtained by the metric introduced in Equation (15.57) with those obtained using the metric from Equation (15.62).

15.4.4.1 Comparing power cost metrics

If we use Equation (15.61) to determine the power for a hop, then the SER per hop can be as high as ε, thereby increasing the total end-to-end SER. The metric $\mathcal{KC}$ will therefore underestimate the amount of power required to transmit a packet along a path with an error that does not exceed ε. Examining the expressions in Equation (15.57), we conclude that

$$\mathcal{PC}(\varepsilon; \mathbf{X}) = \sum_{j=1}^{N} a_j \left(\log(b/\varepsilon) + \log\left(\sum_{i=1}^{N} a_i/a_j \right) \right) \tag{15.63}$$

$$\geq \mathcal{KC}(\varepsilon; \mathbf{X}) \tag{15.64}$$

with equality holding if and only if we are considering a one-hop path. If we treat $p_j = a_j / \sum_{i=1}^{N} a_i$ as a probability distribution, observe that

$$\mathcal{PC}(\varepsilon; \mathbf{X}) = \left(\sum_{j=1}^{N} a_j \right) \left(\log(b/\varepsilon) + \sum_{i=1}^{N} p_i \log 1/p_i \right)$$

$$\leq \mathcal{KC}(\varepsilon; \mathbf{X}) \left(1 + \frac{\log N}{\log(b/\varepsilon)} \right) \tag{15.65}$$

with equality holding when $p_1 = p_2 = \cdots = p_N$. When examining long paths with equidistant hops, using a simple additive cost function will underestimate the power cost by a factor that grows logarithmically with the number of hops compared to a metric that keeps the end-to-end error bounded.

More details on adaptive routing in wireless networks can be found in [1296–1323].

References

1. Adachi, F. (2002) Evolution towards broadband wireless systems, *The 5th International Symposium on Wireless Personal Multimedia Communications*, 27–30 October 2002. **1**, 19–26.

2. Jun-Zhao, Sun, Sauvola, J. and Howie, D. (2001) Features in future: 4G visions from a technical perspective, IEEE Global Telecommunications Conference, 25–29 November 2001, **6**, 3533–3537.

3. Axiotis, D. I., Lazarakis, F. I., Vlahodimitropoulos, C. and Chatzikonstantinou, A. (2002) 4G system level simulation parameters for evaluating the interoperability of MTMR in UMTS and HIPERLAN/2, *4th International Workshop on Mobile and Wireless Communications Networks*, 9–11 September 2002, pp. 559–563.

4. Mihovska, A., Wijting, C., Prasad, R., Ponnekanti, S., Awad, Y. and Nakamura, M. (2002) A novel flexible technology for intelligent base station architecture support for 4G systems. *The 5th International Symposium on Wireless Personal Multimedia Communications*, 27–30 October 2002. **2**, 601–605.

5. Kitazawa, D., Chen, L., Kayama, H. and Umeda, N. (2002) Downlink packet-scheduling considering transmission power and QoS in CDMA packet cellular systems. *4th International Workshop on Mobile and Wireless Communications Networks*, 9–11 September 2002, pp. 183–187.

6. Dell'Uomo, L. and Scarrone, E. (2002) An all-IP solution for QoS mobility management and AAA in the 4G mobile networks. *The 5th International Symposium on Wireless Personal Multimedia Communications*, 27–30 October 2002. **2**, 591–595.

7. Wallenius, E. R. (2002) End-to-end in-band protocol based service quality and transport QoS control framework for wireless 3/4G services. *The 5th International Symposium on Wireless Personal Multimedia Communications*, 27–30 October 2002. **2**, 531–533.

Advanced Wireless Communications. S. Glisic
© 2004 John Wiley & Sons, Ltd. ISBN: 0-470-86776-0

8. Benzaid, M., Minet, P. and Al Agha, K. (2002) Integrating fast mobility in the OLSR routing protocol. *4th International Workshop on Mobile and Wireless Communications Networks*, 9–11 September 2002, pp. 217–221.

9. Kambourakis, G., Rouskas, A. and Gritzalis, S. (2002) Using SSL/TLS in authentication and key agreement procedures of future mobile networks. *4th International Workshop on Mobile and Wireless Communications Networks*, 9–11 September 2002, pp. 152–156.

10. van der Schaar, M. and Meehan, J. (2002) Robust transmission of MPEG-4 scalable video over 4G wireless networks. *Proceedings of the 2002 International Conference on Image Processing*, 24–28 June 2002. **3**, 757–760.

11. Qing-Hui, Zeng, Jian-Ping, Wu, Yi-Lin, Zeng, Ji-Long, Wang and Rong-Hua, Qin (2002) Research on controlling congestion in wireless mobile Internet via satellite, based on multi-information and fuzzy identification technologies. *Proceedings of the 2002 International Conference on Machine Learning and Cybernetics*, 4–5 November 2002. **4**, 1697–1701.

12. 5th International Symposium on Wireless Personal Multimedia Communications. *Proceedings of* (Cat. No.02EX568), *the 5th International Symposium on Wireless Personal Multimedia Communications*, 27–30 October 2002. **1**.

13. Sukuvaara, T., Mahonen, P., and Saarinen, T. (1999) Wireless Internet and multimedia services support through two-layer LMDS system. *1999 IEEE International Workshop on Mobile Multimedia Communications*, (MoMuC '99) 15–17 November 1999, pp 202–207.

14. Martin, C. C., Winters, J. H. and Sollenberger, N. R. (2000) Multiple-input multiple-output (MIMO) radio channel measurements. *Proceedings of the 2000 IEEE Sensor Array and Multichannel Signal Processing Workshop*, 16–17 March 2000, pp. 45–46.

15. Pereira, J. M. (2000) Fourth generation: now it is personal! *The 11th IEEE International Symposium on Personal, Indoor and Mobile Radio Communications*, (PIMRC 2000) 18–21 September 2000, **2**, 1009–1016.

16. Otsu, T., Umeda, N. and Yamao, Y. (2001) System architecture for mobile communications systems beyond IMT-2000. *IEEE Global Telecommunications Conference* (GLOBECOM '01) 25–29 November 2001, **1**, 538–542.

17. Yi Han, Zhang, Makrakis, D., Primak, S. and Yun Bo, Huang (2002) Dynamic support of service differentiation in wireless networks IEEE Canadian Conference on Electrical and Computer Engineering (CCECE 2002) 12–15 May 2002, **3**, 1325–1330.

18. Jun-Zhao, Sun and Sauvola, J. (2002) Mobility and mobility management: a conceptual framework. *10th IEEE International Conference on Networks* (ICON 2002) 27–30 August 2002, pp. 205–210.

19. Vassiliou, V., Owen, H. L., Barlow, D. A., Grimminger, J., Huth, H.-P. and Sokol, J. (2002). A radio access network for next generation wireless networks based on multi-protocol label switching and hierarchical Mobile IP. *Proceedings of IEEE 56th Vehicular Technology Conference, 2002* (VTC 2002) 24–28 September 2002, **2**, 782–786.

20. Nicopolitidis, P., Papadimitriou, G. I., Obaidat, M. S. and Pomportsis, A. S. (2002) 3G wireless systems and beyond: a review. *9th International Conference on Electronics, Circuits and Systems*, 15–18 September 2002, **3**, 1047–1050.

21. 2002 IEEE Wireless Communications and Networking Conference Record, WCNC 2002 (Cat. No.02TH8609), *IEEE Wireless Communications and Networking Conference* (WCNC2002) 17–21 March 2002, **1**.

22. Borras-Chia, J. (2002) Video services over 4G wireless networks: not necessarily streaming. *2002 IEEE Wireless Communications and Networking Conference* (WCNC2002) 17–21 March 2002, **1**, 18–22.

23. Evans, B. G. and Baughan, K. (2000) Visions of 4G, *Electronics and Communication Engineering Journal*, **12**(6), 293–303.

24. Kim, J. and Jamalipour, A. (2001) Traffic management and QoS provisioning in future wireless IP networks. *IEEE Personal Communications*, (see also *IEEE Wireless Communications*), **8**(5), 46–55.

25. Aghvami, A. H., Le, T. H. and Olaziregi, N. (2001) Mode switching and QoS issues in software radio. *IEEE Personal Communications*, (see also *IEEE Wireless Communications*), **8**(5), 38–44.

26. Kanter, T. (2001) An open service architecture for adaptive personal mobile communication. *IEEE Personal Communications*, (see also *IEEE Wireless Communications*), **8**(6), 8–17.

27. Sampath, H., Talwar, S., Tellado, J., Erceg, V. and Paulraj, A. (2002) A fourth-generation MIMO-OFDM broadband wireless system: design, performance, and field trial results. *IEEE Communications Magazine*, **40**(9), 143–149.

28. Kellerer, W. and Vogel, H.-J. (2002). A communication gateway for infrastructure-independent 4G wireless access. *IEEE Communications Magazine*, **40**(3), 126–131.

29. Huang, V. and Weihua Zhuang (2002) QoS-oriented access control for 4G mobile multimedia CDMA communications. *IEEE Communications Magazine*, **40**(3), 118–125.

30. Smulders, P. (2002) Exploiting the 60 GHz band for local wireless multimedia access: prospects and future directions. *IEEE Communications Magazine*, **40**(1), 140–147.

31. Raivio, Y. (2001) 4G-hype or reality. *Second International Conference on 3G Mobile Communication Technologies* (Conf. Publ. No. 477), 26–28 March 2001, pp. 346–350.

32. Becchetti, L., Delli Priscoli, F., Inzerilli, T., Mahonen, P. and Munoz, L. (2001) Enhancing IP service provision over heterogeneous wireless networks: a path toward 4G. *IEEE Communications Magazine*, **39**(8), 74–81.

33. Abe, T., Fujii, H. and Tomisato, S. (2002) A hybrid MIMO system using spatial correlation. *The 5th International Symposium on Wireless Personal Multimedia Communications*, 27–30 October 2002, **3**, 1346–1350.

34. Lincke-Salecker, S. and Hood, C. S. (2003) A supernet: engineering traffic across network boundaries. *36th Annual Simulation Symposium*, 30 March–2 April 2003, pp. 117–124.

35. Yamao, Y., Suda, H., Umeda, N. and Nakajima, N. (2000) Radio access network design concept for the fourth generation mobile communication system. *Proceedings of IEEE 51st Vehicular Technology Conference* (VTC 2000) 15–18 May 2000, **3**, 2285–2289.

36. Ozturk, E. and Atkin, G. E. (2001) Multi-scale DS-CDMA for 4G wireless systems. *IEEE Global Telecommunications Conference* (GLOBECOM '01) 25–29 November 2001, **6**, 3353–3357.

37. Dell'Uomo, L. and Scarrone, E. (2001) The mobility management and authentication/authorization mechanisms in mobile networks beyond 3G. *12th IEEE International Symposium on Personal, Indoor and Mobile Radio Communications*, 30 September–3 October 2001, **1**, C-44–C-48.

38. Kumar, K. J., Manoj, B. S. and Murthy, C. S. R. (2002) On the use of multiple hops in next generation cellular architectures. *10th IEEE International Conference on Networks*, (ICON 2002) 27–30 August 2002, pp. 283–288.

39. Wang, S. S., Green, M. and Malkawi, M. (2002) Mobile positioning and location services. *IEEE Radio and Wireless Conference* (RAWCON 2002) 11–14 August 2002, pp. 9–12.

40. Motegi, M., Kayama, H. and Umeda, N. (2002) Adaptive battery conservation management using packet QoS classifications for multimedia mobile packet communications. *Proceedings of IEEE 56th Vehicular Technology Conference* (VTC 2002) 24–28 September 2002, **2**, 834–838.

41. Ying, Li, Shibua, Zhu, Pinyi, Ren and Gang, Hu (2002) Path toward next generation wireless internet-cellular mobile 4G, WLAN/WPAN and IPv6 backbone. *Proceedings of 2002 IEEE Region 10 Conference on Computers, Communications, Control and Power Engineering* (TENCOM '02) October 28–31 2002, **2**, 1146–1149.

42. Qiu, R. C., Wenwu, Zhu and Ya-Qin, Zhang (2002) Third-generation and beyond (3.5G) wireless networks and its applications. *IEEE International Symposium on Circuits and Systems* (ISCAS 2002) 26–29 May 2002, **1**, I-41–I-44.

43. Bornholdt, C., Sartorius, B., Slovak, J., Mohrle, M., Eggemann, R., Rohde, D. and Grosskopf, G. (2002) 60 GHz millimeter-wave broadband wireless access demonstrator for the next-generation mobile internet. *Optical Fiber Communication Conference and Exhibit* (OFC 2002) 17–22 March 2002, pp. 148–149.

44. Jianhua, He, Zongkai, Yang, Daiqin, Yang, Zuoyin, Tang and Chun Tung, Chou (2002) Investigation of JPEG2000 image transmission over next generation wireless networks. *5th IEEE International Conference on High Speed Networks and Multimedia Communications*, 3–5 July 2002, pp. 71–77.

45. Baccarelli, E. and Biagi, M. (2003) Error resistant space–time coding for emerging 4G-WLANs. *IEEE Wireless Communications and Networking* (WCNC 2003) 16–20 March 2003, **1**, 72–77.

46. Mohr, W. (2002) WWRF – the Wireless World Research Forum. *Electronics and Communication Engineering Journal*, **14**(6), 283–291.

47. Otsu, T., Okajima, I., Umeda, N. and Yamao, Y. (2001) Network architecture for mobile communications systems beyond IMT-2000. *IEEE Personal Communications* (see also *IEEE Wireless Communications*), **8**(5), 31–37.

48. Bria, A., Gessler, F., Queseth, O., Stridh, R., Unbehaun, M., Jiang, Wu, Zander, J. and Flament, M. (2001) 4th-generation wireless infrastructures: scenarios and research challenges. *IEEE Personal Communications* (see also *IEEE Wireless Communications*), **8**(6), 25–31.

49. Fitzek, F., Kopsel, A., Wolisz, A., Krishnam, M. and Reisslein, M. (2002) Providing application-level QoS in 3G/4G wireless systems: a comprehensive framework based on multirate CDMA. *IEEE Wireless Communications* (see also *IEEE Personal Communications*), **9**(2), 42–47.

50. Classon, B., Blankenship, K. and Desai, V. (2002) Channel coding for 4G systems with adaptive modulation and coding. *IEEE Wireless Communications* (see also *IEEE Personal Communications*), **9**(2), 8–13.

51. Yile, Guo and Chaskar, H. (2002) Class-based quality of service over air interfaces in 4G mobile networks. *IEEE Communications Magazine*, **40**(3), 132–137.

52. Win, M. Z. and Scholtz, R. A. (2002) Characterization of ultra-wide bandwidth wireless indoor channels: a communication-theoretic view. *IEEE Journal on Selected Areas in Communications*, **20**(9), 1613–1627.

53. Cassioli, D., Win, M. Z. and Molisch, A. F. (2002) The ultra-wide bandwidth indoor channel: from statistical model to simulations. *IEEE Journal on Selected Areas in Communications*, **20**(6), 1247–1257.

54. Hocquenghem, A. (1959) Codes correcteurs d'erreurs Chiffres (Paris), **2**, 147–156.

55. Bose, R. and Ray-Chaudhuri, D. (1960) On a class of error correcting binary group codes. *Information and Control*, **3**, 68–79.

56. Bose, R. and Ray-Chaudhuri, D. (1960) Further results on error correcting binary group codes. *Information and Control*, **3**, 279–290.

57. Peterson, W. (1960) Encoding and error correction procedures for the Bose–Chaudhuri codes. *IEEE Transactions On Information Theory*, **IT-6**, 459–470.

58. Gorenstein, D. and Zierler, N. (1961) A class of cyclic linear error-correcting codes in $p > n$ synbols. *Journal of the Society of Industrial and Applied Mathematics*, **9**, 107–214.

59. Reed, I. and Solomon, G. (1960) Polynomial codes over certain finite fields. *Journal of the Society of Industrial and Applied Mathematics*, **8**, 300–304.

60. Berlekamp, E. (1965) On decoding binary Bose–Chaudhuri–Hocquenghem codes. *IEEE Transactions On Information Theory*, **11**, 577–579.

61. Berlekamp, E. (1968) *Algebraic Coding Theory* McGraw-Hill, New York, USA.

62. Massey, J. (1965) Step-by-step decoding of the Bose–Chaudhuri–Hocquenghem codes, *IEEE Transactions On Information Theory*, **11**, 580–585.

63. Massey, J. (1969) Shift-register synthesis and BCH decoding. *IEEE Transactions On Information Theory*, **IT-15**, 122–127.

64. Peterson, W. and Weldon Jr, E. (1972) *Error Correcting Codes*, 2nd edition, MIT Press, Cambridge, MA, USA.

65. Sklar, B. (1988) *Digital Communications–Fundamentals and Applications*: Prentice Hall, Englewood Cliffs, NJ, USA.

66. Clark Jr, G. and Cain, J. (1981) *Error Correction Coding for Digital Communications*, Plenum Press, New York, USA.

67. Blahut, R. (1983) *Theory and Practice of Error Control Codes*. Addison Wesley, Reading, MA, USA.

68. Lin, S. and Constello Jr, D. (1982) *Error Control Coding: Fundamentals and Applications*, Prentice Hall, Englewood Cliffs, NJ, USA.

69. Michelson, A. and Levesque, A. (1985) *Error Control Techniques for Digital Communication*, John Wiley & Sons, New York, USA.

70. Hanzo, L. *et al.* (2002) *Turbo coding, Turbo Equalization and Space–Time Coding* John Wiley & Sons, Inc., New York.

71. Benedetto, S. *et al.* (1999) *Principles of Digital Transmissions With Wireless Applications*, Kluwer, New York.

72. Elias, P. (1955) Coding for noisy channels, *IRE Convention Record*, **4**, 37–47.

73. Wozencraft, J. (1957) Sequential decoding for reliable communication, IRE Natl. Convention rec., **5**(2), 11–25.

74. Wozencraft, J. and Reiffen, B. (1961) *Sequential Decoding*, MIT Press, Cambridge, MA, USA.

75. Fano, R. (1963) A heuristic discussion of probabilistic coding. *IEEE Transactions On Information Theory*, **IT-9**, 64–74.

76. Massey, J. (1963) *Threshold Decoding*, MIT Press, Cambridge, MA, USA.

77. Viterbi, A. (1967) Error bounds for convolutional codes and an asymptotically optimum decoding algorithm. *IEEE Transactions On Information Theory*, **IT-13**, 260–269.

78. Forney, G. (1973) The Viterbi algorithm. *Proceedings of the IEEE*, **61**, 268–278.

79. Heller, J. and Jacobs, I. (1971) Viterbi decoding for satellite and space communication. *IEEE Transactions on Communication Technology*, **COM-19**, 835–848.

80. Bahl, L. R., Cocke, J., Jelinek, F. and Raviv, J. (1974) Optimal Decoding of Linear Codes for Minimising Symbol Error Rate. *IEEE Transactions On Information Theory*, **20**, 284–287.

81. Viterbi, A. J. (1965) Optimum detection and signal selection for partially coherent binary communication, *IEEE Transactions On Information Theory*, **IT-11**, 239–246.

82. Viterbi, A. J. (1966) *Principles of Coherent Communications*. McGraw Hill, New York.

83. Viterbi, A. J. (1967) Error bounds for convolutional codes and an asymptotically optimum decoding algorithm. *IEEE Transactions On Information Theory*, **IT-13**, 260–269.

84. Viterbi, A. J. and Omura, J. K. (1979) *Principles of Digital Communication and Coding*. McGraw Hill, New York.

85. Clark, G. C. and Cain, J. B. (1981) *Error Correction Coding for Digital Communications*. Plenum Press, New York.

86. Lin, S. and Costello, D. J. (1983) *Error Control Coding: Fundamentals and Applications*. Prentice Hall, Englewood Cliffs, NJ, USA.

87. Forney, Jr, G. D. (1966) *Concatenated Codes*. MIT Press, Cambridge, MA, USA.

88. Forney, Jr, G. D. (1970) Convolutional Codes I: Algebraic structure. *IEEE Transactions On Information Theory*, **IT-16**, 720–738.

89. Forney, Jr, G. D. (1973) The Viterbi algorithm. *IEEE Proceedings*, **61**, 268–278.

90. Forney, Jr, G. D. (1974a) Convolutional Codes II: Maximum-likelihood decoding, *Information and Control*, **25**, 223–265.

91. Forney, Jr, G. D. (1974b) Convolutional Codes III: Sequential decoding, *Information and Control*, **25**, 267–297.

92. Wozencraft, J. M. (1957) Sequential decoding for reliable communications, *Tech. Report 325*, RLE MIT, Cambridge, MA.

93. Fano, R. M. (1963) A heuristic discussion of probabilistic decoding. *IEEE Transactions On Information Theory*, **IT-9**, 64–74.

94. Jelinek, F. (1969) Fast sequential decoding algorithm using a stack. *IBM Journal of Research and Development*, **13**, 675–685.

95. Zigangirov, K. S. (1966) Some sequential decoding procedures, *Problemy Peredachi Informacii*, **2**, 13–25.

96. Wozencraft, J. M. and Jacobs, I. M. (1965) *Principles of Communication Engineering*. John Wiley & Sons, New York.

97. Savage, J. E. (1966) Sequential decoding. The computational problem. *Bell Systems Technical Journal*, **45**, 149–176.

98. Anderson, J. B. and Mohan, S. (1991) *Source and Channel Coding: An Algorithmic Approach*. Kluwer Academic Press, Boston, MA, USA.

99. Cain, J. B., Clark, G. C. and Geist, J. M. (1979) Punctured convolutional codes of rate $(n - 1)/n$ and simplified maximum likelihood decoding. *IEEE Transactions On Information Theory*, **IT-25**, 97–100.

100. Yasuda, Y., Hirata, Y., Nakamura, K. and Otani, S. (1983) Development of variable-rate Viterbi decoder and its performance characteristics, in *Proceedings Sixth International Conference on Digital Satellite Communications*, Phoenix, AZ, September 1983, pp. XII-24–XII-31.

101. Yasuda, Y. Kashiki, K. and Hirata, Y. (1984) High rate punctured convolutional codes for soft decision Viterbi decoding. *IEEE Transactions On Communications*, **COM-32**, 315–319.

102. Berrou, C. and Glavieux, A. (1996) Near optimum error-correcting coding and decoding: Turbo codes. *IEEE Transactions On Communications*, **COM-44**, 1261–1271.

103. Benedetto, S. and Montorsi, G. (1996a) Unveiling turbocodes: Some results on parallel concatenated coding schemes. *IEEE Transactions On Information Theory*, **COM-44**, 591–600.

104. Divsalar, D. and Pollara, F. (1995) Turbo codes for deep-space communications. *TDA Progress Report 42–120*, pp. 29–39, Jet Propulsion Laboratory, Pasadena, California.

105. Divsalar, D. and McEliece, R. J. (1996) Effective free distance of turbo codes, *Electronics Letters*, **32**(5).

106. Perez, L. C., Seghers, J. and Costello, D. J. (1996) A distance spectrum interpretation of turbo codes. *IEEE Transactions On Information Theory*, **IT-42**, 1698–1709.

107. Benedetto, S. and Montorsi, G. (1996b) Design of parallel concatenated convolutional codes. *IEEE Transactions On Communications*, **IT-43**, 409–428.

108. Bahl, L. R., Cocke, J., Jelinek, F. and Raviv, J. (1974) Optimal decoding of linear codes for minimizing symbol error rate. *IEEE Transactions On Information Theory*, **IT-20**, 284–287.

109. Benedetto, S., Divsalar, D. and Hagenauer, J. (eds). (1998d) Concatenated coding techniques and iterative decoding: Sailing toward channel capacity. *IEEE Journal on Selected Areas in Communications*, **16**(2).

110. Benedetto, S., Divsalar, D., Montorsi, G. and Pollara, F. (1998c) Soft-input soft-output modules for the construction and distributed iterative decoding of code networks. *European Transactions on TeleCommunications*, **9**, 155–172.

111. Benedetto, S., Divsalar, D., Montorsi, G. and Pollara, F. (1998b) Serial concatenation of interleaved codes: Performance analysis, design, and iterative decoding. *IEEE Transactions On Information Theory*, **44**, 909–926.

112. Hagenauer, J., Offer, E. and Papke, L. (1996) Iterative decoding of binary block and convolutional codes. *IEEE Transactions On Information Theory*, **IT-42**, 429–445.

113. Berrou, C., Glavieux, A. and Thitimajshima, P. (1993) Near Shannon limit error correcting codes: Turbo codes. *IEEE International Conference on Communications* (ICC'93), Geneva, Switzerland, pp. 1064–1070.

114. Lin, S. and Costello, D. (1982) *Error Control Coding: Fundamentals and Applications* Prentice Hall, Englewod Cliffs, NJ.

115. Gordon, N., Vucetic, B., Musicki, D. and Du, J. *Joint error control and speech coding for 4.8 kbps digital voice transmission over satellite mobile channels*. Technical Report, Sydney University, Sydney, Australia.

116. Wu, K., Lin, S. and Miller, M. (1982) A hybrid ARQ scheme using multiple shortened cyclic codes, in *Proceedings GLOBECOM*, Miami, FL, pp. C8.61–C8.65.

117. Chase, D. (1985) Code combining – A maximum likelihood decoding approach for combining an arbitrary number of noisy packets. *IEEE Transactions On Communications*, **COM-33**, 385–393.

118. Sovetov, B. and Stah, V. (1982) *Design of adaptive transmission systems*, Energoizdal, Leningrad, in Russian.

119. Sullivan, D. (1971) A generalization of Gallagher's adaptive error control scheme. *IEEE Transactions On Information Theory*, **IT-17**, 727–735.

120. Mandelbaum, D. (1974) An adaptive-feedback coding scheme using incremental redundancy. *IEEE Transactions On Information Theory*, **IT-20**, 388–389.

121. Vucetic, B., Drajic, D. and Perisic, D. (1988) An algorithm for adaptive error control system synthesis, in *ISIT 1985*, Brighton, England; also in *Proceedings IEE*, Part F, 85–94.

122. Mandelbaum, D. M. (1975) On forward error correction with adaptive decoding. *IEEE Transactions On Information Theory*, **IT-21**, 230–233.

123. Kallel, S. and Haccoun, D. (1988) Sequential decoding with ARQ code combining: A robust hybrid FEC/ARQ system. *IEEE Transactions On Communications*, **26**, 773–780.

124. Drukarev, A. and Costello, Jr, D. J. (1983) Hybrid ARQ control using sequential decoding. *IEEE Transactions On Information Theory*, **IT-29**, 521–535.

125. Drukarev, A. and Costello, Jr, D. J. (1982) A comparison of block and convolutional codes in ARQ error control schemes. *IEEE Transactions On Communications*, **COM-30**, 2449–2455.

126. Lugand, L. and Costello, Jr, D. J. (1982) A comparison of three hybrid ARQ schemes on a non-stationary channel, in *Proceedings GLOBECOM*, Miami, FL, pp. C8.4.1–C8.4.5.

127. Hagenauer, J. and Lutz, E. (1987) Forward error correction coding for fading compensation in mobile satellite channels. *IEEE Journal on Selected Areas in Communications*, **SAC-5**, 215–225.

128. Hagenauer, J. (1988) Rate-compatible punctured convolutional codes (RCPC codes) and their applications. *IEEE Transactions On Communications*, **36**, 389–400.

129. Ungerboeck, G. (1982) Channel coding with multilevel/phase signals. *IEEE Transactions On Information Theory*, **IT-28**, 56–67.

130. Ungerboeck, G. (1987) Trellis-coded modulation with redundant signal sets – Part I: Introduction. *IEEE Communications Magazine*, **25**, 5–11; and Trellis-coded modulation with redundant signal sets – Part 11: State of the art, *Ibidem*, pp. 12–21.

131. Forney, Jr., G. D. (1989) Multidimensional constellation – Part II: Voronoi constellations. *IEEE Journal on Selected Areas in Communications*, **7**, 941–958.

132. Forney, Jr, G. D., Gallagher, R. G., Lang, G. R., Longstaff, F. M. and Qureshi, S. H. (1984) Efficient modulation for band-limited channels. *IEEE Journal on Selected Areas in Communications*, **SAC-2**, 632–647.

133. Forney, Jr, G. D. and Ungerboeck, G. (1998) Modulation and coding for linear Gaussian channels. *IEEE Transactions On Information Theory*, pp. 2384–2415.

134. Forney, Jr, G. D. and Wei, L.-F. (1989) Multidimensional constellation – Part I: Introduction, figures of merit, and generalized cross constellations. *IEEE Journal On Selected Areas in Communications*, **7**, 877–892.

135. Biglieri, E. (1984) High-level modulation and coding for nonlinear satellite channels. *IEEE Transactions On Communications*, **COM-32**, 616–626.

136. Biglieri, E. (1992) Parallel demodulation of multidimensional signals. *IEEE Transactions on Communications*, **40**(10), 1581–1587.

137. Biglieri, E., Divsalar, D., McLane, P. J. and Simon, M. K. (1991) *Introduction to Trellis-Coded Modulation with Applications*. Macmillan, New York.

138. Imai, H. and Hirakawa, S. (1977) A new multilevel coding method using error-correcting codes. *IEEE Transactions On Information Theory*, **IT-23**(3), 371–377.

139. Wei, L.-F. (1987) Trellis-coded modulation with multidimensional constellations. *IEEE Transactions On Information Theory*, **IT-33**, 483–501.

140. Cavers, J. K. and Ho, P. (1992) Analysis of the error performance of trellis-coded modulations in Rayleigh-fading channels. *IEEE Transactions On Communications* **40**(1), 74–83.

141. Liu, Y. J., Oka, I. and Biglieri, E. (1990) Error probability for digital transmission over nonlinear channels with applications to TCM. *IEEE Transactions On Information Theory*, **IT 36**, 1101–1110.

142. Zehavi, E. (1992) 8-PSK trellis-codes for a Rayleigh channel. *IEEE Transactions On Communications*, **40**, 873–884.

143. Zehavi, E. and Wolf, J. K. (1987) On the performance evaluation of trellis codes. *IEEE Transactions On Information Theory*, **IT 33**(2), 196–201.

144. Benedetto, S., Mondin, M. and Montorsi, G. (1994) Performance evaluation of trellis-coded modulation schemes. *IEEE Proceedings*, **82**, 833–855.

145. Trott, M. D., Benedetto, S., Garello, R. and Mondin, M. (1996) Rotational invariance of trellis codes–Part I: Encoders and precoders. *IEEE Transactions On Information Theory*, **42**, 751–765.

146. Sundberg, C.-E. W. and Seshadri, N. (1993) Coded modulation for fading channels: An overview. *European Transactions on Telecommunications* **4**(3), 309–324.

147. Pottie, G. J. and Taylor, D. P. (1989) Multilevel codes based on partitioning, *IEEE Transactions On Information Theory*, **IT-35**, 87–98.

148. Pellizzoni, R., Sandri, A., Spalvieri, A. and Biglieri, E. (1997) Analysis and implementation of an adjustable-rate multilevel coded modulation system. *IEE Proceedings on Communications*, **144**, 1–5.

149. Caire, G., Taricco, G. and Biglieri, E. (1998) Bit-interleaved coded modulation. *IEEE Transactions On Information Theory*, **44**, 927–946.

150. Divsalar, D. and Simon, M. K. (1988) The design of trellis coded MPSK for fading channel: Performance criteria. *IEEE Transactions On Communications*, **36**, 1004–1012.

151. Divsalar, D. and Simon, M. K. (1988) The design of trellis coded MPSK for fading channel: Set partitioning for optimum code design. *IEEE Transactions On Communications*, **36**, 1013–1021.

152. Robertson, P., Worz, T. (1998) Bandwidth-efficient Turbo Trellis-coded Modulation Using Punctured Component Codes. *IEEE Journal on Selected Areas in Communications*, **16**, 206–218.

153. Zehavi, E. (1992) 8-PSK trellis-codes for a Rayleigh fading channel. *IEEE Transactions on Communications*, **40**, 873–883.

154. Li, X. and Ritcey, J. A. (1997) Bit-interleaved coded modulation with iterative decoding. *IEEE Communications Letters*, **1**.

155. Li, X. and Ritcey, J. A. (1999) Trellis-coded Modulation with Bit Interleaving and Iterative Decoding. *IEEE Journal on Selected Areas in Communications*, **17**.

156. Li, X. and Ritcey, J. A. (1998) Bit-interleaved coded modulation with iterative decoding – Approaching turbo-TCM performance without code concatenation, in *Proceedings of CISS 1998*, (Princeton University, USA).

157. Ng, S. X., Liew, T. H., Yang, L. L. and Hanzo, L. (2001) Comparative Study of TCM, TTCM, BICM and BICM-ID schemes. *IEEE Vehicular Technology Conference*, p. 265 (CDROM).

158. Ng, S. X., Wong, C. H. and Hanzo, L. (2001) Burst-by-Burst Adaptive Decision Feedback Equalized TCM, TTCM, BICM and BICM-ID. *International Conference on Communications* (ICC), pp. 3031–3035.

159. Ungerboeck, G. (1987) Trellis-coded modulation with redundant signal sets. Part 1 and 2. *IEEE Communications Magazine*, **25**, 5–21.

160. Pietrobon, S. S., Ungerboeck, G., Perez, L. C. and Costello, D. J. (1994) Rotationally invariant nonlinear trellis codes for two-dimensional modulation. *IEEE Transactions On Information Theory*, **IT-40**, 1773–1791.

161. Schlegel, C. (1997) Trellis Coded Modulation, in *Trellis Coding*, IEEE Press, New York, USA, pp. 43–89.

162. Cavers, J. K. and Ho, P. (1992) Analysis of the Error Performance of Trellis-coded Modulations in Rayleigh-fading Channels. *IEEE Transactions On Communications*, **40**, 74–83.

163. Pietrobon, S. S., Deng, R. H., Lafanechere, A., Ungerboeck, G. and Costello, D. J. (1990) Trellis-coded Multidimensional Phase Modulation. *IEEE Transactions on Information Theory*, **36**, 63–89.

164. Wei, L. F. (1987) Trellis-coded modulation with multidimensional constellations. *IEEE Transactions On Information Theory*, **IT-33**, 483–501.

165. Caire, G., Tariceo, G. and Biglieri, E. (1998) Bit-Interleaved Coded Modulation. *IEEE Transactions On Information Theory*, **44**, 927–946,

166. Steele, R. and Webb, W. (1991) Variable rate QAM for data transmission over Rayleigh fading channels, in *Proceedings of Wireless' 91*, Calgary, Alberta, pp. 1–14.

167. Hanzo, L., Wong, C. H. and Yee, M. S. (2002) *Adaptive Wireless Transceivers*. John Wiley & Sons, Inc., New York, USA and IEEE Press. (For detailed contents, please refer to http://www-mobile.ecs.soton.ac.uk.).

168. Benedetto, S. and Biglieri, E. (1999) *Principles of Digital Transmission With Wireless Applications*, Kluwer, New York.

169. Proakis, G. (1995) Digital Communications, 3rd editon, McGraw Hill, New York.

170. Simon, M. K. *et al.* (1995) *Digital Communication Techniques – Signal Design and Detection*, Prentice Hall, Englewood Cliffs, NJ.

171. Sampei, S., Komaki, S. and Morinaga, N. (1994) Adaptive Modulation/TDMA Scheme for Large Capacity Personal Multi-Media Communication Systems. *IEIEE Transactions on Communications*, **E77-B**, 1096–1103.

172. Goldsmith, A. J. and Chua, S. (1997) Variable-rate variable-power MQAM for fading channels. *IEEE Transactions On Communications*, **45**, 1218–1230.

173. Wong, C. and Hanzo, L. (2000) Upper-bound performance of a wideband burst-by-burst adaptive modem. *IEEE Transactions On Communications*, **48**, 367–369.

174. Matsuoka, H., Sampei, S., Morinaga, N. and Kamio, Y. (1996) Adaptive Modulation System with Variable Coding Rate Concatenated Code for High Quality Multi-Media Communications Systems, in *Proceedings of IEEE VTC'96*, Atlanta, USA, 28 April–1 May, **1**, 487–491.

175. Lau, V. and Macleod, M. (1998) Variable rate adaptive trellis coded QAM for high bandwidth efficiency applications in Rayleigh fading channels, in *proceedings of IEEE Vehicular Technology Conference* (VTC'98), Ottawa, Canada, 18–21 May, pp. 348–352.

176. Goldsmith, A. J. and Chua, S. (1998) Adaptive Coded Modulation for Fading Channels. *IEEE Transactions On Communications*, **46**, 595–602.

177. Won, C. H., Liew, T. H. and Hanzo, L. (1999) Burst-by-Burst Turbo Coded Wideband Adaptive Modulation with Blind Modem Mode Detection, in *Proceedings of 4th ACTS Mobile Communications Summit*, Sorrento, Italy, pp. 303–308.

178. Goeckel, D. (1999) Adaptive Coding for Fading Channels using Outdated Fading Estimates. *IEEE Transactions On Communications*, **47**, 844–855.

179. Choi, B. J., Munster, M., Yang, L. L. and Hanzo, L. (2001) Performance of Rake receiver assisted adaptive-modulation based CDMA over frequency selective slow Rayleigh fading channel. *Electronics Letters*, **37**, 247–249.

180. Alamouti, S. M. and Kallel, S. (1994) Adaptive Trellis-coded Multiple-phase-shift Keying Rayleigh Fading Channels. *IEEE Transactions On Communications*, **42**, 2305–2341.

181. Al-Semari, S. and Fuja, T. (1997) I-Q TCM: Reliable communication over the Rayleigh fading channel close to the cutoff rate. *IEEE Transactions On Information Theory*, **43**, 250–262.

182. Webb, W. and Steele, R. (1995) Variable rate QANI for mobile radio. *IEEE Transactions on Communications*, **43**, 2223–2230.

183. Torrance, J. and Hanzo, L. (1996) Performance upper bound of adaptive QAM in slow Rayleigh-fading environments, in *Proceedings of IEEE ICCS'96/ISPACS'96*, Singapore, 25–29 November, pp. 1653–1657.

184. Torrance, J. M., Hanzo, L. and Keller, T. (1999) Interference aspects of adaptive modems over slow Rayleigh fading channels. *IEEE Transactions on Vehicular Technology*, **48**, 1527–1545.

185. Matsuoka, H., Sampei, S., Morinaga, N. and Kamio, Y. (1996) Adaptive modulation systems with variable coding rate concatenated code for high quality multi-media communication systems, in *Proceedings of IEEE VTC'96*, Atlanta, USA, 28 April–1 May, pp. 487–491.

186. Chua, S. G. and Goldsmith, A. J. (1996) Variable-rate variable-power mQAM for fading channels, in *Proceedings of IEEE VTC'96*, Atlanta, USA, 28 April–1 May, pp. 815–819.

187. Torrance, J. and Hanzo, L. (1996) Optimisation of switching levels for adaptive modulation in a slow Rayleigh fading channel. *Electronics Letters*, **32**, 1167–1169.

188. Torrance, J. and Hanzo, L. (1996) On the upper bound performance of adaptive QAM in a slow Rayleigh fading. *IEE Electronics Letters*, 169–171.

189. Lau, V. and Macleod, M. (2001) Variable-rate adaptive trellis coded QAM for flat-fading channels. *IEEE Transactions On Communications*, **49**, 1550–1560.

190. Lau, V. and Marie, S. (1999) Variable rate adaptive modulation for DS-CDMA. *IEEE Transactions On Communications*, **47**, 577–589.

191. Chua, S. and Goldsmith, A. (1998) Adaptive Coded Modulation for Fading Channels. *IEEE Transactions On Communications*, **46**, 595–602.

192. Liu, X., Ormeci, P., Wesel, R. and Goeckel, D. (2001) Bandwidth-efficient, low-latency adaptive coded modulation schemes for time-varying channels, in *Proceedings of IEEE International Conference on Communications*, Helsinki, Finland.

193. Torrance, J. and Hanzo, L. (1996) Demodulation level selection in adaptive modulation. *Electronics Letters*, **32**, 1751–1752.

194. Nanda, S., Balachandran, K. and Kumar, S. (2000) Adaptation techniques in wireless packet data services. *IEEE Communications Magazine*, **38**, 54–64.

195. Forney, Jr, G. D. (1988a) Coset codes – Part I: Introduction and geometrical classification. *IEEE Transactions On Information Theory*, **34**, 1123–1151.

196. Forney, Jr, G. D. (1988b) Coset codes – Part II: Binary lattices and related codes. *IEEE Transactions On Information Theory*, **34**, 1152–1187.

197. Alamouti, S. (1998) A simple transmit diversity technique for wireless communications, *IEEE Journal on Selected Areas in Communications*, **18**(7), 1451–1458.

198. Tarokh, V., Seshadri, N. and Calderbank, A. R. (1998) Space–time codes for high data rate wireless communications: Performance criterion and code construction, *IEEE Transactions On Information Theory*, **44**, 744–765.

199. Furuskar, A., Mazur, S., Muller, F. and Olofsson, H. (1999) EDGE: Enhanced data rates for GSM and TDMA/136 evolution, *IEEE Personal Communications Magazine*, **6**, 56–66.

200. Naguib, A., Seshadri, N. and Calderbank, A. R. (2000) Increasing data rate over wireless channels, *IEEE Signal Processing Magazine*, **17**, 76–92.

201. Tarokh, V., Jafarkhani, H. and Calderbank, A. R. (1999) Space–time block codes from orthogonal designs, *IEEE Transactions On Information Theory*, **45**, 1456–1467.

202. Stoica, P. and Lindskog, E. (2001) *Space–time block coding for channels with intersymbol interference*, presented at the thirty-fifth Asilomar Conference on Signals, Systems and Computing, October 2001.

203. Tarokh, V., Naguib, A., Seshadri, N. and Calderbank, A. R. (1999) Space–time codes for high data rate wireless communication: Performance criteria in the presence of channel estimation errors, mobility and multiple paths, *IEEE Transactions On Communications*, **48**, 199–207.

204. Naguib, A., Tarokh, V., Seshadri, N. and Calderbank, A. R. (1998) A space–time coding modem for high-data-rate wireless communications, *IEEE Journal on Selected Areas in Communications*, **16**, 1459–1477.

205. Seshadri, N. and Winters, J. (1993) Two signaling schemes for improving the error performance of frequency-division-duplex (FDD) transmission systems using transmitter antenna diversity, in *Proceedings of Vehicular Technology Conference*, pp. 508–511.

206. Foschini, G. J. (1996) Layered Space–time architecture for wireless communication in a fading environment when using multi-element antennas, *Bell Labs Technical Journal*, **1**, 41–59.

207. Tarokh, V., Jafarkhani, H. and Calderbank, A. R. (1999) Space–time block coding for wireless communications: Performance results, *IEEE Journal on Selected Areas in Communications*, **17**, 451–460.

208. Franz, V. and Anderson, J. (1998) Concatenated decoding with a reduced-search BCJR algorithm, *IEEE Journal on Selected Areas in Communication*, **16**, 186–195.

209. Bahl, L., Cocke J., Jelinek F. and Raviv, J. (1974) Optimal decoding of linear codes for minimizing symbol error rate, *IEEE Transaction on Information Theory*, **IT-20**, 284–287.

210. Fragouli, C., Al-Dhahir, N., Diggavi, S. and Turin, W. (2002) Prefiltered space–time M-BCJR equalizer for frequency-selective channels, *IEEE Transactions On Communications*, **50**, 742–753.

211. Van Etten, W. (1976) Maximum likelihood receiver for multiple channel transmission systems, *IEEE Transactions On Communications*, **COM-24**, 276–283.

212. Naguib, A. and Seshadri, N. (2000) MLSE and equalization of space–time coded signals, in *Proceedings of Vehicular Technology Conference*, May 2000, 1688–1693.

213. Al-Dhahir, N. (2001) FIR channel-shortening equalizers for MIMO ISI channels, *IEEE Transactions On Communications*, **50**, 213–218.

214. Bauch, G. and Al-Dhahir, N. (2000) Iterative equalization and decoding with channel shortening filters for space–time coded modulation, in *Proceedings of Vehicular Techology Conference*, 2000, pp. 1575–1582.

215. Al-Dhahir, N., Naguib, A. and Calderbank, A. R. (2001) Finite-length MIMO decision-feedback equalization for space–time block-coded signals over multipath fading channels, *IEEE Transactions on Vehicular Technology*, **50**, 1176–1181.

216. Eyuboglu, M. and Qureshi, S. (1999) Reduced-state sequence estimation for coded modulation of inter-symbol interference channels, *IEEE Journal on Selected Areas in Communications*, **17**, 989–999.

217. Weinstein, S. and Ebert, P. (1971) Data transmission by frequency-division multiplexing using the discrete Fourier transform, *IEEE Transactions On Communications*, **COM-19**, 628–634.

218. Zhou, S. and Giannakis, G. (2001) Space–time coded transmissions with maximum diversity gains over frequency-selective multipath fading channels, in *Proceedings of Globecom*, November 2001, pp. 440–444.

219. Sari, H., Karam G. and Jeanclaude, I. (1995) Transmission techniques for digital terrestrial TV broadcasting, *IEEE Communications Magazine*, **33**, 100–109.

220. Clark, M. V. (1998) Adaptive frequency-domain equalization and diversity combining for broadband wireless communications, *IEEE Journal on Selected Areas in Communications*, **16**, 1385–1395.

221. Al-Dhahir, N. (2001) Single-carrier frequency-domain equalization for space–time block-coded transmissions over frequency-selective fading channels, *IEEE Communications Letters*, **5**, 304–306.

222. Kaleh, G. (1995) Channel equalization for block transmission systems, *IEEE Journal on Selected Areas in Communications*, **13**, 110–121.

223. SC FDE PHY Layer System Proposal for Sub 11 GHz BWA [Online]. Available: http://www.ieee802.org/16/tg3/contrib/802 163p–0131r2.pdf

224. Diggavi, S., Al-Dhahir, N., Stamoulis A. and Calderbank, A. R. (2002) Differential space–time transmission for frequency-selective channels, *IEEE Communication Letters*, **6**(6), 253–255.

225. Fragouli, C., Al-Dhahir, N. and Turin, W. (2002) Reduced-complexity training schemes for multiple-antenna broadband transmissions, in *Proceedings of WCNC*, **1**, 78–83.

226. Fragouli, C., Al-Dhahir, N. and Turin, W. (2002) Finite-alphabet constant-amplitude training sequences for multiple-antenna broadband transmissions, *International Control Conference*, **1**, 6–10.

227. Baro, S., Bauchs, G. and Hansmann, A. (2000) Improved codes for space–time trellis coded modulation, *IEEE Communications Letters*, **4**, 20–22.

228. Hammons, R. and Gammal, H. E. (2000) On the theory of space–time codes for PSK modulation, *IEEE Transactions On Information Theory*, **46**, 524–542.

229. Grimm, J., Fitz, M. P. and Krogmeier, J. V. (1998) Further results in space–time coding for Rayleigh fading in *Proceedings of 1998 Allerton Conference*, 1998, pp. 391–400.

230. Grimm, J. (1998) *Transmitter diversity code design for achieving full diversity on Rayleigh channels*, Ph.D. dissertation, Purdue University, West Lafayette, IN.

231. Blum, R. S. (2002) Some analytical tools for designing space–time convolutional codes, *IEEE Transactions On Communications*, **50**, 1593–1599.

232. Guey, J.-C., Fitz, M. P., Bell, M. R. and Kuo, W.-Y. (1999) Signal design for transmitter diversity wireless communication systems over Rayleigh fading channels, *IEEE Transactions On Communications*, **47**, 527–537.

233. Seshadri, N. and Winters, J. H. (1994) Two signaling schemes for improving the error performance of frequency division duplex (FDD) transmission system using antenna diversity, *International Journal of Wireless Infomation Networks*, **1**, 49–60.

234. Raleigh, G. and Cioffi, J. M. (1998) Spatio-temporal coding for wireless communication, *IEEE Transactions On Communications*, **46**, 357–366.

235. Jafarkhani, H. and Seshadri, N. (2003) Super-orthogonal space–time trellis codes, *IEEE Transactions On Information Theory*, **49**(4), 937–950.

236. Genyuan Wang and Xiang-Gen Xia (2002) An orthogonal Space–time coding for CPM systems, *Proceedings of 2002 IEEE International Symposium on Information Theory*, pp. 107–112.

237. Agrawal, A., Ginis, G. and Cioffi, J. M. (2002) Channel diagonalization through orthogonal space–time coding, *IEEE International Conference on Communications* (ICC 2002), 28 April–2 May 2002, **3**, 1621–1624.

238. Sharma, N. and Papadias, C. B. (2002) Improved quasi-orthogonal codes, *IEEE Wireless Communications and Networking Conference* (WCNC2002) 17–21 March 2002, **1**, 169–171.

239. Rouquette, S., Merigeault, S. and Gosse, K. (2002) Orthogonal full diversity space–time block coding based on transmit channel state information for 4 Tx antennas, *IEEE International Conference on Communications* (ICC 2002), 28 April–2 May 2002, **1**, 558–562.

240. Jongren, G., Skoglund, M. and Ottersten, B. (2002) Combining beamforming and orthogonal space–time block coding, *IEEE Transactions On Information Theory*, **48**(3), 611–627.

241. Larsson, E. G., Ganesan, G., Stoica, P. and Wing-Hin Wong (2002) On the performance of orthogonal space–time block coding with quantized feedback, *IEEE Communications Letters*, **6**(11), 487–489.

242. Hsiao-feng, Lu, Kumar, P. V. and Habong, Chung (2002) On orthogonal designs and space–time codes, *Proceedings of 2002 IEEE International Symposium on Information Theory*, 30 June–5 July 2002, pp. 418–423.

243. Gozali, R. and Woerner, B. D. (2002) The impact of channel estimation errors on space–time trellis codes paired with iterative equalization/decoding, *IEEE 55th Vehicular Technology Conference* (VTC Spring 2002), 6–9 May 2002, **2**, 826–831.

244. Garg, P., Mallik, R. K. and Gupta, H. M. (2002) Performance analysis of space–time coding with imperfect channel estimation, *IEEE International Conference on Personal Wireless Communications*, December 15–17 2002, pp. 71–76.

245. Yan, Q. and Blum, R. S. (2002) Improved space–time convolutional codes for quasi-static slow fading channels, *IEEE Transactions on Wireless Communications*, **1**(4), 563–571.

246. Yan, Q. and Blum, R. S. (2000) Optimum space–time convolutional codes, *2000 IEEE Wireless Communications and Networking Conference* (WCNC), 23–28 September 2000, **3**, 1351–1355.

247. Sellathurai, M. and Haykin, S. (2001) Random space–time codes with iterative decoders for BLAST architectures, *Proceedings of 2001 IEEE International Symposium on Information Theory*, 24–29 June 2001, pp. 105–110.

248. Zhou, G., Wang, Y., Zhang, Z. and Chugg, K. M. (2001) On space–time convolutional codes for PSK modulation, *IEEE International Conference on Communications, 2001* (ICC 2001), 11–14 June 2001, **4**, 1122–1126.

249. Youjian Liu, Fitz, M. P. and Takeshita, O. Y. (2000) QPSK space–time turbo codes, *2000 IEEE International Conference on Communications* (ICC 2000), 18–22 June 2000, **1**, 292–296.

250. Jayaweera, S. K. and Poor, H. V. (2002) Turbo (iterative) decoding of a unitary space–time code with a convolutional code, *IEEE 55th Vehicular Technology Conference* (VTC Spring 2002) 6–9 May 2002, **2**, 1020–1024.

251. McCloud, M. L., Brehler, M. and Varanasi, M. K. (2002) Signal design and convolutional coding for noncoherent space–time communication on the block-Rayleigh-fading channel, *IEEE Transactions On Information Theory*, **48**(5), 1186–1194.

252. El Gamal, H. and Hammons, A. R. Jr (2002) On the Design and Performance of Algebraic space–time codes for BPSK and QPSK Modulation, *IEEE Transactions On Communications*, **50**(6).

253. Damen, M. O. and Beaulieu, N. C. (2003) On two high-rate algebraic space–time codes, *IEEE Transactions On Information Theory*, **49**(4), 1059–1063.

254. El Gamal, H. and Hammons, A. R., Jr (2000) 3 Algebraic designs for coherent and differentially coherent space–time codes, *2000 IEEE Wireless Communications and Networking Conference* (WCNC), 23–28 September 2000, **1**, 30–35.

255. El Gamal, H., Hammons, A. R., Jr and Stefanov, A. (2001) Algebraic space–time overlays for convolutionally coded systems, *12th IEEE InternationalSymposium on Personal, Indoor and Mobile Radio Communications*, 30 September–3 October 2001, **1**, C-149–C-154.

256. Damen, M. O., Abed-Meraim, K. and Belfiore, J.-C. (2002) Diagonal algebraic space–time block codes, *IEEE Transactions On Information Theory*, **48**(3), 628–636.

257. Damen, M. O., Tewfik, A. and Belflore, J. C. (2002) A construction of a space–time code based on number theory, *IEEE Transactions On Information Theory*, **48**(3), 753–760.

258. Gamal, H. E. and Hammons, A. R., Jr (2003) On the design of algebraic space–time codes for MIMO block-fading channels, *IEEE Transactions On Information Theory*, **49**(1), 151–163.

259. Hammons, A. R., Jr and El Gamal, H. (2000) Further results on the algebraic design of space–time codes, *Proceedings of IEEE International Symposium on Information Theory*, 25–30 June 2000, p. 339.

260. El Gamal, H. and Damen, M. O. (2002) An algebraic number theoretic framework for space–time coding, *Proceedings of the 2002 IEEE International Symposium on Information Theory*, pp. 132–137.

261. Tarokh, V. and Jafarkani, H (2000) A differential detection scheme for transmit diversity, *IEEE Journal on Selected Areas in Communications*, **18**(7).

262. Yi Yao and Howlader, M. (2002) Multiple symbol double differential space–time coded OFDM, *IEEE 55th Vehicular Technology Conference* (VTC Spring 2002) 6–9 May 2002, **2**, 1050–1054.

263. Diggavi, S. N., Al-Dhahir, N., Stamoulis, A. and Calderbank, A. R. (2002) Differential space–time coding for frequency-selective channels, *IEEE Communications Letters*, **6**(6), 253–255.

264. Jianhua Liu, Jian Li, Hongbin Li and Larsson, E. G. (2001) Differential space-code modulation for interference suppression, *IEEE Transactions on Signal Processing* (see also *IEEE Transactions on Acoustics, Speech, and Signal Processing*), **49**(8), 1786–1795.

265. Zhiqiang Liu, Giannakis, G. B. and Hughes, B. L. (2001) Double differential space–time block coding for time-selective fading channels, *IEEE Transactions On Communications*, **49**(9), 1529–1539.

266. Ingram, M. A., Kuo-Hui Li, van Nguyen, A. and Pratt, T. (2000) Beamforming, Doppler compensation, and differential space–time coding, *Proceedings of the 2000 IEEE Sensor Array and Multichannel Signal Processing Workshop*, 16–17 March 2000, pp. 158–162.

267. Hughes, B. L. (2000) Differential space–time modulation, *IEEE Transactions On Information Theory*, **46**(7), 2567–2578.

268. Lampe, L. H.-J., Schober, R. and Fischer, R. F. H. (2003) Coded differential space–time modulation for flat fading channels, *IEEE Transactions on Wireless Communications*, **2**(3), 582–590.

269. Steiner, A., Peleg, M. and Shamai, S. (2002) Iterative decoding of space–time differentially coded unitary matrix modulation, *IEEE Transactions on Signal Processing* (see also *IEEE Transactions on Acoustics, Speech, and Signal* Processing), **50**(10), 2385–2395.

270. Zhiqiang Liu and Giannakis, G. B. (2003) Block differentially encoded OFDM with maximum multipath diversity, *IEEE Transactions on Wireless Communications*, **2**(3), 420–423.

271. Steiner, A., Peleg, M. and Shamai, S. (2003) SVD iterative detection of turbo-coded multiantenna unitary differential modulation, *IEEE Transactions On Communications*, **51**(3), 441–452.

272. Meixia Tao and Cheng, R. S. (2003) Trellis-coded differential unitary space–time modulation over flat fading channels, *IEEE Transactions On Communications*, **51**(4), 587–596.

273. Jafarkhani, H. and Tarokh, V. (2001) Multiple transmit antenna differential detection from generalized orthogonal designs, *IEEE Transactions On Information Theory*, **47**(6), 2626–2631.

274. Kun Wang and Hongya Ge (2001) New differential transmission scheme with transmit diversity for DS-CDMA systems, *IEEE 54th Vehicular Technology Conference* (VTC 2001 Fall), **1**, 232–236.

275. Sellathurai, M. and Haykin, S. (2001) Further results on diagonal-layered space–time architecture, *IEEE 53rd Vehicular Technology Conference* (VTC 2001 Spring), 6–9 May 2001, **3**, 1958–1962.

276. Zhiqiang Liu and Giannakis, G. B. (2001) Layered space–time coding for high data rate transmissions, *IEEE Military Communications Conference* (MILCOM 2001), 28–31 October 2001, **2**, 1295–1299.

277. Matache, A., Wesel, R. D. and Jun Shi (2002) Trellis coding for diagonally layered space–time systems, *IEEE International Conference on Communications* (ICC 2002), 28 April–2 May 2002, **3**.

278. Gamal, H. E. and Hammons, A. R., Jr (2001) A new approach to layered space–time coding and signal processing, *IEEE Transactions On Information Theory*, **47**(6), 2321–2334.

279. El Gamal, H. (2002) On the design of layered space–time systems for autocoding, *IEEE Transactions On Communications*, **50**(9), 1451–1461.

280. Sellathurai, M. and Haykin, S. (2002) Turbo-BLAST for wireless communications: theory and experiments, *IEEE Transactions on Signal Processing* (see

also *IEEE Transactions on Acoustics, Speech, and Signal, Processing*) **50**(10), 2538–2546.

281. Wubben, D., Bohnke, R., Rinas, J., Kuhn, V. and Kammeyer, K. D. (2001) Efficient algorithm for decoding layered space–time codes, *Electronics Letters*, **37**(22), 1348–1350.

282. Yan Xin and Giannakis, G. B. (2002) High-rate space–time layered OFDM, *IEEE Communications Letters*, **6**(5), 187–189.

283. Sellathurai, M. and Haykin, S. (2003) T-BLAST for wireless communications: first experimental results, *IEEE Transactions on Vehicular Technology*, **52**(3), 530–535.

284. Tarokh, V., Naguib, A., Seshadri, N. and Calderbank, A. R. (1999) Combined array processing and space–time coding, *IEEE Transactions On Information Theory*, **45**(4), 1121–1128.

285. Foschini, G. J., Chizhik, D., Gans, M. J., Papadias, C. and Valenzuela, R. A. (2003) Analysis and performance of some basic space–time architectures, *IEEE Journal on Selected Areas in Communications*, **21**(3), 303–320.

286. Yi Gong and Ben Letaief, K. (2002) Concatenated space–time block coding with trellis coded modulation in fading channels, *IEEE Transactions on Wireless* Communications, **1**(4), 580–590.

287. Ungerboeck, G. (1982) Channel Coding with Multilevel Phase Signal, *IEEE Transactions On Information Theory*, **28**, 55–67.

288. Liew, T. H. and Hanzo, L. (2002) Space–time codes and concatenated channel codes for wireless communications, *Proceedings of the IEEE*, **90**(2), 187–219.

289. Dongzhe Cui and Haimovich, A. M. (2001) Performance of parallel concatenated space–time codes, *IEEE Communications Letters*, **5**(6), 236–238.

290. Goulet, L. and Leib, H. (2003) Serially concatenated space–time codes with iterative decoding and performance limits of block-fading channels, *IEEE Journal on Selected Areas in Communications*, **21**(5), 765–773.

291. Schlegel, C. and Grant, A. (2001) Concatenated space–time coding, *12th IEEE International Symposium on Personal, Indoor and Mobile Radio Communications*, 30 September–3 October 2001, **1**, C-139–C-143.

292. Ying Li, Jun-hong Hui and Xin-mei Wang (2002) Non-full rank space–time trellis codes for serially concatenated system, *IEEE Communications Letters*, **6**(9), 397–399.

293. Xiaotong Lin and Blum, R. S. (2000) Improved space–time codes using serial concatenation, *IEEE Communications Letters*, **4**(7), 221–223.

294. Steiner, A., Peleg, M. and Shamai, S. (2003) SVD iterative detection of turbo-coded multiantenna unitary differential modulation, *IEEE Transactions On Communications*, **51**(3), 441–452.

295. Foschini, G. J. and Gans, M. J. (1998) On limits of wireless communications in a fading environment when using multiple antennas, *Wireless Personal Communications*, **6**, 311–335.

296. Telatar, I. E. (1999) Capacity of multi-antenna Gaussian channels, *European Transactions on Telecommunications*, **10**(6), 585–595.

297. Stege, M., Zillmann, P. and Fettweis, G. (2002) MIMO channel estimation with dimension reduction, *The 5th International Symposium on Wireless Personal Multimedia Communications*, 27–30 October 2002, **2**, 417–421.

298. Qinfang Sun, Cox, D. C., Huang, H. C. and Lozano, A. (2002) Estimation of continuous flat fading MIMO channels, *IEEE Transactions on Wireless Communications*, **1**(4), 549–553.

299. Tugnait, J. K. (1997) Blind spatio-temporal equalization and impulse response estimation for MIMO channels using a Godard cost function, *IEEE Transactions on Signal Processing* (see also *IEEE Transactions on Acoustics, Speech, and Signal Processing*), **45**(1), 268–271.

300. Bai, W., He, C., Jiang, L. G. and Li, X. X. (2003) Robust channel estimation in MIMO-OFDM systems, *Electronics Letters*, **39**(2), 242–244.

301. Tugnait, J. K. (2001) Blind estimation and equalization of MIMO channels via multidelay whitening, *IEEE Journal on Selected Areas in Communications*, **19**(8), 1507–1519.

302. Zhi Ding and Li Qiu (2003) Blind MIMO channel identification from second order statistics using rank deficient channel convolution matrix, *IEEE Transactions on Signal Processing* (see also *IEEE Transactions on Acoustics, Speech and Signal Processing*), **51**(2), 535–544.

303. Haidong Zhu, Farhang-Boroujeny, B. and Schlegel, C. (2003) Pilot embedding for joint channel estimation and data detection in MIMO communication systems, *IEEE Communications Letters*, **7**(1), 30–32.

304. Komninakis, C., Fragouli, C., Sayed, A. H. and Wesel, R. D. (2002) Multi-input multi-output fading channel tracking and equalization using Kalman estimation, *IEEE Transactions on Signal Processing* (see also *IEEE Transactions on Acoustics, Speech, and Signal Processing*), **50**(5), 1065–1076.

305. Xavier, J. M. F., Barroso, V. A. N. and Moura, J. M. F. (1998) Closed-form blind channel identification and source separation in SDMA systems through correlative coding, *IEEE Journal on Selected Areas in Communications*, **16**(8), 1506–1517.

306. Tugnait, J. K. and Bin Huang (2000) Multistep linear predictors-based blind identification and equalization of multiple-input multiple-output channels, *IEEE Transactions on Signal Processing* (see also *IEEE Transactions on Acoustics, Speech, and Signal Processing*), **48**(1), 26–38.

307. Yang, H., Yuan, F. and Vucetic, B. (2002) Performance of space–time trellis codes in frequency selective WCDMA systems, *Proceedings of IEEE 56th Vehicular Technology Conference* (VTC 2002-Fall), 24–28 September 2002, **1**, 233–237.

308. Sun, Q., Cox, D. C., Huang, H. C. and Lozano, A. (2002) Estimation of Continuous Flat Fading MIMO Channels, *IEEE Transactions on Wireless Communications*, **1**(4).

309. Mu Qin and Blum, R. S. (2003) Properties of space–time codes for frequency selective channels and trellis code designs, *IEEE International Conference on Communications* (ICC' 03), **4**, 2286–2290.

310. Shengli Zhou and Giannakis, G. B. (2003) Single-carrier space–time block-coded transmissions over frequency-selective fading channels, *IEEE Transactions on Information Theory*, **49**(1), 164–179.

311. Bahceci, I. and Duman, T. M. (2002) Combined turbo coding and unitary space–time modulation, *IEEE Transactions On Communications*, **50**(8), 1244–1249.

312. Rouquette-Leveil, S. and Gosse, K. (2002) Space–time coding options for OFDM-based WLANs, *IEEE 55th Vehicular Technology Conference* (VTC Spring 2002), 6–9 May 2002, **2**, 904–908.

313. Yue, J. and Gibson, J. D. (2002) Performance of OFDM systems with space–time coding, *2002 IEEE Wireless Communications and Networking Conference* (WCNC2002), 17–21 March 2002, **1**, 280–284.

314. Ben Lu and Xiaodong Wang (2000) Iterative receivers for multiuser space–time coding systems, *IEEE Journal on Selected Areas in Communications*, **18**(11), 2322–2335.

315. Larsson, E. G., Stoica, P. and Li, J. (2002) On maximum-likelihood detection and decoding for space–time coding systems, *IEEE Transactions on Signal Processing* (see also *IEEE Transactions on IEEE Transactions on Acoustics, Speech, and Signal Processing*), **50**(4), 937–944.

316. Youjian Liu, Fitz, M. P. and Takeshita, O. Y. (2002) A rank criterion for QAM space–time codes, *IEEE Transactions On Information Theory*, **48**(12), 3062–3079.

317. Gore, D. A. and Paulraj, A. J. (2002) MIMO antenna subset selection with space–time coding, *IEEE Transactions on Signal Processing* (see also *IEEE Transactions on Acoustics, Speech, and Signal Processing*), **50**(10), 2580–2588.

318. Shengli Zhou, Muquet, B. and Giannakis, G. B. (2002) Subspace-based (semi-) blind channel estimation for block precoded space–time OFDM, *IEEE Transactions on Signal Processing* (see also *IEEE Transactions on Acoustics, Speech, and Signal Processing*), **50**(5), 1215–1228.

319. Goulet, L. and Leib, H. (2003) Serially concatenated space–time codes with iterative decoding and performance limits of block-fading channels, *IEEE Journal on Selected Areas in Communications*, **21**(5), 765–773.

320. Wing Hin Wong and Larsson, E. G. (2003) Orthogonal space–time block coding with antenna selection and power allocation, *Electronics Letters*, **39**(4), 379–381.

321. Caire, G. and Colavolpe, G. (2003) On low-complexity space–time coding for quasi-static channels, *IEEE Transactions On Information Theory*, **49**(6), 1400–1416.

322. Byoungjo Choi and Hanzo, L. (2003) Optimum mode-switching-assisted constant-power single- and multicarrier adaptive modulation, *IEEE Transactions on Vehicular Technology*, **52**(3), 536–560.

323. Ghrayeb, A. and Duman, T. M. (2003) Performance analysis of MIMO systems with antenna selection over quasi-static fading channels, *IEEE Transactions on Vehicular Technology*, **52**(2), 281–288.

324. El Gamal, H. and Damen, M. O. (2003) Universal space–time coding, *IEEE Transactions On Information Theory*, **49**(5), 1097–1119.

325. Xiaoxia Zhang and Fitz, M. P. (2003) Space–time code design with continuous phase modulation, *IEEE Journal on Selected Areas in Communications*, **21**(5), 783–792.

326. Lampe, L. H.-J., Schober, R. and Fischer, R. F. H. (2003) Coded differential space–time modulation for flat fading channels, *IEEE Transactions on Wireless Communications*, **2**(3), 582–590.

327. Jafarkhani, H. and Seshadri, N. (2003) Super-orthogonal space–time trellis codes, *IEEE Transactions On Information Theory*, **49**(4), 937–950.

328. Il-Min Kim and Tarokh, V. (2003) Variable-rate space–time block codes in *M*-ary PSK systems, *IEEE Journal on Selected Areas in Communications*, **21**(3), 362–373.

329. Gesbert, D., Shafi, M., Da-shan Shiu, Smith, P. J. and Naguib, A. (2003) From theory to practice: an overview of MIMO space–time coded wireless systems, *IEEE Journal on Selected Areas in Communications*, **21**(3), 281–302.

330. Banister, B. C. and Zeidler, J. R. (2003) Feedback assisted transmission subspace tracking for MIMO systems, *IEEE Journal on Selected Areas in Communications*, **21**(3), 452–463.

331. Petre, F., Leus, G., Deneire, L., Engels, M., Moonen, M. and De Man, H. (2003) Space–time block coding for single-carrier block transmission DS-CDMA downlink, *IEEE Journal on Selected Areas in Communications*, **21**(3), 350–361.

332. Yan Xin, Zhengdao Wang and Giannakis, G. B. (2003) Space–time diversity systems based on linear constellation precoding, *IEEE Transactions on Wireless Communications*, **2**(2), 294–309.

333. Wallace, J. W., Jensen, M. A., Swindlehurst, A. L. and Jeffs, B. D. (2003) Experimental characterization of the MIMO wireless channel: data acquisition and analysis, *IEEE Transactions on Wireless Communications*, **2**(2), 335–343.

334. Gamal, H. E. and Hammons, A. R., Jr (2003) On the design of algebraic space–time codes for MIMO block-fading channels, *IEEE Transactions On Information Theory*, **49**(1), 151–163.

335. Steiner, A., Peleg, M. and Shamai, S. (2003) SVD iterative detection of turbo-coded multiantenna unitary differential modulation, *IEEE Transactions On Communications*, **51**(3), 441–452.

336. Larsson, E. G. (2003) Unitary nonuniform space–time constellations for the broadcast channel, *IEEE Communications Letters*, **7**(1), 21–23.

337. Meixia Tao and Cheng, R. S. (2003) Trellis-coded differential unitary space–time modulation over flat fading channels, *IEEE Transactions On Communications*, **51**(4), 587–596.

338. Young-Hak Kim and Kaveh, M. (2003) Coordinate-interleaved space–time coding with rotated constellation, *The 57th IEEE Vehicular Technology Conference* (VTC 2003-Spring), April 22–25 2003, **1**, 732–735.

339. Le Nir, V., Mard, M. and Le Gouable, R. (2003) Space–time block coding applied to turbo coded multicarrier CDMA, *The 57th IEEE Vehicular Technology Conference* (VTC 2003-Spring), April 22–25 2003, **1**, 577–581.

340. Yang, J., Sun, Y., Senior, J. M. and Pem, N. (2003) Channel estimation for wireless communications using space–time block coding techniques, *Proceedings of the 2003 International Symposium on Circuits and Systems* (ISCAS '03), May 25–28 2003, **2**, 220–223.

341. Blum, R. S. MIMO capacity with interference, *IEEE Journal on Selected Areas in Communications*, **21**(5), 793–801.

342. Mietzner, J., Hoeher, P. A. and Sandell, M. (2003) Compatible improvement of the GSM/EDGE system by means of space–time coding techniques, *IEEE Transactions on Wireless Communications*, **24**(5), 690–702.

343. Sellathurai, M. and Haykin, S. (2003) T-BLAST for wireless communications: first experimental results, *IEEE Transactions on Vehicular Technology*, **52**(3), 530–535.

344. Shengli Zhou and Giannakis, G. B. (2003) Optimal transmitter eigen-beamforming and space–time block coding based on channel correlations, *IEEE Transactions On Information Theory*, **49**(7), 1673–1690.

345. Debbah, M., Hachem, W., Loubaton, P. and de Courville, M. (2003). MMSE analysis of certain large isometric random precoded systems, *IEEE Transactions On Information Theory*, **49**(5), 1293–1311.

346. Uysal, M. and Georghiades, C. N. (2003) An efficient implementation of a maximum-likelihood detector for space–time block coded systems, *IEEE Transactions on Communications*, **51**(4), 521–524.

347. Baccarelli, E. and Biagi, M. (2003) Error resistant space–time coding for emerging 4G-WLANs, *2003 IEEE Wireless Communications and Networking* (WCNC 2003), 16–20 March 2003, **1**, 72–77.

348. Hochwald, B. M. and ten Brink, S. (2003) Achieving near-capacity on a multiple-antenna channel, *IEEE Transactions On Communications*, **51**(3), 389–399.

349. Banister, B. C. and Zeidler, J. R. (2003) A simple gradient sign algorithm for transmit antenna weight adaptation with feedback, *IEEE Transactions on Signal Processing* (see also *IEEE Transactions on Acoustics, Speech, and Signal Processing*), **51**(5), 1156–1171.

350. Larsson, E. G., Stoica, P. and Jian Li (2003) Orthogonal space–time block codes: maximum likelihood detection for unknown channels and unstructured interferences, *IEEE Transactions on Signal Processing* (see also *IEEE Transactions on Acoustics, Speech, and Signal Processing*), **51**(2), 362–372.

351. Hyundong Shin and Jae Hong Lee (2003) Effect of keyholes on the symbol error rate of space–time block codes, *IEEE Communications Letters*, **7**(1), 27–29.

352. Shengli Zhou and Giannakis, G. B. (2003) Single-carrier space–time block-coded transmissions over frequency-selective fading channels, *IEEE Transactions on Information Theory*, **49**(1), 164–179.

353. Andersen, J. B. (2000) Array gain and capacity for known random channels with multiple element arrays at both ends, *IEEE Journal on Selected Areas in Communications*, **18**(11), 2172–2178.

354. Geman, S. (1980) A limit theorem for the norm of random matrices, *Annals of Probability*, **8**, 252–261.

355. Silverstein, J. W. (1985) The smallest eigenvalue of a large dimensional Wishart matrix, *Annals of Probability*, **13**, 1364–1368.

356. Raleigh, G. G. and Cioffi, J. M. (1998) Spatio-temporal coding for wireless communication, *IEEE Transactions On Communications*, **46**, 357–366.

357. Winters, J. H. (1987) On the capacity of radio communication systems with diversity in a rayleigh fading environment, *IEEE Journal on Selected Areas in Communications*, **5**, 871–878.

358. Glisic, S. (2003) *Adaptive WCDMA–Theory and Practice*, John Wiley & Sons, Chichester.

359. Sarwate, S. V. and Pursley, M. B. (1980) Crosscorrelation Properties of Pseudorandom and Related Sequences, *Proceedings of the IEEE*, **68**, 593–613.

360. Mezger, K. and Bouwens, R. J. (1972) An Ordered Table of Primitive Polynomials over GF(2) of Degrees 2 Through 19 for Use with Linear Maximal Sequence Generators, TM107, Cooley Electronics Laboratory, University of Michigan, Ann Arbor (AD 746876).

361. Golomb, S. W. (1982) *Shift Register Sequences*, Aegean Park Press, Laguna Hills, CA.

362. Lindholm, J. H. (1968) An Analysis of the Pseudo-randomness Properties of Subsequences of long m-sequences, *IEEE Transactions On Information Theory*, **14**(4), 569–576.

363. Massey, J. L. (1969) Shift-Register Synthesis and BCH Decoding, *IEEE Transactions On Information Theory*, **15**(1), 122–127.

364. Groth, E. J. (1971) Generation of Binary Sequences with Controllable Complexity, *IEEE Transactions On Information Theory*, **17**(3), 288–296.

365. Antweiler, M. and Bömer, L. (1992) Complex sequence over GF(p^m) with a two-level autocorrelation function and large linear span, *IEEE Transactions On Information Theory*, **38**, 120–130.

366. Games, R. (1986) The geometry of m-sequences: three-valued crosscorrelations and quadrics in finite projective geometry, *SIAM Journal of Algebraic Discrete Methods*, **7**, 43–52.

367. Schoulz, R. and Welch, L. (1984) GMW sequences, *IEEE Transactions On Information Theory*, **IT-30**, 548–553.

368. Simon, M., Omura, J., Scholtz, R. and Levitt, B. (1985) *Spread-Spectrum Communications*. Computer Science Press, New York.

369. Welch, L. R. (1974) Lower bounds on the maximum correlation of signals, *IEEE Transactions On Information Theory*, **IT-20**, 397–399.

370. Pursley, M. B. (1977) Performance evaluation for phase-coded spread-spectrum multiple-access communication–Part I: System analysis, *IEEE Transactions on Communications*, **COM-25**, 795–799.

371. Pursley, M. B. and Roefs, H. F. A. (1979) Numerical evaluation of correlation parameters for optimal phases of binary shift-register sequences, *IEEE Transactions On Communications*, **COM-27**, 1597–1604.

372. Pursley, M. B. and Sarwate, D. V. (1976) Bounds on aperiodic crosscorrelation for binary sequences, *Electronics Letters*, **12**, 304–305.

373. Pursley, M. B. and Sarwate, D. V. (1977) Evaluation of correlations parameters for periodic sequences, *IEEE Transactions On Information Theory*, **IT-23**, 508–513.

374. Pursley, M. B. and Sarwate, D. V. (1977) Performance evaluation for phase-coded spread-spectrum multiple access communication–Part II: Code sequence analysis, *IEEE Transactions On Communications*, **COM-25**, 800–803.

375. Roefs, H. F. A. and Pursley, M. B. (1977) Correlation parameters of random binary sequences, *Electronics Letters*, **13**, 488–489.

376. Sarwate, D. V. (1979) Bounds on crosscorrelation and autocorrelation of sequences, *IEEE Transactions On Information Theory*, **IT-25**, 720–724.

377. Sarwate, D. V. and Pursley, M. B. (1977) New correlation identities for periodic sequences, *Electronics Letters*, **13**(2), 48–49.

378. Scholtz, R. A. and Welch, L. R. (1978) Group characters: Sequences with good correlation properties, *IEEE Transactions On Information Theory*, **IT-24**, 537–545.

379. Glisic, S. and Vucetic, B. (1997) *CDMA for Wireless Communication* Kluwer AP, Boston.

380. Gold, R. (1966) Characteristic linear sequences and their coset functions, *SIAM Journal of Applied Mathematics*, **14**, 980–985.

381. Gold, R. (1966) *Study of correlation properties of binary sequences*, AF Avionics Lab., Wright-Patterson AFB, OH, Technical Report AFAL-TR-66-234, (AD 488858).

382. Gold, R. (1967) Optimal binary sequences for spread spectrum multiplexing, *IEEE Transactions On Information Theory*, **IT-13**, 619–621.

383. Gold, R. (1968) Maximal recursive sequences with 3-valued recursive crosscorrelation functions, *IEEE Transactions On Information Theory*, **IT-14**, 154–156.

384. Golay, M. (1961) Complementary Series, *IRE Transactions on Information Theory*, **IT 7**, 82–87.

385. Deng, X. and Fan, P. (1999) New binary sequences with good aperiodic autocorrelations obtained by evolutionary algorithm, *IEEE Communications Letters*, **3**(10), 288–290.

386. Deng, H. (1996) Synthesis of binary sequences with good autocorrelation and crosscorrelation properties by simulated annealing, *IEEE Transaction on Aerospace Electronics Systems*, **32**(1), 98–107.

387. Hu, F., Fan, P. Z., Darnell, M. and Jin, F. (1997) Binary sequences with good aperiodic autocorrelation functions obtained by neural network search, *Electronics Letters*, **33**(8), 688–690.

388. Mertens, S. (1996) Exhaustive search for low-autocorrelation binary sequences, *Journal of Physics A: Mathematical and General*, **29**(18), 473–481.

389. Kocabas, S. E. and Atalar, A. (2003) Binary sequences with low aperiodic autocorrelation for synchronization purposes, *IEEE Communications Letters*, **7**(1), 36–38.

390. Zhang Guohua and Zhou Quan (2002) Pseudonoise codes constructed by Legendre sequence, *Electronics Letters*, **38**(8), 376–377.

391. Walther, U. and Ferrweis, G. P. (2001) PN-generators embedded in high performance signal processors, *The 2001 IEEE International Symposium on Circuits and Systems* (ISCAS 2001), 6–9 May 2001, **4**, 45–48.

392. Leon, D., Balkir, S., Hoffman, M. W. and Perez, L. C. (2001) Robust chaotic PN sequence generation techniques, *The 2001 IEEE International Symposium on Circuits and Systems* (ISCAS 2001), 6–9 May 2001, **4**, 53–56.

393. Fujisaki, H. (2001) Optimum binary spreading sequences of Markov chains, *Electronics Letters*, **37**(20), 1234–1235.

394. Hongtao Zhang, Jichang Guo, Huiyun Wang, Runtao Ding and Wai-Kai Chen (2000) Oversampled chaotic map binary sequences: definition, performance and realization, *The 2000 IEEE Asia-Pacific Conference on Circuits and Systems* (IEEE APCCAS 2000), 4–6 December 2000, pp. 618–621.

395. Giardina, C. and Rudrapatna, A. N. (2000) Quasi-Walsh PN sequences and their applications in robust CDMA communication systems, *2000 IEEE International Conference on Personal Wireless Communications*, 17–20 December 2000, pp. 24–27.

396. Abraham, V. D. and Rao, C. D. V. P. (2000) Optimization of spreading code and estimation of channel capacity in multi-code CDMA downlink, *2000 IEEE International Conference on Personal Wireless Communications*, 17–20 December 2000, pp. 469–473.

397. Leon, D., Balkir, S., Hoffman, M. and Perez, L.C. (2000) Fully programmable, scalable chaos-based PN sequence generation, *Electronics Letters*, **36**(16), 1371–1372.

398. Yerdu, S. (1986) Optimum multiuser asymptotic efficiency, *IEEE Transactions On Communications*, **COM-34**, 890–897.

399. Verdu, S. (1986) Minimum probability of error for asynchronous Gaussian multiple-access channels, *IEEE Transactions On Information Theory*, **IT-32**, 85–96.

400. Varanasi, M. and Aazhang, B. (1990) Multistage detection in asynchronous code division multiple access communications, *IEEE Transactions On Communications*, **38**, 509–519.

401. Varanasi, M. and Aazhang, B. (1991) Near optimum detection in synchronous code division multiple access systems, *IEEE Transactions On Communications*, **39**, 725–736.

402. Varanasi, M. (1993) Noncoherent detection in a synchronous multiuser channels, *IEEE Transactions On Information Theory*, **37**(1), 157–176.

403. Zvonar, Z. (1993) *Multiuser detection for Rayleigh fading channel*, Ph.D. thesis, Department of Electrical and Computer Engineering, Northeastern University, Boston, Massachusetts, USA.

404. Xie, Z. *et al.* (1990) Multiuser signal detection using sequential decoding, *IEEE Transactions on Communications*, **38**, 578–583.

405. Aazhang, B. *et al.* (1992) Neural networks for multiuser detection in code division multiple access communications, *IEEE Transactions On Communications*, **COM-40**, 1212–1222.

406. Lupas, R. and Verdu, S. (1989) Linear multiuser detectors for synchronous code division multiple access channels, *IEEE Transactions On Information Theory*, **35**, 123–136.

407. Lupas, R. and Verdu, S. (1990) Near–Far resistance of multiuser detectors in synchronous channels, *IEEE Transactions On Communications*, **38**(4), 496–508.

408. Varanasi, M. and Aazhang, B. (1991) Optimally near–far resistant multiuser detection in differentially coherent synchronous channels, *IEEE Transactions On Information Theory*, **39**, 1006–1018.

409. Xie, Z. *et al.* (1990) A family of suboptimum detectors for coherent multiuser communications, *IEEE ISAC*, **8**(4), 683–690.

410. Xie, Z. *et al.* (1993) Joint signal detection and parameter estimation in multiuser communications, *IEEE Transactions On Communications*, **41**(7), 1208–1216.

411. Zvonar, Z. *et al.* (1992) Optimum detection in synchronous multiple-access multipath Rayleigh fading channels, *Proceedings of the 26th Annual Conference on Information Sciences and Systems*, Princeton University, March 1992.

412. Wijayasuriya, S. S. H., Norton, G. H. and McGeehan, J. P. (1992) Sliding Window Decorrelating Algorithm for DS-CDMA Receivers, *Electronics Letters*, **28**, 1596–1598.

413. Wijayasuriya, S. S. H., McGeehan, J. P. and Norton, G. H. (1993) RAKE Decorrelating Receiver for DS-CDMA Mobile Radio Networks, *Electronics Letters*, **29**, 395–396.

414. Tan, P. H. and Rasmussen, L. (2000) Linear interference cancellation in CDMA based on iterative techniques for linear equation systems, *IEEE Transactions On Communications*, **48**(12).

415. Kapur, A. and Varanasi, M. K. (2003) Multiuser detection for overloaded CDMA systems, *IEEE Transactions On Information Theory*, **49**(7), 1728–1742.

416. Buzzi, S., Lops, M. and Poor, H. V. (2003) Blind adaptive joint multiuser detection and equalization in dispersive differentially encoded CDMA channels, *IEEE Transactions on Signal Processing*, **51**(7), 1880–1893.

417. Lim, H. S., Rao, M. V. C., Tan, A. W. C. and Chuah, H. T. (2003) Multiuser detection for DS-CDMA systems using evolutionary programming, *IEEE Communications Letters*, **7**(3), 101–103.

418. Liping Sun and Guangrui Hu (2003) A new sign algorithm for interference suppression in DS-CDMA systems, *IEEE Communications Letters*, **7**(5), 233–235.

419. Abe, T. and Matsumoto, T. (2003) Space–time turbo equalization in frequency-selective MIMO channels, *IEEE Transactions on Vehicular Technology*, **52**(3), 469–475.

420. Yen, K. and Hanzo, L. (2003) Antenna-diversity-assisted genetic-algorithm-based multiuser detection schemes for synchronous CDMA systems, *IEEE Transactions On Communications*, **51**(3), 366–370.

421. Jong-Hun Rhee, Moo-Yeon Woo and Dong-Ku Kim (2003) Multichannel joint detection of multicarrier 16-QAM DS/CDMA system for high-speed data transmission, *IEEE Transactions on Vehicular Technology*, **52**(1), 37–47.

422. Yin, G. G., Krishnamurthy, V. and Ion, C. (2003) Iterate-averaging sign algorithms for adaptive filtering with applications to blind multiuser detection, *IEEE Transactions On Information Theory*, **49**(3), 657–671.

423. de Lamare, R. C. and Sampaio-Neto, R. (2003) Adaptive MBER decision feedback multiuser receivers in frequency selective fading channels, *IEEE Communications Letters*, **7**(2), 73–75.

424. Al-Bayati, A. K. S., Prakriya, S. and Prasad, S. (2003) Block modulus precoding for blind multiuser detection of DS-CDMA signals, *IEEE Transactions On Communications*, **51**(1), 52–56.

425. Weihua Ye and Varshney, P. K. (2003) An equicorrelation-based multiuser communication scheme for DS-CDMA systems, *IEEE Transactions On Communications*, **51**(1), 43–47.

426. Kuei-Chiang Lai and Shynk, J. J. (2003) Performance evaluation of a generalized linear SIC for DS/CDMA signals, *IEEE Transactions on Signal Processing*, **51**(6), 1604–1614.

427. Kafle, P. L. and Sesay, A. B. (2003) Iterative semi-blind multiuser detection for coded mc-cdma uplink system, *IEEE Transactions On Communications*, **51**(7), 1034–1039.

428. Zhiyu Xu and Cheng, R. S. (2003) A robust rank estimation algorithm of group-blind MMSE multiuser detectors for CDMA systems, *IEEE Transactions on Communications*, **51**(4), 547–552.

429. Wei Zha and Blostein, S. D. (2003) Soft-decision multistage multiuser interference cancellation, *IEEE Transactions on Vehicular Technology*, **52**(2), 380–389.

430. Ee-Lin Kuan and Hanzo, L. (2003) Burst-by-burst adaptive multiuser detection cdma: a framework for existing and future wireless standards, *Proceedings of the IEEE*, **91**(2), 278–302.

431. Brunel, L. and Boutros, J. J. (2003) Lattice decoding for joint detection in direct-sequence CDMA systems, *IEEE Transactions On Information Theory*, **49**(4), 1030–1037.

432. Horlin, F. and Vandendorpe, L. (2003) CA-CDMA: channel-adapted CDMA for MAI/ISI-free burst transmission, *IEEE Transactions On Communications*, **51**(2), 275–283.

433. Buzzi, S. and Lops, M. (2003) Performance analysis for the improved linear multiuser detectors in BPSK-modulated DS-CDMA systems, *IEEE Transactions On Communications*, **51**(1), 37–42.

434. Huaiyu Dai and Poor, H. V. (2002) Iterative space–time processing for multiuser detection in multipath CDMA channels, *IEEE Transactions on Signal Processing*, **50**(9), 2116–2127.

435. Hongbin Li and Jian Li (2002) Differential and coherent decorrelating multiuser receivers for space-time-coded CDMA systems, *Signal Processing*, **50**(10), 2529–2537.

436. Reynolds, D., Xiaodong Wang and Poor, H. V. (2002) Blind adaptive space–time multiuser detection with multiple transmitter and receiver antennas, *IEEE Transactions on Signal Processing*, **50**(6), 1261–1276.

437. Nafie, M. and Tewfik, A. H. (2002) A flexible receiver for CDMA multiuser communications, *IEEE Transactions on Signal Processing*, **50**(7), 1747–1758.

438. Brown, T. and Kaveh, M. (1995) A decorrelating detector for use with antenna arrays, *International Journal of Wireless Information Networks*, **2**(4), 239–246.

439. Jung, P. and Blanz, J. (1995) Joint detection with coherent receiver antenna diversity in CDMA mobile radio systems, *IEEE Transactions on Vehicular Technology*, **44**(1), 76–88.

440. Jung, P., Blanz, J., Nasshan, M. and Baier, P. W. (1994) Simulation of the uplink of JD–CDMA mobile radio systems with coherent receiver antenna diversity, *Wireless Personal Communications*, **1**(2), 61–89.

441. Madhow, U. and Honing, M. L. (1994) MMSE interference suppression for direct sequence spread-spectrum CDMA, *IEEE Transactions On Communications*, **42**(12), 3178–3188.

442. Klein, A., Kaleh, G. K. and Baier, P. W. (1996) Zero forcing and minimum mean-square-error equalization for multiuser detection in code-division multiple access channels, *IEEE Transactions on Vehicular Technology*, **45**(2), 276–287.

443. Berstein, X. and Haimovich, A. M. (1996) Space–time optimum combining for CDMA communications, *Wireless Personal Communications*, **3**(1–2), 73–89.

444. Gray, S. D., Preisig, J. C. and Brady, D. (1997) Multiuser detection in a horizontal underwater acoustic channel using array observations, *IEEE Transactions on Signal Processing*, **45**(1), 148–160.

445. Poor, H. V. and Verdú, S. (1997) Probability of error in MMSE multiuser detection, *IEEE Transactions On Information Theory*, **43**(3), 858–871.

446. Rapajic, P. B. and Vucetic, B. S. (1995) Linear adaptive transmitter–receiver structures for asynchronous SCMA systems, *European Transactions on Telecommunications*, **6**(1), 21–27.

447. Miller, S. L. (1995) An adaptive direct sequence code-division multiple-access receiver for multiuser interference rejection, *IEEE Transactions On Communications*, **43**, 1746–1755.

448. Rapajic, P. B. and Vucetic, B. S. (1994) Adaptive receiver structures for asynchronous CDMA systems, *IEEE Journal on Selected Areas in Communications*, **12**(4), 685–697.

449. Lee, K. B. (1996) Orthogonalization based adaptive interference suppression for direct sequence code-division multiple-access systems, *IEEE Transactions On Communications*, **44**(9), 1082–1085.

450. Miller, S. L. (1996) Training analysis of adaptive interference suppression for direct sequence code-division multiple-access systems, *IEEE Transactions on Communications*, **44**(4), 488–495.

451. Honig, M. (1998) Adaptive linear interference suppression for packet DS-CDMA, *European Transactions on Telecommunications*, **9**(2), 173–181.

452. Hong, M., Madhow, U. and Verdú, S. (1995) Blind adaptive multiuser detection, *IEEE Transactions On Information Theory*, **41**(3), 944–960.

453. Park, S. C. and Dohery, J. F. (1997) Generalized projection algorithm for blind interference suppression in DS/CDMA communications, *IEEE Transactions on Circuits and Systems–Part II Analog and Digital Signal Processing*, **44**(6), 453–460.

454. Schodorf, J. B. and Williams, D. B. (1997) A constrained optimisation approach to multiuser detection, *IEEE Transactions on Signal Processing*, **45**(1), 258–262.

455. Wang, X. and Poor, H. V. (1998) Blind equalization and multiuser detection in dispersive CDMA channels, *IEEE Transactions On Communications*, **46**(1), 91–103.

456. Iltis, R. A. (1998) Performance of constrained and unconstrained adaptive multiuser detectors for quasi-synchronous CDMA, *IEEE Transactions On Communications*, **46**(1), 135–143.

457. Madhow, U. (1997) Blind adaptive interference suppression for the near–far resistant acquisition and demodulation of direct-sequence CDMA, *IEEE Transactions on Signal Processing*, **45**(1), 124–136.

458. Mowbray, R. S., Pringle, R. D. and Grant, P. M. (1992) Increased CDMA system capacity through adaptive cochannel interference regeneration and cancellation, *IEE Proceedings I*, **139**, 515–524.

459. Kohno, R., Imai, H., Hatori, M. and Pasupathy, S. (1990) Combination of an adaptive array antenna and a canceller of interference for direct sequence spread-spectrum multiple-access system, *IEEE Journal on Selected Areas in Communications*, **8**(4), 675–682.

460. Nelson, L. B. and Poor, H. V. (1996) Iterative multiuser receivers for CDMA channels: An EM-based approach, *IEEE Transactions On Communications*, **44**(12), 1700–1710.

461. Soong, A. C. K. and Krzymien, W. A. (1996) A novel CDMA multiuser interference cancellation receiver with reference symbol aided estimation of channel parameters, *IEEE Journal on Selected Areas in Communications*, **14**(8), 1536–1547.

462. Wang, X. and Poor, H. V. (1998) Blind multiuser detection: A subspace approach, *IEEE Transactions On Information Theory*, **44**(2), 677–690.

463. Juntti, M. and Glisic, S. (1997) Advanced CDMA for wireless communications, In Glisic, S. G. and Leppänen P. A. (eds) *Wireless Communications: TDMA Versus DCMA*, Kluwer, pp. 447–490.

464. Zihua Guo and Ben Letaief, K. (2003) A low complexity reduced–rank MMSE receiver for DS/CDMA communications, *IEEE Transactions on Wireless Communications*, **2**(1), 59–68.

465. Zhiyu Xu and Cheng, R. S. (2003) A robust rank estimation algorithm of group-blind MMSE multiuser detectors for CDMA systems, *IEEE Transactions On Communications*, **51**(4), 547–552.

466. Mantravadi, A. and Veeravalli, V. V. (2002) MMSE detection in asynchronous CDMA systems: an equivalence result, *IEEE Transactions On Information Theory*, **48**(12), 3128–3137.

467. Woodward, G., Ratasuk, R., Honig, M. L. and Rapajic, P. B. (2002) Minimum mean-squared error multiuser decision-feedback detectors for DS–CDMA, *IEEE Transactions on Communications*, **50**(12), 2104–2112.

468. Ju Ho Lee and Hyung-Myung Kim (2002) Partial zero-forcing adaptive MMSE receiver for DS-CDMA uplink in multicell environments, *IEEE Transactions on Vehicular Technology*, **51**(5), 1066–1071.

469. Schober, R., Gerstacker, W. H. and Lampe, A. (2002) Noncoherent MMSE interference suppression for DS-CDMA, *IEEE Transactions On Communications*, **50**(4), 577–587.

470. Buzzi, S., Lops, M. and Tulino, A. M. (2002) A generalized minimum-mean-output-energy strategy for CDMA systems with improper MAI, *IEEE Transactions on Information Theory*, **48**(3), 761–767.

471. Iltis, R. (1994) An EKF-based joint estimator for interference, multipath, and code delay in a DS spread-spectrum receiver, *IEEE Transactions On Communications* **42**, 1288–1299.

472. Kay, S. (1993) *Fundamentals of Statistical Signal Processing–Estimation Theory*, Prentice Hall.

473. Bensley, J. S. and Aazhang, B. (1996) Subspace-Based Channel Estimation for Code Division Multiple Access Communications and Systems, *IEEE Transactions on Communications*, **44**(8), 1009–1020.

474. Xie, Z., Rushforth, C., Short, R. and Moon, T. (1993) Joint signal detection and parameter estimation in multiuser communications, *IEEE Transactions On Communications*, **41**, 1208–1216.

475. Aazhang, B., Paris, B. and Orsak, G. (1992) Neural networks for multiuser detection in code-division multiple-access communications, *IEEE Transactions On Communications*, **40**, 1212–1222.

476. Iltis, R. A. and Mailaender, L. (1994)– An adaptive multiuser detector with joint amplitude and delay estimation, *IEEE Journal on Selected Areas in Communications*, Vol. **12** (5), 774–785.

477. Iltis, R. (1990) Joint estimation of PN code delay and multipath using the extended Kalman filter, *IEEE Transactions On Communications*, **38**, 1677–1685.

478. Iltis, R. and Fuxjaeger, A. (1991) A digital DS spread-spectrum receiver with joint channel and Doppler shift estimation, *IEEE Transactions On Communications*, **39**, 1255–1267.

479. Clark, A. P. and Harun, R. (1986) Assesment of Kalman-Filter Channel Estimators for an HF Radio Link, *IEE Proceedings*, **133** (pt.F), 513–521.

480. Hatzinakos, D. and Nikias, C. L. (1989) Estimation of multipath channel response in frequency selective channels, *IEEE Journal on Selected Areas in Communications*, **SAC-7**, 12–19.

481. Shalvi, O. and Weinstein, E. (1990) New criteria for blind deconvolation of nonminimum phase systems (channels), *IEEE Transactions On Information Theory*, **IT-36**, 312–321.

482. Huang, H. C. (1996) *Combined multipath processing, array processing, and multiuser detection for DS-CDMA channels*, Ph.D. Thesis, Princeton University, Princeton, NJ, USA.

483. Latva-aho, M. (1998) *Advanced receivers for wideband CDMA systems*, Ph.D. Thesis, University of Oulu, Finland.

484. Miller, S. Y. (1989) *Detection and estimation in multiple-access channels*, Ph.D. Thesis, Princeton University, Princeton, NJ, USA.

485. Miller, S. Y. and Schwartz, S. C. (1995) Integrated spatial-temporal detectors for asynchronous Gaussian multiple-access channels, *IEEE Transactions On Communications*, **43**, 396–411.

486. Zvonar, Z. (1996) Combined multiuser detection and diversity reception for wireless CDMA systems, *IEEE Transactions on Vehicular Technology*, **45**(1), 205–211.

487. Fanucci, L. *et al.* (2001) VLSI Implementation of CDMA Blind Adaptive Interference-Mitigating Detector, *IEEE JSAC*, **19**(2), 179–190.

488. De Gaudenzi, R. *et al.* (1998) Design of a Low Complexity Adaptive Interference Mitigating Detector for DS/SS Receiver in CDMA Radio Networks, *IEEE Transactions On Communications*, **46**(1), 125–134.

489. Amleh, K. and Hongbin Li (2003) An algebraic approach to blind carrier offset and code timing estimation for DS-CDMA systems, *IEEE Signal Processing Letters*, **10**(2), 32–34.

490. Manikas, A. and Sethi, M. (2003) A space–time channel estimator and single-user receiver for code-reuse in DS-CDMA systems, *IEEE Transactions on Signal Processing*, **51**(1), 39–51.

491. Bin Xu, Chenyang Yang and Shiyi Mao (2002) An improved blind adaptive multiuser detector in multipath CDMA channels based on subspace estimation, *IEEE 55th Vehicular Technology Conference* (VTC Spring 2002), 6–9 May 2002, **1**, 285–288.

492. Fock, G., Schulz-Rittich, P., Schenke, A. and Meyr, H. (2002) Low complexity high resolution Subspace-based delay estimation for DS-CDMA, *IEEE International Conference on Communications* (ICC 2002), 28 April–2 May 2002, **1**, 31–35.

493. Xiaojun Wu, Qinye Yin and Jianguo Zhang (2002) Subspace-based estimation method of uplink FIR channel in MC-CDMA system without cyclic prefix over frequency-selective fading channel, *IEEE International Symposium on Circuits and Systems* (ISCAS 2002), 26–29 May 2002, **1**, I-213–I-216.

494. Yugang Ma, Li, K. H., Kot, A. C. and Ye, G. (2002) A blind code timing estimator and its implementation for DS-CDMA signals in unknown colored noise, *IEEE Transactions on Vehicular Technology*, **51**(6), 1600–1607.

495. Zhengyuan Xu (2002) Asymptotic performance of subspace methods for synchronous multirate CDMA systems, *IEEE Transactions on Signal Processing*, **50**(8), 2015–2026.

496. Lei Huang, Fu-Chun Zheng and Faulkner, M. (2002) Blind adaptive channel estimation for dual-rate DS/CDMA signals, *IEEE Communications Letters*, **6**(4), 129–131.

497. Affes, S. N., Hansen, H. and Mermelstein, P. (2002) Interference subspace rejection: a framework for multiuser detection in wideband CDMA, *IEEE Journal on Selected Areas in Communications*, **20**(2), 287–302.

498. Zhengyuan Xu (2002) Perturbation analysis for subspace decomposition with applications in Subspace-based algorithms, *IEEE Transactions on Signal Processing*, **50**(11), 2820–2830.

499. Tugnait, J. (1994) Blind Estimation of Digital Communication Channel Impulse Response, *IEEE Transactions On Communications*, **42**(2/3/4), 1606–1616.

500. Muirhead, R. (1982) *Aspects of Multivariate Statistical Theory*, Jon Wiley & Sons, New York.

501. Wen-Rong Wu and Yih-Ming Tsuie (2002) An LMS-based decision feedback equalizer for IS-136 receivers, *IEEE Transactions on Vehicular Technology*, **51**(1), 130–143.

502. Zerguine, A., Shafi, A. and Bettayeb, M. (2001) Multilayer perceptron-based DFE with lattice structure, *IEEE Transactions on Neural Networks*, **12**(3), 532–545.

503. Ki Yong Lee (1996) Complex fuzzy adaptive filter with LMS algorithm, *IEEE Transactions on Signal Processing*, **44**(2), 424–427.

504. Reuter, M. and Zeidler, J. R. (1999) Nonlinear effects in LMS adaptive equalizers, *IEEE Transactions on Signal Processing*, **47**(6), 1570–1579.

505. Xiaohua Li and Fan, H. (2000) Linear prediction methods for blind fractionally spaced equalization, *IEEE Transactions on Signal Processing*, **48**(6), 1667–1675.

506. Gi Hun Lee, Jinho Choi, Rae-Hong Park, Iickho Song, Jae Hyuk Park and Byung-Uk Lee (1994) Modification of the reference signal for fast convergence in LMS-based adaptive equalizers, *IEEE Transactions on Consumer Electronics*, **40**(3), 645–654.

507. Mao-Ching Chiu and Chi-chao Chao (1996) Analysis of LMS-adaptive MLSE equalization on multipath fading channels, *IEEE Transactions On Communications*, **44**(12), 1684–1692.

508. Raghunath, K. J. and Parhi, K. K. (1993). Parallel adaptive decision feedback equalizers, *IEEE Transactions on Signal Processing*, **41**(5), 1956–1961.

509. Shanbhag, N. R. and Parhi, K. K. (1995) Pipelined adaptive DFE architectures using relaxed look-ahead, *IEEE Transactions on Signal Processing*, **43**(6), 1368–1385.

510. Al-Dhahir, N. and Cioffi, J. M. (1996) Efficient computation of the delay-optimized finite length MMSE-DFE, *IEEE Transactions on Signal Processing*, **44**(5), 1288–1292.

511. Al-Dhahir, N. and Cioffi, J. M. (1997) Mismatched finite-complexity MMSE decision feedback equalizers, *IEEE Transactions on Signal Processing*, **45**(4), 935–944.

512. Chen, S., Mulgrew, B. and Hanzo, L. (2000) Asymptotic Bayesian decision feedback equalizer using a set of hyperplanes, *IEEE Transactions on Signal Processing*, **48**(12), 3493–3500.

513. Afkhamie, K. H., Zhi-Quan Luo and Kon Max Wong (2001) Interior point least squares estimation: transient convergence analysis and application to MMSE decision-feedback equalization, *IEEE Transactions on Signal Processing*, **49**(7), 1543–1555.

514. Liavas, A. P. (2002) On the robustness of the finite-length MMSE-DFE with respect to channel and second-order statistics estimation errors, *IEEE Transactions on Signal Processing*, **50**(11), 2866–2874.

515. Berberidis, K. and Karaivazoglou, P. (2002) An efficient block adaptive decision feedback equalizer implemented in the frequency domain, *IEEE Transactions on Signal Processing*, **50**(9), 2273–2285.

516. Sheng Chen (2002) Importance of sampling simulation for evaluating lower-bound symbol error rate of the Bayesian DFE with multilevel signaling schemes, *IEEE Transactions on Signal Processing*, **50**(5), 1229–1236.

517. Hunsoo Choo, Muhammad, K. and Roy, K. (2003) Two's complement computation sharing multiplier and its applications to high performance DFE, *IEEE Transactions on Signal Processing*, **51**(2), 458–469.

518. Monsen, P. (1977) Theoretical and Measured Performance of a DFE Modem on a Fading Multipath Channel, *IEEE Transactions On Communications*, **25**(10), 1144–1153.

519. Cantoni, A. and Butler, P. (1976) Stability of Decision Feedback Inverses, *IEEE Transactions On Communications*, **24**(9), 970–977.

520. Ehrman, L. and Monsen, P. (1977) Troposcatter Test Results for a High-Speed Decision-Feedback Equalizer Modem, *IEEE Transactions On Communications*, **25**(12), 1499–1504.

521. Taylor, D. and Shafi, M. (1984) Decision Feedback Equalization for Multipath Induced Interference in Digital Microwave LOS Links, *IEEE Transactions on Communications*, **32**(3), 267–279.

522. Falconer, D., Sheikh, A., Eleftheriou, E. and Tobis, M. (1985) Comparison of DFE and MLSE Receiver Performance on HF Channels, *IEEE Transactions On Communications*, **33**(5), 484–486.

523. Fuyun Ling and Proakis, J. (1985) Adaptive Lattice Decision-Feedback Equalizers–Their Performance and Application to Time-Variant Multipath Channels, *IEEE Transactions On Communications*, **33**(4), 348–356.

524. Leclert, A. and Vandamme, P. (1985) Decision Feedback Equalization of Dispersive Radio Channels, *IEEE Transactions On Communications*, **33**(7), 676–684.

525. Shafi, M. and Moore, D. (1986) Further Results on Adaptive Equalizer Improvements for 16 QAM and 64 QAM Digital Radio, *IEEE Transactions On Communications*, **34**(1), 59–66.

526. Kennedy, R. and Anderson, B. (1987) Recovery Times of Decision Feedback Equalizers on Noiseless Channels, *IEEE Transactions On Communications*, **35**(10), 1012–1021.

527. Kennedy, R., Anderson, B. and Bitmead, R. (1987) Tight Bounds on the Error Probabilities of Decision Feedback Equalizers, *IEEE Transactions On Communications*, **35**(10), 1022–1028.

528. Kennedy, R. and Anderson, B. (1987) Error Recovery of Decision Feedback Equalizers on Exponential Impulse Response Channels, *IEEE Transactions On Communications*, **35**(8), 846–848.

529. Kennedy, R. A., Anderson, B. D. O. and Bitmead, R. R. (1989) Channels leading to rapid error recovery for decision feedback Equalizers, *IEEE Transactions On Communications*, **37**(11), 1126–1135.

530. Zhou, K., Proakis, J. G. and Ling, F. (1990) Decision-feedback equalization of time-dispersive channels with coded modulation, *IEEE Transactions On Communications*, **38**(1), 18–24.

531. Altekar, S. A. and Beaulieu, N. C. (1991) On the tightness of two error bounds for decision feedback Equalizers, *IEEE Transactions On Information Theory*, **37**(3), 638–639.

532. Lin, D. W. (1991) Minimum mean-squared error decision-feedback equalization for digital subscriber line transmission with possibly correlated line codes, *IEEE Transactions On Communications*, **39**(8), 1197–1206.

533. Williamson, D., Kennedy, R. A. and Pulford, G. W. (1992) Block decision feedback equalization, *IEEE Transactions On Communications*, **40**(2), 255–264.

534. Pahlavan, K., Howard, S. J. and Sexton, T. A. (1993) Decision feedback equalization of the indoor radio channel, *IEEE Transactions On Communications*, **41**(1), 164–170.

535. Jaekyun Moon and Sian She (1994) Constrained-complexity equalizer design for fixed delay tree search with decision feedback, *IEEE Transactions on Magnetics*, **30**(5), 2762–2768.

536. Stojanovic, M., Proakis, J. G. and Catipovic, J. A. (1995) Analysis of the impact of channel estimation errors on the performance of a decision-feedback equalizer in fading multipath channels, *IEEE Transactions On Communications*, **43**(2, 3, 4), 877–886.

537. McEwen, P. A. and Kenney, J. G. (1995) Allpass forward equalizer for decision feedback equalization, *IEEE Transactions on Magnetics*, **31**(6), 3045–3047.

538. Russell, M. and Jan W. M. Bergmans (1995) A technique to reduce error propagation in *M*-ary decision feedback equalization, *IEEE Transactions On Communications*, **43**(12), 2878.

539. Cioffi, J. M., Dudevoir, G. P., Vedat Eyuboglu, M. and Forney, G. D., Jr (1995) MMSE decision-feedback equalizers and coding. I. Equalization results, *IEEE Transactions On Communications*, **43**(10), 2582–2594.

540. Sheng Chen, McLaughlin, S., Mulgrew, B. and Grant, P. M. (1995) Adaptive Bayesian decision feedback equalizer for dispersive mobile radio Channels, *IEEE Transactions On Communications*, **43**(5), 1937–1946.

541. Mathew, G., Farhang-Boroujeny, B. and Wood, R. W. (1997) Design of multilevel decision feedback Equalizers, *IEEE Transactions on Magnetics*, **33**(6), 4528–4542.

542. Porat, B. and Friedlander, B. (1991) Blind equalization of digital communication channels using high-order moments, *IEEE Transactions on Signal Processing*, **39**(2), 522–526.

543. Vembu, S., Verdu, S., Kennedy, R. A. and Sethares, W. (1994) Convex cost functions in blind equalization, *IEEE Transactions on Signal Processing*, **42**(8), 1952–1960.

544. Ye Li and Zhi Ding (1995) Convergence analysis of finite length blind adaptive equalizers, *IEEE Transactions on Signal Processing*, **43**(9), 2120–2129.

545. Ye Li and Liu, K. J. R. (1996) Static and dynamic convergence behavior of adaptive blind Equalizers, *IEEE Transactions on Signal Processing*, **44**(11), 2736–2745.

546. Zhi Ding (1997) On convergence analysis of fractionally spaced adaptive blind equalizers, *IEEE Transactions on Signal Processing*, **45**(3), 650–657.

547. Shtrom, V. and Fan, H. (1998) New class of zero-forcing cost functions in blind equalization, *IEEE Transactions on Signal Processing*, **46**(10), 2674–2683.

548. Giannakis, G. B. and Tepedelenlioglu, C. (1999) Direct blind equalizers of multiple FIR channels: a deterministic approach, *IEEE Transactions on Signal Processing*, **47**(1), 62–74.

549. Tugnait, J. K. and Bin Huang (1999) Second-order statistics-based blind equalization of IIR single-input multiple-output channels with common zeros, *IEEE Transactions on Signal Processing*, **47**(1), 147–157.

550. Mannerkoski, J. and Koivunen, V. (2000) Autocorrelation properties of channel encoded sequences–applicability to blind equalization, *IEEE Transactions on Signal Processing*, **48**(12), 3501–3507.

551. Junyu Mai and Sayed, A. H. (2000) A feedback approach to the steady-state performance of fractionally spaced blind adaptive equalizers, *IEEE Transactions on Signal Processing*, **48**(1), 80–91.

552. Borah, D. K., Kennedy, R. A., Zhi Ding and Fijalkow, I. (2001) Sampling and pre-filtering effects on blind equalizer design, *IEEE Transactions on Signal Processing*, **49**(1), 209–218.

553. Lopez-Valcarce, R. and Dasgupta, S. (2001) Blind channel equalization with colored sources based on second-order statistics: a linear prediction approach, *IEEE Transactions on Signal Processing*, **49**(9), 2050–2059.

554. Upez-Valcarce, R. and Dasgupta, S. (2001) Blind equalization of nonlinear channels from second-order statistics, *IEEE Transactions on Signal Processing*, **49**(12), 3084–3097.

555. Luo, Z.-Q. T., Mei Meng, Wong, K. M. and Jian-Kang Zhang (2002) A fractionally spaced blind equalizer based on linear programming, *IEEE Transactions on Signal Processing*, **50**(7), 1650–1660.

556. Prakriya, S. (2002) Eigenanalysis-based blind methods for identification, equalization, and inversion of linear time-invariant channels, *IEEE Transactions on Signal Processing*, **50**(7), 1525–1532.

557. Benveniste, A. and Goursat, M. (1984) Blind Equalizers, *IEEE Transactions On Communications*, **32**(8), 871–883.

558. Mathis, H. and Douglas, S. C. (2003) Bussgang blind deconvolution for impulsive signals *IEEE Transactions on Signal Processing*, **51**(7), 1905–1915.

559. Karaoguz, J. and Ardalan, S. H. (1991) Use of blind equalization for teletext broadcast systems, *IEEE Transactions on Broadcasting*, **37**(2), 44–54.

560. Ding, Z., Kennedy, R. A., Anderson, B. D. O. and Johnson, C. R., Jr (1993) Local convergence of the Sato blind equalizer and generalizations under constraints, *IEEE Transactions On Information Theory*, **39**(1), 129–144.

561. Verdu, S., Anderson, B. D. O. and Kennedy, R. A. (1993) Blind equalization without gain identification, *IEEE Transactions On Information Theory*, **39**(1), 292–297.

562. Tugnait, J. K. (1994) Blind estimation of digital communication channel impulse response, *IEEE Transactions On Communications*, **42**(2,3,4), 1606–1616.

563. Tugnait, J. K. (1996) Blind equalization and estimation of FIR communications channels using fractional sampling, *IEEE Transactions On Communications*, **44**(3), 324–336.

564. Dogancay, K. and Kennedy, R. A. (1999) Least squares approach to blind channel equalization, *IEEE Transactions On Communications*, **47**(11), 1678–1687.

565. Won Lee and Hill, F. (1977) A Maximum-Likelihood Sequence Estimator with Decision-Feedback Equalization, *IEEE Transactions On Communications*, **25**(9), 971–979.

566. Falconer, D., Sheikh, A., Eleftheriou, E. and Tobis, M. (1985) Comparison of DFE and MLSE Receiver Performance on HF Channels, *IEEE Transactions On Communications*, **33**(5), 484–486.

567. Sheen, W.-H. and Stuber, G. L. (1991) MLSE equalization and decoding for multipath-fading channels, *IEEE Transactions On Communications*, **39**(10), 1455–1464.

568. Mao-Ching Chiu and Chi-chao Chao (1996) Analysis of LMS-adaptive MLSE equalization on multipath fading channels, *IEEE Transactions On Communications*, **44**(12), 1684–1692.

569. Yonghai Gu and Tho Le-Ngoc (1996) Adaptive combined DFE/MLSE techniques for ISI channels, *IEEE Transactions On Communications*, **44**(7), 847–857.

570. Hamied, K. A. and Stuber, G. L. (1996) An adaptive truncated MLSE receiver for Japanese personal digital cellular, *IEEE Transactions on Vehicular Technology*, **45**(1), 41–50.

571. Jiunn-Tsair Chen, Paulraj, A. and Reddy, U. (1999) Multichannel maximum-likelihood sequence estimation (MLSE) equalizer for GSM using a parametric channel model, *IEEE Transactions On Communications*, **47**(1), 53–63.

572. Jiunn-Tsair Chen and Yeong-Cheng Wang (2001) Adaptive MLSE equalizers with parametric tracking for multipath fast-fading channels, *IEEE Transactions On Communications*, **49**(4), 655–663.

573. Gerstacker, W. (2002) Equalization concepts for EDGE, *IEEE Transactions on Communications* **1**(1).

574. Dual-Hallen, A. (1989) Delayed Decision-Feedback Sequence Estimation, *IEEE Transactions On Communications* **37**(5).

575. Eyuboglu, M. V. (1988) Reduced-State Sequence Estimation with Set Partitioning and Decision Feedback, *IEEE Transactions On Communications*, **36**(1).

576. Vaidis, T. and Weber, C. L. (1998) Block adaptive techniques for channel identification and data demodulation over band-limited channels, *IEEE Transactions On Communications*, **46**(2), 232–243.

577. Sayed, A. H. and Kailath, T. (1994) A state-space approach to adaptive RLS filtering, *IEEE Signal Processing Magazine*, July.

578. Kennedy, R. A. (1989) Design and optimization of nonlinear mapping in decision feedback equalization, in *Proceedings of the 35th Conference on Decision and Control*, Kobe, Japan, December 1996, pp. 1888–1889.

579. Qureshi, S. U. H. (1985) Adaptive equalization, *Proceedings of IEEE*, **53**, 1349–1387.

580. Qureshi, S. (1985) Adaptive equalization, *Proceedings of IEEE*, **73**, 1349–1387.

581. Proakis, J. G. (1991) Adaptive equalization for TDMA digital mobile radio, *IEEE Transactions on Vehicular Technology*, **40**, 333–341.

582. Magee, F. R. Jr and Proakis, J. G. (1973) Adaptive maximum likelihood estimation for digital signaling in the presence of intersymbol interference, *IEEE Transactions On Information Theory*, **IT–19**, 120–124.

583. Ungerboeck, G. (1974) Adaptive maximum likelihood receivers for carrier modulated data transmission systems, *IEEE Transactions On Communications*, **COM–22**, 624–636.

584. Qureshi, S. and Newhall, E. E. (1973) An adaptive receiver for data transmission over time dispersive channels, *IEEE Transactions On Information Theory*, **IT–19**, 448–457.

585. Qureshi, U. H. (1973) *An Adaptive Decision–Feedback Receiver Using Maximum Likelihood Sequence Estimation.* ICC, Seattle, WA.

586. Raheli, R., Polydoros, A. and Tzou, C. K. (1995) Per survivor processing: A general approach to MLSE in uncertain environments, *IEEE Transactions On Communications*, **43**.

587. Seshadri, N. (1994) Joint data and channel estimation using blind trellis search techniques, *IEEE Transactions On Communications*, **42**, 1000–1011.

588. Forney, G. D. Jr (1972) Maximum likelihood sequence estimation of digital sequences in the presence of intersymbol interference, *IEEE Transactions On Information Theory*, **IT–18**, 363–378.

589. D'Avella, R., Moreno, L. and Sant'Agostino, M. (1989) An adaptive MLSE receiver for TDMA digital mobile radio, *IEEE Journal on Selected Areas in Communications*, **7**, 122–129.

590. Chevillat, P. R. and Eleftheriou, E. (1989) Decoding of trellis–encoded signals in the presence of intersymbol interference and noise, *IEEE Transactions On Communications*, **37**, 669–676.

591. Heller, J. A. and Jacobs, I. M. (1971) Viterbi decoding for satellite and space communication, *IEEE Transactions On Communications*, **COM–19**, 835–848.

592. Gosh, M. and Weber, C. L. (1992) Maximum likelihood blind equalization, *Optical Engineering* **31**(6), 1224–1229.

593. Sato, Y. (1975) A method of self–recovering equalization for multilevel amplitude modulation systems, *IEEE Transactions On Communications*, **COM–23**, 679–682.

594. Bee Leong Yeap, Choong Hin Wong and Hanzo, L. (2003) Reduced complexity in-phase/quadrature-phase *M*-QAM turbo equalization using iterative channel estimation, *IEEE Transactions on Wireless Communications*, **2**(1), 2–10.

595. Nelson, J., Singer, A. and Koetter, R. (2003) Linear turbo equalization for parallel ISI Channels, *IEEE Transactions On Communications*, **51**(6), 860–864.

596. Xiaodong Wang and Rong Chen (2001) Blind turbo equalization in Gaussian and impulsive noise, *IEEE Transactions on Vehicular Technology*, **50**(4), 1092–1105.

597. Yee, M. S., Liew, T. H. and Hanzo, L. (2001) Burst-by-burst adaptive turbo-coded radial basis function-assisted decision feedback equalization, *IEEE Transactions On Communications*, **49**(11), 1935–1945.

598. Bee Leong Yeap, Tong Hooi Liew, Hamorsky, J. and Hanzo, L. (2002) Comparative study of turbo equalization schemes using convolutional turbo, and block-turbo codes, *IEEE Transactions on Wireless Communications*, **1**(2), 266–273.

599. Mong-Suan Yee, Yeap, B. L. and Hanzo, L. (2003) Radial basis function-assisted turbo equalization, *IEEE Transactions On Communications*, **51**(4), 664–675.

600. Okada, T. and Iwanami, Y. (2002) Turbo equalizer detection for GFSK digital FM signals, *IEEE International Conference on Communications* (ICC 2002), 28 April–2 May 2002, **5**, 2952–2956.

601. Dejonghe, A. and Vandendorpe, L. (2002) Turbo-equalization for multilevel modulation: an efficient low-complexity scheme, *IEEE International Conference on Communications* (ICC 2002), 28 April–2 May 2002, **3**, 1863–1867.

602. Tuchler, M., Otnes, R. and Schmidbauer, A. (2002) Performance of soft iterative channel estimation in turbo equalization, *IEEE International Conference on Communications* (ICC 2002), 28 April–2 May 2002, **3**, 1858–1862.

603. Laot, C., Glavieux, A. and Labat, J. (2001) Turbo equalization: adaptive equalization and channel decoding jointly optimized, *IEEE Journal on Selected Areas in Communications*, **19**(9), 1744–1752.

604. Omidi, M. J., Gulak, P. G. and Pasupathy, S. (1998) Parallel structures for joint channel estimation and data detection over fading channels, *IEEE Journal on Selected Areas in Communications*, **16**(9), 1616–1629.

605. Rong Chen, Xiaodong Wang and Liu, J. S. (2000) Adaptive joint detection and decoding in flat-fading channels via mixture Kalman filtering, *IEEE Transactions On Information Theory*, **46**(6), 2079–2094.

606. Giridhar, K., Shynk, J. J., Iltis, R. A. and Mathur, A. (1996) Adaptive MAPSD algorithms for symbol and timing recovery of mobile radio TDMA signals, *IEEE Transactions on Communications*, **44**(8), 976–987.

607. Rollins, M. E. and Simmons, S. J. (1997) Simplified per-survivor Kalman processing in fast frequency-selective fading channels, *IEEE Transactions On Communications*, **45**, 514–553.

608. Anderson, B. D. O. and Moore, J. B. (1979) *Optimal Filtering*. Prentice Hall, Englewood Cliffs, NJ.

609. Raheli, R., Polydoros, A. and Tzou, C. (1995) Per-survivor processing: A general approach to MLSE in uncertain environments, *IEEE Transactions On Communications*, **43**, 354–364.

610. Rao, P. and Bayoumi, M. A. (1991) An algorithm specific VLSI parallel architecture for Kalman filter, in *VLSI Signal Processing IV. IEEE*, Piscataway, NJ pp. 264–273.

611. Bayoumi, M., Rao, P. and Alhalabi, B. (1992) VLSI parallel architecture for Kalman filter – an algorithm specific approach, *Journal of VLSI Signal Processing*, **4**(2, 3), 147–163.

612. Grewal, M. S. and Andrews, A. P. (1993) *Kalman Filtering, Theory and Practice.* Prentice Hall, Englewood Cliffs, NJ.

613. Young-Hoon Kim and Shamsunder, S. (1998) Adaptive algorithms for channel equalization with soft decision feedback, *IEEE Journal on Selected Areas in Communications*, **16**(9), 1660–1669.

614. Jie Zhu, Xi-Ren Cao and Ruey-Wen Liu (1999) A blind fractionally spaced equalizer using higher order statistics, *IEEE Transactions on Circuits and Systems II: Analog and Digital Signal Processing*, **46**(6), 755–764.

615. Chong-Yung Chi and Chii-Horng Chen (2001) Cumulant-based inverse filter criteria for MIMO blind deconvolution: properties, algorithms, and application to DS/CDMA systems in multipath, *IEEE Transactions on Signal Processing*, **49**(7), 1282–1299.

616. Behbahani, A. R. S. and Asjadi, H. (1996) Blind equalization based on third-order cumulant for 4-level and 8-level PAM, *Record of the 5th IEEE International Conference on Universal Personal Communications*, 29 September–2 October 1996 **1**, 136–140.

617. Hatzinakos, D. and Nikias, C. L. (1991) Blind equalization using a tricepstrum-based algorithm, *IEEE Transactions On Communications*, **39**(5), 669–682.

618. Tugnait, J. K. (1995) Blind equalization and estimation of digital communication FIR channels using cumulant matching, *IEEE Transactions On Communications*, **43**(2, 3, 4), 1240–1245.

619. Chih-Chun Feng and Chong-Yung Chi (1999) Performance of cumulant based inverse filters for blind deconvolution, *IEEE Transactions on Signal Processing*, **47**(7), 1922–1935.

620. Papoulis, A. (1991) *Probability, Random Variables, and Stochastic Processes*, 3rd edition, McGraw Hill, New York.

621. Marcos, S., Cherif, S. and Jaidane, M. (1995) Blind cancellation of intersymbol interference in decision feedback equalizers, in *Proceedings of ICASSP*, pp. 1073–1076.

622. Kamel, R. E. and Bar-Ness, Y. (1993) Blind decision feedback equalization using the decorrelation criterion, in *Proceedings of GLOBECOM*, pp. 87–91.

623. Kennedy, R. A. (1993) Blind adaptation of decision feedback equalizers: Gross convergence properties, *International Journal of Adaptive Control Signal Processes*, **7**, 497–523.

624. Papadias, C. B. and Paulraj, A. (1995) Decision–feedback equalization and identification of linear channels using blind algorithms of the Bussgang type, in *Proceedings*

of Asilomar Coni Signals, Systems, Computers, Pacific Grove, CA, 1995, pp. 335–340.

625. Haykin, S. (Ed.) *Blind Deconvolution* Prentice Hall, Englewood Cliffs, NJ.

626. Haykin, S. (1996) *Adaptive Filter Theory*, 3rd edition, Prentice Hall, Englewood Cliffs, NJ.

627. Benveniste, A., Goursat, M. and Ruget, G. (1980) Robust identification of a nonminimum phase system: Blind adjustment of a linear equalizer in data Communications, *IEEE Transactions on Automatic Control*, **AC–25**, 385–399.

628. Comon, P. (1994) Independent component analysis, a new concept?, *Signal Processing*, **36**, 287–314.

629. Moreau, E. and Macchi, O. (1993) New self-adaptive algorithms for source separation based on contrast functions, in *Proceedings of the IEEE Signal Processing Workshop on Higher-Order Statistics*, June 1993, pp. 215–219.

630. Kim, Y.-H. and Shamsunder, S. (1998) Multichannel algorithms for simultaneous equalization and interference suppression, in *Wireless Personal Communications*, Kluwer, Norwell, MA pp. 219–237.

631. Li, Y. G., Cimini, L. J. Jr, and Sollenberger, N. R. (1998) Robust channel estimation for OFDM systems with rapid dispersive fading channels, *IEEE Transactions On Communications*, **46**, 902–915.

632. Jia-Chin Lin (2003) Maximum-likelihood frame timing instant and frequency offset estimation for OFDM communication over a fast Rayleigh-fading channel, *IEEE Transactions on Vehicular Technology*, **52**(4), 1049–1062.

633. Chengyang Li and Roy, S. (2003) Subspace-based blind channel estimation for OFDM by exploiting virtual carriers, *IEEE Transactions on Wireless Communications*, **2**(1), 141–150.

634. Seog Geun Kang, Yong Min Ha and Eon Kyeong Joo (2003) A comparative investigation on channel estimation algorithms for OFDM in mobile Communications, *IEEE Transactions on Broadcasting*, **49**(2), 142–149.

635. Zheng Yuanjin (2003) A novel channel estimation and tracking method for wireless OFDM systems based on pilots and Kalman filtering, *IEEE Transactions on Consumer Electronics*, **49**(2), 275–283.

636. Xiaobo Zhou and Xiaodong Wang (2003) Channel estimation for OFDM systems using adaptive radial basis function networks, *IEEE Transactions on Vehicular Technology*, **52**(1), 48–59.

637. Deneire, L., Vandenameele, P., van der Perre, L., Gyselinckx, B. and Engels, M. (2003) A low-complexity ML channel estimator for OFDM, *IEEE Transactions On Communications*, **51**(2), 135–140.

638. Muquet, B., de Courville, M. and Duhamel, P. (2002) Subspace-based blind and semi-blind channel estimation for OFDM systems, *IEEE Transactions on Signal Processing*, **50**(7), 1699–1712.

639. Luise, M., Marselli, M. and Reggiannini, R. (2002) Low-complexity blind carrier frequency recovery for OFDM signals over frequency-selective radio channels, *IEEE Transactions On Communications*, **50**(7), 1182–1188.

640. Coleri, S., Ergen, M., Puri, A. and Bahai, A. (2002) Channel estimation techniques based on pilot arrangement in OFDM systems, *IEEE Transactions on Broadcasting*, **48**(3), 223–229.

641. Landstrom, D., Wilson, S. K., van de Beek, J.-J., Odling, P. and Borjesson, P. O. (2002) Symbol time offset estimation in coherent OFDM systems, *IEEE Transactions On Communications*, **50**(4), 545–549.

642. May, T., Rohling, H. and Engels, V. (1998) Performance analysis of Viterbi decoding for 64-DAPSK and 64-QAM modulated OFDM signals, *IEEE Transactions On Communications*, **46**(2), 182–190.

643. Shäfer, R. (1995) Terrestrial transmission of DTVB signals–The European specification. in *Proceedings of the International Broadcasting Convention*, Amsterdam, The Netherlands, 1995, pp. 79–84.

644. Bagels, V. and Rohling, H. (1995) Multilevel differential modulation techniques (64-DAPSK) for multicarrier transmission systems, *European Transactions on Telecommunications*, **6**, 633–640.

645. Rohling, H. and Engels, V. (1995) Differential amplitude phase shift keying (DAPSK)– A new modulation method for DTVB, in *Proceedings of the International Broadcasting Convention*, Amsterdam, The Netherlands, 1995, pp. 102–108.

646. Monnier, R., Rault, J. B. and de Couasnon, T. (1992) Digital television broadcasting with high spectral efficiency, in *Proceedings of the International Broadcasting Convention*, Amsterdam, The Netherlands, 1992, pp. 380–384.

647. Marti, B., Bernard, P., Lodge, N. and Shäfer, R. (1993) European activies on digital television broadcasting–From company to cooperative projects, *EBU Technical Review*, pp. 22–29.

648. Chow, Y. C., Nix, A. R. and McGeehan, J. P. (1992) Analysis of 16-APSK modulation in AWGN and Rayleigh fading channel, *Electronics Letters*, **28**, 1608–1610.

649. Giallorenci, T. R. and Wilson, S. G. (1995) Noncoherent demodulation techniques for trellis-coded M-DPSK signals, *IEEE Transactions On Communications*, **43**, 2370–2380.

650. Simon, M. K. and Divsalar, D. (1988) The performance of trellis coded multilevel DPSK on a fading mobile satellite channel, *IEEE Transactions on Vehicular Technology*, **37**, 78–91.

651. Divsalar, D. and Simon, M. K. (1990) Multiple-symbol differential detection of MPSK, *IEEE Transactions On Communications*, **38**, 300–308.

652. Divsalar, D., Simon, M. K. and Shahshahani, M. (1990) The performance of trellis-coded MDPSK with multiple symbol detection, *IEEE Transactions On Communications*, **38**, 1391–1403.

653. Hanzo, L. *et al.* (2002) *Adaptive Wireless Transceivers.* John Wiley & Sons, Ltd, Chichester.

654. Zhiqiang Liu and Giannakis, G. B. (2003) Block differentially encoded OFDM with maximum multipath diversity, *IEEE Transactions on Wireless Communications*, **2**(3), 420–423.

655. Zhiqiang Liu, Yan Xin and Giannakis, G. B. (2002) Space–time-frequency coded OFDM over frequency-selective fading channels, *IEEE Transactions on Signal Processing*, **50**(10), 2465–2476.

656. Stamoulis, A., Zhiqiang, L. and Giannakis, G. B. (2002) Space–time block-coded OFDMA with linear precoding for multirate services, *IEEE Transactions on Signal Processing*, **50**(1), 119–129.

657. Kuo-Hui Li and Ingram, M. A. (2002) Space–time block-coded OFDM systems with RF beamformers for high-speed indoor wireless communications, *IEEE Transactions On Communications*, **50**(12), 1899–1901.

658. Lu, B., Xiaodong Wang and Ye Li (2002) Iterative receivers for space–time block-coded OFDM systems in dispersive fading channels, *IEEE Transactions on Wireless Communications*, **1**(2), 213–225.

659. Ye Li (2002) Simplified channel estimation for OFDM systems with multiple transmit antennas, *IEEE Transactions on Wireless Communications*, **1**(1), 67–75.

660. Lu, B., Xiaodong Wang and Narayanan, K. R. (2002) LDPC-based space–time coded OFDM systems over correlated fading channels: Performance analysis and receiver design, *IEEE Transactions On Communications*, **50**(1), 74–88.

661. Blum, R. S., Ye Geoffrey Li, Winters, J. H. and Qing Yan (2001) Improved space–time coding for MIMO-OFDM wireless communications, *IEEE Transactions on Communications*, **49**(11), 1873–1878.

662. Li, Y. G., Seshadri, N. and Ariyavisitakul, S. (1999) Channel estimation for OFDM systems with transmitter diversity in mobile wireless channels, *IEEE Journal on Selected Areas in Communications*, **17**, 461–471.

663. Blum, R. S., Ye Geoffrey Li, Winters, J. H. and Qing Yan (2001) Improved space–time coding for MIMO-OFDM wireless Communications, *IEEE Transactions On Communications*, **49**(11), 1873–1878.

664. Li, Y. G., Winters, J. H. and Sollenberger, N. R. (2001) Signal detection for MIMO-OFDM wireless Communications, *IEEE International Conference on Communications*, June 2001.

665. Al-Dhahir, N., Uysal, M. and Georghiades, C. N. (2001) Three space–time block-coding schemes for frequency-selective fading channels with application to EDGE, *IEEE 54th Vehicular Technology Conference*, 7–11 October 2001, **3**, 1834–1838.

666. Younis, W. and Al-Dhahir, N. (2002) Joint prefiltering and MLSE equalization of space–time-coded transmissions over frequency-selective channels, *IEEE Transactions on Vehicular Technology*, **51**(1), 144–154.

667. Al-Dhahir, N. (2002) Overview and comparison of equalization schemes for space–time-coded signals with application to EDGE, *IEEE Transactions on Signal Processing*, **50**(10), 2477–2488.

668. Franz, V. and Anderson, J. (1998) Concatenated decoding with a reduced-search BCJR algorithm, *IEEE Journal on Selected Areas in Communications*, **16**, 186–195.

669. Bahl, L., Cocke, J., Jelinek, F. and Raviv, J. (1974) Optimal decoding of linear codes for minimizing symbol error rate, *IEEE Transactions On Information Theory*, **IT-20**, 284–287.

670. Fragouli, C., Al-Dhahir, N., Diggavi, S. and Turin, W. (2002) Prefiltered space–time M-BCJR equalizer for frequency-selective channels, *IEEE Transactions on Communications*, **50**, 742–753.

671. Naguib, A. and Seshadri, N. (2000) MLSE and equalization of space–time coded signals, in *Proceedings of Vehicular Technology Conference*, May 2000, pp. 1688–1693.

672. Al-Dhahir, N. (2001) FIR channel-shortening equalizers for MIMO ISI channels, *IEEE Transactions On Communications*, **50**, 213–218.

673. Duel-Halien, A. and Heegard, C. (1989) Delayed decision-feedback sequence estimation, *IEEE Transactions On Communications*, **36**, 428–436.

674. Lindskog, E. and Paulraj, A. (2000) A transmit diversity scheme for delay spread channels, in *Proceedings of the International Conference on Communications*, June 2000, pp. 307–311.

675. Pollet, T., Van Bladel, M. and Moeneclaey, M. (1995) BER sensitivity of OFDM systems to carrier frequency offset and wiener phase noise, *IEEE Transactions On Communications*, **44**, 191–193.

676. Sari, H., Karam, G. and Jeanclaude, I. (1995) Transmission techniques for digital terrestrial TV broadcasting, *IEEE Communications Magazine*, **33**, 100–109.

677. Clark, M. V. (1998) Adaptive frequency-domain equalization and diversity combining for broadband wireless Communications, *IEEE Journal on Selected Areas in Communications*, **16**, 1385–1395.

678. Al-Dhahir, N. (2001) Single-carrier frequency-domain equalization for space–time block-coded transmissions over frequency-selective fading channels, *IEEE Communications Letters*, **5**, 304–306.

679. Kaleh, G. (1995) Channel equalization for block transmission systems, *IEEE Journal on Selected Areas in Communications*, **13**, 110–121.

680. SC FDE PHY Layer Sys. Proposal for Sub 11 GHz BWA (Online). Available: http://www.ieee802.org/16/tg3/contrib/802 163p-0131r2.pdf.

681. Crazier, S., Falconer, D. and Mahmoud, S. (1991) Least sum of squared errors (LSSE) channel estimation, *Proceedings of the Institute of Electronic Engineers F*, pp. 371–378.

682. Chu, D. (1972) Polyphase codes with good periodic correlation properties, *IEEE Transactions On Information Theory*, **IT-18**, 531–532.

683. Fragouli, C., Al-Dhahir, N. and Turin, W. (2002) Reduced-complexity training schemes for multiple-antenna broadband transmissions, in *Proceedings of WCNC*, **1**, 78–83.

684. Fragouli, C., Al-Dhahir, N. and Turin, W. (2002) Finite-alphabet constant-amplitude training sequences for multiple-antenna broadband transmissions, in *Proceedings of the International Control Conference*, **1**, 6–10.

685. Chong, L. L. and Milstein, L. B. (2000) Error rate of a multicarrier CDMA system with imperfect channel estimates, *IEEE International Conference on Communications* (ICC 2000), 18-22 June 2000, **2**, 934–938.

686. Kondo, S. and Milstein, L. B. (1996) On the performance of multicarrier DS CDMA systems, *IEEE Transactions On Communications*, **44**, 238–246.

687. Fazel, K. and Fettweis, G. P. (Eds.) (1997) *Multi-Carrier Spread Spectrum*. Kluwer, Boston, MA.

688. Sourour, E. and Nakagawa, M. (1996) Performance of orthogonal multicarrier CDMA in a multipath fading channel, *IEEE Transactions On Communications*, **44**, 356–367.

689. Kondo, S. and Milstein, L. B. (1993) On the use of multicarrier direct sequence spread spectrum systems, in *Proceedings of IEEE MILCOM*, Boston, MA, October 1993, pp. 52–56.

690. TIA/TR45.5. (1998, July). *The cdma2000 ITU-R RTT candidate submission (0.18)* (Online). Available: http://www.itu.Int/imt/2-radio-dev/proposals/index.html.

691. Eng, T. and Milstein, L. B. (1994) Comparison of hybrid FDMA/CDMA systems in frequency selective Rayleigh fading, *IEEE Journal on Selected Areas in Communications*, **12**, 938–951.

692. Dongwook Lee and Milstein, L. B. (1999) Comparison of multicarrier DS-CDMA broadcast systems in a multipath fading channel, *IEEE Transactions On Communications*, **47**(12), 1897–1904.

693. Hagenauer, J. (1988) Rate-compatible punctured convolutional codes (RCPC Codes) and their applications, *IEEE Transactions On Communications*, **36**, 389–400.

694. Balachandran, K., Kadaba, S. R. and Nanda, S. (1999) Channel quality estimation and rate adaptation for cellular mobile radio, *IEEE Journal on Selected Areas in Communications*, **17**, 1244–1256.

695. Frenger, P., Orten, P., Ottosson, T. and Svensson, A. (1999) Rate-compatible convolutional codes for multirate DS-CDMA systems, *IEEE Transactions On Communications*, **47**, 828–836.

696. Jumi Lee, Iickho Song, So Ryoung Park, and Seokho Yoon (2001) Analysis of an adaptive rate convolutionally coded multicarrier DS/CDMA system, *IEEE Transactions on Vehicular Technology*, **50**(4), 1014–1023.

697. Weiping Xu and Milstein, L. B. (2001) On the use of Interference suppression to reduce Intermodulation distortion in multicarrier CDMA systems, *IEEE Transactions On Communications*, **49**(1), 130–141.

698. Lin Fang and Milstein, L. B. (2001) Performance of successive Interference cancellation in convolutionally coded multicarrier DS/CDMA systems, *IEEE Transactions On Communications*, **49**(12), 2062–2067.

699. Miller, S. L. and Rainbolt, B. J. (2000) MMSE detection of multicarrier CDMA, *IEEE Journal on Selected Areas in Communications*, **18**(11), 2356–2362.

700. Kalofonos, D. N., Stojanovic, M. and Proakis, J. G. (1998) On the performance of adaptive MMSE detectors for a MC-CDMA system in fast fading Rayleigh channels, *The Ninth IEEE International Symposium on Personal, Indoor and Mobile Radio Communications*, 8–11 September 1998, **3**, 1309–1313.

701. Kalofonos, D. N., Stojanovic, M. and Proakis, J. G. (2003) Performance of adaptive MC-CDMA detectors in rapidly fading Rayleigh channels, *IEEE Transactions on Wireless Communications*, **2**(2), 229–239.

702. Jinghong Ma and Tugnait, J. K. (2002) Blind detection of multirate asynchronous CDMA signals in multipath channels, *IEEE Transactions on Signal Processing*, **50**(9), 2258–2272.

703. Kunjie Wang, Pingping Zong and Bar-Ness, Y. (2001) A reduced complexity partial sampling MMSE receiver for asynchronous MC-CDMA systems, *IEEE Global Telecommunications Conference*, November 2001, **2**, 728–732.

704. Pingping Zong, Kunjie Wang and Bar-Ness, Y. (2001) Partial sampling MMSE Interference suppression in asynchronous multicarrier CDMA system, *IEEE Journal on Selected Areas in Communications*, **19**(8), 1605–1613.

705. Weiping Xu and Milstein, L. B. (1998) MMSE Interference suppression for multicarrier DS-CDMA in frequency selective channels, *IEEE Global Telecommunications Conference*, 1998, 8–12 November 1998, **1**, 259–264.

706. Petre, F., Vandenameele, P., Bourdoux, A., Gyselinckx, B., Engels, M., Moonen, M. and DeMan, H. (2000) Combined MMSE/pcPIC multiuser detection for MC-CDMA, *IEEE 51st Vehicular Technology Conference Proceedings*, Tokyo, 15–18 May 2000, **2**, 770–774.

707. Petre, F., Engels, M., Moonen, M., Gyselinckx, B. and De Man, H. (2001) Adaptive MMSE/pcPIC-MMSE multiuser detector for MC-CDMA satellite system, *IEEE International Conference on Communications*, 11–14 June 2001, **9**, 2640–2644.

708. Hyung-Yun Kong and Chang-Hee Lee (1999) Design of MC-CDMA system based on non-linear MMSE, *Proceedings of the IEEE Region 10 Conference*, 15–17 September 1999, **1**, 53–56.

709. Pingping Zong, Kunjie Wang and Bar-Ness, Y. (2001) A novel partial sampling MMSE receiver for uplink multicarrier CDMA, *IEEE VTS 53rd Vehicular Technology Conference*, 6–9 May 2001, **1**, 741–745.

710. Helard, J.-F., Baudais, J.-Y. and Citerne, J. (2000) Linear MMSE detection technique for MC-CDMA, *Electronics Letters*, **36**(7), 665–666.

711. Namgoong, J., Wong, T. F. and Lehnert, J. S. (1999) Subspace MMSE receiver for multicarrier CDMA, *IEEE Wireless Communications and Networking Conference* (WCNC 1999), 21–24, September 1999, **1**, 90–94.

712. Zigang Yang, Lu, B. and Xiaodong Wang (2001) Blind Bayesian multiuser receiver for space–time coded MC-CDMA system over frequency-selective fading channel, *IEEE Global Telecommunications Conference*, 25–29 November 2001, **2**, 781–785.

713. Zigang Yang, Ben Lu and Xiaodong Wang (2001) Bayesian Monte Carlo multiuser receiver for space–time coded multicarrier CDMA systems, *IEEE Journal on Selected Areas in Communications*, **19**(8), 1625–1637.

714. Lie-Liang Yang and Hanzo, L. (2002) Broadband MC DS-CDMA using space–time and frequency-domain spreading, *IEEE 56th Vehicular Technology Conference*, 24–28 September 2002, **3**, 1632–1636.

715. Gelfand, A. and Smith, A. (1990) Sampling-based approaches to calculating marginal densities, *Journal of the American Statistical Association*, **85**, 398–409.

716. Geman, S. and Geman, D. (1984) Stochastic relaxation, Gibbs distribution, and the Bayesian restoration of images, *IEEE Transactions on Pattern Analysis and Machine Intelligence*, **PAMI-6**, 721–741.

717. Chan, K. (1993) Asymptotic behavior of the Gibbs sampler, *Journal of the American Statistical Association*, **88**, 320–326.

718. Liu, J., Wong, W. and Hong, A. (1995) Covariance structure and convergence rate of the Gibbs sampler with various scans, *Journal of the Royal Statistical Society Series B*, **57**, 157–169.

719. Robert, C. and Casella, G. (1999) *Monte Carlo Statistical Methods*. Springer-Verlag, New York.

720. Wang, X. and Chen, R. (2000) Adaptive Bayesian multiuser detection for synchronous CDMA with Gaussian and impulsive noise, *IEEE Transactions on Signal Processing*, **48**, 2013–2028.

721. Win, M. Z. and Scholtz, R. A. (1998) Impulse radio: How it works, *IEEE Communications Letters*, **2**, 36–38.

722. Scholtz, R. A. (1993) Multiple access with time-hopping impulse modulation, in *Proceedings of MILCOM*, October 1993, pp. 447–450.

723. Win, M. Z. and Scholtz, R. A. (1998) On the energy capture of ultrawide bandwidth signals in dense multipath environments, *IEEE Communications Letters*, **2**(9), 245–247.

724. Win, M. Z., Scholtz, R. A. and Barnes, M. A. (1997) Ultra-wide bandwidth signal propagation for indoor wireless communications, in *Proceedings of the IEEE International Conference on Communications*, Montreal, Canada, June 1997, pp. 56–60.

725. Cramer, R. J.-M., Scholtz, R. A. and Win, M. Z. (2002) Evaluation of an ultra-wideband propagation channel, *IEEE Transactions on Antennas and Propagation*, **50**(5), 561–570.

726. Schantz, H. G. and Fullerton, L. (2001) The diamond dipole: A Gaussian impulse antenna, *Proceedings of the IEEE AP-S International Symposium*, July 8–13, 2001.

727. Scholtz, R. A. and Win, M. Z. (1997) Impulse radio, in *Wireless Communications: TDMA versus CDMA*, S. G. Glisic and P. A. Leppänen (Eds.) Kluwer, Norwell, MA.

728. Hashemi, H. (1993) The indoor radio propagation channel, *Proceedings of the IEEE*, **81**, 943–968.

729. Liberti, J. C. and Rappaport, T. S. (1999) *Smart Antennas for Wireless Communications: IS-95 and Third Generation CDMA Applications*. Prentice Hall, Englewood Cliffs, NJ.

730. Saleh, A. A. M. and Valenzuela, R. A. (1987) A statistical model for indoor multipath propagation, *IEEE Journal on Selected Areas in Communications*, **5**, 128–137.

731. Spencer, Q., Rice, M., Jeffs, B. and Jensen, M. (1997) A statistical model for the angle-of-arrival in indoor multipath propagation, in *Proceedings of the IEEE Vehicular Technology Conference*, 1415–1419.

732. Spencer, Q., Jeffs, B., Jensen, M. and Swindlehurst, A. (2000) Modeling the statistical time and angle of arrival characteristics of an indoor multipath channel, *IEEE Journal on Selected Areas in Communications*, **18**, 347–360.

733. Vaughan, R. G. and Scott, N. L. (1999) Super-resolution of pulsed multipath channels for delay spread characterization, *IEEE Transactions On Communications*, **47**, 343–347.

734. Yano, S. M. (2002) Investigating the ultra-wideband indoor wireless channel, *IEEE 55th Vehicular Technology Conference* (VTC Spring 2002) 6–9 May 2002, **3**, 1200–1204.

735. Rappaport, T. S. (1966) *Wireless Communications: Principles and Practices*, Prentice Hall.

736. Cassioli, D. *et al.* (2001) A statistical model for the UWB indoor channel, *IEEE VTC2001*, **2**, 1159–1163.

737. Ghassemzadeh, S. S., Jana, R., Rice, C. W., Turin, W. and Tarokh, V. (2002) A statistical path loss model for in-home UWB channels, *IEEE Conference on Ultra Wideband Systems and Technologies*, 21–23 May 2002, pp. 59–64.

738. Turin, W., Jana, R., Ghassemzadeh, S. S., Rice, C. W. and Tarokh, T. (2002) Autoregressive modelling of an indoor UWB channel, *IEEE Conference on Ultra Wideband Systems and Technologies*, 21–23 May 2002, pp. 71–74.

739. Ghassemzadeh, S. S. and Tarokh, V. (2003) UWB path loss characterization in residential environments, *IEEE Radio Frequency Integrated Circuits (RFIC) Symposium*, June 8–10 2003, pp. 501–504.

740. Ghassemzadeh, S. S. and Tarokh, V. (2003) UWB path loss characterization in residential environments, *IEEE MTT-S International Microwave Symposium Digest*, 8–13 June 2003, **1**, 365–368.

741. Ramirez-Mireles, F. (2001) On the performance of ultra-wide-band signals in Gaussian noise and dense multipath, *IEEE Transactions on Vehicular Technology*, **50**(1), 244–249.

742. Gagliardi, R. M. (1988) *Introduction to Telecommunications Engineering*. John Wiley & Sons, Inc., New York, pp. 357–437.

743. Win, M. Z., Ramírez-Mireles, F., Scholtz, R. A. and Barnes, M. A. (1997) Ultra-wide bandwidth (UWB) signal propagation for outdoor wireless communications, in *Proceedings of the IEEE VTC Conference*, May 1997, pp. 251–255.

744. Ramírez-Mireles, F. and Scholtz, R. A. (1998) Time-shift-keyed equicorrelated signal sets for impulse radio *M*-ary modulation, in *Proceedings of the IEEE Wireless Conference*, pp. 404–408.

745. Ramírez-Mireles, F. and Scholtz, R. A. (1997) Performance of equicorrelated ultra-wideband pulse-position-modulated signals in the indoor wireless impulse radio channel, in *Proceedings of the IEEE PACRIM Conference*, pp. 640–644.

746. Ramírez-Mireles, F. (2001) Performance of ultrawideband SSMA using time hopping and *M*-ary PPM, *IEEE Journal on Selected Areas in Communications*, **19**(6), 1186–1196.

747. Ramírez-Mireles, F. (1998) *Multiple-access with ultra-wideband impulse radio modulation using spread spectrum time-hopping and block waveform pulse-position-modulated signals*, Ph.D. dissertation, Communication Sciences Institute, Electrical Engineering Department, University of Southern California.

748. Georghiades, C. N. (1988) On PPM sequences with good autocorrelation properties, *IEEE Transactions On Information Theory*, **34**, 571–576.

749. Gagliardi, R., Robbins, J. and Taylor, H. (1987) Acquisition sequences in PPM communications, *IEEE Transactions On Information Theory*, **33**, 738–744.

750. Golomb, S. W. (1991) Construction of signals with favorable correlation properties, in *Surveys in Combinatorics*, Cambridge University Press, Cambridge.

751. Forouzan, A. R., Nasiri-Kenari, M. and Salehi, J. A. (2002) Performance analysis of time-hopping spread-spectrum multiple-access systems: uncoded and coded schemes, *IEEE Transactions on Wireless Communications*, **1**, 671–681.

752. Viterbi, A. J. (1990) Very low-rate convolutional codes for maximum theoretical performance of spread-spectrum multiple-access channels, *IEEE Journal on Selected Areas in Communications*, **8**, 641–649.

753. Viterbi, A. (1995) *CDMA: Principles of Spread-Spectrum Communication*. Addison Wesley, Reading, MA.

754. Shaft, P. D. (1977) Low-rate convolutional code application in spread-spectrum communications, *IEEE Transactions On Communications*, **COM-25**, 815–821.

755. Win, M. Z., Qiu, X., Scholtz, R. A. and Li, V. O. K. (1999) ATM-based TH-SSMA network for multimedia PCS, *IEEE Journal on Selected Areas in Communications*, **17**, 824–836.

756. Le Martret, C. J. and Giannakis, G. B. (2002) All-digital impulse radio with multiuser detection for wireless cellular systems, *IEEE Transactions On Communications*, **50**, 1440–1450.

757. Hussain, M. G. M. (2002) Principles of space–time array processing for ultrawide-band impulse radar and radio communications, *IEEE Transactions on Vehicular Technology*, **51**, 393–403.

758. Hussain, M. G. M. (1988) Performance analysis and advancement of self-steering arrays for nonsinusoidal waves–I, II, *IEEE Transactions on Electromagnetic Compatibility*, **30**, 161–174.

759. Hussain, M. G. M. (1988) A self-steering array for nonsinusoidal waves based on array impulse response measurement, *IEEE Transactions on Electromagnetic Compatibility*, **30**, 154–160.

760. Hussain, M. G. M., Al-Halabi, M. M. M. and Omar, A. A. (1989) Antenna patterns of nonsinusoidal waves with the time variation of a Gaussian pulse–Part III, *IEEE Transactions on Electromagnetic Compatibility*, **31**, 34–47.

761. Balanis, C. A. (1982) Antenna Theory Analysis and Design, Harper and Row, New York, pp. 274–279.

762. Ma, M. T. (1974) Theory and Application of Antenna Arrays, John Wiley & Sons New York, pp. 191–202.

763. Hussain, M. G. M. (1988) Antenna patterns for nonsinusoidal waves with the time variation of a Gaussian pulse–Part I and II, *IEEE Transactions on Electromagnetic Compatibility*, **30**, 504–522.

764. Rajeswaran, A., Somayazulu, V. S. and Foerster, J. R. (2003) RAKE performance for a pulse based UWB system in a realistic UWB indoor channel, *IEEE International Conference on Communications*, **4**, 2879–2883.

765. Canadeo, C. M., Temple, M. A., Baldwin, R. O. and Raines, R. A. (2003) UWB multiple access performance in synchronous and asynchronous networks, *Electronics Letters*, **39**(11), 880–882.

766. Miller, L. E. (2003) Autocorrelation functions for Hermite-polynomial ultra-wideband pulses, *Electronics Letters*, **39**(11), 870–871.

767. Porcino, D. and Hirt, W. (2003) Ultra-wideband radio technology: potential and challenges ahead, *IEEE Communications Magazine*, **41**(7), 66–74.

768. Aiello, G. R. and Rogerson, G. D. (2003) Ultra-wideband wireless systems, *IEEE Microwave Magazine*, **4**(2), 36–47.

769. Hyun-Jin Park, Mi-Jeong Kim, Yoon-Jae So, Young-Hwan You and Hyoung-Kyu Song (2003) UWB communication system for home entertainment network, *IEEE Transactions on Consumer Electronics*, **49**(2), 302–311.

770. Durisi, G. and Benedetto, S. (2003) Performance evaluation of TH-PPM UWB systems in the presence of multiuser interference, *IEEE Communications Letters*, **7**(5), 224–226.

771. Parr, B., ByungLok Cho, Wallace, K. and Zhi Ding (2003) A novel ultra-wideband pulse design algorithm, *IEEE Communications Letters*, **7**(5), 219–221.

772. Won Namgoong (2003) A channelized digital ultrawideband receiver, *IEEE Transactions on Wireless Communications*, **2**(3), 502–510.

773. Nakagawa, M., Honggang Zhang and Sato, H. (2003) Ubiquitous homelinks based on IEEE 1394 and ultra wideband solutions, *IEEE Communications Magazine*, **41**(4), 74–82.

774. Saberinia, E. and Tewfik, A. H. (2003) Single and multi-carrier UWB communications, *Seventh International Symposium on Signal Processing and Its Applications*, July 1–4, 2003, **2**, 343–346.

775. Saberinia, E. and Tewfik, A. H. (2003) Receiver structures for multi-carrier UWB systems, *Seventh International Symposium on Signal Processing and Its Applications*, July 1–4, 2003, **1**, 313–316.

776. Pidre Mosquera, J. M. and Isasa, M. V. (2003) Planar resistively loaded UWB dipoles analysis and comparison, *IEEE Society International Conference on Antennas and Propagation*, June 22–27 2003, **3**, 636–639.

777. Taniguchi, T. and Kobayashi, T. (2003) An omnidirectional and low-vswr antenna for the FCC-approved UWB frequency band, *IEEE Society International Conference on Antennas and Propagation*, June 22–27 2003, **3**, 460–463.

778. Schantz, H. G. (2003) UWB magnetic antennas, *IEEE Society International Conference on Antennas and Propagation*, June 22–27 2003, **3**, 604–607.

779. Ogawa, T., Tomiki, A. and Kobayashi, T. (2003) Development of two kinds of UWB sources for propagation, EMC and other experimental studies: impulse radio and direct-sequence spread spectrum, *IEEE Society International Conference on Antennas and Propagation*, June 22–27 2003, **3**, 273–276.

780. Xianming Qing, Wah Chia, M. Y. and Xuanhui Wu (2003) Wide-slot antenna for uwb applications, *IEEE Society International Conference on Antennas and Propagation*, June 22–27 2003, **1**, 834–837.

781. Kerkhoff, A. and Hao Ling (2003) Design of a planar monopole antenna for use with ultra-wideband (uwb) having a band-notched characteristic, *IEEE Society International Conference on Antennas and Propagation*, June 22–27 2003, **1**, 830–833.

782. Zwierzchowski, S. and Jazayeri, P. (2003) A systems and network analysis approach to antenna design for uwb communications, *IEEE Society International Conference on Antennas and Propagation*, June 22–27 2003, **1**, 826–829

783. Zhi Ning Chen, Xuan Hui Wu, Ning Yang and Chia, M. Y. W (2003) Design considerations for antennas in uwb wireless communication systems, *IEEE Society International Conference on Antennas and Propagation*, June 22–27 2003, **1**, 822–825.

784. Kwan-ho Lee, Chi-Chih Chen, Teixeira, F. L. and Lee, R. (2003) Numerical study of a uwb dual-polarized feed design for enhanced tapered chambers, *IEEE Society International Conference on Antennas and Propagation*, June 22–27 2003, **1**, 265–268.

785. Ghassemzadeh, S. S. and Tarokh, V. (2003) UWB path loss characterization in residential environments, *IEEE Radio Frequency Integrated Circuits (RFIC) Symposium*, June 8–10 2003, pp. 501–504.

786. Aiello, G. R. (2003) Challenges for ultra-wideband (UWB) CMOS integration, *IEEE Radio Frequency Integrated Circuits (RFIC) Symposium*, June 8–10 2003, pp. 497–500.

787. Woo Cheol Chung and Dong Sam Ha (2003) On the performance of bi-phase modulated uwb signals in a multipath channel, *IEEE Vehicular Technology Conference*, April 22–25 2003, **3**, 1654–1658.

788. Piazzo, L. and Romme, J. (2003) Spectrum control by means of the th code in uwb systems, *IEEE Vehicular Technology Conference*, April 22–25 2003, **3**, 1649–1653.

789. Guangrong Yue, Lijia Ge and Shaoqian Li (2003) Performance of uwb time-hopping spread-spectrum impulse radio in multipath environments, *IEEE Vehicular Technology Conference*, April 22–25 2003, **3**, 1644–1648.

790. Terri, M., Hong, A., Guibe, G. and Legrand, F. (2003) Major characteristics of UWB indoor transmission for simulation, *IEEE Vehicular Technology Conference*, April 22–25, 2003 **1**, 19–23.

791. Yongfu Huang, Xiangning Fan, Jiang Wang and Guangguo Bi (2003) Analysis of the energy dynamic of UWB signal in multi-path environments, *IEEE Vehicular Technology Conference*, April 22–25 2003, **1**, 15–18.

792. Ray-Rong Lao, Jeon-Hwan Tarng and Chiuder Hsiao (2003) Transmission coefficients measurement of building materials for UWB systems in 3–10 GHz, *IEEE Vehicular Technology Conference*, April 22–25 2003, **1**, 11–14.

793. Alvarez, A., Valera, G., Lobeira, M., Torres, R. and Garcia, J. L. (2003) New channel impulse response model for UWB indoor system simulations, *IEEE Vehicular Technology Conference*, April 22–25 2003, **1**, 1–5.

794. Nakache, Y.-P. and Molisch, A. F. (2003) Spectral shape of UWB signals influence of modulation format, multiple access scheme and pulse shape, *IEEE Vehicular Technology Conference*, April 22–25 2003, **4**, 2510–2514.

795. Saberinia, E. and Tewfik, A. H. (2003) Generating UWB-OFDM signal using sigma-delta modulator, *IEEE Vehicular Technology Conference*, April 22–25 2003, **2**, 1425–1429.

796. Durisi, G. and Benedetto, S. (2003) Performance evaluation and comparison of different modulation schemes for UWB multi access systems, *IEEE International Conference on Communications*, 11–15 May 2003, **3**, 2187–2191.

797. Nassar, C. R., Fang Zhu and Zhiqiang Wu (2003) Direct sequence spreading UWB systems: frequency domain processing for enhanced performance and

throughput, *IEEE International Conference on Communications*, 11–15 May 2003, **3**, 2180–2186.

798. Baccarelli, E. and Biagi, M. (2003) An adaptive codec for multi-user interference mitigation for UWB-based WLANs, *IEEE International Conference on Communications*, 11–15 May 2003, **3**, 2020–2024.

799. Cassioli, D., Win, M. Z., Vatalaro, F. and Molisch, A. F. (2003) Effects of spreading bandwidth on the performance of UWB RAKE receivers, *IEEE International Conference on Communications*, 11–15 May 2003, **5**, 3545–3549.

800. Kusuma, J., Maravic, I. and Vetterli, M. (2003) Sampling with finite rate of innovation: Channel and timing estimation for UWB and GPS, *IEEE International Conference on Communications*, 11–15 May 2003, **5**, 3540–3544.

801. Yamamoto, N. and Ohtsuki, T. (2003) Adaptive internally turbo-coded ultra wideband-impulse radio (AITC-UWB-IR) system, *IEEE International Conference on Communications*, 11–15 May 2003, **5**, 3535–3539.

802. Weisenhorn, M. and Hirt, W. (2003) Performance of binary antipodal signaling over the indoor UWB MIMO channel, *IEEE International Conference on Communications*, 11–15 May 2003, **4**, 2872–2878.

803. Saberinia, E. and Tewfik, A. H. (2003) *N*-tone sigma-delta uwb-ofdm transmitter and receiver, *IEEE International Conference on Acoustics, Speech, and Signal Processing*, (ICASSP'03), April 6–10 2003, **4**, IV-129–IV-132.

804. Huaning Niu, Ritcey, J. A. and Hai Liu (2003) Performance of uwb RAKE receivers with imperfect tap weights, *IEEE International Conference on Acoustics, Speech, and Signal Processing*(ICASSP'03), April 6–10, 2003, **4**, IV-125–IV-128.

805. Mo, S. S., Gelman, A. D. and Gopal, J. (2003) Frame synchronization in UWB using multiple SYNC words to eliminate line frequencies, *IEEE Wireless Communications and Networking Conference*, 16–20 March 2003, **2**, 773–778.

806. Canadeo, C. M., Temple, M. A., Baldwin, R. O. and Raines, R. A. (2003) Code selection for enhancing UWB multiple access communication performance using TH-PPM and DS-BPSK modulations, *IEEE Wireless Communications and Networking Conference*, 16–20 March 2003, **2**, 678–682.

807. Boubaker, N. and Letaief, K. B. (2003) A low complexity MMSE-RAKE receiver in a realistic UWB channel and in the presence of NBI, *IEEE Wireless Communications and Networking Conference*, 16–20 March 2003, **1**, 233–237.

808. Siwiak, K., Bertoni, H. and Yano, S. M. (2003) Relation between multipath and wave propagation attenuation, *Electronics Letters*, **39**(1), 142–143.

809. Giannakis, G. B. and Halford, S. D. (1997) Asymptotically optimal blind fractionally spaced channel estimation and performance analysis, *IEEE Transactions on Signal Processing*, **45**(7), 1815–1830.

810. Tsatsanis, M. K. and Giannakis, G. B. (1997) Blind estimation of direct sequence spread spectrum signals in multipath, *IEEE Transactions on Signal Processing*, **45**(7), 1241–1252.

811. Shengli Zhou, Muquet, B. and Giannakis, G. B. (2002) Subspace-based (semi-) blind channel estimation for block precoded space–time OFDM, *IEEE Transactions on Signal Processing*, **50**(5), 1215–1228.

812. Hongbin Li, Xuguang Lu and Giannakis, G. B. (2002) Capon multiuser receiver for CDMA systems with space–time coding, *IEEE Transactions on Signal Processing*, **50**(5), 1193–1204.

813. Dogandzic, A. and Nehorai, A. (2002) Finite-length MIMO equalization using canonical correlation analysis, *IEEE Transactions on Signal Processing*, **50**(4), 984–989.

814. Manikas, A. and Sethi, M. (2003) A space–time channel estimator and single-user receiver for code-reuse DS-CDMA systems, *IEEE Transactions on Signal Processing*, **51**(1), 39–51.

815. Lang Tong, van der Veen, A.-J., Dewilde, P. and Youngchul Sung (2003) Blind decorrelating RAKE receivers for long-code WCDMA, *IEEE Transactions on Signal Processing*, **51**(6), 1642–1655.

816. van der Veen, A.-J., Talwar, S. and Paulraj, A. (1997) A subspace approach to blind space–time signal processing for wireless communication systems, *IEEE Transactions on Signal Processing*, **45**(1), 173–190.

817. Tugnait, J. K. (1995) On blind identifiability of multipath channels using fractional sampling and second-order cyclostationary statistics, *IEEE Transactions On Information Theory*, **41**(1), 308–311.

818. Tugnait, J. K. (1996) Blind equalization and estimation of FIR communications channels using fractional sampling, *IEEE Transactions On Communications*, **44**(3), 324–336.

819. Jie Zhu, Xi-Ren Cao and Ruey-Wen Liu (1999) A blind fractionally spaced equalizer using higher order statistics, *IEEE Transactions on Circuits and Systems II: Analog and Digital Signal Processing*, **46**(6), 755–764.

820. Kadous, T. A. and Sayeed, A. M. (2000) Decentralized multiuser detection for time-varying multipath channels, *IEEE Transactions On Communications*, **48**(11), 1840–1852.

821. Yung-Fang Chen, Zoltowski, M. D., Ramos, J., Chatterjee, C. and Roychowdhury, V. P. (2000) Reduced-dimension blind space–time 2-D RAKE receivers for DS-CDMA communication systems, *IEEE Transactions on Signal Processing*, **48**(6), 1521–1536.

822. Gupta, R. and Hero, A. O. (2000) Power versus performance tradeoffs for reduced resolution LMS adaptive filters, *IEEE Transactions on Signal Processing*, **48**(10), 2772–2784.

823. Joon Ho Cho and Lehnert, J. S. (2001) Blind adaptive multiuser detection for DS/SSMA communications with generalized random spreading, *IEEE Transactions On Communications*, **49**(6), 1082–1091.

824. Bugallo, M. F., Miguez, J. and Castedo, L. (2001) A maximum likelihood approach to blind multiuser interference cancellation, *IEEE Transactions on Signal Processing*, **49**(6), 1228–1239.

825. Swindlehurst, A. L. and Leus, G. (2002) Blind and semi-blind equalization for generalized space–time block codes, *IEEE Transactions on Signal Processing*, **50**(10), 2489–2498.

826. Damen, M. O., Safavi, A. and Abed-Meraim, K. (2003) On CDMA with space–time codes over multipath fading channels, *IEEE Transactions on Wireless Communications*, **2**(1), 11–19.

827. Roy, R. and Kailath, T. (1989) ESPRIT-estimation of signal parameters via rotational invariance techniques, *IEEE Transactions on Acoustics Speech and Signal Processing*, **37**(7), 984–995.

828. Rao, B. D. and Hari, K. V. S. (1989) Performance analysis of ESPRIT and TAM in determining the direction of arrival of plane waves in noise, *IEEE Transactions on Acoustics, Speech and Signal Processing*, **37**(12), 1990–1995.

829. Soon, V. C. and Huang, Y. F. (1992) An analysis of ESPRIT under random sensor uncertainties, *IEEE Transactions on Signal Processing*, **40**(9), 2353–2358.

830. van der Veen, A. J., Ober, P. B. and Deprettere, E. F. (1992) Azimuth and elevation computation in high resolution DOA estimation, *IEEE Transactions on Signal Processing*, **40**(7), 1828–1832.

831. Rao, B. D. and Hari, K. V. S. (1993) Weighted subspace methods and spatial smoothing: analysis and comparison, *IEEE Transactions on Signal Processing*, **41**(2), 788–803.

832. Swindlehurst, A. L. and Kailath, T. (1993) A performance analysis of subspace-based methods in the presence of model error II–Multidimensional algorithms, *IEEE Transactions on Signal Processing*, **41**(9), 2882–2890.

833. Yuan-Hwang Chen and Yih-Sheng Lin (1994) A modified cumulant matrix for DOA estimation, *IEEE Transactions on Signal Processing*, **42**(11), 3287–3291.

834. Gansman, J. A., Zoltowski, M. D. and Krogmeier, J. V. (1996) Multidimensional multirate DOA estimation in beamspace, *IEEE Transactions on Signal Processing*, **44**(11), 2780–2792.

835. Yuen, N. and Friedlander, B. (1996) Asymptotic performance analysis of ESPRIT, higher order ESPRIT, and virtual ESPRIT algorithms, *IEEE Transactions on Signal Processing*, **44**(10), 2537–2550.

836. Kautz, G. M. and Zoltowski, M. D. (1996) Beamspace DOA estimation featuring multirate eigenvector Processing, *IEEE Transactions on Signal Processing*, **44**(7), 1765–1778.

837. Swindlehurst, A. L., Stoica, P. and Jansson, M. (2001) Exploiting arrays with multiple invariances using MUSIC and MODE, *IEEE Transactions on Signal Processing*, **49**(11), 2511–2521.

838. Tichavsky, P., Wong, K. T. and Zoltowski, M. D. (2001) Near-field/far-field azimuth and elevation angle estimation using a single vector hydrophone, *IEEE Transactions on Signal Processing*, **49**(11), 2498–2510.

839. Besson, O., Stoica, P. and Kamiya, Y. (2002) Direction finding in the presence of an intermittent interference, *IEEE Transactions on Signal Processing*, **50**(7), 1554–1564.

840. Rao, B. D. and Hari, K. V. S. (1989) Performance analysis of Root-Music, *IEEE Transactions on Acoustics, Speech and Signal Processing*, **37**(12), 1939–1949.

841. Li, F., Vaccaro, R. J. and Tufts, D. W. (1991) Performance analysis of the state-space realization (TAM) and ESPRIT algorithms for DOA estimation, *IEEE Transactions on Antennas and Propagation*, **39**(3), 418–423.

842. Li, F. and Vaccaro, R. J. (1992) Sensitivity analysis of DOA estimation algorithms to sensor errors, *IEEE Transactions on Aerospace and Electronic Systems*, **28**(3), 708–717.

843. Li, J. and Compton, R. T., Jr. (1992) Two-dimensional angle and polarization estimation using the ESPRIT algorithm, *IEEE Transactions on Antennas and Propagation*, **40**(5), 550–555.

844. Swindlehurst, A. (1992) DOA identifiability for rotationally invariant arrays, *IEEE Transactions on Signal Processing*, **40**(7), 1825–1828.

845. Li, F., Liu, H. and Vaccaro, R. J. (1993) Performance analysis for DOA estimation algorithms: unification, simplification, and observations, *IEEE Transactions on Aerospace and Electronic Systems*, **29**(4), 1170–1184.

846. Li, J. (1993) Direction and polarization estimation using arrays with small loops and short dipoles, *IEEE Transactions on Antennas and Propagation*, **41**(3), 379–387.

847. Hamza, R. and Buckley, K. (1994) Resolution enhanced ESPRIT, *IEEE Transactions on Signal Processing*, **42**(3), 688–691.

848. Fu Li and Yang Lu (1994) Bias analysis for ESPRIT-type estimation algorithms, *IEEE Transactions on Antennas and Propagation*, **42**(3), 418–423.

849. Fuhl, J., Rossi, J.-P. and Bonek, E. (1997) High-resolution 3-D direction-of-arrival determination for urban mobile radio, *IEEE Transactions on Antennas and Propagation*, **45**(4), 672–682.

850. Haardt, M. (1997) Structured least squares to improve the performance of ESPRIT-type algorithms, *IEEE Transactions on Signal Processing*, **45**(3), 792–799.

851. Kwok-Chiang Ho, Kah-Chye Tan and Tan, B. T. G. (1997) Efficient method for estimating directions-of-arrival of partially polarized signals with electromagnetic vector sensors, *IEEE Transactions on Signal Processing*, **45**(10), 2485–2498.

852. Wong, K. T. and Zoltowski, M. D. (1997) Uni-vector-sensor ESPRIT for multisource azimuth, elevation, and polarization estimation, *IEEE Transactions on Antennas and Propagation*, **45**(10), 1467–1474.

853. Tsung-Hsien Liu and Mendel, J. M. (1998) Azimuth and elevation direction finding using arbitrary array geometries, *IEEE Transactions on Signal Processing*, **46**(7), 2061–2065.

854. Strobach, P. (1998) Fast recursive subspace adaptive ESPRIT algorithms, *IEEE Transactions on Signal Processing*, **46**(9), 2413–2430.

855. Lindmark, B., Lundgren, S., Sanford, J. R. and Beckman, C. (1998) Dual-polarized array for signal-processing applications in wireless communications, *IEEE Transactions on Antennas and Propagation*, **46**(6), 758–763.

856. Hongyi Wang and Liu, K. J. R. (1998) 2-D spatial smoothing for multipath coherent signal separation, *IEEE Transactions on Aerospace and Electronic Systems*, **34**(2), 391–405.

857. Gershman, A. B. and Haardt, M. (1999) Improving the performance of Unitary ESPRIT via pseudo-noise resampling, *IEEE Transactions on Signal Processing*, **47**(8), 2305–2308.

858. Astely, D., Swindlehurst, A. L. and Ottersten, B. (1999) Spatial signature estimation for uniform linear arrays with unknown receiver gains and phases, *IEEE Transactions on Signal Processing*, **47**(8), 2128–2138.

859. Lemma, A. N., van der Veen, A.-J. and Deprettere, E.F. (1999) Multiresolution ESPRIT algorithm, *IEEE Transactions on Signal Processing*, **47**(6), 1722–1726.

860. Jansson, M., Goransson, B. and Ottersten, B. (1999) A subspace method for direction of arrival estimation of uncorrelated emitter signals, *IEEE Transactions on Signal Processing*, **47**(4), 945–956.

861. Hasan, M. A., Azimi-Sadjadi, M. R. and Hasan, A. A. (2000) Rational invariant subspace approximations with applications, *IEEE Transactions on Signal Processing*, **48**(11), 3032–3041.

862. Zoltowski, M. D. and Wong, K. T. (2000) ESPRIT-based 2-D direction finding with a sparse uniform array of electromagnetic vector sensors, *IEEE Transactions on Signal Processing*, **48**(8), 2195–2204.

863. Thoma, R. S., Hampicke, D., Richter, A., Sommerkorn, G., Schneider, A., Trautwein, U. and Wirnitzer, W. (2000) Identification of time-variant directional mobile radio channels, *IEEE Transactions on Instrumentation and Measurement*, **49**(2), 357–364.

864. Blanz, J. J., Papathanassiou, A., Haardt, M., Furio, I. and Baier, P. W. (2000) Smart antennas for combined DOA and joint channel estimation in time-slotted CDMA mobile radio systems with joint detection, *IEEE Transactions on Vehicular Technology*, **49**(2), 293–306.

865. Shahbazpanahi, S., Valaee, S. and Bastani, M. H. (2001) Distributed source localization using ESPRIT algorithm, *IEEE Transactions on Signal Processing*, **49**(10), 2169–2178.

866. Strobach, P. (2001) Total least squares phased averaging and 3-D ESPRIT for joint azimuth-elevation-carrier estimation, *IEEE Transactions on Signal Processing*, **49**(1), 54–62.

867. Lemma, A. N., van der Veen, A.-J. and Deprettere, E. F. (2003) Analysis of joint angle-frequency estimation using ESPRIT, *IEEE Transactions on Signal Processing*, **51**(5), 1264–1283.

868. Lagunas, M. A., Vidal, J. and Perez-Neira, A. I. (2000) Joint array combining and MLSE for single-user receivers in multipath Gaussian multiuser channels, *IEEE Journal on Selected Areas in Communications*, **18**(11), 2252–2259.

869. Bottomley, G. E. and Jamal, K. (1995) Adaptive arrays and MLSE equalization, *IEEE 45th Vehicular Technology Conference*, 25–28 July 1995, **1**, 50–54.

870. Fujii, M. (1996) Joint processing of an adaptive array and an MLSE for multipath channels, *IEEE Global Telecommunications Conference*, (GLOBECOM '96), 18–22 November 1996, **1**, 560–564.

871. Fujii, M. (1997) Joint processing of an adaptive array and an MLSE for frequency-selective fading channels, *IEEE International Conference on Communications*, Montreal, 8–12 June 1997, **2**, 636–640.

872. Monzingo, R. A. and Miller, T. W. (1980) *Introduction to Adaptive Arrays*, John Wiley & Sons, Inc. New York.

873. Steele, R. (1992) *Mobile Radio Communications*, Pentech, New York.

874. Poor, H. V. and Verdu, S. (1988) Single-user detectors for multiuser channels, *IEEE Transactions On Communications*, **36**, 50–60.

875. Miller, Y. and Schwartz, S. C. (1995) Integrated spatial-temporal detectors for aynchronous Gaussian multiple-access channels, *IEEE Transactions on Communications*, **43**.

876. Suard, B., Xu, G., Liu, H. and Kailath, T. (1998) Uplink channel capacity of space-division multiple-access schemes, *IEEE Transactions On Information Theory*, **44**, 1468–1476.

877. Magee, F. R. (1975) A comparison of compromise Viterbi algorithm and standard equalization techniques over band limited channels, *IEEE Transactions On Communications*, **COM-23**, 361–367.

878. Pipon, F., Chevalier, P., Vila, P. and Monott, J. J. (1997) Joint spatial and temporal equalization for channels with ISI and ICI: Theoretical and experimental results for a base-station, in *Proceedings of the IEEE Workshop on Signal Advances in Wireless Communications* (SPAWC'97) Paris, April 1997, pp. 309–312.

879. Proakis, J. (1995) *Digital Communications*. McGraw Hill, New York.

880. Liberti, J. C. and Rappaport, T. S. (1999) *Smart Antennas for Wireless Communications: IS-95 and Third Generation CDMA Applications*, Prentice Hall, Englewood Cliffs, NJ.

881. Escartin, M. and Ranta, P. A. (1997) Interference rejection with a small antenna array at the mobile scattering environment, in *Proceedings of the IEEE Signal Processing Workshop on Signal Processing Advances in Wireless Communications*, Paris, April 1997, pp. 165–168.

882. Bottomley, G. and Jamal, K. (1995) Adaptive arrays and MLSE equalization, in *Proceeding of the 45th IEEE Vehicular Technology Conference*, Chicago, pp. 50–54.

883. Pedersen, K., Mogensen, P. and Fleury, B. (2000) A stochastic model of the temporal and azimuthal dispersion seen at the base station in outdoor propagation environments, *IEEE Transactions on Vehicular Technology*, **49**(2).

884. European TeleCommunications Standard (1975), p. 12.

885. ETSI, (1998) *Evaluation report for ETSI UMTS terrestrial radio access (UTRA) ITU-R RTT candidate*, Technical Report ETSI-SMG2.

886. Chien, C., Srivastava, M. B., Jain, R., Lettieri, P., Aggarwal, V. and Sternowski, R. (1999) Adaptive radio for multimedia wireless links, *IEEE Journal on Selected Areas in Communications*, **17**(5), 793–813.

887. Seskar, I. P. and Mandayam, N. B. (1999) A software radio architecture for linear multiuser detection, *IEEE Journal on Selected Areas in Communications*, **17**(5), 814–823.

888. Hanzo, L. and Streit, J. (1999) Adaptive low-rate wireless videophone schemes, *IEEE Transactions on Circuits and Systems for Video Technology*, **5**(4), 305–318.

889. Brown, C. and Feher, J. (1996) A reconfigurable modem for increased network capacity and video, voice, and data transmission over GSM PCS, *IEEE Transactions on Circuits and Systems for Video Technology*, **6**(2), 215–224.

890. Streit, J. and Hanzo, L. (1997) Dual-mode vector-quantized low-rate cordless videophone systems for indoors and outdoors applications, *IEEE Transactions on Vehicular Technology*, **46**(2), 340–357.

891. Brown, E. R. (1998) RF-MEMS switches for reconfigurable integrated circuits, *IEEE Transactions on Microwave Theory and Techniques*, **46**(11), 1868–1880.

892. Von Herzen, B. (1998) Signal processing at 250 MHz using high-performance FPGAs, *IEEE Transactions on Very Large Scale Integration (VLSI) Systems*, **6**(2), 238–246.

893. Chi-Kuang Chen, Po-Chih Tseng, Yung-Chil Chang and Liang-Gee Chen (2001) A digital signal processor with programmable correlator array architecture for third generation wireless communication system, *IEEE Transactions on Circuits and Systems II: Analog and Digital Signal Processing*, **48**(12), 1110–1120.

894. Minnis, B. J. and Moore, P. A. (2003) A highly digitized multimode receiver architecture for 3G mobiles, *IEEE Transactions on Vehicular Technology*, **52**(3), 637–653.

895. Papapolymerou, J., Lange, K. L., Goldsmith, C. L., Malczewski, A. and Kleber, J. (2003) Reconfigurable double-stub tuners using MEMS switches for intelligent RF front-ends, *IEEE Transactions on Microwave Theory and Techniques*, **51**(1), 271–278.

896. Palicot, J. and Roland, C. (2000) A two step architecture for an adaptive receiver, *First International Conference on 3G Mobile Communication Technologies* (IEE Conf. Publ. No. 471), 27–29 March 2000, pp. 301–305.

897. Miyamoto, R. Y., Yongxi Qian and Itoh, T. (2001) A reconfigurable active retrodirective/direct conversion receiver array for wireless sensor systems, *IEEE MTT-S International Microwave Symposium Digest*, 20–25 May 2001, **2**, 1119–1122.

898. Jae Ho Jung and Deuk Su Lyu (2002) An architecture of a reconfigurable transceiver based on digital IF for WCDMA and IS-95 base stations, *The 5th International Symposium on Wireless Personal Multimedia Communications*, 27–30 October 2002, **2**, 831–834.

899. Bian, Y. Q., Nix, A. R. and McGeehan, J. P. (2002) Base station 2-dimensional reconfigurable receiver architecture for DS-CDMA, *International Zurich Seminar on Broadband Communications, Access, Transmission, Networking*, 19–21 February 2002, pp. 36-1–36-6.

900. Veljanovski, R., Singh, J. and Faulkner, M. (2003) Design and implementation of reconfigurable filter, *Electronics Letters*, **39**(10), 813–814.

901. Palicot, J. and Roland, C. (2003) A new concept for wireless reconfigurable receivers, *IEEE Communications Magazine*, **41**(7), 124–132.

902. Baines, R. and Pulley, D. (2003) A total cost approach to evaluating different reconfigurable architectures for baseband processing in wireless receivers, *IEEE Communications Magazine*, **41**(1), 105–113.

903. Tao Long and Shanbhag, N. R. (1999) Low-power CDMA multiuser receiver architectures, *IEEE Workshop on Signal Processing Systems* (SiPS 99), 20–22 October 1999, pp. 493–502.

904. Swanchara, S. and Athanas, P. (1999) A methodical approach for stream-oriented configurable signal processing, *Proceedings of the 32nd Annual Hawaii International Conference on System Sciences* (HICSS-32), 5–8 January 1999, **Track 3**, 6 pp.

905. Tschanz, J. and Shanbhag, N. R. (1999) A low-power, reconfigurable adaptive equalizer architecture, *Conference Record of the Thirty-Third Asilomar Conference on Signals, Systems and Computers*, 24–27 October 1999, **2**, 1391–1395.

906. Becker, J., Pionteck, T. and Glesner, M. (2000) An application-tailored dynamically reconfigurable hardware architecture for digital baseband processing, *Proceedings of the 13th Symposium on Integrated Circuits and Systems Design*, 18–24 September 2000, pp. 341–346.

907. Rice, M., Dick, C. and Harris, F. (2001) Maximum likelihood carrier phase synchronization in FPGA-based software defined radios, *Proceedings of the IEEE International Conference on Acoustics, Speech and Signal Processing* (ICASSP, 01) 7–11 May 2001, **2**, 889–892.

908. Chadha, K. and Cavallaro, J. R. (2001) A reconfigurable Viterbi decoder architecture, *Conference Record of the Thirty-Fifth Asilomar Conference on Signals, Systems and Computers*, 4–7 November 2001, **1**, 66–71.

909. Srikanteswara, S., Neel, J., Reed, J. H. and Athanas, P. (2001) Soft radio implementations for 3G and future high data rate systems, *IEEE Global Telecommunications Conference*, 25–29 November 2001, **6**, 3370–3374.

910. Sumanen, L. and Halonen, K. (2002) Dual-mode pipeline A/D converter for direct conversion receivers, *Electronics Letters*, **38**(19), 1101–1103.

911. Hentschel, T., Henker, M. and Fettweis, G. (1999) The digital front-end of software radio terminals, *IEEE Personal Communications*, **6**(4), 40–46.

912. Colsell, S. and Edwards, R. (2001) A comparative study of reconfigurable digital and analogue technologies for future mobile communication systems, *Second International Conference on 3G Mobile Communication Technologies* (IEE Conf. Publ. No. 477), 26–28 March 2001, pp. 302–305.

913. Bucknell, P. and Pitchers, S. (2000) Overcoming the challenges of lower layer protocol reconfiguration for software radio based mobile terminals, *First International Conference on 3G Mobile Communication Technologies* (IEE Conf. Publ. No. 471), 27–29 March 2000, pp. 331–335.

914. Moessner, K. and Tafazolli, R. (2000) Terminal reconfigurability–the software download aspect, *First International Conference on 3G Mobile Communication Technologies* (IEE Conf. Publ. No. 471) 27–29 March 2000, pp. 326–330.

915. Chao, C. Y. and Ilyas, M. (1989) Fast reconfigurable communication networks, *Eighth Annual International Phoenix Conference on Computers and Communications*, 22–24 March 1989, pp. 248–252.

916. Ward, C. R. (1990) A multi-mode programmable communication radio receiver element, *Proceedings of the Tactical Communications Conference*, 24–26 April 1990, **1**, 179–204.

917. Luecke, J. and Jordan, M. (1990) Programmable digital receiver architecture for high data rate and multichannel communications applications, *IEEE Military Communications Conference* (MILCOM '90) 30 September–3 October 1990, **3**, 1256–1260.

918. Ning Zhang and Brodersen, R. W. (2000) Architectural evaluation of flexible digital signal processing for wireless receivers, *Thirty-Fourth Asilomar Conference on Signals, Systems and Computers*, 29 October–1 November 2000, **1**, 78–83.

919. Veljanovski, R., Singh, J. and Faulkner, M. (2002) DSP and ASIC implementation of a channel filter for a 3G UTRA-TDD system, *The 13th IEEE International Symposium on Personal, Indoor and Mobile Radio Communications*, 15–18 September 2002, **3**, 1447–1451.

920. Oswald, M. T., Hagness, S. C., Van Veen, B. D. and Popovic, Z. (2002) Reconfigurable single-feed antennas for diversity wireless communications, *IEEE Antennas and Propagation Society International Symposium*, 16–21 June 2002, **1**, 469–472.

921. Veljanovski, R., Stojcevski, A., Singh, J., Zayegh, A. and Faulkner, M. (2002) Reconfigurable architecture for UTRA-TDD system, *Electronics Letters*, **38**(25), 1732–1733.

922. Gupta, C. N., Evans, J. B. and Minden, G. J. (1993) Reconfigurable ATM transmitter/receiver implementation, *Electronics Letters*, **29**(24), 2139–2140.

923. Latva-aho, M., Juntti, M. and Oppermann, I. (1998) Reconfigurable adaptive RAKE receiver for wideband CDMA systems, 48th *IEEE Vehicular Technology Conference* (VTC 98), 18–21 May 1998, **3**, 1740–1744.

924. Haruyama, S. and Morelos-Zaragoza, R. (2001) A software defined radio platform with direct conversion: SOPRANO, *IEEE VTS 54th Vehicular Technology Conference*, 7–11 October 2001, **3**, 1558–1560.

925. Miranda, H. C., Pinto, P. C. and Silva, S. B. (2003) A self-reconfigurable receiver architecture for software radio systems, *Radio and Wireless Conference* (RAWCON '03) 10–13 August 2003, pp. 241–244.

926. Luecke, J. and Jordan, M. (1990) Programmable digital communications receiver architecture for high data rate avionics and ground applications, *IEEE/AIAA/NASA 9th Digital Avionics Systems Conference*, 15–18 October 1990, pp. 552–556.

927. Suzuki, F., Koizumi, H., Nishino, K. and Yasuura, H. (1997) A method for reconfigurable multimedia equipment development using inverse problem, *IEEE Pacific Rim Conference on Communications, Computers and Signal Processing*, 20–22 August 1997, **1**, 74–80.

928. Goel, M., Appadwedula, S., Shambhag, N. R., Ramchandran, K. and Jones, D. L. (1999) A low-power multimedia communication system for indoor wireless applications, *IEEE Workshop on Signal Processing Systems*, 20–22 October 1999, pp. 473–482.

929. Fu-Yen Kuo and Chung-Wei Ku (2000) Software radio based reconfigurable correlator/FIR filter for CDMA/TDMA receiver, *The 2000 IEEE International Symposium on Circuits and Systems* (ISCAS 2000) Geneva, 28–31 May 2000, **1**, 112–115.

930. Jondral, F., Wiesler, A. and Machauer, R. (2000) A software defined radio structure for 2nd and 3rd generation mobile communications standards, *IEEE Sixth International Symposium on Spread Spectrum Techniques and Applications*, 6–8 September 2000, **2**, 637–640.

931. Tarver, B., Christensen, E. and Miller, A. (2001) Software defined radios (SDR) platform and application programming interfaces (API), *IEEE Military Communications Conference*, 28–31 October 2001, **1**, 153–157.

932. Yanjun Hu and Jinkang Zhu (2001) Active dynamic multiuser detection for WCDMA communication systems, *IEEE VTS 54th Vehicular Technology Conference*, **1**, pp. 497–501.

933. Darbel, N., Rasse, Y., Bastidas-Garcia, O., Faux, G., Jubelin, B. and Carrie, M. (2002) Reconfigurable low power cell search engine for UMTS-FDD mobile terminals, *IEEE Workshop on Signal Processing Systems*, 16–18 October 2002, pp. 171–176.

934. Veljanovski, R., Singh, J. and Faulkner, M. (2002) A proposed reconfigurable digital filter for a mobile station receiver, *IEEE Global Telecommunications Conference*, 17–21 November 2002, pp. 524–528.

935. Stojcevski, A., Singh, J. and Zayegh, A. (2002) A reconfigurable analog-to-digital converter for UTRA-TDD mobile terminal receiver, *The 2002 45th Midwest Symposium on Circuits and Systems*, 4–7 August 2002, pp. 613–616.

936. Velianovski, R., Singh, J. and Faulkner, M. (2002) A low-power reconfigurable digital pulse-shaping filter for an utra-tdd mobile terminal receiver, *The 2002 45th Midwest Symposium on Circuits and Systems*, 4–7 August 2002, pp. 1–4.

937. Miyamoto, R. Y., Leong, K. M. K. H., Seong-Sik Jeon, Yuanxun Wang and Itoh, T. (2002) An adaptive multi-functional array for wireless sensor systems, *IEEE MTT-S International Microwave Symposium Digest*, 2–7 June 2002, pp. 1369–1372.

938. Minnis, B. and Moore, P. (2002) A reconfigurable receiver architecture for 3G mobiles, *IEEE Radio Frequency Integrated Circuits (RFIC) Symposium*, 2–4 June 2002, pp. 187–190.

939. Xinyu Xu, Ke Wu and Bosisio, R. G. (2003) Software defined radio receiver based on six-port technology, *IEEE MTT-S International Microwave Symposium*, 8–13 June 2003, **2**, 1059–1062.

940. Veljanovski, R., Stojccvski, A., Singh, J., Faulkner, M. and Zayegh, A. (2003) A highly efficient reconfigurable architecture for an utra-tdd mobile station receiver, *International Symposium on Circuits and Systems* (ISCAS '03), 25–28 May 2003, pp. 45–48.

941. Arnold, J., Caldow, A. and Harman, K. (2003) A reconfigurable 100 Mchip/s spread spectrum receiver, *IEEE International Conference on Acoustics, Speech and Signal Processing* (ICASSP '03), 6–10 April 2003, **2**, pp. 445–448.

942. Bianco, A., Dassatti, A., Martina, M., Molino, A. and Vacca, F. (2003) A reconfigurable, power-scalable RAKE receiver IP for W-CDMA, *Asia and South Pacific Design Automation Conference*, 21–24 January 2003, pp. 499–502.

943. Palico, J. and Roland, C. (2003) FFT: a basic function for a reconfigurable receiver, *10th International Conference on TeleCommunications*, February 23–March 1 2003, pp. 898–902.

944. Stojcevski, A., Singh, J. and Zayegh, A. (2003) Reconfigurable ADC for 3 -G UTRA-TDD mobile receiver, *Southwest Symposium on Mixed-Signal Design*, 23–25 February 2003, pp. 27–31.

945. Mc Cormick, A. C., Grant, P. M., Thompson, J. S., Arslan, T. and Erdogan, A. T. (2002) Low power receiver architectures for multi-carrier CDMA, *IEE Proceedings Circuits, Devices and Systems*, **149**(4), 227–233.

946. Hanzo, L. and Streit, J. (1995) Adaptive low-rate wireless videophone schemes, *IEEE Transactions on Circuits and Systems for Video Technology*, **5**(4), 305–318.

947. Dick, C. and Harris, F. J. (1999) Configurable logic for digital communications: some signal processing perspectives, *IEEE Communications Magazine*, **37**(8), 107–111.

948. Blaickner, A. and Grunbacher, H. (2000) On reconfigurable methods and gate array based solutions of fast forward error correction systems for software radio and set-top-box applications, *IEEE Transactions on Consumer Electronics*, **46**(4), 994–998.

949. Chi-Kuang Chen, Po-Chih Tseng, Yung-Chil Chang and Liang-Gee Chen (2001) A digital signal processor with programmable correlator array architecture for third generation wireless communication systems, *IEEE Transactions on Circuits and Systems II: Analog and Digital Signal Processing*, **48**(12), 1110–1120.

950. Grayver, E. and Daneshrad, B. (2000) A reconfigurable 8 GOP ASIC architecture for high-speed data communications, *IEEE Journal on Selected Areas in Communications*, **18**(11), 2161–2171.

951. TMS320C6000 power consumption summary (Online) Available: http://www-s.ti.com/sc/psheets/spra486b/spra486b.pdf.

952. (Online). Available: http://brass.cs.berkeley.edu/reproc.html.

953. (Online). Available: http://www.acm.org/crossroads/xrds5-3//rcconcept.html.

954. Athanas, P. M. and Silverman, H. F. (1993) Processor reconfiguration through instruction-set metamorphosis, *Computer*, **26**, 11–18.

955. Hauck, S., Fry, T. W., Hosler, M. M. and Kao, J. P. (1997) The Chimaera reconfigurable functional unit, in *Proceedings of the IEEE Symposium on FPGAs for Custom Computing Machines*, pp. 87–96.

956. DeHon, A. (1999) Balancing interconnect and computation in a reconfigurable computing array (or, why you don't really want 100% LUT utilization), in *Proceedings of FPGA'99*, pp. 69–78.

957. Zhang, H., Prabhu, V., George, V., Wan, M., Benes, M. and Abnous, A. (2000) A 1 V heterogeneous reconfigurable processor IC for baseband wireless applications, in *Proceedings of ISSCC 2000*.

958. Virtex power estimator 1.5 (Online). Available: http://www.xilinx.com/cgi-bin/powerweb.pl.

959. Virtex 2.5 V programmable gate arrays, XILINX databook, 1999. rev. 1.9.

960. TMS320C6201, TMS320C6201B digital signal processors (Online). Available: http://www.ti.com/sc/docs/products/dsp/tms320c6201.html.

961. TMS320C62X assembly benchmarks (Online). Available: http://www.ti.com/sc/docs/products/dsp/c6000/62bench.htm.

962. (Online). Available: http://www.xilinx.com/ipcenter/catalog/logicore/docs/mult_vgen_v1_0.pdf.

963. 106 Ms/s digital filter. Graychip (Online). Available: http://www.graychip.com/GC2011/GC2011.html.

964. Master, P. (1999) Adaptive computing processors for next generation wireless, *Wireless Design and Development*.

965. Haykin, S. (1995) *Adaptive Filter Theory*. Prentice Hall, Englewood Cliffs, NJ.

966. Grayver, E. and Daneshrad, B. (1998) Direct digital frequency synthesis using a modified CORDIC, in *Proceedings of IEEE ISCAS*.

967. Tan, L. K. *et al.* (1995) An 800-MHz quadrature digital synthesizer, *IEEE Journal on Solid-State Circuits*, **30**, 1463–1473.

968. Volder, J. E. (1959) The CORDIC trigonometric computing technique, *IRE Transactions on Electronics and Computers*, **EC-8**, 330–334.

969. Oppenheim, A. V. and Schafer, R. W. (1989) *Discrete-Time Signal Processing*. Prentice Hall, Englewood Cliffs, NJ.

970. Gunn, J. E., Barron, K. S. and Ruczczyk, W. (1999) A low-power DSP core-based software radio architecture, *IEEE Journal on Selected Areas in Communications*, **17**(4), 574–590.

971. Mitola, J., III (2000) SDR architecture refinement for JTRS, *IEEE Military Communications Conference Proceedings*, 22–25 October 2000, **1**, 214–218.

972. Mitola, J. (1995) The software radio architecture, *IEEE Communications Magazine*, **33**, 26–38.

973. Raytheon TI System (RTIS), Purdue University, and University of Michigan; DARPA Small Unit Operation (SUO) Software Radio and Algorithms Development Program: Open Review 2 Presentation, November 18, 1997.

974. Raytheon TI System (RTIS), Purdue University, and University of Michigan; DARPA Small Unit Operation (SUO) Software Radio and Algorithms Development Program: Open Review 1 Presentation, September 17, 1997.

975. Walden, R. H. (1999) Analog-to-digital converter survey and analysis, *IEEE Journal on Selected Areas in Communications*, **JSAC-17**, 539–550.

976. Candy, J. and Temes, G. (1992) *Oversampling Delta-Sigma Data Converters: Theory, Design, and Simulation*. IEEE Press, New York.

977. Texas Instruments Website at http://www.ti.com/sc/docs/news/1996/96037a.htm#support.

978. Lee, W. *et al.* (1997) A 1 V DSP for wireless communication, in *ISSCC Digital Technical Papers*, pp. 92.

979. Milstein, L. B. (1988) Interference rejection techniques in spread spectrum communication, *Proceedings of the IEEE*, **76**.

980. Skidmore, W., Snajder, M., Indoo, P. and Klose, D. R. (2000) A NATO-interoperable, tactical command and control system for land forces, *EUROCOMM 2000, Information Systems for Enhanced Public Safety and Security*. IEEE/AFCEA, 17 May 2000, pp. 278–282.

981. Cook, P. G. and Bonser, W. (1999) Architectural overview of the SPEAKeasy system, *IEEE Journal on Selected Areas in Communications*, **17**(4), 650–661.

982. Ivers, A. and Smith, D. (1997) A practical approach to the implementation of multiple radio configurations utilizing reconfigurable hardware and software building blocks, *MILCOM 97 Proceedings*, 2–5 November 1997, **3**, 1327–1332.

983. Place, J., Kerr, D. and Schaefer, D. (2000) Joint Tactical Radio System, *21st Century Military Communications Conference Proceedings* (MILCOM 2000), 22–25 October 2000, **1**, 209–213.

984. Tarver, B., Christensen, E. and Miller, A. (2000) The Wireless Information Transfer System (WITS) architecture for the Digital Modular Radio (DMR) software defined radio (SDR), *21st Century Military Communications Conference* (MILCOM 2000), 22–25 October 2000, **1**, 226–230.

985. Ward, C. R. (1990) A multi-mode programmable communication radio receiver element, *Proceedings of the Tactical Communications Conference*, 24–26 April 1990, **1**, 179–204.

986. Gudaitis, M. S. and Hinman, R. D. (1997) Tactical software radio concept, *MILCOM 97 Proceedings*, 2–5 November 1997, **3**, 1207–1211.

987. Vidano, R. (1997) SPEAKeasy II-an IPT approach to software programmable radio development, *MILCOM 97 Proceedings*, 2–5 November 1997, **3**, 1212–1215.

988. Wiesler, A. and Jondral, F. K. (2002) A software radio for second- and third-generation mobile systems, *IEEE Transactions on Vehicular Technology*, **51**(4), 738–748.

989. Fox, P. W. (2000) Software defined radios–Motorola's Wireless Information Transfer System (WITS), *EUROCOMM 2000 Information Systems for Enhanced Public Safety and Security*. IEEE/AFCEA, 17 May 2000, pp. 43–46.

990. Reichhart, S. P., Youmans, B. and Dygert, R. (1999) The software radio development system, *IEEE Personal Communications*, **6**(4), 20–24.

991. Sanchez, R., Sparks, C., Malinimohan, K., Roberts, J., Plumb, R. and Petr, D. (1999) The rapidly deployable radio network, *IEEE Journal on Selected Areas in Communications*, **17**(4), 689–703.

992. Diamond, D. B., Beale, F. T. and Robertson, M. R. (1995) Keys to the digital battlefield: automated requirements analysis for Force, *XXI IEEE Military Communications Conference* (MILCOM'95) 5–8 November 1995, **3**, 1103–1107.

993. Bard, J. D. and Mears, T. J., II (2000) A CORBA-based DAMA modem for the Digital Modular Radio (DMR), *MILCOM 2000, 21st Century Military Communications Conference Proceedings*, 22–25 October 2000, **1**, 236–240.

994. Christensen, E., Miller, A. and Wing, E. (2000) Waveform Application Development Process for software defined radios, *MILCOM 2000, 21st Century Military Communications Conference Proceedings*, 22–25 October 2000, **1**, 231–235.

995. Tarver, B., Christensen, E., Miller, A. and Wing, E. R. (2001) Digital modular radio (DMR) as a maritime/fixed Joint Tactical Radio System (JTRS), *IEEE Military Communications Conference* (MILCOM 2001), 28–31 October 2001, **1**, 163–167.

996. Eyermann, P. A. and Powell, M. A. (2001) Maturing the software communications architecture for JTRS, *IEEE Military Communications Conference* (MILCOM 2001), 28–31 October 2001, **1**, 158–162.

997. Tarver, B., Christensen, E. and Miller, A. (2001) Software defined radios (SDR) platform and application programming interfaces (API), *IEEE Military Communications Conference* (MILCOM 2001) 28–31 October 2001, **1**, 153–157.

998. Roggow, S. W. (1990) Software architecture for a miniature radio system, *IEEE Military Communications Conference*, 30 September–3 October 1990, **3**, 1054–1057.

999. Bartholomew, R. G. (1996) An open software architecture in a DoD standard GPS receiver, *IEEE Position Location and Navigation Symposium*, 22–26 April 1996, pp. 205–212.

1000. Cowen-Hirsch, R., Shrum, D., Davis, B., Stewart, D. and Kontson, K. (2000) Software radio: evolution or revolution in spectrum management, *MILCOM 2000, 21st Century Military Communications Conference Proceedings*, 22–25 October 2000, **1**, 8–14.

1001. Chuprun, S., Bergstrom, C. and Fette, B. (2000) SDR strategies for information warfare and assurance, *MILCOM 2000, 21st Century Military Communications Conference Proceedings*, 22–25 October 2000, **2**, 1219–1223.

1002. Pearson, D. M. (2001) SDR (systems defined radio): how do we get there from here? *IEEE Military Communications Conference* (MILCOM 2001), 28–31 October 2001, **1**, 571–575.

1003. Gifford, S., Kleider, J. E. and Chuprun, S. (2001) Broadband OFDM using 16-bit precision on a SDR platform, *IEEE Military Communications Conference* (MILCOM 2001), 28–31 October 2001, **1**, 180–184.

1004. Patti, J. J., Husnay, R. M. and Pintar, J. (1999) A smart software radio: concept development and demonstration, *IEEE Journal on Selected Areas in Communications*, **17**(4), 631–649.

1005. Razavilar, J., Rashid-Farrokhi, F. and Liu, K. J. R. (1999) Software radio architecture with smart antennas: a tutorial on algorithms and complexity, *IEEE Journal on Selected Areas in Communications*, **17**(4), 662–676.

1006. Gebauer, T. and Gockler, H. (1995) Channel-individual adaptive beamforming for mobile satellite communications, *IEEE Journal on Selected Areas in Communications*, **13**(2), 439–448.

1007. Winters, J. H. (1998) Smart antennas for wireless systems, *IEEE Personal Communications Magazine*, **5**, 23–27.

1008. Widrow, B. and Stearns, S. (1985) *Adaptive Signal Processing*. Prentice Hall, Englewood Cliffs, NJ.

1009. Donoho, D. L. (1981) On minimum entropy deconvolution, in *Applied Time Series Analysis II*. Academicess, New York, pp. 565–608.

1010. Giannakis, G. B. and Mendel, J. M. (1989) Identification of nonminimum phase systems using via higher order statistics, *IEEE Transactions on Acoustics, Speech and Signal Processing*, **37**, 360–377.

1011. Nikias, C. L. (1988) ARMA bispectrum approach to nonminimum phase system identification, *IEEE Transactions on Acoustics, Speech and Signal Processing*, **36**, 513–525.

1012. Tugnait, J. K. (1987) Identification of linear stochastic systems via second and fourth-order cumulant matching, *IEEE Transactions On Information Theory*, **IT-33**, 393–407.

1013. Benveniste, A., Goursat, M. and Ruget, G. (1980) Robust identification of a nonminimum phase system: Blind adjustment of a linear equalizer in data communications, *IEEE Transactions on Automatic Control*, **AC-25**, 385–399.

1014. Chen, Y. and Nikias, C. L. (1992) Blind equalization with criterion with memory nonlinearity, *Optical Engineering*, **31**, 1200–1210.

1015. Godard, N. (1980) Self-recovering equalization and carrier tracking in two-dimensional data communication systems, *IEEE Transactions On Communications*, **COM-28**, 1867–1875.

1016. Gustafsson, F. and Wahlberg, B. (1995) Blind equalization by direct examination of the input sequence, *IEEE Transactions On Communications*, **43**, 2213–2222.

1017. Picchi, G. and Prati, G. (1987) Blind equalization and carrier recovery using a 'stop-and-go' decision-directed algorithm, *IEEE Transactions On Communications*, **COM-35**, 877–887.

1018. Sato, Y. (1975) A method of self-recovering equalization for multi-level amplitude modulation, *IEEE Transactions On Communications*, **COM-23**, 679–682.

1019. Shalvi, O. and Weinstein, E. (1990) New criteria for blind deconvolution of nonminimum phase systems (channels), *IEEE Transactions On Information Theory*, **36**, 312–321.

1020. Treichler, J. R. and Agee, B. G. (1983) A new approach to multipath correction of constant modulus signals, *IEEE Transactions on Acoustics, Speech and Signal Processing*, **ASSP-31**, 349–372.

1021. Treichler, J. R. and Larimore, M. G. (1985) New processing techniques based on the constant modulus adaptive algorithm, *IEEE Transactions on Acoustics, Speech and Signal Processing*, **ASSP-33**, 420–431.

1022. Verdu, S., Anderson, B. D. O. and Kennedy, R. A. (1993) Blind equalization without gain identification, *IEEE Transactions On Information Theory*, **39**, 292–297.

1023. Tugnait, J. K. (1995) On fractionally-spaced blind adaptive equalization under symbol timing offsets using Godard and related equalizers, in *Proceedings of IEEE ICASSP'95*, May 1995, **3**, 1976–1979.

1024. Petersen, B. R. and Falconer, D. D. (1994) Suppression of adjacent-channel, co-channel, and intersymbol interference by equalizers and linear combiners, *IEEE Transactions On Communications*, **42**, 3109–3118.

1025. Yang, J. and Roy, S. (1994) On joint transmitter and receiver optimization for multiple-input-multiple output (MIMO) transmission systems, *IEEE ICASSP'94*, Adelaide, Australia, April 1994, pp. 317–320.

1026. Treichler, J. R. and Larimore, M. G. (1985) The tone capture properties of CMA-based interference suppressors, *IEEE Transactions on Acoustics, Speech and Signal Processing*, **ASSP-33**, 946–958.

1027. Weinstein, E., Swami, A., Giannakis, G. and Shamsunder, S. (1994) Multi-channel ARMA processes, *IEEE Transactions on Signal Processing*, **42**, 898–913.

1028. Weinstein, E., Oppenheim, A. V., Feder, M. and Buck, J. R. (1994) Iterative and sequential algorithms for multisensor signal enhancement, *IEEE Transactions on Signal Processing*, **42**, 846–859.

1029. Yellin, D. and Weinstein, E. (1994) Criteria for multichannel signal separation, *IEEE Transactions on Signal Processing*, **42**, 2158–2168.

1030. Naguib, A. Paulraj, A. and Kailath, T. (1994) Capacity improvement with base-station antenna arrays in cellular CDMA, *IEEE Transactions on Vehicular Technology*, **43**.

1031. Moulines, E., Duhamel, P., Cardoso, J. and Mayrargue, S. (1995) Subspace methods for the blind identification of multichannel FIR filters, *IEEE Transactions on Signal Processing*, **43**, 516–525.

1032. Guerin, R. (1988) Queueing-blocking system with two arrivals stream and guard channels, *IEEE Transactions On Communications*, **36**, 153–163.

1033. Rashid-Farrokhi, F., Tassiulas, L. and Liu, K. J. R. (1996) Joint optimal power control and beamforming for wireless networks with antenna arrays, in *Proceedings of the IEEE Global Communications Conference (GLOBECOM' 96)*, London, England, November 1996, pp. I-555–I-559.

1034. Farrokhi, F., Tassiulas, L. and Liu, K. J. R. (1998) Joint optimal power control and beamforming in wireless networks with antenna arrays, *IEEE Transactions On Communications*, **46**, 1313–1324.

1035. Zander, J. (1992) Distributed cochannel interference control in cellular radio systems, *IEEE Transactions on Vehicular Technology*, **41**, 305–311.

1036. Wu, J., Sheng, W.-X., Chan, K.-P., Chung, W.-K., Cheng, K.-K. M. and Wu, K.-L. (2002) Smart antenna system implementation based on digital beam-forming and software radio technologies, *IEEE MTT-S International Microwave Symposium Digest*, 2–7 June 2002, **1**, 323–326.

1037. Green, P. J. and Taylor, D. P. (2002) Smart antenna software radio test system, *The First IEEE International Workshop on Electronic Design, Test and Applications*, 29–31 January 2002, pp. 68–72.

1038. Perez-Neira, A., Mestre, X. and Fonollosa, J. R. (2001) Smart antennas in software radio base stations, *IEEE Communications Magazine*, **39**(2), 166–173.

1039. Kohno, R. (1997) Software antenna and its communication theory for mobile radio communications, *IEEE International Conference on Personal Wireless Communications*, 17–19 December 1997, pp. 227–233.

1040. Xu, B., Vu, T. B. and Jonas, G. (1998) Implementation of a smart antenna using TMS320C80 DSPs for mobile communications, *Fourth International Conference on Signal Processing* (ICSP '98), 12–16 October 1998, **1**, 355–358.

1041. Ousterhout, J. K. (1994) *Tcl and the Tk Toolkit*, Addison Wesley, Reading, MA.

1042. Turlett, T., Bentzen, H. J. and Tennenhouse, D. (1999) Toward the software realization of a GSM base station, *IEEE Journal on Selected Areas in Communications*, **17**(4), 603–612.

1043. GSM 06.10–European digital cellular telecommunications system (phase 2); Full rate speech transcoding, *ETS 300 580-2, European Telecommunication Standard*, September 1994.

1044. Murota, K. and Hirade, K. (1981) GMSK modulation for digital radio telephony, *IEEE Transactions On Communications*, **COM-29**, 1044–1050.

1045. Vaidyanathan, P. P. (1990) Multirate digital filters, filter banks, polyphase networks, and applications: A tutorial, *Proceedings of the IEEE*, **78**.

1046. Karn, P. (1995) Convolutional decoders for amateur packet radio, in *Proceedings of ARRL'1995 Digital Communications Conference*.

1047. 3GPP technical specifications – 3G TR 25.944: Channel coding and multiplexing examples, Release 1999.

1048. 3GPP technical specifications – 3G TS 25.211: Physical channels and mapping of transport channels onto physical channels (FDD), Release 1999.

1049. (Online). http://www.spec.org.

1050. Adachi, F., Sawahashi, M. and Suda, H. (1998) Wideband DS-CDMA for Next-Generation Mobile Communications Systems, *IEEE Communications Magazine*, 56–69.

1051. Chapman, K., Hardy, P., Miller, A. and George, M. (2000) CDMA Matched Filter Implementation in Virtex Devices, Xilinx XAPP212(v1.0).

1052. Esmailzadeh, R. and Nakagawa, M. (1993) Pre-RAKE Diversity Combination for Direct Sequence Spread Spectrum Mobile Communications Systems, *IEEE Transactions On Communications*, **E76-B** (8), 1008–1015.

1053. Reves, X., Gelonch, A. and Casadevall, F. (1999) Reconfigurable Hardware Platform for Software Radio Applications (SHaRe) in Mobile Communications Environments, *Proceedings of the ACTS Mobile Communication Summit*, Sorrento, Italy, June 1999.

1054. XILINX Virtex™2.5 V Field Programmable Gate Arrays (2000).

1055. XILINX XC4000E XC4000X Series Field Programmable Gate Arrays (1999).

1056. Cummings, M. and Haruyama, S. (1999) FPGA in the Software Radio, *IEEE Communications Magazine*, **37**(2), 108–112.

1057. Taylor, C. (1997) Using Software Radio in 3rd Generation Communications System, *ACTS Mobile Communications Summit*, Aalborg, Denmark, October 1997.

1058. Reves, X., Gelonch, A. and Casadevall, F. (2001) Software radio implementation of a DS-CDMA indoor subsystem based on FPGA devices, *12th IEEE International*

Symposium on Personal, Indoor and Mobile Radio Communications, 30 September– 3 October 2001, **1**, D-86–D-90.

1059. Glisic, S., Nikolic, Z. and Dimitrijevic, B. (1999) Adaptive self reconfigurable interference suression schemes for CDMA wireless networks, *IEEE Transactions On Communications*, **47**(4), 598–607.

1060. Milstein L. B. and Wang J. (1995) Interference suression for CDMA overlays of narrowband waveforms, in S. G. Glisic and P. A. Leänen (Eds.) *Code Division Multiple Access Communications*, Kluwer Academic Publishers, 147–160.

1061. Grieco, D. (1994) The capacity achievable with a broadband CDMA microcell underlay to an existing cellular macrosystem, *IEEE JSAC*, **12**(4), 744–750.

1062. Milstein, L. B., Schilling, D. L., Pickholtz, R. L., Erzeg, V., Kullback, M., Kanterakis, E., Fishman, D., Biederman, W. H. and Salerno D. (1992) On the feasibility of a CDMA overlay for personal communications networks, *IEEE Journal on Selected Areas in Communications*, **10**, 655–668.

1063. Milstein, L. B. (1988) Interference rejection techniques in spread spectrum communications. *Proceedings of the IEEE*, **76**, 657–971.

1064. Ketchum, J. W. and Proakis, J. G. (1982) Adaptive algorithm for estimating and suressing narrowband interference in PN spread spectrum systems. *IEEE Transactions On Communications*, **COM-30**, 913–924.

1065. Milstein, L. B. and Iltis, R. A. (1986) Signal processing for interference rejection in spread spectrum communications. *IEEE ASSP magazine*, pp. 18–31.

1066. Gupta, A. K. (1985) On suression of sinusoidal signal in broad band noise. *IEEE Transactions on Acoustics Speech and Signal Processing*, **ASSP-33**(4).

1067. Li, L. and Milstein, L. B. (1983) Rejection of pulsed cw interference in PN spread spectrum signals using complex adaptive filters, *IEEE Transactions On Communications*, **COM-31**, 10–20.

1068. Li, L. and Milstein, L. B. (1982) Rejection of narrowband interference in PN spread spectrum signals using transversal filters, *IEEE Transactions On Communications*, **COM-30**, 925–928.

1069. Masry, E. (1985) Closed form analytical results for the rejection of narrow band interference in PN spread spectrum systems – part II: Linear interpolator filters, *IEEE Transactions On Communications*, **COM-33**, 10–19.

1070. Masry, E. (1984) Closed form analytical results for the rejection of narrow band interference in PN spread spectrum systems – part I: Linear prediction filters, *IEEE Transactions On Communications*, **COM-32**, 888–896.

1071. Vijyan, R. and Poor, H. V. (1991) Nonlinear techniques for interference suppression in spread spectrum systems, *IEEE Transactions On Communications*, **COM-38**, 1060–1065.

1072. Garth, L. and Poor, H. V. (1992) Narrowband interference suppression in impulsive channels, *IEEE Transactions on Aerospace and Electronic Systems*, **28**, 15–34.

1073. Poor, H. V. (1988) *An Introduction to Signal Detection and Estimation*, Springer-Verlag.

1074. Masreliez, C. J. (1975) Aroximate non-Gaussian filtering with linear state and observation relations, *IEEE Transactions on Automatic Control*, pp. 107–110.

1075. Hasan, M. A. *et al.* (1994) A narrowband interference canceller with adjustable centre weight, *IEEE Transactions On Communications*, **42**(2, 3, 4), 877–880.

1076. Doherty, J. F. (1994) Linearly constrained direct-sequence spread-spectrum interference rejection, *IEEE Transactions On Communications*, **42**(2, 3, 4), 865–871.

1077. Glisic, S. G. *et al.* (1995) Rejection of a FH signal in a DS spread-spectrum system using complex adaptive filters, *IEEE Transactions On Communications*, **43**(5), 1982–1991.

1078. Glisic, S. G. *et al.* (1995) Rejection of a frequency sweeping signal in a DS spread-spectrum system using complex adaptive filters, *IEEE Transactions On Communications*, **43**(1), 136–145.

1079. Ansari, A. *et al.* (1994) Performance study of maximum likelihood receivers and transversal filters for the detection of direct-sequence spread-spectrum signal in narrowband interference, *IEEE Transactions On Communications*, **42**(2, 3, 4), 1939–1946.

1080. Rusch, L. A. *et al.* (1994) A narrowband interference suppression in CDMA spread spectrum communications, *IEEE Transactions On Communications*, **42**(2, 3, 4), 1969–1979.

1081. Glisic, S., Nikolic, Z., Pokrajac, D. and Leanen, P. (1999) Performance Enhancement of DS Spread Spectrum Systems: Two Dimensional Interference Suppression, *IEEE Transactions On Communications*, **47**(10), 1549–1560.

1082. Grieco, D. M. (1994) The Capacity Achievable with a Broadband CDMA Microcellular Underlay to an Existing Cellular Macrosystem, *IEEE JSAC*, **12**(4), 744–750.

1083. Milstein, L. B. *et al.* (1992) On the feasibility of a CDMA overlay for personal communication networks. *IEEE JSAC*, **10**(4), 655–667.

1084. Glisic, S. (2003) *Adaptive WCDMA*, John Wiley & Sons, Ltd, Chichester.

1085. Glisic, S. and Vucetic, B. (1997) *Spread Spectrum CDMA Systems for Wireless Communications*, Artech House, London.

1086. Lee., H. B. (1975) Accuracy Limitations of Hyperbolic Multilateration Systems, *IEEE Transactions Aerospace and Electronic Systems*, **AES-11**(1), 16–29.

1087. Zhao, Y. (1997) *Vehicle Location and Navigation Systems*, Artech House, Norwood, MA.

1088. 3GPP TS 25.305, Stage 2 Functional Specification of UE Positioning, March 2002.

1089. 3GPP TS 25.331, Radio Resource Control (RRC) Protocol Specification, March 2002.

1090. Zhao, Y. (2000) Mobile Phone Location Determination and Its Impact on Intelligent Transportation Systems, *IEEE Transactions on Intelligent Transport System,* **1**(1), 55–67.

1091. TIA/EIA/IS-801-1, Position Determination Service Standard for Dual-Mode Spread Spectrum Systems-Addendum, March 2001.

1092. Beadles, J. T. (1996) Land Based Radio Navigation Systems, http:/www.nav-cen.uscg.mil/loran.loranff.html.

1093. Rappaport, T. S. (1996) *Wireless Communications: Principles & Practice*, Prentice Hall, Upper Saddle River, NJ.

1094. Ivanov, N. and Salischev, V. (1992) The GLONASS System – An Overview, *Journal of Navigation*, **45**(2), 1992, 75–182.

1095. Beser, J. and Balendra, A. (1993) Integrated GPS/GLONASS Navigation Results, *Proceeding of ION GPS-93*, Salt Lake City, UT, September 1993, pp. 171–183.

1096. Kennedy, J. and Sullivan, M. C. (1995) Direction Finding and Smart Antennas Using Software Radio Architecture, *IEEE Communications Magazine*, **3**(5), 62–68.

1097. Oh, S. K. and Un, C. K. (1992) Simple Computational Methods of the AP Algorithm for Maximum Likelihood Localization of Multiple Radiating Sources, *IEEE Transactions on Signal Processing*, **40**(11), 2848–2854.

1098. Ziskind, I. and Wax, M. (1988) Maximum Likelihood Localization of Multiple Sources by Alternating Projection, *IEEE Transactions on Acoustics, Speech and Signal Processing*, **36**(10), 1553–1560.

1099. Bancroft, S. (1985) An Algebraic Solution of the GPS Equations, *IEEE Transactions on Aerospace and Electronic Systems*, **AES-21**(7).

1100. McGillom, C. D. and Rappaport, T. S. (1989) A Beacon Navigation Method for Autonomous Vehicles, *IEEE Transactions on Vehicular Technology*, **38**(3), 132–139.

1101. Raner, W. H. and Schmidt, M. O. (1969) *Fundamentals of Surveying*, Van Nostrand, New York.

1102. Schell, S. V. and Gardner, W. A. (1993) High Resolution Direction Finding, in *Handbook of Statistics: Volume 10*, Elsevier, pp. 755–817.

1103. Schmidt, R. O. (1996) Multiple Emitter Location and Signal Parameter Estimation, *IEEE Transactions on Antennas and Propagation*, **AP-34**(3), 27G–80.

1104. Rao, B. D. and Had, K. V. (1989) Performance Analysis of Root-Music, *IEEE Transactions on Acoustics, Speech and Signal Processing*, **37**(12), 1939–1949.

1105. Roy, R. and Kailath, T. (1989) ESPRIT – Estimation of Signal Parameters via Rotational Invariance Techniques, *IEEE Transactions on Acoustics, Speech and Signal Processing*, **29**(4), 984–995.

1106. Fang, B. T. (1990) Simple Solutions for Hyperbolic and Related Position Fixes, *IEEE Transactions on Aerospace and Electronic Systems*, **26**(5), 748–753.

1107. Smith, O. J. and Abel, J. S. (1987) The Spherical Interpolation Method of Source Localization, *IEEE Journal of Oceanic Engineering*, **OE-12**(1), 246–252.

1108. Chan, Y. I. and Ho, K. C. (1994) A Simple and Efficient Estimator for Hyperbolic Location, *IEEE Transactions on Signal Processing*, **42**(8), 1905–1915.

1109. Foy, W. H, (1976) Position–Location Solutions by Taylor-Series Estimation, *IEEE Transactions on Aerospace and Electronic Systems*, **AES-12**(2), 187–193.

1110. Stoica, P. and Nehorai, A. (1969) MUSIC, Maximum Likelihood, and Cramer–Rao Bound, *Proceedings of the IEEE*, **57**(8), 1408–1418.

1111. Paulraj, A., Roy, A. R. and Kailath, T. (1986) A Subspaced Rotation Approach to Signal Parameter Estimation, *Proceedings of the IEEE*, **74**, 1044–1045.

1112. Xu, G. and Liu., H. (1995) An Effective Transmission Beamforming Scheme for Frequency-Division-Duplex Digital Wireless Communication Systems, Proceedings of *IEEE ICASSP*, pp. 1729–1732.

1113. Li, J. *et al.* (1995) Computationally Efficient Angle Estimation for Signals with Known Waveforms, *IEEE Transactions on Signal Processing*, **43**(9), 2154–2163.

1114. Knapp, C. H. and Carter, G. C. (1976) The Generalized Correlation Method for Estimation of Time Delay, *IEEE Transactions on Acoustics, Speech and Signal Processing*, **ASSP-24**(4), 320–327.

1115. Hahn, W. R. and Tretter, S. A. (1973) Optimum Processing for Delay Vector Estimation in Passive Signal Arrays, *IEEE Transactions On Information Theory*, **IT-19**(5), 608–619.

1116. Gardner, W. A. and Chen, C.-K. (1992) Signal-Selective Time-Difference-of-Arrival Estimation for Passive Location of Man-Made Signal Sources in Highly Corruptive Environments, Part I: Theory and Method, *IEEE Transactions on Signal Processing*, **40**(5), 1168–1184.

1117. Gardner, W. A. and Chen, C.-K. (1992) Signal-Selective Time-Difference-of-Arrival Estimation for Passive Location of Man-Made Signal Sources in Highly Corruptive Environments, Part II: Algorithms and Performance, *IEEE Transactions on Signal Processing*, **40**(5), 1185–1197.

1118. Hahn, W. R. (1975) Optimum Signal Processing for Passive Sonar Range and Bearing Estimation, *Journal of the Acoustical Society of America*, **58**(1), 201–207.

1119. Carter, G. C. (1981) Time Delay Estimation for Passive Sonar Signal Estimation, *IEEE Transactions on Acoustics, Speech and Signal Processing*, **ASSP-29**(3), 463–470.

1120. Abel, J. S. and Smith, J. O. (1989) Source Range and Depth Estimation from Multipath Range Difference Measurements, *IEEE Transactions on Acoustics, Speech and Signal Processing*, **37**(8), 1157–65.

1121. Foy, W. H. (1976) Position–Location Solutions by Taylor-Series Estimation, *IEEE Transactions on Aerospace and Electronic Systems*, **AES-12**(2), 187–193.

1122. Torrieri, D. J. (1984) Statistical Theory of Passive Location Systems, *IEEE Transactions on Aerospace and Electronic* Systems, **AE5-20**(2), 183–198.

1123. Nicholson, D. L. (1976) Multipath and Ducting Tolerant Location Techniques for Automatic Vehicle Location Systems, Proceedings of the *IEEE Vehicular Technology Conference*, Washington, DC, 24–16 March 1976, pp. 151–154.

1124. Smith, O. J. and Abel, J. S. (1987) The Spherical Interpolation Method of Source Localization, *IEEE Journal of Oceanic Engineering*, **OE-12**(1), 246–252.

1125. Smith, J. O. and Abel, J. S. (1987) Closed-Form Least-Squares Source Location Estimation from Range Difference Measurements, *IEEE Transactions on Acoustics, Speech and Signal Processing*, **ASSP-35**(8), 1223–1225.

1126. Friedlander, B. (1987) A Passive Localization Algorithm and Its Accurate Analysis, *IEEE Journal of Oceanic Engineering*, **OE-12**(1), 234–244.

1127. Abel, J. S. (1990) A Divide and Conquer Approach to Least-Squares Estimation, *IEEE Transactions on Aerospace and Electronic Systems*, **26**(2), 423–427.

1128. Scholtz, R. (1980) Frame Synchronization Techniques, *IEEE Transactions On Communications*, **28**(8), 1204–1213.

1129. Sekimoto, T. and Kaneko, H. (1962) Group Synchronization for Digital Transmission Systems, *IEEE Transactions On Communications*, **10**(4), 381–390.

1130. Epstein, R. (1965) An Automatic Synchronization Technique, *IEEE Transactions On Communications*, **13**(4), 547–550.

1131. Hawkes, T., Dupraz, J., Girault, J. and Anglade, P. (1968). Construction and Performace of a PCM Frame Synchronizer with Self-Varying Threshold, *IEEE Transactions On Communications*, **16**(1), 142–148.

1132. Massey, J. (1972) Optimum Frame Synchronization, *IEEE Transactions On Communications*, **20**(2), 115–119.

1133. Lin, P., Yi-Bing Lin and Chlamtac, I. (2003) Modeling frame synchronization for UMTS high-speed downlink packet access, *IEEE Transactions on Vehicular Technology*, **52**(1), 132–141.

1134. Zae Yong Choi and Lee, Y. H. (2002) Frame synchronization in the presence of frequency offset, *IEEE Transactions On Communications*, **50**(7), 1062–1065.

1135. Boukerche, A., Sungbum Hong and Jacob, T. (2002) An efficient synchronization scheme of multimedia streams in wireless and mobile systems, *IEEE Transactions on Parallel and Distributed Systems*, **13**(9).

1136. Newton, N. J. (2002) Data synchronization and noisy environments, *IEEE Transactions On Information Theory*, **48**(8), 2253–2262.

1137. Sai-Weng Lei and Lau, V. K. N. (2002) Performance analysis of adaptive interleaving for OFDM systems, *IEEE Transactions on Vehicular Technology*, **51**(3), 435–444.

1138. van Wijngaarden, A. J. and Morita, H. (2001) Partial-prefix synchronizable codes, *IEEE Transactions On Information Theory*, **47**(5), 1839–1848.

1139. Lampow-Maundy, H. and Fair, I. J. (2000) Frame synchronization in guided scrambling line codes, *IEEE Transactions On Communications*, **48**(12), 1992–1996.

1140. de Lind van Wijngaarden, A. J. and Willink, T. J. (2000) Frame synchronization using distributed sequences, *IEEE Transactions On Communications*, **48**(12), 2127–2138.

1141. Meng-Han Hsieh and Che-Ho Wei (1999) A low-complexity frame synchronization and frequency offset compensation scheme for OFDM systems over fading channels, *IEEE Transactions on Vehicular Technology*, **48**(5), 1596–1609.

1142. Lindsey, W. C. *et al.* (1985) Network Synchronization, *Proceedings of the IEEE*, **73**, 1445–1467.

1143. Buckwalter, J. F., Heath, T. H. and York, R. A. (2003) Synchronization design of a coupled phase-locked loop, *IEEE Transactions on Microwave Theory and Techniques*, **51**(3), 952–960.

1144. Abdel-Ghaffar, H. S. (2002) Analysis of synchronization algorithms with time-out control over networks with exponentially symmetric delays, *IEEE Transactions On Communications*, **50**(10), 1652–1661.

1145. Xiao Fan Wang and Guanrong Chen (2002) Synchronization in scale-free dynamical networks: robustness and fragility, *IEEE Transactions on Circuits and Systems I: Fundamental Theory and Applications*, **49**(1), 54–62.

1146. Walsh, G. C., Hong Ye and Bushnell, L. G. (2002) Stability analysis of networked control systems, *IEEE Transactions on Control Systems Technology*, **10**(3), 438–446.

1147. Mazzenga, F., Vatalaro, F. and Wheatley, C. E. III (2002) Performance evaluation of a network synchronization technique for CDMA cellular communications, *IEEE Transactions on Wireless Communications*, **1**(2), 322–332

1148. Shengming Jiang, Jianqiang Rao, Dajiang He, Xinhua Ling and Chi Chung Ko (2002) A simple distributed PRMA for MANETs, *IEEE Transactions on Vehicular Technology*, **51**(2), pp. 293–305.

1149. Komatsu, F., Torikai, H. and Saito, T. (2001) On a network of chaotic oscillators by intermittently coupled capacitors, *IEEE Transactions on Circuits and Systems I: Fundamental Theory and Applications*, **48**(2), 226–232.

1150. Lynch, J. J. York, R. A. (2001) Synchronization of oscillators coupled through narrow-band networks, *IEEE Transactions on Microwave Theory and Techniques* **49**(2), 237–249.

1151. Chai Wah Wu (2001) Synchronization in arrays of coupled nonlinear systems: passivity, circle criterion, and observer design, *IEEE Transactions on Circuits and Systems I: Fundamental Theory and Applications*, **48**(10), 1257–1261.

1152. Chuang, J. C.-I. (1994) Autonomous time synchronization among radio ports in wireless personal communications, *IEEE Transactions on Vehicular Technology*, **43**(1), 27–32.

1153. Berthaud, J.-M. (2000) Time synchronization over networks using convex closures, *IEEE/ACM Transactions on Networking*, **8**, 265–277.

1154. Algans, A., Pedersen, K. I. and Mogensen, P. E. (2002) Experimental analysis of the joint statistical properties of azimuth spread, delay spread, and shadow fading, *IEEE Journal on Selected Areas in Communications*, **20**.

1155. Pedersen, K. I., Mogensen, P. E. and Fleury, B. H. (2000) A stochastic model of the temporal and azimuthal dispersion seen at the base station in outdoor propagation environments, *IEEE Transactions on Vehicular Technology*, **49**, 437–447.

1156. Frederiksen, F., Mogensen, P., Pedersen, K. I. and Leth-Espensen, P. (1998) A software testbed for performance evaluation of adaptive antennas in FH GSM and wideband-CDMA, in *Proceedings of the 3rd ACTS Mobile Communication Summit*, Rhodes, Greece, June 1998, **2**, 430–435.

1157. Cox, D. C., Murray, R. and Norris, A. (1994) 800 MHz attenuation measured in and around sub-urban houses, *AT&T Bell Labs Technical Journal*, **673**.

1158. Bernhardt, R. C. (1987) Macroscopic diversity in frequency reuse systems, *IEEE Journal on Selected Areas in Communications*, **5**, 862–878.

1159. Hata, M. (1980) Empirical formula for propagation loss in land mobile radio service, *IEEE Transactions on Vehicular Technology*, **VT-29**, 317–325 (1997).

1160. Bertoni, H. L. (2000) *Radio Propagation for Modern Wireless Systems*. Prentice Hall, Englewood Cliffs, NJ.

1161. Martin, U. (1997) A directional radio channel model for densely built-up urban areas, in *Proceedings of the European Personal Mobile Communications Conference (EPMCC)*, Bonn, Germany, October 1997, pp. 237–244.

1162. Correia, L. (2001) *Wireless Flexible Personalised Communications–Cost 259 Final Report*. John Wiley & Sons, Inc., New York 2001.

1163. Winters, J. (1984) Optimum combining in digital mobile radio with cochannel interference, *IEEE Transactions on Vehicular Technology*, **VT-33**.

1164. Pedersen, K. I., Mogensen, P. E. and Fleury, B. H. (1998) Spatial channel characteristics in outdoor environments and their impact on BS antenna system performance, in *Proceedings of the IEEE Vehicular Technology Conference (VTC'98)*, Ottawa, Canada, May 1998, pp. 719–724.

1165. Pedersen, K. I. and Mogensen, K. I. (1999) Evaluation of vector-RAKE receivers using different antenna array configurations and combining schemes, *International Journal of Wireless Information Networks*, **6**, 181–194.

1166. Veen, V. and Buckley, K. M. (1988) Beamforming: A versatile approach to spatial filtering, *IEEE Acoustics, Speech and Signal Processing (ASSP) Magazine*, pp. 4–24.

1167. Liberti, J. and Rappaport, T. (1996) A geometrically based model for line-of-sight multipath radio channels, in *Proceedings of the IEEE Vehicular Technology Conference (VTC 96)*, May 1996, pp. 844–848.

1168. Lu, M., Lo, T. and Litva, J. (1997) A physical spatio-temporal model of multipath propagation channels, in *Proceedings of the IEEE Vehicular Technology Conference (VTC 97)*, May 1997, pp. 810–814.

1169. Nørklit, O. and Andersen, J. B. (1998) Diffuse channel model and experimential results for antenna arrays in mobile environments, *IEEE Transactions on Antenna Propagation*, **96**, 834–840.

1170. Cheon, C., Bertoni H. L. and Liang, G. (2000) *Monte Carlo simulation of delay and angle spread in different building environments*, presented at the IEEE Vehicular Technology Conference (VTC '00), Boston, MA.

1171. Gudmundson, M. (1992) Correlation model for shadow fading in mobile radio systems, *IEEE Electronics Letters*, **27**, 2126–2145.

1172. Mawira, A. (1992) Models for the spatial correlation functions of the log-normal component of the variability of VHF/UHF field strength in urban environments, in *Proceedings of Personal, Indoor and Mobile Radio Communications (PIMRC'92)*, pp. 436–440.

1173. Gehring, A., Steinbauer, M., Gaspard, I. and Grigat, M. (2001) *Empirical channel stationarity in urban environments*, presented at the European Personal Mobile Communications Conference (EPMCC), Vienna, February 2001.

1174. Sørensen, T. B. (1998) *Correlation model for shadow fading in a small urban macro cell*, presented at the Personal, Indoor and Mobile Radio Communications Conference (PIMRC '98), Boston.

1175. Algorithms and Antenna Array Recommendations, Public deliverable from European ACTS, TSUNAMI II Project, Deliverable code: AC020/AUC/A1.2/DR/P/005/bl, May 1997.

1176. Information Technologies and Sciences–Digital Land Mobile Radio Communications, Commission of the European Communities, COST 207, 1998.

1177. Bello, P. A. (1963) Characterization of randomly time-variant linear channels, *IEEE Transactions On Communications Systems*, **CS-11**, 360–393.

1178. Fessler, J. A. and Hero, A. (1994) Space-alternating generalised expectation-maximization algorithm, *IEEE Transactions on Signal Processing*, **42**, 2664–2677.

1179. Fleury, B. H., Tschudin, M., Heddergott, R., Dahlhaus, D. and Pedersen, K. I. (1999) Channel parameter estimation in mobile radio environments using the SAGE algorithm, *IEEE Journal on Selected Areas in Communications*, **17**, 434–450.

1180. Pedersen, K. I., Fleury, B. H. and Mogensen, P. E. (1997) High resolution of electromagnetic waves in time-varying radio channels, in *Proceedings of the International Symposium on Personal, Indoor and Mobile Radio Communications (PIMRC '97)*, Helsinki, Finland, September 1997, pp. 650–654.

1181. Pedersen, K. I., Mogensen, P. E., Fleury, B. H., Frederiksen, F. and Olesen, K. (1997) Analysis of time, azimuth and Doppler dispersion in outdoor radio channels, in *Proceedings of the ACTS Mobile Communication Summit '97,* Aalborg, Denmark, October 1997, pp. 308–313.

1182. Greenstein, L. J., Erceg, V., Yeh, Y. S. and Clark, M. V. (1997) A new path-gain/delay-spread propagation model for digital cellular channels, *IEEE Transactions on Vehicular Technology*, **46**, 477–485.

1183. Toeltsch, M., Laurila, J., Kalliola, K., Molisch, A. F., Vainikainen, P. and Bonek, E. (2002) Statistical characterization of urban spatial radio channels, *IEEE Journal On Selected Areas In Communications*, **20**(3), 539–549.

1184. Roy, R., Paulraj, A. and Kailath, T. (1986) ESPRIT–A subspace rotation approach to estimation of parameters of cisoids in noise, *IEEE Transactions on Acoustics, Speech Signal Processing*, **32**, 1340–1342.

1185. Zoltowski, M., Haardt, M. and Mathews, C. (1994) Closed-form 2-D angle estimation with rectangular arrays in element space or beamspace via unitary ESPRIT, *IEEE Transactions on Signal Processing*, **44**, 316–328.

1186. Haardt, M. and Nossek, J. (1995) Unitary ESPRIT: How to obtain an increased estimation accuracy with a reduced computational burden, *IEEE Transactions on Signal Processing*, **43**, 1232–1242.

1187. Swales, S. C., Beach, M., Edwards, D. and McGeehan, J. P. (1990) The performance enhancement of multibeam adaptive base-station antennas for cellular land mobile ratio systems, *IEEE Transactions on Vehicular Technology*, **39**, 56–67.

1188. Ertel, R. B., Cardieri, P., Sowerby, K. W., Rappaport, T. S. and Reed, J. H. (1998) Overview of spatial channel models for antenna array communications systems, *IEEE Personal Communications*, **5**(1), 10–22.

1189. Martin, U., Fuhl, J., Gaspard, I., Haardt, M., Kuchar, A., Math, C., Molisch, A. F. and Thomä, R. (1999) Model scenarios for direction-selective adaptive antennas in cellular mobile communication systems–Scanning the literature, *Wireless Personal Communications Magzine, (Special Issue on Space Division Multiple Access)*, **11**(1), 109–129.

1190. Pedersen, K., Mogensen, P. and Fleury, B. (2000) A stochastic model of the temporal and azimuthal dispersion seen at the base station in outdoor propagation environments, *IEEE Transactions on Vehicular Technology*, **49**(2), 437–447.

1191. Fuhl, J., Rossi, J.-P. and Bonek, E. (1997) High resolution 3-D direction-of-arrival determination for urban mobile radio, *IEEE Transactions on Antenna Propagation*, **4**, 672–682.

1192. Kuchar, A., Rossi, J.-P. and Bonek, E. (2000) Directional macro-cell channel characterization from urban measurements, *IEEE Transactions on Antenna Propagation*, **48**, 137–146.

1193. Kalliola, K., Laitinen, H., Vaskelainen, L. and Vainikainen, P. (2000) Real-time 3D spatial-temporal dual-polarized measurement of wideband radio channel at mobile station, *IEEE Transactions on Instrumentation and Measurment*, **49**, 439–448.

1194. Kivinen, J., Korhonen, T., Aikio, P., Gruber, R., Vainikainen, P. and Häggman, S.-G. (1999) Wideband radio channel measurement system at 2 GHz, *IEEE Transactions on Instrumentation and Measurment*, **48**, 39–44.

1195. Kermoal, J. P., Schumacher, L., Pedersen, K. I., Mogensen, P. E. and Frederiksen, F. (2002) A Stochastic MIMO radio channel model with experimental validation, *IEEE Journal on Selected Areas in Communications*, **20**, 1211–1226.

1196. Yu, K., Bengtsson, M., Ottersten, B., McNamara, D., Karlsson, P. and Beach, M. (2001) Second order statistics of NLOS indoor MIMO channels based on 5.2 GHz measurements, in *Proceedings of GLOBECOM'01*, San Antonio, Texas, USA, November 2001, pp. 156–160.

1197. Golub, G. H. and Van Loan, C. F. (1996) *Matrix Computations*, 3rd edition, The Johns Hopkins University Press, Baltimore, MD.

1198. Petrus, P., Reed, J. H. and Rappaport, T. S. (1997) Effects of directional antennas at the base station on the Doppler spectrum, *IEEE Communications Letters*, **1**, 40–42.

1199. Frederiksen, F., Mogensen, P., Pedersen, K. I. and Leth-Espensen, P. (1998) A Software testbed for performance evaluation of adaptive antennas in FH GSM and wideband-CDMA, in *Proceedings of the 3rd ACTS Mobile Communication Summit*, Rhodes, Greece, June 1998, **2**, 430–435.

1200. Andersen, J. B. (2000) Array gain and capacity for known random channels with multiple element arrays at both ends, *IEEE Journal on Selected Areas in Communications*, **18**, 2172–2178.

1201. Zhao, X., Kivinen, J., Vainikainen, P. and Skog, K. (2002) Propagation characteristics for wideband outdoor mobile communications at 5.3 GHz, *IEEE Journal on Selected Areas in Communications*, **20**, 507–514.

1202. Lee, W. C. Y. (1985) Estimate of local average power of a mobile radio signal, *IEEE Transactions on Vehicular Technology*, **34**, 22–27.

1203. Green, E. (1990) Radio link design for micro-cellular systems, *British Telecom Technology Journal*, **8**, 85–96.

1204. Dersch, U. and Zollinger, E. (1994) Physical characteristics of urban micro-cellular propagation, *IEEE Transactions on Antenna Propagation*, **42**, 1528–1539.

1205. Kivinen, J., Zhao, X. and Vainikainen, P. (2001) Empirical characterization of wideband indoor radio channel at 5.3 GHz, *IEEE Transactions on Antenna Propagation*, **49**, 1192–1203.

1206. Karasawa, Y. and Ivai, H. (2000) Formulation of spatial correlation statistic in Nakagami–Rice fading environments, *IEEE Transactions on Antenna Propagation*, **48**, 12–18.

1207. Suzuki, H. (1977) A statistical model for urban radio propagation, *IEEE Transactions On Communications*, **25**, 673–680.

1208. Takada, J., Jiye Fu, Houtao Zhu and Kobayashi, T. (2002) Spatio-temporal channel characterization in a suburban non line-of-sight microcellular environment, *IEEE Journal On Selected Areas In Communications*, **20**(3), 532–538.

1209. Masui, H., Takahashi, K., Takahashi, S., Kage, K. and Kobayashi, T. (1999) Delay profile measurement system for microwave broadband transmission and analysis of delay characteristics in an urban environment, *IEICE Transactions on Electronics*, **E82-C**, 1287–1292.

1210. Molisch, A. F., Steinbauer, M., Toeltsch, M., Bonek, E. and Thomä, R. S. (2002) Capacity of MIMO systems based on measured wireless channels, *IEEE Journal on Selected Areas in Communications*, **20**, 561–569.

1211. Parks-Gornet, J. and Imam, I. N. (1989) Using rank factorization in calculating the Moore–Penrose generalized inverse, *IEEE Southeastcon '89, Energy and Information Technologies in the Southeast*, 9–12 April 1989, **2**, 427–431.

1212. Tokarzewski, J., (1998) System zeros analysis via the Moore–Penrose pseudoinverse and SVD of the first nonzero Markov parameter, *IEEE Transactions on Automatic Control*, **43**(9), 1285–1291.

1213. Withers, L. P., Jr. (1993) A parallel algorithm for generalized inverses of matrices, with applications to optimum beamforming, *IEEE International Conference on Acoustics, Speech and Signal Processing* (ICASSP-93), 27–30 April 1993, **1**, 369–372.

1214. Shu Wang, and Xilang Zhou (1999) Extending ESPRIT algorithm by using virtual array and Moore–Penrose general inverse techniques, *IEEE Southeastcon '99*, 25–28 March 1999, pp. 315–318.

1215. Steinbauer, M., Molisch, A. F. and Bonek, E. (2001) The double-directional radio channel, *IEEE Antennas and Propagation Magazine*, pp. 51–63.

1216. Rubio, M. L., Garcia-Armada, A., Torres, R. P. and Garcia, J. L. (2002) Channel modeling and characterization at 17 GHz for indoor broadband WLAN, *IEEE Journal on Selected Areas in Communications*, **20**.

1217. Hashemi, H. (1993) The indoor radio propagation channel, *Proceedings of the IEEE*, **81** (7), 943–968.

1218. Bury, K. (1999) *Statistical Distributions in Engineering*, Cambridge University Press, Cambridge UK.

1219. Torres, R. P., Valle, L., Domingo, M., Loredo, S. and Diez, M. C., (1999) CIN-DOOR: An engineering tool for planning and design of wireless systems in enclosed spaces, *IEEE Antennas and Propagation Magazine*, **41**, 11–22.

1220. Loredo, S., Torres, R. P., Domingo, M., Valle, L. and Pérez, J. R. (2000) Measurements and predictions of the local mean power and small-scale fading statistics in indoor wireless environments, *Microwave Optical Technology Letters*, **24**, 329–331.

1221. Torres, R. R., Loredo, S., Valle, L. and Domingo, M. (2001) An accurate and efficient method based on ray-tracing for the prediction of local flat-fading statistic in picocell radio channels, *IEEE Journal on Selected Areas in Communications*, **18**, 170–178.

1222. Loredo, S., Valle, L. and Torres, R. P. (2001) Accuracy analysis of GO/UTD radio channel modeling in indoor scenarios at 1.8 and 2.5 GHz, *IEEE Antennas and Propagation Magazine*, **43**.

1223. Bohdanowicz, A. (2000) *Wide band indoor and outdoor radio channel measurements at 17 GHz*, Ubicom Technical Report/2000/2.

1224. Bohdanowicz, A., Janssen, G. J. M. and Pietrzyk, S. (1999) WideBand indoor and outdoor multipath channel measurements at 17 GHz, in *Proceedings of the IEEE Vehicular Technology Conference*, **4**, 1998–2003.

1225. Talbi, L. and Delisle, G. Y. (1996) Experimental characterization of EHF multipath indoor radio channels, *IEEE Journal on Selected Areas in Communications*, **14**, 431–439.

1226. Rappaport, T. S. and McGillem, C. D. (1988) UHF fading in factories, *IEEE Journal on Selected Areas in Communications*, **7**, 40–48.

1227. Rappaport, T. S. (1989) Indoor radio communications for factories of the future, *IEEE Communications Magazine*, pp. 15–24.

1228. Abou-Raddy, A. F. and Elnoubi, S. M. (1997) Propagation measurements and channel modeling for indoor mobile radio at 10 GHz, in *Proceedings of the IEEE Vehicular Technology Conference*, **3**, 1395–1399.

1229. Saleh, A. A. M. and Valenzuela, R. A. (1987) A statistical model for indoor multipath propagation, *IEEE Journal on Selected Areas in Communications*, **SAC-5**.

1230. Hao Xu, Kukshya, V. and Rappaport, T. S. (2002) Spatial and temporal characteristics of 60 GHz indoor channels, *IEEE Journal On Selected Areas In Communications*, **20**(3), 620–630.

1231. Rappaport, T. S. (1996) *Wireless Communications: Principles and Practice*. Prentice Hall, Englewood Cliffs, NJ.

1232. Durgin, G. and Rappaport, T. S. (2000) Theory of multipath shape factors for small-scale fading wireless channels, *IEEE Transactions on Antenna Propagation*, **48**, 682–693.

1233. Manabe, T., Miura, Y. and Ihara, T. (1996) Effects of antenna directivity and polarization on indoor multipath propagation characteristics at 60 GHz, *IEEE Journal on Selected Areas in Communications*, **14**, 441–448.

1234. Sato, K., Manabe, T., Ihara, T., Saito, H., Ito, S., Tanaka, T., Sugai, K., Ohmi, N., Murakami, Y., Shibayama, M., Konishi, Y. and Kimura, T. (1999) Measurements of reflection and transmission characteristics of interior structures of office building in the 60 GHz band, *IEEE Transactions on Antenna Propagation*, **45**, 1783–1792.

1235. Langen, B., Lober, G. and Herzig, W. (1994) Reflection and transmission behavior of building materials at 60 GHz, in *Proceedings of IEEE PIMRC '94*, The Hague, The Netherlands, September 1994, pp. 505–509.

1236. Rossi, J. P., Barbot, J. P. and Levy, A. J. (1997) Theory and measurement of the angle of arrival and time delay of UHF radiowave using a ring array, *IEEE Transactions on Antenna Propagation*, **45**, 876–884.

1237. de Jong, Y. L. C. and Herben, M. H. A. J. (1999) High-resolution angle-of-arrival measurement of the mobile radio channel, *IEEE Transactions on Antenna Propagation*, **47**, 1677–1687.

1238. Droste, H. and Kadel, G. (1995) Measurement and analysis of wideband indoor propagation characteristics at 17 GHz and 60 GHz, in *Proceedings of IEE Antennas Propagation Conference (Publication 407)*, April 1995, pp. 288–291.

1239. Holloway, C. L., Perini, P. L., Delyzer, R. R. and Allen, K. C. (1997) Analysis of composite walls and their effects on short-path propagation modeling, *IEEE Transactions on Vehicular Technology*, **46**,730–738.

1240. Honcharenko, W. and Bertoni, H. (1994) Transmission and reflection characteristics at concrete block walls in the UHF bands proposed for future PCS, *IEEE Transactions on Antenna Propagation*, **42**, 232–239.

1241. Smulders, P. and Correia, L. (1997) Characterization of propagation in 60 GHz radio channels, *Electronics Communications and Engineering Journal*, **9**(2) 73–80.

1242. Xu, H., Rappaport, T. S., Boyle, R. J. and Schaffner, J. (2000) Measurements and modeling for 38-GHz point-to-multipoint radiowave propagation, *IEEE Journal on Selected Areas in Communications*, **18**, 310–321.

1243. Xu, H., Kukshya, V. and Rappaport, T. (2000) Spatial and temporal characterization of 60 GHz channels, in *Proceedings of the IEEE VTC'2000*, Boston, MA, September 24–28, 2000.

1244. Cassioli, D., Win, M. Z., Molisch, A. F. (2002) The ultra-wide bandwidth indoor channel: from statistical model to simulations, *IEEE Journal On Selected Areas In Communications*, **20**(6), 1247–1257.

1245. Greenstein, L. J., Erceg, V., Yeh, Y. S. and Clark, M. V. (1997) A new path-gain/delay-spread propagation model for digital cellular channels, *IEEE Transactions on Vehicular Technology*, **46**, 477–485.

1246. Erceg, V., Greenstein, L. J., Tjandra, S. Y., Parkoff, S. R., Gupta, A., Kulic, B., Julius, A. A. and Bianchi, R. (1999) An empirically based path loss model for wireless channels in suburban environments, *IEEE Journal on Selected Areas in Communications*, **17**, 1205–1211.

1247. Failli, E. (Ed.), (1989), *Digital Land Mobile Radio. Final Report of COST 207*. Commission of the European Union, Luxemburg.

1248. Nakagami, M. (1960) The *m*-distribution – A general formula of intensity distribution of rapid fading, in *Statistical Method in Radio Wave Propagation*, W. C. Hoffman (Ed.), Pergamon, Oxford, UK, pp. 3–36.

1249. Braun, W. R. and Dersch, U. (1991) A physical mobile radio channel model, *IEEE Transactions on Vehicular Technology*, **40**, 472–482.

1250. Yin, M. and Li, V. O. K., (1990) Unslotted CDMA with Fixed Packet Lengths, *IEEE Journal on Selected Areas in Communications*, **8**(4), 529–541.

1251. Toshimitsu, K., Yamazato, T., Katayama, M. and Ogawa, A. (1994) A Novel Spread Slotted ALOHA System with CLSP, *IEEE Journal on Selected Areas in Communications*, **12**(4), 665–672.

1252. Sato, T., Okada, H., Yamazato, T., Katayama, M. and Ogawa, A. (1996) Throughput Analysis of DS/SSMA Unslotted ALOHA System with Fixed Packet Length, *IEEE Journal on Selected Areas in Communications*, **14**(4), 750–756.

1253. Phan, V. V. and Glisic, S. (2001) Estimation of Implementation Losses in MAC Protocols in Wireless CDMA Networks, *International Journal on Wireless Information Networks*, **8**(3), 115–132.

1254. Viterbi, A. J. (1995) *Principle of Spread Spectrum Communication*, Addison Wesley Inc.

1255. Prasad, R. (1996) *CDMA for Wireless Personal Communications*, Artech House Inc.

1256. Steel, R. (1992) *Mobile Radio Communications*, John Wiley & Sons, Inc. and IEEE Press.

1257. Rappaport, T. S. (1996) *Wireless Communications Principles & Practice*, Prentice–Hall Inc.

1258. Glisic, S. (2002) *Adaptive WCDMA Theory & Practice*, John Wiley & Sons Inc., New York.

1259. Chien, C., Srivastava, M. B., Jain, R., Lettieri, P., Aggrawal, V. and Sternowski, R. (1999) Adaptive Radio for Multimedia Wireless Links, *IEEE Journal on Selected Areas in Communications*, **17**(5), 793–813.

1260. Hara, S., Ogino, A., Araki, M., Okada M. and Morinaga, N. (1996) Throughput Performance of SAW-ARQ Protocol with Adaptive Packet Length in Mobile Packet Data Transmission, *IEEE Transactions on Vehicular Technology*, **45**(3), 561–569.

1261. Lettieri, P. and Srivastava, M. B. (1998) Adaptive frame length control for improving wireless link range and energy efficiency, *IEEE InfoCom98*, **2**, 564–571.

1262. Siew, C. K. and Goodman, D. J. (1989) Packet Data Transmission Over Mobile Radio Channels, *IEEE Transactions on Vehicular Technology*, **38**(2), 95–101.

1263. Zorzi, M., Rao, R. R. and Milstein, L. B. (1996) A Markov Model for Block Errors on Fading Channels, *IEEE PIMRC96*, 1074–1078.

1264. Zorzi, M., Rao, R. R. and Milstein, L. B. (1998) Error Statistics in Data Transmission over Fading Channel, *IEEE Transactions On Communications*, **46**(11), 1468–1477.

1265. Zorzi, M. (1998) Performance of FEC and ARQ Error Control in Bursty Channel under Delay Constraints, *IEEE VTC98*, 1390–1394.

1266. Pahlavan, K. and Levesque, A. H. (1994) Wireless Data Communications, *Proceedings of the IEEE*, Invited Paper, **82**(9), 1398–1430.

1267. Jakes, W. C. (1995) Microwave Mobile Communications, IEEE Press.

1268. Adachi, F., Feeney, M. T. and Parsons, J. D. (1988) Effects of Correlated Fading on Level Crossing Rates and Average Fade Durations with Predetection Diversity Reception, *IEE Proceedings*, **135**(1), 11–17.

1269. Dong, X. and Beaulieu, N. C. (2002) Average Level Crossing Rate and Fade Duration of Maximal Ratio Diversity in Unbalanced and Correlated Channels, *IEEE WCNC02*, **2**, 762–767.

1270. Rice, S. O. (1958) Distribution of the Duration of Fades in Radio Transmissions: Gaussian Noise Model, *Bell Systems Technical Journal*, **37**, 581–635.

1271. Bodtmann, W. F. and Arnold, H. W. (1982) Fade-Duration Statistics of Rayleigh Distributed Waves, *IEEE Transactions On Communications*, **30**(3), 549–553.

1272. Durgin, G. D. and Rappaport, T. S. (1999) Level-Crossing Rates and Average Fade Duration for Wireless Channels with Spatially Complicated Multipath, *IEEE Globe-Com99*, pp. 427–431.

1273. Turin, W. and Nobelen, R. V. (1998) Hidden Markov Modelling of Flat Fading Channels, *IEEE Journal on Selected Areas in Communications*, **16**(9), 1809–1817.

1274. Zorzi, M., Rao, R. R. and Milstein, L. B. (1995) On the Accuracy of a First-Order Markov Model for Data Transmission on Fading Channels, *IEEE ICUPC95*, pp. 211–215.

1275. Wang, H. S. and Chang, P. C. (1996) On Verifying the First-Order Markovian Assumption for a Rayleigh Fading Channel Model, *IEEE Transactions on Vehicular Technology*, **45**(2), 353–357.

1276. Tan, C. C. and Beaulieu, N. C. (2000) On First-Order Markov Modelling for the Rayleigh Fading Channel, *IEEE Transactions On Communications*, **48**(2), 2032–2040.

1277. Omnet++ Discrete Event Simulation Tool, http://www.hit.bme.hu/phd/vargaa/omnetpp.htm.

1278. Gross, D. and Harris, C. M. (1998) Fundamentals of Queuing Theory, John Wiley & Sons Inc., New York.

1279. Ross, K. W. and Tsang, D. H. K. (1989) The Stochastic Knapsack Problem, *IEEE Transactions On Communications*, **37**(7), 740–747.

1280. Gentile, C., Haerri, J. and VanDyck, R. E. (2002) Kinetic minimum-power routing and clustering in mobile adhoc networks, *IEEE 56th Vehicular Technology Conference*, 24–28 September 2002, **3**, 1328–1332.

1281. Rodoplu, V. and Meng, T. H. (1999) Minimum energy mobile wireless networks, *IEEE Journal on Selected Areas in Communications*, **17**(8), 1333–1344.

1282. Shahani, A. R., Schaeffer, D. K. and Lee, T. H. (1997) A 12 mW wide dynamic range CMOS front-end for a portable GPS receiver, in *Proceedings of the IEEE International Solid-State Circuits Conference* February 1997, **40**, 368–369.

1283. Lynch, N. A. (1996) *Distributed Algorithms.* Morgan Kaufmann, San Mateo, CA, pp. 51–80.

1284. Gray, P. and Meyer, R. (1995) Future directions in silicon ICs for RF personal communications, in *Proceedings of CICC'95*, pp. 83–90.

1285. Rudell, J. C. *et al.* (1997) A 1.9-GHz wide-band IF double conversion CMOS receiver for cordless telephone applications, *IEEE Journal on Solid-State Circuits*, **32**, 2071–2088.

1286. Parkingson, B. W. and Spilker, J. J. (1996) *Global Positioning System: Theory and Applications Vol. I.* American Institute of Aeronautics and Astronautics, Inc. Washington, DC.

1287. Pursley, M. B., Russell, H. B. and Staples, P. E. (1999) Routing for multimedia traffic in wireless frequency-hop communication networks, *IEEE Journal on Selected Areas in Communications*, **17**(5), 784–792.

1288. Pursley, M. B. and Russell, H. B. (1993) Routing in frequency-hop packet radio networks with partial-band jamming, *IEEE Transactions On Communications* **41**(7), 1117–1124.

1289. Pursley, M. B. and Russell, H. B. (1994) Network protocols for frequency-hop packet radios with decoder side information, *IEEE Journal on Selected Areas in Communications*, **12**(4), 612–621.

1290. Pursley, M. B. and Russell, H. B., (1995) Recognizing and responding to reduced-quality links in FH packet radio networks, in *Proceedings of the 1995 IEEE Military Communications Conference*, pp. 75–79.

1291. Steenstrup, M. (1995) *Routing in Communications Networks.* Prentice Hall, Englewood Cliffs, NJ.

1292. Manohar, R. and Scaglione, A. (2003) Power optimal routing in wireless networks, *IEEE International Conference on Communications* (ICC '03), **4**, 2979–2984.

1293. Manohar, R. and Scaglione, A. (2003) Power Optimal Routing in Wireless Networks, Cornell Computer Systems Technical Report CSL-TR-2003-1028, January 2003. Available at http://www.csl.cornell.edu/publications.html.

1294. Rappaport, T. (1999) *Wireless Communications*, Prentice Hall.

1295. Narayanaswamy, S., Kawadia, V., Sreenivas, R. S. and Kumar, P. R. (2002) Power Control in Ad Hoc Networks: Theory, Architecture, Algorithm and Implementation of the COMPOW protocol, *Proceedings of European Wireless 2002, Next Generation Wireless Networks: Technologies, Protocols, Services and Applications*, Florence, Italy, 25–28 February, 2002, pp. 156–162.

1296. Chang, J.-H. and Tassiulas, L. (2000) Energy conserving routing in wireless ad hoc networks, in *Proceedings of IEEE INFOCOM 2000*, **1**, 22–31.

1297. Hou, T. and Li, V. (1986) Transmission Range Control in Multihop Packet Radio Networks, *IEEE Transactions On Communications*, **34**(1), 38–44.

1298. Sousa, E. S. and Silvester, J. A. (1990) Optimum transmission ranges in a direct-sequence spread-spectrum multihop packet radio network, *IEEE Journal on Selected Areas in Communications*, **8**(5), 762–771.

1299. Jonnavithula, S. and Billinton, R. (1996) Minimum cost analysis of feeder routing in distribution system planning, *IEEE Transactions on Power Delivery*, **11**(4), 1935–1940.

1300. Zadeh, A. N., Jabbari, B., Pickholtz, R. and Vojcic, B. (2002) Self-organizing packet radio ad hoc networks with overlay (SOPRANO), *IEEE Communications Magazine*, **40**(6), 149–157.

1301. Stojmenovic, I. and Datta, S. (2002) Power and cost aware localized routing with guaranteed delivery in wireless networks, *Seventh International Symposium on Computers and Communications* (ISCC 2002), 1–4 July 2002, pp. 31–36.

1302. Stojmenovic, I. and Lin, X. (2001) Power-aware localized routing in wireless networks, *IEEE Transactions on Parallel and Distributed Systems*, **12**(11), 1122–1133.

1303. Maleki, M., Dantu, K. and Pedram, M. (2002) Power-aware source routing protocol for mobile ad hoc networks, *Proceedings of the 2002 International Symposium on Low Power Electronics and Design*, (ISLPED '02), 12–14 August 2002, pp. 72–75.

1304. Jung-Hee Ryu and Dong-Ho Cho (2001) A new routing scheme concerning energy conservation in wireless home ad hoc networks, *IEEE Transactions on Consumer Electronics*, **47**(1), 1–5.

1305. Neely, M. J., Modiano, E. and Rohrs, C. E. (2003) Power allocation and routing in multibeam satellites with time-varying channels, *IEEE /ACM Transactions on Networking*, **11**(1), 138–152.

1306. Mitlin, V. (2003) Optimal mac packet size in networks without cut-through routing, *IEEE Transactions on Wireless Communications*, **2**(5), 901–910.

1307. Pollini, G. P. and Meier-Hellstern, K. S. (1995) Efficient routing of information between interconnected cellular mobile switching centers, *IEEE /ACM Transactions on Networking*, **3**(6), 765–774.

1308. Ramjee, R., La Porta, T. F., Kurose, J. and Towsley, D. (1998) Performance evaluation of connection rerouting schemes for ATM-based wireless networks, *IEEE /ACM Transactions on Networking*, **6**(3), 249–261.

1309. Bigioi, P., Cucos, A., Corcoran, P., Chahil, C. and Lusted, K. (1999) Transparent, dynamically configurable RF network suitable for home automation applications, *IEEE Transactions on Consumer Electronics*, **45**(3), 474–480.

1310. Binh Vien Dao, Duato, J. and Yalamanchili, S. (1999) Dynamically configurable message flow control for fault-tolerant routing, *IEEE Transactions on Parallel and Distributed Systems*, **10**(1), 7–22.

1311. Stojmenovic, I. and Xu Lin (2001) Loop-free hybrid single-path/flooding routing algorithms with guaranteed delivery for wireless networks, *IEEE Transactions on Parallel and Distributed Systems*, **12**(10), 1023–1032.

1312. Wong, V. W. S., Lewis, M. E. and Leung, V. C. M. (2001) Stochastic control of path optimization for inter-switch handoffs in wireless ATM networks, *IEEE /ACM Transactions on Networking*, **9**(3), 336–350.

1313. Jung-Hee Ryu and Dong-Ho Cho (2001) A new routing scheme concerning energy conservation in wireless home ad hoc networks, *IEEE Transactions on Consumer Electronics*, **47**(1), 1–5.

1314. Mohorcic, M., Werner, M., Svigelj, A. and Kandus, G. (2002) Adaptive routing for packet-oriented intersatellite link networks: performance in various traffic scenarios, *IEEE Transactions on Wireless Communications*, **1**(4), 808–818.

1315. Garcia Nocetti, F. and Stojmenovic, I. and Jingyuan Zhang (2002) Addressing and routing in hexagonal networks with applications for tracking mobile users and connection rerouting in cellular networks, *IEEE Transactions on Parallel and Distributed Systems*, **13**(9), 963–971.

1316. Sunghyun Choi and Shin, K. G. (2002) Adaptive bandwidth reservation and admission control in QoS-sensitive cellular networks, *IEEE Transactions on Parallel and Distributed Systems*, **13**(9), 882–897.

1317. Jie Wu (2002) Extended dominating-set-based routing in ad hoc wireless networks with unidirectional links, *IEEE Transactions on Parallel and Distributed Systems*, **13**(9), 866–881.

1318. Schurgers, C., Kulkarni, G. and Srivastava, M. B. (2002) Distributed on-demand address assignment in wireless sensor networks, *IEEE Transactions on Parallel and Distributed Systems*, **13**(10), 1056–1065.

1319. Agglou, G. and Tafazolli, R. (2002) Determining the optimal configuration for the relative distance microdiscovery ad hoc routing protocol, *IEEE Transactions on Vehicular Technology*, **51**(2), 354–370.

1320. Beraldi, R. and Baldoni, R. (2003) A caching scheme for routing in mobile ad hoc networks and its application to ZRP, *IEEE Transactions on Computers*, **52**(8), 1051–1062.

1321. Duggirala, R., Gupta, R., Qing-An Zeng and Agrawal, D. P. (2003) Performance enhancements of ad hoc networks with localized route repair, *IEEE Transactions on Computers*, **52**(7), 854–861.

1322. Toumpis, S. and Goldsmith, A. J. (2003) Capacity regions for wireless ad hoc networks, *IEEE Transactions on Wireless Communications*, **2**, 736–748.

1323. Jen-Yi Pan, Wei-Tsong Lee and Nen-Fu Huang (2003) Providing multicast short message services over self-routing mobile cellular backbone network, *IEEE Transactions on Vehicular Technology*, **52**(1), 240–253.

Index

Advanced Wireless Communications. S. Glisic

© 2004 John Wiley & Sons, Ltd. ISBN: 0-470-86776-0